The UK Pesticide Guide 2001

Editor – R. Whitehead BA, MSc

BRITISH
CROP
PROTECTION
COUNCIL

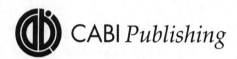

CABI *Publishing*

CABI *Publishing* (a division of CAB International) is one of the world's foremost publishers of databases, books and journals in agriculture, and applied life sciences. It has a worldwide reputation for producing high quality, value-added information, drawing on its links with the scientific community. From its Headquarters in Wallingford, UK, CABI *Publishing* runs a worldwide operation, distributing books, journals and electronic products to customers in over 150 countries, and selling its products through a network of international agents. For further information, please contact CABI *Publishing*, CAB International, Wallingford, Oxon OX10 8DE, UK.

The British Crop Protection Council (BCPC) is a self-supporting limited company with charity status, which was formed in 1968 to promote the knowledge and understanding of the science and practice of crop protection. The corporate members include government departments and research councils; advisory services; associations concerned with the farming industry; agrochemical manufacturers; agricultural engineering, agricultural contracting and distribution services; universities; scientific societies; organisations concerned with the environment; and some experienced independent members. Further details available from BCPC, 49 Downing Street, Farnham GU9 7PH.

ISBN 0 85199 497 0

Printed and bound in the UK by Biddles Ltd, Guildford and King's Lynn

Contents

Editor's Note

We aim to make *The UK Pesticide Guide* a comprehensive and up-to-date listing of pesticides or adjuvants available in the UK market place and registered for professional users. The pace of change in the pesticide industry makes this a formidable challenge.

Not all approved products are available, while some 'old' products following a merger or take-over remain in the supply chain for at least a season. That is why some products are 'double entered'. We try to include all that are available, but heavy reliance is placed on suppliers to notify the Editor about which products will be on sale. Some are better at this than others, but suppliers give an undertaking that their products listed in the *Guide* will be on sale during the season to which the edition refers. *No product is included unless the supplier has requested it, on these terms.* Suppliers are given the opportunity to do this every year, and inevitably this leads to omissions. Readers are asked to notify the Editor of any such omissions that they notice, and, perhaps more importantly, notify the supplier also.

Accuracy is a prime objective. Much rests on the goodwill and co-operation of manufacturers and suppliers in ensuring the accuracy of the information given about their products. They are asked to check this every year, and we are very grateful for the commitment and diligence with which the majority approach this task. Sadly some do not, and in these cases, although we make strenuous efforts to avoid it, some information may not be fully updated. Because of this, the *Guide* must *never* be regarded as a substitute for the product label.

New active ingredients in this edition include the fungicides triticonazole, metconazole, and trifloxystrobin, the sulfonylurea herbicide DE570, and pymetrozine, a new azomethine insecticide.

This fourteenth edition contains over 500 product changes, about equally divided between additions and deletions. Many of these result from commercial changes, some of which are still in progress. At the time of going to press it has not been possible to include the Ministry registration numbers of some BASF products acquired following the take-over of Cyanamid. They are included in the interests of completeness, and on the assurance given by the company that the products will be registered by the time of publication. Similarly the formation of Syngenta was being ratified shortly before printing and therefore no products are entered under this name. Full lists are, however, shown for Novartis and Zeneca in the certainty that most of these products will be available throughout 2001.

R. Whitehead
Editor

Disclaimer

Every effort has been made to ensure that the information in this book is complete and correct at the time of going to press but the Editor and the publishers do not accept liability for any error or omission in the content, or for any loss, damage or other accident arising from the use of products listed herein. Omission of a product does not necessarily mean that it is not approved or that it is not available for use.

The information has been selected from official sources and from suppliers' labels and product manuals of pesticides approved under the Control of Pesticides Regulations 1986 and available for pest, disease and weed control applications in the United Kingdom.

It is essential to follow the instructions on the approved label before handling, storing or using any crop-protection product. Approved 'off-label' uses are made entirely at the risk of the user.

The contents of this publication are based on information received up to October 2000.

Changes Since 2000 Edition

Pesticides and adjuvants added or deleted since the 2000 edition are listed below. Every effort has been made to ensure the accuracy of this list, but late notifications to the Editor are not included.

Products Added

Products new to this edition are shown here. In addition, products that were listed in the previous edition whose MAFF Approval number *and* supplier have changed are included. Where only the MAFF number has changed, the product is not listed.

Product	MAFF No	Supplier	Product	MAFF No	Supplier
Actipron	ADJ0013	Batsons	Biosyl	ADJ0385	Intracrop
Adagio	10057	Interfarm	Biothene	ADJ0360	Intracrop
Agricola Lens	10257	Agricola	Bison 83 WG	10063	Nufarm Whyte
Agroxone	09947	Headland	Borocil K	09924	Aventis Environ.
Aliette 80 WG	09156	Hortichem			
Alpha Protugan Plus	08799	Makhteshim	Boxer	09819	Dow
Alphabird	10146	Killgerm	Bumper 250 EC	09039	Makhteshim
Alphathrin	10025	Nufarm Whyte	Calidan	09980	Aventis
Amazon	10266	Aventis	Camber	09901	Headland
Amigo	ADJ0397	Intracrop	Capricorn	10035	Aventis Environ.
Ardent	09968	Aventis			
Arsenal	10208	BASF	Capture	09982	Aventis
Arsenal 50	BASF	BASF	Caramba	09864	Cyanamid
Asulox	09969	Aventis	Caramba	10213	BASF
Atol	10235	Nufarm Whyte	Cardel Egret	09703	Cardel
Avenge 2	10210	BASF	Cat G	ADJ0348	Greenhill
Aventis Betanal Progress OF	09722	Aventis	CDA Vanquish	09927	Aventis Environ.
Aventis Cerone	09972	Aventis			
Aventis Challenge	09721	Aventis	Cercobin Liquid	09984	Aventis
Aventis Cheetah Super	09730	Aventis	Cerone	09985	Aventis
Aventis Compass	10040	Aventis	Charger	09986	Aventis
Aventis Decis	09710	Aventis	Chess	09817	Novartis
Aventis Eagle	09765	Aventis	Chiltern Blues	10071	Chiltern
Aventis Foil	09709	Aventis	Chiltern Gold 98	ADJ0302	Chiltern
Aventis Galtak 50 SC	09711	Aventis	Chiltern Hundreds	10072	Chiltern
Aventis Harvest	09810	Aventis	Clortosip 500	09320	Sipcam
Aventis Merlin	09719	Aventis	Clovotox	09928	Aventis Environ.
Aventis Mocap 10G	09973	Aventis			
Aventis Rovral Flo	09974	Aventis	Commando	10215	BASF
Aventis Tattoo	09811	Aventis	Compass	10041	Aventis
Aventis Temik 10G	09749	Aventis	Comrade	10181	United Phosphorus
Aventis Tigress Ultra	09725	Aventis			
Aztec	10211	BASF	Condox	09429	Zeneca
Azural	09582	Monsanto	Contest	10216	BASF
Bacara	09976	Aventis	Crossfire 480	09929	Aventis Environ.
Ballad	09775	Headland			
Banlene Super	10053	Aventis	Cutback	ADJ0351	Greenhill
Barclay Barbarian	09865	Barclay	Cutinol Plus	ADJ0395	Intracrop
Barclay Fentin Flow 532	09434	Barclay	Dagger	10218	BASF
Barclay Gallup 360	09127	Barclay	Dancer FL	10048	Sipcam
Barclay Gallup Biograde 360	09840	Barclay	Decoy Wetex	09707	Bayer
Barclay Gallup Biograde Amenity	10203	Barclay	DFF + IPU WDG	10204	Aventis
			Dicotox Extra	09930	Aventis Environ.
Barclay Hurler 200	09917	Barclay			
Bavistin FL	00218	BASF	Dimilin 25-WP	08902	Hortichem
Biofilm	ADJ0359	Intracrop	Dithane 945	09897	Interfarm

Product	MAFF No	Supplier	Product	MAFF No	Supplier
Dithianon Flowable	10219	BASF	Karate Ready to Use Rat Bait	H6745	DiverseyLever
Diuron 80 WP	09931	Aventis Environ.	Karate Ready to Use Rodenticide Sachets	H6746	DiverseyLever
Dixie 6	10052	Greencrop	Kerb 50 W	10166	SumiAgro
Dockmaster	10033	Nufarm Whyte	Kerb 50 W	10200	Interfarm
Doff New Formula Metaldehyde Slug Killer Mini Pellets	09772	Doff Portland	Kerb 80 EDF	10201	Interfarm
			Kerb Flo	10157	SumiAgro
Dormone	09932	Aventis Environ.	Kerb Flo	10160	Interfarm
			Kerb Pro Flo	10158	SumiAgro
DP 911 WSB	09867	DuPont	Killgerm Fenitrothion 40 WP	H4858	Killgerm
DP 928	09632	DuPont	Killgerm Fenitrothion 50 EC	H4722	Killgerm
Drat Rat Bait	H6743	B H & B	Kruga 5EC	09863	Interfarm
Easel	09878	Nufarm Whyte	Landgold Deltaland	09906	Landgold
Elan	ADJ0392	Intracrop	Landgold Difenoconazole	09964	Landgold
Elvaron Multi	10080	Bayer	Landgold Epoxiconale FM	08806	Landgold
Encore	BASF	BASF	Landgold Epoxiconazole	09821	Landgold
Euroagkem Pace	ADJ0305	Euroagkem	Landgold Strobilurin 250	09595	Landgold
Exit Wetex	10149	Bayer	Landgold Strobilurin KE	09908	Landgold
Farmatin 560	09877	Aventis	Legumex Extra	08676	Aventis
Fastac	10220	BASF	Lo Dose G	ADJ0349	Intracrop
Flamenco	09913	Aventis	Logic	ADJ0288	Microcide
Freeway	09933	Aventis Environ.	Lorsban WG	10139	Dow
			Marathon	09959	Me2
Fubol Gold WG	10184	Novartis	Marquise	08738	Makhteshim
Galben M	07220	Sipcam	Masai	10223	BASF
Genesis	09993	Aventis	MCPA 25%	07998	Nufarm Whyte
Glimark	ADJ0371	Nufarm Whyte	Me2 Aducksbackside	10065	Me2
GLY-480	09837	Powaspray	Me2 Aldee	09781	Me2
Granit	09995	Aventis	Me2 Azoxystrobin	09654	Me2
Greencrop Boulevard	09960	Greencrop	Me2 Cymoxeb	09486	Me2
Greencrop Glenroe	09903	Greencrop	Me2 Exodus	09786	Me2
Greencrop Gweedore	09882	Greencrop	Me2 KME	09594	Me2
Greencrop Rosette	09648	Greencrop	Me2 Lambda	09712	Me2
Greencrop Saffron FL	10244	Greencrop	Me2 Tebuconazole	09751	Me2
Greencrop Valentia	10197	Greencrop	Menace 80 EDF	10165	Interfarm
Hallmark with Zeon Technology	09809	Zeneca	Meothrin	BASF	BASF
			Merit	BASF	BASF
Harlequin 500 SC	09779	Makhteshim	Meteor	BASF	BASF
Headland Cedar	09948	Headland	Micene DF	09957	Sipcam
Headland Guard 2000	ADJ0369	Headland	Mildothane Liquid	10002	Aventis
Helmsman	09934	Aventis Environ.	Mildothane Turf Liquid	09935	Aventis Environ.
Herboxone	10032	Headland	Mircam	09888	Nufarm Whyte
Huron	10148	Bayer	Mircam Plus	09884	Nufarm Whyte
Hycamba Plus	10180	Agrichem	Mithras 80 EDF	10164	Interfarm
Hymec Triple	09949	Agrichem	Mocap 10G	10003	Aventis
I T Asulam	10186	I T Agro	Nemolt	10226	Fargro
I T Cyanazine	10179	I T Agro	NWA CTU 500	10173	Nufarm Whyte
Ingot	09997	Aventis	Optica	09963	Headland
Insegar WG	09789	Novartis	Oxytril CM	10005	Aventis
Intracrop Green Oil	ADJ0262	Intracrop	Panther	10008	Aventis
Intracrop Rapeze	ADJ0390	Intracrop	Pentil	ADJ0340	Nufarm Whyte
Invader	BASF	BASF	Plenum	09816	Novartis
Isomec	09881	Nufarm Whyte	PP Captan 80-WG	08971	Tomen
Javelin	09998	Aventis	Precis	10159	SumiAgro
Javelin Gold	09999	Aventis	Quintozene WP	09936	Aventis Environ.
Jockey	10076	Aventis			
Judge	10000	Aventis	Ravine	10095	Aventis
K & S Chlorofume	08722	K&S Fumigation	Redeem Flo	10236	Interfarm
			Regulox K	09937	Aventis

Product	MAFF No	Supplier
Environ.		
Reward 5EC	09862	Interfarm
Rizolex	09673	Hortichem
Rizolex Flowable	09358	Hortichem
Robust	10176	Aventis
Rover	09848	Sipcam
Rovral Flo	10013	Aventis
Rovral Green	09938	Aventis
Environ.		
Rovral Liquid FS	10014	Aventis
Rovral WP	10015	Aventis
Satellite	BASF	BASF
Sectacide 50 EC	H4939	Killgerm
Sequel	09886	Hortichem
SHL Granular Feed and Weed	09926	Sinclair
SHL Granular Feed, Weed and Mosskiller	09925	Sinclair
sHYlin	10030	Agrichem
Siltex	ADJ0398	Intracrop
Sipcam Echo 75	08302	Sipcam
Sirocco	09939	Aventis
Environ.		
Snooker	10018	Aventis
Spasor	09945	Aventis
Environ.		
Spasor Biactive	09940	Aventis
Environ.		
Spearhead	09941	Aventis
Environ.		
Sphinx	07607	Aventis
Standon Bentazone S	10124	Standon
Standon Kresoxim Super	09794	Standon
Standon Kresoxim-Epoxiconazole	09281	Standon
Standon Metazachlor-Q	09676	Standon
Standon Propyzamide 400 SC	10255	Standon
Standon Rimsulfuron	09955	Standon

Product	MAFF No	Supplier
Stomp 400 SC	10229	BASF
Strate	10020	Aventis
Sumi-Alpha	BASF	BASF
Super Mosstox	09942	Aventis
Environ.		
Supertox 30	09946	Aventis
Environ.		
Surpass 5EC	09861	Interfarm
suSCon Indigo	09902	Fargro
T 25	ADJ0352	Intracrop
Talon Rat & Mouse Bait (Cut Wheat)	H6709	Killgerm
Talon Rat & Mouse Bait (Whole Wheat)	H6710	Killgerm
Tecto Superflowable Turf Fungicide	09528	Vitax
Temik 10G	10021	Aventis
Terpitz	09634	Me2
Tolkan Liquid	10023	Aventis
Tolugan Extra	09393	Makhteshim
Top Up G	ADJ0357	Greenhill
Torpedo II	ADJ0393	Newman
Totril	10026	Aventis
Tricur	10281	Bayer
Trifolex-Tra	BASF	BASF
Trump	BASF	BASF
Twist	10125	Novartis
Upgrade	10029	Aventis
Vitaflo Extra	07048	Uniroyal
Vitesse	10042	Aventis
Environ.		
Voodoo	09414	Sipcam
Warfarin 0.5% Concentrate	H6815	B H & B
Warfarin Ready Mixed Bait	H6816	B H & B
Xanadu	09943	Aventis
Environ.		
Zapper	09944	Aventis
Environ.		
Zenon	09193	Bayer

Products Deleted

The appearance of a product name in the following list may not necessarily mean that it is no longer available. In cases of doubt, refer to the supplier.

Product	MAFF No	Supplier
Actellifog	08078	Hortichem
Adagio	07832	PBI
Advizor	06571	DuPont
Afugan	07301	Promark
Agroxone 50	08345	Mirfield
Agrys	08083	Ciba Agric.
Aitken's Lawn Sand Plus	04542	Aitken
Alfacron 10 WP	02832	Ciba Agric.
Aliette	05648	Hortichem
Alpha Glyphogan	05784	Makhteshim

Product	MAFF No	Supplier
Alto 100 SL	05065	Sandoz
Alto 100 SL	08350	Novartis
Alto Combi	05066	Sandoz
Alto Combi	08465	Novartis
Alto Elite	05069	Sandoz
Alto Major	06979	Sandoz
Amazon TP	07681	Ciba Agric.
Amazon TP	08384	Novartis
Angle	08066	Ciba Agric.
Aplan	06121	Sandoz

Product	MAFF No	Supplier	Product	MAFF No	Supplier
Apron Combi FS	07203	Ciba Agric.	Deloxil	07313	AgrEvo
Arelon 500	08100	AgrEvo	Dipterex 80	00711	Bayer
Argylene	03386	Fargro	Disyston P 10	00715	Bayer
Armillatox	06234	Armillatox	Dithane 945	00719	PBI
Ashlade 700 5C	07046	Ashlade	Doff Agricultural Slug Killer	06058	Doff Portland
Ashlade 700 CCC	06473	Ashlade	with Animal Repellent		
Ashlade Mancarb FL	07977	Ashlade	Dosaflo	00754	Sandoz
Ashlade Mancarb Plus	08160	Ashlade	DUK 110	06266	DuPont
Ashlade Maneb Flowable	06477	Ashlade	DUK 51	06764	DuPont
Ashlade SMC Flowable	06494	Ashlade	DUK 880	04121	DuPont
Atlas 2,4-D	07699	Atlas	Early Impact	06659	Zeneca
Atlas MCPA	07717	Atlas	EF1166	07966	Dow
Atlas Protrum K	07723	Atlas	Enforcer	07866	Miracle
Atlas Tecgran 100	07730	Atlas	Euroagkem Taktic	ADJ0210	Euroagkem
Atlas Tecnazene 6% Dust	07731	Atlas	Exit	07632	Bayer
Atol	07347	Ashlade	Favour 600 SC	05842	Ciba Agric.
Aura 750 EC	05705	Ciba Agric.	Ferrax	05284	Bayer
Autumn Kite	07119	AgrEvo	Folio 575 SC	05843	Ciba Agric.
Avadex BW Granular	00174	Monsanto	Folio 575 SC	08406	Novartis
Barleyquat B	06001	Mandops	Fongarid 25 WP	03595	Ciba Agric.
Beret Gold	07531	Ciba Agric.	Fonofos Seed Treatment	06664	Zeneca
Bettaquat B	06004	Mandops	Forester	00914	Vitax
Bolda FL	07653	Atlas	Foundation	07465	Sandoz
Bolero	07436	Ciba Agric.	Fubol 58 WP	00927	Ciba Agric.
Boscor	07416	Ciba Agric.	Fubol 58 WP	08534	Novartis
Brasoran 50 WP	08394	Novartis	Fubol 75 WP	03462	Ciba Agric.
Bravo 720	09104	Zeneca	Fumite Lindane 10	00933	Hortichem
Bravocarb	09105	Zeneca	Fumite Lindane 40	00934	Hortichem
Bygran F	00365	Wheatley	Fumite Lindane Pellets	00937	Hortichem
Bygran S	00366	Wheatley	Fumite Propoxur Smoke	09290	Hortichem
Campbell's CMPP	02918	MTM Agrochem.	Fusarex Granules	06668	Zeneca
			Fusion	04908	DuPont
Campbell's Field Marshal	08594	MTM Agrochem.	Gamma-Col	06670	Zeneca
			Gesagard 50 WP	00981	Ciba Agric.
Campbell's Grassland Herbicide	06157	MTM Agrochem.	Gesaprim 500 SC	05845	Ciba Agric.
			Gesatop 500 SC	05846	Ciba Agric.
Camppex	08266	United Phosphorus	Glint 500 EC	04126	Ciba Agric.
			Glyfos	07109	Cheminova
Carbetamex	06186	RP Agric.	Graphic	07585	Ciba Agric.
Childion	03821	Hortichem	Hallmark	06434	Zeneca
Cirsium	08729	Mirfield	Halo	06520	Zeneca
Clortosip	06126	Sipcam	Harlequin 500 SC	05847	Ciba Agric.
Cogito	07384	Ciba Agric.	Harlequin 500 SC	08416	Novartis
Competitor	07310	AgrEvo	Hawk	08030	Ciba Agric.
Condox	06519	Zeneca	Headland Guard	ADJ0073	Headland
Coopex Maxi Smoke Generators	H5131	AgrEvo Environ.	Headland Kor Flo	08019	Headland
			Headland Quell	08317	Headland
Coopex Mini Smoke Generators	H5130	AgrEvo Environ.	Headland Quilt	ADJ0188	Headland
			Headland Redeem Flo	08340	Headland
Coopex WP	H5096	AgrEvo Environ.	Headland Zebra Flo	07442	Headland
Cudgel	06648	Zeneca	Headland Zebra WP	07441	Headland
Cymag	06651	Zeneca	Hispor 45 WP	01050	Ciba Agric.
Cyren	08358	Cheminova	Hostaquick	07326	AgrEvo
Daconil Turf	07929	Miracle	Hyquat 70	03364	Agrichem
Danadim Dimethoate 40	07351	Cheminova	Impact Excel	06680	Zeneca
Darlingtons Diazinon Granules	08400	Sylvan	Intracrop Balance	08037	Intracrop
			Intracrop BLA	ADJ0125	Intracrop
Darlingtons Dichlorvos	05699	Sylvan	Intracrop MCCC	08506	Intracrop
Decade 500 EC	05757	Ciba Agric.	Intrepid	07819	Miracle
Decisquick	07312	AgrEvo	ISK 375	09103	Zeneca
Decoy	06535	Bayer	Isoproturon 500	06718	AgrEvo

Product	MAFF No	Supplier	Product	MAFF No	Supplier
Joust	08122	Unicrop	Pacer	06690	Zeneca
Justice	07963	DuPont	Palette	06691	Zeneca
Karate Ready-to-Use Rat & Mouse Bait	05321	Lever Industrial	Patrol	08661	Zeneca
			Pirimor	06694	Zeneca
Karate Ready-to-Use Rodenticide Sachets	05890	Lever Industrial	Plover	07232	Ciba Agric.
			Pointer	06695	Zeneca
Karmex	01128	DuPont	Poise	08276	Unicrop
Kerb 50 W	02986	PBI	Polyram DF	08234	BASF
Kerb Flo	04521	PBI	Portman Mandate 80	06320	Portman
Kerb Pro Flo	08679	PBI	Portman Pirimicarb	06922	Portman
Killgerm Seconal	04715	Killgerm	Portman Sysdim 40	06902	Portman
Kor DF	08979	Headland	PP 375	07898	Zeneca
Kruga 5EC	08756	Headland	PP Captan 80 WG	06696	Zeneca
Landgold Triallate 480	08505	Landgold	Prebane SC	07634	Ciba Agric.
Lentagran WP	07556	Sandoz	Prophet 500 EC	07447	Ciba Agric.
Levington Octave	07505	Levington	Prospa	07899	Zeneca
Lo-Gran 20 WG	05993	Ciba Agric.	Pybuthrin 33	H5106	AgrEvo Environ.
Longlife Cleanrun 2	07851	Miracle			
Longlife Renovator 2	08379	Miracle	Quickstep	07389	Sandoz
Luxan Mancozeb Flowable	06812	Luxan	Quickstep	08557	Novartis
			Radar	06747	Zeneca
Mallard	07934	Ciba Agric.	Rampart	05166	Sipcam
Manex II	07637	Agrichem	Rapier	05314	MTM Agrochem.
Mantis 250 EC	06240	Ciba Agric.			
Mantle 425 EC	05715	Ciba Agric.	Recoil	04039	Sandoz
Manzate 200 PI	07209	DuPont	Regulex	07997	Zeneca
Mavrik Aquaflow	08347	Novartis	Reward 5EC	08757	Headland
MCC 25 EC	09115	Chiltern	Ridomil mbc 60 WP	01804	Ciba Agric.
Menace 80 EDF	09255	Headland	Ridomil Plus 50 WP	01803	Ciba Agric.
Metaphor	08017	United Phosphorus	Ripost Pepite	06485	Sandoz
			Roundup Amenity	08721	Monsanto
Methyl Bromide 100%	01336	Bromine & Chem.	Roundup Four 80	03176	Monsanto
			Roundup Pro	04146	Monsanto
Methyl Bromide 98	01335	Bromine & Chem.	Salvo	07092	Zeneca
			Sambarin 312.5 SC	05809	Ciba Agric.
Midstream	07739	Miracle	Sambarin 312.5 SC	08439	Novartis
Mirage Super 600EC	08531	Makhteshim	Sapecron 240 EC	01861	Ciba Agric.
Miros DF	04966	Sipcam	Sapecron 240 EC	08440	Novartis
Mistral	06943	Ciba Agric.	Selective Weedkiller	06579	Reabrook
Mithras 80 EDF	09219	Headland	Semeron	08441	Novartis
Moddus	07830	Ciba Agric.	Semeron 25WP	01916	Ciba Agric.
MON 240	04538	Monsanto	Sequel	07624	Promark
Moot	06990	Sandoz	Sheen	07828	Ciba Agric.
MSS 2,4-DP+MCPA	01396	Mirfield	SHL Turf Feed and Weed + Mosskiller	04438	Sinclair
MSS Chlormequat 70	03937	Mirfield			
MSS Ethosan	08558	Mirfield	Sipcam UK Rover 500	04165	Sipcam
MSS Mircam	01415	Mirfield	Skirmish	08079	Ciba Agric.
MSS Mircam Plus	01416	Mirfield	Skirmish	08444	Novartis
MSS Mircarb	08788	Mirfield	Sovereign	08152	Ciba Agric.
MSS Mirquat	08166	Mirfield	Sparkle 45 WP	04968	Ciba Agric.
MSS Optica	04973	Mirfield	Spraying Oil	ADJ9999	Hortichem
MTM Trifluralin	05313	MTM Agrochem.	Standon Pirimicarb H	05669	Standon
			Standon Spiroxamin 500	08916	Standon
Neporex 2SG	06985	Ciba Agric.	Standon Tridemorph 750	05667	Standon
Nimrod-T	07865	Miracle	Stefes Banlene Super	07691	Stefes
Novosol FC	06566	Ashlade	Stefes Cypermethrin 2	05719	Stefes
Nuvan 500 EC	03861	Ciba Agric.	Stefes Docklene Super	07696	Stefes
Octolan	06256	Sandoz	Stefes IPU 500	08102	Stefes
Opogard 500 SC	05850	Ciba Agric.	Stefes Legumex Extra	07841	Stefes
Option	07951	DuPont	Stefes Leyclene	08173	Stefes
Osprey 58 WP	05717	Ciba Agric.	Stefes Mancozeb DF	08010	Stefes

Product	MAFF No	Supplier	Product	MAFF No	Supplier
Stefes Mancozeb WP	07655	Stefes	Tripart Faber	05505	Tripart
Stefes Phenoxylene 50	07612	Stefes	Tripart Legion	06113	Tripart
Stefes Poraz	07528	Stefes	Tripart Mini Slug Pellets	02207	Tripart
Stefes Pride	05616	Stefes	Tripart Pugil	06153	Tripart
Stefes Toluron	05779	Stefes	Turfclear WDG	07490	Levington
Sting CT	04754	Monsanto	Twin	07374	Headland
Storite SS	08702	Banks	Unicrop Atrazine FL	08045	Unicrop
Strada	08824	Sipcam	Unicrop Fenitrothion 50	02267	Unicrop
Super-Tin 4L	02995	Chiltern	Unicrop Mancozeb 80	07451	Unicrop
Surpass 5EC	08758	Headland	Unicrop Maneb 80	06926	Unicrop
Swipe P	08150	Ciba Agric.	Unicrop Simazine FL	08032	Unicrop
Systhane 6 Flo	07334	Promark	Valiant	09251	Stefes
Teal	06117	Ciba Agric.	Vassgro Cypermethrin Insecticide	03240	Vass
Tecto Flowable Turf Fungicide	06273	Vitax			
Tern	07933	Ciba Agric.	Vassgro Non Ionic	ADJ0190	Vass
Thiovit	05572	Sandoz	Vassgro Spreader	ADJ0035	Vass
Tilt 250 EC	02138	Ciba Agric.	Walkover Mosskiller	04662	Allen
Tipoff	05878	Unicrop	Weed and Brushkiller (New Formulation)	07072	Vitax
Topas 100 EC	03231	Ciba Agric.			
Topas C 50 WP	03232	Ciba Agric.	Yaltox	02371	Bayer
Topik	07763	Ciba Agric.	Zolone Liquid	06206	Hortichem
Touchdown LA	07747	Miracle	Zulu	07829	Ciba Agric.

SECTION 1
INTRODUCTORY INFORMATION

About the UK Pesticide Guide

Purpose

The primary aim of this book is to provide a practical guide to what pesticides, plant growth regulators and adjuvants the farmer or grower can realistically and legally obtain in the UK, and to indicate the purposes for which they may be used. It is designed to help in the identification of products appropriate for a particular problem, and a Crop/Pest Guide is included to facilitate this. In addition to uses recommended on product labels, details are provided of those uses which do not appear on labels but which have been granted off-label approval. Such uses are not endorsed by the manufacturer, and are undertaken entirely at the risk of the user.

As well as identifying the products available, the book provides guidance on how to use them safely and effectively, but without giving details of doses, volumes, spray schedules or approved tank mixtures. Other sections provide essential background information on a whole range of pesticide-related issues including legislation, codes of practice, poisons and treatment of poisoning, products for use in special situations, and weed and crop growth stage keys. While we have tried to cover all other important factors, this book does **not** provide a full statement of product recommendations. **Before using any pesticide product it is essential that the user should read the label carefully and comply strictly with the instructions it contains**.

Scope

Individual profiles are provided of some 350 individual active ingredients, and a further 180 profiles of mixtures containing one or more of these active ingredients. Each profile has a list of available approved products for use in agriculture, horticulture (including amenity horticulture), forestry and areas near water, and the supplier from whom each may be obtained. Within these fields all types of pesticide covered by the Control of Pesticides Regulations or the Plant Protection Products Regulations are included. This embraces acaricides, algicides, fungicides, herbicides, insecticides, lumbricides, molluscicides, nematicides and rodenticides, together with plant growth regulators used as straw shorteners, sprout inhibitors and for various horticultural purposes. The total number of pesticide products covered in the *Guide* is about 1400.

Products are only included if requested by the supplier and supported by evidence of approval. This is intended to ensure that only marketed products are listed.

In addition, the *Guide* includes information on more than 100 authorised adjuvants which, although not themselves pesticides, may be added to pesticides to improve their effectiveness in use.

The *Guide* does **not** cover home garden, domestic, food storage, public health or animal health uses.

Sources of Information

The information for this edition has been drawn from these authoritative sources:
- approved labels and product manuals received from the suppliers of pesticides up to October 2000
- the MAFF/HSE publication *Pesticides 2000*
- entries in *The Pesticides Register* and *Pesticides Monitor*, published monthly by MAFF and HSE, listing new UK approvals (including off-label approvals), up to and including the issue for August/September 2000
- MAFF lists of approval expiries

Criteria for Inclusion

To be included in the Guide, a product must meet the following conditions:
- it must have extant MAFF/HSE Approval under UK pesticides legislation
- information on the approved uses must have been provided by the supplier
- the product must be expected to be on the UK market during the currency of this edition

Active ingredients that have been banned in the period since 1987 are listed, but no further details are given.

When a company changes its name, whether by merger, take-over or joint venture, it is obliged to re-register its products in the name of the new company, and new Ministry registration numbers are assigned. Where stocks of the previously registered products remain legally available, both old and new numbers are included in the *Guide* and remain until approval for the former lapses or stocks are notified as exhausted.

Products that have been withdrawn from the market and whose approval will finally lapse during 2001 are identified in each profile. After the indicated date, sale, storage or use of the product *bearing that approval number* becomes illegal. Where there is a direct replacement product, it is indicated.

How to Use the Guide

The book consists of five main sections:
- Introductory Information
- Adjuvants
- Crop/Pest Guide
- Pesticide Profiles
- Appendices

Introductory Information

This section summarises legislation covering approval, storage, sale and use of pesticides in the UK. Chemicals subject to the Poisons Laws are listed and there is a summary of first aid measures if pesticide poisoning should be suspected. The section also provides lists of products approved for use in or near water, in forestry and as seed treatments. Products specifically approved for aerial application are tabulated.

Adjuvants

Adjuvants are listed in a table ordered alphabetically by product name. For each product, details are shown of the main supplier, the authorisation number and the mode of action (e.g. wetter, sticker, etc.) as shown on the label. A brief statement of the uses of the adjuvant is given Protective clothing requirements and label precautions are listed using the codes from Appendix 4.

Crop/Pest Guide

This section (p. 71) enables the user to identify which active ingredients are approved for a particular crop/pest combination. The crops are grouped as shown in the Crop/Pest Guide Index. For convenience, some crops and targets have been grouped into generic units (for example, 'cereals' are handled as a group, not individually). Therefore indications from the Crop/Pest Guide must always be followed up by reference to the specific entry in the pesticide profiles section because a product may not be approved on all crops in the group. Chemicals indicated as having uses in cereals, for example, may only be approved for use on winter wheat and barley, not for other cereals. Because of differences in the wording of product labels it may sometimes be necessary to refer to broader categories of organism in addition to specific organisms (e.g. annual grasses and annual weeds in addition to annual meadow grass).

Pesticide Profiles

Each active ingredient has a separate numbered profile entry, as does each mixture of active ingredients. The entries are arranged in alphabetical order using the common names approved by the British Standards Institution, and used in *The Pesticide Manual*. Where an active ingredient is available only in mixtures, this is stated. The ingredients of the mixtures are themselves ordered alphabetically and the entries appear at the correct point for the first named ingredient.

Within each profile entry a table lists the products approved and available on the market, in the following style:

Product name	Main supplier	Active ingredient content	Formulation type	Registration No.*
6 Scotts Octave	Scotts	46% w/w	WP	09275
7 Sporgon 50 WP	Sylvan	46% w/w	WP	03829
8 Sportak 45 EW	AgrEvo	450 g/l	EW	07996

*Normally refers to registration with MAFF. In a few cases where products registered with the Health and Safety Executive are used in agriculture, HSE numbers are quoted, e.g. H0462

Many of the *product names* are registered trade-marks but no special indication of this is given. Individual products are indexed under their **entry number** in the Index of Proprietary Names at the back of the book. The *main supplier* indicates the marketing outlet through which the product may be purchased. Full addresses and telephone/fax numbers of suppliers are listed in Appendix I. Some website and e-mail addresses are also given. For mixtures, the *active ingredient contents* are given in the same order as they appear in the profile heading. The *formulation types* are listed in full in the Key to Abbreviations and Acronyms (Appendix 5).

The **Uses** section lists all approved uses (on-label and off-label) notified by suppliers to the Editor by October 2000, giving both the principal target organisms and the recommended crops or situations. Where there is an important condition of use, or the approval is off-label, this is shown in parentheses, e.g. *(off-label)*. Numbers in square brackets refer to the numbered products in the table above. Thus a typical use approved for Sporgon 50 WP (product 7) but not for Scotts Octave or Sportak 45 EW (products 6 and 8) appears as

Dry bubble in **mushrooms** [7]

Below the **Uses** paragraph, **Notes** are listed under the following headings. Unless otherwise stated any reference to dose is made in terms of product rather than active ingredient. Where notes refer to particular products, rather than to the entry generally, this is indicated by numbers in square brackets as described above.

Efficacy	Factors important in making the most effective use of the treatment. Certain factors, such as the need to apply chemicals uniformly, the need to spray at appropriate volume rates and the need to prevent settling out of active ingredient in spray tanks, are assumed to be important for all treatments and are not emphasised in individual profiles.
Crop safety/ Restrictions	Factors important in minimising the risk of crop damage including statutory conditions relating to the maximum permitted number of applications, or the maximum total dose. Recommended timing of treatment in relation to crop growth stage is included, with reference to standardised growth stage keys where possible (GS numbers – see Appendix 3). Where a statutory 'latest time of application' has been established this is given in the section 'latest application/Harvest interval (HI)'.
Special precautions/ Environmental safety	Notes are included in this section where products are subject to the Poisons Law, where Maximum Exposure Limits apply and where the label warns about organophosphorus and/or anticholinesterase compounds. Where the label specifies a Hazard Class this is noted, together with the associated risk phrases. Any other special operator precautions are specified.
	Where any of the products in the profile are subject to Category A or Category B buffer zone restrictions under the LERAP scheme the relevant information is given.
	Other environmental hazards are also noted here, including potential dangers to livestock, game, wildlife, bees and fish. The need to avoid drift onto neighbouring crops and to wash out equipment thoroughly after use are important with all pesticide treatments, but may receive special mention here if of particular significance.
Personal protective equipment/ Label precautions	The COSHH regulations require that wherever there is a label recommendation for use of personal protective equipment (PPE) it should be preceded by the phrase '*Engineering control of operator exposure must be used where reasonably practicable in addition to the following personal protective equipment:*' and followed by '*However, engineering controls may replace personal protective equipment if a COSHH assessment shows that they provide an equal or higher standard of protection.*'
	These phrases are not repeated in the profiles but an indication is given of the items of PPE which are mentioned on the labels of the various products. It is not possible to show the items required for different operations e.g. handling concentrate, spraying with hand lance, cleaning equipment etc. However, a

list of all the PPE mentioned on the label is shown as a series of letters on the first line of this section. These letter codes are explained in Appendix 4A.

Other label precautions not concerned with choice of PPE are given on the second line as a list of numbers referring to the key to standard phrases in Appendix 4B. **The letters referring to PPE and the numbered precautions are given for information only and should not be used for the purpose of making a COSHH assessment without reference to the current product label**.

Withholding period Any requirements relating to pesticides used on grazing land with potentially harmful effects on livestock are given here.

Latest application/ Harvest interval The 'Latest time of application' given in this section is laid down as a statutory condition of use for individual crops. Where quoted in terms of the period which must elapse between the last application and harvesting for human or animal consumption this 'Harvest Interval' is also given.

Approval Includes notes on approval for aerial or ULV application and 'Off-label' approvals, giving references to the official approval document numbers (OLA numbers), copies of which can be obtained from ADAS or NFU and must be consulted before the treatment is used.

Where an active ingredient has been accepted by the Brewers' and Licensed Retailers' Association (BLRA) for use on malting barley and/or hops, this is indicated here.

Where a product has been withdrawn by its manufacturer, and its approval will finally expire in 2001, the expiry date is shown here, together with the registration number of its direct replacement, if any.

Maximum residue level (MRL) If MRLs have been set for any/all of the active ingredients in the profile, they are listed here, or else a cross-reference is given to the "parent" profile where they may be found. All crops and foods for which an MRL has been set are listed. Products in the profile are not necessarily approved for use on these crops in the UK.

Definitions

The descriptions used in this *Guide* for the crops or situations in which products are approved for use are those used on the approved product labels. These can sometimes be ambiguous or imprecise, especially where simply limiting use to a certain crop is not relevant, leading to misunderstandings among advisors and users. To clarify matters for all concerned, the Pesticides Safety Directorate (PSD) has commenced a periodic issue of definitions drawn up after wide consultation with the industry. Further guidance can be obtained from PSD.

The definitions already published are:

Ornamental plant production: All ornamental plants that are grown for sale or are produced for replanting into their final growing position (e.g. *flowers, house plants, nursery stock, bulbs grown in containers or in the ground*).

Managed amenity turf: Frequently mown, intensively managed turf (e.g. *public parks, golf courses, sports fields*).

Amenity grassland: Areas of semi-natural or planted grassland with minimal management (e.g. *railway and motorway embankments, airfields and grassland nature reserves*). These areas may be managed for their botanical interest, and the relevant authority should be contacted before using pesticides in such locations.

Amenity vegetation: Areas of semi-natural or planted herbaceous plants, trees and shrubs.

Land not intended to bear vegetation: Soil or man-made surfaces where it is intended that no, or minimal, vegetation will be grown for several years (e.g. *pavements, tennis courts, industrial areas, railway ballast*). It does *not* include the land between rows of crops.

Green cover on land temporarily removed from production: Fields covered by natural regeneration or by a planted green cover crop which will not be consumed by humans or livestock, but which will be growing harvested crops in other years (e.g. *green cover on setaside*).

Forest nursery: Areas where young trees are raised outside for subsequent forest planting.

Forest: Groups of trees being grown in their final positions. Covers all woodland grown for whatever objective, including commercial timber production, amenity and recreation, conservation and landscaping, ancient traditional coppice and farm forestry, and trees from natural regeneration, colonisation or coppicing. Also includes restocking of established woodlands and new planting on both improved and unimproved land.

Farm forestry: Groups of trees established on arable land or improved grassland including those planted for short rotation coppicing.

Indoors (for rodenticide use)**:** Situations where the bait is placed within a building or other enclosed structure, and where the target is living or feeding predominantly within that building or structure.

These phrases will be phased into product labels over a period of time, and others will be published in due course.

Pesticide Legislation

Anyone who advertises, sells, supplies, stores or uses a pesticide is affected by legislation, including those who use pesticides in their own homes, gardens and allotments. There are numerous statutory controls but the major legal instruments are outlined below.

The Food and Environment Protection Act 1985 (FEPA) and Control of Pesticides Regulations 1986 (COPR)

FEPA introduced statutory powers to control pesticides with the aims of protecting human beings, creatures and plants, safeguarding the environment, ensuring safe, effective and humane methods of controlling pests and making pesticide information available to the public. Control of pesticides is achieved by COPR, which lay down the Approvals required before any pesticide may be sold, stored, supplied, advertised or used, and allow for the general requirements set out in various Consents, which specify the conditions subject to which approval is given. Consent A relates to advertisement, Consent B to sale, supply and storage, Consent C (i) to use and Consent C (ii) to aerial application of pesticides. The conditions of the Consents may be changed from time to time. Details are given in the MAFF/HSE Reference Book *Pesticides 200x* (revised annually) and are updated in the monthly publication *Pesticides Monitor*.

The controls currently in force include the following:

- Only approved products may be sold, supplied, stored, advertised or used
- Only products specifically approved for the purpose may be applied from the air
- A recognised Storeman's Certificate of Competence is required by anyone who stores for sale or supply pesticides approved for agricultural use
- A recognised Certificate of Competence is required by anyone who gives advice when selling or supplying pesticides approved for agricultural use. Users of pesticides must comply with the Conditions of Approval relating to use
- A recognised Certificate of Competence is required for all contractors and for persons born after 31 December 1964 applying pesticides approved for agricultural use (unless working under direct supervision of a certificate holder). Proposals are now in place that *every* user of agricultural pesticides will have to hold a Certificate of Competence regardless of age or supervision. A suitable transition period will allow this requirement to be enacted
- Only those adjuvants authorised by MAFF may be used (see Section 2).
- Regarding tank-mixes, 'no person shall combine or mix for use two or more pesticides which are anti-cholinesterase compounds unless the approved label of at least one of the pesticide products states that the mixture may be made; and no person shall combine or mix for use two or more pesticides if all the conditions of the approval relating to this use cannot be complied with'

The 'Authorisation' Directive

European Council Directive 91/414/EEC, known as the 'Authorisation' Directive, is intended to harmonise national arrangements for the authorisation of plant protection products within the European Union. It became effective on 25 July 1993. Under the provisions of the Directive, individual Member States are responsible for authorisation within their own territory of products containing active substances that appear in a list agreed at Community level. This list, known as Annex I, is being created over a period of time by review of existing active ingredients and authorisation of new ones.

The process of reviewing existing active ingredients is taking considerably longer than originally anticipated. Out of the original 'first list' of 90 substances, several have been withdrawn from the market and, by October 2000, eight substances had been included in Annex I. A second priority list of 148 active substances was published in February 2000, and initial arrangements for the third phase of the programme, covering the remaining 426 substances, was also published during 2000. New guidelines are being drawn up to expedite the process, but, even so, completion of the programme will be several years later than the original target date of 2003.

Individual Member States are amending their national arrangements and legislation in order to meet the requirements of Directive 91/414/EEC. In the UK this has been achieved by a series of Plant Protection Products Regulations (PPPR), under which, over a period of time, all agricultural and horticultural pesticides will come to be regulated. Meanwhile existing product approvals are being maintained under COPR, and new ones are granted for products containing active ingredients already on the market by 25 July 1993. Products containing new active substances not on the market at this date may be granted provisional approval under PPPR in advance of Annex I listing of their active ingredients. As active ingredients are placed on Annex 1 of the Directive, products containing only those active ingredients will be regulated solely under PPPR. Active ingredients in this Guide that have been included in Annex I are identified in the Approval section of the profile.

The Directive also provides for a system of mutual recognition of products registered in other Member States. Annex I listing, and a relevant approval in the Member State on which the mutual recognition is to be based, are essential pre-requisites. Informal discussions are also in progress to develop proposals for a 'voluntary mutual recognition scheme' for minor uses.

The Control of Substances Hazardous to Health Regulations 1988 (COSHH)

The COSHH regulations, which came into force on 1 October 1989, were made under the Health and Safety at Work Act 1974 and are also important as a means of regulating the use of pesticides. The regulations cover virtually all substances hazardous to health, including those pesticides classed as Very toxic, Toxic, Harmful, Irritant or Corrosive, other chemicals used in farming or industry and substances with occupational exposure limits. They also cover harmful micro-organisms, dusts and any other material, mixture or compound used at work which can harm people's health.

The original Regulations, together with all subsequent amendments, have been consolidated into a single set of regulations: The Control of Substances Hazardous to Health Regulations 1994 (COSHH 1994).

The basic principle underlying the COSHH regulations is that the risks associated with the use of any substance hazardous to health must be assessed before it is used and the appropriate measures taken to control the risk. The emphasis is changed from that pertaining under the Poisonous Substances in Agriculture Regulations 1984 (now repealed), whereby the principal method of ensuring safety was the use of protective clothing, to the prevention or control of exposure to hazardous substances by a combination of measures. In order of preference the measures should be:

a. substitution with a less hazardous chemical or product
b. technical or engineering controls (e.g. the use of closed handling systems etc.)
c. operational controls (e.g. operators located in cabs fitted with air-filtration systems etc.)
d. use of personal protective equipment (PPE), which includes protective clothing

Consideration must be given as to whether it is necessary to use a pesticide at all in a given situation and, if so, the product posing the least risk to humans, animals and the environment must be selected. Where other measures do not provide adequate control of exposure and the use of PPE is necessary, the items stipulated on the product label must be used as a minimum. It is essential that equipment is properly maintained and the correct procedures adopted. Where necessary, the exposure of workers must be monitored, health checks carried out and employees must be instructed and trained in precautionary techniques. Adequate records of all operations involving pesticide application must be made and retained for at least 3 years.

Certificates of Competence – the roles of BASIS and NPTC

COPR, COSHH and other legislation places certain obligations on those who handle and use pesticides. Minimum standards are laid down for the transport, storage and use of pesticides and the law requires those who act as storekeepers, sellers and advisors to hold recognised Certificates of Competence.

BASIS is an independent Registration Scheme for the pesticide industry, recognised under COPR. It is responsible for organising training courses and examinations to enable such staff to obtain a Certificate of Competence.

In addition, BASIS undertakes annual assessment of pesticide supply stores, enabling distributors, contractors and seedsmen to meet their obligations under the Code of Practice for Suppliers of Pesticides. Further information can be obtained from BASIS (see Appendix 2).
Certain spray operators also require Certificates of Competence under the Control of Pesticides Regulations. These certificates are awarded by the National Proficiency Tests Council (NPTC) to candidates who pass an assessment carried out by an approved NPTC or Scottish Skills Testing Service Assessor in one or more of a series of competence modules. Holders are required to produce their certificate on demand for inspection to any person authorised to enforce COPR. Further information can be obtained from NPTC (see Appendix 2).

Maximum Residue Levels

A small number of pesticides are liable to leave residues in foodstuffs, even when used correctly. Where residues can occur, statutory limits, known as Maximum Residue Levels (MRLs), have been established. MRLs provide a check that products have been used as directed; they are not safety limits. However, they do take account of consumer safety because they are set at levels that ensure that normal dietary intake of residues presents no risk to health. Wide safety margins are built in and eating food containing residues above the MRL does not automatically imply a risk to health. Nevertheless, it is an offence to put into circulation any produce where the MRL is exceeded.

The UK has set statutory MRLs since 1988. The European Union intends eventually to introduce MRLs for all pesticide/commodity combinations. These are being introduced initially by a series of priority lists but will subsequently be covered by the review programme under Directive 91/414.

MRLs apply to imported as well as to home-produced foodstuffs. Details of those that have been set have been published in *The Pesticides (Maximum Residue Levels in Crops, Food and Feeding Stuffs) Regulations 1994*, and successive amendments to these Regulations, increasing the number of MRLs from 5,000 to over 11,000. All these regulations have now been consolidated into the *Pesticides (Maximum Residue Levels in Crops, Foods and Feeding Stuffs) (England and Wales) Regulations 1999*, which came into force on 1 February 2000 and are available from the Stationery Office Publications Centre (Tel: 0870 600 5522). The consolidation of the Regulations will make the legislation more user friendly and accessible to those who need to ensure that MRLs are respected.

The MRLs applicable to the active ingredients included in this Guide are shown in the relevant pesticide profiles (Section 4). However, it is essential to consult the Regulations for full details and definitions, and for information on chemicals not currently marketed in Britain.

MRLs are shown by crop in the pesticide profiles (Section 4). All crops for which MRLs have been set are listed regardless of whether they are grown in UK or whether the products listed are approved for use on them. To make the information more manageable, crops have been grouped according to the structure used in the Regulations and reproduced below. Where the same MRL applies to all crops in a group, only the group name is given. Thus *pome fruits 0.5* indicates that a MRL of 0.5 mg/kg has been set for all crops in the pome fruit sub-group (apples, pears, quinces). *Apples, pears 0.5* indicates that either a MRL has not been set for quinces, or that a different level has been set and appears elsewhere in the list. *Fruits 0.5* indicates the level set for the entire fruits group.

Group	Sub-group	Crops/products
FRUITS	Citrus fruits	Grapefruit, lemons, limes, mandarins (including clementines & similar hybrids), oranges, pomelos, others
	Tree nuts	Almonds, brazil nuts, cashew nuts, chestnuts, coconuts, hazelnuts, macadamia nuts, pecans, pine nuts, pistachios, walnuts, others
	Pome fruits	Apples, pears, quinces, others
	Stone fruits	Apricots, cherries, peaches, nectarines, plums, others
	Berries & small fruit	Grapes, strawberrries, cane fruits (blackberries, loganberries, raspberries, others), bilberries, cranberries, currants, gooseberries, wild berries, others
	Miscellaneous fruit	Avocados, bananas, dates, figs, kiwi fruit, kumquats, litchis, mangoes, olives, passion fruit, pineapples, pomegranates, others
VEGETABLES	Root & tuber vegetables	Beetroot, carrots, celeriac, horseradish, Jerusalem artichokes, parsnips, parsley root, radishes, salsify, sweet potatoes, swedes, turnips, yams, others
	Bulb vegetables	Garlic, onions, shallots, spring onions, others
	Fruiting vegetables	Tomatoes, peppers, aubergines, cucumbers, gherkins, courgettes, melons, squashes, watermelons, sweet corn, others
	Brassica vegetables	Broccoli, cauliflowers, Brussels sprouts, head cabbages, Chinese cabbage, kale, kohlrabi, others
	Leafy vegetables/herbs	Lettuces (cress, lamb's lettuce, lettuce, scarole), spinach, beet leaves, watercress, witloof, herbs (chervil, chives, parsley, celery leaves), others
	Legume vegetables	beans (with pods), beans (without pods), peas (with pods), peas (without pods), others
	Stem vegetables	Asparagus, cardoons, celery, fennel, globe artichokes, leeks, rhubarb, others
	Fungi	wild and cultivated mushrooms
PULSES		Beans, lentils, peas, others
OILSEEDS		Linseed, peanuts, poppy seed, sesame seed, sunflower seed, rape seed, soya bean, mustard seed, cotton seed, others
POTATOES		Early potatoes, ware potatoes
TEA		
HOPS		
CEREALS		Wheat, rye, barley, oats, triticale, maize, rice, others
ANIMAL PRODUCTS		Meat, fat and preparations of meat, milk, dairy produce, eggs

Approval (On-label and Off-label)

Only officially approved pesticides may be marketed and used in the UK. Approvals are normally granted only in relation to individual products and for specified uses. It is an offence to use non-approved products or to use approved products in a manner that does not comply with the statutory conditions of use, except where the crop or situation is the subject of an off-label extension of use.

Statutory Conditions of Use

Statutory conditions have been laid down for the use of individual products and may include:

- field of use (e.g. agriculture, horticulture, etc.)
- crop or situations for which treatment is permitted
- maximum individual dose
- maximum number of treatments or the maximum total dose
- maximum area or quantity which may be treated
- latest time of application or harvest interval
- operator protection or training requirements
- environmental protection requirements
- any other specific restrictions relating to particular pesticides.

All products must now display these statutory conditions of use in a 'statutory box' on the label.

Types of Approval

There are three categories of approval that may be granted, under the Control of Pesticides Regulations, to products containing active ingredients that were already on the market by 25 July 1993:

- *Full* (granted for an unstipulated period)
- *Provisional* (granted for a stipulated period)
- *Experimental Permit* (granted for the purposes of testing and developing new products, formulations or uses). Products with only an Experimental Permit may not be advertised or sold and do not appear in this *Guide*.

Products containing new active substances, and those containing older ingredients once they have been listed in Annex I to Directive 91/414 (see previous section), will be granted approval under the Plant Protection Products Regulations. Again, there are three categories, similar to those listed above:

- *Standard approval* (granted for a period not exceeding ten years). Only applicable to products whose active substances are listed in Annex I.
- *Provisional approval* (granted for a period not exceeding three years, but renewable). Applicable where Annex I listing of the active substance is awaited.
- *Approval for research and development* to enable field experiments or tests to be carried out with active substances not otherwise approved.

The official list of approved products, including all the above categories except those for experimental purposes, is the MAFF/HSE Reference Book 500, *Pesticides 200x*. Details of new approvals, amendments to existing approvals and off-label approvals (see below) are published monthly in *Pesticides Monitor*.

Withdrawal of Approval

Product approvals may be reviewed, amended, suspended or revoked at any time. Revocation may occur for various reasons, such as commercial withdrawal or failure by the approval holder to meet data requirements. Where there are no safety concerns, a new approval will normally be issued for up to two years to allow the using up of stocks by persons other than the approval holder. This is known as the phased revocation procedure. Where safety considerations make it necessary, however, immediate revocation may occur.

Approval of Commodity Substances

Some chemicals have minor uses as pesticides but are predominantly used for non-pesticidal purposes. Approval is granted to such commodity substances for use only. They may not be sold, supplied, stored or advertised as pesticides unless specific approval has been granted. 18 such substances have been approved for certain specified uses as laid down in Annex D of *Pesticides 1998*. Of these, carbon dioxide, formaldehyde, methyl bromide, sodium chloride, sodium hypochlorite, strychnine hydrochloride and sulphuric acid are approved for agricultural or horticultural use and are included in this Guide.

Off-label Extension of Use

Products may legally be used in a manner not covered by the printed label in several ways:
- in accordance with the "Off-label Arrangements" (see below)
- in accordance with a specific off-label approval (SOLA). SOLAs are uses for which approval has been sought by individuals or organisations other than the manufacturers. The Notices of Approval are published by MAFF and are widely available from ADAS or NFU offices. Users of SOLAs must first obtain a copy of the relevant Notice of Approval and comply strictly with the conditions laid down therein
- in tank mixture with other approved pesticides in accordance with Consent C(i) made under FEPA. Full details of Consent C(i) are given in Annex A of *Pesticides 200x* but there are two essential requirements for tank mixes. Firstly, all the condititons of approval of all the components of a mixture must be complied with. Secondly, no person may mix or combine pesticides which are anticholinesterase compounds unless allowed by the label of at least one of the pesticides in the mixture
- in conjunction with authorised adjuvants
- in reduced spray volume under certain conditions
- the use of certain herbicides on specified set-aside areas subject to restrictions which differ between Scotland and the rest of the UK
- by mutual recognition of a use fully approved in another Member State of the European Union and authorised by PSD

Although approved, off-label uses are not endorsed by manufacturers and such treatments are made entirely at the risk of the user.

The Off-label Arrangements

Since 1 January 1990, arrangements have been in place allowing many approved products to be used for additional minor uses, without the need for Specific Off-Label Approvals (SOLAs). The arrangements are regularly reviewed to ensure they stay in line with current science and regulations. They were revised in December 1994, and again in December 1999, to come into force on 1 January 2000. They are now called *The Long Term Arrangements for Extension of Use (2000)*, and will remain in force until they are reviewed again, which must take place before the end of December 2004.

The scheme is founded on the principle that the risks to the operator, consumer, wildlife and the environment are acceptable and no greater than those which arise from the approved uses of a product, which appear on its label. All off-label approvals, including those covered by these arrangements, undergo a full safety assessment by the Pesticide Safety Directorate, in consultation with the Advisory Committee on Pesticides. There is, however, no test of efficacy, and **use of these extensions of use, as with any off-label use, is done at the user's choosing, and the commercial risk is entirely theirs.**

The arrangements are set out in full at Annex C of *Pesticides 200x*. *Readers should note that the December 1999 review resulted in a number of amendments to the arrangements. It is important that users are fully aware of these new arrangements before proceeding with pesticide uses that were permitted under the previous arrangements.* The following is a **summary** of the main points and uses.

Specific restrictions for extension of use

Certain restrictions are necessary to ensure that the extension of use does not increase the risk to the operator, the consumer or the environment:

- The arrangements apply to label and SOLA recommendations, but ONLY for the use of products approved for use as Agricultural and/or Horticultural pesticides.
- Safety precautions and statutory conditions relating to use (shown in the statutory box on the label) MUST be observed. If extrapolation from a SOLA is to be used, all conditions specified on the SOLA Notice of Approval MUST be observed in addition to the safety precautions and statutory conditions on the specified product's label.
- Pesticides must only be used in the same situation (i.e. outdoor or protected) as that specified on the product label or SOLA Notice. Approvals for use on tomatoes, cucumbers, lettuces, chrysanthemums and mushrooms include protected crops unless otherwise stated. Use on other protected crops may only occur when the product label specifically allows use under protection on the crop on which the extrapolation is based. Similarly, pesticides approved only for use in protected situations must not be used outdoors. **If a label or SOLA Notice does not specify a situation, then extrapolation only to an outdoor use is permitted.**
- The application method must be as stated on the product label and in accordance with relevant Codes of Practice and COSHH requirements. Where application of a pesticide under these arrangements is planned to be made by hand-held equipment users MUST ensure that hand-held use is appropriate for the current label or SOLA Notice conditions. **Note:** unless otherwise stated, spray applications to protected crops include hand-held uses.

In cases where hand-held application is not appropriate to the use on which extrapolation is to be made, hand-held application may nevertheless be permitted provided certain conditions specified in the arrangements are met (consult Annex C of *Pesticides 200x*). When planning to apply a pesticide under these arrangements by broadcast air-assisted sprayer, only those products with a specific label or SOLA recommendation for such use may be used. Any associated buffer zone restrictions must also be observed.

- Use in or near water (including coastal waters), or by aerial application, is not permitted under the arrangements.
- Rodenticides and other vertebrate control agents are not included, nor is use on land not intended for cropping.
- Pesticides classed as harmful, dangerous, extremely dangerous or high risk to bees must not be used off-label on any flowering crop. Where use is recommended on-label on flowering peas, cereals or oilseed rape this relates only to those crops. If the label or SOLA Notice specifies an aquatic buffer zone, users must conduct a LERAP assessment for the extension of use.
- The user of a pesticide under these arrangements must take all reasonable precautions to safeguard wildlife and the environment. The arrangements apply as follows:

I Non-edible crops and plants

Subject to the specific restrictions set out above:

- Pesticides approved for use on any growing crop may be used on commercial agricultural and horticultural holdings and forest nurseries on: (i) ornamental crops including hardy nursery stock, plants, bulbs, flowers and seed crops where neither the seed nor any part of the plant is to be consumed by humans or animals; (ii) forest nursery crops prior to final planting out.
- In addition, pesticides approved for use on any growing *edible* crop (except seed treatments) may be used on non-ornamental crops grown for seed subject to the same consumption restrictions as above (but not including seed crops of potatoes, cereals, oilseeds, peas, beans and other pulses).
- Pesticides (except seed treatments) approved for use on oilseed rape may be used on hemp grown for fibre.
- Herbicides approved for use on cereals, grass and maize may be used on commercial agricultural and horticultural holdings on *Miscanthus* spp., (Elephant grass), but not after the crop is 1 m high. The crops and its products must not be used for food or feed.

II Farm forestry and rotational cropping

Subject to the specific restrictions set out above:
- Herbicides approved for use on cereals may be used in the first 5 years of establishment in

farm forestry on land previously under arable cultivation or improved grassland, and reclaimed brownfield sites.
- Herbicides approved for use on cereals, oilseed rape, sugar beet, potatoes, peas and beans may be used in the first year of regrowth after cutting in coppices established on land previously under arable cultivation or improved grassland, or reclaimed brownfield sites.

III Nursery fruit crops

Subject to the specific restrictions set out above:

- Pesticides approved for use on any crop for human or animal consumption may be used on commercial agricultural and horticultural holdings on nursery fruit trees, nursery grape vines prior to final planting out, bushes, canes and non-fruiting strawberry plants provided any fruit harvested within 12 months is destroyed. Applications must NOT be made if fruit is present.

IV Hops

Subject to the specific restrictions set out above:

- On commercial and horticultural holdings pesticides may be used on mature stock or mother plants kept specifically for propagation, hop propagules prior to final planting out, and first year "nursery hops" in their final planting position but not taken to harvest in that year. Treated hops must not be harvested for human or animal consumption (including idling) within 12 months of treatment.

V Crops used partly or wholly for consumption by humans or livestock

Subject to the specific restrictions set out above:

- Pesticides may be used on commercial agricultural and horticultural holdings on crops in the first column below if they have been approved for use on the crop(s) opposite them in the second column. **However, it is the responsibility of the user to ensure that the proposed use does not result in any statutory UK MRL being exceeded. These extrapolations may NOT be used where the MRL for the crop in column 1 is set at the limit of determination, or is lower than the MRL for the crop in column 2, or where an MRL has been established for the crop in column 1 but not for the crop in column 2.**

Minor use	Crops on which use is approved	Additional special conditions
Arable crops		
Poppy (grown for oilseed), sesame	Sunflower	
Mustard, linseed, evening primrose, honesty	Oilseed rape	
Borage (grown for oilseed), canary flower e.g. *Echium vulgare, E. plantagineum* (grown for oilseed)	Oilseed rape	Seed treatments are not permitted
Rye, triticale	Barley	Treatments applied before first spikelet of inflorescence just visible
Rye, triticale	Wheat	
Grass seed crop	Grass for grazing or fodder	

Minor use	Crops on which use is approved	Additional special conditions
Grass seed crop	Wheat, barley, oats, rye, triticale	Treated crops must not be grazed or cut for fodder until 90 days after treatment. Seed treatments are not permitted. Use of chlormequat containing products is not permitted
Lupins	Combining peas or field beans	
Fruit crops		
Almond, chestnut, hazelnut, walnut	Apple, cherry or plum	For herbicides used on the orchard floor ONLY
Quince, crab apple	Apple or pear	
Almond, chestnut, hazelnut, walnut	Products approved for use on two of the following: almond, chestnut, hazelnut, walnut	
Nectarine, apricot	Peach	
Blackberry, dewberry, *Rubus* species (e.g. tayberry, loganberry)	Raspberry	
Whitecurrant, bilberry, cranberry	Blackcurrant or redcurrant	
Redcurrant	Blackcurrant	
Vegetable crops		
Parsley root	Carrot or radish	
Fodder beet, mangel	Sugar beet	
Horseradish	Carrot or radish	
Parsnip	Carrot	
Salsify	Carrot or celeriac	
Swede	Turnip	
Turnip	Swede	
Garlic, shallot	Bulb onion	
Aubergine	Tomato	
Squash, pumpkin, marrow, watermelon	Melon	
Broccoli	Calabrese	
Calabrese	Broccoli	
Roscoff cauliflower	Cauliflower	
Collards	Kale	
Lamb's lettuce, frisée/frise, radicchio, cress, scarole	Lettuce	
Leaf herbs and edible flowers*	Lettuce or spinach or parsley or sage or mint or tarragon	

* This extension of use applies to the following leaf herbs and edible flowers: angelica, balm, basil, bay, borage, burnet (salad), caraway, camomile, chervil, chives, clary, coriander, dill, fennel, fenugreek, feverfew, hyssop, land cress, lovage, marjoram, marigold, mint, nasturtium, nettle, oregano, parsley, rocket, rosemary, rue, sage, savory, sorrel, tarragon, thyme, verbena (lemon), woodruff.

Minor use	Crops on which use is approved	Additional special conditions
Beet leaves, red chard, white chard, yellow chard	Spinach	
Edible podded peas (e.g. mange-tout, sugar snap)	Edible podded beans	
Runner beans	Dwarf French beans	
Rhubarb, cardoon	Celery	
Edible fungi other than mushroom (e.g. oyster mushroom)	Mushroom	

For applications in store on crops PARTLY OR WHOLLY FOR HUMAN OR ANIMAL CONSUMPTION, the following extensions of use apply:

Minor use	Crops on which use is approved
Rye, barley, oats, buckwheat, millet, sorghum, triticale	Wheat
Dried peas	Dried beans
Dried beans	Dried peas
Mustard, sunflower, honesty, sesame, evening primrose, poppy (grown for oilseed), borage (grown for oilseed), canary flower e.g. *Echium vulgare/E. plantaginium* (grown for oilseed)	Oilseed rape

VI Clarifications

Under these arrangements the following crops are considered synonymous or equivalent and as such, uses on crops in Column 1 can be read across to uses in Column 2

Column 1	Column 2: equivalent
Hazelnut	Cobnuts, filberts
French bean	Navy bean
Vining pea	Picking pea, shelling pea, non-edible podded pea
Linseed	Linola, flax
Wheat	Durum wheat

Guidance on the Use of Pesticides

The information given in the *UK Pesticide Guide* provides some of the answers needed to assess health risks, including the hazard classification and the level of operator protection required. However, the Guide cannot provide all the details needed for a complete hazard assessment, which must be based on the product label itself and, where necessary, the Health and Safety Data Sheet and other official literature.

Detailed guidance on how to comply with the Regulations is available from several sources:

1. *Pesticides: Code of practice for the safe use of pesticides on farms and holdings*, 1998. (MAFF Publication PB3528)

The Code of Practice is available from MAFF Publications and has been substantially revised and updated from the previous (1990) edition. It gives comprehensive guidance on how to comply with COPR and the COSHH regulations under the headings: *User training and certification, Planning and preparation, Working with pesticides, Disposal of pesticide waste* and *Keeping records*. In particular it details what is involved in making a COSHH Assessment. The principal source of information for making such an assessment is the approved product label. In most cases the label provides all the necessary information but in certain circumstances other sources must be consulted and these are listed in the Code of Practice.

2. Other codes of practice:
* *Code of Practice for the Use of Approved Pesticides in Amenity & Industrial Areas* (ISBN 1 871140 12 9)
* *The Safe Use of Pesticides for Non-Agricultural Purposes – Approved Code of Practice* (ISBN 0 11 885673 1)
* *Code of Practice for Suppliers of Pesticides to Agriculture, Horticulture and Forestry* (MAFF Booklet PB0091)
* *Code of Good Agricultural Practice for the Protection of Soil* (MAFF Booklet PB0617)
* *Code of Good Agricultural Practice for the Protection of Water* (MAFF Booklet PB0587)
* *Code of Good Agricultural Practice for the Protection of Air* (MAFF Booklet PB0618)

3. Other guidance and practical advice:

HSE (from HSE Books – see Appendix 2)
* *A step by step guide to COSHH assessment*, HS(G) 97 (ISBN 0 11 886 379 7)
* *Food and Environment Protection Act 1985/Control of Pesticides Regulations 1986. An open learning course* (ISBN 0 11 8857 43 6)
* *COSHH in agriculture*, AS 28
* *COSHH in forestry*, AS 30
* *Occupational Exposure Limits 1995*, Guidance Note EH 40/95 (ISBN 0 7176 0876 X
* *Biological monitoring of workers exposed to organophosphorus pesticides*, Guidance Note MS 17 (ISBN 0 11 883951 9)

MAFF (from The Stationery Office – see Appendix 2)

* *Pesticides 200x* (published annually)
* *Pesticides Monitor* (monthly) (ISSN 1466-5670)
* *Local Environment Risk Assessments for Pesticides – A Practical Guide* (PB4168)

Crop Protection Association (see Appendix 2)

* *The plain man's guide to pesticides and the COSHH assessment*
* *Amenity Handbook: A guide to the selection and use of amenity pesticides*
* *Pesticides and Water Quality*
* *Sprayer Cleaning*
* *Container Cleaning*
* *Agrochemical Disposal*
* *Hand Protection*
* *Good Agrochemical Storage*
* *Avoiding drift*

British Crop Protection Council (see Appendix 2)

- *The Pesticide Manual* (12th edition) (ISBN 1 901396 12 6)
- *The e-Pesticide Manual* (ISBN 1 901396 23 1)
- *Hand-held and Amenity Sprayers Handbook* (ISBN 1 901396 03 7)
- *Boom and Fruit Sprayers Handbook* (ISBN 1 901396 02 9)
- *Using Pesticides: A Complete Guide to Safe, Effective Spraying* (ISBN 1 901396 01 0)

The Environment Agency (see Appendix 2)

- *The Prevention of Pollution of Controlled Waters by Pesticides* (Leaflet PPG9)
- *The Use of Herbicides in or near Water*
- *Introduction of the Groundwater Regulations*

Other useful references

- *Agrochemicals Accepted by the Brewers' and Licensed Retailers' Association for Use on Barley and Hops* (BLRA Technical Circular No 349, July 2000)

Protection of Bees

Honey Bees

Honey bees are a source of income for their owners and important to farmers and growers as pollinators of their crops. It is irresponsible and unnecessary to use pesticides in such a way that may endanger them. Pesticides vary in their toxicity to bees, but those that present a special hazard are classed as "harmful", "dangerous" or "extremely dangerous" to bees. Products so classified carry a specific warning in the precautions section of the label. These are indicated in this *Guide* in the **personal protective equipment/label precautions** section of the pesticide profile by the numbers E12b, E12c, E12d or E12e, depending on which phrase applies. The phrases are stated in full in Appendix 4.

In July 1996 a new classification was introduced that more accurately reflects the actual risk to honey bees when a product is used. Products assessed by the Pesticides Safety Directorate as posing such a risk are classified as "High risk to bees". Such products carry a new warning in the precautions section of the label, indicated in this Guide by the number E12a.

Where use of a product that is hazardous to bees is contemplated, there are simple guidelines to follow, all of which will be stated on the product label. Most important of these are:

- give local beekeepers as much warning of your intention as possible
- avoid spraying crops in flower or where bees are actively foraging
- keep down flowering weeds

The British Beekeepers' Association (see Appendix 2) can give further advice on protecting honey bees from pesticides.

Wild Bees

Wild bees also play an important role. Bumblebees are useful pollinators of spring flowering crops and fruit trees because they forage in cool, dull weather when honey bees are inactive. They play a particularly important part in pollinating field beans, red and white clover, lucerne and borage. Bumblebees nest and overwinter in field margins and woodland edges. Avoidance of direct or indirect spray contamination of these areas, in addition to the above measures, therefore helps the survival of these valuable insects.

The Campaign Against Illegal Poisoning of Wildlife

The Campaign Against Illegal Poisoning of Wildlife, aimed at protecting some of Britain's rarest birds of prey and wildlife whilst also safeguarding domestic animals, was launched in March 1991 by the Ministry of Agriculture, Fisheries and Food and the Department of the Environment, Transport and the Regions. The main objective is to deter those who may be considering using pesticides illegally.

The Campaign is supported by a range of organisations associated with animal welfare, nature preservation, field sports and game keeping including the RSPB, English Nature, the British Field Sports Society and the Game Conservancy Trust.

The three objectives are:

- To advise farmers, gamekeepers and other land managers on legal ways of controlling pests;
- To advise the public on how to report illegal poisoning incidents and to respect the need for legal alternatives;
- To investigate incidents and prosecute offenders.

A freephone number (0800 321 600) is available to make it easier for the public to report incidents and numerous leaflets, posters, postcards and stickers have been created to publicise the existence of the Campaign. A video has also been produced to illustrate the many talks, demonstrations and exhibitions presented by ADAS Consulting Ltd, on behalf of MAFF.

The Campaign arose from the results of the MAFF Wildlife Incident Investigation Scheme for the investigation of possible cases of illegal poisoning. Under this scheme, all reported incidents are considered and thoroughly investigated where appropriate. Enforcement action is taken wherever sufficient evidence of an offence can be obtained and numerous prosecutions have been made since the start of the Campaign.

Further information about the Campaign is available from: Pesticides Safety Directorate, Research Co-ordination and Environmental Policy Branch, Room 308, Mallard House, 3 Peasholme Green, York YO1 7PX, or from the website: www.pesticides.gov.uk

Chemicals Subject to the Poisons Law

Certain products in this book are subject to the provisions of the Poisons Act 1972, the Poisons List Order 1982 and the Poisons Rules 1982 (copies of all these are obtainable from The Stationery Office). These Rules include general and specific provisions for the storage and sale and supply of listed non-medicine poisons.

The chemicals approved for use in the UK and included in this book are specified under Parts I and II of the Poisons List as follows:

Part I Poisons (sale restricted to registered retail pharmacists and to registered non-pharmacy businesses provided sales do not take place on retail premises)

aluminium phosphide
chloropicrin
magnesium phosphide

methyl bromide
strychnine

Part II Poisons (sale restricted to registered retail pharmacists and listed sellers registered with a local authority)

aldicarb
alphachloralose
carbofuran (a)
chlorfenvinphos (a,b)
demeton-S-methyl
dichlorvos (a, e)
disulfoton (a)
endosulfan
fentin acetate
fentin hydroxide

fonofos (a)
formaldehyde
methomyl (f)
nicotine (c)
oxamyl (a)
paraquat (d)
phorate (a)
sulphuric acid
zinc phosphide

(a) Granular formulations containing up to 12% w/w of this, or a combination of similarly flagged poisons, are exempt
(b Treatments on seeds are exempt
(c) Formulations containing not more than 7.5% of nicotine are exempt
(d) Pellets containing not more than 5% paraquat ion are exempt
(e) Preparations in aerosol dispensers containing not more than 1% a.i. are exempt. Materials impregnated with dichlorvos for slow release are exempt
(f) Solid substances containing not more than 1% w/w a.i. are exempt

Occupational Exposure Limits

A fundamental requirement of the COSHH Regulations is that exposure of employees to substances hazardous to health should be prevented or adequately controlled. Exposure by inhalation is usually the main hazard, and in order to measure the adequacy of control of exposure by this route various substances have been assigned occupational exposure limits.

There are two types of occupational exposure limits defined under COSHH: Occupational Exposure Standards (OES) and Maximum Exposure Limits (MEL). The key difference is that an OES is set at a level at which there is no indication of risk to health; for a MEL a residual risk may exist and the level takes socio-economic factors into account. In practice, MELs have been most often allocated to carcinogens and to other substances for which no threshold of effect can be identified and for which there is no doubt about the seriousness of the effects of exposure.

OESs and MELs are set on the recommendations of the Advisory Committee on Toxic Substances (ACTS). Full details are published by HSE in *EH 40/95 Occupational Exposure Limits 1995* (ISBN 0 7176 0876 X).

As far as pesticides are concerned, OESs and MELs have been set for relatively few active ingredients. This is because pesticide products usually contain other substances in their formulation, including solvents, which may have their own OES/MEL. In practice inhalation of solvent may be at least, or more, important than that of the active ingredient. These factors are taken into account in the approval process of the pesticide product under the Control of Pesticide Regulations. This indicates one of the reasons why a change of pesticide formulation usually necessitates a new approval assessment under COPR.

Poisoning by Pesticides – First Aid Measures

If pesticides are handled in accordance with the required safety precautions, as given on the container label, poisoning should not occur. It is difficult however, to guard completely against the occasional accidental exposure. Thus, if a person handling, or exposed to, pesticides becomes ill, it is a wise precaution to apply first aid measures appropriate to pesticide poisoning even though the cause of illness may eventually prove to have been quite different. An employer has a legal duty to make adequate first aid provision for employees. Regular pesticide users should consider appointing a trained first aider, even if numbers of employees are not large, since there is a specific hazard.

The first essential in a case of suspected poisoning is for the person involved to stop work, to be moved away from any area of possible contamination and for a doctor to be called at once. If no doctor is available the patient should be taken to hospital as quickly as possible. In either event it is most important that the name of the chemical being used should be recorded and preferably the whole product label or leaflet should be shown to the doctor or hospital concerned.

Some pesticides which are unlikely to cause poisoning in normal use are extremely toxic if swallowed accidentally or deliberately. In such cases get the patient to hospital as quickly as possible, with all the information you have.

General Measures

Measures appropriate in all cases of suspected poisoning include:

- Remove any protective or other contaminated clothing (taking care to avoid personal contamination)
- Wash any contaminated areas carefully with water or with soap and water if available
- In cases of eye contamination, flush with plenty of clean water for at least 15 min
- Lay the patient down, keep at rest and under shelter. Cover with one clean blanket or coat etc. Avoid overheating
- Monitor level of consciousness, breathing and pulse-rate
- If consciousness is lost, place the casualty in the recovery position (on his/her side with head down and tongue forward to prevent inhalation of vomit)
- If breathing ceases or weakens commence mouth to mouth resuscitation. Ensure that the mouth is clear of obstructions such as false teeth, that the breathing passages are clear, and that tight clothing around the neck, chest and waist has been loosened. If a poisonous chemical has been swallowed, it is essential that the first aider is protected by the use of a resuscitation device (several types are available on the market)

Specific measures

In case of poisoning with particular chemical groups, the following measures may be taken before transfer to hospital:

Dinitro compounds* (dinoseb, DNOC, dinocap etc.)
Keep the patient at rest and cool.

Organophosphorus and carbamate insecticides
(aldicarb, azinphos-methyl, benfuracarb, carbofuran, chlorfenvinphos, chlorpyrifos, dichlorvos, heptenophos, malathion, methiocarb, mevinphos, propoxur, quinalphos, thiometon, triazophos etc.)
Keep the patient at rest. The patient may suddenly stop breathing so be ready to give artificial respiration.

Organochlorine compounds (aldrin†, dienochlor, endosulfan, gamma-HCH)
If convulsions occur, do not interfere unless the patient is in danger of injury; if so any restraint

---

*Approvals for use of dinoseb, dinoseb acetate. dinoseb amine, dinoterb, and binapacryl were revoked on 22 Jan 1988. Approval for storage of these materials was withdrawn on 30 June 1988. Approvals for supply of DNOC were revoked in Dec 1988 and for its use in Dec 1989.
†Approvals for supply, storage and use of aldrin were revoked in May 1989.

must be gentle. When convulsions cease, place the casualty on his or her side with head down, tongue forward (recovery position).

Paraquat, diquat

Irrigate skin and eye splashes copiously with water. If any chemical has been swallowed, take to hospital for tests.

Cyanide (including sodium cyanide)

Send for medical aid. Remove casualty to fresh air, if necessary using breathing apparatus and protective clothing. Remove casualty's contaminated clothing. Gently brush solid particles from the skin, making sure you protect your own skin from contamination. Wash the skin and eyes copiously with water. Transfer casualty to nearest accident and emergency hospital by the quickest possible means together with the first aid cyanide antidote, if held on the premises.

Reporting of Pesticide Poisoning

Any cases of poisoning by pesticides must be reported without delay to an HM Agricultural Inspector of the Health and Safety Executive. In addition any cases of poisoning by substances named in schedule 2 of The Reporting of Injuries, Diseases and Dangerous Occurrences Regulations 1985, must also be reported to HM Agricultural Inspectorate (this includes organophosphorus chemicals, mercury and some fumigants).

Cases of pesticide poisoning should also be reported to the manufacturer concerned.

Additional information

General advice on the safe use of pesticides is given in a range of Health and Safety Executive leaflets available from HSE Books (see Appendix 2)

A useful booklet produced by GCPF (Global Crop Protection Federation) titled 'Guidelines for emergency measures in cases of pesticide poisoning' is available from GCPF, Avenue Louise 143, B-1050 Brussels, Belgium.

The major agrochemical companies are able to provide authoritative medical advice about their own pesticide products. Detailed advice is also available to doctors from The National Poisons Information Service, New Cross Hospital, London SE14 5ER (020 7635 9191) and from regional centres (see Appendix 2).

Agricultural Pesticides and Water

The Food and Environment Protection Act 1985 (FEPA) places a special obligation on users of pesticides to "safeguard the environment and in particular avoid the pollution of water". Under the Water Resources Act 1991 it is an offence to pollute any controlled waters (watercourses or groundwater), either deliberately or accidentally. Protection of controlled waters from pollution is the responsibility of the Environment Agency (EA).

Users of pesticides therefore have a duty to adopt responsible working practices and, unless they are applying herbicides in or near water, to prevent them getting into water. Guidance on how to achieve this is given in the MAFF Code of Good Agricultural Practice for the Protection of Water.

The duty of care covers not only the way in which a pesticide is sprayed, but also its storage, preparation and disposal of surplus, sprayer washings and the container. Products in this Guide that are a major hazard to fish, other aquatic life or aquatic higher plants carry one of several specific label precautions in their profile, depending on the assessed hazard level.

Advice on any water pollution problems is available from the Environment Agency and the Crop Protection Association (see Appendix 2).

Protecting Surface Waters

Surface waters are particularly vulnerable to contamination. One of the best ways of preventing those pesticides that carry the greatest risk to aquatic wildlife from reaching surface waters is to prohibit their application within a boundary adjacent to the water. Such areas are known as no-spray, or buffer, zones. Certain products in this Guide are restricted in this way and have a legally binding label precaution to make sure the potential exposure of aquatic organisms to pesticides that might harm them is minimised.

Before 1999 the protected zones were measured from the edge of the water. The distances were 2 metres for hand-held or knapsack sprayers, 6 metres for ground crop sprayers, and a variable distance (but often 18 metres) for broadcast air-assisted applications, such as in orchards. The introduction of LERAPs (see below) has changed the method of measuring buffer zones.

"Surface water" includes lakes, ponds, reservoirs, streams, rivers and watercourses (natural or artificial). It also includes temporarily or seasonally dry ditches, which have the potential to carry water at different times of the year.

Buffer zone restrictions do not necessarily apply to all products containing the same active ingredient. Those in formulations that are not likely to contaminate surface water through spray drift do not pose the same risk to aquatic life and are not subject to the restrictions.

Local Environmental Risk Assessments for Pesticides (LERAPs)

Local Environment Risk Assessments for Pesticides (LERAPs) were introduced in March 1999. They give the user of most products currently subject to a buffer zone restriction the option of continuing to comply with the existing buffer zone restriction (using the new method of measurement), or carrying out a LERAP and possibly reducing the size of the buffer zone as a result. In either case there is a new legal obligation for the user to record his decision, including the results of the LERAP.

The scheme has changed the method of measuring the buffer zone. Previously the zone was measured from the edge of the water, but it is now the distance from the top of the bank of the watercourse to the edge of the spray area.

The LERAP provides a mechanism for taking into account other factors that may reduce the risk, such as dose reduction, the use of low drift nozzles, and whether the watercourse is dry or flowing. The previous arrangements applied the same restriction regardless of whether there was actually water present. Now there is a new standard zone of 1 metre from the top of a dry ditch bank.

Other factors to include in a LERAP that may allow a reduction in the buffer zone are:

- The size of the watercourse, because the wider it is, the greater the dilution factor, and the lower the risk of serious pollution.
- The dose applied. The lower the dose, the less is the risk.
- The application equipment. Sprayers and nozzles are to be star-rated over coming months according to their ability to reduce spray drift fallout. Equipment offering the greatest reductions will achieve the highest rating of three stars. Not all sprayers are included; the scheme is restricted to ground crop sprayers only. It does not yet apply to users of orchard or hop sprayers, but proposals to extend the scheme to broadcast air-assisted spray equipment are being considered.

Not all products that had a label buffer zone restriction are included in the LERAP scheme. The option to reduce the buffer zone does not apply to organophosphorus or synthetic pyrethroid insecticides. These groups are classified as Category A products. All other products that had a label buffer zone restriction are classified as Category B. The wording of the buffer zone label precautions has been amended for all products to take account of the new method of measurement, and whether or not the particular product qualifies for inclusion in the LERAP scheme.

Products in this Guide that are in Category A or B are identified in the **Special Precautions/ Environmental Safety** section of the profile. Updates to the list are published regularly by PSD and details can be obtained from their website: *www.pesticides.gov.uk*

The introduction of LERAPs is an important step forward because it demonstrates a willingness to reduce the impact of regulation on users of pesticides by allowing flexibility where local conditions make it safe to do so. This places a legal responsibility on the user to ensure the risk assessment is done either by himself or by the spray operator or by a professional consultant or advisor. It is compulsory to record the LERAP and make it available for inspection by enforcement authorities.

Groundwater Regulations

New Groundwater Regulations were introduced in 1999 to complete the implementation in the UK of the EU Groundwater Directive (Protection of Groundwater Against Pollution Caused by Certain Dangerous Substances - 80/68/EEC). These Regulations help prevent the pollution of groundwater by controlling discharges or disposal of certain substances, including all pesticides.

Groundwater is defined under the Regulations as any water contained in the ground below the water table. Pesticides must not enter groundwater unless it is deemed by the appropriate Agency to be permanently unsuitable for other uses.

The Regulations make it a criminal offence to dispose of pesticides onto land without official authorisation from the Environment Agency (in England and Wales) or the Scottish Environmental Protection Agency (SEPA). Normal use of a pesticide in accordance with product approval does not require authorisation. This includes spraying the washings and rinsings back on the crop provided that, in so doing, the maximum approved dose for that product on that crop is not exceeded.

The Agencies will review all authorisations within a four year interval. It may also grant authorisations for a limited period. The Agencies can serve notice at any time to modify the conditions of an authorisation where necessary to prevent pollution of groundwater.

In practice the best advice to farmers and growers is to plan to use all diluted spray within the crop and to dispose of all washings via the same route making sure that they stay within the conditions of approval of the product.

Use of Herbicides in or near Water

Products in this Guide approved for use in or near water are listed below. Before use of any product in or near water the appropriate water regulatory body (Environment Agency/Local Rivers Purification Authority or, in Scotland, the Scottish Environment Protection Agency) must be consulted. Guidance and definitions of the situation covered by approved labels are given in the MAFF publication *Guidelines for the Use of Herbicides on Weeds in or near Watercourses and Lakes*. Always read the label before use.

Chemical	Product	Weeds controlled
asulam	Asulox	Docks, bracken
2,4-D	Dormone MSS 2,4-D Amine	Water weeds, dicotyledon weeds on banks
dichlobenil	Casoron G Luxan Dichlobenil Granules Sierraron G	Aquatic weeds
diquat	Midstream	Aquatic weeds
fosamine-ammonium	Krenite	Woody weeds
glyphosate	Barclay Gallup 360 Barclay Gallup Amenity Barclay Gallup Biograde 360 Barclay Gallup Biograde Amenity Buggy SG Clinic Glyfos Glyfos Pro-Active Glyper Helosate Helosate MSS MSS Glyfield Rival Roundup Biactive Roundup Biactive Dry Roundup Pro Biactive Spasor Spasor Biactive	Reeds, rushes, sedges, waterlilies, grass weeds on banks
maleic hydrazide	Regulox K	Suppression of grass growth on banks
terbutryn	Clarosan	Aquatic weeds

Use of Pesticides in Forestry

The table below lists those products that have approved uses in forestry.

Chemical	Product	Use
aluminium phosphide	Phostek	Mole and rabbit control
aluminium ammonium sulphate	Guardsman B Guardsman M Liquid Curb Crop Spray	Reduction of damage by deer, hares, rabbits, birds
ammonium sulphamate	Amcide Root-Out	Rhododendron and other woody weed control
asulam	Asulox I T Asulam	Bracken control
atrazine	Atlas Atrazine Atrazol	Weed control in conifers
azaconazole + imazalil	Nectec Paste	Canker, silver leaf
carbosulfan	Marshal/suSCon	Weevil control
chlorpyrifos	Barclay Clinch II Choir Dursban 4 Greencrop Pontoon Lorsban T	Weevil and beetle control in transplant lines and cut logs
copper hydroxide	Spin Out	Root control in forest nursery containers
cycloxydim	Greencrop Valentia Standon Cycloxydim	Annual grasses
2,4-D	Dicotox Extra Easel MSS 2,4-D Ester	Perennial and woody weed control Perennial and woody weed control
2,4-D + dicamba + triclopyr	Nufarm Nu-Shot	Perennial and woody weed control
diflubenzuron	Dimilin 25-WP Dimilin Flo	Caterpillar control on forest trees
diquat + paraquat	PDQ	Pre-planting weed control, pre-emergence control in nurseries, directed sprays or ring-weeding in plantations. Firebreak maintenance
fluazifop-P-butyl	Fusilade 250 EW	Perennial grass weed control

Chemical	Product	Use
fosamine-ammonium	Krenite	Woody weed control in forestry land (not conifer plantations)
glufosinate-ammonium	Aventis Challenge Aventis Harvest Challenge Harvest	Control of annual and perennial weeds (directed sprays)
glyphosate	Agriguard Glyphosate 360 Barbarian Barclay Barbarian Barclay Gallup Barclay Gallup 360 Barclay Gallup Amenity Barclay Gallup Biograde 360 Barclay Gallup Biograde Amenity Barclay Garryowen Buggy SG Clinic Glyfos ProActive Glyper Greencrop Gypsy Helosate Hilite MSS Glyfield Portman Glyphosate 360 Portman Glyphosate Rival Roundup Pro Biactive Stefes Glyphosate Stefes Kickdown Stirrup	Control of annual, perennial and woody weeds (directed sprays and wiper application). Chemical thinning
indolylbutyric acid	Synergol	Rooting cuttings
isoxaben	Flexidor 125 Gallery 125	Pre-emergence weed control in forestry transplants, second year undercuts, forest and woodland plantings
paraquat	Barclay Total Gramoxone 100	Pre-planting weed control, pre-emergence control in nurseries directed sprays or ring-weeding in plantations. Firebreak maintenance
pirimicarb	Phantom	Control of aphids
propaquizafop	Falcon Landgold PQF 100 Standon Propaquizafop	Control of annual and perennial grass weeds

Chemical	Product	Use
propyzamide	Barclay Piza 400 FL Headland Judo Kerb 50 W Kerb 80 EDF Kerb Granules Kerb Pro Flo Kerb Pro Granules Menace 80 EDF	Grass weed control in forest trees
simazine	Alpha Simazine 50SC Gesatop Sipcam Simazine Flowable	Annual weed control in nursery beds and transplant lines
sulphonated cod liver oil	Scuttle	Reduction of damage by deer, rabbits
triclopyr	Chipman Garlon 4 Convoy Garlon 4 Timbrel Triptic 48 EC	Woody weed control
ziram	Aaprotect	Bird and animal repellent

Pesticides Used as Seed Treatments (including treatments on seed potatoes)

Information on the target pests for these products can be found in the relevant pesticide profile in Section 4.

Chemical	Product	Formulation	Crop(s)
aluminium ammonium sulphate	Guardsman SDP Seed Dressing Powder	DS	Field crops, corms, flower bulbs
bitertanol + fuberidazole	Sibutol	FS	Oats, rye, triticale, wheat
bitertanol + fuberidazole + imidacloprid	Sibutol Secur	LS	Oats, wheat
carboxin + imazalil + thiabendazole	Vitaflow Extra	FS	Barley, oats
carbendazim + tecnazene	Hortag Tecnacarb Tripart Arena Plus	DS DS	Potatoes Potatoes
carboxin + thiram	Anchor	FS	Wheat, barley, oats, rye, triticale
ethirimol + flutriafol + thiabendazole	Ferrax	FS	Barley
fludioxonil	Beret Gold	FS	Wheat, barley
fluquinconazole + prochloraz	Jockey	FS	Wheat
fuberidazole + imidacloprid + triadimenol	Bayton Secur	FS	Cereals
fuberidazole + triadimenol	Baytan Flowable	FS	Cereals
guazatine	Panoctine Ravine	LS LS	Wheat, barley, oats Wheat, barley, oats
guazatine + imazalil	Panoctine Plus	LS	Barley, oats
hymexazol	Tachigaren 70WP	WP	Sugar beet
imazalil	Fungazil 100 SL Sphinx Stryper	LS LS LS	Potatoes Barley Barley
imazalil + pencycuron	Monceren IM Monceren IM Flowable	DS FS	Potatoes Potatoes

33

Chemical	Product	Formulation	Crop(s)
imazalil + thiabendazole	Extratect Flowable	FS	Potatoes
imazalil + triticonazole	Robust	LS	Barley
imidacloprid	Gaucho	WS	Sugar beet
imidacloprid + tebuconazole + triazoxide	Raxil Secur	LS	Barley
iprodione	Rovral Liquid FS	FS	Oilseed rape, linseed, potatoes
metalaxyl + thiabendazole + thiram	Apron Combi FS	FS	Beans, peas
pencycuron	Monceren DS	DS	Potatoes
	Monceren Flowable	FS	Potatoes
	Standon Pencycuron DP	DS	Potatoes
prochloraz	Prelude 20 LF	LS	Linseed, flax
tebuconazole + triazoxide	Raxil S	FS	Barley
tecnazene	Various products, see tecnazene entry	Various	Potatoes
tecnazene + thiabendazole	Hytec Super	DS	Potatoes
tefluthrin	Force ST	LS	Sugar and fodder beet
thiabendazole	Hykeep	DS	Potatoes
	Storite Clear Liquid	LS	Potatoes
	Storite Flowable	FS	Potatoes
thiabendazole + thiram	Hy-TL	FS	Peas
	sHYlin	FS	Carrots, flax, linseed, lupins, parsnips
thiram	Agrichem Flowable Thiram	FS	Beans, peas, maize, vegetables, flowers
	Thiraflo		Peas
tolclofos-methyl	Rizolex	DS	Potatoes
	Rizolex Flowable	FS	Potatoes

Aerial Application of Pesticides

Only these products specifically approved for aerial application may be so applied and they may only be applied to specific crops or for specified uses. A list of products approved for application from the air was published as Annex B in *Pesticides 2000*. The list given below is taken mainly from this source with updating from issues of the *Pesticides Monitor* and product labels.

It is emphasised that the list is for guidance only — reference must be made to the product labels for detailed conditions of use which must be complied with. The list does not include those products which have been granted restricted aerial application approval limiting the area which may be treated.

Detailed rules are imposed on aerial application regarding prior notification of the Nature Conservancy, water authorities, bee keepers, Environmental Health Officers, neighbours, hospitals, schools etc. and the conditions under which application may be made. The full conditions are available from MAFF and must be consulted before any aerial application is made.

Chemical	Product	Crops/uses
asulam	Asulox	Grassland, forestry, non-crop land (bracken control)
benalaxyl + mancozeb	Galben M	Potatoes
carbendazim	Ashlade Carbendazim FL	Wheat, oilseed rape
	Carbate Flowable	Wheat, oilseed rape
	Hinge	Wheat, oilseed rape
	Tripart Defensor FL	Wheat, oilseed rape
chlormequat	Agriguard Chlormequat 700	Wheat, oats
	Atlas 3C: 645	Wheat, oats
	Atlas Chlormequat 46	Wheat, oats
	Atlas Chlormequat 700	Wheat, oats
	Atlas Quintacel	Wheat, oats
	Atlas Terbine	Wheat, oats
	Atlas Tricol	Wheat, oats
	Barclay Holdup	Wheat, oats, barley
	Barclays Holdup 600	Wheat, oats, barley
	Barclays Holdup 640	Wheat, oats, barley
	Barclay Take 5	Wheat, oats, rye, triticale
	BASF 3C Chlormequat 720	Wheat, oats, barley

Chemical	Product	Crops/Uses
	Greencrop Carna	Wheat, oats, barley
	Greencrop Coolfin	Wheat, oats
	Mandops Chlormequat 700	Wheat, oats, winter barley
	New 5C Cycocel	Wheat, oats, barley, rye, triticale
	Sigma PCT	Wheat, barley
	Supaquat	Wheat, oats, barley, rye, triticale
	Trio	Wheat, oats, barley, rye, triticale
	Tripart Brevis	Wheat, oats
	Uplift	Wheat, oats, barley
	Whyte Chlormequat 700	Wheat, oats
2-chloroethyl phosphonic acid	Aventis Cerone Barclay Coolmore Cerone Charger	Barley Barley Barley Barley
chlorothalonil	Agriguard Chlorothalonil	Potatoes
	Barclay Corrib 500	Wheat, potatoes, field beans, oilseed rape, brassicas, onions, peas
	Bombardier FL	Potatoes
	Bravo 500	Potatoes
	Clortosip 500	Potatoes
	Flute	Potatoes
	Greencrop Orchid	Potatoes
	Mainstay	Potatoes
	Repulse	Potatoes
	Rover	Potatoes
	Strada	Potatoes
	Tripart Faber	Wheat, potatoes
	Ultrafaber	Potatoes
chlorotoluron	Lentipur CL 500	Wheat, barley, triticale
	Luxan Chlorotoluron 500 Flowable	Wheat, barley

Chemical	Product	Crops/uses
	Tripart Ludorum	Winter wheat, winter barley, durum wheat, triticale
copper oxychloride	Cuprokylt	Potatoes
cymoxanil + mancozeb	Ashlade Solace	Potatoes
	Standon Cymoxanil Extra	Potatoes
diflubenzuron	Dimilin Flo	Forest trees
dimethoate	Barclay Dimethosect	Cereals, peas, ware potatoes, sugarbeet
	BASF Dimethoate 40	Cereals, peas, ware potatoes, sugar beet
	Danadim Dimethoate 40	Cereals, peas, potatoes, beet crops
	Rogor L40	Cereals, peas, potatoes, beet
fenitrothion	Dicofen	Cereals, peas
flamprop-M-isopropyl	Commando	Cereals
iprodione	Landgold Iprodione 250	Oilseed rape, field beans
mancozeb	Agrichem Mancozeb 80	Potatoes
	Ashlade Mancozeb FL	Potatoes, wheat, barley, oats, rye, triticale
	Dithane 945	Potatoes
	Dithane Dry Flowable	Potatoes
	Helm 75 WP	Potatoes
	Mandate 80	Potatoes
	Opie WP	Potatoes
	Penncozeb WDG	Potatoes
mancozeb + metalaxyl	Fubol 75WP	Potatoes
	Osprey 58WP	Potatoes
mancozeb + oxadixyl	Recoil	Potatoes
maneb	Agrichem Maneb 80	Potatoes
	X-Spor SC	Potatoes

Chemical	Product	Crops/uses
metaldehyde	Doff Horticultural Slug Killer Blue Mini Pellets	All crops
	Doff Metaldehyde Slug Killer Mini Pellets	All crops
	Metarex RG	All crops
	Mifaslug	All crops
	Optimol	All crops
	PBI Slug Pellets	All crops
	Superflor 6% Metaldehyde Slug Killer	All crops
methabenzthiazuron	Tribunil	Cereals, grass
methiocarb	Decoy Wetex	All crops
	Draza	All crops
	Exit Wetex	All crops
	Huron	All crops
	Karan	All crops
	Lupus	All crops
	Rivet	All crops
monolinuron	Arresin	Potatoes, dwarf beans, leeks
pirimicarb	Aphox	Cereals, peas, potatoes, sugar beet, beans, leaf brassicas, maize, swedes, oilseed rape, turnips, carrots
	Barclay Pirimisect	Wheat, barley, triticale, oats, rye
	Phantom	Cereals, peas, potatoes, sugar beet, beans, leaf brassicas, maize, sweetcorn, swedes, oilseed rape, turnips, carrots
	Portman Pirimicarb	Wheat, barley, oats
propiconazole	Barclay Bolt	Wheat, barley, oats
	Mantis	Wheat, barley, rye, oats
	Standon Propiconazole	Wheat, barley, rye, oats
	Tilt	Wheat, barley, rye, oats,

		triticale
Chemical	**Product**	**Crops/uses**
sulphur	Thiovit	Sugar beet
triadimefon	Bayleton	Sugar beet, cereals, brassicas (including turnips and swedes)
tri-allate	Avadex Excel 15G	Winter wheat, barley, winter beans, peas
zineb	Unicrop Zineb	Potatoes

SECTION 2
ADJUVANTS

Adjuvants

Adjuvants are not themselves classed as pesticides and there is considerable misunderstanding over the extent to which they are legally controlled under the Food and Environment Protection Act. An adjuvant is a substance other than water which enhances the effectiveness of a pesticide with which it is mixed. Consent C(i)5 under the Control of Pesticides Regulations allows that an adjuvant can be used with a pesticide only if that adjuvant is authorised and on a list published from time to time in *Pesticides Monitor*. An authorised adjuvant has an *adjuvant number* and may have specific requirements about the circumstances in which it may be used.

Adjuvant product labels must be consulted for full details of authorised use, but the table below provides a summary of the label information to indicate the area of use of the adjuvant. Label precautions refer to the keys given in Appendix 4, A and B, and may include warnings about products harmful or dangerous to fish. The table includes all adjuvants notified by suppliers as available in 2001.

Product	Supplier	Adj. No.	Type
Actipron	Batsons	A0013	adjuvant
Contains	97% highly refined mineral oil		
Use with	Amazon, Asulox, Basagran SG, Benlate, Betanal E, Betanal E + Goltix WG, Betanal E + Magnum, Checkmate, Checkmate + Goltix WG, Dow Shield + Goltix WG, Goltix WG, Goltix WG + Magnum, Hawk, Laser, Pilot, Roundup, Topik		
Precautions	U08, U19, U20a, E15, E30a, E31a		
Actipron	Bayer	A0013	adjuvant
Contains	97% highly refined mineral oil		
Use with	Amazon, Asulox, Basagran SG, Benlate, Betanal E, Betanal E + Goltix WG, Betanal E + Magnum, Checkmate, Checkmate + Goltix WG, Dow Shield + Goltix WG, Goltix WG, Goltix WG + Magnum, Hawk, Laser, Pilot, Roundup, Topik		
Precautions	U08, U19, U20a, E15, E30a, E31a		
Actirob B	Stefes	A0278	vegetable oil
Contains	842 g/l esterified rapeseed oil		
Use with	Cheetah Super and Wildcat on wheat; Sceptre on sugar beet. For details of use with all approved pesticides on specified crops consult label or supplier		
Activator 90	Newman	A0337	non-ionic surfactant/spreader/wetter
Contains	750 g/l alkylphenyl hydroxypolyoxyethylene and 150 g/l natural fatty acids		
Use with	Any spray for which additional wetter is approved and recommended		
Protective clothing	A, C		
Precautions	M03, R04a, R04b, U02, U04a, U05a, U10, U11, U19, U20b, C03, E01, E13c, E30a, E31a, E34		

Product	Supplier	Adj. No.	Type
Admix-P	Newman	A0301	wetter
Contains	80% w/w polyalkylene oxide modified heptamethyltrisiloxane and a maximum of 20% w/w allyloxypolyethylene glycol methyl ether		
Use with	A wide range of pesticides applied as corm, tuber, onion and other bulb treatments in seed production and in seed potato treatment		
Protective clothing	A, C, H		
Precautions	M03, R03a, R03c, R04a, R04b, R04e, U02, U05a, U08, U19, U20b, C03, E01, E13c, E30a, E31a, E34		
Agral	Zeneca	A0154	non-ionic surfactant/spreader/wetter
Contains	948 g/l alky phenol ethylene oxide		
Use with	Any spray for which additional wetter is approved and recommended		
Protective clothing	A, C		
Precautions	R04a, R04b, R07d, U05a, U08, U20c, C03, E01, E15, E26, E30a, E31a, E34		
Agropen	Intracrop	A0143	vegetable oil
Contains	95% refined rapeseed oil		
Use with	Pesticides which have a recommendation for the addition of a wetter/spreader		
Protective clothing	A, C		
Precautions	U02, U05a, U08, U19, U20b, C03, E01, E13c, E30a, E31a, E34		
Amigo	Intracrop	A0397	spreader/wetter
Contains	85% w/w rapeseed oil		
Use with	Any pesticide for which use of a wetter/spreader is recommended. Also for use with recommended doses of approved pesticides in non-crop situaitons and on a range of specified fruit and vegetable crops		
Protective clothing	A, C		
Precautions	R04a, U05a, U08, U19, U20b, C03, E01, E13c, E30a, E31a, E34		
Anoint	Intracrop	A0236	spreader/vegetable oil/wetter
Contains	95% rapeseed oil		
Use with	Pesticides that have a recommendation for the addition of a wetter/spreader		
Protective clothing	A, C		
Precautions	R04a, R04b, U05a, U08, U19, U20b, C03, E01, E13c, E30a, E31a, E34		
Aquator	Zeneca	A0265	surfactant
Contains	A blend of surfactant and ammonium sulphate		
Use with	Approved formulations of glyphosate-trimesium		
Protective clothing	C		

Product	Supplier	Adj. No.	Type
Precautions	U05a, U08, U19, U20b, C03, E01, E13c, E30a, E31c		

Arma — Interagro — A0306 — penetrant

Contains	500 g/l alkoxylated fatty amine + 500 g/l polyoxyethylene monolaurate
Use with	Cereal growth regulators, cereal herbicides, cereal fungicides, oilseed rape fungicides and a wide range of other pesticides on specified crops
Precautions	U05a, C03, E01, E13c, E30a, E34

AS Elite — Monsanto — A0230 — extender

Contains	40% w/w ammonium sulphate
Use with	Roundup GT, Roundup Elite
Precautions	U20a, E30a, E31b

Atlas Adherbe — Nufarm Whyte — A0023 — mineral oil

Contains	83% highly refined paraffinic oil and surfactants
Use with	Recommended rates of approved pesticides up to specified growth stages of a wide range of agricultural arable crops, and in non-crop situations. Crops beyond specified growth stages, and grassland, should only be used with half recommended rates of the pesticide or less. See label for details of growth stage restrictions and specific recommendations for use with herbicides on beet crops.
Protective clothing	A, C
Precautions	R04a, R04b, R04c, U05a, U08, U19, U20b, C03, E01, E13c, E26, E30a, E31b, E37

Atlas Adjuvant Oil — Nufarm Whyte — A0021 — mineral oil

Contains	95% w/w highly refined mineral oil and surfactants
Use with	Recommended rates of approved pesticides up to specified growth stages of a wide range of agricultural arable crops, and in non-crop situations. Crops beyond specified growth stages, and grassland, should only be used with half recommended rates of the pesticide or less. See label for details of growth stage restrictions and specific recommendations for use with herbicides on beet crops.
Precautions	R04a, R04b, R04c, U05a, U08, U19, U20a, C03, E01, E13c, E26, E30a, E31b

Axiom — Batsons — A0136 — mineral oil

Contains	96.8% w/w paraffinic base oil
Use with	Asulox, Benlate, Betanal E, Checkmate, Gesaprim 500 SC, Goltix WG, Goltix WG/Betanal E, Goltix WG/Checkmate, Goltix WG/Dow Shield, Goltix WG/Fusilade 5, Laser, Pilot and Roundup
Precautions	U09a, U19, U20a, E15, E30a, E31a

Banka — Interagro — A0245 — spreader/sticker/wetter

Contains	29.2% w/w alkyl pyrrolidones

Product	Supplier	Adj. No.	Type

Use with	Potato fungicides and a wide range of other pesticides on specified crops
Protective clothing	A, C
Precautions	R04b, R04d, R07d, U02, U05a, U11, U19, U20a, C03, E01, E13c, E30a, E34

Barclay Actol	Barclay	A0126	mineral oil
Contains	99% highly refined paraffinic oil
Use with	Any approved pesticide for which the addition of a spraying oil is recommended
Precautions	U19, U20b, E13c, E30a, E31b

Barclay Clinger	Barclay	A0198	sticker/wetter
Contains	96% w/v poly-1-p-menthene
Use with	Barclay Gallup, Barclay Gallup Amenity and a wide range of approved fungicides and insecticides. See label for details
Precautions	U19, U20b, E13c, E30a, E31a

Bazuka	Euroagkem	A0241	spreader/wetter
Contains	800 g/l polyalkyleneoxide modified heptamethyltrisiloxane
Use with	All fungicides applied to spring and winter sown wheat, spring and winter sown barley, spring and winter sown oats, triticale, rye and durum wheat where the pesticide label recommends the addition of wetters and spreaders
Protective clothing	A, C
Precautions	M03, R03a, R03c, R04e, U02, U05a, U08, U19, U20b, C03, E01, E13c, E30a, E31a, E34

Biodew	Intracrop	A0253	non-ionic surfactant/spreader/wetter
Contains	85% w/w alkyl polyglycol ether and fatty acids
Use with	Pesticides where a wetter/spreader is recommended
Protective clothing	A, C
Precautions	M03, R03c, R04a, R07d, U05a, U08, U19, U20c, C03, E01, E13c, E30a, E31a, E34

Bioduo	Intracrop	A0299	wetter
Contains	85% alkyl polyglycol ether and fatty acids
Use with	A wide range of pesticides used in grassland, agriculture and horticulture and with pesticides used in non-crop situations
Protective clothing	A, C
Precautions	M03, R03c, R04a, R04b, R07d, U05a, U08, U20c, C03, E01, E13c, E30a, E31a, E34

Product	Supplier	Adj. No.	Type
Biofilm	Intracrop	A0359	sticker/wetter
Contains	96% poly-l-p-menthene		
Use with	All approved fungicides and insecticides on edible crops and all approved formulations of glyphosate used pre-harvest on wheat, barley, oilseed rape, stubble, and in non-crop situations and grassland destruction		
Precautions	U19, U20c, E13c, E30a, E32a		
Biosyl	Intracrop	A0385	spreader/sticker/wetter
Contains	1% w/w polyoxyethylen-alpha-methyl-omega-[3-(1,3,3,3-tetramethyl-3-trimethylsiloxy)-disiloxanyl] propylether and 32.67% w/w ethylene oxide condensate		
Use with	Recommended rates of approved pesticides on a range of specified fruit and vegetable crops and on managed amenity turf		
Protective clothing	A, C		
Precautions	R04a, R04b, U05a, U08, U20c, C03, E01, E15, E31a		
Biothene	Intracrop	A0360	sticker/wetter
Contains	96% poly-l-p-menthene		
Use with	All approved fungicides and insecticides on edible crops and all approved formulations of glyphosate used pre-harvest on wheat, barley, oilseed rape, stubble, and in non-crop situations and grassland destruction		
Precautions	U19, U20c, E13c, E30a, E32a		
Bond	Newman	A0184	drift retardant/extender/sticker/wetter
Contains	450 g/l synthetic latex and 100 g/l alkylphenylhydroxy-polyoxyethylene		
Use with	Blight sprays on potatoes and other pesticides on a wide range of crops. See label or contact supplier for details		
Protective clothing	A, C		
Precautions	M03, R04a, R04b, U05a, U08, U19, U20a, C03, E01, E13c, E30a, E31a		
Cat G	Greenhill	A0348	wetter
Contains	431 g/l polyethoxylated tallow amine		
Use with	All approved formulations of glyphosate used pre-harvest in wheat, barley and oats, post-harvest on stubbles and in non-crop areas. In addition may be used with all approved herbicides up to their full approved dose on a range of specified fruit and vegetable crops		
Protective clothing	A, C		
Precautions	M03, R04a, R04b, U05a, U08, U13, U19, U20b, C03, E01, E13c, E30a, E31a, E34, E37		
Celect	Fargro	A0334	extender/sticker/wetter
Contains	95% rapeseed oil		

Product	Supplier	Adj. No.	Type
Use with	Pesticides in horticulture		
Protective clothing	A, C		
Precautions	U19, U20b, E30a, E31b		

Chiltern Gold 98	Chiltern	A0302	adjuvant/vegetable oil
Contains	97.5% w/w methylated rapeseed oil		
Use with	A wide range of pesticides used in agriculture, horticulture, forestry and amenity situations. See label for details of crops and dose rates.		
Protective clothing	A, H		
Precautions	R04e, U05a, U19, U20b, C03, E01, E13c, E26, E30a, E31b		

Codacide Oil	Microcide	A0011	vegetable oil
Contains	95% emulsifiable vegetable oil		
Use with	All approved pesticides and tank mixes. See label for details		
Protective clothing	A, C		
Precautions	U19, U20b, E13c, E30a, E31b		

Companion	Allied Colloids	A0239	spreader/sticker/wetter
Contains	25% w/w polyacrylamide		
Use with	Approved herbicides on cereals and any pesticide on non-food crops. See label for restrictions on use with wettable powders, suspension concentrates and water dispersible granules, and other usage limitations		
Precautions	U05a, U08, C03, E01, E13c, E30a		

Comulin	Batsons	A0164	mineral oil
Contains	96.8% w/w refined paraffinic oil		
Use with	Amazon, Asulox, Basagran, Benlate, Checkmate, Gesaprim 500 SC, Goltix WG, Goltix WG/Checkmate, Goltix WG/Dow Shield, Goltix WG/Fusilade 5, Goltix WG/Laser/Checkmate, Hawk, Laser, Pilot, Roundup, Topik		
Precautions	U08, U19, U20a, E15, E30a, E31a		

Contact	Interagro	A0191	mineral oil
Contains	92% w/w paraffinic oil		
Use with	Atrazine, Beetomax + metamitron, fenoxaprop-p-ethyl, metamitron, phenmedipham and other approved pesticides which have a label recommendation for use with authorised adjuvant mineral oils		
Protective clothing	A, C		
Precautions	R04a, R04b, U02, U05a, U08, U19, U20a, C03, E01, E13c, E30a, E31b		

Cropoil	Chiltern	A0137	mineral oil
Contains	99% highly refined mineral oil		

Product	Supplier	Adj. No.	Type

Use with	Approved pesticides for use on certain specified crops (cereals, combinable break crops, field and dwarf beans, peas, oilseed rape, brassicas, potatoes, carrots, parsnips, sugar beet, fodder beet, mangels, red beet, maize, sweetcorn, onions, leeks, horticultural crops, forestry, amenity grassland)
Protective clothing	A, C
Precautions	R04, R04e, U19, U20b, E13c, E26, E30a, E31b

Cutback	Greenhill	A0351	wetter
Contains	800 g/l polyoxyethylene tallow amine
Use with	Specified herbicides when used at half or less than their approved dose in a wide range of arable crops, forestry and amenity situations (see label for details) and also with approved formulations of diquat and glyphosate in specified situations up to the full approved dose of the herbicide
Protective clothing	A, C
Precautions	M03, R03c, R04a, R04b, R07d, U02, U05a, U08, U13, U19, U20b, C03, E01, E13b, E16e, E30a, E31a, E34

Cutinol	Greenhill	A0151	vegetable oil
Contains	95% emulsifiable rapeseed oil
Use with	All approved pesticides for which addition of a wetter is recommended
Precautions	R04a, R04b, U05a, U08, U20a, C03, E01, E13c, E30a, E31a

Cutinol Plus	Intracrop	A0395	spreader/wetter
Contains	85% w/w rapeseed oil
Use with	Any pesticide for which use of a wetter/spreader is recommended. Also for use with recommended doses of approved pesticides in non-crop situaitons and on a range of specified fruit anf vegetable crops
Protective clothing	A, C
Precautions	R04a, U05a, U08, U19, U20b, C03, E01, E13c, E30a, E31a, E34

Cutonic Foliar Booster	Lambson	A0228	non-ionic surfactant/wetter
Contains	100% w/w polyether-polymethylsiloxane-copolymer
Use with	All pesticides authorised for use on all cereals and grassland crops where the use of a wetting/spreading/penetrating surfactant is recommended. Can be used with micronutrients and fertilisers when applied as a foliar treatment
Protective clothing	A, C
Precautions	M03, R03a, R03c, R04d, R04e, U02, U05a, U08, U19, U20b, C03, E01, E13c, E26, E30a, E31b, E34

Designer	Newman	A0322	drift retardant/extender/sticker/wetter
Contains	8.44% w/w organosilicone wetter and 50% w/w synthetic latex

Product	Supplier	Adj. No.	Type

Use with	A wide range of fungicides, insecticides and trace elements for cereals and specified agricultural and horticultural crops		
Protective clothing	A, C		
Precautions	M03, R04a, R04b, R04e, U02, U05a, U08, U19, U20b, C03, E01, E13c, E30a, E31a, E34		

Du Pont Adjuvant	DuPont	A0119	non-ionic surfactant/wetter
Contains	900 g/l ethylene oxide condensate		
Use with	Any spray for which a non-ionic wetter is recommended		
Protective clothing	A, C		
Precautions	R04a, R04b, R07d, U05a, U08, U20c, C03, E01, E13c, E30a, E31a		

Elan	Intracrop	A0392	spreader/sticker/wetter
Contains	20% w/w polyoxyethylen-alpha-methyl-omega-[3-(1,3,3,3-tetramethyl-3-trimethylsiloxy)-disiloxanyl] propylether		
Use with	Approved pesticides used at full dose on a range of specified fruit and vegetable crops, and on managed amenity turf		
Protective clothing	A, C, H		
Precautions	U05a, U08, U20c, C03, E01, E13c, E15, E30a, E31a, E37		

Emerald	Intracrop	A0031	anti-transpirant/extender
Contains	96% di-1-p-menthene		
Use with	Alone or with most approved insecticides, fungicides or growth regulators. See label for details. Do not use in mixture with adjuvant oils or surfactants		
Precautions	U19, U20b, E13c, E30a, E31a		

Enhance	Greenhill	A0147	non-ionic surfactant/spreader/wetter
Contains	900 g/l phenol ethylene oxide condensate		
Use with	Any spray for which additional wetter is approved and recommended		
Precautions	R03c, R04a, R04b, R07d, U05a, U08, U13, U19, U20a, C03, E01, E13c, E30a, E31a		

Enhance Low Foam	Greenhill	A0148	non-ionic surfactant/spreader/wetter
Contains	900 g/l alkyl phenol ethylene oxide condensate with silicone anti-foaming agent		
Use with	Any spray for which wetter is approved and recommended through low volume spraying systems or where low foam is required		
Protective clothing	A, C		
Precautions	R04a, R04b, U05a, U08, U20a, C03, E01, E13c, E30a, E31a		

Ethokem	Greenhill	A0146	cationic surfactant
Contains	870 g/l polyoxyethylene tallow amine		

Product	Supplier	Adj. No.	Type

Use with	A wide range of fungicides, herbicides, insecticides, growth regulators and micronutrients. See leaflet for details
Precautions	R04a, R04b, U05a, U08, U20a, C03, E01, E13c, E30a, E31a

Ethokem C 12	Greenhill	A0149	cationic surfactant

Contains	870 g/l bis-2 hydroxyethyl coco-amine
Use with	All approved foliar acting herbicides for use on edible and non-edible crops
Protective clothing	A, C
Precautions	M03, R03c, R04a, R04b, R07d, U02, U05a, U08, U13, U19, U20b, C03, E01, E13c, E30a, E31a, E34, E37

Ethowet	Greenhill	A0319	wetter

Contains	471 g/l polyethoxylated tallow amine
Use with	All approved formulations of glyphosate in stubble and pre-harvest treatments of cereals and in non-crop areas; also with all approved pesticides on a wide range of specified fruit and vegetable crops
Precautions	M03, R04a, R04b, U05a, U08, U13, U19, U20b, C03, E01, E13c, E30a, E31a, E34, E37

Euroagkem Pace	Euroagkem	A0305	spreader/wetter

Contains	42% w/w propionic acid
Use with	Chlormequat, glyphosate and a wide range of herbicides for use cereals and other agricultural and horticultural crops. Must not be used with any pesticide to be applied in or near water. See label for details
Protective clothing	A, C
Precautions	M05, R05b, U02, U04a, U05a, U11, U12, U13, U14, U15, U19, U20b, C03, E01, E15, E30a, E31a

Euroagkem Pen-e-trate	Euroagkem	A0216	acidifier/non-ionic surfactant

Contains	350 g/l propionic acid
Use with	Pesticides that recommend the use of a wetter/spreader
Protective clothing	A, C
Precautions	R04a, R04b, R05, R05b, U02, U05a, U08, U19, U20a, C03, E01, E30a, E31a

Exell	Truchem	A0017	spreader/wetter

Contains	64% polyethoxylated tallow amine
Use with	Diquat, glyphosate
Protective clothing	A, C
Precautions	M03, R04a, R04b, U05a, U08, U19, U20a, C03, E01, E13c, E30a, E31a, E34

Product	Supplier	Adj. No.	Type
Felix	Intracrop	A0178	spreader/wetter
Contains	60% ethylene oxide condensate		
Use with	Mecoprop, 2,4-D in cereals and amenity turf, and a range of grass weedkillers in agriculture. See label for details		
Protective clothing	A, C		
Precautions	R04a, R04b, R07d, U05a, U08, U20a, C03, E01, E15, E30a, E31a		
Frigate	Unicrop	A0325	sticker/wetter
Contains	800 g/l tallow amine ethoxylate		
Use with	Roundup formulations		
Protective clothing	A, C		
Precautions	M03, R03c, R04a, R04b, R04c, R07d, U05a, U09a, U19, U20a, C03, E01, E13b, E30a, E31a		
Galion	Intracrop	A0162	spreader/wetter
Contains	60% polyoxyalkylene glycol		
Use with	Herbicides in grassland and amenity turf and with grass weedkillers in cereals, oilseed rape, sugar beet and other crops		
Protective clothing	A, C		
Precautions	R04a, R04b, R07d, U05a, U08, U20b, C03, E01, E15, E30a, E31a		
Glimark	Nufarm Whyte	A0371	mineral oil
Contains	140 g/l highly refined mineral oil and surfactants		
Use with	Glyphosate		
Protective clothing	A, C, D, H, M		
Precautions	R04a, R04b, R04c, U05a, U08, U20b, C03, E13c, E26, E31a, E37		
Grip	Newman	A0211	drift retardant/extender/sticker/wetter
Contains	450 g/l synthetic latex and 100 g/l alkylphenyl hydroxypolyoxyethylene		
Use with	Blight sprays on potatoes and other pesticides on a wide range of crops. See label or contact supplier for details		
Protective clothing	A, C		
Precautions	M03, R04a, R04b, U02, U05a, U08, U19, U20b, C03, E01, E13c, E30a, E31a, E34		
GS 800	Greenhill	A0150	cationic surfactant
Contains	800 g/l polyoxyethylene tallow amine		
Use with	A wide range of pesticides. See label for details		
Protective clothing	A, C		
Precautions	R03c, R04a, R04b, R07d, U05a, U08, U13, U19, U20a, C03, E01, E13c, E30a, E31a		

Product	Supplier	Adj. No.	Type
Guide	Newman	A0208	acidifier/drift retardant/penetrant
Contains	350 g/l modified soya lecithin, 100 g/l alkylphenyl hydroxypolyoxyethylene and 350 g/l propionic acid		
Use with	All plant growth regulators for cereals and oilseed rape, dimethoate and chlorpyrifos for wheat bulb fly control. For other uses see label or contact supplier for details		
Protective clothing	A, C		
Precautions	M03, R04b, R04d, U02, U04a, U05a, U10, U11, U19, U20a, C03, E01, E13c, E30a, E31a		
Headland Fortune	Headland	A0277	penetrant/spreader/vegetable oil/wetter
Contains	75% w/w mixed methylated fatty acid esters of seed oil and N-butanol		
Use with	Herbicides and fungicides in a wide range of crops. See label for details		
Protective clothing	A, C		
Precautions	R04, R04e, U02, U05a, U14, U20a, C03, E01, E13c, E26, E30a, E31b, E34		
Headland Guard 2000	Headland	A0369	extender/sticker/wetter
Contains	52% w/w synthetic latex solution		
Use with	All approved pesticides that recommend the use of a sticking, wetting or extending agent; also with all approved pesticides for use on crops not destined for human or animal consumption; also with all approved pesticides on beans, peas, edible podded peas, oilseed rape, linseed, sugar beet, cereals, maize, Brussels sprouts, potatoes, cauliflowers		
Precautions	U05a, U08, U19, U20b, C03, E01, E13b, E26, E31b, E34, E37		
Headland Inflo XL	Headland	A0329	wetter
Contains	85% w/w polyalkylene oxide modified heptamethyl siloxane		
Use with	Any pesticide for use on amenity grassland or managed amenity turf (except any product applied in or near water) where the use of a wetting, spreading and penetrating surfactant is recommended to improve foliar coverage		
Protective clothing	A, C		
Precautions	M03, R03a, R03c, R04d, R04e, U02, U05a, U08, U19, U20b, C03, E01, E13c, E26, E30a, E31b, E34		
Headland Intake	Headland	A0074	penetrant
Contains	450 g/l propionic acid		
Use with	All approved pesticides on any crop not intended for human or animal consumption, and with all approved pesticides on beans, peas, edible podded peas, oilseed rape, linseed, sugar beet, cereals (except triazole fungicides), maize, Brussels sprouts, potatoes cauliflowers. See label for detailed advice on timing on these crops		
Protective clothing	A, C		

Product	Supplier	Adj. No.	Type
Precautions	R05, R05b, U02, U05a, U10, U11, U14, U15, U19, C03, E01, E26, E30a, E31b, E34		

Headland Rhino	Headland	A0328	wetter
Contains	85% w/w polyalkylene oxide modified heptamethyl siloxane		
Use with	Any herbicide, systemic fungicide, systemic insecticide or plant growth regulator (except any product applied in or near water) where the use of a wetting, spreading and penetrating surfactant is recommended to improve foliar coverage		
Protective clothing	A, C		
Precautions	M03, R03a, R03c, R04d, R04e, U02, U05a, U08, U19, U20b, C03, E01, E13c, E26, E30a, E31b, E34		

Headland Sure-Fire	Headland	A0214	vegetable oil/wetter
Contains	874 g/l rapeseed oil		
Use with	Any pesticide where the use of a wetting agent is recommended		
Protective clothing	A, C		
Precautions	R04a, R04b, U05a, U08, U20b, C03, E01, E13c, E26, E30a, E31a		

Holdtite	Greenhill	A0180	extender/sticker
Contains	52% w/w synthetic latex and 20% alkyl phenol ethylene oxide condensate		
Use with	A wide range of agricultural chemicals/micronutrients		
Protective clothing	A, C		
Precautions	R04a, R04b, R04e, U02, U05a, U08, U19, U20b, C03, E01, E13b, E26, E30a, E31a		

Hyspray	Fine	A0020	cationic surfactant
Contains	800 g/l polyethoxylated tallow amine		
Use with	All approved formulations of glyphosate		
Protective clothing	A, C		
Precautions	R03c, R04a, R04b, R07d, U02, U05a, U08, U13, U19, U20a, C03, E01, E13b, E30a, E31b, E34		

Intracrop Archer	Intracrop	A0242	spreader/wetter
Contains	800 g/l polyalkyleneoxide modified heptamethyltrisiloxane		
Use with	All fungicides used in cereals where the addition of wetters/spreaders is recommended on the pesticide label		
Protective clothing	A, C		
Precautions	M03, R03a, R03c, R04e, U02, U05a, U08, U19, U20b, C03, E01, E13c, E30a, E31a, E34		

Product	Supplier	Adj. No.	Type
Intracrop Bla-Tex	Intracrop	A0173	extender/sticker
Contains	450 g/l synthetic latex		
Use with	Any approved pesticide where the addition of a sticker/extender is required and recommended		
Protective clothing	A, C		
Precautions	M03, R04a, R04b, U02, U05a, U08, U19, U20a, C03, E01, E13c, E30a, E31a		
Intracrop Green Oil	Intracrop	A0262	wetter
Contains	950 g/l refined rapeseed oil		
Use with	All approved pesticides on grassland and edible crops when used at half or less than the approved dose, and with approved pesticides up to their full approved dose on a range of specified fruit and vegetable crops		
Precautions	U08, U20b, E13c, E30a, E31b, E37		
Intracrop Non-Ionic Wetter		Intracrop	A0009	spreader/wetter
Contains	900 g/l alkyl phenol ethoxylate		
Use with	Pesticides which have a recommendation for use with a wetter or emulsified oil		
Protective clothing	A, C		
Precautions	R04a, R04b, R07d, U02, U05a, U08, U19, U20a, C03, E01, E13c, E30a, E31a		
Intracrop Quiver	Intracrop	A0237	spreader/wetter
Contains	800 g/l polyalkyleneoxide modified heptamethyltrisiloxane		
Use with	All approved fungicides applied to wheat, barley, oats, triticale, rye and durum wheat		
Protective clothing	A, C		
Precautions	M03, R03a, R03c, R04e, U02, U05a, U08, U19, U20a, C03, E01, E13c, E30a, E31a, E34		
Intracrop Rapeze	Intracrop	A0390	adjuvant/vegetable oil
Contains	842 g/l methylated rapeseed oil		
Use with	All approved pesticides on edible and non-edible crops when used at half or less than the approved dose, and with approved pesticides up to their full approved dose on a range of specified fruit and vegetable crops		
Precautions	U19, U20b, E13b, E30a, E31b		
Intracrop Rapide Beta	Intracrop	A0175	acidifier/non-ionic surfactant
Contains	350 g/l propionic acid and 100 g/l alkylphenyl-hydroxypolyoxyethylene		

Product	Supplier	Adj. No.	Type
Use with	Any spray where the addition of a wetter/spreader or self-emulsifying oil is required and recommended on the pesticide label		
Protective clothing	A, C		
Precautions	M03, R05, R05b, U02, U04a, U05a, U08, U10, U11, U19, U20a, C03, E01, E13c, E30a, E31a, E34		
Iona	Intracrop	A0232	spreader/wetter
Contains	921 g/l alkyl phenol ethoxylate		
Use with	Any approved pesticide for which the addition of a wetter/spreader is recommended		
Protective clothing	A, C		
Precautions	R04a, R04b, U05a, U08, U19, U20a, C03, E01, E13c, E30a, E32a, E34		
Iona Low Foam	Intracrop	A0255	spreader/wetter
Contains	921 g/l alkyl phenol ethoxylate and silicone antifoam agent		
Use with	Pesticides that have a recommendation for the use of a wetter/spreader		
Protective clothing	A, C		
Precautions	R04a, R04b, U05a, U08, U19, U20b, C03, E01, E13c, E30a, E31a, E34		
Jogral	Intracrop	A0226	cationic surfactant
Contains	800 g/l tallow amine ethoxylate		
Use with	Glyphosate		
Protective clothing	A, C		
Precautions	M03, R03c, R07d, U08, U19, U20a, C03, E01, E13b, E28, E31b, E32a, E34		
Kandu	Zeneca	A0264	surfactant
Contains	A blend of surfactant and ammonium sulphate		
Use with	Approved formulations of glyphosate-trimesium		
Protective clothing	C		
Precautions	U05a, U08, U19, U20b, C03, E01, E13c, E30a, E31c		
Klipper	Service Chemicals	A0260	spreader/wetter
Contains	600 g/l ethylene oxide condensate		
Use with	Approved salt formulations of mecoprop alone or in mixtures with 2,4-D on managed amenity turf		
Protective clothing	A, C		
Precautions	R04a, R04b, R07d, U05a, U08, U20c, C03, E01, E15, E30a, E31a		

Product	Supplier	Adj. No.	Type
LI-700	Newman	A0176	acidifier/drift retardant/penetrant

Contains: 350 g/l modified soya lecithin, 100 g/l alkylphenyl hydroxypolyoxyethylene and 350 g/l propionic acid

Use with: All plant growth regulators for cereals and oilseed rape, dimethoate and chlorpyrifos for wheat bulb fly control. For other uses see label or contact supplier for details

Protective clothing: A, C

Precautions: M03, R04b, R04d, U02, U04a, U05a, U10, U11, U19, U20a, C03, E01, E13c, E30a, E31a

---

Product	Supplier	Adj. No.	Type
Libsorb	Allied Colloids	A0251	spreader/wetter

Contains: alkyl alcohol ethoxylate

Use with: Any spray for which additional wetter is approved and recommended

Protective clothing: A, C

Precautions: R04a, R04b, U05a, U08, U20a, C03, E01, E13c, E30a, E31a

---

Product	Supplier	Adj. No.	Type
Lightning	Rigby Taylor	A0280	mineral oil

Contains: 4.84% w/w paraffinic oil

Use with: Roundup Pro Biactive

Precautions: U08, U19, U20a, E15, E30a, E31a

Product	Supplier	Adj. No.	Type
Lo Dose G	Intracrop	A0349	wetter

Contains: 431 g/l polyoxyethylated tallow amine

Use with: All approved formulations of glyphosate used pre-harvest in wheat, barley and oats, post-harvest on stubbles and in non-crop areas. In addition may be used with all approved herbicides up to their full approved dose on a range of specified fruit and vegetable crops

Protective clothing: A, C

Precautions: M03, R04a, R04b, U05a, U08, U13, U19, U20b, C03, E01, E13c, E30a, E31a, E34, E37

Product	Supplier	Adj. No.	Type
Logic	Microcide	A0288	vegetable oil

Contains: 95% w/w rape seed or soya oil

Use with: All approved pesticides on edible and non-edible crops

Protective clothing: A, C

Precautions: U19, U20b, E30a, E31b

Product	Supplier	Adj. No.	Type
Luxan Non-Ionic Wetter	Luxan	A0139	non-ionic surfactant/spreader/wetter

Contains: 900 g/l alkyl phenol ethylene oxide condensate

Use with: Any spray for which additional wetter/spreader is approved and recommended

Protective clothing: A, C

Product	Supplier	Adj. No.	Type
Precautions	R04a, R04b, R07d, U05a, U08, U20b, C03, E01, E13c, E30a, E31b, E34		

Luxan Oil-H   Luxan   A0321   mineral oil

Contains	800 g/l mineral oil
Use with	Approved formulations of phenmedipham on sugar beet, fodder beet and mangels
Precautions	C03, E15, E30a, E31b, E34, U05a, U08, U19

Lyrex   Batsons   A0177   mineral oil

Contains	96.8% w/w paraffinic base oil
Use with	Asulox, Basagran, Benlate, Checkmate, Gesaprim 500 SC, Goltix WG, Goltix WG/Checkmate, Goltix WG/Dow Shield, Goltix WG/Fusilade 5, Goltix WG/Checkmate/Laser, Laser, Pilot, Roundup

Lyrol   Batsons   A0165   mineral oil

Contains	96.8% w/w refined paraffinic oil
Use with	Asulox, Basagran, Benlate, Checkmate, Gesaprim 500 SC, Goltix WG, Goltix WG/Checkmate, Goltix WG/Dow Shield, Goltix WG/Fusilade 5, Goltix WG/Laser/Checkmate, Laser, Pilot and Roundup
Precautions	U08, U19, U20a, E13c, E30a, E31a

Mangard   Mandops   A0132   adjuvant/anti-transpirant

Contains	96% w/w di-1-p-menthene
Use with	Glyphosate and a wide range of other herbicides, fungicides and insecticides – see label for details
Precautions	U20c, E15, E30a, E31a

Minder   Stoller   A0087   vegetable oil

Contains	94% vegetable oil
Use with	Glyphosate
Precautions	U20c, E01, E15, E30a, E32a

Mixture B   Service Chemicals   A0161   non-ionic surfactant/ spreader/wetter

Contains	500 g/l nonyl phenol ethylene oxide condensate and 500 g/l primary alcohol ethylene oxide condensate
Use with	Glyphosate in forestry, amenity and non-crop situations
Protective clothing	A, C
Precautions	M03, R03a, R03c, R04a, R04b, U05a, U09a, C03, E01, E13c, E30a, E31a, E34

New Ethokem   Greenhill   A0318   wetter

Contains	471 g/l polyethoxylated tallow amine

Product	Supplier	Adj. No.	Type

Use with	All approved formulations of glyphosate in stubble and pre-harvest treatments of cereals and in non-crop areas; also with all approved pesticides on a wide range of specified agricultural and horticultural crops
Precautions	U05a, U08, U20a, C03, E01, E13c, E30a, E31a

Newman Cropspray 11E Newman   A0195   mineral oil

Contains	99% highly refined mineral oil
Use with	Approved herbicides for which the addition of an adjuvant oil is recommended
Precautions	R03, R04g, U19, U20a, E13c, E26, E30a, E31b

Newman's T-80   Newman   A0192   spreader/wetter

Contains	78% w/w polyoxyethylene tallow amine
Use with	Glyphosate
Protective clothing	A, C
Precautions	M04, R03c, R04a, R04b, U05a, U08, U13, U19, U20a, C03, E01, E13c, E16e, E17, E30a, E31a, E34

Nion   Service Chemicals   A0258   spreader/wetter

Contains	90% (900 g/l) ethylene oxide condensate
Use with	Pesticides which have a recommendation for use with a wetter/spreader
Protective clothing	A, C
Precautions	R04a, R04b, U05a, U08, U19, U20b, C03, E01, E13c, E30a, E31a

Nu Film 17   Intracrop   A0055   sticker/wetter

Contains	96% di-1-p-menthene
Use with	A range of fungicides, insecticides and foliar nutrients. See label for details
Precautions	U19, U20c, C02 (30 d), E13c, E30a, E32a

Nu Film P   Intracrop   A0039   sticker/wetter

Contains	96% poly-1-p-menthene
Use with	Glyphosate and many other pesticides and growth regulators for which a protectant is recommended. Do not use in mixture with adjuvant oils or surfactants
Precautions	U19, U20b, E13c, E30a, E31a

Output   Zeneca   A0163   mineral oil/surfactant

Contains	60% mineral oil and 40% surfactants
Use with	Grasp

Product	Supplier	Adj. No.	Type
Protective clothing	A, C		
Precautions	R04b, U05a, U08, U19, U20a, C03, E01, E13c, E30a, E31b		
Partna	Zeneca	A0246	mineral oil/surfactant
Contains	A blend of surfactants and mineral oils		
Use with	Formulations of fluazifop-P-butyl		
Protective clothing	A		
Precautions	U05a, U08, U19, U20b, C03, E01, E13c, E30a, E31c		
Pentil	Nufarm Whyte	A0340	non-ionic surfactant/wetter
Contains	450 g/l propionic acid and alkyphenylhydroxypolyoxyethylene		
Use with	Approved pesticides applied in amenity and forestry situations and on a range of specified arable and vegetable crops		
Protective clothing	A, C		
Precautions	M03, U02, U04a, U05a, U10, U11, U19, U20b, C03, E01, E13c, E30a, E31a		
Perm-E8	Intracrop	A0304	penetrant/surfactant
Contains	42% w/w propionic acid		
Use with	Approved chlormequat formulations and other pesticides on a range of agricultural and horticultural crops		
Protective clothing	A, C		
Precautions	M05, R05, R05b, U02, U04a, U05a, U10, U11, U13, U14, U15, U19, U20b, C03, E01, E30a, E31a		
Phase	Newman	A0279	vegetable oil
Contains	75% w/w methylated mixed fatty acid esters of rapeseed oil and 25% w/w proprietary surfactant blend		
Use with	Herbicides and fungicides in sugar beet up to 6 leaves; triazole and morpholine fungicides in winter and spring sown cereals. For other uses see label or contact supplier for details		
Protective clothing	A, H		
Precautions	U02, U05a, U14, U20b, C03, E01, E13c, E30a, E31a, E34		
Phase II	Newman	A0331	
Contains	95% w/w esterified rapeseed oil		
Use with	Pesticides approved for use in sugar beet, oilseed rape, cereals (for grass weed control), and other specified agricultural and horticultural crops		
Precautions	U19, U20b, E13c, E26, E30a, E31b		
Pin-o-Film	Intracrop	A0209	sticker/wetter
Contains	96% di-1-p-menthene		

Product	Supplier	Adj. No.	Type

Use with Approved pesticides on agricultural horticultural and forestry crops. Do not use in tank mixture with adjuvant oils or surfactants

Precautions U19, U20b, C02 (30 d), E13c, E30a, E32a

Planet Intracrop A0224 non-ionic surfactant/spreader/wetter

Contains 85% alkyl polyglycol ether and fatty acid

Use with Any spray for which additional wetter is recommended

Precautions R03c, R04a, R07d, U05a, U08, U19, U20b, E01, E13c, E30a, E31a, E32a

---

Prima Newman A0310 mineral oil

Contains 99% highly refined mineral oil

Use with All pesticides on all crops when used up to 50% of maximum approved dose for that use (mixtures with Roundup must only be used for treatment of stubbles); all approved pesticides at full dose on listed crops (see label). For other uses see label or contact supplier for details

Precautions R03, R04g, U19, U20b, E13c, E26, E30a, E31b

Q-900 Non-ionic Wetter Greenhill A0261 spreader/wetter

Contains 90% w/w (921 g/l) alkyl phenol ethoxylate

Use with Pesticides which have a recommendation for the use of a wetter/spreader

Protective clothing A, C

Precautions R04a, R04b, U05a, U08, U19, U20b, C03, E01, E13c, E30a, E31a, E34

Quadrangle Quad-Fast Quadrangle A0054 coating agent/surfactant

Contains di-1-p-menthene

Use with Glyphosate and other pesticides. Growth regulators and nutrients for which a coating agent is approved and recommended

Precautions U20b, E15, E30a, E31a

Ranger Newman A0296 acidifier/drift retardant/penetrant

Contains 350 g/l modified soya lecithin, 100 g/l alkylphenyl hydroxypolyoxyethylene and 350 g/l propionic acid

Use with All plant growth regulators for cereals and oilseed rape, dimethoate and chlorpyrifos for wheat bulb fly control. For other uses see label or contact supplier for details

Protective clothing A, C

Precautions M03, R04b, R04d, U02, U04a, U05a, U10, U11, U19, U20a, C03, E01, E13c, E30a, E31a

---

Rapide Intracrop A0116 penetrant/surfactant

Contains 40% propionic acid

Product	Supplier	Adj. No.	Type
Use with	A wide range of pesticides and growth regulators, especially chlormequat		
Protective clothing	A, C		
Precautions	R04a, R04b, R05, R05b, U02, U05a, U08, U19, U20a, C03, E01, E30a, E31a		

Retain	Euroagkem	A0263	mineral oil/non-ionic surfactant
Contains	60% w/w mineral oil and 40% w/w non-ionic surfactants		
Use with	Tralkoxydim		
Protective clothing	A, C		
Precautions	R04b, U05a, U08, U19, C03, E01, E13c, E30a, E31b		

Ryda	Interagro	A0168	cationic surfactant/wetter
Contains	800 g/l polyethoxylated tallow amine		
Use with	Glyphosate		
Protective clothing	A, C		
Precautions	M03, R03c, R04a, R04b, R07d, U02, U05a, U08, U13, U19, U20a, C03, E01, E13c, E30a, E31b, E34		

SAS 90	Intracrop	A0311	spreader/wetter
Contains	100% polyoxyethylen-alpha-methyl-omega-[3-(1,3,3,3-tetramethyl-3-trimethylsiloxy)-disiloxanyl] propylether		
Use with	A wide range of pesticides on specified crops in agriculture and horticulture, and on amenity vegetation and land not intended for cropping		
Precautions	U02, U08, U19, U20b, E13c, E30a, E31a, E34		

Scout	Newman	A0207	acidifier/drift retardant/non-ionic surfactant/penetrant
Contains	350 g/l modified soya lecithin, 100 g/l alkylphenyl hydroxypolyoxyethylene and 350 g/l propionic acid		
Use with	All plant growth regulators for cereals and oilseed rape, dimethoate and chlorpyrifos for wheat bulb fly control. For other uses see label or contact supplier for details		
Protective clothing	A, C		
Precautions	M03, R04b, R04d, U02, U04a, U05a, U10, U11, U19, U20b, C03, E01, E13c, E30a, E31a		

Signal	Intracrop	A0308	extender/sticker
Contains	450 g/l synthetic latex		
Use with	All approved pesticides where addition of a sticker/extender is required and recommended		
Protective clothing	A, C		

Product	Supplier	Adj. No.	Type
Precautions	M03, R04a, R04b, U02, U05a, U08, U19, U20b, C03, E01, E13c, E30a, E31a		
Siltex	Intracrop	A0398	spreader/sticker/wetter
Contains	27% w/w polyoxyethylen-alpha-methyl-omega-[3-(1,3,3,3-tetramethyl-3-trimethylsiloxy)-disiloxanyl] propylether and 7.3% w/w cocoiminodipropionate		
Use with	Approved pesticides used at full dose on a range of specified fruit and vegetable crops, and on managed amenity turf		
Protective clothing	A, C		
Precautions	U05a, U08, U20c, C03, E01, E13c, E30a, E31a, E37		
Silwet L-77	Newman	A0193	drift retardant/spreader/wetter
Contains	80% w/w polyalkylene oxide modified heptamethyltrisiloxane and a maximum of 20 % w/w allyloxypolyethylene glycol methyl ether		
Use with	All approved fungicides on winter and spring sown cereals; all approved pesticides applied at 50% or less of their full approved dose. A wide range of other uses - see label or contact supplier for details		
Protective clothing	A, C		
Precautions	M03, U02, U05a, U08, U19, U20a, C03, E01, E13c, E30a, E31a, E34		
Slippa	Interagro	A0206	spreader
Contains	655 g/l polyalkyleneoxide modified heptamethyltrisiloxane + non-ionic wetters		
Use with	Cereal fungicides and a wide range of other pesticides and trace elements on specified crops		
Protective clothing	A, C		
Precautions	M03, R03a, R03c, R04b, R04d, R04e, U02, U05a, U08, U11, U19, U20a, C03, E01, E13c, E30a, E31a, E34		
Slyx	Euroagkem	A0250	spreader/vegetable oil/wetter
Contains	95% refined rapeseed oil		
Use with	Pesticides that have a recommendation for the addition of a wetter or spreader		
Protective clothing	A, C		
Precautions	M03, R04a, U04a, U05a, U08, U20b, C03, E01, E13c, E30a, E31b, E34		
SM99	Newman	A0134	mineral oil
Contains	99% w/w highly refined paraffinic oil		
Use with	Approved herbicides for which the addition of an adjuvant oil is recommended		
Precautions	R03, R04g, U19, U20a, E13c, E15, E30a, E31b		

Product	Supplier	Adj. No.	Type
Solar	Intracrop	A0225	activator/non-ionic surfactant/spreader
Contains	75% polypropoxypropanol and 15% alkyl polyglycol ether		
Use with	Foliar applied plant growth regulators		
Protective clothing	A, C		
Precautions	M03, R04a, R04b, U04a, U05a, U08, U20a, C03, E01, E13c, E30a, E31b, E32a, E34		
Sprayfast	Mandops	A0131	extender/sticker/wetter
Contains	di-1-p-menthene and nonyl phenol ethylene oxide condensate		
Use with	Glyphosate and other pesticides, growth regulators or nutrients for which a coating agent is approved and recommended		
Precautions	U20b, E13c, E30a, E31a		
Sprayfix	Newman	A0145	drift retardant/extender/sticker/wetter
Contains	450 g/l synthetic latex and 100 g/l alkylphenyl hydroxypolyoxyethylene		
Use with	Blight sprays on potatoes and other pesticides on a wide range of crops. See label or contact supplier for details		
Protective clothing	A, C		
Precautions	M03, R04a, R04b, U02, U05a, U08, U19, U20a, C03, E01, E13c, E30a, E31a		
Spraymac	Newman	A0144	acidifier/non-ionic surfactant
Contains	350 g/l propionic acid and 100 g/l alkylphenyl hydroxypolyoxyethylene		
Use with	All plant growth regulators for cereals and oilseed rape and any spray for which the addition of a wetter/spreader is recommended. See label or contact supplier for details		
Protective clothing	A, C		
Precautions	M03, R04a, R04b, U02, U04a, U05a, U10, U11, U19, U20a, C03, E01, E13c, E30a, E32a		
Sprayprover	Fine	A0238	mineral oil
Contains	95% highly refined mineral oil		
Use with	Any spray for which the addition of a mineral oil is approved and recommended		
Protective clothing	A, C		
Precautions	R04a, R04b, U05a, U08, U19, U20a, C03, E01, E13c, E30a, E31b		
Spreader	PBI	A0189	non-ionic surfactant/spreader/wetter
Contains	nonylphenol ethylene oxide condensate		
Use with	Any spray for which additional wetter is approved and recommended		
Precautions	U20b, E15, E30a, E31a, E34		

Product	Supplier	Adj. No.	Type
Squat	Greenhill	A0179	penetrant
Contains	40% w/w propionic acid		
Use with	Agricultural herbicides, systemic fungicides and desiccants where rapid penetration is required		
Protective clothing	A, C		
Precautions	R04a, R04b, R05b, U02, U05a, U08, U19, U20b, C03, E01, E13c, E26, E30a, E31a		
Stamina	Interagro	A0202	penetrant
Contains	100% w/w alkoxylated fatty amine		
Use with	Glyphosate and a wide range of other pesticides on specified crops		
Protective clothing	A, C		
Precautions	M03, R03c, R04b, U02, U05a, U20a, C03, E01, E13b, E30a, E34		
Stefes Abet	Stefes	A0259	spreader/wetter
Contains	90% w/w ethylene oxide condensate		
Use with	All approved pesticides which have a recommendation for use with a wetting or spreading agent		
Stefes CAT 80	Stefes	A0182	cationic surfactant
Contains	800 g/l (78% w/w) polyoxyethylene tallow amine		
Use with	Glyphosate		
Protective clothing	A, C		
Precautions	M03, R03, U05a, U08, U13, U19, U20a, C03, E01, E13c, E30a, E31a, E34		
Stefes Cover	Stefes	A0187	spreader/wetter
Contains	921 g/l (90% w/w) alkyl phenol ethoxylate		
Use with	Any spray for which additional wetter or spreader is approved and recommended		
Protective clothing	A, C		
Precautions	R04a, R04b, U05a, U08, U19, U20a, C03, E01, E13c, E30a, E31a, E34		
Stefes Eros	Stefes	A0257	vegetable oil
Contains	95% w/w methylated vegetable oil		
Use with	All approved herbicides for cereals, all approved pesticides for forestry and managed amenity turf; all approved formulations of glyphosate; all approved pesticides on all edible and non-edible crops when used at less than 50% of approved dose		
Stefes Oberon	Stefes	A0292	wetter
Contains	500 g/l alkoxylated fatty amine and 500 g/l polyoxyethylene monolaurate		

Product	Supplier	Adj. No.	Type
Use with	All approved herbicides on cereals; all approved fungicides and insecticides on cereals up to GS 52 or at any time when used at less than 50% approved dose; all approved pesticides for forestry and managed amenity turf. Consult label or supplier for details of other uses		

Stefes Spread and Seal Stefes   A0122   anti-transpirant/extender/sticker/wetter

Contains	di-1-p-menthene and nonyl phenol ethylene oxide condensate
Use with	Glyphosate and other pesticides, growth regulators and nutrients for which a coating agent is approved and recommended
Precautions	U20b, C02 (4 wk), E13c, E30a, E31a

Stik-It   Quadrangle   A0071   non-ionic surfactant/spreader/wetter

Use with	A wide range of fungicides and insecticides for which additional wetter is approved and recommended
Precautions	U20b, E15, E30a, E31a, E34

Sward   Service Chemicals   A0215   spreader/wetter

Contains	15.2% w/w polyalkylene oxide modified heptomethyltrisiloxane
Use with	Specified pesticides in amenity situations
Protective clothing	A, C
Precautions	R04, R04d, R04e, U05a, U08, U20c, C03, E01, E13c, E30a, E31a

Swirl   Cyanamid   A0167   mineral oil

Contains	590 g/l highly refined mineral oil
Use with	Aztec, Commando
Precautions	R07d, U08, U20a, E13b, E30a, E31b, E34

T 25   Intracrop   A0352   spreader/wetter

Contains	800 g/l polyoxyethylated tallow amine
Use with	Specified herbicides when used at half or less than their approved dose in a wide range of arable crops, forestry and amenity situations (see label for details) and also with approved formulations of diquat and glyphosate in specified situations up to the full approved dose of the herbicide
Protective clothing	A, C
Precautions	M03, R03c, R04a, R04b, R07d, U02, U05a, U08, U13, U19, U20b, C03, E01, E13b, E16e, E30a, E31a, E34

Talzene   Intracrop   A0233   wetter

Contains	800 g/l polyoxyethylene tallow amine
Use with	Glyphosate
Protective clothing	A, C

Product	Supplier	Adj. No.	Type
Precautions	M03, R03c, R04a, R04b, U05a, U08, U13, U19, U20b, C03, E01, E13c, E30a, E31a, E34		

Tech E	Greenhill	A0275	wetter
Contains	471 g/l polyethoxylated tallow amine		
Use with	All approved formulations of glyphosate in stubble and pre-harvest treatments of cereals and in non-crop areas; also with all approved pesticides on a wide range of specified agricultural and horticultural crops		
Precautions	U05a, U08, U20a, C03, E01, E13c, E30a, E31a		

Tech G	Greenhill	A0274	wetter
Contains	431 g/l polyethoxylated tallow amine		
Use with	All approved formulations of glyphosate in stubble and pre-harvest treatments of cereals and in non-crop areas; also with all approved pesticides on a wide range of specified agricultural and horticultural crops		
Precautions	U05a, U08, U20a, C03, E01, E13c, E30a, E31a		

Techsol Spreader	Greenhill	A0196	spreader/wetter
Contains	23% w/w (235 gm/litre) alkyl phenol ethoxylate		
Use with	Pesticides which have a recommendation for the use of a wetter/ spreader		
Protective clothing	A, C		
Precautions	R04a, R04b, U02, U05a, U08, U19, U20b, C03, E01, E13c, E30a, E31a, E34		

Toil	Interagro	A0248	vegetable oil
Contains	95% w/w methylated vegetable oil		
Use with	Sugar beet herbicides, oilseed rape herbicides, cereal graminicides and a wide range of other pesticides on specified crops		
Protective clothing	A		
Precautions	R04b, U02, U05a, U08, U20a, C03, E01, E15, E30a, E31a		

Top Up G	Greenhill	A0357	wetter
Contains	431 g/l polyoxyethylated tallow amine		
Use with	All approved formulations of glyphosate used pre-harvest in wheat, barley and oats, post-harvest on stubbles and in non-crop areas. In addition may be used with all approved herbicides up to their full approved dose on a range of specified fruit and vegetable crops		
Protective clothing	A, C		
Precautions	M03, R04a, R04b, U05a, U08, U13, U19, U20b, C03, E01, E13c, E30a, E31a, E34, E37		

Product	Supplier	Adj. No.	Type
Torpedo	Newman	A0336	penetrant/wetter
Contains	400 g/l alkoxylated tallow amine, 375 g/l alkylphenylhydroxypolyoxyethylene and 75 g/l natural fatty acids		
Use with	Sulfonylurea herbicides and their approved tank mixes, cereal fungicides, oilseed rape fungicides and a wide range of other pesticides on specified crops		
Protective clothing	A, C		
Precautions	M03, R03c, R04a, R04b, R04d, U02, U04a, U05a, U10, U11, U19, U20b, C03, E01, E13b, E16e, E17, E30a, E31a, E34		
Torpedo II	Newman	A0393	wetter
Contains	215.5 g/litre alkoxylated tallow amine, 375 g/litre alkylphenyl hydroxypolyoxyethylene and 75 g/litre natural fatty acids		
Use with	All pesticides on edible and non-edible crops when used half or less than the full approved dose and with products up to their full approved dose on a wide range of specified fruit and vegetable crops.		
Protective clothing	A, C		
Precautions	R04a, R04b, U05a, U10, U11, U12, U20c, C03, E01, E13c, E30a, E31a		
Tripart Acer	Tripart	A0097	acidifier/penetrant
Contains	750 g/l soyal phospholipids		
Use with	Chlormequat and a wide range of systemic pesticides and trace elements. Contact supplier for details		
Protective clothing	A, C		
Precautions	M03, R04a, R04b, U02, U05a, U08, U19, U20b, C03, E01, E13c, E30a, E31a		
Tripart Lentus	Tripart	A0117	extender/sticker
Contains	450 g/l synthetic latex		
Use with	A wide range of contact herbicides, fungicides and insecticides. Contact supplier for further details		
Protective clothing	A, C		
Precautions	R04a, R04b, U05a, U08, U19, U20a, C03, E01, E13c, E30a, E31a		
Tripart Minax	Tripart	A0108	non-ionic surfactant/spreader/wetter
Contains	900 g/l alkyl alcohol ethoxylate		
Use with	Any spray for which additional wetter or spreader is approved and recommended		
Precautions	E30a, E31a		
Tripart Orbis	Tripart	A0204	mineral oil
Contains	99% w/w highly refined paraffinic oil		

Product	Supplier	Adj. No.	Type
Use with	A wide range of approved pesticides on sugar beet, cereals, oilseed rape, maize, sweetcorn, fodder beet, mangels and red beet (including aerial use where recommended). See label for details		
Precautions	U19, U20a, E07, E15, E26, E30a, E31b		
Wayfarer	Hortichem	A0045	cationic surfactant/spreader/wetter
Contains	80% tallow amine ethoxylate		
Use with	Glyphosate in agriculture		
Protective clothing	A, C		
Precautions	R04a, R04b, R07d, U05a, U08, U20b, C03, E01, E15, E30a, E31a		

SECTION 3
CROP/PEST GUIDE

Crop/Pest Guide Index

Important Note: The Crop/Pest Guide Index refers to pages on which the subject can be located.

Crop/Pest Guide

Important note: For convenience, some crops and pests or targets have been brought together into generic groups in this guide, eg "cereals", "annual grasses". It is essential to check the profile entry in Section 4 *and* the label to ensure that a product is approved for a specific crop/pest combination, eg winter wheat/blackgrass.

Arable crops

Arable crops

Pests	Slugs and snails	metaldehyde

Arable general

Diseases	Soil-borne diseases	dazomet, methyl bromide with amyl acetate, methyl bromide with chloropicrin
Pests	Birds	aluminium ammonium sulphate *(seed treatment)*, ziram
	Damaging mammals	aluminium ammonium sulphate *(seed treatment)*, ziram
	Nematodes	dazomet, methyl bromide with amyl acetate, methyl bromide with chloropicrin
	Slugs and snails	aluminium sulphate, methiocarb
	Soil pests	dazomet, methyl bromide with amyl acetate, methyl bromide with chloropicrin
Weeds	Annual and perennial weeds	glyphosate (agricultural uses) *(wiper application)*, glyphosate (agricultural uses)
	Annual dicotyledons	diquat + paraquat, glufosinate-ammonium, glyphosate (agricultural uses), paraquat *(minimum cultivation)*
	Annual grasses	diquat + paraquat, fluazifop-P-butyl, glufosinate-ammonium, glyphosate (agricultural uses), paraquat *(minimum cultivation)*
	Annual weeds	amitrole, glyphosate (agricultural uses)
	Barren brome	fluazifop-P-butyl, paraquat *(stubble treatment)*
	Couch	amitrole
	Creeping bent	paraquat *(minimum cultivation)*
	Creeping thistle	amitrole
	Docks	amitrole
	General weed control	dazomet, methyl bromide with amyl acetate, methyl bromide with chloropicrin
	Green cover	diquat + paraquat, paraquat

	Perennial dicotyledons	glufosinate-ammonium
	Perennial grasses	diquat + paraquat, glufosinate-ammonium
	Perennial weeds	amitrole
	Volunteer cereals	diquat + paraquat, fluazifop-P-butyl, glyphosate (agricultural uses), paraquat *(minimum cultivation)*
	Wild oats	fluazifop-P-butyl

Beans, field/broad

Diseases	Ascochyta	benomyl, metalaxyl + thiabendazole + thiram *(seed treatment)*, thiabendazole + thiram *(seed treatment)*
	Chocolate spot	benomyl, chlorothalonil, chlorothalonil + cyproconazole, cyproconazole *(with chlorothalonil)*, iprodione, tebuconazole, vinclozolin
	Damping off	metalaxyl + thiabendazole + thiram *(seed treatment)*, thiabendazole + thiram *(seed treatment)*, thiram *(seed treatment)*
	Downy mildew	chlorothalonil + metalaxyl, fosetyl-aluminium, metalaxyl + thiabendazole + thiram *(seed treatment)*
	Rust	chlorothalonil + cyproconazole, cyproconazole, fenpropimorph, tebuconazole
Pests	Aphids	dimethoate, disulfoton, fatty acids, nicotine, pirimicarb
	Birds	aluminium ammonium sulphate
	Caterpillars	nicotine
	Damaging mammals	aluminium ammonium sulphate
	Leaf miners	oxamyl *(off-label)*
	Mealy bugs	fatty acids
	Pea and bean weevils	alpha-cypermethrin, cypermethrin, deltamethrin, lambda-cyhalothrin, lambda-cyhalothrin + pirimicarb, zeta-cypermethrin
	Red spider mites	fatty acids
	Scale insects	fatty acids
	Whitefly	fatty acids
Weeds	Annual dicotyledons	bentazone, chlorpropham + fenuron, cyanazine *(Scotland only)*, cyanazine, cyanazine + pendimethalin, fomesafen + terbutryn, propyzamide, simazine, simazine + trietazine, terbuthylazine + terbutryn, terbutryn + trietazine, trifluralin
	Annual grasses	chlorpropham + fenuron, cyanazine, cyanazine *(Scotland only)*, cycloxydim, fluazifop-P-butyl, fluazifop-P-butyl *(off-label)*, propaquizafop, propyzamide, simazine, terbuthylazine + terbutryn, trifluralin
	Annual meadow grass	cyanazine + pendimethalin, terbutryn + trietazine

	Annual weeds	glyphosate (agricultural uses)
	Barren brome	sethoxydim
	Blackgrass	cycloxydim, fluazifop-P-butyl, sethoxydim, tri-allate
	Chickweed	chlorpropham + fenuron
	Cleavers	pendimethalin *(off-label)*
	Couch	cycloxydim, glyphosate (agricultural uses), sethoxydim
	Creeping bent	cycloxydim, glyphosate (agricultural uses), sethoxydim
	Perennial grasses	cycloxydim, fluazifop-P-butyl *(off-label)*, fluazifop-P-butyl, glyphosate (agricultural uses), propaquizafop, propyzamide, sethoxydim, tri-allate
	Rough meadow grass	cyanazine + pendimethalin
	Volunteer cereals	cycloxydim, fluazifop-P-butyl, glyphosate (agricultural uses), propyzamide, sethoxydim
	Wild oats	cycloxydim, fluazifop-P-butyl, propyzamide, sethoxydim, tri-allate
Crop control	Pre-harvest desiccation	diquat *(stock or pigeon feed only)*, glufosinate-ammonium, glyphosate (agricultural uses)
Plant growth regulation	Increasing yield	chlormequat with di-1-p-menthene
	Lodging control	chlormequat with di-1-p-menthene

Beet crops

Diseases	Black leg	hymexazol
	Damping off	hymexazol
	Disease control/ foliar feed	carbendazim + prochloraz *(off-label)*
	Powdery mildew	carbendazim + flusilazole, copper sulphate + sulphur, cyproconazole, propiconazole, sulphur, triadimefon, triadimenol
	Ramularia leaf spots	cyproconazole, propiconazole, propiconazole *(reduction)*
	Rust	carbendazim + flusilazole, cyproconazole, fenpropimorph *(off-label)*, propiconazole
	Seed-borne diseases	thiram *(seed soak)*
Pests	Aphids	aldicarb, carbosulfan, deltamethrin + pirimicarb, dimethoate *(excluding Myzus persicae)*, disulfoton, imidacloprid *(seed treatment)*, lambda-cyhalothrin + pirimicarb, oxamyl, pirimicarb, triazamate *(including Myzus persicae)*
	Birds	aluminium ammonium sulphate
	Caterpillars	cypermethrin

FOR FULL DETAILS ALWAYS CHECK THE PROFILE AND THE PRODUCT LABEL

	Cutworms	cypermethrin, gamma-HCH, lambda-cyhalothrin, lambda-cyhalothrin + pirimicarb, methiocarb *(reduction)*, zeta-cypermethrin
	Damaging mammals	aluminium ammonium sulphate
	Docking disorder vectors	aldicarb, benfuracarb, carbofuran, oxamyl
	Flea beetles	carbofuran, carbosulfan, deltamethrin, gamma-HCH, imidacloprid *(seed treatment)*, lambda-cyhalothrin, lambda-cyhalothrin + pirimicarb
	Leaf miners	aldicarb, benfuracarb, dimethoate, lambda-cyhalothrin, lambda-cyhalothrin + pirimicarb
	Leatherjackets	chlorpyrifos, gamma-HCH, methiocarb *(reduction)*
	Mangold fly	carbofuran, carbosulfan, dimethoate, disulfoton, imidacloprid *(seed treatment)*, lambda-cyhalothrin, oxamyl
	Millipedes	aldicarb, benfuracarb, carbofuran, carbosulfan, gamma-HCH, imidacloprid *(seed treatment)*, methiocarb *(reduction)*, oxamyl, tefluthrin *(seed treatment)*
	Nematodes	aldicarb, carbofuran, carbosulfan
	Pygmy beetle	aldicarb, benfuracarb, carbofuran, carbosulfan, chlorpyrifos, gamma-HCH, imidacloprid *(seed treatment)*, oxamyl, tefluthrin *(seed treatment)*
	Slugs and snails	methiocarb
	Springtails	benfuracarb, carbofuran, carbosulfan, gamma-HCH, imidacloprid *(seed treatment)*, tefluthrin *(seed treatment)*
	Symphylids	benfuracarb, carbosulfan, gamma-HCH, imidacloprid *(seed treatment)*, tefluthrin *(seed treatment)*
	Tortrix moths	carbofuran
	Virus vectors	deltamethrin + pirimicarb
	Wireworms	carbofuran, carbosulfan, gamma-HCH
Weeds	Annual dicotyledons	chloridazon, chloridazon + ethofumesate, clopyralid, desmedipham + ethofumesate + phenmedipham, desmedipham + phenmedipham, diquat, ethofumesate, ethofumesate + metamitron + phenmedipham, ethofumesate + phenmedipham, glufosinate-ammonium, lenacil, lenacil + phenmedipham, metamitron, paraquat, phenmedipham, propyzamide, trifluralin, triflusulfuron-methyl
	Annual grasses	cycloxydim, fluazifop-P-butyl, glufosinate-ammonium, metamitron, paraquat, propaquizafop, propyzamide, quizalofop-P-ethyl, trifluralin
	Annual meadow grass	chloridazon, chloridazon + ethofumesate, desmedipham + ethofumesate + phenmedipham, ethofumesate, ethofumesate + metamitron + phenmedipham, ethofumesate + phenmedipham, lenacil, metamitron
	Annual weeds	glyphosate (agricultural uses)
	Barren brome	sethoxydim

	Blackgrass	chloridazon + ethofumesate, cycloxydim, ethofumesate, ethofumesate + phenmedipham, sethoxydim, tri-allate
	Corn marigold	clopyralid
	Couch	cycloxydim, quizalofop-P-ethyl, sethoxydim
	Creeping bent	cycloxydim, sethoxydim
	Creeping thistle	clopyralid
	Fat-hen	metamitron
	Mayweeds	clopyralid
	Perennial dicotyledons	clopyralid
	Perennial grasses	cycloxydim, fluazifop-P-butyl, propaquizafop, propyzamide, quizalofop-P-ethyl, sethoxydim, tri-allate
	Polygonums	sodium chloride (commodity substance)
	Volunteer cereals	cycloxydim, fluazifop-P-butyl, glyphosate (agricultural uses), paraquat, propyzamide, quizalofop-P-ethyl, sethoxydim
	Volunteer potatoes	sodium chloride (commodity substance)
	Volunteer sugar beet	glyphosate (agricultural uses) *(wiper application)*
	Weed beet	glyphosate (agricultural uses) *(wiper application)*
	Wild oats	cycloxydim, fluazifop-P-butyl, propyzamide, sethoxydim, tri-allate
Plant growth regulation	Increasing yield	sulphur

Cereals

Diseases	Alternaria	iprodione
	Blue mould	fuberidazole + imidacloprid + triadimenol *(seed treatment)*, fuberidazole + triadimenol *(seed treatment)*
	Botrytis	carbendazim + chlorothalonil + maneb, fenpropidin + tebuconazole, iprodione

Brown foot rot and ear blight	bitertanol + fuberidazole + imidacloprid *(seed treatment - reduction)*, bromuconazole, carbendazim + chlorothalonil + maneb, carboxin + imazalil + thiabendazole *(partial control)*, carboxin + thiram *(seed treatment)*, epoxiconazole *(reduction)*, epoxiconazole + fenpropimorph *(reduction)*, epoxiconazole + fenpropimorph + kresoxim-methyl *(reduction)*, epoxiconazole + kresoxim-methyl *(reduction)*, fenpropidin + tebuconazole, fludioxonil *(seed treatment)*, fuberidazole + imidacloprid + triadimenol *(seed treatment)*, fuberidazole + imidacloprid + triadimenol *(seed treatment - reduction)*, fuberidazole + triadimenol *(seed treatment)*, guazatine *(seed treatment - reduction)*, guazatine + imazalil *(seed treatment - reduction)*, imazalil *(seed treatment)*, imazalil + triticonazole *(moderate control)*, metconazole *(reduction)*, spiroxamine + tebuconazole, tebuconazole, tebuconazole + triadimenol
Covered smut	carboxin + imazalil + thiabendazole, fludioxonil *(seed treatment)*, fuberidazole + imidacloprid + triadimenol *(seed treatment)*, fuberidazole + triadimenol *(seed treatment)*, imazalil + triticonazole
Crown rust	chlorothalonil + tetraconazole, cyproconazole, fuberidazole + triadimenol *(seed treatment)*, propiconazole, tebuconazole *(reduction)*, tebuconazole + triadimenol, tetraconazole
Eyespot	bromuconazole *(reduction)*, carbendazim, carbendazim + chlorothalonil + maneb, carbendazim + flusilazole, carbendazim + maneb, carbendazim + prochloraz, carbendazim + propiconazole, chlorothalonil + cyproconazole, cyproconazole, cyproconazole + cyprodinil, cyproconazole + prochloraz, cyproconazole + quinoxyfen *(reduction)*, cyproconazole + tridemorph, cyprodinil, epoxiconazole *(reduction)*, epoxiconazole + fenpropimorph *(reduction)*, epoxiconazole + fenpropimorph + kresoxim-methyl *(reduction)*, epoxiconazole + kresoxim-methyl *(reduction)*, fenpropidin + prochloraz, fenpropimorph + prochloraz, flusilazole, maneb, prochloraz, prochloraz + tebuconazole, propiconazole *(with carbendazim)*, propiconazole *(low levels only)*
Foot rot	fludioxonil *(seed treatment)*, guazatine + imazalil *(seed treatment)*
Fusarium root rot	bitertanol + fuberidazole *(seed treatment)*, bitertanol + fuberidazole + imidacloprid *(seed treatment)*, fluquinconazole + prochloraz *(seed treatment)*
Late ear diseases	azoxystrobin, azoxystrobin + fenpropimorph, carbendazim, chlorothalonil, cyproconazole, cyproconazole + propiconazole, prochloraz + tebuconazole
Leaf stripe	carboxin + imazalil + thiabendazole, carboxin + thiram *(seed treatment - reduction)*, fludioxonil *(seed treatment - reduction)*, fuberidazole + triadimenol *(seed treatment)*, guazatine + imazalil *(seed treatment)*, imazalil *(seed treatment)*, imazalil + triticonazole, imidacloprid + tebuconazole + triazoxide *(seed treatment)*, tebuconazole + triazoxide *(seed treatment)*

Loose smut   bitertanol + fuberidazole *(seed treatment - reduction)*, bitertanol + fuberidazole + imidacloprid *(seed treatment - reduction)*, carboxin + imazalil + thiabendazole, carboxin + imazalil + thiabendazole *(partial control)*, carboxin + thiram *(seed treatment - reduction)*, fuberidazole + imidacloprid + triadimenol *(seed treatment)*, fuberidazole + triadimenol *(seed treatment)*, imazalil + triticonazole, imidacloprid + tebuconazole + triazoxide *(seed treatment)*, tebuconazole + triazoxide *(seed treatment)*

Net blotch   azoxystrobin, azoxystrobin + fenpropimorph, bromuconazole, carbendazim + chlorothalonil + maneb, carboxin + imazalil + thiabendazole *(moderate control)*, chlorothalonil + cyproconazole, chlorothalonil + tetraconazole, cyproconazole, cyproconazole + cyprodinil, cyproconazole + prochloraz, cyproconazole + propiconazole, cyproconazole + tridemorph, cyprodinil, epoxiconazole, epoxiconazole + fenpropimorph, epoxiconazole + fenpropimorph + kresoxim-methyl, epoxiconazole + kresoxim-methyl, fenpropidin + prochloraz, fenpropidin + propiconazole, fenpropidin + propiconazole + tebuconazole, fenpropidin + tebuconazole, fenpropimorph + flusilazole, fenpropimorph + prochloraz, fenpropimorph + propiconazole, flusilazole, fuberidazole + imidacloprid + triadimenol *(seed treatment - seed-borne only)*, guazatine + imazalil *(seed treatment)*, imazalil *(seed treatment)*, imazalil + triticonazole *(moderate control)*, imidacloprid + tebuconazole + triazoxide, iprodione, mancozeb, maneb, metconazole *(reduction)*, prochloraz, prochloraz + tebuconazole, propiconazole, propiconazole + tebuconazole, spiroxamine + tebuconazole, tebuconazole, tebuconazole + triadimenol, tebuconazole + triazoxide *(seed treatment)*, trifloxystrobin

Powdery mildew   azoxystrobin, azoxystrobin + fenpropimorph, bromuconazole, carbendazim *(partial control)*, carbendazim + chlorothalonil + maneb, carbendazim + flusilazole, carbendazim + maneb, carbendazim + prochloraz, carbendazim + propiconazole, chlorothalonil + cyproconazole, chlorothalonil + fluquinconazole *(reduction)*, chlorothalonil + tetraconazole, copper oxychloride + maneb + sulphur, cyproconazole, cyproconazole + cyprodinil, cyproconazole + prochloraz, cyproconazole + propiconazole, cyproconazole + quinoxyfen, cyproconazole + tridemorph, cyprodinil, epoxiconazole, epoxiconazole + fenpropimorph, epoxiconazole + fenpropimorph + kresoxim-methyl, epoxiconazole + kresoxim-methyl, fenbuconazole *(reduction)*, fenbuconazole + propiconazole *(moderate control)*, fenpropidin, fenpropidin + fenpropimorph, fenpropidin + prochloraz, fenpropidin + propiconazole, fenpropidin + propiconazole + tebuconazole, fenpropidin + tebuconazole, fenpropimorph, fenpropimorph + flusilazole, fenpropimorph + kresoxim-methyl, fenpropimorph + prochloraz, fenpropimorph + propiconazole, fenpropimorph + quinoxyfen, fluquinconazole, fluquinconazole + prochloraz, fluquinconazole + prochloraz *(moderate control)*, flusilazole, fuberidazole + imidacloprid + triadimenol *(seed treatment)*, fuberidazole + triadimenol *(seed treatment)*, maneb, metconazole *(moderate control)*, metconazole, prochloraz, prochloraz *(protection)*, prochloraz + tebuconazole, propiconazole, propiconazole + tebuconazole, quinoxyfen, spiroxamine, spiroxamine + tebuconazole, sulphur, tebuconazole, tebuconazole + triadimenol, tetraconazole, triadimefon, triadimefon *(off-label - research/breeding)*, triadimenol

Pyrenophora leaf spot	fuberidazole + imidacloprid + triadimenol *(seed treatment)*, fuberidazole + triadimenol *(seed treatment)*, guazatine + imazalil *(seed treatment)*
Rhynchosporium	azoxystrobin *(off-label)*, azoxystrobin, azoxystrobin + fenpropimorph, bromuconazole, carbendazim, carbendazim + chlorothalonil + maneb, chlorothalonil, chlorothalonil + cyproconazole, chlorothalonil + tetraconazole, copper oxychloride + maneb + sulphur, cyproconazole, cyproconazole + cyprodinil, cyproconazole + prochloraz, cyproconazole + propiconazole, cyproconazole + tridemorph, cyprodinil, epoxiconazole, epoxiconazole + fenpropimorph, epoxiconazole + fenpropimorph + kresoxim-methyl, epoxiconazole + kresoxim-methyl, fenbuconazole, fenpropidin, fenpropidin + fenpropimorph *(moderate control)*, fenpropidin + prochloraz, fenpropidin + propiconazole, fenpropidin + propiconazole + tebuconazole, fenpropidin + tebuconazole, fenpropimorph, fenpropimorph + flusilazole, fenpropimorph + kresoxim-methyl, fenpropimorph + prochloraz, fenpropimorph + propiconazole, fluquinconazole + prochloraz, flusilazole, mancozeb, maneb, metconazole, prochloraz, prochloraz + propiconazole, prochloraz + tebuconazole, propiconazole, propiconazole + tebuconazole, spiroxamine *(reduction)*, spiroxamine + tebuconazole, tebuconazole, tebuconazole + triadimenol, triadimenol, trifloxystrobin
Rust	azoxystrobin, azoxystrobin *(off-label)*, azoxystrobin + fenpropimorph, bromuconazole, carbendazim + chlorothalonil + maneb, carbendazim + flusilazole, carbendazim + maneb, carbendazim + propiconazole, chlorothalonil + cyproconazole, chlorothalonil + fluquinconazole, chlorothalonil + tetraconazole, cyproconazole, cyproconazole + cyprodinil, cyproconazole + prochloraz, cyproconazole + propiconazole, cyproconazole + quinoxyfen, cyproconazole + tridemorph, difenoconazole, epoxiconazole, epoxiconazole + fenpropimorph, epoxiconazole + fenpropimorph + kresoxim-methyl, epoxiconazole + kresoxim-methyl, fenbuconazole, fenbuconazole + propiconazole, fenpropidin, fenpropidin + fenpropimorph, fenpropidin + prochloraz, fenpropidin + propiconazole, fenpropidin + propiconazole + tebuconazole, fenpropidin + tebuconazole, fenpropimorph, fenpropimorph + flusilazole, fenpropimorph + prochloraz, fenpropimorph + propiconazole, fluquinconazole, fluquinconazole + prochloraz, flusilazole, fuberidazole + imidacloprid + triadimenol *(seed treatment)*, fuberidazole + triadimenol *(seed treatment)*, mancozeb, maneb, metconazole, prochloraz + tebuconazole, propiconazole, propiconazole + tebuconazole, spiroxamine, spiroxamine + tebuconazole, tebuconazole, tebuconazole + triadimenol, tetraconazole, triadimefon *(off-label - research/breeding)*, triadimefon, triadimenol, trifloxystrobin

	Septoria diseases	azoxystrobin, azoxystrobin + fenpropimorph, bitertanol + fuberidazole *(seed treatment - reduction)*, bromuconazole, carbendazim *(partial control)*, carbendazim + chlorothalonil + maneb, carbendazim + flusilazole, carbendazim + maneb, carbendazim + prochloraz, carbendazim + propiconazole, carboxin + imazalil + thiabendazole, carboxin + thiram *(seed treatment)*, chlorothalonil, chlorothalonil + cyproconazole, chlorothalonil + fluquinconazole, chlorothalonil + tetraconazole, copper oxychloride + maneb + sulphur, cyproconazole, cyproconazole + cyprodinil, cyproconazole + prochloraz, cyproconazole + propiconazole, cyproconazole + quinoxyfen, cyproconazole + tridemorph, difenoconazole, epoxiconazole, epoxiconazole + fenpropimorph, epoxiconazole + fenpropimorph + kresoxim-methyl, epoxiconazole + kresoxim-methyl, fenbuconazole, fenbuconazole + propiconazole, fenpropidin, fenpropidin + prochloraz, fenpropidin + propiconazole, fenpropidin + propiconazole + tebuconazole, fenpropidin + tebuconazole, fenpropimorph + flusilazole, fenpropimorph + kresoxim-methyl *(reduction)*, fenpropimorph + prochloraz, fenpropimorph + propiconazole, fluquinconazole, fluquinconazole + prochloraz *(seed treatment)*, fluquinconazole + prochloraz, flusilazole, fuberidazole + imidacloprid + triadimenol *(seed treatment - reduction)*, fuberidazole + imidacloprid + triadimenol *(seed treatment)*, fuberidazole + triadimenol *(seed treatment)*, guazatine *(seed treatment)*, iprodione, mancozeb, maneb, prochloraz, prochloraz + propiconazole, prochloraz + tebuconazole, propiconazole, propiconazole + tebuconazole, spiroxamine + tebuconazole, tebuconazole, tebuconazole + triadimenol, tetraconazole, triadimenol
	Snow mould	fludioxonil *(seed treatment)*
	Snow rot	fuberidazole + imidacloprid + triadimenol *(seed treatment - reduction)*, propiconazole, triadimefon, triadimenol
	Sooty moulds	bromuconazole, carbendazim, carbendazim + chlorothalonil + maneb, carbendazim + maneb, chlorothalonil + tetraconazole, cyproconazole + tridemorph, epoxiconazole *(reduction)*, epoxiconazole + fenpropimorph *(reduction)*, epoxiconazole + fenpropimorph + kresoxim-methyl *(reduction)*, epoxiconazole + kresoxim-methyl *(reduction)*, fenpropidin + tebuconazole, fenpropimorph + propiconazole, mancozeb, maneb, propiconazole, spiroxamine + tebuconazole, tebuconazole, tebuconazole + triadimenol, tetraconazole
	Stinking smut	bitertanol + fuberidazole *(seed treatment)*, bitertanol + fuberidazole + imidacloprid *(seed treatment)*, carboxin + thiram *(seed treatment)*, fludioxonil *(seed treatment)*, fluquinconazole + prochloraz *(seed treatment)*, fuberidazole + imidacloprid + triadimenol *(seed treatment)*, fuberidazole + triadimenol *(seed treatment)*
	Take-all	fluquinconazole + prochloraz *(reduction (seed treatment))*
Pests	Aphids	alpha-cypermethrin, bifenthrin, chlorpyrifos, cypermethrin *(summer)*, cypermethrin, deltamethrin, deltamethrin + pirimicarb *(on ears)*, dimethoate, esfenvalerate, lambda-cyhalothrin, lambda-cyhalothrin + pirimicarb, pirimicarb, tau-fluvalinate, zeta-cypermethrin
	Barley yellow dwarf vectors	cypermethrin, deltamethrin

Barley yellow dwarf virus vectors	cypermethrin, deltamethrin, lambda-cyhalothrin, tau-fluvalinate, zeta-cypermethrin
Birds	aluminium ammonium sulphate
Cutworms	gamma-HCH
Damaging mammals	aluminium ammonium sulphate, sulphonated cod liver oil
Frit fly	chlorpyrifos
Leatherjackets	chlorpyrifos, fenitrothion, gamma-HCH, methiocarb *(reduction)*
Saddle gall midge	fenitrothion
Slugs and snails	metaldehyde *(seed admixture)*, methiocarb, thiodicarb *(seed admixture)*, thiodicarb
Thrips	chlorpyrifos, fenitrothion
Virus vectors	bifenthrin, bitertanol + fuberidazole + imidacloprid *(seed treatment)*, fuberidazole + imidacloprid + triadimenol *(seed treatment)*, imidacloprid + tebuconazole + triazoxide *(seed treatment)*
Wheat bulb fly	chlorpyrifos, dimethoate, tefluthrin
Wheat-blossom midges	chlorpyrifos, fenitrothion
Wireworms	bitertanol + fuberidazole + imidacloprid *(reduction of damage)*, fuberidazole + imidacloprid + triadimenol *(reduction of damage)*, gamma-HCH, imidacloprid + tebuconazole + triazoxide *(reduction of damage)*, tefluthrin
Yellow cereal fly	alpha-cypermethrin, cypermethrin, deltamethrin, lambda-cyhalothrin

Weeds	Annual dicotyledons	2,4-D, 2,4-DB + MCPA, amidosulfuron, benazolin + bromoxynil + ioxynil, bromoxynil, bromoxynil + diflufenican + ioxynil, bromoxynil + ioxynil, bromoxynil + ioxynil + mecoprop-P, bromoxynil + ioxynil + triasulfuron, carfentrazone-ethyl + flupyrsulfuron-methyl, carfentrazone-ethyl + isoproturon, carfentrazone-ethyl + mecoprop-P, carfentrazone-ethyl + metsulfuron-methyl, carfentrazone-ethyl + thifensulfuron-methyl, chlorotoluron, chlorotoluron + isoproturon, cinidon-ethyl, clodinafop-propargyl + diflufenican, clodinafop-propargyl + trifluralin, clopyralid, cyanazine, cyanazine + pendimethalin, cyanazine + terbuthylazine, DE 570, dicamba + MCPA + mecoprop, dicamba + MCPA + mecoprop-P, dicamba + mecoprop-P, dichlorprop, dichlorprop + MCPA, dichlorprop-P + MCPA + mecoprop-P, diflufenican + flurtamone, diflufenican + flurtamone + isoproturon, diflufenican + isoproturon, diflufenican + terbuthylazine, diflufenican + trifluralin, flupyrsulfuron-methyl, flupyrsulfuron-methyl + metsulfuron-methyl, flupyrsulfuron-methyl + thifensulfuron-methyl, fluroxypyr, imazamethabenz-methyl, isoproturon, isoproturon + pendimethalin, isoproturon + simazine, linuron, MCPA, MCPA + MCPB, mecoprop-P, methabenzthiazuron, metoxuron, metsulfuron-methyl, metsulfuron-methyl + thifensulfuron-methyl, metsulfuron-methyl + tribenuron-methyl, pendimethalin, pendimethalin + simazine, terbutryn, terbutryn + trifluralin, thifensulfuron-methyl + tribenuron-methyl, triasulfuron, tribenuron-methyl, trifluralin
	Annual grasses	chlorotoluron, chlorotoluron + isoproturon, cyanazine, diflufenican + isoproturon, isoproturon, isoproturon + pendimethalin, isoproturon + simazine, metoxuron, pendimethalin, terbutryn, terbutryn + trifluralin, trifluralin
	Annual meadow grass	carfentrazone-ethyl + isoproturon, cyanazine + pendimethalin, cyanazine + terbuthylazine, diflufenican + flurtamone, diflufenican + flurtamone + isoproturon, diflufenican + terbuthylazine, diflufenican + trifluralin, fenoxaprop-P-ethyl + isoproturon, isoproturon *(off-label)*, linuron, methabenzthiazuron, pendimethalin, terbutryn, terbutryn + trifluralin
	Annual weeds	glyphosate (agricultural uses)
	Barren brome	metoxuron
	Black bindweed	dichlorprop, dichlorprop + MCPA, fluroxypyr, linuron
	Blackgrass	carfentrazone-ethyl + flupyrsulfuron-methyl, chlorotoluron, chlorotoluron + isoproturon, clodinafop-propargyl, clodinafop-propargyl + diflufenican, clodinafop-propargyl + trifluralin, diclofop-methyl + fenoxaprop-P-ethyl, diflufenican + flurtamone, diflufenican + flurtamone + isoproturon, diflufenican + isoproturon, fenoxaprop-P-ethyl, fenoxaprop-P-ethyl + isoproturon, flupyrsulfuron-methyl, flupyrsulfuron-methyl + metsulfuron-methyl, flupyrsulfuron-methyl + thifensulfuron-methyl, imazamethabenz-methyl, isoproturon, isoproturon + pendimethalin, isoproturon + simazine, methabenzthiazuron, metoxuron, pendimethalin, terbutryn, terbutryn + trifluralin, tralkoxydim, tri-allate
	Canary grass	fenoxaprop-P-ethyl, fenoxaprop-P-ethyl + isoproturon, tralkoxydim *(qualified minor use)*

Chickweed

bromoxynil + diflufenican + ioxynil, carfentrazone-ethyl + isoproturon, carfentrazone-ethyl + mecoprop-P, cinidon-ethyl, cyanazine + pendimethalin, DE 570, dicamba + MCPA + mecoprop, dicamba + MCPA + mecoprop-P, dicamba + mecoprop-P, dichlorprop-P + MCPA + mecoprop-P, fenoxaprop-P-ethyl + isoproturon, fluroxypyr, linuron, mecoprop-P, metsulfuron-methyl, metsulfuron-methyl + thifensulfuron-methyl, metsulfuron-methyl + tribenuron-methyl, terbutryn + trifluralin, thifensulfuron-methyl + tribenuron-methyl, triasulfuron, tribenuron-methyl

Cleavers

amidosulfuron, carfentrazone-ethyl, carfentrazone-ethyl + isoproturon, carfentrazone-ethyl + mecoprop-P, carfentrazone-ethyl + thifensulfuron-methyl, cinidon-ethyl, DE 570, dicamba + MCPA + mecoprop, dicamba + MCPA + mecoprop-P, dicamba + mecoprop-P, dichlorprop-P + MCPA + mecoprop-P, fluroxypyr, mecoprop-P, metsulfuron-methyl + thifensulfuron-methyl, pendimethalin

Corn marigold

clopyralid, fenoxaprop-P-ethyl + isoproturon, linuron

Couch

amitrole *(direct drilling)*, glyphosate (agricultural uses)

Creeping bent

glyphosate (agricultural uses)

Creeping thistle

clopyralid

Docks

dicamba + MCPA + mecoprop, fluroxypyr

Fat-hen

linuron, MCPA

Field pansy

cyanazine + pendimethalin, dichlorprop-P + MCPA + mecoprop-P, metsulfuron-methyl + thifensulfuron-methyl

Hemp-nettle

dichlorprop + MCPA, fenoxaprop-P-ethyl + isoproturon, fluroxypyr, MCPA

Loose silky bent

carfentrazone-ethyl + isoproturon, diflufenican + flurtamone, diflufenican + flurtamone + isoproturon, fenoxaprop-P-ethyl, imazamethabenz-methyl

Mayweeds

bromoxynil + diflufenican + ioxynil, carfentrazone-ethyl + isoproturon, carfentrazone-ethyl + thifensulfuron-methyl, clopyralid, cyanazine + pendimethalin, dicamba + MCPA + mecoprop, dicamba + MCPA + mecoprop-P, dicamba + mecoprop-P, dichlorprop-P + MCPA + mecoprop-P, fenoxaprop-P-ethyl + isoproturon, metsulfuron-methyl, metsulfuron-methyl + thifensulfuron-methyl, metsulfuron-methyl + tribenuron-methyl, terbutryn + trifluralin, thifensulfuron-methyl + tribenuron-methyl, triasulfuron, tribenuron-methyl

Perennial dicotyledons

2,4-D, 2,4-DB + MCPA, clopyralid, dicamba + MCPA + mecoprop, dicamba + MCPA + mecoprop-P, dicamba + mecoprop-P, dichlorprop, dichlorprop + MCPA, MCPA, MCPA + MCPB, mecoprop-P

Perennial grasses

glyphosate *(agricultural uses)*, imazamethabenz-methyl, pendimethalin + simazine, tralkoxydim *(qualified minor use)*, tri-allate

Perennial ryegrass

terbutryn

	Polygonums	2,4-DB + MCPA, bromoxynil + diflufenican + ioxynil, carfentrazone-ethyl + thifensulfuron-methyl, dicamba + MCPA + mecoprop, dicamba + MCPA + mecoprop-P, dicamba + mecoprop-P, dichlorprop, dichlorprop + MCPA, linuron, metsulfuron-methyl + thifensulfuron-methyl
	Rough meadow grass	carfentrazone-ethyl + isoproturon, chlorotoluron, chlorotoluron + isoproturon, clodinafop-propargyl, clodinafop-propargyl + diflufenican, clodinafop-propargyl + trifluralin, cyanazine + pendimethalin, fenoxaprop-P-ethyl, isoproturon, methabenzthiazuron, terbutryn, tralkoxydim
	Ryegrass	diclofop-methyl + fenoxaprop-P-ethyl, tralkoxydim
	Speedwells	bromoxynil + diflufenican + ioxynil, carfentrazone-ethyl, carfentrazone-ethyl + mecoprop-P, carfentrazone-ethyl + thifensulfuron-methyl, cinidon-ethyl, metsulfuron-methyl + thifensulfuron-methyl, pendimethalin, terbutryn + trifluralin
	Volunteer cereals	glyphosate (agricultural uses)
	Volunteer oilseed rape	DE 570, diflufenican + flurtamone, imazamethabenz-methyl
	Volunteer potatoes	fluroxypyr
	Wild oats	chlorotoluron, chlorotoluron + isoproturon, clodinafop-propargyl, clodinafop-propargyl + diflufenican, clodinafop-propargyl + trifluralin, diclofop-methyl + fenoxaprop-P-ethyl, difenzoquat, diflufenican + isoproturon, fenoxaprop-P-ethyl, fenoxaprop-P-ethyl + isoproturon, flamprop-M-isopropyl, glyphosate (agricultural uses) *(wiper application)*, imazamethabenz-methyl, isoproturon, isoproturon + pendimethalin, pendimethalin, tralkoxydim, tri-allate
Crop control	Pre-harvest desiccation	diquat *(stockfeed only)*, glufosinate-ammonium *(not seed crops)*, glyphosate (agricultural uses)
Plant growth regulation	Increasing yield	2-chloroethylphosphonic acid + mepiquat chloride (*low lodging situations*), chlormequat, chlormequat + choline chloride, chlormequat + choline chloride + imazaquin, sulphur
	Lodging control	2-chloroethylphosphonic acid, 2-chloroethylphosphonic acid + mepiquat chloride, chlormequat, chlormequat + 2-chloroethylphosphonic acid, chlormequat + 2-chloroethylphosphonic acid + imazaquin, chlormequat + 2-chloroethylphosphonic acid + mepiquat chloride, chlormequat + choline chloride, chlormequat + choline chloride + imazaquin, chlormequat + mepiquat chloride, fuberidazole + imidacloprid + triadimenol *(seed treatment - reduction)*, trinexapac-ethyl

Cereals undersown

Weeds	Annual dicotyledons	2,4-D, 2,4-DB, 2,4-DB + linuron + MCPA, 2,4-DB + MCPA, 2,4-DB + MCPA *(red or white clover)*, benazolin + 2,4-DB + MCPA, bentazone + MCPA + MCPB, bromoxynil + ioxynil, dicamba + MCPA + mecoprop-P *(grass only)*, dicamba + MCPA + mecoprop-P, dichlorprop, MCPA, MCPA *(red clover or grass)*, MCPA + MCPB, MCPB
	Black bindweed	dichlorprop

	Chickweed	benazolin + 2,4-DB + MCPA, dicamba + MCPA + mecoprop-P *(grass only)*, dicamba + MCPA + mecoprop-P
	Cleavers	benazolin + 2,4-DB + MCPA, dicamba + MCPA + mecoprop-P
	Fat-hen	MCPA *(red clover or grass)*, MCPA
	Hemp-nettle	MCPA *(red clover or grass)*, MCPA
	Mayweeds	dicamba + MCPA + mecoprop-P, dicamba + MCPA + mecoprop-P *(grass only)*
	Perennial dicotyledons	2,4-D, 2,4-DB, 2,4-DB + MCPA *(red or white clover)*, 2,4-DB + MCPA, benazolin + 2,4-DB + MCPA, dicamba + MCPA + mecoprop-P *(grass only)*, dicamba + MCPA + mecoprop-P, dichlorprop, MCPA *(red clover or grass)*, MCPA + MCPB, MCPB
	Polygonums	2,4-DB + MCPA, benazolin + 2,4-DB + MCPA, dicamba + MCPA + mecoprop-P *(grass only)*, dicamba + MCPA + mecoprop-P, dichlorprop

Clovers

Weeds	Annual dicotyledons	2,4-DB + MCPA, benazolin + 2,4-DB + MCPA, MCPA + MCPB, MCPB, propyzamide
	Annual grasses	propyzamide
	Chickweed	benazolin + 2,4-DB + MCPA
	Cleavers	benazolin + 2,4-DB + MCPA
	Docks	asulam *(off-label)*
	Perennial dicotyledons	2,4-DB + MCPA, benazolin + 2,4-DB + MCPA, MCPA + MCPB, MCPB
	Perennial grasses	propyzamide
	Polygonums	2,4-DB + MCPA, benazolin + 2,4-DB + MCPA
Crop control	Pre-harvest desiccation	diquat

Fodder brassica seed crops

Pests	Aphids	dimethoate *(excluding Myzus persicae)*
	Leaf miners	dimethoate
Weeds	Annual dicotyledons	benazolin + clopyralid *(off-label)*, propyzamide
	Annual grasses	propyzamide
	Chickweed	benazolin + clopyralid *(off-label)*
	Cleavers	benazolin + clopyralid *(off-label)*
	Mayweeds	benazolin + clopyralid *(off-label)*
	Perennial grasses	propyzamide

Fodder brassicas

Diseases	Alternaria	iprodione *(seed treatment)*, iprodione
	Botrytis	chlorothalonil
	Ring spot	tebuconazole *(off-label)*
Pests	Aphids	alpha-cypermethrin, deltamethrin + pirimicarb, nicotine, pirimicarb
	Cabbage root fly	carbofuran
	Cabbage stem weevil	carbofuran
	Caterpillars	alpha-cypermethrin, cypermethrin, deltamethrin, deltamethrin + pirimicarb, nicotine
	Flea beetles	alpha-cypermethrin, carbofuran
	General insect control	dimethoate *(off-label)*
Weeds	Annual dicotyledons	chlorthal-dimethyl, clopyralid, cyanazine *(off-label)*, propachlor, sodium monochloroacetate, tebutam, trifluralin
	Annual grasses	cyanazine *(off-label)*, fluazifop-P-butyl *(stockfeed only)*, fluazifop-P-butyl *(off-label)*, propachlor, tebutam, trifluralin
	Corn marigold	clopyralid
	Creeping thistle	clopyralid
	Mayweeds	clopyralid
	Perennial dicotyledons	clopyralid
	Perennial grasses	fluazifop-P-butyl *(stockfeed only)*, fluazifop-P-butyl *(off-label)*
	Volunteer cereals	fluazifop-P-butyl *(stockfeed only)*, tebutam
	Wild oats	fluazifop-P-butyl *(stockfeed only)*

Grass seed crops

Diseases	Crown rust	propiconazole
	Drechslera leaf spot	propiconazole
	Powdery mildew	propiconazole
	Rhynchosporium	propiconazole
Pests	Aphids	deltamethrin *(off-label)*, dimethoate
Weeds	Annual dicotyledons	2,4-D, 2,4-D + mecoprop-P, bromoxynil + ethofumesate + ioxynil, clopyralid *(off-label)*, dicamba + MCPA + mecoprop, dicamba + MCPA + mecoprop-P, MCPA, mecoprop-P

FOR FULL DETAILS ALWAYS CHECK THE PROFILE AND THE PRODUCT LABEL

	Annual grasses	ethofumesate
	Annual meadow grass	bromoxynil + ethofumesate + ioxynil
	Blackgrass	ethofumesate
	Chickweed	dicamba + MCPA + mecoprop, dicamba + MCPA + mecoprop-P, ethofumesate, mecoprop-P
	Cleavers	dicamba + MCPA + mecoprop, dicamba + MCPA + mecoprop-P, ethofumesate, mecoprop-P
	Docks	dicamba + MCPA + mecoprop, dicamba + MCPA + mecoprop-P
	Fat-hen	MCPA
	Hemp-nettle	MCPA
	Mayweeds	dicamba + MCPA + mecoprop, dicamba + MCPA + mecoprop-P
	Perennial dicotyledons	2,4-D, 2,4-D + mecoprop-P, clopyralid *(off-label)*, dicamba + MCPA + mecoprop, dicamba + MCPA + mecoprop-P, MCPA, mecoprop-P
	Polygonums	dicamba + MCPA + mecoprop, dicamba + MCPA + mecoprop-P
	Ryegrass	chlorpropham *(off-label)*
	Volunteer cereals	ethofumesate
	Wild oats	difenzoquat
Plant growth regulation	Lodging control	trinexapac-ethyl

Linseed/flax

Diseases	Alternaria	iprodione *(seed treatment)*
	Botrytis	tebuconazole *(reduction)*, thiabendazole + thiram *(seed treatment)*
	Damping off	thiabendazole + thiram *(seed treatment)*
	Fusarium root rot	thiabendazole + thiram *(seed treatment)*
	Powdery mildew	tebuconazole
	Seed-borne diseases	prochloraz *(seed treatment)*
Weeds	Annual dicotyledons	amidosulfuron, bentazone, bromoxynil, bromoxynil + clopyralid, clopyralid, MCPA, metsulfuron-methyl, trifluralin
	Annual grasses	cycloxydim, fluazifop-P-butyl, propaquizafop, quizalofop-P-ethyl, sethoxydim, trifluralin
	Annual weeds	glyphosate (agricultural uses)
	Barren brome	sethoxydim

	Blackgrass	cycloxydim, sethoxydim
	Chickweed	metsulfuron-methyl
	Cleavers	amidosulfuron
	Corn marigold	clopyralid
	Couch	cycloxydim, glyphosate (agricultural uses), quizalofop-P-ethyl
	Creeping bent	cycloxydim
	Creeping thistle	clopyralid
	Fat-hen	MCPA
	Hemp-nettle	MCPA
	Mayweeds	clopyralid, metsulfuron-methyl
	Perennial dicotyledons	clopyralid, MCPA
	Perennial grasses	cycloxydim, fluazifop-P-butyl, propaquizafop, quizalofop-P-ethyl
	Volunteer cereals	cycloxydim, fluazifop-P-butyl, glyphosate (agricultural uses), quizalofop-P-ethyl, sethoxydim
	Wild oats	cycloxydim, fluazifop-P-butyl, sethoxydim
Crop control	Pre-harvest desiccation	diquat, glufosinate-ammonium, glyphosate (agricultural uses)

Lucerne

Weeds	Annual dicotyledons	2,4-DB, chlorpropham, paraquat *(off-label)*, propyzamide
	Annual grasses	chlorpropham, fluazifop-P-butyl *(off-label)*, paraquat *(off-label)*, propyzamide
	Chickweed	chlorpropham
	Perennial dicotyledons	2,4-DB
	Perennial grasses	propyzamide
	Polygonums	chlorpropham

Lupins

Diseases	Damping off	thiabendazole + thiram *(seed treatment)*
Weeds	Annual dicotyledons	terbuthylazine + terbutryn
	Annual grasses	sethoxydim, terbuthylazine + terbutryn
	Barren brome	sethoxydim
	Blackgrass	sethoxydim

Volunteer cereals	sethoxydim
Wild oats	sethoxydim

Maize/sweetcorn

Diseases	Damping off	thiram *(seed treatment)*
Pests	Aphids	nicotine, pirimicarb *(off-label)*, pirimicarb
	Caterpillars	nicotine
	Cutworms	gamma-HCH
	Frit fly	aldicarb *(off-label)*, carbofuran, chlorpyrifos, fenitrothion, lambda-cyhalothrin *(off-label)*
	Leatherjackets	gamma-HCH
	Wireworms	gamma-HCH
Weeds	Annual dicotyledons	atrazine, bromoxynil *(off-label)*, bromoxynil, bromoxynil + prosulfuron, bromoxynil + terbuthylazine, clopyralid, cyanazine, cyanazine + pendimethalin, fluroxypyr, pendimethalin *(off-label)*, pendimethalin, rimsulfuron, simazine
	Annual grasses	atrazine, cyanazine, pendimethalin *(off-label)*, simazine
	Annual meadow grass	cyanazine + pendimethalin, pendimethalin
	Black bindweed	bromoxynil + prosulfuron, fluroxypyr
	Black nightshade	pyridate
	Chickweed	bromoxynil + prosulfuron, fluroxypyr
	Cleavers	fluroxypyr, pyridate
	Corn marigold	clopyralid
	Creeping thistle	clopyralid
	Docks	fluroxypyr
	Fat-hen	pyridate
	Hemp-nettle	bromoxynil + prosulfuron, fluroxypyr
	Mayweeds	bromoxynil + prosulfuron, clopyralid
	Perennial dicotyledons	clopyralid
	Polygonums	bromoxynil + prosulfuron
	Rough meadow grass	cyanazine + pendimethalin
	Speedwells	pendimethalin
	Volunteer oilseed rape	rimsulfuron
	Volunteer potatoes	fluroxypyr

	Wild oats	difenzoquat

Miscellaneous field crops

Diseases	Botrytis	iprodione + thiophanate-methyl *(off-label)*
	Fungus diseases	propiconazole *(off-label)*
	Sclerotinia	iprodione + thiophanate-methyl *(off-label)*
Pests	Aphids	pirimicarb *(off-label)*
	Flea beetles	deltamethrin *(off-label)*
	General insect control	lambda-cyhalothrin *(off-label)*
	Leaf miners	oxamyl *(off-label)*
	Pollen beetles	deltamethrin *(off-label)*
Weeds	Annual dicotyledons	benazolin + clopyralid *(off-label)*, bentazone *(off-label)*, cyanazine *(off-label)*, MCPA + MCPB, metazachlor *(off-label)*, pendimethalin *(off-label)*, pendimethalin, trifluralin *(off-label)*
	Annual grasses	pendimethalin *(off-label)*, pendimethalin, sethoxydim *(off-label)*, trifluralin *(off-label)*
	Annual meadow grass	pendimethalin
	Blackgrass	pendimethalin, tri-allate
	Chickweed	benazolin + clopyralid *(off-label)*
	Cleavers	benazolin + clopyralid *(off-label)*, pendimethalin
	Creeping thistle	clopyralid *(off-label)*
	Mayweeds	benazolin + clopyralid *(off-label)*, clopyralid *(off-label)*
	Perennial dicotyledons	MCPA + MCPB
	Perennial grasses	sethoxydim *(off-label)*, tri-allate
	Speedwells	pendimethalin
	Wild oats	pendimethalin, tri-allate
Crop control	Pre-harvest desiccation	diquat *(off-label)*

Oilseed rape

Diseases	Alternaria	carbendazim + iprodione, carbendazim + prochloraz, carbendazim + tebuconazole, carbendazim + vinclozolin *(reduction)*, difenoconazole, iprodione *(seed treatment)*, iprodione, iprodione + thiophanate-methyl, metconazole, prochloraz, propiconazole, tebuconazole, vinclozolin
	Black scurf and stem canker	difenoconazole, iprodione + thiophanate-methyl, tebuconazole

FOR FULL DETAILS ALWAYS CHECK THE PROFILE AND THE PRODUCT LABEL

	Botrytis	carbendazim *(reduction)*, carbendazim + iprodione, carbendazim + prochloraz, carbendazim + vinclozolin *(reduction)*, chlorothalonil, iprodione, iprodione + thiophanate-methyl, prochloraz, thiophanate-methyl, vinclozolin
	Canker	carbendazim + flusilazole *(reduction)*, carbendazim + prochloraz, carbendazim + tebuconazole, prochloraz
	Damping off	thiram *(seed treatment)*
	Downy mildew	chlorothalonil, chlorothalonil + metalaxyl, mancozeb
	Light leaf spot	carbendazim *(reduction)*, carbendazim, carbendazim + flusilazole, carbendazim + iprodione, carbendazim + prochloraz, carbendazim + tebuconazole, carbendazim + vinclozolin *(reduction)*, cyproconazole, difenoconazole, flusilazole, iprodione + thiophanate-methyl, prochloraz, propiconazole *(reduction)*, propiconazole, tebuconazole, thiophanate-methyl
	Phoma leaf spot	carbendazim + flusilazole, carbendazim + tebuconazole, cyproconazole, prochloraz + propiconazole, tebuconazole
	Ring spot	carbendazim + tebuconazole *(reduction)*, tebuconazole *(reduction)*
	Sclerotinia	metconazole *(reduction)*
	Sclerotinia stem rot	carbendazim + iprodione, carbendazim + prochloraz, carbendazim + tebuconazole, carbendazim + vinclozolin, iprodione, iprodione + thiophanate-methyl, prochloraz, tebuconazole, thiophanate-methyl, vinclozolin
	White leaf spot	carbendazim + prochloraz, prochloraz
Pests	Aphids	deltamethrin, deltamethrin + pirimicarb, lambda-cyhalothrin, lambda-cyhalothrin + pirimicarb, pirimicarb, tau-fluvalinate
	Birds	aluminium ammonium sulphate
	Cabbage root fly	carbofuran
	Cabbage seed weevil	alpha-cypermethrin, deltamethrin, lambda-cyhalothrin
	Cabbage stem flea beetle	alpha-cypermethrin, bifenthrin, carbofuran, cypermethrin, deltamethrin *(in store)*, deltamethrin, lambda-cyhalothrin, lambda-cyhalothrin + pirimicarb, zeta-cypermethrin
	Cabbage stem weevil	deltamethrin
	Damaging mammals	aluminium ammonium sulphate, sulphonated cod liver oil
	Flea beetles	lambda-cyhalothrin, lambda-cyhalothrin + pirimicarb
	Pod midge	alpha-cypermethrin, cypermethrin, deltamethrin, lambda-cyhalothrin, lambda-cyhalothrin + pirimicarb, zeta-cypermethrin
	Pollen beetles	aipha-cypermethrin, cypermethrin, deltamethrin, lambda-cyhalothrin, lambda-cyhalothrin + pirimicarb, tau-fluvalinate, zeta-cypermethrin

	Rape winter stem weevil	alpha-cypermethrin, bifenthrin, carbofuran, cypermethrin, deltamethrin *(in store)*, deltamethrin, zeta-cypermethrin
	Seed weevil	cypermethrin, lambda-cyhalothrin, lambda-cyhalothrin + pirimicarb, zeta-cypermethrin
	Slugs and snails	methiocarb, thiodicarb
	Virus vectors	lambda-cyhalothrin
Weeds	Annual dicotyledons	benazolin + clopyralid *(off-label)*, benazolin + clopyralid, chlorthal-dimethyl, clopyralid, clopyralid + propyzamide, cyanazine, metazachlor, metazachlor + quinmerac, napropamide, propachlor, propyzamide, tebutam, trifluralin
	Annual grasses	clopyralid + propyzamide, cyanazine, cycloxydim, fluazifop-P-butyl, napropamide, propachlor, propaquizafop, propyzamide, quizalofop-P-ethyl, sethoxydim, tebutam, trifluralin
	Annual meadow grass	metazachlor, metazachlor + quinmerac
	Annual weeds	glyphosate (agricultural uses)
	Barren brome	clopyralid + propyzamide, sethoxydim
	Blackgrass	cycloxydim, metazachlor, metazachlor + quinmerac, sethoxydim
	Canary grass	sethoxydim
	Chickweed	benazolin, benazolin + clopyralid
	Cleavers	benazolin, benazolin + clopyralid, metazachlor + quinmerac, napropamide
	Corn marigold	clopyralid
	Couch	cycloxydim, glyphosate (agricultural uses), quizalofop-P-ethyl
	Creeping bent	cycloxydim, glyphosate (agricultural uses)
	Creeping thistle	clopyralid
	Groundsel	napropamide
	Groundsel, triazine resistant	metazachlor
	Mayweeds	benazolin + clopyralid, clopyralid, clopyralid + propyzamide
	Perennial dicotyledons	clopyralid, glyphosate (agricultural uses)
	Perennial grasses	cycloxydim, fluazifop-P-butyl, glyphosate (agricultural uses), propaquizafop, propyzamide, quizalofop-P-ethyl
	Volunteer cereals	cycloxydim, fluazifop-P-butyl, glyphosate (agricultural uses), propyzamide, quizalofop-P-ethyl, sethoxydim, tebutam
	Wild oats	cycloxydim, fluazifop-P-butyl, propyzamide, sethoxydim

FOR FULL DETAILS ALWAYS CHECK THE PROFILE AND THE PRODUCT LABEL

Crop control	Pre-harvest desiccation	diquat, glufosinate-ammonium, glyphosate (agricultural uses)
Plant growth regulation	Increasing yield	chlormequat with di-1-p-menthene, sulphur
	Lodging control	chlormequat with di-1-p-menthene

Peas

Diseases	Ascochyta	carbendazim + cymoxanil + oxadixyl + thiram *(seed treatment)*, chlorothalonil, metalaxyl + thiabendazole + thiram *(seed treatment)*, thiabendazole + thiram *(seed treatment)*, vinclozolin
	Botrytis	chlorothalonil, chlorothalonil + cyproconazole, vinclozolin
	Damping off	carbendazim + cymoxanil + oxadixyl + thiram *(seed treatment)*, metalaxyl + thiabendazole + thiram *(seed treatment)*, thiabendazole + thiram *(seed treatment)*, thiram *(seed treatment)*
	Downy mildew	carbendazim + cymoxanil + oxadixyl + thiram *(seed treatment)*, fosetyl-aluminium *(off-label)*, metalaxyl + thiabendazole + thiram *(seed treatment)*
	Mycosphaerella	chlorothalonil, vinclozolin
	Powdery mildew	triadimefon *(off-label - research/breeding)*
	Rust	triadimefon *(off-label - research/breeding)*
Pests	Aphids	alpha-cypermethrin, cypermethrin, deltamethrin, deltamethrin + pirimicarb, dimethoate, fatty acids, fenitrothion, lambda-cyhalothrin, lambda-cyhalothrin + pirimicarb, nicotine, pirimicarb, zeta-cypermethrin
	Birds	aluminium ammonium sulphate
	Caterpillars	nicotine
	Damaging mammals	aluminium ammonium sulphate
	General insect control	chlorpyrifos *(off-label)*
	Mealy bugs	fatty acids
	Pea and bean weevils	alpha-cypermethrin, cypermethrin, deltamethrin, fenitrothion, lambda-cyhalothrin, lambda-cyhalothrin + pirimicarb, zeta-cypermethrin
	Pea midge	dimethoate, fenitrothion, lambda-cyhalothrin + pirimicarb
	Pea moth	alpha-cypermethrin, cypermethrin, deltamethrin, fenitrothion, lambda-cyhalothrin, lambda-cyhalothrin + pirimicarb, zeta-cypermethrin
	Red spider mites	fatty acids
	Scale insects	fatty acids
	Slugs and snails	methiocarb
	Thrips	dimethoate, fenitrothion

	Whitefly	fatty acids
Weeds	Annual dicotyledons	bentazone, bentazone + MCPB, bentazone + pendimethalin, chlorpropham + fenuron, cyanazine, cyanazine + pendimethalin, fomesafen + terbutryn *(spring sown)*, MCPA + MCPB, MCPB, pendimethalin, prometryn, simazine + trietazine, terbuthylazine + terbutryn, terbutryn + trietazine, trifluralin *(off-label)*
	Annual grasses	chlorpropham + fenuron, cyanazine, cycloxydim, fluazifop-P-butyl, pendimethalin, prometryn, propaquizafop, sethoxydim, terbuthylazine + terbutryn, trifluralin *(off-label)*
	Annual meadow grass	cyanazine + pendimethalin, pendimethalin, terbutryn + trietazine
	Annual weeds	glyphosate (agricultural uses)
	Blackgrass	cycloxydim, pendimethalin, sethoxydim, tri-allate
	Chickweed	bentazone + pendimethalin, chlorpropham + fenuron
	Cleavers	bentazone + pendimethalin, pendimethalin
	Couch	cycloxydim, glyphosate (agricultural uses)
	Creeping bent	cycloxydim, glyphosate (agricultural uses)
	Mayweeds	bentazone + pendimethalin
	Perennial dicotyledons	MCPA + MCPB, MCPB
	Perennial grasses	cycloxydim, fluazifop-P-butyl, glyphosate (agricultural uses), propaquizafop, tri-allate
	Polygonums	bentazone + pendimethalin
	Rough meadow grass	cyanazine + pendimethalin
	Speedwells	pendimethalin
	Volunteer cereals	cycloxydim, fluazifop-P-butyl, glyphosate (agricultural uses), sethoxydim
	Wild oats	cycloxydim, fluazifop-P-butyl, pendimethalin, sethoxydim, tri-allate
Crop control	Pre-harvest desiccation	diquat, glufosinate-ammonium *(not seed crops)*, glyphosate (agricultural uses), sulphuric acid (commodity substance)
Plant growth regulation	Increasing yield	chlormequat with di-1-p-menthene
	Lodging control	chlormequat with di-1-p-menthene

Potatoes

Diseases	Black scurf and stem canker	imazalil + pencycuron *(tuber treatment - reduction)*, imazalil + pencycuron *(tuber treatment)*, imazalil + thiabendazole *(tuber treatment - reduction)*, iprodione *(seed treatment)*, pencycuron *(tuber treatment)*, tolclofos-methyl *(seed tuber treatment)*, tolclofos-methyl *(tuber treatment)*
	Blight	benalaxyl + mancozeb, Bordeaux mixture, chlorothalonil, chlorothalonil + mancozeb, chlorothalonil + metalaxyl, chlorothalonil + propamocarb hydrochloride, copper oxychloride, copper sulphate + sulphur, cymoxanil + mancozeb, cymoxanil + mancozeb + oxadixyl, dimethomorph + mancozeb, fentin acetate + maneb, fentin hydroxide, ferbam + maneb + zineb, fluazinam, mancozeb, mancozeb + metalaxyl, mancozeb + metalaxyl-M, mancozeb + ofurace *(off-label)*, mancozeb + ofurace, mancozeb + oxadixyl, mancozeb + propamocarb hydrochloride, maneb, zineb
	Dry rot	carbendazim + tecnazene, imazalil, imazalil + thiabendazole *(tuber treatment - reduction)*, tecnazene, tecnazene + thiabendazole, thiabendazole *(tuber treatment - post-harvest)*, thiabendazole *(post-harvest)*
	Gangrene	2-aminobutane, carbendazim + tecnazene, imazalil, imazalil + thiabendazole *(tuber treatment - reduction)*, tecnazene + thiabendazole, thiabendazole *(post-harvest)*, thiabendazole *(tuber treatment - post-harvest)*
	Rhizoctonia	tolclofos-methyl *(off-label - tuber treatment)*
	Silver scurf	2-aminobutane, carbendazim + tecnazene, imazalil, imazalil + pencycuron *(tuber treatment - reduction)*, imazalil + thiabendazole *(tuber treatment - reduction)*, tecnazene + thiabendazole, thiabendazole *(tuber treatment - post-harvest)*, thiabendazole *(post-harvest)*
	Skin spot	2-aminobutane, carbendazim + tecnazene, imazalil, imazalil + thiabendazole *(tuber treatment - reduction)*, tecnazene + thiabendazole, thiabendazole *(tuber treatment - post-harvest)*, thiabendazole *(post-harvest)*
Pests	Aphids	aldicarb, deltamethrin + pirimicarb, dimethoate *(excluding Myzus persicae)*, disulfoton, lambda-cyhalothrin, lambda-cyhalothrin + pirimicarb, nicotine, oxamyl, phorate, pirimicarb, pymetrozine
	Capsids	phorate
	Caterpillars	cypermethrin, nicotine
	Colorado beetle	chlorfenvinphos *(off-label)*, chlorpyrifos *(off-label)*, deltamethrin *(off-label - Statutory Notice)*
	Cutworms	chlorpyrifos, cypermethrin, lambda-cyhalothrin + pirimicarb, zeta-cypermethrin
	Leafhoppers	phorate
	Leatherjackets	methiocarb *(reduction)*
	Nematodes	1,3-dichloropropene, aldicarb, oxamyl

	Potato cyst nematode	1,3-dichloropropene, aldicarb, carbofuran, ethoprophos, fosthiazate, oxamyl
	Potato virus vectors	nicotine
	Slugs and snails	methiocarb, thiodicarb
	Spraing vectors	aldicarb, oxamyl
	Virus vectors	deltamethrin + pirimicarb
	Wireworms	ethoprophos, phorate *(damage reduction)*
Weeds	Annual dicotyledons	bentazone, cyanazine, cyanazine + pendimethalin, diquat, diquat + paraquat, glufosinate-ammonium, linuron, metribuzin, monolinuron, paraquat, pendimethalin, prometryn, rimsulfuron, terbuthylazine + terbutryn, terbutryn + trietazine
	Annual grasses	cyanazine, cycloxydim, diquat + paraquat, glufosinate-ammonium, metribuzin, monolinuron, paraquat, pendimethalin, prometryn, propaquizafop, terbuthylazine + terbutryn
	Annual meadow grass	cyanazine + pendimethalin, linuron, monolinuron, pendimethalin, terbutryn + trietazine
	Black bindweed	linuron
	Blackgrass	cycloxydim, pendimethalin, sethoxydim
	Chickweed	linuron
	Cleavers	pendimethalin
	Corn marigold	linuron
	Couch	cycloxydim, sethoxydim
	Creeping bent	cycloxydim, sethoxydim
	Fat-hen	linuron, monolinuron
	Perennial grasses	cycloxydim, diquat + paraquat, propaquizafop, sethoxydim
	Polygonums	linuron, monolinuron
	Rough meadow grass	cyanazine + pendimethalin
	Speedwells	pendimethalin
	Volunteer cereals	cycloxydim, diquat + paraquat, paraquat, sethoxydim
	Volunteer oilseed rape	metribuzin, rimsulfuron
	Wild oats	cycloxydim, pendimethalin, sethoxydim
Crop control	Haulm destruction	sulphuric acid (commodity substance)
	Pre-harvest desiccation	diquat, glufosinate-ammonium *(not seed crops)*
Plant growth regulation	Increasing yield	sulphur

	Sprout suppression	chlorpropham *(fog)*, chlorpropham, maleic hydrazide, tecnazene, tecnazene + thiabendazole
	Volunteer suppression	maleic hydrazide

Seed brassicas/mustard

Diseases	Alternaria	iprodione *(seed treatment)*, iprodione
	Botrytis	iprodione
Pests	Cabbage seed weevil	deltamethrin
	Cabbage stem weevil	deltamethrin
	Pod midge	deltamethrin
	Pollen beetles	deltamethrin
Weeds	Annual dicotyledons	benazolin + clopyralid *(off-label)*, chlorthal-dimethyl, propachlor, trifluralin
	Annual grasses	propachlor, quizalofop-P-ethyl, sethoxydim, trifluralin
	Annual weeds	glyphosate (agricultural uses)
	Barren brome	sethoxydim
	Blackgrass	sethoxydim
	Chickweed	benazolin + clopyralid *(off-label)*
	Cleavers	benazolin + clopyralid *(off-label)*
	Couch	glyphosate (agricultural uses), quizalofop-P-ethyl
	Mayweeds	benazolin + clopyralid *(off-label)*
	Perennial grasses	propaquizafop, quizalofop-P-ethyl
	Volunteer cereals	glyphosate (agricultural uses), quizalofop-P-ethyl, sethoxydim
	Wild oats	sethoxydim
Crop control	Pre-harvest desiccation	glyphosate (agricultural uses)

Field vegetables

Asparagus

Diseases	Fungus diseases	thiabendazole *(off-label)*
Pests	Aphids	nicotine
	Asparagus beetle	cypermethrin *(off-label)*

	Caterpillars	nicotine
Weeds	Annual and perennial weeds	glyphosate (top fruit, horticulture, forestry, amenity etc.) *(off-label)*
	Annual dicotyledons	diuron *(off-label)*, metamitron *(off-label)*, simazine, simazine *(off-label)*, terbacil
	Annual grasses	diuron *(off-label)*, simazine *(off-label)*, simazine, terbacil
	Cleavers	isoxaben *(off-label)*
	Couch	terbacil
	Creeping thistle	clopyralid *(off-label)*
	Fat-hen	isoxaben *(off-label)*
	Perennial grasses	terbacil

Beans, french/runner

Diseases	Anthracnose	carbendazim
	Botrytis	carbendazim, iprodione *(off-label)*, vinclozolin
	Damping off	thiram *(seed treatment)*
	Rust	tebuconazole *(off-label)*
	Sclerotinia	iprodione *(off-label)*
Pests	Aphids	dimethoate, fatty acids, nicotine, pirimicarb
	Caterpillars	lambda-cyhalothrin *(off-label)*, nicotine
	General insect control	chlorpyrifos *(off-label)*, nicotine *(off-label)*
	Mealy bugs	fatty acids
	Red spider mites	fatty acids
	Scale insects	fatty acids
	Whitefly	fatty acids
Weeds	Annual dicotyledons	bentazone *(off-label)*, bentazone, chlorpropham + fenuron *(off-label)*, chlorthal-dimethyl, fomesafen, monolinuron, pendimethalin *(off-label)*, simazine *(off-label)*, trifluralin
	Annual grasses	chlorpropham + fenuron *(off-label)*, cycloxydim, fluazifop-P-butyl *(off-label)*, monolinuron, pendimethalin *(off-label)*, simazine *(off-label)*, trifluralin
	Annual meadow grass	monolinuron
	Blackgrass	cycloxydim
	Chickweed	chlorpropham + fenuron *(off-label)*
	Couch	cycloxydim
	Creeping bent	cycloxydim

FOR FULL DETAILS ALWAYS CHECK THE PROFILE AND THE PRODUCT LABEL

	Fat-hen	monolinuron
	Perennial grasses	cycloxydim, fluazifop-P-butyl *(off-label)*
	Polygonums	monolinuron
	Volunteer cereals	cycloxydim
	Volunteer oilseed rape	fomesafen
	Wild oats	cycloxydim

Brassica seed crops

Diseases	Alternaria	iprodione
	Botrytis	iprodione
Weeds	Annual dicotyledons	propyzamide, trifluralin *(off-label)*
	Annual grasses	propyzamide, trifluralin *(off-label)*

Brassicas

Diseases	Alternaria	chlorothalonil, chlorothalonil + metalaxyl *(moderate control)*, difenoconazole, iprodione *(seed treatment)*, iprodione, tebuconazole, triadimenol
	Botrytis	chlorothalonil
	Damping off	etridiazole *(seedlings and transplants)*, fosetyl-aluminium *(off-label)*, thiram *(seed treatment)*
	Damping off and wirestem	tolclofos-methyl
	Downy mildew	chlorothalonil, chlorothalonil + metalaxyl, copper oxychloride + metalaxyl *(off-label)*, dichlofluanid, fosetyl-aluminium *(off-label)*, propamocarb hydrochloride
	Light leaf spot	benomyl, tebuconazole, triadimenol
	Phytophthora	propamocarb hydrochloride
	Powdery mildew	copper sulphate + sulphur, tebuconazole, triadimefon, triadimefon *(off-label - research/breeding)*, triadimenol
	Pythium	propamocarb hydrochloride
	Rhizoctonia	chlorothalonil
	Ring spot	benomyl, chlorothalonil, chlorothalonil + metalaxyl *(reduction)*, difenoconazole, tebuconazole *(off-label)*, tebuconazole, triadimenol
	Rust	triadimefon *(off-label - research/breeding)*
	Spear rot	copper oxychloride *(off-label)*
	White blister	chlorothalonil + metalaxyl, mancozeb + metalaxyl-M *(off-label)*

	Wirestem	quintozene
Pests	Aphids	alpha-cypermethrin, bifenthrin, chlorpyrifos, chlorpyrifos + disulfoton, cypermethrin, deltamethrin *(off-label)*, deltamethrin + pirimicarb, dimethoate, disulfoton, fatty acids, lambda-cyhalothrin + pirimicarb, nicotine, pirimicarb *(off-label)*, pirimicarb, triazamate, triazamate *(not Savoys)*
	Birds	aluminium ammonium sulphate
	Cabbage root fly	carbofuran, carbosulfan, chlorfenvinphos, chlorpyrifos, chlorpyrifos + disulfoton
	Cabbage stem flea beetle	cypermethrin
	Cabbage stem weevil	carbofuran, carbosulfan
	Caterpillars	alpha-cypermethrin, Bacillus thuringiensis, bifenthrin, chlorpyrifos, cypermethrin, deltamethrin *(off-label)*, deltamethrin, deltamethrin + pirimicarb, diflubenzuron, lambda-cyhalothrin, lambda-cyhalothrin + pirimicarb, nicotine
	Cutworms	chlorpyrifos
	Damaging mammals	aluminium ammonium sulphate, sulphonated cod liver oil
	Flea beetles	alpha-cypermethrin, carbofuran, carbosulfan, deltamethrin
	General insect control	chlorpyrifos *(off-label)*, dimethoate *(off-label)*
	Leaf miners	dichlorvos *(off-label)*, nicotine
	Leatherjackets	chlorpyrifos
	Mealy bugs	fatty acids
	Pollen beetles	alpha-cypermethrin *(off-label)*, lambda-cyhalothrin + pirimicarb
	Red spider mites	fatty acids
	Scale insects	fatty acids
	Slugs and snails	metaldehyde *(seed admixture)*, methiocarb
	Western flower thrips	dichlorvos *(off-label)*
	Whitefly	bifenthrin, chlorpyrifos, fatty acids, lambda-cyhalothrin, lambda-cyhalothrin + pirimicarb
Weeds	Annual dicotyledons	chlorthal-dimethyl, clopyralid, cyanazine *(off-label)*, metazachlor, pendimethalin, propachlor *(off-label)*, propachlor, sodium monochloroacetate *(off-label)*, sodium monochloroacetate, tebutam, trifluralin
	Annual grasses	cyanazine *(off-label)*, cycloxydim, fluazifop-P-butyl *(off-label)*, pendimethalin, propachlor *(off-label)*, propachlor, tebutam, trifluralin
	Annual meadow grass	metazachlor

	Black nightshade	pyridate
	Blackgrass	cycloxydim, metazachlor
	Cleavers	pyridate
	Corn marigold	clopyralid
	Couch	cycloxydim
	Creeping bent	cycloxydim
	Creeping thistle	clopyralid
	Fat-hen	pyridate
	Groundsel, triazine resistant	metazachlor
	Mayweeds	clopyralid
	Perennial dicotyledons	clopyralid
	Perennial grasses	cycloxydim, fluazifop-P-butyl *(off-label)*
	Volunteer cereals	cycloxydim, tebutam
	Wild oats	cycloxydim
Plant growth regulation	Increasing yield	sulphur

Carrots/parsnips/parsley

Diseases	Alternaria	iprodione + thiophanate-methyl *(off-label)*, tebuconazole *(off-label)*
	Canker	tebuconazole *(off-label)*
	Cavity spot	metalaxyl-M
	Crown rot	fenpropimorph *(off-label)*, iprodione + thiophanate-methyl *(off-label)*
	Damping off	thiabendazole + thiram *(seed treatment)*, thiram *(seed treatment)*
	Fungus diseases	triadimenol *(off-label)*
	Powdery mildew	sulphur *(off-label)*, triadimefon
	Seed-borne diseases	thiram *(seed soak)*
Pests	Aphids	aldicarb, carbofuran, carbosulfan, dimethoate, disulfoton, lambda-cyhalothrin + pirimicarb, nicotine, pirimicarb *(off-label)*, pirimicarb
	Birds	aluminium ammonium sulphate
	Carrot fly	carbofuran, carbosulfan, chlorfenvinphos, disulfoton, lambda-cyhalothrin *(off-label)*, tefluthrin *(off-label - seed treatment)*
	Caterpillars	nicotine

	Cutworms	chlorpyrifos, lambda-cyhalothrin, lambda-cyhalothrin + pirimicarb
	Damaging mammals	aluminium ammonium sulphate
	General insect control	deltamethrin *(off-label)*
	Nematodes	aldicarb, carbofuran, carbosulfan
Weeds	Annual dicotyledons	chlorpropham, chlorpropham + pentanochlor, ioxynil *(off-label)*, isoxaben *(off-label)*, linuron, metoxuron, pendimethalin, pentanochlor, prometryn, trifluralin
	Annual grasses	chlorpropham, cycloxydim, fluazifop-P-butyl *(off-label)*, fluazifop-P-butyl, metoxuron, pendimethalin, prometryn, propaquizafop, quizalofop-P-ethyl, trifluralin
	Annual meadow grass	chlorpropham + pentanochlor, linuron, pentanochlor
	Annual weeds	metribuzin *(off-label)*
	Black bindweed	linuron
	Blackgrass	cycloxydim
	Chickweed	chlorpropham, linuron
	Corn marigold	linuron
	Couch	cycloxydim, quizalofop-P-ethyl
	Creeping bent	cycloxydim
	Docks	asulam *(off-label)*
	Fat-hen	linuron
	Mayweeds	metoxuron
	Perennial grasses	cycloxydim, fluazifop-P-butyl *(off-label)*, fluazifop-P-butyl, propaquizafop, quizalofop-P-ethyl
	Polygonums	chlorpropham, linuron
	Volunteer cereals	cycloxydim, fluazifop-P-butyl, quizalofop-P-ethyl
	Wild oats	cycloxydim, fluazifop-P-butyl

Celery

Diseases	Celery leaf spot	Bordeaux mixture, carbendazim *(off-label)*, chlorothalonil *(qualified minor use)*, chlorothalonil, copper ammonium carbonate, copper oxychloride
	Damping off	etridiazole *(seedlings and transplants)*
	Seed-borne diseases	thiram *(seed soak)*
	Septoria diseases	chlorothalonil *(qualified minor use)*, chlorothalonil
Pests	Aphids	nicotine, pirimicarb

	Carrot fly	lambda-cyhalothrin *(off-label)*
	Caterpillars	cypermethrin, lambda-cyhalothrin *(off-label)*, nicotine
	General insect control	deltamethrin *(off-label)*
	Leaf miners	nicotine
Weeds	Annual dicotyledons	chlorpropham, chlorpropham + pentanochlor, linuron, pentanochlor, prometryn
	Annual grasses	chlorpropham, prometryn
	Annual meadow grass	chlorpropham + pentanochlor, linuron, pentanochlor
	Black bindweed	linuron
	Chickweed	chlorpropham, linuron
	Fat-hen	linuron
	Polygonums	chlorpropham, linuron
Plant growth regulation	Increasing yield	gibberellins

Cucurbits

Diseases	Black spot	benomyl *(off-label)*
	Botrytis	iprodione *(off-label)*
	Gummosis	benomyl *(off-label)*
	Powdery mildew	bupirimate, bupirimate *(off-label)*, fenarimol *(off-label)*, imazalil *(off-label)*
	Sclerotinia	iprodione *(off-label)*
Pests	Aphids	fatty acids, nicotine, pirimicarb *(off-label)*
	Caterpillars	nicotine
	Leaf miners	deltamethrin *(off-label - fog)*, oxamyl *(off-label)*
	Mealy bugs	fatty acids
	Red spider mites	fatty acids
	Scale insects	fatty acids
	Western flower thrips	deltamethrin *(off-label - fog)*
	Whitefly	fatty acids
Weeds	Annual dicotyledons	chlorthal-dimethyl *(off-label)*

Field vegetables general

Diseases	Soil-borne diseases	dazomet

Pests	Aphids	nicotine, rotenone
	Capsids	nicotine
	Leafhoppers	nicotine
	Nematodes	dazomet
	Sawflies	nicotine
	Slugs and snails	aluminium sulphate
	Soil pests	dazomet
	Thrips	nicotine
Weeds	Annual dicotyledons	diquat *(between row treatment)*, diquat, diquat + paraquat, glufosinate-ammonium, paraquat *(stale seedbed/inter-row)*, paraquat
	Annual grasses	diquat + paraquat, glufosinate-ammonium, paraquat *(stale seedbed/inter-row)*, paraquat
	Annual weeds	ammonium sulphamate *(pre-planting)*
	Barley cover crops	fluazifop-P-butyl
	General weed control	dazomet
	Perennial grasses	diquat + paraquat
	Perennial weeds	ammonium sulphamate *(pre-planting)*
	Volunteer cereals	diquat + paraquat, paraquat *(stale seedbed/inter-row)*, paraquat
Plant growth regulation	Increasing yield	sulphur

Herb crops

Diseases	Alternaria	iprodione + thiophanate-methyl *(off-label)*, tebuconazole *(off-label)*
	Botrytis	propamocarb hydrochloride *(off-label)*
	Crown rot	fenpropimorph *(off-label)*, iprodione + thiophanate-methyl *(off-label)*
	Downy mildew	fosetyl-aluminium *(off-label)*, propamocarb hydrochloride *(off-label)*
	Fungus diseases	triadimenol *(off-label)*
	Powdery mildew	sulphur *(off-label)*, tebuconazole *(off-label)*
	Ring spot	prochloraz *(off-label)*
	Rust	tebuconazole *(off-label)*
	White blister	metalaxyl-M *(off-label)*, propamocarb hydrochloride *(off-label)*
Pests	Aphids	pirimicarb *(off-label)*

FOR FULL DETAILS ALWAYS CHECK THE PROFILE AND THE PRODUCT LABEL

	Bean seed flies	tefluthrin *(off-label - seed treatment)*
	Flea beetles	deltamethrin *(off-label)*
	Leafhoppers	deltamethrin *(off-label)*
Weeds	Annual dicotyledons	chlorpropham + pentanochlor *(off-label)*, chlorthal-dimethyl, clopyralid *(off-label)*, ethofumesate *(off-label)*, metamitron *(off-label)*, paraquat *(off-label)*, pendimethalin *(off-label)*, pentanochlor *(off-label)*, prometryn *(off-label)*, propachlor *(off-label)*, propachlor, simazine *(off-label)*, terbacil *(off-label)*, trifluralin *(off-label)*
	Annual grasses	ethofumesate *(off-label)*, paraquat *(off-label)*, pendimethalin *(off-label)*, prometryn *(off-label)*, propachlor *(off-label)*, propachlor, simazine *(off-label)*, terbacil *(off-label)*, trifluralin *(off-label)*
	Annual meadow grass	chlorpropham + pentanochlor *(off-label)*, pentanochlor *(off-label)*
	Couch	terbacil *(off-label)*
	Docks	asulam *(off-label)*
	General weed control	monolinuron *(off-label)*
	Perennial dicotyledons	clopyralid *(off-label)*
	Perennial grasses	terbacil *(off-label)*

Lettuce, outdoor

Diseases	Big vein	carbendazim *(off-label)*
	Botrytis	iprodione, iprodione *(off-label)*, propamocarb hydrochloride *(off-label)*, quintozene, thiram
	Damping off	thiram *(seed treatment)*
	Downy mildew	fosetyl-aluminium *(off-label)*, mancozeb, metalaxyl + thiram, propamocarb hydrochloride *(off-label)*, zineb
	Rhizoctonia	quintozene, tolclofos-methyl
	Ring spot	prochloraz *(off-label)*
	Sclerotinia	iprodione *(off-label)*, quintozene
Pests	Aphids	cypermethrin, deltamethrin *(off-label)*, fatty acids, imidacloprid *(off-label - seed treatment)*, lambda-cyhalothrin + pirimicarb, malathion *(indoors)*, nicotine, pirimicarb
	Caterpillars	cypermethrin, deltamethrin *(off-label)*, deltamethrin
	Cutworms	cypermethrin, lambda-cyhalothrin, lambda-cyhalothrin + pirimicarb
	Damaging mammals	sulphonated cod liver oil
	Mealy bugs	fatty acids

	Red spider mites	fatty acids
	Scale insects	fatty acids
	Whitefly	fatty acids
Weeds	Annual dicotyledons	chlorpropham, paraquat *(off-label)*, pendimethalin *(off-label)*, propachlor *(off-label)*, propyzamide, trifluralin
	Annual grasses	chlorpropham, paraquat *(off-label)*, pendimethalin *(off-label)*, propachlor *(off-label)*, propyzamide, trifluralin
	Chickweed	chlorpropham
	Perennial grasses	propyzamide
	Polygonums	chlorpropham

Miscellaneous field vegetables

Diseases	Botrytis	iprodione *(off-label)*
	Celery leaf spot	chlorothalonil *(off-label)*
	Crown rot	fenpropimorph *(off-label)*
	Damping off	thiram *(off-label - seed treatment)*
	Fungus diseases	triadimenol *(off-label)*
	Phytophthora	fosetyl-aluminium *(off-label)*
	Sclerotinia	iprodione *(off-label)*, quintozene *(forcing)*
	Septoria diseases	chlorothalonil *(off-label)*
Pests	Aphids	chlorpyrifos *(off-label)*, deltamethrin *(off-label)*, nicotine, pirimicarb *(off-label)*
	Carrot fly	chlorfenvinphos *(off-label)*, lambda-cyhalothrin *(off-label)*
	Caterpillars	deltamethrin *(off-label)*, nicotine
	Cutworms	chlorpyrifos *(off-label)*, lambda-cyhalothrin *(off-label)*
	Flea beetles	deltamethrin *(off-label)*
	General insect control	deltamethrin *(off-label)*, dimethoate *(off-label)*
	Leafhoppers	deltamethrin *(off-label)*
Weeds	Annual dicotyledons	bentazone *(off-label)*, chlorpropham + pentanochlor, linuron *(off-label)*, pendimethalin *(off-label)*, pentanochlor, prometryn *(off-label)*, trifluralin *(off-label)*
	Annual grasses	cycloxydim *(off-label)*, linuron *(off-label)*, pendimethalin *(off-label)*, prometryn *(off-label)*, trifluralin *(off-label)*
	Annual meadow grass	chlorpropham + pentanochlor, pentanochlor
	Perennial grasses	cycloxydim *(off-label)*

Onions/leeks

Diseases	Bacterial blight	copper oxychloride *(off-label)*
	Botrytis	chlorothalonil, iprodione
	Cladosporium leaf blotch	propiconazole
	Cladosporium leaf spot	propiconazole *(off-label)*
	Collar rot	iprodione
	Damping off	thiram *(seed treatment)*
	Downy mildew	chlorothalonil + metalaxyl *(qualified minor use)*, mancozeb *(off-label)*
	Fungus diseases	ferbam + maneb + zineb *(off-label)*
	Phytophthora	propamocarb hydrochloride
	Pythium	propamocarb hydrochloride
	Rhynchosporium	chlorothalonil
	Rust	chlorothalonil, cyproconazole, fenpropimorph, propiconazole *(qualified minor use)*, propiconazole, tebuconazole, triadimefon
	White blister	mancozeb + metalaxyl-M *(off-label)*
	White rot	tebuconazole *(off-label)*
	White tip	chlorothalonil + metalaxyl *(qualified minor use)*, propamocarb hydrochloride *(off-label)*
Pests	Aphids	nicotine
	Bean seed flies	tefluthrin *(off-label - seed treatment)*
	Caterpillars	nicotine
	Cutworms	chlorpyrifos
	General insect control	deltamethrin *(off-label)*, dimethoate *(off-label)*
	Leaf miners	deltamethrin *(off-label - fog)*, oxamyl *(off-label)*
	Onion fly	carbofuran, tefluthrin *(off-label - seed treatment)*
	Stem nematodes	aldicarb *(off-label)*, aldicarb, carbofuran, oxamyl *(off-label)*
	Western flower thrips	deltamethrin *(off-label - fog)*, dichlorvos *(off-label)*

Weeds	Annual dicotyledons	bentazone *(off-label)*, chloridazon *(off-label)*, chloridazon + propachlor *(off-label - post-emergence)*, chloridazon + propachlor, chlorpropham, chlorpropham + fenuron, chlorpropham + tar acids + fenuron, chlorthal-dimethyl, clopyralid, cyanazine, ethofumesate *(off-label)*, fluroxypyr *(off-label)*, ioxynil *(off-label)*, ioxynil, linuron *(off-label)*, monolinuron, pendimethalin *(off-label)*, pendimethalin, prometryn *(off-label)*, prometryn, propachlor, sodium monochloroacetate
	Annual grasses	chloridazon + propachlor *(off-label - post-emergence)*, chloridazon + propachlor, chlorpropham, chlorpropham + fenuron, chlorpropham + tar acids + fenuron, cyanazine, cycloxydim, ethofumesate *(off-label)*, fluazifop-P-butyl *(off-label)*, fluazifop-P-butyl, linuron *(off-label)*, monolinuron, pendimethalin *(off-label)*, pendimethalin, prometryn *(off-label)*, prometryn, propachlor, propaquizafop
	Annual meadow grass	chloridazon *(off-label)*, monolinuron
	Annual weeds	glyphosate (top fruit, horticulture, forestry, amenity etc.)
	Black nightshade	pyridate
	Blackgrass	cycloxydim
	Chickweed	chlorpropham, chlorpropham + fenuron, chlorpropham + tar acids + fenuron
	Cleavers	pyridate
	Corn marigold	clopyralid
	Couch	cycloxydim
	Creeping bent	cycloxydim
	Creeping thistle	clopyralid
	Fat-hen	monolinuron, pyridate
	Mayweeds	clopyralid
	Perennial dicotyledons	clopyralid
	Perennial grasses	cycloxydim, fluazifop-P-butyl *(off-label)*, fluazifop-P-butyl, propaquizafop
	Polygonums	chlorpropham, monolinuron
	Volunteer cereals	cycloxydim, fluazifop-P-butyl, glyphosate (top fruit, horticulture, forestry, amenity etc.)
	Volunteer potatoes	fluroxypyr *(off-label)*
	Wild oats	cycloxydim, fluazifop-P-butyl
Plant growth regulation	Sprout suppression	maleic hydrazide

Peas, mange-tout

Diseases	Botrytis	iprodione *(off-label)*
	Sclerotinia	iprodione *(off-label)*
Pests	Aphids	nicotine
	General insect control	chlorpyrifos *(off-label)*
Weeds	Annual dicotyledons	simazine *(off-label)*, simazine + trietazine *(off-label)*, terbuthylazine + terbutryn
	Annual grasses	simazine *(off-label)*, terbuthylazine + terbutryn

Red beet

Diseases	Aphanomyces cochlioides	hymexazol *(off-label)*
	Botrytis	iprodione *(off-label - dip)*
	Canker	iprodione *(off-label - dip)*
	Rust	fenpropimorph *(off-label)*
	Seed-borne diseases	thiram *(seed soak)*
Pests	Aphids	dimethoate *(excluding Myzus persicae)*, nicotine, pirimicarb *(off-label)*
	Caterpillars	cypermethrin, nicotine
	Cutworms	cypermethrin
	Flea beetles	carbofuran
	Leaf miners	dimethoate, lambda-cyhalothrin *(off-label)*
Weeds	Annual dicotyledons	clopyralid, ethofumesate, ethofumesate + phenmedipham, lenacil, metamitron, phenmedipham, triflusulfuron-methyl *(off-label)*
	Annual grasses	fluazifop-P-butyl *(off-label)*, metamitron, quizalofop-P-ethyl
	Annual meadow grass	ethofumesate, ethofumesate + phenmedipham, lenacil, metamitron
	Blackgrass	ethofumesate, ethofumesate + phenmedipham, tri-allate
	Corn marigold	clopyralid
	Couch	quizalofop-P-ethyl
	Creeping thistle	clopyralid
	Fat-hen	metamitron
	Mayweeds	clopyralid
	Perennial dicotyledons	clopyralid

	Perennial grasses	fluazifop-P-butyl *(off-label)*, quizalofop-P-ethyl, tri-allate
	Volunteer cereals	propaquizafop *(off-label)*, quizalofop-P-ethyl
	Wild oats	tri-allate

Rhubarb

Diseases	Botrytis	benomyl *(off-label)*, iprodione *(off-label)*
	Fungus diseases	benomyl *(off-label)*
Weeds	Annual dicotyledons	chlorpropham + tar acids + fenuron, propyzamide *(outdoor)*, propyzamide, simazine *(off-label)*, simazine
	Annual grasses	chlorpropham + tar acids + fenuron, propyzamide *(outdoor)*, propyzamide, simazine *(off-label)*, simazine
	Annual weeds	dichlobenil *(off-label)*
	Chickweed	chlorpropham + tar acids + fenuron
	Perennial grasses	dichlobenil *(off-label)*, propyzamide *(outdoor)*, propyzamide
Plant growth regulation	Increasing yield	gibberellins

Root brassicas

Diseases	Alternaria	iprodione *(seed treatment)*, tebuconazole *(off-label)*
	Damping off	thiram *(seed treatment)*
	Downy mildew	propamocarb hydrochloride *(off-label)*
	Light leaf spot	tebuconazole *(off-label)*
	Powdery mildew	copper sulphate + sulphur, sulphur, tebuconazole, triadimefon, triadimenol
	Ring spot	tebuconazole *(off-label)*
	White blister	mancozeb + metalaxyl-M *(off-label)*
Pests	Aphids	carbofuran, deltamethrin + pirimicarb, nicotine, pirimicarb *(off-label)*, pirimicarb
	Cabbage root fly	carbofuran, carbosulfan, chlorfenvinphos *(off-label)*, chlorpyrifos *(off-label)*
	Cabbage stem weevil	carbofuran, carbosulfan
	Caterpillars	deltamethrin, deltamethrin + pirimicarb, nicotine
	Flea beetles	carbofuran, carbosulfan, deltamethrin
	General insect control	dimethoate *(off-label)*
	Turnip root fly	carbofuran

Weeds	Annual dicotyledons	chlorthal-dimethyl, clopyralid *(off-label)*, clopyralid, metazachlor *(off-label)*, metazachlor, prometryn *(off-label)*, propachlor *(off-label)*, propachlor, sodium monochloroacetate *(off-label)*, tebutam, trifluralin *(off-label)*, trifluralin
	Annual grasses	cycloxydim, fluazifop-P-butyl *(stockfeed only)*, fluazifop-P-butyl *(off-label)*, metazachlor *(off-label)*, prometryn *(off-label)*, propachlor *(off-label)*, propachlor, propaquizafop, tebutam, trifluralin *(off-label)*, trifluralin
	Annual meadow grass	metazachlor
	Annual weeds	glyphosate (agricultural uses)
	Blackgrass	cycloxydim, metazachlor
	Corn marigold	clopyralid
	Couch	cycloxydim
	Creeping bent	cycloxydim
	Creeping thistle	clopyralid
	Groundsel, triazine resistant	metazachlor
	Mayweeds	clopyralid
	Perennial dicotyledons	clopyralid
	Perennial grasses	cycloxydim, fluazifop-P-butyl *(stockfeed only)*, fluazifop-P-butyl *(off-label)*, propaquizafop
	Volunteer cereals	cycloxydim, fluazifop-P-butyl *(stockfeed only)*, glyphosate (agricultural uses), tebutam
	Wild oats	cycloxydim, fluazifop-P-butyl *(stockfeed only)*
Plant growth regulation	Increasing yield	sulphur

Spinach

Diseases	Downy mildew	copper oxychloride + metalaxyl *(off-label)*, fosetyl-aluminium *(off-label)*
Pests	Aphids	dimethoate, nicotine, pirimicarb *(off-label)*
	Bean seed flies	tefluthrin *(off-label - seed treatment)*
	Caterpillars	nicotine
	Leaf miners	dimethoate
	Western flower thrips	dichlorvos *(off-label)*
Weeds	Annual dicotyledons	chlorpropham + fenuron, clopyralid *(off-label)*, phenmedipham *(off-label)*
	Annual grasses	chlorpropham + fenuron

| | Chickweed | chlorpropham + fenuron |

Tomatoes, outdoor

Diseases	Blight	copper ammonium carbonate, copper oxychloride
	Botrytis	iprodione, pyrimethanil *(off-label)*
	Leaf mould	copper ammonium carbonate
	Soil-borne diseases	metam-sodium *(Jersey only)*
Pests	Nematodes	metam-sodium *(Jersey only)*
	Soil pests	metam-sodium *(Jersey only)*
Weeds	General weed control	metam-sodium *(Jersey only)*

Watercress

Diseases	Downy mildew	fosetyl-aluminium *(off-label)*, propamocarb hydrochloride *(off-label)*
	Phytophthora	etridiazole *(off-label)*, fosetyl-aluminium *(off-label)*, propamocarb hydrochloride *(off-label)*
	Pythium	copper oxychloride *(off-label)*, etridiazole *(off-label)*, fosetyl-aluminium *(off-label)*, metalaxyl-M *(off-label)*, propamocarb hydrochloride *(off-label)*
	Rhizoctonia	benomyl *(off-label)*, copper oxychloride *(off-label)*
Pests	Aphids	dimethoate *(off-label)*, fatty acids *(off-label)*
	Flea beetles	malathion *(off-label)*
	Midges	malathion *(off-label)*
	Mustard beetle	malathion *(off-label)*

Flowers and ornamentals

Annuals/biennials

Weeds	Annual dicotyledons	chlorpropham, pentanochlor
	Annual grasses	chlorpropham
	Annual meadow grass	pentanochlor
	Chickweed	chlorpropham
	Polygonums	chlorpropham

Bedding plants

Diseases	Botrytis	carbendazim *(off-label)*, quintozene
	Damping off	furalaxyl
	Phytophthora	propamocarb hydrochloride
	Powdery mildew	carbendazim *(off-label)*
	Pythium	propamocarb hydrochloride
	Rhizoctonia	quintozene
	Sclerotinia	quintozene
Pests	Aphids	imidacloprid
	Sciarid flies	imidacloprid
	Vine weevil	imidacloprid
	Whitefly	imidacloprid
Weeds	Annual dicotyledons	pentanochlor
	Annual grasses	sethoxydim *(off-label)*
	Annual meadow grass	pentanochlor
	Perennial grasses	sethoxydim *(off-label)*
Plant growth regulation	Increasing flowering	paclobutrazol
	Stem shortening	chlormequat, chlormequat + choline chloride, daminozide, paclobutrazol

Bulbs/corms

Diseases	Botrytis	carbendazim *(off-label - spray and dip)*, thiram, zineb
	Fire	dichlofluanid, mancozeb, thiram, zineb
	Fungus diseases	formaldehyde (commodity substance) *(dip)*
	Fusarium diseases	carbendazim *(off-label - drench)*, carbendazim *(off-label - dip)*, prochloraz *(off-label)*, thiabendazole
	Ink disease	chlorothalonil *(qualified minor use)*, chlorothalonil
	Penicillium rot	carbendazim *(off-label)*, prochloraz *(off-label)*
	Phytophthora	etridiazole, propamocarb hydrochloride
	Powdery mildew	carbendazim *(off-label)*
	Pythium	etridiazole, propamocarb hydrochloride
	Rhizoctonia	quintozene
	Sclerotinia	carbendazim *(off-label - spray and dip)*, quintozene

	Stagonospora	carbendazim *(off-label)*
Pests	Aphids	malathion *(indoors)*, nicotine
	Birds	aluminium ammonium sulphate *(seed treatment)*
	Bulb scale mite	endosulfan
	Capsids	nicotine
	Damaging mammals	aluminium ammonium sulphate *(seed treatment)*
	General insect control	chlorpyrifos *(off-label)*
	Leaf miners	nicotine
	Leafhoppers	malathion *(indoors)*, nicotine
	Mealy bugs	malathion *(indoors)*
	Sawflies	nicotine
	Stem and bulb nematodes	1,3-dichloropropene
	Thrips	malathion *(indoors)*, nicotine
	Whitefly	malathion *(indoors)*
Weeds	Annual dicotyledons	bentazone, chlorpropham, chlorpropham + linuron, chlorpropham + pentanochlor, chlorpropham + tar acids + fenuron, cyanazine, diquat, diquat + paraquat, paraquat, pentanochlor
	Annual grasses	chlorpropham, chlorpropham + linuron, chlorpropham + tar acids + fenuron, cyanazine, cycloxydim, diquat + paraquat, paraquat, sethoxydim *(off-label)*
	Annual meadow grass	chlorpropham + pentanochlor, pentanochlor
	Blackgrass	cycloxydim
	Chickweed	chlorpropham, chlorpropham + tar acids + fenuron
	Couch	cycloxydim
	Creeping bent	cycloxydim, paraquat
	Perennial grasses	cycloxydim, diquat + paraquat, sethoxydim *(off-label)*
	Polygonums	chlorpropham
	Volunteer cereals	cycloxydim
	Wild oats	cycloxydim
Crop control	Pre-harvest desiccation	sulphuric acid (commodity substance)
Plant growth regulation	Increasing flowering	paclobutrazol
	Stem shortening	2-chloroethylphosphonic acid, paclobutrazol

FOR FULL DETAILS ALWAYS CHECK THE PROFILE AND THE PRODUCT LABEL

Camellias

| Plant growth regulation | Stem shortening | chlormequat, chlormequat + choline chloride |

Carnations

Diseases	Rust	oxycarboxin, thiram, zineb
Pests	Aphids	malathion *(indoors)*
	Leafhoppers	malathion *(indoors)*
	Mealy bugs	malathion *(indoors)*
	Thrips	malathion *(indoors)*
	Whitefly	malathion *(indoors)*
Weeds	Annual dicotyledons	pentanochlor
	Annual meadow grass	pentanochlor

Chrysanthemums

Diseases	Botrytis	carbendazim *(off-label)*, quintozene, thiram
Diseases	Powdery mildew	bupirimate, carbendazim *(off-label)*, copper ammonium carbonate, dinocap
	Ray blight	mancozeb
	Rhizoctonia	quintozene
	Rust	oxycarboxin, propiconazole *(off-label)*, thiram
	Sclerotinia	quintozene
	Soil-borne diseases	metam-sodium
Pests	Aphids	malathion *(indoors)*, nicotine
	Leaf miners	nicotine
	Leafhoppers	malathion *(indoors)*
	Mealy bugs	malathion *(indoors)*
	Nematodes	metam-sodium
	Soil pests	metam-sodium
	Thrips	malathion *(indoors)*
	Whitefly	malathion *(indoors)*
Weeds	Annual dicotyledons	chlorpropham, chlorpropham + pentanochlor, pentanochlor
	Annual grasses	chlorpropham

118

	Annual meadow grass	chlorpropham + pentanochlor, pentanochlor
	Chickweed	chlorpropham
	General weed control	metam-sodium
	Polygonums	chlorpropham
Plant growth regulation	Stem shortening	daminozide

Container-grown stock

Diseases	Fungus diseases	prochloraz
	Phytophthora	etridiazole, propamocarb hydrochloride
Pests	Vine weevil	chlorpyrifos
Weeds	Annual dicotyledons	isoxaben, isoxaben + trifluralin, napropamide, oxadiazon
	Annual grasses	napropamide, oxadiazon
	Annual meadow grass	isoxaben + trifluralin
	Cleavers	napropamide
	Fat-hen	isoxaben + trifluralin
	Groundsel	napropamide
Plant growth regulation	Root control	copper hydroxide
	Stem shortening	chlormequat, chlormequat + choline chloride

Dahlias

Diseases	Botrytis	quintozene
	Rhizoctonia	quintozene
	Sclerotinia	quintozene
Pests	Aphids	malathion *(indoors)*
	Leafhoppers	malathion *(indoors)*
	Mealy bugs	malathion *(indoors)*
	Thrips	malathion *(indoors)*
	Whitefly	malathion *(indoors)*

Flowers and ornamentals

Pests	Slugs and snails	metaldehyde

Flowers and ornamentals general

Diseases	Alternaria	iprodione *(seed treatment)*
	Black root rot	carbendazim *(off-label)*
	Black spot	bupirimate + triforine
	Botrytis	chlorothalonil, dichlofluanid *(off-label)*, prochloraz, thiram *(except Hydrangea)*
	Damping off	copper ammonium carbonate, tolclofos-methyl
	Disease control/ foliar feed	difenoconazole *(off-label)*
	Foot rot	tolclofos-methyl
	Phytophthora	propamocarb hydrochloride
	Powdery mildew	bupirimate + triforine, pyrifenox
	Pythium	propamocarb hydrochloride
	Root rot	tolclofos-methyl
	Rust	oxycarboxin
	Soil-borne diseases	metam-sodium
	Verticillium wilt	chloropicrin *(soil fumigation)*
Pests	Aphids	cypermethrin, deltamethrin, dimethoate, nicotine, permethrin, pirimicarb, pymetrozine, pyrethrins + resmethrin *(protected crops)*, rotenone, Verticillium lecanii
	Birds	ziram
	Browntail moth	diflubenzuron, teflubenzuron
	Capsids	cypermethrin, deltamethrin, nicotine
	Caterpillars	Bacillus thuringiensis, cypermethrin, diflubenzuron, nicotine, permethrin, teflubenzuron
	Cutworms	cypermethrin
	Damaging mammals	bone oil, ziram
	General insect control	aldicarb *(off-label)*
	Leaf miners	abamectin, cypermethrin *(off-label)*, dimethoate, nicotine, oxamyl *(off-label)*
	Leafhoppers	nicotine
	Nematodes	metam-sodium
	Red spider mites	abamectin, bifenthrin, fenazaquin
	Sawflies	nicotine
	Scale insects	deltamethrin

	Sciarid flies	chlorpyrifos
	Slugs and snails	aluminium sulphate, methiocarb
	Soil pests	metam-sodium
	Stem nematodes	aldicarb *(off-label)*
	Thrips	cypermethrin, deltamethrin, nicotine
	Tortrix moths	diflubenzuron
	Vine weevil	chlorpyrifos
	Western flower thrips	abamectin
	Whitefly	cypermethrin, permethrin, pyrethrins + resmethrin *(protected crops)*, teflubenzuron
	Winter moth	diflubenzuron
	Woolly aphid	nicotine
Weeds	Annual and perennial weeds	glyphosate (top fruit, horticulture, forestry, amenity etc.) *(wiper application)*, glyphosate (top fruit, horticulture, forestry, amenity etc.)
	Annual dicotyledons	chlorpropham + fenuron, chlorthal-dimethyl, diquat + paraquat, metazachlor, napropamide, paraquat, propachlor, propyzamide, trifluralin *(off-label)*
	Annual grasses	chlorpropham + fenuron, diquat + paraquat, fluazifop-P-butyl *(off-label)*, napropamide, paraquat, propachlor, propyzamide, trifluralin *(off-label)*
	Annual meadow grass	metazachlor
	Annual weeds	ammonium sulphamate *(pre-planting)*, ammonium sulphamate
	Blackgrass	metazachlor
	Chickweed	chlorpropham + fenuron
	Cleavers	napropamide
	Creeping bent	paraquat
	General weed control	metam-sodium
	Groundsel	napropamide
	Groundsel, triazine resistant	metazachlor
	Perennial grasses	diquat + paraquat, fluazifop-P-butyl *(off-label)*, propyzamide
	Perennial weeds	ammonium sulphamate *(pre-planting)*, ammonium sulphamate
	Total vegetation control	glyphosate (top fruit, horticulture, forestry, amenity etc.)

FOR FULL DETAILS ALWAYS CHECK THE PROFILE AND THE PRODUCT LABEL

Plant growth regulation	Rooting of cuttings	2-(1-naphthyl)acetic acid, 4-indol-3-yl-butyric acid, 4-indol-3-yl-butyric acid + 2-(1-naphthyl)acetic acid with dichlorophen, indol-3-ylacetic acid
	Stem shortening	chlormequat, chlormequat + choline chloride, daminozide

Hardy ornamental nursery stock

Diseases	Fungus diseases	prochloraz
	Phytophthora	etridiazole, fosetyl-aluminium, furalaxyl, propamocarb hydrochloride
	Powdery mildew	pyrifenox
	Pythium	furalaxyl, propamocarb hydrochloride
Pests	Aphids	imidacloprid, nicotine
	Browntail moth	diflubenzuron
	Capsids	deltamethrin
	Caterpillars	diflubenzuron
	Scale insects	deltamethrin
	Sciarid flies	imidacloprid
	Thrips	deltamethrin
	Tortrix moths	diflubenzuron
	Vine weevil	carbofuran, imidacloprid
	Whitefly	imidacloprid
	Winter moth	diflubenzuron
Weeds	Annual dicotyledons	chlorpropham + tar acids + fenuron, diuron, diuron + paraquat, glufosinate-ammonium, isoxaben, isoxaben + trifluralin, metazachlor, pentanochlor, simazine
	Annual grasses	chlorpropham + tar acids + fenuron, diuron, diuron + paraquat, glufosinate-ammonium, simazine
	Annual meadow grass	isoxaben + trifluralin, metazachlor, pentanochlor
	Blackgrass	metazachlor
	Chickweed	chlorpropham + tar acids + fenuron
	Fat-hen	isoxaben + trifluralin
	Groundsel, triazine resistant	metazachlor
	Perennial dicotyledons	diuron + paraquat, glufosinate-ammonium
	Perennial grasses	diuron + paraquat, glufosinate-ammonium

Hedges

Pests	Browntail moth	diflubenzuron
	Caterpillars	diflubenzuron
	Tortrix moths	diflubenzuron
	Winter moth	diflubenzuron
Weeds	Annual and perennial weeds	glyphosate (top fruit, horticulture, forestry, amenity etc.)
Plant growth regulation	Growth retardation	maleic hydrazide

Perennials

Weeds	Annual dicotyledons	pentanochlor, propachlor
	Annual grasses	propachlor, sethoxydim *(off-label)*
	Annual meadow grass	pentanochlor
	Perennial grasses	sethoxydim *(off-label)*

Roses

Diseases	Black spot	captan, dichlofluanid, mancozeb, myclobutanil
	Powdery mildew	bupirimate, carbendazim, dichlofluanid, dinocap, dodemorph, fenarimol, myclobutanil
	Rust	bupirimate + triforine, mancozeb, myclobutanil, oxycarboxin, penconazole
Pests	Aphids	dimethoate, malathion *(indoors)*
	Leaf miners	dimethoate
	Leafhoppers	malathion *(indoors)*
	Mealy bugs	malathion *(indoors)*
	Red spider mites	fenpropathrin
	Scale insects	malathion *(indoors)*
	Slug sawflies	rotenone
	Thrips	malathion *(indoors)*
	Whitefly	malathion *(indoors)*
Weeds	Annual dicotyledons	atrazine, pentanochlor, propyzamide, simazine
	Annual grasses	atrazine, propyzamide, simazine

	Annual meadow grass	pentanochlor
	Annual weeds	dichlobenil
	Perennial dicotyledons	dichlobenil
	Perennial grasses	dichlobenil, propyzamide
Plant growth regulation	Increasing flowering	paclobutrazol
	Stem shortening	paclobutrazol

Standing ground

Weeds	Annual dicotyledons	chlorpropham + tar acids + fenuron
	Annual grasses	chlorpropham + tar acids + fenuron
	Chickweed	chlorpropham + tar acids + fenuron

Trees/shrubs

Diseases	Canker	azaconazole + imazalil, octhilinone
	Powdery mildew	penconazole
	Pruning wounds	octhilinone
	Scab	penconazole
	Silver leaf	azaconazole + imazalil, octhilinone
Pests	Aphids	dimethoate, fatty acids
	Browntail moth	diflubenzuron
	Capsids	deltamethrin
	Caterpillars	diflubenzuron
	Leaf miners	dimethoate
	Mealy bugs	fatty acids
	Red spider mites	fatty acids
	Scale insects	deltamethrin, fatty acids
	Thrips	deltamethrin
	Tortrix moths	diflubenzuron
	Whitefly	fatty acids
	Winter moth	diflubenzuron
Weeds	Annual and perennial weeds	glyphosate (top fruit, horticulture, forestry, amenity etc.) *(wiper application)*, glyphosate (top fruit, horticulture, forestry, amenity etc.)

	Annual dicotyledons	carbetamide + diflufenican + oxadiazon, clopyralid, diuron, diuron + glyphosate, diuron + paraquat, glufosinate-ammonium, isoxaben, isoxaben + trifluralin, napropamide, oxadiazon, paraquat, propachlor, propyzamide, simazine
	Annual grasses	cycloxydim *(off-label)*, diuron, diuron + glyphosate, diuron + paraquat, glufosinate-ammonium, napropamide, oxadiazon, paraquat, propachlor, propyzamide, sethoxydim *(off-label)*, simazine
	Annual meadow grass	carbetamide + diflufenican + oxadiazon, diuron, isoxaben + trifluralin
	Annual weeds	ammonium sulphamate, dichlobenil, diuron
	Bindweeds	oxadiazon
	Bracken	asulam, glyphosate (top fruit, horticulture, forestry, amenity etc.)
	Cleavers	napropamide, oxadiazon
	Corn marigold	clopyralid
	Creeping bent	paraquat
	Creeping thistle	clopyralid
	Fat-hen	isoxaben + trifluralin
	Groundsel	isoxaben + trifluralin, napropamide
	Horsetails	propyzamide
	Mayweeds	clopyralid
	Perennial dicotyledons	ammonium sulphamate, carbetamide + diflufenican + oxadiazon, clopyralid, dichlobenil, diuron + glyphosate, diuron + paraquat, glufosinate-ammonium
	Perennial grasses	cycloxydim *(off-label)*, dichlobenil, diuron + glyphosate, diuron + paraquat, glufosinate-ammonium, propyzamide, sethoxydim *(off-label)*
	Polygonums	oxadiazon
	Sedges	propyzamide
	Total vegetation control	glyphosate (top fruit, horticulture, forestry, amenity etc.)
Plant growth regulation	Sucker inhibition	maleic hydrazide

Water lily

Diseases	Crown rot	carbendazim + metalaxyl *(off-label)*

Forestry

Cut logs/timber

Pests	Ambrosia beetle	chlorpyrifos
	Black pine beetle	permethrin *(off-label)*
	Elm bark beetle	chlorpyrifos
	Larch shoot beetle	chlorpyrifos
	Pine shoot beetle	chlorpyrifos
	Pine weevil	permethrin *(off-label)*

Farm woodland

Diseases	Dutch elm disease	metam-sodium *(off-label)*, thiabendazole *(off-label - injection)*
Weeds	Annual dicotyledons	cyanazine *(off-label)*, lenacil *(off-label)*, metazachlor *(off-label)*, metazachlor, napropamide *(off-label)*, pendimethalin *(off-label)*, propyzamide
	Annual grasses	cyanazine *(off-label)*, cycloxydim, fluazifop-P-butyl, metazachlor, napropamide *(off-label)*, pendimethalin *(off-label)*, propaquizafop, propyzamide
	Annual meadow grass	lenacil *(off-label)*
	Perennial grasses	cycloxydim, fluazifop-P-butyl, propaquizafop, propyzamide
	Volunteer cereals	fluazifop-P-butyl
	Wild oats	fluazifop-P-butyl

Forest nursery beds

Pests	Aphids	pirimicarb
	Birds	aluminium ammonium sulphate
	Black pine beetle	permethrin *(off-label)*
	Damaging mammals	aluminium ammonium sulphate
	Large pine weevil	carbosulfan *(off-label - growing medium treatment)*
	Pine weevil	chlorpyrifos, permethrin *(off-label)*
Weeds	Annual dicotyledons	napropamide, paraquat *(stale seedbed)*, paraquat, pentanochlor, simazine
	Annual grasses	napropamide, paraquat *(stale seedbed)*, simazine
	Annual meadow grass	pentanochlor

Cleavers	napropamide
Groundsel	napropamide

Forestry plantations

Diseases	Canker	azaconazole + imazalil
	Silver leaf	azaconazole + imazalil
Pests	Birds	aluminium ammonium sulphate, ziram
	Black pine beetle	chlorpyrifos, permethrin *(off-label)*
	Caterpillars	diflubenzuron
	Damaging mammals	aluminium ammonium sulphate, aluminium phosphide, sulphonated cod liver oil, ziram
	Grey squirrels	warfarin
	Large pine weevil	carbosulfan, chlorpyrifos
	Pine weevil	chlorpyrifos, permethrin *(off-label)*
	Rodents	aluminium phosphide
	Vine weevil	chlorpyrifos
	Winter moth	diflubenzuron
Weeds	Annual and perennial weeds	glyphosate (top fruit, horticulture, forestry, amenity etc.)
	Annual dicotyledons	2,4-D, 2,4-D + dicamba + triclopyr, atrazine *(off-label)*, atrazine, clopyralid *(off-label)*, cyanazine *(off-label)*, diquat + paraquat, isoxaben, metazachlor, paraquat, propyzamide, simazine
	Annual grasses	atrazine *(off-label)*, atrazine, cyanazine *(off-label)*, cycloxydim, diquat + paraquat, metazachlor, paraquat, propyzamide, simazine
	Annual weeds	ammonium sulphamate
	Bracken	asuiam *(off-label)*, asulam, glyphosate (top fruit, horticulture, forestry, amenity etc.), imazapyr *(site preparation)*
	Brambles	2,4-D + dicamba + triclopyr, glyphosate (top fruit, horticulture, forestry, amenity etc.), triclopyr
	Broom	triclopyr
	Couch	atrazine, glyphosate (top fruit, horticulture, forestry, amenity etc.)
	Creeping bent	paraquat
	Docks	2,4-D + dicamba + triclopyr, triclopyr
	Firebreak desiccation	paraquat
	Gorse	2,4-D + dicamba + triclopyr, triclopyr

	Heather	2,4-D, glyphosate (top fruit, horticulture, forestry, amenity etc.)
	Horsetails	propyzamide
	Japanese knotweed	2,4-D + dicamba + triclopyr
	Perennial dicotyledons	2,4-D, 2,4-D + dicamba + triclopyr, ammonium sulphamate, clopyralid *(off-label)*, glyphosate (top fruit, horticulture, forestry, amenity etc.) *(wiper application)*, triclopyr
	Perennial grasses	ammonium sulphamate, atrazine, cycloxydim, diquat + paraquat, glyphosate (top fruit, horticulture, forestry, amenity etc.), propyzamide
	Rhododendrons	2,4-D + dicamba + triclopyr, ammonium sulphamate, glyphosate (top fruit, horticulture, forestry, amenity etc.), triclopyr
	Rushes	glyphosate (top fruit, horticulture, forestry, amenity etc.)
	Sedges	propyzamide
	Stinging nettle	2,4-D + dicamba + triclopyr, triclopyr
	Total vegetation control	imazapyr *(site preparation)*
	Volunteer cereals	diquat + paraquat
	Woody weeds	2,4-D, 2,4-D + dicamba + triclopyr, ammonium sulphamate, fosamine-ammonium *(off-label)*, fosamine-ammonium, glyphosate (top fruit, horticulture, forestry, amenity etc.), triclopyr
Crop control	Chemical thinning	glyphosate (top fruit, horticulture, forestry, amenity etc.)

Transplant lines

Pests	Black pine beetle	chlorpyrifos
	Pine weevil	chlorpyrifos
	Vine weevil	chlorpyrifos
Weeds	Annual dicotyledons	paraquat, simazine
	Annual grasses	paraquat, simazine
	Creeping bent	paraquat

Fruit and hops

Apples/pears

Diseases	Blossom wilt	vinclozolin
	Botrytis fruit rot	thiram

	Canker	Bordeaux mixture, carbendazim *(partial control)*, carbendazim, copper oxychloride, octhilinone, thiophanate-methyl
	Collar rot	copper oxychloride + metalaxyl, fosetyl-aluminium
	Crown rot	fosetyl-aluminium
	Gloeosporium	captan, carbendazim, thiram
	Phytophthora	mancozeb + metalaxyl-M *(off-label)*
	Powdery mildew	bupirimate, captan + penconazole, carbendazim, dinocap, fenarimol, fenbuconazole *(reduction)*, kresoxim-methyl *(reduction)*, myclobutanil, penconazole, pyrifenox, sulphur, thiophanate-methyl, triadimefon
	Pruning wounds	octhilinone
	Scab	Bordeaux mixture, captan, captan + penconazole, carbendazim, carbendazim *(off-label)*, dithianon, dodine, fenarimol, fenbuconazole, kresoxim-methyl, mancozeb, myclobutanil, pyrifenox, pyrimethanil, sulphur, thiophanate-methyl, thiram
	Silver leaf	octhilinone
	Storage rots	carbendazim *(off-label - dip)*, carbendazim, carbendazim + metalaxyl, thiophanate-methyl
Pests	Aphids	chlorpyrifos, cypermethrin, deltamethrin, dimethoate, fenitrothion, nicotine, pirimicarb, tar oils
	Apple blossom weevil	chlorpyrifos, fenitrothion
	Bryobia mites	dimethoate
	Capsids	carbaryl, chlorpyrifos, cypermethrin, deltamethrin, dimethoate, fenitrothion, nicotine
	Caterpillars	chlorpyrifos, cypermethrin, deltamethrin, diflubenzuron, fenitrothion, fenpropathrin, nicotine
	Codling moth	carbaryl, chlorpyrifos, cypermethrin, deltamethrin, diflubenzuron, fenitrothion
	Earwigs	carbaryl
	Red spider mites	amitraz, bifenthrin, chlorpyrifos, clofentezine, dicofol, dimethoate, fenazaquin, fenpropathrin, fenpyroximate, tebufenpyrad, tetradifon
	Rust mite	amitraz, diflubenzuron
	Sawflies	chlorpyrifos, cypermethrin, deltamethrin, dimethoate, fenitrothion
	Scale insects	tar oils
	Slug sawflies	rotenone
	Suckers	amitraz, chlorpyrifos, cypermethrin, deltamethrin, diflubenzuron, dimethoate, fenitrothion, lambda-cyhalothrin, tar oils
	Tortrix moths	carbaryl, chlorpyrifos, cypermethrin, deltamethrin, diflubenzuron, fenitrothion, fenoxycarb

FOR FULL DETAILS ALWAYS CHECK THE PROFILE AND THE PRODUCT LABEL

	Winter moth	carbaryl, chlorpyrifos, cypermethrin, diflubenzuron, fenitrothion, tar oils
	Woolly aphid	chlorpyrifos
Weeds	Annual and perennial weeds	glyphosate (top fruit, horticulture, forestry, amenity etc.)
	Annual dicotyledons	2,4-D, 2,4-D + dichlorprop + MCPA + mecoprop-P, dicamba + MCPA + mecoprop-P, diquat + paraquat, diuron, glufosinate-ammonium, isoxaben, napropamide, oxadiazon, pendimethalin, pentanochlor, propyzamide, simazine, sodium monochloroacetate *(directed spray)*, sodium monochloroacetate
	Annual grasses	diquat + paraquat, diuron, glufosinate-ammonium, napropamide, oxadiazon, pendimethalin, propyzamide, simazine
	Annual meadow grass	pentanochlor
	Barren brome	amitrole
	Bindweeds	oxadiazon
	Chickweed	2,4-D + dichlorprop + MCPA + mecoprop-P
	Cleavers	2,4-D + dichlorprop + MCPA + mecoprop-P, napropamide, oxadiazon
	Couch	amitrole
	Docks	asulam, dicamba + MCPA + mecoprop-P
	General weed control	amitrole
	Groundsel	napropamide
	Perennial dicotyledons	2,4-D, 2,4-D + dichlorprop + MCPA + mecoprop-P, dicamba + MCPA + mecoprop-P, glufosinate-ammonium
	Perennial grasses	diquat + paraquat, glufosinate-ammonium, propyzamide
	Perennial weeds	amitrole
	Polygonums	oxadiazon
	Volunteer cereals	diquat + paraquat
Crop control	Sucker control	glyphosate (top fruit, horticulture, forestry, amenity etc.)
Plant growth regulation	Controlling vigour	paclobutrazol
	Fruit ripening	2-chloroethylphosphonic acid *(off-label)*
	Fruit thinning	carbaryl
	Increasing fruit set	gibberellins, paclobutrazol
	Reducing fruit russeting	gibberellins

Apricots/peaches/nectarines

Diseases	Leaf curl	Bordeaux mixture, copper ammonium carbonate, copper oxychloride
Pests	Aphids	tar oils
	Red spider mites	tetradifon *(protected crops)*
	Scale insects	tar oils
	Winter moth	tar oils
Weeds	Docks	asulam *(off-label)*
	General weed control	amitrole *(off-label)*

Blueberries/cranberries

Diseases	Botrytis	pyrimethanil *(off-label)*
Weeds	Annual dicotyledons	propachlor *(off-label)*
	Annual grasses	propachlor *(off-label)*
	Docks	asulam *(off-label)*

Cane fruit

Diseases	Botrytis	chlorothalonil, dichlofluanid, fenhexamid, iprodione, pyrimethanil *(off-label)*, thiram, tolylfluanid
	Cane blight	dichlofluanid
	Cane spot	Bordeaux mixture, chlorothalonil, copper ammonium carbonate, copper oxychloride, dichlofluanid, thiram
	Phytophthora	metalaxyl-M *(off-label - soil drench)*
	Powdery mildew	bupirimate, chlorothalonil, dichlofluanid, fenarimol, myclobutanil *(off-label - protected crops)*, triadimefon *(off-label)*
	Purple blotch	copper oxychloride
	Root rot	mancozeb + oxadixyl *(off-label)*
	Spur blight	Bordeaux mixture, thiram
Pests	Aphids	chlorpyrifos, dimethoate, pirimicarb, tar oils
	Birds	aluminium ammonium sulphate
	Blackberry mite	endosulfan
	Capsids	dimethoate
	Caterpillars	Bacillus thuringiensis

	Damaging mammals	aluminium ammonium sulphate
	Nematodes	1,3-dichloropropene
	Raspberry beetle	chlorpyrifos, deltamethrin, fenitrothion, rotenone
	Raspberry cane midge	chlorpyrifos, fenitrothion
	Red spider mites	chlorpyrifos, clofentezine *(off-label - protected crops)*, clofentezine *(off-label)*, dimethoate, tetradifon
	Scale insects	tar oils
	Virus vectors	1,3-dichloropropene
	Winter moth	tar oils
Weeds	Annual dicotyledons	atrazine, bromacil, chlorthal-dimethyl, diquat + paraquat, glufosinate-ammonium, isoxaben, napropamide, oxadiazon, paraquat, pendimethalin, propyzamide *(England only)*, propyzamide, simazine, trifluralin
	Annual grasses	atrazine, bromacil, diquat + paraquat, fluazifop-P-butyl, glufosinate-ammonium, napropamide, oxadiazon, paraquat, pendimethalin, propyzamide *(England only)*, propyzamide, simazine, trifluralin
	Barren brome	sethoxydim
	Bindweeds	oxadiazon
	Blackgrass	sethoxydim
	Cleavers	napropamide, oxadiazon
	Couch	bromacil, sethoxydim
	Creeping bent	paraquat, sethoxydim
	Docks	asulam *(off-label)*
	Groundsel	napropamide
	Perennial dicotyledons	bromacil, glufosinate-ammonium
	Perennial grasses	bromacil, diquat + paraquat, fluazifop-P-butyl, glufosinate-ammonium, propyzamide *(England only)*, propyzamide, sethoxydim
	Perennial ryegrass	paraquat
	Polygonums	oxadiazon
	Rough meadow grass	paraquat
	Volunteer cereals	diquat + paraquat, fluazifop-P-butyl, sethoxydim
	Wild oats	fluazifop-P-butyl, sethoxydim
Crop control	Sucker control	sodium monochloroacetate *(off-label)*

Currants

Diseases	Botrytis	chlorothalonil, dichlofluanid, fenhexamid, pyrimethanil *(off-label)*, tolylfluanid
	Currant leaf spot	Bordeaux mixture, chlorothalonil *(qualified minor use)*, chlorothalonil, copper ammonium carbonate, dodine, mancozeb
	Leaf spots	pyrifenox
	Powdery mildew	bupirimate, chlorothalonil, fenarimol, myclobutanil, penconazole, pyrifenox, triadimefon *(off-label)*
	Rust	copper oxychloride, thiram
Pests	Aphids	chlorpyrifos, dimethoate, pirimicarb, tar oils
	Big-bud mite	endosulfan
	Capsids	chlorpyrifos, fenitrothion
	Caterpillars	chlorpyrifos, fenpropathrin
	Gall mite	sulphur
	Red spider mites	bifenthrin, chlorpyrifos, dimethoate, fenpropathrin, tetradifon
	Sawflies	fenitrothion
	Scale insects	tar oils
	Slugs and snails	methiocarb
	Winter moth	diflubenzuron, tar oils
Weeds	Annual and perennial weeds	glyphosate (top fruit, horticulture, forestry, amenity etc.) *(off-label)*
	Annual dicotyledons	chlorpropham, chlorthal-dimethyl, diquat + paraquat, diuron *(off-label)*, glufosinate-ammonium, isoxaben, MCPB, napropamide, oxadiazon, paraquat, pendimethalin, pentanochlor, propachlor *(off-label)*, propyzamide, simazine, sodium monochloroacetate *(directed spray)*, sodium monochloroacetate
	Annual grasses	chlorpropham, diquat + paraquat, diuron *(off-label)*, fluazifop-P-butyl, glufosinate-ammonium, napropamide, oxadiazon, paraquat, pendimethalin, propachlor *(off-label)*, propyzamide, simazine
	Annual meadow grass	pentanochlor
	Bindweeds	oxadiazon
	Chickweed	chlorpropham
	Cleavers	napropamide, oxadiazon
	Creeping bent	paraquat
	Docks	asulam *(off-label)*, asulam
	Groundsel	napropamide

	Perennial dicotyledons	glufosinate-ammonium, MCPB
	Perennial grasses	diquat + paraquat, fluazifop-P-butyl, glufosinate-ammonium, propyzamide
	Perennial ryegrass	paraquat
	Polygonums	chlorpropham, oxadiazon
	Rough meadow grass	paraquat
	Volunteer cereals	diquat + paraquat, fluazifop-P-butyl
	Wild oats	fluazifop-P-butyl

Fruit and hops general

Diseases	Powdery mildew	fenpropimorph *(off-label)*
Pests	Aphids	nicotine, rotenone
	Birds	aluminium ammonium sulphate
	Capsids	nicotine
	Damaging mammals	aluminium ammonium sulphate
	Leafhoppers	nicotine
	Sawflies	nicotine
	Slugs and snails	aluminium sulphate, methiocarb
	Woolly aphid	nicotine
Weeds	Annual dicotyledons	diquat + paraquat, glufosinate-ammonium, trifluralin *(off-label)*
	Annual grasses	diquat + paraquat, glufosinate-ammonium, trifluralin *(off-label)*
	Annual weeds	dichlobenil
	Perennial dicotyledons	dichlobenil, glufosinate-ammonium
	Perennial grasses	dichlobenil, diquat + paraquat, glufosinate-ammonium
	Volunteer cereals	diquat + paraquat
Plant growth regulation	Increasing yield	sulphur

Fruit nursery stock

Weeds	Annual dicotyledons	metazachlor, trifluralin *(off-label)*
	Annual grasses	trifluralin *(off-label)*

Annual meadow grass	metazachlor	
Blackgrass	metazachlor	
Groundsel, triazine resistant	metazachlor	

Gooseberries

Diseases	Botrytis	chlorothalonil, dichlofluanid, fenhexamid, pyrimethanil *(off-label)*, tolylfluanid
	Currant leaf spot	chlorothalonil, dodine, mancozeb
	Leaf spots	pyrifenox
	Powdery mildew	bupirimate, chlorothalonil, fenarimol, myclobutanil, pyrifenox, sulphur, triadimefon *(off-label)*
	Septoria diseases	chlorothalonil *(qualified minor use)*, chlorothalonil
Pests	Aphids	chlorpyrifos, dimethoate, pirimicarb, tar oils
	Bryobia mites	lambda-cyhalothrin *(off-label)*
	Capsids	chlorpyrifos, dimethoate, fenitrothion, lambda-cyhalothrin *(off-label)*
	Caterpillars	chlorpyrifos, lambda-cyhalothrin *(off-label)*, nicotine
	Red spider mites	chlorpyrifos, dimethoate, lambda-cyhalothrin *(off-label)*, tetradifon
	Sawflies	fenitrothion, nicotine, rotenone
	Scale insects	tar oils
	Winter moth	tar oils
Weeds	Annual dicotyledons	chlorpropham, chlorthal-dimethyl, diquat + paraquat, isoxaben, napropamide, oxadiazon, paraquat, pendimethalin, pentanochlor, propachlor *(off-label)*, propyzamide, simazine, sodium monochloroacetate *(directed spray)*, sodium monochloroacetate
	Annual grasses	chlorpropham, diquat + paraquat, fluazifop-P-butyl, napropamide, oxadiazon, paraquat, pendimethalin, propyzamide, simazine
	Annual meadow grass	pentanochlor
	Bindweeds	oxadiazon
	Chickweed	chlorpropham
	Cleavers	napropamide, oxadiazon
	Creeping bent	paraquat
	Docks	asulam *(off-label)*
	Groundsel	napropamide
	Perennial grasses	diquat + paraquat, fluazifop-P-butyl, propyzamide

Perennial ryegrass	paraquat
Polygonums	chlorpropham, oxadiazon
Rough meadow grass	paraquat
Volunteer cereals	diquat + paraquat, fluazifop-P-butyl
Wild oats	fluazifop-P-butyl

Grapevines

Diseases	Botrytis	chlorothalonil, dichlofluanid, fenhexamid, iprodione *(off-label)*, pyrimethanil *(off-label)*
	Downy mildew	chlorothalonil, copper oxychloride, copper oxychloride + metalaxyl *(off-label)*, mancozeb *(off-label)*
	Powdery mildew	copper sulphate + sulphur, dinocap *(off-label)*, sulphur, triadimefon *(off-label)*
Pests	Red spider mites	tetradifon
	Scale insects	tar oils
Weeds	Annual and perennial weeds	glyphosate (top fruit, horticulture, forestry, amenity etc.) *(off-label)*
	Annual dicotyledons	glufosinate-ammonium, isoxaben, oxadiazon, paraquat, simazine *(off-label)*
	Annual grasses	glufosinate-ammonium, oxadiazon, paraquat
	Bindweeds	oxadiazon
	Cleavers	oxadiazon
	Perennial dicotyledons	glufosinate-ammonium
	Perennial grasses	glufosinate-ammonium
	Polygonums	oxadiazon

Hops

Diseases	Downy mildew	Bordeaux mixture, chlorothalonil, copper oxychloride, copper oxychloride + metalaxyl, copper sulphate + sulphur, fosetyl-aluminium, zineb
	Powdery mildew	bupirimate, copper sulphate + sulphur, fenpropimorph *(off-label)*, myclobutanil *(off-label)*, penconazole, sulphur, triadimefon
Pests	Aphids	bifenthrin, cypermethrin, deltamethrin, fenpropathrin, imidacloprid, lambda-cyhalothrin, tebufenpyrad
	Caterpillars	fenpropathrin
	Nematodes	1,3-dichloropropene
	Red spider mites	bifenthrin, dicofol, fenpropathrin, lambda-cyhalothrin, tebufenpyrad, tetradifon

	Virus vectors	1,3-dichloropropene
Weeds	Annual dicotyledons	diquat + paraquat, isoxaben, oxadiazon, paraquat, pendimethalin, simazine
	Annual grasses	diquat + paraquat, fluazifop-P-butyl, oxadiazon, paraquat, pendimethalin, simazine
	Bindweeds	oxadiazon
	Blackgrass	fluazifop-P-butyl
	Cleavers	oxadiazon
	Creeping bent	paraquat
	Docks	asulam
	Perennial grasses	diquat + paraquat, fluazifop-P-butyl
	Perennial ryegrass	paraquat
	Polygonums	oxadiazon
	Rough meadow grass	paraquat
	Volunteer cereals	diquat + paraquat, fluazifop-P-butyl
	Wild oats	fluazifop-P-butyl
Crop control	Chemical stripping	anthracene oil, diquat, diquat + paraquat, paraquat, sodium monochloroacetate *(off-label)*, sodium monochloroacetate

Plums/cherries/damsons

Diseases	Bacterial canker	Bordeaux mixture, copper oxychloride
	Blossom wilt	myclobutanil *(off-label)*
	Canker	octhilinone
	Pruning wounds	octhilinone
	Rust	myclobutanil *(off-label)*
	Silver leaf	octhilinone
Pests	Aphids	chlorpyrifos, cypermethrin, deltamethrin, dimethoate, fenitrothion, pirimicarb *(off-label)*, pirimicarb, tar oils
	Caterpillars	cypermethrin, deltamethrin, fenitrothion
	Plum fruit moth	deltamethrin, diflubenzuron
	Red spider mites	chlorpyrifos, clofentezine, dimethoate, tetradifon
	Rust mite	diflubenzuron
	Sawflies	deltamethrin, dimethoate
	Scale insects	tar oils
	Tortrix moths	chlorpyrifos, deltamethrin, diflubenzuron

	Winter moth	chlorpyrifos, diflubenzuron, tar oils
Weeds	Annual and perennial weeds	glyphosate (top fruit, horticulture, forestry, amenity etc.)
	Annual dicotyledons	diquat + paraquat, glufosinate-ammonium, isoxaben, napropamide, pendimethalin, pentanochlor, propyzamide, sodium monochloroacetate *(directed spray)*, sodium monochloroacetate
	Annual grasses	diquat + paraquat, glufosinate-ammonium, napropamide, pendimethalin, propyzamide
	Annual meadow grass	pentanochlor
	Cleavers	napropamide
	Docks	asulam *(off-label)*, asulam
	General weed control	amitrole *(off-label)*
	Groundsel	napropamide
	Perennial dicotyledons	glufosinate-ammonium
	Perennial grasses	diquat + paraquat, glufosinate-ammonium, propyzamide
	Volunteer cereals	diquat + paraquat
Crop control	Sucker control	glyphosate (top fruit, horticulture, forestry, amenity etc.)
Plant growth regulation	Controlling vigour	paclobutrazol
	Increasing fruit set	paclobutrazol

Quinces

Weeds	Docks	asulam *(off-label)*
	General weed control	amitrole *(off-label)*

Strawberries

Diseases	Botrytis	captan, carbendazim, chlorothalonil *(outdoor crops only)*, chlorothalonil *(off-label)*, chlorothalonil, dichlofluanid, fenhexamid, iprodione, pyrimethanil, thiram, tolylfluanid
	Crown rot	chloropicrin *(soil fumigation)*, fosetyl-aluminium *(off-label)*
	Powdery mildew	bupirimate, carbendazim, dinocap, fenarimol, myclobutanil, sulphur, triadimefon *(off-label)*
	Red core	chloropicrin *(soil fumigation)*, copper oxychloride + metalaxyl, fosetyl-aluminium *(off-label - spring treatment)*, fosetyl-aluminium
	Verticillium wilt	chloropicrin *(soil fumigation)*
Pests	Aphids	chlorpyrifos, dimethoate, malathion, nicotine, pirimicarb

	Birds	aluminium ammonium sulphate
	Caterpillars	Bacillus thuringiensis, nicotine
	Chafer grubs	gamma-HCH
	Damaging mammals	aluminium ammonium sulphate
	Leatherjackets	gamma-HCH
	Nematodes	1,3-dichloropropene, chloropicrin *(soil fumigation)*
	Red spider mites	bifenthrin, chlorpyrifos, clofentezine *(off-label - protected crops)*, dicofol, dimethoate, fenbutatin oxide, fenpropathrin, tebufenpyrad, tetradifon
	Slugs and snails	methiocarb
	Stem nematodes	1,3-dichloropropene
	Strawberry blossom weevil	chlorpyrifos
	Strawberry seed beetle	methiocarb
	Tarsonemid mites	dicofol, endosulfan
	Tortrix moths	chlorpyrifos, fenitrothion
	Vine weevil	carbofuran, chlorpyrifos
	Virus vectors	1,3-dichloropropene
	Wireworms	gamma-HCH
Weeds	Annual dicotyledons	chlorpropham, chlorpropham + fenuron, chlorthal-dimethyl, clopyralid, diquat + paraquat, glufosinate-ammonium, isoxaben, napropamide, pendimethalin, phenmedipham, propachlor *(off-label)*, propachlor, propyzamide, simazine, trifluralin
	Annual grasses	chlorpropham, chlorpropham + fenuron, cycloxydim, diquat + paraquat, ethofumesate *(off-label)*, fluazifop-P-butyl, glufosinate-ammonium, napropamide, pendimethalin, propachlor *(off-label)*, propachlor, propyzamide, simazine, trifluralin
	Blackgrass	cycloxydim
	Chickweed	chlorpropham, chlorpropham + fenuron
	Cleavers	napropamide
	Clover	ethofumesate *(off-label)*
	Corn marigold	clopyralid
	Couch	cycloxydim
	Creeping bent	cycloxydim
	Creeping thistle	clopyralid
	Docks	asulam *(off-label)*
	Groundsel	napropamide

	Mayweeds	clopyralid
	Perennial dicotyledons	clopyralid, glufosinate-ammonium
	Perennial grasses	cycloxydim, diquat + paraquat, fluazifop-P-butyl, glufosinate-ammonium, propyzamide
	Polygonums	chlorpropham
	Volunteer cereals	cycloxydim, diquat + paraquat, fluazifop-P-butyl
	Wild oats	cycloxydim, fluazifop-P-butyl
Crop control	Runner desiccation	paraquat

Tree fruit

Diseases	Soil-borne diseases	methyl bromide with chloropicrin
	Verticillium wilt	chloropicrin *(soil fumigation)*
Pests	Aphids	fatty acids, nicotine, rotenone
	Birds	aluminium ammonium sulphate, ziram
	Capsids	nicotine
	Damaging mammals	aluminium ammonium sulphate, ziram
	Mealy bugs	fatty acids
	Nematodes	methyl bromide with chloropicrin
	Red spider mites	fatty acids
	Sawflies	nicotine
	Scale insects	fatty acids
	Soil pests	methyl bromide with chloropicrin
	Whitefly	fatty acids
Weeds	Annual and perennial weeds	glyphosate (top fruit, horticulture, forestry, amenity etc.)
	Annual dicotyledons	glufosinate-ammonium, paraquat
	Annual grasses	glufosinate-ammonium, paraquat
	Annual weeds	dichlobenil
	Creeping bent	paraquat
	General weed control	methyl bromide with chloropicrin
	Perennial dicotyledons	dichlobenil, glufosinate-ammonium, glyphosate (top fruit, horticulture, forestry, amenity etc.) *(wiper application)*
	Perennial grasses	dichlobenil, glufosinate-ammonium
	Perennial ryegrass	paraquat

| | Rough meadow grass | paraquat |
| Plant growth regulation | Increasing yield | sulphur |

Tree nuts

Diseases	Bacterial blight	copper oxychloride *(off-label)*
	Bacterial canker	copper oxychloride *(off-label)*
	Botrytis	iprodione *(off-label)*
Pests	Aphids	lambda-cyhalothrin *(off-label)*
	Caterpillars	lambda-cyhalothrin *(off-label)*
Weeds	Annual and perennial weeds	glyphosate (top fruit, horticulture, forestry, amenity etc.) *(off-label)*
	Annual dicotyledons	glufosinate-ammonium
	Annual grasses	glufosinate-ammonium
	Perennial dicotyledons	glufosinate-ammonium
	Perennial grasses	glufosinate-ammonium

Grain/crop store uses

Food storage

| Pests | Food storage pests | methyl bromide with amyl acetate |

Stored cabbages

Diseases	Alternaria	iprodione
	Botrytis	iprodione
	Phytophthora	carbendazim + metalaxyl

Stored grain/rapeseed/linseed

| Pests | General insect control | magnesium phosphide |
| | Grain storage pests | chlorpyrifos-methyl, deltamethrin, etrimfos, fenitrothion, fenitrothion + permethrin + resmethrin, methyl bromide, phenothrin + tetramethrin, pirimiphos-methyl |

Stored products

Pests	Stored product pests	aluminium phosphide

Grassland

Grassland

Diseases	Crown rust	propiconazole, triadimefon
	Damping off	thiram *(seed treatment)*
	Drechslera leaf spot	propiconazole
	Powdery mildew	propiconazole, sulphur, triadimefon
	Rhynchosporium	propiconazole, triadimefon
Pests	Aphids	pirimicarb
	Birds	aluminium ammonium sulphate
	Cutworms	gamma-HCH
	Damaging mammals	aluminium ammonium sulphate, strychnine hydrochloride (commodity substance) *(areas of restricted public access)*
	Frit fly	chlorpyrifos
	Leatherjackets	chlorpyrifos, gamma-HCH
	Wireworms	gamma-HCH
Weeds	Annual and perennial weeds	glyphosate (agricultural uses)
	Annual dicotyledons	2,4-D, 2,4-D + dicamba + triclopyr, 2,4-D + mecoprop-P, 2,4-DB + linuron + MCPA, 2,4-DB + MCPA, benazolin + 2,4-DB + MCPA, benazolin + bromoxynil + ioxynil, bentazone + MCPA + MCPB, bromoxynil + ioxynil + mecoprop-P, clopyralid, dicamba + MCPA + mecoprop, dicamba + MCPA + mecoprop-P, dicamba + mecoprop-P, fluroxypyr, MCPA, MCPA + MCPB, mecoprop-P, paraquat *(sward destruction/direct drilling)*
	Annual grasses	ethofumesate, paraquat *(sward destruction/direct drilling)*
	Annual weeds	glyphosate (agricultural uses)
	Black bindweed	fluroxypyr
	Blackgrass	ethofumesate
	Bracken	asulam, glyphosate (agricultural uses)
	Brambles	2,4-D + dicamba + triclopyr, clopyralid + triclopyr, triclopyr
	Broom	clopyralid + triclopyr, triclopyr
	Buttercups	MCPA

.	Chickweed	benazolin + 2,4-DB + MCPA, dicamba, dicamba + MCPA + mecoprop, dicamba + MCPA + mecoprop-P, dicamba + mecoprop-P, ethofumesate, fluroxypyr, mecoprop-P
	Cleavers	benazolin + 2,4-DB + MCPA, dicamba + MCPA + mecoprop, dicamba + MCPA + mecoprop-P, dicamba + mecoprop-P, ethofumesate, fluroxypyr, mecoprop-P
	Corn marigold	clopyralid
	Couch	glyphosate (agricultural uses)
	Creeping bent	paraquat *(sward destruction/direct drilling)*
	Creeping thistle	clopyralid *(off-label - weed wiper)*, clopyralid, clopyralid + 2,4-D + MCPA
	Destruction of short term leys	glyphosate (agricultural uses)
	Docks	2,4-D + dicamba + triclopyr, asulam, clopyralid + fluroxypyr + triclopyr, clopyralid + triclopyr, dicamba, dicamba + MCPA + mecoprop, dicamba + MCPA + mecoprop-P, dicamba + mecoprop-P, fluroxypyr, fluroxypyr + triclopyr, MCPA, thifensulfuron-methyl, triclopyr
	Fat-hen	MCPA
	Gorse	2,4-D + dicamba + triclopyr, clopyralid + triclopyr, triclopyr
	Hemp-nettle	fluroxypyr, MCPA
	Japanese knotweed	2,4-D + dicamba + triclopyr
	Mayweeds	clopyralid, dicamba + MCPA + mecoprop, dicamba + MCPA + mecoprop-P, dicamba + mecoprop-P
	Perennial dicotyledons	2,4-D, 2,4-D + dicamba + triclopyr, 2,4-D + mecoprop-P, 2,4-DB + MCPA, benazolin + 2,4-DB + MCPA, clopyralid, clopyralid + fluroxypyr + triclopyr, clopyralid + triclopyr *(off-label - weed wiper)*, clopyralid + triclopyr, dicamba + MCPA + mecoprop, dicamba + MCPA + mecoprop-P, dicamba + mecoprop-P, glyphosate (agricultural uses) *(wiper application)*, MCPA, MCPA + MCPB, mecoprop-P, triclopyr
	Perennial ryegrass	paraquat *(sward destruction/direct drilling)*
	Polygonums	benazolin + 2,4-DB + MCPA, dicamba + MCPA + mecoprop, dicamba + MCPA + mecoprop-P, dicamba + mecoprop-P
	Ragwort	MCPA
	Rhododendrons	2,4-D + dicamba + triclopyr
	Rough meadow grass	paraquat *(sward destruction/direct drilling)*
	Rushes	glyphosate (agricultural uses), MCPA, triclopyr
	Stinging nettle	2,4-D + dicamba + triclopyr, clopyralid + fluroxypyr + triclopyr, clopyralid + triclopyr, triclopyr
	Total vegetation control	diquat + paraquat, glufosinate-ammonium, glyphosate (agricultural uses)

FOR FULL DETAILS ALWAYS CHECK THE PROFILE AND THE PRODUCT LABEL

	Volunteer cereals	ethofumesate
	Volunteer potatoes	fluroxypyr
	Woody weeds	2,4-D + dicamba + triclopyr, triclopyr
Plant growth regulation	Increasing yield	sulphur

Leys

Pests	Frit fly	cypermethrin
	Leatherjackets	methiocarb *(reduction)*
	Slugs and snails	methiocarb
Weeds	Annual dicotyledons	2,4-DB + MCPA, benazolin + 2,4-DB + MCPA, bromoxynil + ethofumesate + ioxynil, dicamba + MCPA + mecoprop, dicamba + MCPA + mecoprop-P, dicamba + mecoprop-P, fluroxypyr, MCPA, MCPA + MCPB, MCPB, mecoprop-P
	Annual grasses	ethofumesate
	Annual meadow grass	bromoxynil + ethofumesate + ioxynil
	Black bindweed	fluroxypyr
	Blackgrass	ethofumesate
	Chickweed	benazolin + 2,4-DB + MCPA, dicamba, dicamba + MCPA + mecoprop, dicamba + MCPA + mecoprop-P, ethofumesate, fluroxypyr, mecoprop-P
	Cleavers	benazolin + 2,4-DB + MCPA, dicamba + MCPA + mecoprop, dicamba + MCPA + mecoprop-P, ethofumesate, fluroxypyr, mecoprop-P
	Docks	dicamba, dicamba + MCPA + mecoprop, dicamba + MCPA + mecoprop-P, dicamba + mecoprop-P, fluroxypyr
	Fat-hen	MCPA
	Hemp-nettle	fluroxypyr, MCPA
	Mayweeds	dicamba + MCPA + mecoprop, dicamba + MCPA + mecoprop-P
	Perennial dicotyledons	2,4-DB + MCPA, benazolin + 2,4-DB + MCPA, dicamba + MCPA + mecoprop, dicamba + MCPA + mecoprop-P, dicamba + mecoprop-P, MCPA, MCPA + MCPB, MCPB, mecoprop-P
	Polygonums	2,4-DB + MCPA, benazolin + 2,4-DB + MCPA, dicamba + MCPA + mecoprop, dicamba + MCPA + mecoprop-P
	Volunteer cereals	ethofumesate
	Volunteer potatoes	fluroxypyr

Non-crop pest control

Farm buildings/yards

Diseases	Fungus diseases	formaldehyde (commodity substance)
Pests	Ants	fenitrothion
	Beetles	dichlorvos, fenitrothion
	Birds	alphachloralose
	Bugs	fenitrothion
	Cockroaches	fenitrothion
	Crickets	fenitrothion
	Damaging mammals	bone oil
	Earwigs	fenitrothion
	Fleas	fenitrothion
	Flies	azamethiphos, cyromazine, dichlorvos, diflubenzuron, fenitrothion, methomyl + (Z)-9-tricosene, phenothrin, phenothrin + tetramethrin, pyrethrins, tetramethrin
	Grey squirrels	warfarin
	Hide beetle	alpha-cypermethrin
	Mealworms	alpha-cypermethrin
	Mites	fenitrothion
	Mosquitoes	dichlorvos, phenothrin + tetramethrin
	Moths	fenitrothion
	Poultry ectoparasites	dichlorvos
	Poultry house pests	alpha-cypermethrin, fenitrothion
	Rodents	alphachloralose *(indoor use)*, brodifacoum *(indoor use only)*, brodifacoum, bromadiolone, calciferol + difenacoum, chlorophacinone, coumatetralyl, difenacoum, diphacinone, warfarin, zinc phosphide
	Silverfish	fenitrothion
	Wasps	phenothrin + tetramethrin
Weeds	Annual and perennial weeds	glyphosate (top fruit, horticulture, forestry, amenity etc.)
	Annual weeds	diuron
	Bracken	imazapyr
	Perennial weeds	diuron
	Total vegetation control	imazapyr

FOR FULL DETAILS ALWAYS CHECK THE PROFILE AND THE PRODUCT LABEL

Farmland

Pests	Damaging mammals	aluminium phosphide, bone oil
	Rodents	aluminium phosphide
Weeds	Volunteer potatoes	dichlobenil *(blight prevention)*

Food storage areas

Pests	Ants	fenitrothion
	Beetles	fenitrothion
	Bugs	fenitrothion
	Cockroaches	fenitrothion
	Crickets	fenitrothion
	Earwigs	fenitrothion
	Fleas	fenitrothion
	Flies	fenitrothion
	Food storage pests	methyl bromide
	Mites	fenitrothion
	Moths	fenitrothion
	Silverfish	fenitrothion

Manure heaps

Pests	Flies	cyromazine, diflubenzuron

Miscellaneous situations

Pests	Birds	carbon dioxide (commodity substance), paraffin oil (commodity substance) *(egg treatment)*
	Damaging mammals	strychnine hydrochloride (commodity substance) *(areas of restricted public access)*
	Rodents	carbon dioxide (commodity substance)
	Wasps	resmethrin + tetramethrin

Refuse tips

Pests	Ants	fenitrothion
	Beetles	fenitrothion
	Bugs	fenitrothion

Cockroaches	fenitrothion
Crickets	fenitrothion
Earwigs	fenitrothion
Fleas	fenitrothion
Flies	diflubenzuron, fenitrothion
Mites	fenitrothion
Moths	fenitrothion
Silverfish	fenitrothion

Stored plant material

Pests	General insect control	magnesium phosphide, methyl bromide

Protected crops

Aubergines

Diseases	Botrytis	dichlofluanid *(off-label)*, pyrimethanil *(off-label)*
	Phytophthora	propamocarb hydrochloride
	Powdery mildew	sulphur *(off-label)*
	Pythium	propamocarb hydrochloride
Pests	Ants	pirimiphos-methyl
	Aphids	nicotine, permethrin, pirimiphos-methyl, Verticillium lecanii
	Capsids	pirimiphos-methyl
	Earwigs	pirimiphos-methyl
	Leaf miners	deltamethrin *(off-label - fog)*, oxamyl *(off-label)*, pirimiphos-methyl
	Leafhoppers	nicotine
	Red spider mites	pirimiphos-methyl
	Sawflies	pirimiphos-methyl
	Sciarid flies	permethrin
	Thrips	nicotine, pirimiphos-methyl
	Tomato fruitworm	permethrin
	Western flower thrips	deltamethrin *(off-label - fog)*

	Whitefly	buprofezin, nicotine, permethrin, pirimiphos-methyl, Verticillium lecanii

Beans

Pests	Aphids	Verticillium lecanii
	Red spider mites	tetradifon
	Whitefly	Verticillium lecanii

Chicory

Pests	Leaf miners	cypermethrin *(off-label)*

Cucumbers

Diseases	Botrytis	chlorothalonil, iprodione
	Damping off	etridiazole *(seedlings and transplants)*
	Damping off and foot rot	propamocarb hydrochloride *(off-label)*
	Downy mildew	copper oxychloride + metalaxyl *(off-label)*
	Phytophthora	propamocarb hydrochloride
	Powdery mildew	bupirimate, chlorothalonil, copper ammonium carbonate, imazalil
	Pythium	propamocarb hydrochloride
	Rhizoctonia	quintozene
	Root diseases	carbendazim *(off-label)*
Pests	Ants	pirimiphos-methyl
	Aphids	deltamethrin, fatty acids, nicotine, permethrin, pirimicarb, pirimiphos-methyl, pymetrozine, pyrethrins + resmethrin, Verticillium lecanii
	Capsids	pirimiphos-methyl
	Caterpillars	Bacillus thuringiensis, deltamethrin
	Earwigs	pirimiphos-methyl
	Leaf miners	cypermethrin *(off-label)*, deltamethrin *(off-label - fog)*, pirimiphos-methyl
	Leafhoppers	nicotine
	Mealy bugs	deltamethrin, fatty acids, petroleum oil
	Red spider mites	abamectin, fatty acids, fenbutatin oxide, petroleum oil, pirimiphos-methyl, tetradifon
	Sawflies	pirimiphos-methyl
	Scale insects	deltamethrin, fatty acids, petroleum oil

	Sciarid flies	permethrin
	Thrips	deltamethrin, nicotine, pirimiphos-methyl
	Tomato fruitworm	permethrin
	Western flower thrips	abamectin, deltamethrin *(off-label - fog)*, dichlorvos
	Whitefly	buprofezin, cypermethrin, deltamethrin, fatty acids, nicotine, permethrin, pirimiphos-methyl, pyrethrins + resmethrin, Verticillium lecanii

Herbs

Pests	Aphids	nicotine
	Leaf miners	dichlorvos *(off-label)*
	Leafhoppers	nicotine
	Thrips	nicotine
	Western flower thrips	dichlorvos *(off-label)*
	Whitefly	nicotine
Weeds	Annual dicotyledons	prometryn *(off-label)*, terbacil *(off-label)*
	Annual grasses	prometryn *(off-label)*, terbacil *(off-label)*
	Couch	terbacil *(off-label)*
	Perennial grasses	terbacil *(off-label)*

Lettuce

Diseases	Big vein	carbendazim *(off-label)*
	Botrytis	dicloran, iprodione, quintozene, thiram
	Downy mildew	fosetyl-aluminium, mancozeb, metalaxyl + thiram, thiram, zineb
	Rhizoctonia	dicloran, quintozene, tolclofos-methyl
	Sclerotinia	quintozene
Pests	Aphids	nicotine, pirimicarb
	Flea beetles	deltamethrin *(off-label)*
	Leaf miners	abamectin, cypermethrin *(off-label)*
	Leafhoppers	deltamethrin *(off-label)*, nicotine
	Sciarid flies	permethrin
	Thrips	nicotine
	Western flower thrips	abamectin

	Whitefly	cypermethrin, nicotine, permethrin, Verticillium lecanii
Weeds	Annual dicotyledons	chlorpropham with cetrimide, pendimethalin *(off-label)*
	Annual grasses	chlorpropham with cetrimide, pendimethalin *(off-label)*
	Chickweed	chlorpropham with cetrimide
	Polygonums	chlorpropham with cetrimide

Mushrooms

Diseases	Bacterial blotch	sodium hypochlorite (commodity substance)
	Cobweb	prochloraz, pyrifenox *(off-label)*
	Dry bubble	chlorothalonil, prochloraz
	Fungus diseases	formaldehyde (commodity substance) *(spray or fumigant)*
	Mycogone	carbendazim
	Trichoderma	carbendazim *(off-label - spawn treatment)*
	Wet bubble	carbendazim, chlorothalonil, prochloraz
Pests	Mushroom flies	pyrethrins + resmethrin *(off-label)*
	Sciarid flies	diflubenzuron, methoprene, permethrin, pyrethrins + resmethrin *(off-label)*

Mustard and cress

Diseases	Damping off	etridiazole *(seedlings and transplants)*

Peppers

Diseases	Botrytis	chlorothalonil, dichlofluanid *(off-label)*
	Damping off and foot rot	propamocarb hydrochloride *(off-label)*
	Phytophthora	propamocarb hydrochloride
	Powdery mildew	fenarimol *(off-label)*, sulphur *(off-label)*
	Pythium	propamocarb hydrochloride
Pests	Ants	pirimiphos-methyl
	Aphids	deltamethrin, fatty acids, nicotine, permethrin, pirimicarb, pirimiphos-methyl, Verticillium lecanii
	Capsids	pirimiphos-methyl
	Caterpillars	Bacillus thuringiensis, deltamethrin
	Earwigs	pirimiphos-methyl
	Leaf miners	deltamethrin *(off-label - fog)*, oxamyl *(off-label)*, pirimiphos-methyl

Leafhoppers	nicotine	
Mealy bugs	deltamethrin, fatty acids	
Red spider mites	fatty acids, fenbutatin oxide *(off-label)*, pirimiphos-methyl, tetradifon	
Sawflies	pirimiphos-methyl	
Scale insects	deltamethrin, fatty acids	
Sciarid flies	permethrin	
Thrips	nicotine, pirimiphos-methyl	
Tomato fruitworm	permethrin	
Tomato moth	diflubenzuron *(off-label)*	
Western flower thrips	deltamethrin *(off-label - fog)*	
Whitefly	buprofezin, deltamethrin, fatty acids, nicotine, permethrin, pirimiphos-methyl, Verticillium lecanii	

Pot plants

Diseases	Botrytis	carbendazim *(off-label)*, carbendazim, iprodione, quintozene
	Disease control/ foliar feed	difenoconazole *(off-label)*
	Phytophthora	fosetyl-aluminium, furalaxyl, propamocarb hydrochloride
	Powdery mildew	bupirimate, carbendazim *(off-label)*, carbendazim
	Pythium	furalaxyl, propamocarb hydrochloride
	Rhizoctonia	quintozene
	Root rot	fosetyl-aluminium
	Rust	mancozeb, oxycarboxin
	Sclerotinia	quintozene
	Soil-borne diseases	metam-sodium
Pests	Aphids	deltamethrin, imidacloprid, malathion *(indoors)*, nicotine
	Caterpillars	deltamethrin
	Leaf miners	abamectin
	Leafhoppers	malathion *(indoors)*
	Mealy bugs	deltamethrin, malathion *(indoors)*, petroleum oil
	Nematodes	metam-sodium
	Red spider mites	abamectin, petroleum oil, tetradifon
	Scale insects	deltamethrin, petroleum oil
	Sciarid flies	imidacloprid

	Soil pests	metam-sodium
	Thrips	malathion *(indoors)*
	Vine weevil	carbofuran, imidacloprid
	Western flower thrips	abamectin
	Whitefly	deltamethrin, imidacloprid, malathion *(indoors)*
Weeds	General weed control	metam-sodium
Plant growth regulation	Flower induction	2-chloroethylphosphonic acid
	Improving colour	paclobutrazol
	Increasing branching	2-chloroethylphosphonic acid
	Increasing flowering	paclobutrazol
	Stem shortening	chlormequat, chlormequat + choline chloride, daminozide, paclobutrazol

Protected asparagus

Pests	Aphids	nicotine

Protected crops general

Diseases	Black root rot	carbendazim *(off-label)*
	Botrytis	chlorothalonil
	Disease control/ foliar feed	carbendazim + prochloraz *(off-label)*
	Fungus diseases	formaldehyde (commodity substance) *(spray, dip or fumigant)*
	Phytophthora	fosetyl-aluminium
	Powdery mildew	imazalil
	Rhizoctonia	tolclofos-methyl *(off-label)*
	Root rot	fosetyl-aluminium
	Soil-borne diseases	dazomet, formaldehyde (commodity substance) *(drench)*, metam-sodium, methyl bromide with amyl acetate, methyl bromide with chloropicrin
Pests	Ants	pirimiphos-methyl
	Aphids	deltamethrin, nicotine, permethrin, pirimiphos-methyl, pymetrozine, rotenone
	Capsids	nicotine, pirimiphos-methyl
	Caterpillars	Bacillus thuringiensis, deltamethrin, nicotine

	Earwigs	pirimiphos-methyl
	Leaf miners	nicotine, pirimiphos-methyl
	Leafhoppers	nicotine
	Mealy bugs	deltamethrin
	Nematodes	dazomet, metam-sodium, methyl bromide with amyl acetate, methyl bromide with chloropicrin
	Red spider mites	fenbutatin oxide, pirimiphos-methyl, tetradifon
	Sawflies	nicotine, pirimiphos-methyl
	Scale insects	deltamethrin
	Sciarid flies	permethrin
	Slugs and snails	aluminium sulphate, methiocarb *(off-label)*
	Soil pests	dazomet, metam-sodium, methyl bromide with amyl acetate, methyl bromide with chloropicrin
	Thrips	nicotine, pirimiphos-methyl
	Western flower thrips	dichlorvos *(off-label)*
	Whitefly	buprofezin, cypermethrin, deltamethrin, nicotine, permethrin, pirimiphos-methyl, Verticillium lecanii
Weeds	Annual dicotyledons	chlorpropham + fenuron, isoxaben *(off-label)*, paraquat *(off-label)*
	Annual grasses	chlorpropham + fenuron, paraquat *(off-label)*
	Chickweed	chlorpropham + fenuron
	General weed control	dazomet, metam-sodium, methyl bromide with amyl acetate, methyl bromide with chloropicrin

Protected cucurbits

Diseases	Downy mildew	copper oxychloride + metalaxyl *(off-label)*
Pests	Aphids	nicotine, pirimicarb *(off-label)*
	Leaf miners	cypermethrin *(off-label)*
	Leafhoppers	nicotine
	Red spider mites	tetradifon
	Thrips	nicotine
	Whitefly	nicotine
Weeds	Annual dicotyledons	isoxaben *(off-label)*

Protected cut flowers

Diseases	Botrytis	quintozene
	Powdery mildew	imazalil
	Rhizoctonia	quintozene
	Rust	azoxystrobin *(off-label)*, mancozeb
	Sclerotinia	quintozene
	Soil-borne diseases	metam-sodium
Pests	Aphids	dimethoate, nicotine, pirimicarb, Verticillium lecanii
	Leaf miners	dimethoate
	Nematodes	metam-sodium
	Red spider mites	tebufenpyrad
	Soil pests	metam-sodium
	Whitefly	Verticillium lecanii
Weeds	General weed control	metam-sodium
Plant growth regulation	Basal bud stimulation	2-chloroethylphosphonic acid

Protected grapevines

Pests	Mealy bugs	petroleum oil
	Red spider mites	petroleum oil, tetradifon
	Scale insects	petroleum oil

Protected onions/leeks/garlic

Pests	Aphids	nicotine
	General insect control	chlorpyrifos *(off-label)*
	Leafhoppers	nicotine
	Stem nematodes	oxamyl *(off-label)*
	Thrips	nicotine
	Whitefly	nicotine

Protected potatoes

Pests	Aphids	nicotine
	Leafhoppers	nicotine

	Thrips	nicotine
	Whitefly	nicotine

Protected radishes

Diseases	Rhizoctonia	tolclofos-methyl *(off-label)*
	White blister	mancozeb + metalaxyl-M *(off-label)*

Tomatoes

Diseases	Blight	Bordeaux mixture *(outdoor)*, chlorothalonil, copper sulphate + sulphur
	Botrytis	chlorothalonil, dichlofluanid, iprodione, pyrimethanil *(off-label)*, quintozene, thiram
	Damping off	copper oxychloride, etridiazole *(seedlings and transplants)*, etridiazole
	Damping off and foot rot	propamocarb hydrochloride *(off-label)*
	Didymella stem rot	carbendazim *(off-label)*
	Foot rot	copper oxychloride
	Leaf mould	chlorothalonil, copper ammonium carbonate, dichlofluanid
	Phytophthora	copper oxychloride, etridiazole, propamocarb hydrochloride
	Powdery mildew	bupirimate *(off-label)*, fenarimol *(off-label)*, sulphur *(off-label)*
	Pythium	propamocarb hydrochloride
	Rhizoctonia	quintozene
	Root diseases	etridiazole *(off-label)*
	Sclerotinia	quintozene
	Soil-borne diseases	metam-sodium
Pests	Ants	pirimiphos-methyl
	Aphids	deltamethrin, dimethoate, fatty acids, malathion *(indoors)*, nicotine, permethrin, pirimicarb, pirimiphos-methyl, pyrethrins + resmethrin, Verticillium lecanii
	Capsids	pirimiphos-methyl
	Caterpillars	Bacillus thuringiensis, deltamethrin *(off-label)*, deltamethrin
	Earwigs	pirimiphos-methyl
	Leaf miners	abamectin, deltamethrin *(off-label - fog)*, deltamethrin *(off-label)*, oxamyl *(off-label)*, pirimiphos-methyl
	Leafhoppers	malathion *(indoors)*, nicotine

FOR FULL DETAILS ALWAYS CHECK THE PROFILE AND THE PRODUCT LABEL

	Mealy bugs	deltamethrin, fatty acids, malathion *(indoors)*, petroleum oil
	Nematodes	metam-sodium
	Red spider mites	abamectin, fatty acids, fenbutatin oxide, petroleum oil, pirimiphos-methyl, tetradifon
	Sawflies	pirimiphos-methyl
	Scale insects	deltamethrin, fatty acids, petroleum oil
	Sciarid flies	permethrin
	Soil pests	metam-sodium
	Thrips	malathion *(indoors)*, nicotine, pirimiphos-methyl
	Tomato fruitworm	permethrin
	Western flower thrips	abamectin, deltamethrin *(off-label - fog)*
	Whitefly	buprofezin, deltamethrin, fatty acids, malathion *(indoors)*, nicotine, permethrin, pirimiphos-methyl, pyrethrins + resmethrin, Verticillium lecanii
Weeds	Annual dicotyledons	pentanochlor
	Annual meadow grass	pentanochlor
	General weed control	metam-sodium
Plant growth regulation	Fruit ripening	2-chloroethylphosphonic acid
	Increasing fruit set	(2-naphthyloxy)acetic acid

Total vegetation control

Land temporarily removed from production

Weeds	Annual and perennial weeds	glyphosate (agricultural uses)
	Docks	thifensulfuron-methyl *(in green cover)*
	Green cover	cycloxydim, diquat + paraquat, fluazifop-P-butyl, glufosinate-ammonium, glyphosate (agricultural uses), metsulfuron-methyl, paraquat
	Perennial dicotyledons	glufosinate-ammonium
	Perennial grasses	glufosinate-ammonium

Non-crop areas

Diseases	Fungus diseases	dichlorophen
Pests	Grey squirrels	warfarin
Weeds	Algae	dichlorophen
	Annual and perennial weeds	glyphosate (agricultural uses), glyphosate (top fruit, horticulture, forestry, amenity etc.), glyphosate + oxadiazon *(soil surface treatment)*
	Annual dicotyledons	2,4-D + dicamba + triclopyr, chlorpropham + tar acids + fenuron, diquat + paraquat, diuron, diuron + glyphosate, glufosinate-ammonium, glyphosate (agricultural uses), paraquat, picloram
	Annual grasses	chlorpropham + tar acids + fenuron, diquat + paraquat, diuron, diuron + glyphosate, glufosinate-ammonium, glyphosate (agricultural uses), paraquat
	Annual meadow grass	diuron
	Annual weeds	diuron
	Bracken	asulam, glyphosate (top fruit, horticulture, forestry, amenity etc.), imazapyr, picloram
	Brambles	2,4-D + dicamba + triclopyr, triclopyr *(directed treatment)*, triclopyr
	Broom	triclopyr *(directed treatment)*, triclopyr
	Chickweed	chlorpropham + tar acids + fenuron
	Couch	amitrole, glyphosate (agricultural uses)
	Creeping bent	amitrole, paraquat
	Creeping thistle	amitrole
	Docks	2,4-D + dicamba + triclopyr, amitrole, asulam, triclopyr
	Gorse	2,4-D + dicamba + triclopyr, triclopyr *(directed treatment)*, triclopyr
	Green cover	glyphosate (agricultural uses)
	Japanese knotweed	2,4-D + dicamba + triclopyr, glyphosate (agricultural uses), picloram
	Mosses	dichlorophen
	Perennial dicotyledons	2,4-D + dicamba + triclopyr, diuron + glyphosate, glufosinate-ammonium, MCPA, picloram, triclopyr
	Perennial grasses	amitrole, diquat + paraquat, diuron + glyphosate, glufosinate-ammonium
	Perennial ryegrass	paraquat
	Perennial weeds	diuron, glyphosate (agricultural uses)
	Rhododendrons	2,4-D + dicamba + triclopyr, triclopyr

FOR FULL DETAILS ALWAYS CHECK THE PROFILE AND THE PRODUCT LABEL

	Rough meadow grass	paraquat
	Rushes	triclopyr
	Stinging nettle	2,4-D + dicamba + triclopyr, MCPA, triclopyr
	Total vegetation control	amitrole, amitrole + 2,4-D + diuron, bromacil, bromacil + diuron, bromacil + picloram, dichlobenil, diuron + paraquat, glyphosate (top fruit, horticulture, forestry, amenity etc.), glyphosate + oxadiazon *(soil surface treatment)*, imazapyr, sodium chlorate
	Volunteer cereals	diquat + paraquat
	Woody weeds	2,4-D + dicamba + triclopyr, fosamine-ammonium, picloram, triclopyr *(directed treatment)*, triclopyr
Plant growth regulation	Growth suppression	maleic hydrazide

Stubbles

Weeds	Annual and perennial weeds	Glyphosate (agricultural uses)
	Annual dicotyledons	diquat + paraquat, glufosinate-ammonium, glyphosate (agricultural uses), paraquat
	Annual grasses	diquat + paraquat, glufosinate-ammonium, glyphosate (agricultural uses), paraquat
	Annual weeds	Amitrole
	Barren brome	Paraquat
	Couch	amitrole, glyphosate (agricultural uses)
	Creeping bent	Glyphosate (agricultural uses), paraquat
	Creeping thistle	Amitrole
	Docks	Amitrole
	Perennial grasses	diquat + paraquat, glyphosate (agricultural uses)
	Volunteer cereals	diquat + paraquat, glyphosate (agricultural uses), paraquat
	Volunteer oilseed rape	Glyphosate (agricultural uses)
	Volunteer potatoes	amitrole *(barley stubble)*, glyphosate (agricultural uses)
	Wild oats	Paraquat

Weeds in or near water

Weeds	Aquatic weeds	2,4-D, dichlobenil, diquat, glyphosate (top fruit, horticulture, forestry, amenity etc.), terbutryn
	Perennial dicotyledons	2,4-D
	Perennial grasses	glyphosate (top fruit, horticulture, forestry, amenity etc.)

	Rushes	glyphosate (top fruit, horticulture, forestry, amenity etc.)
	Sedges	glyphosate (top fruit, horticulture, forestry, amenity etc.)
	Waterlilies	glyphosate (top fruit, horticulture, forestry, amenity etc.)
	Woody weeds	fosamine-ammonium
Plant growth regulation	Growth suppression	maleic hydrazide

Turf/amenity grass

Amenity grass

Diseases	Brown patch	Iprodione
	Dollar spot	Iprodione
	Fusarium patch	Iprodione
	Grey snow mould	Iprodione
	Melting out	Iprodione
	Red thread	Iprodione
Pests	Frit fly	Chlorpyrifos
	Leatherjackets	Chlorpyrifos
Weeds	Annual and perennial weeds	glyphosate (top fruit, horticulture, forestry, amenity etc.) *(wiper application)*, glyphosate (top fruit, horticulture, forestry, amenity etc.)
	Annual dicotyledons	2,4-D + picloram, bifenox + MCPA + mecoprop-P, dicamba + maleic hydrazide + MCPA, dicamba + MCPA + mecoprop-P, dichlorprop + MCPA, fluroxypyr + triclopyr, isoxaben, MCPA, MCPA + mecoprop-P
	Brambles	2,4-D + picloram, clopyralid + triclopyr
	Broom	clopyralid + triclopyr
	Buttercups	dichlorprop + MCPA, MCPA
	Clovers	dichlorprop + MCPA
	Creeping thistle	2,4-D + picloram
	Daisies	dichlorprop + MCPA
	Docks	2,4-D + picloram, asulam *(not fine turf)*, clopyralid + triclopyr, dicamba + MCPA + mecoprop-P
	Gorse	clopyralid + triclopyr
	Japanese knotweed	2,4-D + picloram

	Perennial dicotyledons	2,4-D + picloram, bifenox + MCPA + mecoprop-P, clopyralid + triclopyr, dicamba + maleic hydrazide + MCPA, dicamba + MCPA + mecoprop-P, dichlorprop + MCPA, fluroxypyr + triclopyr, MCPA, MCPA + mecoprop-P
	Ragwort	2,4-D + picloram
	Stinging nettle	clopyralid + triclopyr
	Total vegetation control	glyphosate (top fruit, horticulture, forestry, amenity etc.)
	Woody weeds	2,4-D + picloram, fluroxypyr + triclopyr
Plant growth regulation	Growth retardation	dicamba + maleic hydrazide + MCPA, trinexapac-ethyl
	Growth suppression	maleic hydrazide, mefluidide

Turf/amenity grass

Diseases	Anthracnose	carbendazim + chlorothalonil, carbendazim + iprodione, chlorothalonil
	Brown patch	Iprodione
	Cladosporium leaf spot	carbendazim + iprodione
	Dollar spot	carbendazim, carbendazim + chlorothalonil, chlorothalonil, fenarimol, iprodione, quintozene, thiabendazole, thiophanate-methyl
	Fairy rings	Triforine
	Fungus diseases	Dichlorophen
	Fusarium patch	carbendazim, carbendazim + chlorothalonil, carbendazim + epoxiconazole, carbendazim + iprodione, chlorothalonil, fenarimol, iprodione, quintozene, thiabendazole, thiophanate-methyl
	Grey snow mould	Iprodione
	Melting out	Iprodione
	Red thread	carbendazim + chlorothalonil, carbendazim + iprodione, chlorothalonil, dichlorophen, fenarimol, iprodione, quintozene, thiabendazole, thiophanate-methyl
Pests	Damaging mammals	aluminium ammonium sulphate *(anti-fouling)*, aluminium ammonium sulphate
	Earthworms	carbendazim, carbendazim + chlorothalonil, gamma-HCH + thiophanate-methyl, thiophanate-methyl
	Frit fly	Chlorpyrifos
	Leatherjackets	chlorpyrifos, gamma-HCH + thiophanate-methyl

Weeds	Annual and perennial weeds	glyphosate (top fruit, horticulture, forestry, amenity etc.)
	Annual dicotyledons	2,4-D, 2,4-D + dicamba, 2,4-D + mecoprop-P, 2,4-D + picloram, bifenox + MCPA + mecoprop-P, clopyralid + diflufenican + MCPA, clopyralid + fluroxypyr + MCPA, dicamba + dichlorprop + ferrous sulphate + MCPA, dicamba + dichlorprop + MCPA, dicamba + maleic hydrazide + MCPA, dicamba + MCPA + mecoprop-P, dichlorprop + ferrous sulphate + MCPA, dichlorprop + MCPA, diquat + paraquat *(sward destruction/turf renovation)*, fluroxypyr + mecoprop-P, ioxynil, isoxaben, MCPA, mecoprop-P, picloram
	Annual grasses	diquat + paraquat *(sward destruction/turf renovation)*, ethofumesate
	Annual weeds	Diuron
	Blackgrass	Ethofumesate
	Bracken	Picloram
	Brambles	2,4-D + picloram
	Buttercups	Dichlorprop + ferrous sulphate + MCPA, dichlorprop + MCPA, MCPA
	Chickweed	dicamba + MCPA + mecoprop-P, ethofumesate, mecoprop-P
	Cleavers	dicamba + MCPA + mecoprop-P, ethofumesate, mecoprop-P
	Clovers	mecoprop-P
	Creeping thistle	2,4-D + picloram, MCPA
	Daisies	MCPA
	Docks	2,4-D + picloram, asulam, dicamba + MCPA + mecoprop-P, MCPA
	Japanese knotweed	2,4-D + picloram, picloram
	Mayweeds	dicamba + MCPA + mecoprop-P
	Mosses	dicamba + dichlorprop + ferrous sulphate + MCPA, dichlorophen, dichlorophen + ferrous sulphate, dichlorprop + ferrous sulphate + MCPA, ferrous sulphate
	Perennial dicotyledons	2,4-D, 2,4-D + dicamba, 2,4-D + mecoprop-P, 2,4-D + picloram, bifenox + MCPA + mecoprop-P, clopyralid + diflufenican + MCPA, dicamba + dichlorprop + ferrous sulphate + MCPA, dicamba + dichlorprop + MCPA, dicamba + maleic hydrazide + MCPA, dicamba + MCPA + mecoprop-P, dichlorprop + ferrous sulphate + MCPA, dichlorprop + MCPA, fluroxypyr + mecoprop-P, MCPA, mecoprop-P, picloram
	Perennial grasses	diquat + paraquat *(sward destruction/turf renovation)*
	Perennial weeds	Diuron
	Polygonums	dicamba + MCPA + mecoprop-P
	Ragwort	2,4-D + picloram

	Slender speedwell	chlorthal-dimethyl, fluroxypyr + mecoprop-P
	Total vegetation control	glyphosate (top fruit, horticulture, forestry, amenity etc.)
	Volunteer cereals	Ethofumesate
	Weed grasses	amitrole *(off-label)*
	Woody weeds	2,4-D + picloram, picloram
Plant growth regulation	Growth retardation	dicamba + maleic hydrazide + MCPA, trinexapac-ethyl
	Growth suppression	maleic hydrazide, mefluidide
	Increasing yield	Sulphur

SECTION 4
PESTICIDE PROFILES

1   abamectin

A selective acaricide and insecticide for use in ornamentals

Products

Dynamec	Novartis BCM	19 g/l	EC	08701

Uses

Leaf miners in *flowers, ornamentals, protected flowers, protected lettuce, protected tomatoes*. Two-spotted spider mite in *flowers, ornamentals, protected cucumbers, protected flowers, protected tomatoes*. Western flower thrips in *flowers, ornamentals, protected cucumbers, protected flowers, protected lettuce, protected tomatoes*.

Efficacy

- Treat at first sign of infestation. Repeat sprays may be required
- For effective control total cover of all plant surfaces is essential, but avoid run-off
- Mites quickly become immobilised but 3-5 d may be required for maximum mortality

Crop safety/Restrictions

- Number of treatments 6 on protected tomatoes and cucumbers (only 4 of which can be made when flowers or fruit present); 4 on protected lettuce; not restricted on flowers but rotation with other products advised
- Maximum concentration must not exceed 50 ml per 100 l water
- Do not mix with wetters, stickers or other adjuvants
- On tomato or cucumber crops that are in flower, or have started to set fruit, treat only between 1 Mar and 31 Oct. Seedling tomatoes or cucumbers that have not started to flower or set fruit may be treated at any time. Do not treat cherry tomatoes
- Apply to lettuce only between 1 Mar and 31 Oct
- Do not use on ferns (*Adiantum* spp) or Shasta daisies
- Some spotting or staining may occur on carnation, kalanchoe and begonia foliage
- Consult manufacturer for list of plant varieties tested for safety
- There is insufficient evidence to support product compatibility with integrated and biological pest control programmes

Special precautions/Environmental safety

- Harmful if swallowed, in contact with skin and by inhalation
- Irritating to eyes
- Flammable
- Not to be used on food crops
- Extremely dangerous to fish or other aquatic life. Do not contaminate surface waters or ditches with chemical or used container
- High risk to bees. Do not apply to crops in flower or to those in which bees are actively foraging. Do not apply when flowering weeds are present
- Where bumble bees are used in tomatoes as pollinators, keep them out for 24 h after treatment
- Unprotected persons must be kept out of treated areas until the spray has dried

Personal protective equipment/Label precautions

- A, C, D, H, K, M
- M04, R03a, R03b, R03c, R04a, U02, U05a, U09a, U19, U20a, C03, E01, E12a, E13a, E26, E30b, E31b, E34

2   aldicarb

A soil-applied, systemic carbamate insecticide and nematicide

Products

1	Aventis Temik 10G	Aventis	10% w/w	GR	09749
2	Landgold Aldicarb 10G	Landgold	10% w/w	GR	06036
3	Me2 Aldee	Me2	10% w/w	GR	09781
4	Standon Aldicarb 10G	Standon	10% w/w	GR	05915
5	Temik 10G	Aventis	10% w/w	GR	10021
6	Temik 10G	RP Agric.	10% w/w	GR	06210

FOR FULL CONDITIONS OF USE ALWAYS READ THE PRODUCT LABEL

Uses

Aphids in **carrots, parsnips, potatoes, sugar beet** [1-6]. Docking disorder vectors in **sugar beet** [1, 5, 6]. Free-living nematodes in **carrots, parsnips, sugar beet** [1-6]. Free-living nematodes in **potatoes** [1, 5, 6]. Frit fly in **sweetcorn** *(off-label)* [5, 6]. Insect pests in **ornamentals** *(off-label)* [5, 6]. Leaf miners in **sugar beet** [1-6]. Millipedes in **sugar beet** [1-6]. Potato cyst nematode in **potatoes** [1-6]. Pygmy beetle in **sugar beet** [1-6]. Spraing vectors in **potatoes** [1-6]. Stem nematodes in **bulb onions** [1-6]. Stem nematodes in **leeks** *(off-label)*, **ornamentals** *(off-label)* [5, 6].

Efficacy

- Must be incorporated into soil by physical means. See label for details of application rates, timing, suitable applicators and techniques of incorporation
- Persistence and activity may be reduced in very wet soils or where pH exceeds 8.0. Do not use within 14 d of liming
- Use in potatoes reduces incidence of spraing disease

Crop safety/Restrictions

- Maximum number of treatments 1 per crop
- No edible crops other than those listed (see label) should be planted into treated soil for at least 8 wk after application

Special precautions/Environmental safety

- Aldicarb is subject to the Poisons Rules 1982 and the Poisons Act 1972. See notes in Section 1
- This product contains an anticholinesterase carbamate compound. Do not use if under medical advice not to work with such compounds
- Toxic in contact with skin, by inhalation and if swallowed
- Keep in original container, tightly closed, in a safe place, under lock and key
- Dangerous to game, wild birds and animals. Cover granules completely and immediately after application. Bury spillages. Failure to bury granules immediately and completely is hazardous to wildlife
- Dangerous to fish or other aquatic life. Do not contaminate surface waters or ditches with chemical or used container
- Keep unprotected persons out of treated glasshouses for at least 1 d
- Option to return empty container as instructed by supplier [6]

Personal protective equipment/Label precautions

- B, H, K, M [1-6]; C [1-6] (or D+E); A [1, 4-6]; D [2, 3]
- M02, M04, R02a, R02b, R02c, U02, U04a, U05a, U09a, U13, U19, U20a, C03, E01, E10a, E13b, E30b, E32a, E34 [1-6]; E06a [1-6] (13 wk); E02 [2-4] (24 h); E33 [1, 5, 6]

Withholding period

- Keep all livestock out of treated areas for at least 13 wk. Bury or remove spillages

Latest application/Harvest interval

- At planting, sowing, drilling or transplanting
- HI potatoes 8 wk; carrots, parsnips 12 wk; spring sown bulb onions do not harvest until mature bulb stage

Approval

- Off-label approval unlimited for use on outdoor and protected sweetcorn (OLA 2771/96)[6], (OLA 1930/00)[5]; unlimited for use on a range of outdoor and protected ornamentals - see OLA notice for details (OLA 1325/95)[6], (OLA 1932/00)[5]; unstipulated for use on outdoor leeks (OLA 2440/99)[6], (OLA 1928/00)[5]

Maximum Residue Level (mg residue/kg food)

- citrus, pecans, cauliflower, Brussels sprouts 0.2; tree nuts (except pecans), pome fruit, stone fruit, berries (except strawberries), cane fruit, miscellaneous fruit (except bananas), root and tuber vegetables (except beetroot, carrots, parsnips), bulb vegetables, fruiting vegetables (except tomatoes), leaf brassicas, leaf vegetables, fresh herbs, legumes, stem vegetables (except leeks), mushrooms, pulses, oilseeds (except linseed, rapeseed, cotton seed), tea, cereals 0.05; products of animal origin 0.01

3	**aldrin**

A persistent organochlorine insecticide, all approvals for which were revoked in 1989. Aldrin is an environmental hazard and is banned under the EC Directive 70/117/EEC

4	**alphachloralose**

A narcotic glucofuranose rodenticide used to kill mice

Products

1	Alphabird	Killgerm	100%	CB	10146
2	Alphamouse	Killgerm	100%	CB	H6692

Uses

Birds in **farm buildings** [1]. Mice in **farm buildings** *(indoor use)* [2].

Efficacy
- To be used only by professionals under licence to lay stupefying baits [1]
- Place prepared bait in shallow trays where mouse activity observed, or spread over selected area for bird control
- Place baits where mice are active, in runs or harbourages. Best results obtained where temperature is below 15°C
- Leave in position for 7-10 days only. If mouse activity remains continue treatment with a rodenticide containing a different active ingredient

Special precautions/Environmental safety
- Alphachloralose is subject to the Poisons Rules 1982 and Poisons Act 1972. See notes in Section 1
- May be used only by persons instructed or trained in the use of alphachloralose
- Harmful by inhalation and if swallowed
- Do not use outdoors. Products must be used in situations where baits are placed within a building or other enclosed structure, and the target is living or feeding predominantly within that building or structure
- Wash out all mixing equipment thoroughly at the end of every operation
- Prevent access to baits by children, birds and non-target animals
- Do not prepare or lay baits where food, feed or water could become contaminated
- A suitable warning dye must always be used [2]
- Remove all remains of bait and bait containers after treatment and burn or bury. Do not dispose of in refuse sacks or on open rubbish tips
- Search for and burn or bury all rodent bodies. Do not place in refuse bins or on rubbish tips
- Search for, and humanely kill any feral pigeons, house sparrows and other birds and burn or bury. Do not place in refuse bins or on rubbish tips [1]

Personal protective equipment/Label precautions
- A, D, H
- M04, R03b, R03c, U02, U05a, U13, U20b, C03, E01, E15, E30b, E32a, E34, V01b, V02, V03a, V04a

5	**alpha-cypermethrin**

A contact and ingested pyrethroid insecticide for use in arable crops

Products

1	Alphathrin	Nufarm Whyte	100 g/l	EC	10025
2	Contest	BASF	15% w/w	WG	10216
3	Contest	Cyanamid	15% w/w	WG	09024
4	Fastac	Cyanamid	100 g/l	EC	07008
5	Fastac	BASF	100 g/l	EC	10220
6	I T Alpha-cypermethrin	I T Agro	100 g/l	EC	08274
7	I T Alpha-cypermethrin 100 EC	I T Agro	100 g/l	EC	09372
8	Littac	Sorex	15 g/l	SC	H5176
9	Standon Alpha-C10	Standon	100 g/l	EC	08823

Uses

Blossom beetles in *broccoli (off-label)*, *calabrese (off-label)*, *cauliflowers (off-label)* [4]. Brassica pod midge in *winter oilseed rape* [1-7, 9]. Brassica pod midge in *spring oilseed rape* [6, 7]. Cabbage aphid in *broccoli, brussels sprouts, cabbages, calabrese, cauliflowers, kale* [2-5, 9]. Cabbage seed weevil in *spring oilseed rape, winter oilseed rape* [1-7, 9]. Cabbage stem flea beetle in *winter oilseed rape* [1-7, 9]. Caterpillars in *broccoli, brussels sprouts, cabbages, calabrese, cauliflowers, kale* [1-7, 9]. Cereal aphids in *barley, wheat* [1, 9]. Cereal aphids in *spring barley, spring wheat, winter barley, winter wheat* [2-7]. Flea beetles in *broccoli, brussels sprouts, cabbages, calabrese, cauliflowers, kale* [1-7, 9]. Hide beetle in *poultry houses* [8]. Lesser mealworm in *poultry houses* [8]. Pea and bean weevils in *broad beans, combining peas, spring field beans, vining peas, winter field beans* [1]. Pea aphid in *combining peas, vining peas* [1]. Pea moth in *combining peas, vining peas* [1]. Pollen beetles in *spring oilseed rape, winter oilseed rape* [1-7, 9]. Poultry house pests in *poultry houses* [8]. Rape winter stem weevil in *winter oilseed rape* [1-5, 9]. Yellow cereal fly in *winter barley, winter wheat* [2, 3].

Efficacy

- For cabbage stem flea beetle control spray oilseed rape when adult or larval damage first seen and about 1 mth later
- For flowering pests on oilseed rape apply at any time during flowering, on pollen beetle best results achieved at green to yellow bud stage (GS 3,3-3,7), on seed weevil between 20 pods set stage and 80% petal fall (GS 4,7-5,8)
- Spray cereals in autumn for control of cereal aphids, in spring/summer for grain aphids. (See label for details)
- For flea beetle, caterpillar and cabbage aphid control on brassicas apply when the pest or damage first seen or as a preventive spray. Repeat if necessary
- For flea beetle and caterpillar control in peas and beans apply when pest attack first seen and repeat as necessary
- For lesser mealworm control in poultry houses apply a coarse, low-pressure spray as routine treatment after clean-out and before each new crop. Spray vertical surfaces and ensure an overlap onto ceilings. It is not necessary to treat the floor [8]

Crop safety/Restrictions

- Maximum number of treatments 4 per crop on edible brassicas, 3 per crop on winter oilseed rape, peas, 2 on beans, spring oilseed rape (only 1 after yellow bud stage - GS 3,7)
- Apply up to 2 sprays on cereals in autumn and spring, 1 in summer between 1 Apr and 31 Aug. See label for details of rates
- Only 1 aphicide treatment may be applied in cereals between 1 Apr and 31 Aug in any one year and spray volume must not be reduced in this period

Special precautions/Environmental safety

- Harmful in contact with skin or if swallowed [4-7, 9]
- Irritating to skin. Risk of serious damage to eyes [4-7, 9]
- Flammable [4-7, 9]
- Irritant. May cause sensitisation by skin contact [2, 3]
- Dangerous to bees. Do not apply at flowering stage except as directed on oilseed rape and cereals. Keep down flowering weeds in all crops [2-7, 9]
- Extremely dangerous to fish or other aquatic life. Do not contaminate surface waters or ditches with chemical or used container
- LERAP Category A (except 8)
- When used as recommended treatment presents low hazard to foraging bees in oilseed rape and has minimal effect on beneficial insects
- For summer cereal application do not spray within 6 m from edge of crop and do not reduce volume when used after 31 Mar
- Do not apply to a cereal crop if any product containing a pyrethroid or dimethoate has been applied after the start of ear emergence (GS 51)
- Do not apply directly to poultry; collect eggs before application [8]

Personal protective equipment/Label precautions

- A, H [1-9]; C [1, 4-9]

• M03 [1, 4-7, 9]; M05 [1]; R03a, R03c, R04b, R04d, R07d [1, 4-7, 9]; R04, R04e [2, 3]; U05a [1-7, 9]; U04a [1, 4-7, 9]; U02, U19 [1, 4-9]; U10 [1-7]; U20b [1-8]; U11 [1, 4-7]; U20a [9]; U09a [8]; C03 [1-7, 9]; C04, C05, C07, C09 [8]; E12c, E16c, E16d, E26, E31b [1-7, 9]; E01, E13a, E30a [1-9]; E34 [1, 4-7, 9]; E04, E05, E32a [8]

Latest application/Harvest interval

• Before the end of flowering for oilseed rape, before 31 Mar in year of harvest for cereals (autumn and spring application), before early dough stage (GS 83) for cereals (summer application).
• HI vining peas 1 d, brassicas 7 d, combining peas, broad beans, field beans 11 d

Approval

• Off-label approval unlimited for use on cauliflowers, calabrese, broccoli (OLA 1750/96)[4]
• Accepted by BLRA for use pre-harvest on malting barley

---

6   aluminium ammonium sulphate

An inorganic bird and animal repellent

Products

1	Guardsman B	Chiltern	83 g/l	SL	05494
2	Guardsman M	Chiltern	83 g/l	SL	05495
3	Guardsman STP Seed Dressing Powder	Sphere	88% w/w	DS	03606
4	Liquid Curb Crop Spray	Sphere	83 g/l	SC	03164
5	Rezist	Barrettine	88% w/w	WP	08576

Uses

Birds in **brassicas, broad beans, bush fruit, cane fruit, carrots, forest nursery beds, forestry plantations, grassland, oilseed rape, peas, spring field beans, strawberries, sugar beet, top fruit, winter cereals, winter field beans** [1, 2, 4]. Birds in **corms** *(seed treatment)*, **flower bulbs** *(seed treatment)*, **seeds** *(seed treatment)* [3]. Damaging mammals in **brassicas, broad beans, bush fruit, cane fruit, carrots, forest nursery beds, forestry plantations, grassland, oilseed rape, peas, spring field beans, strawberries, sugar beet, top fruit, winter cereals, winter field beans** [1, 2, 4]. Damaging mammals in **corms** *(seed treatment)*, **flower bulbs** *(seed treatment)*, **seeds** *(seed treatment)* [3]. Dogs in **amenity turf** *(anti-fouling)* [5]. Moles in **amenity turf** [5].

Efficacy

• Apply as overall spray to growing crops before damage starts or mix powder with seed depending on type of protection required
• Spray deposit protects growth present at spraying but gives little protection to new growth
• Product must be sprayed onto dry foliage to be effective and must dry completely before dew or frost forms. In winter this may require some wind
• Treatments to deter moles should be applied as a planned programme when moles active. Other wild and domestic animals will be repelled during treatment period [5]
• Treatments for dog fouling should be applied before fouling occurs [5]

Personal protective equipment/Label precautions

• M05, U04b, U10 [1, 2, 4]; U20a [1, 2, 4, 5]; U20b [3]; E15, E30a, E32a [1-5]; E31b [1, 2, 4]; S02, S05 [1-4]

Latest application/Harvest interval

• HI fruit crops 6 wk

---

7   aluminium phosphide

A phosphine generating compound used against vertebrates and grain store pests

Products

1	Luxan Talunex	Luxan	57% w/w	GE	06563
2	Phostek	Killgerm	57% w/w	GE	07921
3	Phostek	Killgerm	57% w/w	GE	05115
4	Phostoxin	Rentokil	56% w/w	GE	09315

FOR FULL CONDITIONS OF USE ALWAYS READ THE PRODUCT LABEL

Uses

Mice in **farmland, forestry** [2]. Moles in **forestry** [1, 2]. Moles in **farmland** [1, 2, 4]. Rabbits in **forestry** [1, 2]. Rabbits in **farmland** [1, 2, 4]. Rats in **farmland, forestry** [1, 2]. Rodents in **farmland** [4]. Stored product pests in **stored products** [3].

Efficacy

- Product releases poisonous hydrogen phosphide gas in contact with moisture
- Place pellets in burrows or runs and seal hole by heeling in or covering with turf. Do not cover pellets with soil. Inspect daily and treat any new or re-opened holes
- Apply pellets by means of Luxan Topex Applicator [1]
- See label for details of fumigation of grain in silos and commodities not stored in bulk [3]

Special precautions/Environmental safety

- Aluminium phosphide is subject to the Poisons Rules 1982 and the Poisons Act 1972. See notes in Section 1
- Only to be used by operators instructed or trained in the use of aluminium phosphide and familiar with the precautionary measures to be taken. See label and HSE Guidance Notes for full precautions
- Very toxic by inhalation, if swallowed and in contact with skin [1-3]
- Very toxic by inhalation and if swallowed [4]
- Harmful in contact with skin [4]
- Product liberates very toxic, highly flammable gas
- Highly flammable
- Only open container outdoors [1, 4], in well ventilated space [2, 3], and for immediate use. Keep away from liquid or water as this causes immediate release of gas. Do not use in wet weather
- Keep in original container, tightly closed, in a safe place, under lock and key
- Do not use within 3 m of human or animal habitation. Before application ensure that no humans or domestic animals are in adjacent buildings or structures [2, 3]
- Dangerous to fish or other aquatic life. Do not contaminate surface waters or ditches with chemical or used container
- Pellets must never be placed or allowed to remain on ground surface
- Do not use adjacent to watercourses
- Take particular care to avoid gassing non-target animals, especially those protected under the Wildlife and Countryside Act (e.g. badgers. polecat, reptiles, natterjack toads, most birds). Do not use in burrows where there is evidence of badger or fox activity, or when burrows might be occupied by birds
- Dust remaining after decomposition is harmless and of no environmental hazard
- Dispose of empty containers as directed on label

Personal protective equipment/Label precautions

- A [1-4]; D, H [4]
- M04 [1-3]; M05 [2, 3]; R01a [1-3]; R01b, R01c, R07c [1-4]; R03a [4]; U05b, U07, U13, U19, U20a [1-4]; U10, U18 [2, 3]; U14 [1]; U01 [4]; C03 [1-4]; E34 [1-3]; E01, E13b, E30b, E32b [1-4]; E02 [2, 3]; E29 [2-4]; E27 [1]; V04a [4]

Withholding period

- Keep livestock out of treated areas

Approval

- Accepted by BLRA for use in stores for malting barley [3]

---

8   aluminium sulphate

An inorganic salt for slug and snail control

Products

Growing Success Slug Killer	Growing Success	99.5% w/w	SG	04386

Uses

Slugs in **field crops, fruit crops, ornamentals, protected crops, vegetables**. Snails in **field crops, fruit crops, ornamentals, protected crops, vegetables**.

Efficacy
- Apply granules to soil surface. Product has contact effect
- Best results achieved during mild, damp weather when slugs and snails active

9   amidosulfuron

A post-emergence sulfonylurea herbicide for cleavers and other broad-leaved weed control in cereals

Products

1	Aventis Eagle	Aventis	75% w/w	WG	09765
2	Barclay Cleave	Barclay	75% w/w	WG	09489
3	Druid	Aventis	50% w/w	WG	08714
4	Eagle	Aventis	75% w/w	WG	07318
5	Landgold Amidosulfuron	Landgold	75% w/w	WG	09021
6	Pursuit	Aventis	75% w/w	WG	07333

Uses

Annual dicotyledons in *durum wheat, oats, rye, spring barley, spring wheat, triticale, winter barley, winter wheat* [1-6]. Annual dicotyledons in *linseed* [1-4, 6]. Cleavers in *durum wheat, oats, rye, spring barley, spring wheat, triticale, winter barley, winter wheat* [1-6]. Cleavers in *linseed* [1-4, 6].

Efficacy
- For best results apply in spring (from 1 Feb) in warm weather when soil moist and weeds growing actively
- Weed kill is slow, especially under cool, dry conditions. Weeds may sometimes only be stunted but will have little or no competitive effect on crop
- May be used on all soil types unless certain sequences are used on linseed. See label
- Spray is rainfast after 1 h
- Cleavers controlled from emergence to flower bud stage. If present at application charlock (up to flower bud), shepherds purse (up to flower bud) and field forget-me-not (up to 6 leaves) will also be controlled

Crop safety/Restrictions
- Maximum number of treatments 1 per crop
- Do not apply to crops undersown or due to be undersown with clover or lucerne
- Broadcast crops should be sprayed post-emergence after plants have a well established root system
- Do not spray crops under stress, suffering drought, waterlogged, grazed, lacking nutrients or if soil compacted
- Do not spray if frost expected
- Do not roll or harrow within 1 wk of spraying
- Only cereals, winter oilseed rape, mustard, winter field beans, vetches may be sown in the same year as treatment
- Certain mixtures or sequences with other sulfonylurea products are permitted under explicitly detailed conditions. See label for details. There are no recommendations for mixtures with metsulfuron-methyl products on linseed
- Cereals or potatoes must be sown as the following crop after use of such mixtures or sequences in cereals. Only cereals may be sown after the use of such sequences in linseed
- Certain mixtures with fungicides are expressly forbidden. See label for details
- If crop fails cereals may be sown after 15 d
- Avoid drift onto neighbouring broad-leaved plants or onto surface waters or ditches

Special precautions/Environmental safety
- Irritating to eyes [3]
- Dangerous to fish or other aquatic life. Do not contaminate surface waters or ditches with chemical or used container
- Take care to wash out sprayers thoroughly. See label for details

Personal protective equipment/Label precautions
- C [3]
- R04a [3]; U20a [1, 2, 4, 5]; U20b [3, 6]; U05a, C03 [3]; E13b, E31a [1-6]; E01 [3]

Latest application/Harvest interval
- Before first spikelets just visible (GS 51) for cereals; before flower buds visible for linseed

Approval
- Accepted by BLRA for use on malting barley

10   2-aminobutane

A fumigant alkylamine fungicide permitted for use only on stored seed potatoes

Products

Hortichem 2-Aminobutane	Hortichem	720 g/l	VP	06147

Uses

Gangrene in **seed potatoes**. Silver scurf in **seed potatoes**. Skin spot in **seed potatoes**.

Efficacy
- Product used for treatment of seed potato tubers by fumigation in appropriate premises (see label)

Crop safety/Restrictions
- Maximum number of treatments 1 per batch. Maximum quantity to be fumigated in a single stack must not exceed 2000 tonnes
- Do not treat immature tubers. Allow period of healing before treating damaged tubers
- Treatment must only be carried out by trained operators in suitable fumigation chambers under licence from the British Technology Group
- Fumigate within 21 d of lifting

Special precautions/Environmental safety
- Harmful in contact with skin
- Irritating to eyes, skin and respiratory system
- Highly flammable. Keep away from sources of ignition. No smoking
- Keep in original container, tightly closed, in a safe place, under lock and key
- Do not empty into drains
- Do not contaminate surface waters or ditches with chemical or used container
- Do not supply treated potatoes for consumption by humans or lactating dairy cows
- Use must be in accordance with approved Code of Practice for the Control of Substances Hazardous to Health: Fumigation Operations

Personal protective equipment/Label precautions
- A, C
- M03, R03b, R04a, R04b, R04c, R07c, U02, U04a, U05a, U11, U13, U19, U20b, C03, E01, E15, E30b, E31a, E34

Approval
- 2-aminobutane was last reviewed in 1990 and is a severely restricted pesticide in UK. It is permitted only for use on seed potatoes

Maximum Residue Level (mg residue/kg food)
- citrus fruits 5; potatoes 1

11   amitraz

An amidine acaricide and insecticide for use in top fruit

Products

Mitac HF	Aventis	200 g/l	EC	07358

Uses

Pear sucker in **pears**. Red spider mites in **apples, pears**. Rust mite in **apples**.

Efficacy
- For red spider mites on apples and pears spray at 60-80% egg hatch and repeat 3 wk later
- For pear sucker control spray when significant numbers of nymphs have hatched but before there is significant contamination of the fruit with honeydew, normally Jun/Jul
- Best results achieved in dry conditions, do not spray if rain imminent

Crop safety/Restrictions
- Maximum total dose 7 l/ha product per yr

Special precautions/Environmental safety
- Harmful in contact with skin and if swallowed
- Harmful to fish. Do not contaminate surface waters or ditches with chemical or used container
- Keep in original container, tightly closed, in a safe place, under lock and key

Personal protective equipment/Label precautions
- A, C, H
- M03, R03a, R03c, U02, U04a, U05a, U08, U13, U19, U20a, C03; C02 (2 wk); E01, E13c, E26, E30b, E31b, E34

Latest application/Harvest interval
- HI apples, pears 2 wk

Maximum Residue Level (mg residue/kg food)
- hops 50; tomatoes 0.5; tea 0.1; all other crops (except tea) 0.02

---

12   amitrole

A translocated, foliar-acting, non-selective triazole herbicide

Products

1	MSS Aminotriazole Technical	Nufarm Whyte	98% w/w	TC	04645
2	Weedazol-TL	Bayer	225 g/l	SL	02979

Uses

Annual weeds in *fallows, headlands, stubbles* [2]. Barren brome in *apples, pears* [2]. Bent grasses in *non-crop areas* [1]. Contaminant grasses in *amenity turf - amitrole resistant (off-label)* [2]. Couch in *non-crop areas* [1]. Couch in *apples, fallows, headlands, pears, stubbles, winter wheat (direct drilling)* [2]. Creeping bent in *non-crop areas* [1]. Creeping thistle in *non-crop areas* [1]. Creeping thistle in *fallows, stubbles* [2]. Docks in *non-crop areas* [1]. Docks in *fallows, headlands, stubbles* [2]. General weed control in *apples, apricots (off-label), cherries (off-label), peaches (off-label), pears, plums (off-label), quinces (off-label)* [2]. Perennial weeds in *apples, fallows, headlands, pears* [2]. Total vegetation control in *non-crop areas* [1]. Volunteer potatoes in *stubbles (barley stubble)* [2].

Efficacy
- In non-crop land may be applied at any time from Apr to Oct. Best results achieved in spring or early summer when weeds growing actively. For coltsfoot, hogweed and horsetail summer and autumn applications are preferred [1]
- In cropland apply when couch in active growth and foliage at least 7.5 cm high [2]
- In fallows and stubble plough 3-6 wk after application to depth of 20 cm, taking care to seal the furrow [2]

Crop safety/Restrictions
- Maximum number of treatments 1 per crop for winter wheat, 1 per yr for stone fruit, 2 per yr for amenity turf
- Keep off suckers or foliage of desirable trees or shrubs [1]
- Do not spray areas into which the roots of adjacent trees or shrubs extend [1]
- Do not spray on sloping ground when rain imminent and run-off may occur [1]
- Allow specified interval between treatment and sowing crops (see label) [2]
- Apply in autumn on land to be used for spring barley [2]
- Do not sow direct-drilled winter wheat less than 2 wk after application [2]
- Apply round base of established fruit trees taking care to avoid contaminating trees, especially where bark is damaged. Keep off trees [2]

Special precautions/Environmental safety
- Harmful to fish. Do not contaminate surface waters or ditches with chemical or used container

Personal protective equipment/Label precautions
- C [2]; A [1, 2]
- U09a [2]; U19 [1, 2]; U08, U20a [1]; C03, E15 [2]; E30a, E31b [1, 2]; E13c [1]

Latest application/Harvest interval
- At least 2 wk before direct drilling for winter wheat; before drilling for field crops; before end Jun or after harvest for apples, pears; end Oct for fallows [2]

Approval
- Off-label approval unlimited for use on amenity turf (OLA 0693/91) and around stone fruit trees (OLA 0333/92) [2]

Maximum Residue Level (mg residue/kg food)
- tea, hops 0.1; fruits, vegetables, pulses, oilseeds, potatoes 0.05

13 amitrole + 2,4-D + diuron

A total herbicide mixture of translocated and residual chemicals

Products

Trik	Nufarm Whyte	26.6:11.0:46.4% w/w	WP	07853

Uses

Total vegetation control in *land not intended to bear vegetation*.

Efficacy
- Apply in spring or late summer/early autumn when weeds are growing actively and have sufficient leaf area to absorb chemical
- Apply maintenance treatment if necessary at lower rate when weeds 7-10 cm high
- Increase rate on areas of peat or high carbon content

Crop safety/Restrictions
- Do not use on ground under which roots of valuable trees or shrubs are growing

Special precautions/Environmental safety
- Harmful to fish. Do not contaminate surface waters or ditches with chemical or used container

Personal protective equipment/Label precautions
- A, C, D, H, M
- R04a, R04b, R04c, U05a, U08, U19, U20a, C03, E01, E07, E13c, E30a, E32a

Withholding period
- Keep livestock out of treated areas until foliage of poisonous weeds such as ragwort has died and become unpalatable

Maximum Residue Level (mg residue/kg food)
- see amitrole entry

14 ammonium sulphamate

A non-selective, inorganic, general purpose herbicide and tree-killer

Products

1	Amcide	B H & B	99.5% w/w	CR	04246
2	Root-Out	Dax	98.5% w/w	CR	03510

Uses

Annual weeds in *forestry, ornamentals* (pre-planting), *vegetables* (pre-planting) [1]. Annual weeds in *ornamentals, trees and shrubs* [2]. Perennial dicotyledons in *forestry* [1]. Perennial dicotyledons in *trees and shrubs* [2]. Perennial grasses in *forestry* [1, 2]. Perennial weeds in *ornamentals* (pre-planting), *vegetables* (pre-planting) [1]. Perennial weeds in *ornamentals* [2]. Rhododendrons in *forestry* [1, 2]. Woody weeds in *forestry* [1, 2].

Efficacy
- Apply as spray to low scrub and herbaceous weeds from Apr to Sep in dry weather when rain unlikely and cultivate after 3-8 wk
- Apply as crystals in frills or notches in trunks of standing trees at any time of year

- Apply as concentrated solution or crystals to stump surfaces within 48 h of cutting. Rhododendrons must be cut level with ground and sprayed to cover cut surface, bark and immediate root area
- Stainless steel or plastic sprayers are recommended. Solutions are corrosive to mild steel, galvanised iron, brass and copper

Crop safety/Restrictions
- Allow 8-12 wk after treatment before replanting
- Keep spray at least 30 cm from growing plants. Low doses may be used under mature trees with undamaged bark

Special precautions/Environmental safety
- Harmful to fish. Do not contaminate surface waters or ditches with chemical or used container

Personal protective equipment/Label precautions
- U15, U20a [2]; U11, U14, U19 [1, 2]; U09b, U20b [1]; E01, E13c, E30a [1, 2]; E32a [1]

15 anthracene oil

A crop desiccant

Products

Sterilite Hop Defoliant   Coventry Chemicals   63.6% w/w   EC   05060

Uses

Chemical stripping in *hops*.

Efficacy
- Spray when hop bines 1.2 m high and direct spray downward at 45° onto area to be defoliated. Repeat as necessary until cones are formed

Crop safety/Restrictions
- Do not spray if temperature is above 21°C or after cones have formed
- Do not drench rootstocks
- Do not spray on windy, wet or frosty days

Special precautions/Environmental safety
- Harmful if swallowed. Irritating to eyes, skin and respiratory system
- Dangerous to fish. Do not contaminate surface waters or ditches with chemical or used container

Personal protective equipment/Label precautions
- A, C, H
- R03c, R04a, R04b, R04c, U05a, U08, U19, U20a, C02, C03, E13a, E25, E26, E29, E30a, E33, S07

Approval
- Accepted by BLRA for use on hops

16 asulam

A translocated carbamate herbicide for control of docks and bracken

Products

1	Asulox	RP Agric.	400 g/l	SL	06124
2	Asulox	Aventis	400 g/l	SL	09969
3	I T Asulam	I T Agro	400 g/l	SL	10186

Uses

Bracken in *forestry plantations* (off-label), *non-crop areas* [1, 2]. Bracken in *forestry, permanent pasture, rough grazing* [1-3]. Bracken in *amenity vegetation* [3]. Docks in *blueberries* (off-label), *cane fruit* (off-label), *cranberries* (off-label), *currants* (off-label), *damsons* (off-label), *gooseberries* (off-label), *mint* (off-label), *nectarines* (off-label), *parsley* (off-label), *quinces* (off-label), *road verges, strawberries* (off-label), *tarragon* (off-label), *waste ground, white clover seed crops* (off-label) [1, 2]. Docks in *amenity grass* (not fine turf), *apples, blackcurrants, cherries, hops, pears, permanent pasture, plums* [1-3].

Efficacy
- Spray bracken when fronds fully expanded but not senescent, usually Jul-Aug; docks in full leaf before flower stem emergence
- Bracken fronds must not be damaged by stock, frost or cutting before treatment
- Do not apply in drought or hot, dry conditions
- Uptake and reliability of bracken control may be improved by use of specified additives - see label. Additives not recommended on forestry land
- To allow adequate translocation do not cut or admit stock for 14 d after spraying bracken or 7 d after spraying docks. Preferably leave undisturbed until late autumn
- Complete bracken control rarely achieved by one treatment. Survivors should be sprayed when they recover to full green frond, which may be in the ensuing year but more likely in the second year following initial application

Crop safety/Restrictions
- Maximum number of treatments 1 per yr
- In forestry areas some young trees may be checked if sprayed directly (see label)
- Allow at least 6 wk between spraying and planting any crop
- Do not use in pasture before mowing for hay
- In fruit crops apply as a directed spray
- Do not treat blackcurrant cuttings, hop sets or weak hills
- Some grasses and herbs will be damaged by full dose. Most sensitive are cocksfoot, Yorkshire fog, timothy, bents, annual meadow-grass, daisies, docks, plaintains, saxifrage
- Apply as spot treatment in parsley, mint and tarragon, not directly to crop

Special precautions/Environmental safety
- Product is approved for use near surface waters. Whilst every care should be taken to avoid contamination of water, any that does occur from normal use should not offer harm to users, consumers of the water, or the environment
- The appropriate water regulatory body must be consulted before application near surface waters

Personal protective equipment/Label precautions
- A, C, D, H [1-3]; M [1-3] (for ULV application)
- U20c [1, 2]; U19, U20b [1-3] (ULV); U08 [3]; E31a [1, 2]; E15, E30a [1-3]; E07 [1-3] (14 d); E26, E31b [3]

Withholding period
- Keep livestock out of treated areas for at least 14 d and until foliage of poisonous weeds such as ragwort has died and become unpalatable

Approval
- May be applied through CDA equipment
- Approved for aerial application on bracken. See notes in Section 1
- Approved for use near surface waters. See notes in Section 1
- Off-label Approval unlimited for use on white clover seed crops (OLA 0931/92) [1], (OLA 1894/00) [2]; unlimited for use on strawberries (OLA 0932/92) [1], (OLA 1896/00) [2]; unlimited for use on cane fruit, currants, gooseberries, strawberries, blueberries, cranberries, damsons, nectarines, quinces (OLA 0930/92) [1], (OLA 1892/00) [2]; unlimited for use in forestry areas (OLA 1001/92) [1], (OLA 1898/00) [2]; unlimited as spot treatment in parsley, mint, tarragon (OLA 0097/93) [1], (OLA 1900/00) [2]
- Accepted by BLRA for use on hops (not applied to hop sets)

---

17   atrazine

A triazine herbicide with residual and foliar activity, with restricted permitted uses

Products

1	Alpha Atrazine 50 SC	Makhteshim	500 g/l	SC	04877
2	Atlas Atrazine	Nufarm Whyte	500 g/l	SC	07702
3	Atrazol	Sipcam	500 g/l	SC	07598
4	DAPT Atrazine 50 SC	DAPT	500 g/l	SC	08031
5	DG90	Sipcam	90% w/w	WG	09310
6	Gesaprim	Novartis	500 g/l	SC	08411
7	MSS Atrazine 50 FL	Nufarm Whyte	500 g/l	SC	01398

Uses

Annual dicotyledons in *maize* [1-7]. Annual dicotyledons in *sweetcorn* [1, 3-7]. Annual dicotyledons in *conifer plantations, forestry* (off-label), *raspberries, roses* [2]. Annual grasses in *maize* [1-4, 6, 7]. Annual grasses in *sweetcorn* [1, 3, 4, 6, 7]. Annual grasses in *conifer plantations, forestry* (off-label), *raspberries, roses* [2]. Couch in *conifer plantations* [2]. Perennial grasses in *conifer plantations* [2].

Efficacy

- May be used pre- or early post-weed emergence in maize and sweetcorn
- Root activity enhanced by rainfall soon after application and reduced on high organic soils. Foliar activity effective on weeds up to 3 cm high
- Not recommended for use on soils with more than 10% organic matter
- In conifers apply as overall spray in Feb-Apr. May be used in first spring after planting
- Apply to raspberries in spring before new cane emergence, not in season of planting
- Application rates vary with crop, soil type and weed problem. See label for details
- Resistant weed strains may develop with repeated use of atrazine or other triazines

Crop safety/Restrictions

- Maximum number of applications (including other atrazine/simazine products) 1 per crop (or lower doses to 3.0 l/ha total) for maize, sweetcorn; 1 per season for conifer plantations, raspberries, roses
- Maximum total dose equivalent to one full dose treatment [1]
- Do not apply to raspberries in season of planting
- On slopes heavy rainfall soon after application may cause surface run-off
- To reduce soil run-off, especially from forest plantations, users are advised to plant grass strips 6 m wide between treated areas and surface waters
- Do not use on Christmas trees
- After annual weed control only maize or sweetcorn should be sown for at least 7 mth after application. Do not sow oats in autumn following spring treatment

Special precautions/Environmental safety

- Irritant [5]
- Dangerous to fish or other aquatic life and aquatic higher plants. Do not contaminate surface waters or ditches with chemical or used container
- LERAP Category B
- Use must be restricted to one product containing atrazine or simazine, and either to a single application at the maximum approved rate or (subject to any existing maximum permitted number of treatments) to several applications at lower doses up to the maximum approved rate for a single application

Personal protective equipment/Label precautions

- A, C, D, H, M [1-7]; B [3, 5, 6]
- R04 [5]; U20c [3, 5, 7]; U20b [1, 2, 4, 6]; U05a, C03 [5]; E32a [7]; E13b, E16a, E16b, E30a [1-7]; E26 [1-5]; E31b [1, 4, 6]; E31a [2, 3, 5]; E01 [5, 6]

Latest application/Harvest interval

- 7 mth before a succeeding crop for maize, sweetcorn; Apr for conifer plantations; before cane emergence for raspberries; before weeds at 3 cm for roses

Approval

- Atrazine was reviewed in 1992 and approvals for aerial use, and use on non-crop land (excluding home garden use) revoked. Restrictions were also placed on the number of applications that could be made to crops
- Off-label approval unstipulated for use in forest situations (OLA 0431/99)[2]

Maximum Residue Level (mg residue/kg food)

- fruits, vegetables, pulses, oilseeds, potatoes, tea, hops 0.1

---

18   azaconazole

A conazole available only in mixture

FOR FULL CONDITIONS OF USE ALWAYS READ THE PRODUCT LABEL

19 azaconazole + imazalil

A fungicide mixture for use in horticulture

Products

Nectec Paste	Hortichem	1:2% w/w	PA	08510

Uses

Canker in **forestry, ornamental trees, shrubs**. Silver leaf in **forestry, ornamental trees, shrubs**.

Efficacy
- Paint pruning cuts immediately. If necessary clean wounds and cut back any loose bark
- Ensure whole of cut area is fully covered beyond wound to surrounding healthy bark
- Cut back established cankers to sound healthy wood before treatment. Work paste into all crevices

Crop safety/Restrictions
- Maximum number of treatments 1 per wound per yr
- Treat only during dry weather and not in frosty conditions
- Use only during dormant periods
- Do not apply to grafting cuts

Special precautions/Environmental safety
- Harmful to fish or other aquatic life. Do not contaminate surface waters or ditches with chemical or used container

Personal protective equipment/Label precautions
- A
- U05a, U20c, C03, E01, E13c, E26, E30a

Approval
- Imazalil included in Annex I under EC Directive 91/414

Maximum Residue Level (mg residue/kg food)
- see imazalil entry

20 azamethiphos

A residual organophosphorus insecticide for fly control

Products

Alfacron 10 WP	Novartis A H	10% w/w	WP	08587

Uses

Flies in **livestock houses**.

Efficacy
- Add water as directed to produce paint consistency and paint onto 2.5% of total wall and ceiling surface area at a minimum of 5 points in building or apply as spray to 30% of surface area

Crop safety/Restrictions
- Only apply to areas out of reach of children and animals

Special precautions/Environmental safety
- This product contains an anticholinesterase organophosphorus compound. Do not use if under medical advice not to work with such compounds
- Irritating to eyes and skin. May cause sensitization by skin contact
- Dangerous to fish. Do not contaminate surface waters or ditches with chemical or used container
- Do not apply directly to livestock and poultry
- Do not apply to surfaces on which food or feed is stored, prepared or eaten. Cover feedstuffs and remove exposed milk and eggs before application

Personal protective equipment/Label precautions
- A, C
- M01, R04a, R04b, R04e, U05a, U10, U13, U14, U15, U19, U20b, C03, C04, C05, C06, C07, E01, E05, E13b, E30a, E32a

21 azoxystrobin

A systemic translaminar and protectant strobilurin fungicide for cereals

Products

1 Amistar	Zeneca	250 g/l	SC	08517
2 Barclay ZX	Barclay	250 g/l	SC	09570
3 Landgold Strobilurin 250	Landgold	250 g/l	SC	09595
4 Me2 Azoxystrobin	Me2	250 g/l	SC	09654
5 Standon Azoxystrobin	Standon	250 g/l	SC	09515

Uses

Brown rust in *rye (off-label)*, *triticale (off-label)* [1, 2]. Brown rust in *spring barley, spring wheat, winter barley, winter wheat* [1-5]. Glume blotch in *spring wheat, winter wheat* [1-5]. Late ear diseases in *spring wheat, winter wheat* [1-5]. Leaf spot in *spring wheat, winter wheat* [1-5]. Net blotch in *spring barley, winter barley* [1-5]. Powdery mildew in *spring barley, spring wheat, winter barley, winter wheat* [1-5]. Rhynchosporium in *rye (off-label)*, *triticale (off-label)* [1, 2]. Rhynchosporium in *spring barley, winter barley* [1-5]. White rust in *protected chrysanthemums (off-label)* [1]. Yellow rust in *spring wheat, winter wheat* [1-5].

Efficacy

- Best results obtained from use as a protectant or during early stages of disease establishment
- Control of developed mildew or Rhynchosporium infections can be improved by appropriate tank mixture. See label for details

Crop safety/Restrictions

- Maximum number of treatments 2 per crop [3]
- Maximum total dose equivalent to three full dose treatments [1, 2, 5]
- To discourage development of resistance do not apply more than two foliar treatments of strobilurin products to the same crop
- Avoid sequential use alone against powdery mildew

Special precautions/Environmental safety

- Dangerous to fish or other aquatic life. Do not contaminate surface waters or ditches with chemical or used container

Personal protective equipment/Label precautions

- U20a [1]; U09a, U19 [1-5]; U02 [1, 2]; U20b [2-5]; E13b, E30a [1-5]; E31c [1, 2]; E26 [2-5]; E31b [3-5]; E24 [4]

Latest application/Harvest interval

- Grain watery ripe (GS 71)

Approval

- Azoxystrobin included in Annex I under EC Directive 91/414
- Off-label approval to Jul 2008 for use on rye, triticale (OLA 1244/99)[1]; to Jul 2008 for use in ornamental plant production, pot chrysanthemums, protected chrysanthemums grown in peat substrate/compost on hard surfaces (OLA 1536/00)[1]
- Accepted by BLRA for use on malting barley

Maximum Residue Level (mg residue/kg food)

- wheat, barley, rye, triticale 0.3

22 azoxystrobin + fenpropimorph

A protectant and eradicant fungicide mixture

Products

1 Amistar Pro	Zeneca	100:280 g/l	SE	08871
2 Barclay ZA	Barclay	100:280 g/l	SE	09462

Uses

Brown rust in *spring barley, spring wheat, winter barley, winter wheat*. Late ear diseases in *spring wheat, winter wheat*. Net blotch in *spring barley, winter barley*. Powdery mildew in **spring barley, spring wheat, winter barley, winter wheat**. Rhynchosporium in **spring barley, winter barley**. Septoria diseases in **spring wheat, winter wheat**. Yellow rust in **spring wheat, winter wheat**.

Efficacy
- Best results obtained from application before infection following a disease risk assessment, or when disease first seen in crop
- Results may be less reliable when used on crops under stress
- Treatments for protection against ear disease should be made at ear emergence

Crop safety/Restrictions
- Maximum total dose equivalent to two full dose treatments
- Active ingredients have different modes of action and therefore development of resistance less likely. However to discourage development of resistance a maximum of two applications of strobilurin fungicides to the same crop is recommended

Special precautions/Environmental safety
- Irritating to skin
- May cause sensitization by skin contact [1]
- Dangerous to fish or other aquatic life. Do not contaminate surface waters or ditches with chemical or used container
- LERAP Category B [2]

Personal protective equipment/Label precautions
- A, H [1, 2]
- M03 [1, 2]; R04e [1]; R04b, U02, U05a, U09a, U14, U15, U19, U20b, C03, E01, E13b, E26, E30a, E31c [1, 2]; E16a, E16b [2]

Latest application/Harvest interval
- Before early milk stage (GS 73)
- HI 5 wk

Approval
- Azoxystrobin included in Annex I under EC Directive 91/414
- Accepted by BLRA for use on malting barley

---

23   Bacillus thuringiensis

A bacterial insecticide for control of caterpillars

Products

Dipel	Biowise	16000 IU/mg	WP	08634

Uses

Caterpillars in **broccoli, brussels sprouts, cabbages, cauliflowers, cucumbers, ornamentals, peppers, protected brassica seedlings, protected cucumbers, protected ornamentals, protected peppers, protected tomatoes, raspberries, strawberries, tomatoes**.

Efficacy
- Product affects gut of larvae and must be eaten to be effective. Caterpillars cease feeding and die in 2-3 d (3-5 d for large caterpillars)
- Apply as soon as larvae appear on crop and repeat every 3-14 d for outdoor crops, every 3 wk under glass
- Addition of a wetter recommended for use on brassicas
- Good coverage is essential, especially of undersides of leaves. Use a drop-leg sprayer in field crops

Crop safety/Restrictions
- No restriction on number of treatments of edible crops
- Do not mix more spray than can be used in a 12 h period

Special precautions/Environmental safety
- Store out of direct sunlight

Personal protective equipment/Label precautions
- A, C, H
- U20c, E01, E15, E30a, E32a

Latest application/Harvest interval
- HI zero for crops

---

24   benalaxyl

A phenylamide (acylalanine) fungicide available only in mixtures

---

25   benalaxyl + mancozeb

A systemic and protectant fungicide mixture

Products

1	Galben M	Sipcam	8:65% w/w	WP	07220
2	Tairel	Sipcam	8:65% w/w	WB	07767

Uses

Blight in *early potatoes, maincrop potatoes*.

Efficacy
- Apply to potatoes at blight warning prior to crop becoming infected and repeat at 10-21 d intervals depending on risk of infection
- Spray irrigated potatoes after irrigation and at 14 d intervals, crops in polythene tunnels at 10 d intervals
- When active potato growth ceases use a non-systemic fungicide to end of season, starting not more than 10 d after last application
- Do not treat potatoes showing active blight infection
- For reduction of downy mildew in oilseed rape apply at seedling to 7-leaf stage, before end Nov as soon as infection seen

Crop safety/Restrictions
- Maximum number of treatments 5 per crop (including other phenylamide-based fungicides) for potatoes, 1 per crop for winter oilseed rape

Special precautions/Environmental safety
- Irritating to eyes
- Dangerous to fish or other aquatic life. Do not contaminate surface waters or ditches with chemical or used container

Personal protective equipment/Label precautions
- A [1, 2]
- R04a, U05a, U08, U20b, C03 [1, 2]; C02 [2] (7 d); E32a [1]; E01, E13b, E29, E30a [1, 2]

Latest application/Harvest interval
- HI 7 d

Approval
- Approved for aerial application on potatoes [1]. See notes in Section 1

Maximum Residue Level (mg residue/kg food)
- (benalaxyl) grapes, onions, tomatoes, peppers 0.2; tea, hops 0.1; fruits (except grapes), vegetables (except onions, tomatoes, peppers), pulses, oilseeds, potatoes, cereals, animal products 0.05. See also mancozeb entry

---

26   benazolin

A translocated arylacetic acid herbicide for oilseed rape

Products

1	Aventis Galtak 50 SC	Aventis	500 g/l	SC	09711
2	Galtak 50 SC	Aventis	500 g/l	SC	07258

Uses

Chickweed in *winter oilseed rape*. Cleavers in *winter oilseed rape*.

Efficacy
- For best results spray during mild moist weather when weeds actively growing

- Weeds shaded by crop or grass weeds will not be completely controlled and may grow away later
- Frost after application will not reduce weed control but it should not be on foliage at time of treatment

Crop safety/Restrictions
- Maximum number of treatments 1 per crop
- Treat from 3 developed crop leaf stage
- Only treat healthy crops and allow time before treatment for recovery from any conditions that reduce wax formation
- Autumn treatment may result in abnormal growth of leaves and stems later in season from which recovery is normally complete
- Do not apply to spring oilseed rape or fodder rape
- Avoid drift outside the target area
- Interval of 14 d must elapse before or after treatment with any other pesticide

Special precautions/Environmental safety
- Irritant, may cause sensitization by skin contact
- Harmful to fish or other aquatic life. Do not contaminate surface waters or ditches with chemical or used container

Personal protective equipment/Label precautions
- A
- M03, R04, R04e, U04a, U05a, U09a, U10, U14, U20a, C03, E01, E07, E13c, E26, E30a, E31b, E34

Withholding period
- Keep livestock out of treated areas until foliage of any poisonous weeds such as ragwort has died and become unpalatable

Latest application/Harvest interval
- Before crop flower buds visible above leaves (GS 3,5)

---

27 benazolin + bromoxynil + ioxynil

A post-emergence, HBN herbicide mixture for cereal crops and grass

Products

Asset	Aventis	50:125:62.5 g/l	EC	07243

Uses

Annual dicotyledons in *barley, durum wheat, newly sown grass, oats, wheat*.

Efficacy
- Commonly used in mixture with mecoprop and other cereal herbicides to extend the range of weeds controlled. Sequential treatments also recommended
- Best results achieved when weeds small and actively growing and crop competitive
- Weeds should be dry when sprayed

Crop safety/Restrictions
- Maximum number of treatments 2 per crop or yr
- Use in cereals from 2-leaf stage to first node detectable (GS 12-31), in grass from 2-leaf stage when crop growing vigorously (other times may be advised for mixtures)
- Use on oats only in spring, on other cereals in autumn or spring
- Do not apply to crops undersown with legumes
- Do not use on crops affected by pests, disease, waterlogging or prolonged frost
- Severe frost within 3-4 wk of spraying may scorch crop
- Do not roll or harrow crops for 3 d before or after spraying
- Do not spray grass seed crops later than 5 wk before heading

Special precautions/Environmental safety
- Harmful if swallowed. Irritating to skin and eyes
- Flammable
- Do not apply by hand-held equipment or at concentrations higher than those recommended

- Harmful to bees. Do not apply to crops in flower or to those in which bees are actively foraging. Do not apply when flowering weeds are present
- Dangerous to fish or other aquatic life. Do not contaminate surface waters or ditches with chemical or used container

Personal protective equipment/Label precautions
- A, C
- M03, R03c, R04a, R04b, R07d, U05a, U08, U13, U19, U20b, C03; C02 (6 wk); E01, E07, E12e, E13b, E26, E30a, E31b, E34

Withholding period
- Keep livestock out of treated areas for at least 6 wk and until foliage of poisonous plants such as ragwort has died and become unpalatable

Latest application/Harvest interval
- 5 wk before heading (seed crops) for newly sown grass; up to GS 31 for barley, durum wheat, oats, wheat
- HI 6 wk (before grazing)

Approval
- Accepted by BLRA for use on malting barley

Maximum Residue Level (mg residue/kg food)
- see ioxynil entry

---

28   benazolin + clopyralid

A post-emergence herbicide mixture for use mainly in winter oilseed rape

Products

Benazalox	Aventis	30:5% w/w	WP	07246

Uses

Annual dicotyledons in **evening primrose** *(off-label)*, **honesty** *(off-label)*, **spring oilseed rape** *(off-label)*, **swede seed crops** *(off-label)*, **white mustard** *(off-label)*, **winter oilseed rape**. Chickweed in **evening primrose** *(off-label)*, **swede seed crops** *(off-label)*, **white mustard** *(off-label)*, **winter oilseed rape**. Cleavers in **evening primrose** *(off-label)*, **swede seed crops** *(off-label)*, **white mustard** *(off-label)*, **winter oilseed rape**. Mayweeds in **evening primrose** *(off-label)*, **swede seed crops** *(off-label)*, **white mustard** *(off-label)*, **winter oilseed rape**.

Efficacy
- Weeds are controlled from cotyledon stage to early flowerbud or up to 15 cm high or across in winter oilseed rape
- For best results apply during mild, moist weather when weeds are growing actively and still visible in young crop. Do not spray if rain expected within 4 h
- Do not spray if frost present on foliage; frost after application will not reduce effectiveness
- For grass weed control various tank mixtures are recommended or sequential treatments, allowing specified interval after application of grass-killer. See label for details

Crop safety/Restrictions
- Maximum number of treatments 1 per crop
- May be used on winter oilseed rape when crop between stages of 3 fully developed leaves to flower buds hidden beneath leaves (GS 1,3-3,1), on spring oilseed rape before green bud
- Oilseed rape may be replanted in the event of crop failure. Wheat, barley or oats may be sown after 1 mth, other crops in the following autumn after ploughing. Winter beans should not be planted in the same year
- Chop and incorporate straw and trash in early autumn to release any clopyralid residues. Ensure remains of plants completely decayed before planting susceptible crops

Special precautions/Environmental safety
- Irritating to eyes and skin
- Harmful to fish or other aquatic life. Do not contaminate surface waters or ditches with chemical or used container

Personal protective equipment/Label precautions
- A, C
- R04a, R04b, U05a, U09a, U13, U19, U20a, C03, E01, E07, E13c, E30a, E32a

---

Withholding period
- Keep livestock out of treated areas until foliage of any poisonous weeds such as ragwort has died and become unpalatable

Latest application/Harvest interval
- Oilseed rape before green bud stage (GS 3,1) (before end Jan if in tank mix with Butisan S), honesty up to 4-5 pairs true leaves

Approval
- Off-label approval unlimited for use on evening primrose, white mustard and swedes grown for seed (OLA 1596, 1597/96), on honesty (OLA 0610/93), on spring oilseed rape (OLA 1598/97)

---

29   benazolin + 2,4-DB + MCPA

A post-emergence herbicide for undersown cereals, grass and clover

Products

1	Legumex Extra	Aventis	27:237:42.8 g/l	SL	08676
2	Setter 33	Dow	50:237:43 g/l	SL	05623

Uses

Annual dicotyledons in **grassland** [1]. Annual dicotyledons in **clovers, seedling leys, undersown barley, undersown oats, undersown wheat** [1, 2]. Annual dicotyledons in **direct-sown seedling clovers** [2]. Chickweed in **grassland** [1]. Chickweed in **clovers, seedling leys, undersown barley, undersown oats, undersown wheat** [1, 2]. Chickweed in **direct-sown seedling clovers** [2]. Cleavers in **grassland** [1]. Cleavers in **clovers, seedling leys, undersown barley, undersown oats, undersown wheat** [1, 2]. Cleavers in **direct-sown seedling clovers** [2]. Knotgrass in **grassland** [1]. Knotgrass in **clovers, seedling leys, undersown barley, undersown oats, undersown wheat** [1, 2]. Knotgrass in **direct-sown seedling clovers** [2]. Perennial dicotyledons in **clovers, grassland, seedling leys, undersown barley, undersown oats, undersown wheat** [1].

Efficacy
- Best results on annual weeds achieved when treated young and in active growth
- Top growth of established perennials may be checked [2]
- Spray perennial weeds when well developed but before flowering. Higher rates may be used against perennials in established grassland [1]
- Do not spray when rain is imminent or during drought

Crop safety/Restrictions
- Maximum number of treatments 1 per crop or yr
- Spray undersown cereals before 1st node detectable (GS 31), grass seedlings should have at least 3 leaves
- Spray winter cereals in spring from leaf sheath erect exceeding 5 cm, spring wheat after 5 leaves unfolded (GS 15), spring barley and oats after 2 leaves unfolded (GS 12)
- Spray clovers after the 1-trifoliate leaf stage and before red clover has more than 3 trifoliate leaves. Do not spray clover seed crops in the yr seed is to be taken
- Spray new swards before end Oct when grasses have at least 3 leaves and clovers at stages as above
- Clovers may be damaged if frost occurs soon after spraying
- Do not spray legumes other than clover
- Do not roll, harrow or cut crops for 3 d before or after spraying

Special precautions/Environmental safety
- Harmful to fish or other aquatic life. Do not contaminate surface waters or ditches with chemical or used container

Personal protective equipment/Label precautions
- A, C [1]
- R03c, R04b, R04d [1]; U08, U20b [1, 2]; E26 [2]; E13c, E30a, E31b [1, 2]; E07 [1, 2] (2 wk)

Withholding period
- Keep livestock out of treated areas for at least 2 wk and until foliage of poisonous weeds such as ragwort has died and become unpalatable

Latest application/Harvest interval
- Before first node detectable (GS 31) for undersown cereals

Approval
- Accepted by BLRA for use on malting barley

---

30   benfuracarb

A soil-applied carbamate insecticide and nematicide for beet crops

Products

Oncol 10G	Nufarm Whyte	10% w/w	GR	08249

Uses

Docking disorder vectors in **fodder beet, mangels, sugar beet**. Leaf miners in **fodder beet, mangels, sugar beet**. Millipedes in **fodder beet, mangels, sugar beet**. Pygmy beetle in **fodder beet, mangels, sugar beet**. Springtails in **fodder beet, mangels, sugar beet**. Symphylids in **fodder beet, mangels, sugar beet**.

Efficacy
- Apply at sowing with suitable applicator so that the granules are mixed with the moving soil closing the seed furrow. See label for recommended applicators and calibration

Crop safety/Restrictions
- Maximum number of treatments 1 per crop

Special precautions/Environmental safety
- Irritating to eyes and skin
- Dangerous to game, wild birds and animals
- Extremely dangerous to fish or other aquatic life. Do not contaminate surface waters or ditches with chemical or used container

Personal protective equipment/Label precautions
- A, B, C, D, E, H, K, M
- M02, R04a, R04b, U02, U05a, U09a, U13, U19, U20a, C03, E01, E10a, E13a, E30a, E32a

Latest application/Harvest interval
- At planting

---

31   benomyl

A systemic, MBC fungicide with protectant and eradicant activity

Products

Benlate Fungicide	DuPont	50% w/w	WP	00229

Uses

Black spot in **courgettes** (off-label). Botrytis in **rhubarb** (off-label). Chocolate spot in **field beans**. Fungus diseases in **rhubarb** (off-label). Gummosis in **courgettes** (off-label). Leaf and pod spot in **field beans**. Light leaf spot in **brussels sprouts**. Rhizoctonia in **watercress** (off-label). Ring spot in **brussels sprouts**.

Efficacy
- Apply as overall spray. One spray effective for some diseases, repeat sprays at 1-4 wk intervals needed for others. Recommended spray programmes vary with disease and crop. See label for details of timing and recommended tank mixes
- Addition of non-ionic wetter recommended for certain uses. See label for details
- To delay appearance of resistant strains in diseases needing more than 2 applications per season, use in programme with fungicide of different mode of action
- Product may lose effectiveness if allowed to become damp during storage

Crop safety/Restrictions
- Maximum number of treatments (including applications of any product containing benomyl, carbendazim or thiophanate-methyl) 2 per crop

Personal protective equipment/Label precautions
- A, C, D, H, M
- U20b, E15, E30a, E32a

Latest application/Harvest interval
- HI 3 wk

Approval
- Following implementation of Directive 98/82/EC, approval for use of benomyl on numerous crops was revoked in 1999
- Off-label approval unlimited for use on protected rhubarb (OLA 1806/96), and outdoor courgettes (OLA 1866/96); unlimited for use on watercress (OLA 0875/97)

---

32   bentazone

A post-emergence contact diazinone herbicide

Products

1	Basagran SG	BASF	87% w/w	SG	08360
2	I T Bentazone	I T Agro	480 g/l	SL	08677
3	I T Bentazone 48	I T Agro	480 g/l	SL	09283
4	Standon Bentazone	Standon	480 g/l	SL	09204
5	Standon Bentazone S	Standon	87% w/w	SG	10124

Uses

Annual dicotyledons in **bulb onions** *(off-label)*, **evening primrose** *(off-label)*, **salad onions** *(off-label)* [1]. Annual dicotyledons in **linseed, narcissi, navy beans, peas, potatoes, runner beans, spring field beans, winter field beans** [1-5]. Annual dicotyledons in **soya beans** *(off-label)* [1, 4]. Annual dicotyledons in **broad beans** [1, 4, 5]. Annual dicotyledons in **french beans** [1, 5]. Annual dicotyledons in **dwarf beans** [2-4]. Annual dicotyledons in **outdoor dwarf beans** *(off-label)* [4].

Efficacy
- Most effective control obtained when weeds are growing actively and less than 5 cm high or across. Good spray cover is essential
- Split dose application may be made in beans, linseed, potatoes and narcissi and generally gives better weed control. See label for details
- Do not apply if rain or frost expected, if unseasonably cold, if foliage wet or in drought
- A minimum of 6 h free from rain is required after application
- Various recommendations are made for use in spray programmes with other herbicides or in tank mixes. See label for details
- Addition of Actipron, Adder or Cropspray 11E adjuvant oils recommended for use in dwarf beans and potatoes to improve fat hen control. Do not use under hot or humid conditions

Crop safety/Restrictions
- Maximum number of treatments varies with crop and product - see labels
- Crops must be treated at correct stage of growth to avoid danger of scorch. See label for details and for varietal tolerances. Apply to broad beans from 3 to 4 leaf pairs only
- Do not use on crops which have been affected by drought, waterlogging, frost or other stress conditions
- Do not spray at temperatures above 21°C. Delay spraying until evening if necessary
- Consult processor before using on crops for processing
- A satisfactory wax test must be carried out before use on peas
- Not all varieties are fully tolerant. Do not use on forage peas and other named varieties of peas, beans and potatoes
- May be used on selected varieties of maincrop and second early potatoes (see label for details), not on seed crops or first earlies
- Do not treat narcissi during flower bud initiation

Special precautions/Environmental safety
- Harmful: If swallowed [2-4]
- Risk of serious damage to eyes [1, 5]
- May cause sensitization by skin contact

Personal protective equipment/Label precautions
- A [1-5]; C [1, 5]
- M03 [2-4]; R04e [1-5]; R04, R04d [1, 5]; R03c [2-4]; U05a, U08, U19, U20b, C03, E01, E15, E30a [1-5]; E32a [1, 5]; E31b, E34 [2-4]; E26 [2-5]

Latest application/Harvest interval
- Before shoots exceed 15 cm high for potatoes and spring field beans (or 6-7 leaf pairs), 4 leaf pairs (6 pairs or 15 cm high with split dose) for broad beans, before flower buds visible for French, navy, runner and winter field beans and linseed, before flower buds can be found enclosed in terminal shoot for peas
- HI onions 3 wk

Approval
- Bentazone included in Annex I under EC Directive 91/414
- Off-label approval unlimited for use on outdoor bulb and salad onions (OLA 1234/97)[1], and evening primrose (OLA 1232/97)[1]; to Apr 2002 for use on soya beans (OLA 0793/97)[1]

33   bentazone + MCPA + MCPB

A post-emergence herbicide for undersown spring cereals and grass

Products

1	Acumen	BASF	200:80:200 g/l	SL	00028
2	Headland Archer	Headland	200:80:200 g/l	SL	08814

Uses

Annual dicotyledons in *newly sown grass, undersown spring barley, undersown spring oats, undersown spring wheat*.

Efficacy
- Best results when weeds small and actively growing provided crop at correct stage
- A minimum of 6 h free from rain is required after treatment
- Do not apply if frost expected, if crop wet or when temperatures at or above 21°C
- On first year grass leys use alone or in mixture with cyanazine on seedling stage weeds before end of Sep

Crop safety/Restrictions
- Maximum number of treatments 1 per crop on undersown cereals; 1 per yr on grassland
- Apply to cereals from 2-fully expanded leaf stage but before first node detectable (GS 12-30) provided clover has reached 1-trifoliate leaf stage
- Do not treat red clover after 3-trifoliate leaf stage
- Apply to newly sown leys after grass has reached 2-leaf stage provided clovers have at least 1 trifoliate leaf and red clover has not passed 3-trifoliate leaf stage. Grasses must have at least 3 leaves before treating with cyanazine mixture
- Do not use on crops suffering from herbicide damage or physical stress
- Do not use on seed crops or on cereals undersown with lucerne
- Do not roll or harrow for 7 d before or after spraying
- Clovers may be scorched and undersown crop checked but effects likely to be outgrown

Special precautions/Environmental safety
- Harmful if swallowed. Irritating to eyes and skin. May cause sensitization by skin contact
- Harmful to fish or other aquatic life. Do not contaminate surface waters or ditches with chemical or used container

Personal protective equipment/Label precautions
- A, C [1, 2]
- M03, R03c, R04a, R04b, R04e [1, 2]; U20a [1]; U05a, U08, U19 [1, 2]; U20b [2]; C03 [1, 2]; E31c [1]; E01, E07, E13c, E30a, E34 [1, 2]; E26 [2]

Withholding period
- Keep livestock out of treated areas for at least 2 wk and until foliage of poisonous weeds such as ragwort has died and become unpalatable

Latest application/Harvest interval
- Before 1st node detectable for undersown cereals (GS 31), 2 wk before grazing for grassland (4 wk if mixed with cyanazine)

Approval
- Bentazone included in Annex I under EC Directive 91/414
- Accepted by BLRA for use on malting barley

---

34   bentazone + MCPB

A post-emergence herbicide mixture for tank mixing with cyanazine

Products

Pulsar	BASF	200:200 g/l	SL	04002

Uses

Annual dicotyledons in *peas*.

Efficacy

- To be used in tank mix with cyanazine
- Best results achieved when weeds small and actively growing provided crop at correct stage. Good spray cover is essential
- May be applied as single treatment or as split dose treatment applying first spray when susceptible weeds are not beyond the 2 leaf stage
- Do not apply if rain or frost expected or if foliage wet. A minimum of 24 h free from rain required after treatment
- Do not apply during drought or unseasonably cold weather

Crop safety/Restrictions

- Maximum number of treatments 1 per crop or 2 per crop with split dose
- Apply only to listed cultivars (see label) from 3 fully expanded leaf (for full dose) or from 2 node stage (for split dose) to before flower buds can be found enclosed in terminal shoot (GS 102 or 103 to before GS 201)
- Do not treat forage pea cultivars or mange-tout peas
- Apply after a satisfactory wax test. Early drilled crops or crops affected by frost or abrasion may not have sufficiently waxy cuticle
- Allow 7 d after spraying before using a grass weed herbicide, or wait 14 d afterwards and test leaf wax before treating
- Do not use as tank mix with any other product than cyanazine, nor after use of TCA
- Do not apply to any crop that may have been subjected to stress conditions, where foliage damaged or under hot, sunny conditions when temperature exceeds 21°C
- Do not apply insecticides within 7 d of treatment

Special precautions/Environmental safety

- Irritating to eyes and skin. May cause sensitization by skin contact
- Harmful to fish or other aquatic life. Do not contaminate surface waters or ditches with chemical or used container

Personal protective equipment/Label precautions

- A, C
- R04a, R04b, R04e, U05a, U08, U19, U20a, C03, E01, E07, E13c, E30a, E31c, E34

Withholding period

- Keep livestock out of treated areas for 14 d and until foliage of any poisonous weeds such as ragwort has died and become unpalatable

Latest application/Harvest interval

- Before enclosed bud stage of crop (GS 10x)

Approval

- Bentazone included in Annex I under EC Directive 91/414

---

35   bentazone + pendimethalin

A contact and residual herbicide mixture for combining peas

Products

Impuls	BASF	480:400 g/l	KL	09720

Uses

Annual dicotyledons in *combining peas*. Chickweed in *combining peas*. Cleavers in *combining peas*. Knotgrass in *combining peas*. Mayweeds in *combining peas*.

Efficacy
- Best results achieved by treating weeds when they are small and ensuring maximum spray coverage
- Moist soil required for maximum efficacy. Loose or cloddy seedbeds must be consolidated before spraying
- Residual control reduced on soils with more than 6% OM
- Minimum 6 hr free from rain required for optimum foliar activity. Do not spray when foliage wet

Crop safety/Restrictions
- Maximum total dose equivalent to one full dose treatment
- Severe crop damage may occur if mixed with any other product. Leave minimum 7 d before, or 14 d after, using a post-emergence grass herbicide. Do not apply insecticides within 7 d of use. Do not apply if TCA has been used
- Crystal violet leaf wax test must be carried out before use. Transient leaf margin scorch may occur after treatment
- Crop damage may occur if heavy rain follows application on stony or gravelly soils
- Do not use on soils prone to waterlogging
- If day time temperatures likely to exceed 21°C spray in the evening
- Do not treat crops under stress from any cause
- Land must be ploughed to 150 mm before any succeeding crop except cereals. In the event of crop failure spring barley may be sown after ploughing

Special precautions/Environmental safety
- Harmful if swallowed
- May cause sensitization by skin contact
- Dangerous to fish or other aquatic life. Do not contaminate surface waters or ditches with chemical or used container
- Equipment must be washed out thoroughly after use. Traces of product may damage susceptible crops sprayed later

Personal protective equipment/Label precautions
- A
- M03, R03c, R04e, U05a, U08, U13, U19, U20b, C03, E01, E13b, E30a, E31c, E34

Latest application/Harvest interval
- Before 3rd node stage (GS 103)

Approval
- Bentazone included in Annex I under EC Directive 91/414

36	**bifenox**

A diphenyl ether herbicide available only in mixtures

37	**bifenox + MCPA + mecoprop-P**

A contact and translocated herbicide mixture for amenity turf

Products

Sirocco	Aventis Environ.	150:100:200 g/l	SC	09939

Uses

Annual dicotyledons in *amenity grass, managed amenity turf*. Perennial dicotyledons in *amenity grass, managed amenity turf*.

Efficacy
- For best results grass and weeds should be actively growing at the time of treatment
- Treatment during the early part of the season is recommended although spraying at any time between Apr and Sep will be effective, except during drought or waterlogged conditions or where there is a risk of frost

---

FOR FULL CONDITIONS OF USE ALWAYS READ THE PRODUCT LABEL

Crop safety/Restrictions
- Maximum number of treatments 1 per yr
- Only use on established turf
- Do not use on turf within one year of sowing, except on swards consisting of perennial rye-grass and/or smooth-stalked meadow-grass, which may be treated 6 mth after sowing
- Do not apply in drought, waterlogging or frost, otherwise damage may occur which may include bleaching or scorch and which may persist
- Great care must be taken not to overlap spray swaths, as persistent damage may result
- Avoid mowing during the 3-4 d preceding or following treatment

Special precautions/Environmental safety
- Harmful if swallowed
- Risk of serious damage to eyes
- Dangerous to fish or other aquatic life. Do not contaminate surface waters or ditches with chemical or used container
- Leave treated clippings in situ. If treated clippings are collected, do not use as mulch either as fresh clippings or following composting
- Avoid spray drift onto all broad-leaved plants outside the target area. Small amounts may cause serious injury to herbaceous plants, vegetables, fruit and glasshouse crops

Personal protective equipment/Label precautions
- A, C, H
- M03, R03c, R04d, U02, U05a, U19, U20b, C03, E01, E13b, E26, E30a, E31b, E34

38   bifenthrin

A contact and residual pyrethroid acaricide/insecticide for use in agricultural and horticultural crops

Products

1	Starion	FMC	100 g/l	EC	09795
2	Talstar	Hortichem	100 g/l	EC	06913

Uses

Aphids in **winter barley, winter oats, winter wheat** [1]. Aphids in **broccoli, brussels sprouts, cabbages, calabrese, cauliflowers** [1, 2]. Cabbage stem flea beetle in **winter oilseed rape** [1]. Caterpillars in **broccoli, brussels sprouts, cabbages, calabrese, cauliflowers** [1, 2]. Damson-hop aphid in **hops** [2]. Fruit tree red spider mite in **apples, pears** [2]. Rape winter stem weevil in **winter oilseed rape** [1]. Two-spotted spider mite in **blackcurrants, hops, ornamentals, strawberries** [2]. Virus vectors in **winter barley, winter oats, winter wheat** [1]. Whitefly in **broccoli, brussels sprouts, cabbages, calabrese, cauliflowers** [1, 2].

Efficacy
- Timing of application varies with crop and pest. See label for details
- Good spray cover of upper and lower plant surfaces essential to achieve effective pest control

Crop safety/Restrictions
- Maximum number of treatments 5 per yr for hops; 2 per yr for all other crops

Special precautions/Environmental safety
- Harmful if swallowed and by inhalation
- Irritating to skin and eyes
- Flammable
- Extremely dangerous to bees. Do not apply to crops in flower or to those in which bees are actively foraging. Do not apply when flowering weeds are present
- Extremely dangerous to fish or other aquatic life. Do not contaminate surface waters or ditches with chemical or used container
- LERAP Category A
- Keep in original container, tightly closed, in a safe place under lock and key

Personal protective equipment/Label precautions
- A, C, H [1, 2]
- M03, R03b, R03c, R04a, R04b, R07d [1, 2]; U19 [2]; U02, U04a, U05a, U08, U20b, C03 [1, 2]; E34 [2]; E01, E12b, E13a, E16c, E16d, E26, E30b, E31b [1, 2]; E17 [1, 2] (18 m)

Latest application/Harvest interval
- Before 31 Mar in yr of harvest for cereals
- HI zero for all other crops

Approval
- Accepted by BLRA for use on hops

---

39   bitertanol

A conazole fungicide available only in mixtures

---

40   bitertanol + fuberidazole

A broad spectrum fungicide mixture for seed treatment in cereals

Products

Sibutol	Bayer	375:23 g/l	FS	07238

Uses

Bunt in **spring wheat** *(seed treatment)*, **winter wheat** *(seed treatment)*. Fusarium root rot in **spring oats** *(seed treatment)*, **spring rye** *(seed treatment)*, **spring wheat** *(seed treatment)*, **triticale** *(seed treatment)*, **winter oats** *(seed treatment)*, **winter rye** *(seed treatment)*, **winter wheat** *(seed treatment)*. Loose smut in **spring wheat** *(seed treatment - reduction)*, **winter wheat** *(seed treatment - reduction)*. Septoria seedling blight in **spring wheat** *(seed treatment - reduction)*, **winter wheat** *(seed treatment - reduction)*.

Efficacy
- Must be applied simultaneously with water in the ratio 1 part product to 2 parts water in a recommended seed treatment machine
- Treated seed should preferably be drilled in the same season
- Control of loose smut may be inadequate for use on seed for multiplication

Crop safety/Restrictions
- Maximum number of treatments 1 per batch of seed
- Do not use on seed with more than 16% moisture content, or on sprouted, cracked or skinned seed

Special precautions/Environmental safety
- Dangerous to fish or other aquatic life. Do not contaminate surface waters or ditches with chemical or used container
- Do not use treated seed as food or feed
- Treated seed harmful to game and wildlife
- Product also supplied in returnable containers. See label for guidance on handling, storage, protective clothing and precautions

Personal protective equipment/Label precautions
- A, H
- U20a, E03, E13b, E30a, E32a, E34, S01, S02, S03, S04b, S05, S06, S07

Latest application/Harvest interval
- Before drilling

Maximum Residue Level (mg residue/kg food)
- (bitertanol) pome fruits, apricots, peaches, nectarines, plums 1; bananas 0.5

41   bitertanol + fuberidazole + imidacloprid

A broad spectrum fungicide and insecticide seed treatment for winter wheat and winter oats

Products

Sibutol Secur	Bayer	140:8.6:87.5 g/l	LS	09131

Uses

Bunt in **winter wheat** *(seed treatment)*. Fusarium root rot in **winter oats** *(seed treatment)*, **winter wheat** *(seed treatment)*. Loose smut in **winter wheat** *(seed treatment - reduction)*. Seedling blight and foot rot in **winter wheat** *(seed treatment - reduction)*. Virus vectors in **winter oats** *(seed treatment)*, **winter wheat** *(seed treatment)*. Wireworms in **winter oats** *(reduction of damage)*, **winter wheat** *(reduction of damage)*.

Efficacy

- Best applied through recommended seed treatment machines
- Evenness of seed cover improved by simultaneous application of equal volumes of product and water or dilution of product with an equal volume of water
- Drill treated seed in the same season. Use minimum 125 kg treated seed per ha. Lower drilling rates and/or early drilling affect duration of BYDV protection needed and may require follow-up aphicide treatment
- In high risk areas where aphid activity is heavy and prolonged a follow-up aphicide treatment may be required
- Protection against foliar air-borne and splash-borne diseases later in the season will require appropriate fungicide follow-up sprays

Crop safety/Restrictions

- Maximum number of treatments 1 per batch of seed
- Slightly delayed and reduced emergence may occur but this is normally outgrown
- Field emergence which is delayed for any reason may be accentuated by treatment
- Do not use on seed with more than 16% moisture content, or on sprouted, cracked or skinned seed

Special precautions/Environmental safety

- Harmful if swallowed
- Dangerous to fish or other aquatic life. Do not contaminate surface waters or ditches with chemical or used container
- Do not use treated seed as food or feed
- Dangerous to birds, game and other wildlife. Treated seed should not be left on the soil surface. Bury spillages
- Treated seed should not be broadcast, but drilled to a depth of 4 cm in a well prepared seedbed
- If seed is left on the soil surface the field should be harrowed and rolled to ensure good incorporation

Personal protective equipment/Label precautions

- A, H
- M03, R03c, U04a, U05a, U13, U20b, C03, E01, E03, E13b, E26, E30a, E32a, E34, S01, S02, S03, S04c, S05, S06, S07

Latest application/Harvest interval

- Before drilling

Maximum Residue Level (mg residue/kg food)

- see bitertanol + fuberidazole entry

42   bone oil

A ready-to-use animal repellent

Products

Renardine	Roebuck Eyot	33.3% w/w	AL	06769

Uses

Badgers in *amenity areas, farmland*. Cats in *agricultural premises, amenity areas, farmland*. Dogs in *agricultural premises, amenity areas, farmland*. Foxes in *amenity areas, farmland*. Moles in *amenity areas, farmland*. Rabbits in *amenity areas, farmland*.

Efficacy

- Soak pieces of stick, rags or sand in product and distribute around area to be protected as directed
- Repeat treatment weekly or after heavy rainfall

Crop safety/Restrictions

- Do not apply directly to animals or crops
- Used as directed product is harmless to animals

---

43   Bordeaux mixture

A protectant copper sulphate/lime complex fungicide

Products

Wetcol 3	Ford Smith	30 g/l (copper)	SC	02360

Uses

Bacterial canker in *cherries*. Blight in *potatoes, tomatoes* *(outdoor)*. Cane spot in *loganberries, raspberries*. Canker in *apples, pears*. Celery leaf spot in *celery*. Currant leaf spot in *blackcurrants*. Downy mildew in *hops*. Leaf curl in *apricots, nectarines, peaches*. Scab in *apples, pears*. Spur blight in *raspberries*.

Efficacy

- Spray interval normally 7-14 d but varies with disease and crop. See label for details
- Commence spraying potatoes before crop meets in row or immediately first blight period occurs
- For canker control spray monthly from Aug to Oct
- For peach leaf curl control spray at leaf fall in autumn and again in Feb
- Spray when crop foliage dry. Do not spray if rain imminent

Crop safety/Restrictions

- Do not use on copper sensitive cultivars, including Doyenne du Comice pears

Special precautions/Environmental safety

- Harmful if swallowed. Irritating to eyes, skin and respiratory system
- Harmful to fish or other aquatic life. Do not contaminate surface waters or ditches with chemical or used container
- Harmful to livestock

Personal protective equipment/Label precautions

- R03c, R04a, R04b, R04c, U13, U15, U20a, E06b, E13c, E30a, E32a

Withholding period

- Keep all livestock out of treated areas for at least 3 wk

Latest application/Harvest interval

- HI 7 d

---

44   brodifacoum

An anticoagulant coumarin rodenticide

Products

1	Klerat	Scotts	0.005% w/w	RB	06869
2	Klerat Wax Blocks	Scotts	0.005% w/w	RB	06827
3	Sorexa Checkatube	Sorex	0.25% w/w	RB	09474
4	Talon Rat & Mouse Bait (Cut Wheat)	Killgerm	0.005% w/w	RB	H6709
5	Talon Rat & Mouse Bait (Whole Wheat)	Killgerm	0.005% w/w	RB	H6710

Uses

Mice in **farm buildings** [1, 2]. Mice in **farm buildings** *(indoor use only)* [3]. Mice in **agricultural premises** [4, 5]. Rats in **farm buildings** [1, 2]. Rats in **agricultural premises** [4, 5].

Efficacy

- Effective against rodents resistant to other commonly used anticoagulants
- A single feed representing a fraction of the pest's normal daily food requirement can be lethal
- Ready-to-use in a baiting programme
- Cover baits by placing in bait boxes, drain pipes or under boards [1, 2], or use bait tubes [3]
- Inspect bait sites frequently and top up as long as there is evidence of feeding

Special precautions/Environmental safety

- For use only by professional pest contractors
- Do not use outdoors. Products must be used in situations where baits are placed within a building or other enclosed structure, and the target is living or feeding predominantly within that building or structure
- Prevent access to baits by children, birds and other animals
- Do not prepare or lay baits where food, feed or water could become contaminated
- Remove all remains of bait or bait containers after use and burn or bury
- Search for and burn or bury all rodent bodies. Do not place in refuse bins or on rubbish tips

Personal protective equipment/Label precautions

- M05 [1-3]; M03 [4, 5]; U20b [1-3]; U13 [1-5]; U20a [4, 5]; C03 [1, 2, 4, 5]; E30b [1-5]; E32a [1, 2, 4, 5]; V01a [1, 2]; V03a, V04a [1-3]; V01b, V02 [3-5]; V03b, V04b [4, 5]

---

45   bromacil

A soil acting uracil herbicide for non-crop areas and cane fruit

Products

Hyvar X	DuPont	80% w/w	WP	01105

Uses

Annual dicotyledons in **blackberries, loganberries, raspberries, tayberries**. Annual grasses in **blackberries, loganberries, raspberries, tayberries**. Couch in **blackberries, loganberries, raspberries, tayberries**. Perennial dicotyledons in **blackberries, loganberries, raspberries, tayberries**. Perennial grasses in **blackberries, loganberries, raspberries, tayberries**. Total vegetation control in **non-crop areas**.

Efficacy

- Apply to established cane fruit as soon as possible in spring after cultivation and before bud break. Avoid further soil disturbance for as long as possible. If inter-row areas are cultivated, only a 30 cm band either side of row need be treated
- Best results achieved when soil is moist at time of application
- For total vegetation control apply to bare ground or standing vegetation. Adequate rainfall is needed to carry chemical into root zone
- Against existing vegetation best results achieved in late winter to early spring but application also satisfactory at other times

Crop safety/Restrictions

- Maximum number of treatments 1 per yr
- May be used on cane fruit established for at least 2 yr
- In Scotland only may be used on newly planted raspberries at a reduced rate immediately after planting followed by light ridging
- Do not use in last 2 yr before grubbing crop to avoid injury to subsequent crops. Carrots, lettuce, beet, leeks and brassicas are extremely sensitive
- When used non-selectively, take care not to apply where chemical can be washed into root zone of desirable plants
- Treated land should not be cropped within 3 full yr of treatment and then only after obtaining advice from manufacturer

Special precautions/Environmental safety

- Irritating to eyes, skin and respiratory system

Personal protective equipment/Label precautions
- A, C
- R04a, R04b, R04c, U05a, U09a, U19, U20a, C03, E01, E15, E30a, E32a

Latest application/Harvest interval
- Apr for cane fruit

---

46   bromacil + diuron

A root-absorbed residual total herbicide mixture

Products

1 Borocil K	Aventis Environ.	0.88:0.88% w/w	GR	09924
2 Borocil K	RP Amenity	0.88:0.88% w/w	GR	05183

Uses

Total vegetation control in *non-crop areas*.

Efficacy
- Spray or apply granules in early stage of weed growth at any time of year, provided adequate moisture to activate chemical is supplied by rainfall
- Use higher rates on adsorptive soils or established weed growth
- Do not apply when ground frozen

Crop safety/Restrictions
- Do not apply on or near trees, shrubs, crops or other desirable plants
- Do not apply where roots of desirable plants may extend or where chemical may be washed into contact with their roots
- Do not use on ground intended for subsequent cultivation

Special precautions/Environmental safety
- Harmful to fish or other aquatic life. Do not contaminate surface waters or ditches with chemical or used container

Personal protective equipment/Label precautions
- U19, U20b, E13c, E30a, E32a

---

47   bromacil + picloram

A persistent residual and translocated herbicide mixture

Products

Hydon	Nomix-Chipman	1.08:0.33% w/w	GR	01088

Uses

Total vegetation control in *non-crop areas*.

Efficacy
- Apply at any time from spring to autumn. Best results achieved in Mar-Apr when weeds 50-75 mm high
- Do not apply in very dry weather as moisture needed to carry chemical to roots
- May be used on high fire risk sites

Crop safety/Restrictions
- Do not apply near crops, cultivated plants and trees
- Do not apply on slopes where run-off to cultivated plants or water courses may occur

Special precautions/Environmental safety
- Harmful to fish or other aquatic life. Do not contaminate surface waters or ditches with chemical or used container

Personal protective equipment/Label precautions
- U19, U20b, E07, E13c, E30a, E32a

Withholding period
- Keep livestock out of treated area until foliage of any poisonous weeds such as ragwort has died and become unpalatable

---

48   bromadiolone

An anti-coagulant coumarin-derivative rodenticide

Products

1	Endorats Premium Rat Killer	Irish Drugs	0.005% w/w	GB	H6725
2	Slaymor	Novartis A H	0.005% w/w	RB	08592
3	Slaymor Bait Bags	Novartis A H	0.005% w/w	RB	08593
4	Tomcat 2	Antec	0.005% w/w	PT	08257
5	Tomcat 2 Blox	Antec	0.005% w/w	BB	08261

Uses

Mice in *farmyards* [2, 3]. Mice in *farm buildings* [2-5]. Rats in *farm buildings* [1-5]. Rats in *farmyards* [2, 3].

Efficacy

- Ready-to-use baits are formulated on a mould-resistant, whole-wheat base
- Use in baiting programme. Place baits in protected situations, sufficient for continuous feeding between treatments
- Chemical is effective against warfarin- and coumatetralyl-resistant rats and mice and does not induce bait shyness
- Use bait bags where loose baiting inconvenient (eg behind ricks, silage clamps etc)

Special precautions/Environmental safety

- Access to baits by children, birds and animals, particularly cats, dogs, pigs and poultry, must be prevented
- Baits must not be placed where food, feed or water could become contaminated
- Remains of bait and bait containers must be removed after treatment and burned or buried
- Rodent bodies must be searched for and burned or buried. They must not be placed in refuse bins or on rubbish tips

Personal protective equipment/Label precautions

- M03 [4, 5]; M04 [2, 3]; U20b [4, 5]; U13 [1-5]; U20a [2, 3]; E15, E32a [4, 5]; E30a [1-5]; E31a [2, 3]; S06 [1]; V01a, V02, V04a [1-5]; V03a [1, 4, 5]

49   bromoxynil

A contact acting HBN herbicide

See also benazolin + bromoxynil + ioxynil

Products

1	Alpha Bromolin 225 EC	Makhteshim	225 g/l	EC	08255
2	Alpha Bromotril P	Makhteshim	250 g/l	LI	07099
3	Barclay Mutiny	Barclay	250 g/l	SC	08933
4	Flagon 400 EC	Makhteshim	400 g/l	LI	08875
5	Greencrop Tassle	Greencrop	240 g/l	SC	09659

Uses

Annual dicotyledons in *linseed* [1, 3]. Annual dicotyledons in *maize, sweetcorn* (off-label) [2]. Annual dicotyledons in *forage maize* [3, 5]. Annual dicotyledons in *barley, oats, wheat* [4].

Efficacy

- Apply to spring sown maize from 2 fully expanded leaves up to 9 fully expanded leaves [2, 3]
- Apply to spring sown linseed from 1 fully expanded leaf to 20 cm tall [1]
- Spray when main weed flush has germinated and the largest are at the 4 leaf stage
- Weed control can be enhanced by using a split treatment spraying each application when the weeds are seedling to 2 true leaves. Apply the second treatment before the crop canopy covers the ground [2, 3]
- Tank mixture with atrazine recommended for enhanced weed control - see label [2, 3]

Crop safety/Restrictions

- Maximum number of treatments 1 per yr
- Maximum total dose equivalent to one full dose treatment [2, 3, 5]

- Foliar scorch, which rapidly disappears without affecting growth, will occur if treatment made in hot weather or during rapid growth
- Do not apply with oils or other adjuvants
- Do not apply during frosty weather, drought, when soil is waterlogged, when rain expected within 4 h or to crops under any stress
- Take particular care to avoid drift onto neighbouring susceptible crops or open water surfaces

Special precautions/Environmental safety
- Harmful if swallowed
- Irritating to eyes
- May cause sensitization by skin contact [4]
- Flammable [4]
- Dangerous to fish or other aquatic life. Do not contaminate surface waters or ditches with chemical or used container
- Harmful to bees. Do not apply to crops in flower or to those in which bees are actively foraging. Do not apply when flowering weeds are present
- Do not apply using hand-held equipment or at concentrations higher than those recommended

Personal protective equipment/Label precautions
- A, C [1-5]; H [1, 4]
- M03 [2-5]; R03c, R04a [1-5]; R04e, R07d [4]; U19 [1-4]; U05a, U08, U13, U20a [1-5]; U02 [1-3, 5]; U14, U15 [1, 4]; C03 [1-5]; C02 [4] (6 wk); E12e, E13b, E30a, E31b, E34 [1-5]; E01 [2-5]; E07 [1, 4] (6 wk); E26 [1, 3-5]

Withholding period
- Keep livestock out of treated areas for at least 6 wk

Latest application/Harvest interval
- Before 10 fully expanded leaf stage of crop for maize [2, 3]; before crop 20 cm tall and before flower buds visible for linseed; before 2nd node detectable (GS 32) for cereals [4]

Approval
- Bromoxynil was reviewed in 1995 and approvals for home garden use, and most hand held applications revoked. There are timing restrictions on grassland, leeks and onions and some other crops
- Off-label unlimited for use on outdoor sweetcorn (OLA 2230/97)[2]
- Accepted by BLRA for use on malting barley

50   bromoxynil + clopyralid

A post-emergence contact and translocated herbicide mixture

Products

Vindex	Dow	240:50 g/l	EC	05470

Uses

Annual dicotyledons in *linseed*.

Efficacy
- Best results achieved when weeds small and growing actively in warm, moist weather
- Vigorous crop competition enhances control of more resistant weeds and prevents those that germinate after treatment becoming a problem
- Do not spray during drought, waterlogging, frost or if rain is imminent
- Weed spectrum can be widened by mixture with bentazone. See label for details

Crop safety/Restrictions
- Maximum number of treatments 1 per crop
- Avoid spraying during drought, waterlogging, frost, extremes of temperature or if rain is imminent or falling
- Spraying in frosty weather or when hard frost occurs within 3-4 wk may result in leaf scorch
- Do not roll or harrow within 7 d before or after spraying
- Do not apply any other product within 7 d of application (10 d if crop is stressed)

- Wash equipment thoroughly with water and an authorised non-ionic wetting agent according to manufacturer's instructions immediately after use. Traces of product could harm susceptible crops sprayed later

Special precautions/Environmental safety
- Harmful in contact with skin or if swallowed. Irritating to skin and eyes
- May cause lung damage if swallowed
- Flammable
- Do not apply by hand-held equipment or at concentrations higher than those recommended
- Dangerous to fish or other aquatic life. Do not contaminate surface waters or ditches with chemical or used container
- Harmful to bees. Do not apply to crops in flower or to those in which bees are actively foraging. Do not apply when flowering weeds are present
- Do not harvest crops for animal or human consumption for at least 6 wk after last application

Personal protective equipment/Label precautions
- A, C
- M03, M05, R03a, R03c, R04a, R04b, R04g, R07d, U05a, U08, U13, U19, U20b, C03; C02 (6 wk); E01, E12e, E13b, E26, E30a, E31b, E34; E07 (6 wk)

Withholding period
- Keep livestock out of treated areas for at least 6 wk after treatment

Latest application/Harvest interval
- Before first flower and before crop exceeds 30 cm tall

---

51   bromoxynil + diflufenican + ioxynil

A selective contact and translocated herbicide for cereals

Products

1 Capture	Aventis	300:50:200 g/l	SC	09982
2 Capture	RP Agric.	300:50:200 g/l	SC	06881

Uses

Annual dicotyledons in *spring barley, spring wheat, triticale, winter barley, winter rye, winter wheat*. Chickweed in *spring barley, spring wheat, triticale, winter barley, winter rye, winter wheat*. Knotgrass in *spring barley, spring wheat, triticale, winter barley, winter rye, winter wheat*. Mayweeds in *spring barley, spring wheat, triticale, winter barley, winter rye, winter wheat*. Speedwells in *spring barley, spring wheat, triticale, winter barley, winter rye, winter wheat*.

Efficacy
- Best results obtained on small weeds and when competition removed early
- Good spray coverage essential for good activity

Crop safety/Restrictions
- Maximum number of treatments 1 per crop
- Do not treat crops undersown or to be undersown
- Do not treat frosted crops or those that are under stress from any cause
- In the event of crop failure winter wheat may be drilled immediately after normal cultivations. Winter barley may be re-drilled after ploughing
- Land must be ploughed and an interval of 12 wk elapse after treatment before planting spring crops of wheat, barley, oilseed rape, peas, field beans, sugar beet, potatoes, carrots, edible brassicas or onions
- Successive treatments of any products containing diflufenican can lead to soil build-up and inversion ploughing must precede sowing any following non-cereal crop. Even where ploughing occurs some crops may be damaged

Special precautions/Environmental safety
- Harmful if swallowed
- Harmful to bees. Do not apply to crops in flower or to those in which bees are actively foraging. Do not apply when flowering weeds are present

- Dangerous to fish or other aquatic life. Do not contaminate surface waters or ditches with chemical or used container
- LERAP Category B
- Do not apply by hand-held equipment or at concentrations higher than those recommended

Personal protective equipment/Label precautions
- A, C, H
- M03, R03c, U05a, U08, U13, U19, U20b, C03; C02 (2 wk); E01, E12e, E13b, E16a, E26, E30a, E31b, E34; E07 (2 wk)

Latest application/Harvest interval
- Before 2nd node detectable (GS 32)

Approval
- Accepted by BLRA for use on malting barley

Maximum Residue Level (mg residue/kg food)
- see ioxynil entry

52   bromoxynil + ethofumesate + ioxynil

A post-emergence herbicide for new grass leys

Products

Leyclene	Aventis	50:200:25 g/l	EC	07263

Uses

Annual dicotyledons in *grass seed crops, seedling leys*. Annual meadow grass in *grass seed crops, seedling leys*.

Efficacy
- Best results when weeds small and growing actively in a vigorous crop, soil moist and further rain within 10 d. Mid Oct to end Dec normally suitable
- Annual meadow grass controlled during early crop establishment. Spray weed grasses before fully tillered
- Do not spray in cold conditions or when heavy rain or frost imminent
- Do not use on soils with more than 10% organic matter
- Do not cut for 14 d after spraying or graze in Jan-Feb after spraying in Oct-Dec
- Ash or trash should be burned, buried or removed before spraying

Crop safety/Restrictions
- Maximum number of treatments 2 per yr
- Apply to healthy ryegrasses or tall fescue after 2-3 leaf stage, to cocksfoot, timothy and meadow fescue at least 60 d after emergence and after 2-3 leaf stage
- Do not use on crops under stress, during periods of very dry weather, prolonged frost or waterlogging
- Do not use where clovers or other legumes are valued components of ley
- Do not roll for 7 d before or after spraying
- Any crop may be sown 5 mth after application following ploughing to at least 15 cm

Special precautions/Environmental safety
- Harmful if swallowed. Irritating to skin and eyes. Flammable
- Do not apply by hand-held equipment or at concentrations higher than those recommended
- Dangerous to fish or other aquatic life. Do not contaminate surface waters or ditches with chemical or used container
- Harmful to bees. Do not apply to crops in flower or to those in which bees are actively foraging. Do not apply when flowering weeds are present

Personal protective equipment/Label precautions
- A, C
- M03, R03c, R04a, R04b, R07d, U05a, U08, U13, U19, U20b, C03; C02 (6 wk); E01, E12e, E13b, E30a, E31b, E34; E07 (6 wk)

Withholding period
- Keep livestock out of treated areas for at least 6 wk after treatment

Latest application/Harvest interval
- 6 wk before cutting or grazing

Maximum Residue Level (mg residue/kg food)
- see ioxynil entry

53   bromoxynil + ioxynil

A contact acting post-emergence HBN herbicide for cereals

Products

1 Alpha Briotril	Makhteshim	240:160 g/l	EC	04876
2 Alpha Briotril 19/19	Makhteshim	190:190 g/l	EC	04740
3 Mextrol Biox	Nufarm Whyte	200:200 g/l	EC	09470
4 Oxytril CM	RP Agric.	200:200 g/l	EC	08667
5 Oxytril CM	Aventis	200:200 g/l	EC	10005

Uses

Annual dicotyledons in *spring barley, spring oats, spring wheat, winter barley, winter oats, winter wheat* [1]. Annual dicotyledons in *barley, oats, wheat* [2-5]. Annual dicotyledons in *rye, triticale, undersown cereals* [3-5].

Efficacy
- Best results achieved on young weeds growing actively in a highly competitive crop
- Do not apply during periods of drought or when rain imminent (some labels say 'if likely within 4 or 6 h')
- Recommended for tank mixture with hormone herbicides to extend weed spectrum. See labels for details

Crop safety/Restrictions
- Maximum number of treatments 1 per crop
- Apply to winter or spring cereals from 1-2 fully expanded leaf stage, but before second node detectable (GS 32)
- Spray oats in spring when danger of frost past. Do not spray winter oats in autumn
- Apply to undersown cereals pre-sowing or pre-emergence of legume provided cover crop is at correct stage. Only spray trefoil pre-sowing
- On crops undersown with grasses alone apply from 2-leaf stage of grass
- Do not spray crops stressed by drought, waterlogging or other factors
- Do not roll or harrow for several days before or after spraying. Number of days specified varies with product. See label for details

Special precautions/Environmental safety
- Harmful if swallowed
- Irritating to eyes
- Flammable [1, 2]
- Do not apply by hand-held equipment or at concentrations higher than those recommended
- LERAP category B [2]
- Dangerous to fish or other aquatic life. Do not contaminate surface waters or ditches with chemical or used container
- Harmful to bees. Do not apply to crops in flower or to those in which bees are actively foraging. Do not apply when flowering weeds are present

Personal protective equipment/Label precautions
- A, C [1-5]
- M03 [1-5]; R07d [1, 2]; R03c, R04a [1-5]; U02, U14, U15, U20a [1, 2]; U05a, U08, U13, U19 [1-5]; U20b [3-5]; C03 [1-5]; C02 [2-5] (6 wk); E16a, E16b [2]; E31b [1, 2]; E01, E12e, E13b, E30a, E34 [1-5]; E07 [1-5] (6 wk); E26 [1]; E32a [3-5]

Withholding period
- Keep livestock out of treated areas for at least 6 wk [2], 14 d [4]

Latest application/Harvest interval
- Before 2nd node detectable stage (GS 31).
- HI (animal consumption) 14 d

Approval
- Accepted by BLRA for use on malting barley

Maximum Residue Level (mg residue/kg food)
- see ioxynil entry

54   bromoxynil + ioxynil + mecoprop-P

A post-emergence contact and translocated herbicide

Products

Swipe P	Novartis	56:56:224 g/l	EC	08452

Uses

Annual dicotyledons in *durum wheat, ryegrass, spring barley, spring oats, spring wheat, triticale, winter barley, winter oats, winter wheat*.

Efficacy
- Best results when weeds small and growing actively in a strongly competitive crop
- Do not spray in rain or when rain imminent. Control may be reduced by rain within 6 h
- Application to wet crops or weeds may reduce control
- To achieve optimum control of large over-wintered weeds or in advanced crops increase water volume to aid spray penetration and cover
- Recommended for tank-mixing with approved MCPA-amine for hemp nettle control

Crop safety/Restrictions
- Maximum number of treatments 1 per crop. The total amount of mecoprop-P applied in a single yr must not exceed the maximum total dose approved for any single product for the crop/situation
- Spray cereals from 3 leaves unfolded (GS 13) to before second node detectable (GS 32) for winter sown cereals, spring wheat and spring barley and before first node detectable (GS 31) for spring oats
- Do not treat durum wheat or winter oats in autumn
- Apply to direct sown ryegrass or cereals undersown with ryegrass from 2-3 leaf stage of grass
- Do not spray crops undersown with legumes or use on winter oats or durum wheat in autumn
- Do not spray crops under stress from frost, waterlogging, drought or other causes
- Do not roll within 5 d after spraying
- Yield of barley may be reduced if frost occurs within 3-4 wk of treatment of low vigour crops on light soils or subject to stress
- Some crop yellowing may follow treatment but yield not normally affected
- Do not mix with manganese sulphate
- Avoid drift onto neighbouring susceptible crops

Special precautions/Environmental safety
- Harmful if swallowed
- Irritating to skin. May cause sensitization by skin contact
- Risk of serious damage to eyes
- Do not apply by hand-held equipment or at concentrations higher than those recommended
- Dangerous to fish or other aquatic life. Do not contaminate surface waters or ditches with chemical or used container
- LERAP Category B
- Harmful to bees. Do not apply to crops in flower or to those in which bees are actively foraging. Do not apply when flowering weeds are present
- Do not harvest crops for animal consumption for at least 6 wk after last application

Personal protective equipment/Label precautions
- A, C, H, M
- M03, R03c, R04b, R04d, R04e, U05a, U08, U13, U14, U15, U19, U20a, C03; C02 (6 wk); E01, E12e, E13b, E16a, E26, E30a, E31b, E34; E07 (6 wk)

Withholding period
- Keep livestock out of treated areas for at least 6 wk and until foliage of poisonous weeds such as ragwort has died and become unpalatable

Latest application/Harvest interval
- Before first node detectable (GS31) for spring oats; before second node detectable (GS32) for durum wheat, spring barley, spring wheat, triticale, winter barley, winter oats, winter wheat
- HI (animal consumption) 6 wk

Approval
- Accepted by BLRA for use on malting barley

Maximum Residue Level (mg residue/kg food)
- see ioxynil entry

55   bromoxynil + ioxynil + triasulfuron

An HBN and sulfonylurea herbicide mixture for cereals

Products

| Teal | Novartis | 190:190 g/l:20% w/w | KK | 08453 |

Uses
Annual dicotyledons in **barley, oats, rye, triticale, wheat**.

Efficacy
- For best results apply during warm, moist weather when weeds growing actively
- Do not spray when rain imminent. Rainfall within 6 h after spraying may reduce weed control

Crop safety/Restrictions
- Maximum number of treatments 1 per crop
- Apply in spring, after 1 Feb, from 3 leaves unfolded but before second node detectable (GS 13-32)
- Do not spray undersown crops or those due to be undersown
- Do not spray winter barley of low vigour, on light soil and subject to stress during frosty weather
- Do not use on oats until risk of frost is over
- Do not use during frosty weather, when frost imminent, or on crops under stress from frost, waterlogging or drought
- Do not spray in tank mixture, or in sequence, with a product containing any other sulfonylurea
- See label for details of restrictions on sequential use of other herbicides and sowing of subsequent crops

Special precautions/Environmental safety
- Harmful if swallowed.
- Irritating to eyes
- Do not apply by hand-held equipment or at concentrations higher than those recommended
- Dangerous to fish or other aquatic life and extremely dangerous to aquatic higher plants. Do not contaminate surface waters or ditches with chemical or used container
- LERAP Category B
- Harmful to bees. Do not apply to crops in flower or to those in which bees are actively foraging. Do not apply when flowering weeds are present
- Take care to wash out sprayers thoroughly. See label for details

Personal protective equipment/Label precautions
- A, C, H
- M03, R03c, R04a, U05a, U08, U13, U14, U15, U19, U20a, C03, E01, E12e, E13b, E16a, E30a, E32a, E34

Withholding period
- Keep livestock out of treated areas for at least 6 wk

Latest application/Harvest interval
- Before second node detectable (GS 32).
- HI (animal consumption) 6 wk

Approval
- Triasulfuron included in Annex I under EC Directive 91/414
- Accepted by BLRA for use on malting barley

Maximum Residue Level (mg residue/kg food)
- see ioxynil entry

56   bromoxynil + prosulfuron

A contact and residual herbicide mixture for maize and sweetcorn

Products

Jester	Novartis	60:3 % w/w	WG	08681

Uses

Annual dicotyledons in *maize, sweetcorn*. Black bindweed in *maize, sweetcorn*. Chickweed in *maize, sweetcorn*. Hemp-nettle in *maize, sweetcorn*. Knotgrass in *maize, sweetcorn*. Mayweeds in *maize, sweetcorn*.

Efficacy
- Apply post-emergence up to when the crop has four unfolded leaves
- To minimise the possible development of resistant weeds, use mixtures or sequences with other herbicides with a different mode of action
- Product must be used with Agral

Crop safety/Restrictions
- Maximum number of treatments 1 per crop
- Apply in cool conditions, or during the evening, to avoid scorch
- Do not treat crops grown for seed production
- Consult before use on crops intended for processing
- Do not use in frosty weather or on crops under stress
- Do not tank mix with organophosphate insecticides or apply in tank mix or sequence with any other sulfonylurea

Special precautions/Environmental safety
- Harmful if swallowed
- Irritating to eyes
- LERAP Category B
- Do not apply by knapsack sprayer or in volumes less than those recommended
- Take special care to avoid drift outside the target area
- Take care to wash out sprayers thoroughly. See label for details

Personal protective equipment/Label precautions
- A, C
- R03c, R04a, U05a, U11, U15, U20c, C03, E01, E13a, E16a, E29, E30a, E31a

Latest application/Harvest interval
- Before 5 crop leaves unfolded for maize, sweetcorn

57   bromoxynil + terbuthylazine

A post-emergence contact and residual herbicide for maize

Products

Alpha Bromotril PT	Makhteshim	200:300 g/l	SC	09435

Uses

Annual dicotyledons in *forage maize*.

Efficacy
- Best results obtained from treatment when main flush of weeds has germinated and largest at 4 leaf stage
- Control can be enhanced by use of split dose treatment but second spray must be applied before crop canopy covers ground
- Soils should be moist at time of treatment and should not be disturbed afterwards
- Residual activity may be reduced on soils with high OM content, or where high amounts of slurry or manure have been applied

Crop safety/Restrictions
- Maximum total dose equivalent to one full dose treatment

- Foliar scorch may occur after treatment in hot conditions or when growth particularly rapid. In such conditions spray in the evening
- Do not apply during frosty weather, drought, when soil is waterlogged or when rain expected within 12 hr
- Do not treat crops under any stress whatever
- Product should be used only where damaging populations of competitive weeds have emerged. Yield may be reduced when used in absence of significant weed populations
- Succeeding crops may not be planted until the spring following application. Mould board plough to 150 mm before doing so

Special precautions/Environmental safety
- Harmful if swallowed
- Irritating to eyes
- May cause sensitization by skin contact
- Harmful to bees. Do not apply to crops in flower or to those in which bees are actively foraging. Do not apply when flowering weeds are present
- Dangerous to fish or other aquatic life. Do not contaminate surface waters or ditches with chemical or used container
- Do not apply by hand-held equipment or at concentrations higher than those recommended

Personal protective equipment/Label precautions
- A, C, H
- M03, R03c, R04a, R04e, U02, U05a, U08, U13, U19, U20a, C03, E01, E12e, E13b, E30a, E31b, E34; E07 (6 wk)

Withholding period
- Keep livestock out of treated areas for at least 6 wk

Latest application/Harvest interval
- Before 9 fully expanded leaf stage

58   bromuconazole

A systemic conazole fungicide for cereals

Products

1 Granit	Aventis	200 g/l	SC	09995
2 Granit	RP Agric.	200 g/l	SC	08268

Uses

Brown rust in *spring barley, spring wheat, winter barley, winter wheat* [2]. Eyespot in *spring barley (reduction), spring wheat (reduction), winter barley (reduction), winter wheat (reduction)* [2]. Fusarium ear blight in *spring wheat, winter wheat* [2]. Glume blotch in *spring wheat, winter wheat* [2]. Net blotch in *spring barley, winter barley* [2]. Powdery mildew in *spring barley, spring wheat, winter barley, winter wheat* [2]. Rhynchosporium in *spring barley, winter barley* [2]. Septoria leaf spot in *spring wheat, winter wheat* [2]. Sooty moulds in *spring wheat, winter wheat* [2]. Yellow rust in *spring barley, spring wheat, winter barley, winter wheat* [2].

Efficacy
- Apply at any time from the tillering stage up to the latest times indicated for each crop
- Best results achieved from applications before disease becomes established
- If disease pressure persists a second treatment may be required to prevent late season attacks
- Best control of ear diseases in wheat obtained from application at or after the ear completely emerged stage
- More consistent control of powdery mildew obtained from tank mixture with fenpropimorph. See label

Crop safety/Restrictions
- Maximum number of treatments 2 per crop
- Only cereals, oilseed rape, field beans, peas, potatoes, linseed or Italian ryegrass may be sown as following crops

Special precautions/Environmental safety
- Dangerous to fish or other aquatic life. Do not contaminate surface waters or ditches with chemical or used container

• LERAP Category B

Personal protective equipment/Label precautions
  • A, C, H [2]
  • U08, U19, U20b, E13b, E16a, E16b, E26, E30a, E31b [2]

Latest application/Harvest interval
  • Varies with dose used and crop - see label

---

59   bupirimate

A systemic pyrimidine fungicide active against powdery mildew

Products

Nimrod	Zeneca	250 g/l	EC	06686

Uses

Powdery mildew in **apples, begonias, blackcurrants, chrysanthemums, courgettes, cucumbers, gooseberries, hops, marrows, pears, pumpkins** (off-label), **raspberries, roses, squashes** (off-label), **strawberries, tomatoes** (off-label).

Efficacy
  • Apply before or at first signs of disease and repeat at 7-14 d intervals. Timing and maximum dose vary with crop. See label for details
  • On apples during periods that favour disease development lower doses applied weekly give better results than higher rates fortnightly
  • Product has negligible effect on *Phytoseiulus* and *Encarsia* and may be used in conjunction with biological control of red spider mite
  • Not effective in protected crops against strains of mildew resistant to bupirimate

Crop safety/Restrictions
  • Maximum number of treatments or maximum total dose depends on crop and variety (see label for details)
  • With apples, hops and ornamentals cultivars may vary in sensitivity to spray. See label for details
  • If necessary to spray cucurbits in winter or early spring spray a few plants 10-14 d before spraying whole crop to test for likelihood of leaf spotting problem
  • On roses some leaf puckering may occur on young soft growth in early spring or under low light intensity. Avoid use of high rates or wetter on such growth
  • Never spray flowering begonias (or buds showing colour) as this can scorch petals
  • Do not mix with other chemicals for application to begonias, cucumbers or gerberas

Special precautions/Environmental safety
  • Irritating to eyes and skin.
  • Flammable
  • Harmful to fish or other aquatic life. Do not contaminate surface waters or ditches with chemical or used container

Personal protective equipment/Label precautions
  • A, C
  • R04a, R04b, R04g, R07d, U05a, U20a, C03, E01, E13c, E30a, E31c, E34

Latest application/Harvest interval
  • HI depends on crop and variety (see label for details)

Approval
  • Off-label approval unlimited for use on protected tomatoes (OLA 0516/96); unstipulated for use on outdoor crops of squashes and pumpkins (OLA 0080/00)
  • Accepted by BLRA for use on hops

---

60   bupirimate + triforine

A systemic protectant and eradicant fungicide for ornamental crops

Products

Nimrod-T	Scotts	62.5:62.5 g/l	EC	09268

FOR FULL CONDITIONS OF USE ALWAYS READ THE PRODUCT LABEL

Uses

Black spot in *ornamentals*. Powdery mildew in *ornamentals*. Rust in *roses*.

Efficacy

- To prevent infection on roses spray in early May and repeat every 10-14 d. If infected with blackspot in previous season commence spraying at bud burst

Crop safety/Restrictions

- Spraying in high glasshouse temperatures may cause temporary leaf damage
- Test varietal susceptibility of roses and other ornamentals by spraying a few plants and allow 14 d for any symptoms to develop

Special precautions/Environmental safety

- Irritating to eyes
- Harmful to fish or other aquatic life. Do not contaminate surface waters or ditches with chemical or used container

Personal protective equipment/Label precautions

- C
- R04a, U05a, U09a, U20b, C03, E01, E13c, E26, E30a, E31b, E34

Approval

- Accepted by BLRA for use on hops

Maximum Residue Level (mg residue/kg food)

- see triforine entry

61	**buprofezin**

A moulting inhibitor, thiadiazine insecticide for whitefly control

Products

Applaud	Zeneca	250 g/l		SC	06900

Uses

Glasshouse whitefly in *aubergines, cucumbers, peppers, protected ornamentals, tomatoes*. Tobacco whitefly in *aubergines, cucumbers, peppers, protected ornamentals, tomatoes*.

Efficacy

- Product has contact, residual and some vapour activity
- Whitefly most susceptible at larval stages but residual effect can also kill nymphs emerging from treated eggs and application to pupae reduces emergence
- Adult whitefly not directly affected. Resistant strains of tobacco whitefly are known and where present control likely to be reduced or ineffective
- Product may be used either in IPM programme in association with *Encarsia formosa* or in All Chemical programme
- In IPM programme apply as single application and allow at least 60 d before re-applying
- In All Chemical programme apply twice at 7-14 d interval and allow at least 60 d before re-applying
- Do not leave spray liquid in sprayer for long periods
- Do not apply as fog or mist

Crop safety/Restrictions

- Maximum number of treatments up to 8 per crop for tomatoes and cucumbers; up to 2 per crop for aubergines and peppers
- Do not apply more than 2 sprays within a 65 d period on tomatoes, or within a 45 d period on cucumbers
- See label for list of ornamentals successfully treated but small scale test advised to check varietal tolerance. This is especially important if spraying flowering ornamentals with buds showing colour
- Do not treat Dieffenbachia or Closmoplictrum
- Do not apply to crops under stress

Personal protective equipment/Label precautions

- U20c, E15, E30a, E32a, E34

Latest application/Harvest interval
- HI edible crops 3 d

62   calciferol

A hypercalcaemic rodenticide available only in mixtures

63   calciferol + difenacoum

A mixture of rodenticides with different modes of action

Products

| Sorexa CD Concentrate | Sorex | 0.1:0.0025% w/w | RB | 03513 |

Uses

Mice in **farm buildings**.

Efficacy
- Lay baits in mouse runs in many locations throughout infested area
- It is important to lay many small baits as mice are sporadic feeders
- Cover baits to protect from moisture in outdoor situations
- Inspect baits frequently and replace or top up with fresh material as necessary

Special precautions/Environmental safety
- Protect baits from access by children, domestic or other animals

Personal protective equipment/Label precautions
- U13, U20a, E30a, E32a, V01a, V02, V03a, V04a

64   captafol

A protectant dicarboximide fungicide approvals for which expired on 31 December 1990. It is banned under the EC Directive 70/117/EEC

65   captan

A protectant dicarboximide fungicide with horticultural uses

Products

1	Alpha Captan 80 WDG	Makhteshim	80% w/w	WG	07096
2	Alpha Captan 83 WP	Makhteshim	83% w/w	WP	04806
3	PP Captan 80-WG	Tomen	80% w/w	WG	08971

Uses

Black spot in **roses** [1-3]. Botrytis in **strawberries** [1, 2]. Gloeosporium in **apples** [3]. Gloeosporium rot in **apples** [1, 2]. Scab in **apples, pears** [1-3].

Efficacy
- For control of scab apply at bud burst and repeat at 10-14 d intervals until danger of scab infection ceased
- For suppression of fruit storage rots apply from late Jul and repeat at 2-3 wk intervals
- For black spot control in roses apply after pruning with 3 further applications at 14 d intervals or spray when spots appear and repeat at 7-10 d intervals
- For grey mould in strawberries spray at first open flower and repeat every 7-10 d [1, 2]
- Do not leave diluted material for more than 2 h. Agitate well before and during spraying

Crop safety/Restrictions
- Maximum number of treatments 12 per yr on apples and pears as pre-harvest sprays
- Product must not be used as dip or drench on apples or pears [3]
- Do not use on apple cultivars Bramley, Monarch, Winston, King Edward, Spartan, Kidd's Orange or Red Delicious or on pear cultivar D'Anjou
- Do not mix with alkaline materials or oils

- Do not use on fruit for processing

Special precautions/Environmental safety
- Harmful [1]
- Irritating to eyes, skin and respiratory system [2]; to eyes only [3]; to respiratory system only [1]
- Risk of serious damage to eyes [1]
- May cause sensitization by skin contact [1, 3]
- Harmful to fish or other aquatic life. Do not contaminate surface waters or ditches with chemical or used container [1-3]
- Powered visor respirator with hood and neck cape must be used when handling concentrate [3]

Personal protective equipment/Label precautions
- A [1-3]; D, E, H [1, 3]; C, K, M [3]
- R04b [2]; R04c [1, 2]; R04a [2, 3]; R03, R04d [1]; R04e [1, 3]; U20b [1, 2]; U05a [1-3]; U09a, U19 [1]; U20c [3]; C03 [1-3]; E32a [1, 2]; E01, E13c, E30a [1-3]; E31b, E34 [3]

Latest application/Harvest interval
- HI apples, pears 14 d; strawberries 7 d

Maximum Residue Level (mg residue/kg food)
- pome fruits, grapes, cane fruits, bilberries, cranberries, currants, gooseberries, tomatoes, peppers, aubergines 3; apricots, peaches, nectarines, plums, lettuce, beans, peas, leeks 2; citrus fruits, bananas, carrots, horseradish, parsnips, parsley root, salsify, swedes, turnips, garlic, onions, shallots, cucumbers, gherkins, courgettes, cauliflowers, Brussels sprouts, head cabbage, celery, rhubarb, mushrooms, potatoes 0.1

66   captan + penconazole

A protectant fungicide for use on apple trees

Products

Topas C 50 WP	Novartis	47.5:2.5% w/w	WB	08459

Uses

Powdery mildew in *apples*. Scab in *apples*.

Efficacy
- Use as a protective spray every 7-14 d from bud burst until extension growth ceases
- High antisporulant activity reduces development of primary mildew and controls spread of secondary mildew
- Does not affect beneficial insects so can be used in integrated control programmes
- Penconazole is recommended alone for mildew control if scab is not a problem

Crop safety/Restrictions
- Maximum number of treatments 10 per yr

Special precautions/Environmental safety
- Irritating to eyes, skin and respiratory system
- Dangerous to fish or other aquatic life. Do not contaminate surface waters or ditches with chemical or used container

Personal protective equipment/Label precautions
- A, C
- R04a, R04b, R04c, U02, U05a, U08, U19, U20b, C03; C02 (14 d); E01, E13b, E30a, E32a

Latest application/Harvest interval
- HI apples 14 d

Maximum Residue Level (mg residue/kg food)
- see captan entry

67   carbaryl

A contact carbamate insecticide, with restricted permitted uses

Products

Thinsec	Zeneca	450 g/l	SC	06710

Uses

Apple capsid in **apples**. Codling moth in **apples**. Earwigs in **apples**. Fruit thinning in **apples**. Tortrix moths in **apples**. Winter moth in **apples**.

Efficacy

- Application rate and timing vary with pest. See label for details
- Effective thinning of apples depends on thorough wetting of foliage and fruitlets. Dose and timing vary with cultivar
- Fruit from treated trees may need picking 5-7 d earlier than from untreated trees

Crop safety/Restrictions

- Spray concentration must not exceed 3.8 l/500 l water
- Carbaryl is dangerous to pollinating insects and particular care must be taken when spraying close to blossom period or when orchard adjacent to other flowering crops
- Do not spray Laxton's Fortune between pink bud stage and end of first wk in Jun
- Do not use on apples between late pink bud and end of first wk in Jun except for purpose of fruit thinning

Special precautions/Environmental safety

- This product contains an anticholinesterase carbamate compound. Do not use if under medical advice not to work with such compounds
- Harmful if swallowed and in contact with skin
- Harmful to fish or other aquatic life. Do not contaminate surface waters or ditches with chemical or used container
- Dangerous to bees. Do not apply to crops in flower or to those in which bees are actively foraging. Do not apply when flowering weeds are present
- Spray equipment may only be used where the operator's normal working position is within a closed cab on a tractor or on a self-propelled sprayer. Hand-held or similar application equipment may not be used
- Low level induction bowls, or a closed transfer system must be used when transferring product from container to spray tank

Personal protective equipment/Label precautions

- A, C, H, K
- M02, M03, R03c, U05a, U08, U20a, C03, E01, E12c, E13c, E30a, E31a, E34

Latest application/Harvest interval

- HI apples 3 wk

Approval

- All approvals for non-agricultural uses of carbaryl were revoked in November 1995 following review
- Approval expiry 31 Jul 2001

Maximum Residue Level (mg residue/kg food)

- apricots, peaches, nectarines, plums, cane fruits, bilberries, cranberries, currants, gooseberries, lettuce 10; citrus fruits, strawberries 7; pome fruits, grapes, bananas, tomatoes, peppers, aubergines, head cabbage, beans, peas 5; cucumbers, gherkins, courgettes, celery, rhubarb 3; carrots, horseradish, parsnips, parsley root, salsify, swedes 2; turnips, garlic, onions, shallots, cauliflowers, Brussels sprouts, leeks, mushrooms 1; potatoes 0.2

---

68   carbendazim

A systemic benzimidazole fungicide with curative and protectant activity

Products

1 Ashlade Carbendazim Flowable	Nufarm Whyte	500 g/l	SC	06213
2 Bavistin DF	BASF	50% w/w	WG	03848
3 Bavistin FL	BASF	500 g/l	SC	00218
4 Carbate Flowable	PBI	500 g/l	SC	08957
5 Derosal Liquid	AgrEvo	500 g/l	SC	07315
6 Derosal WDG	Aventis	80% w/w	WG	07316
7 Greencrop Mooncoin	Greencrop	500 g/l	SC	09392
8 Headland Addstem DF	Headland	50% w/w	WG	08904

FOR FULL CONDITIONS OF USE ALWAYS READ THE PRODUCT LABEL

Products (Continued)

9	Headland Regain	Headland	50% w/w	WG	08675
10	Mascot Systemic	Rigby Taylor	500 g/l	SC	08776
11	Quadrangle Hinge	Quadrangle	500 g/l	SC	04929
12	Stefes C-Flo 2	Stefes	500 g/l	SC	08059
13	Stefes Derosal Liquid	Stefes	500 g/l	SC	07649
14	Stefes Derosal WDG	Stefes	80% w/w	WG	07658
15	Tripart Defensor FL	Tripart	500 g/l	SC	02752
16	Turf Systemic Fungicide	PBI	500 g/l	SC	09349
17	Turfclear	Scotts	500 g/l	SC	07506
18	Twincarb	Vitax	500 g/l	SC	08777
19	UPL Carbendazim 500 FL	United Phosphorus	500 g/l	SC	07472

Uses

Anthracnose in **dwarf beans** [2, 3]. Big vein in **outdoor lettuce** *(off-label)*, **protected lettuce** *(off-label)* [2]. Black root rot in **ornamentals** *(off-label)*, **protected ornamentals** *(off-label)* [2]. Botrytis in **bedding plants** *(off-label)*, **bulbs/corms** *(off-label - spray and dip)*, **chrysanthemums** *(off-label)*, **pot plants** *(off-label)* [2]. Botrytis in **dwarf beans** [2, 3]. Botrytis in **strawberries** [2, 3, 5, 6, 8, 11-15]. Botrytis in **oilseed rape** *(reduction)* [4, 7]. Botrytis in **protected pot plants** [5, 12-14]. Botrytis in **pot plants** [6]. Canker in **apples** *(partial control)* [11]. Canker in **apples** [15]. Celery leaf spot in **celery** *(off-label)* [2]. Didymella in **protected tomatoes** *(off-label)* [2]. Dollar spot in **amenity turf** [16]. Dollar spot in **managed amenity turf** [18]. Dollar spot in **turf** [9, 10, 17]. Eye rot in **apples, pears** [5, 12-14]. Eyespot in **spring barley** [1]. Eyespot in **winter barley** [1-3]. Eyespot in **winter wheat** [1-5, 7, 8, 12, 13, 15, 19]. Eyespot in **winter rye** [2-4, 8]. Eyespot in **wheat** [6, 14]. Fusarium in **bulbs/corms** *(off-label - dip)*, **freesias** *(off-label - drench)* [2]. Fusarium patch in **amenity turf** [16]. Fusarium patch in **managed amenity turf** [18]. Fusarium patch in **turf** [9, 10, 17]. Gloeosporium in **pears** [14]. Gloeosporium in **apples** [2, 8, 11, 14]. Gloeosporium rot in **apples, pears** [5, 12, 13]. Glume blotch in **winter wheat** *(partial control)* [11]. Late ear diseases in **winter wheat** [3]. Light leaf spot in **oilseed rape** [1, 4-7, 11-15]. Light leaf spot in **oilseed rape** *(reduction)* [2, 3, 8]. Mycogone in **mushrooms** [2, 3, 8]. Penicillium rot in **bulbs/corms** *(off-label)* [2]. Powdery mildew in **winter wheat** *(partial control)* [11]. Powdery mildew in **bedding plants** *(off-label)*, **chrysanthemums** *(off-label)*, **corms** *(off-label)*, **pot plants** *(off-label)* [2]. Powdery mildew in **roses** [2, 3, 8]. Powdery mildew in **pears** [5, 12, 13]. Powdery mildew in **protected pot plants** [5, 12-14]. Powdery mildew in **apples** [5, 6, 11-13, 15]. Powdery mildew in **pot plants** [6]. Powdery mildew in **strawberries** [6, 12-15]. Rhynchosporium in **spring barley, winter barley** [1-3]. Root diseases in **inert substrate cucumbers** *(off-label)* [2]. Scab in **pears** *(off-label)* [2]. Scab in **apples** [2, 5, 6, 8, 11-15]. Scab in **pears** [5, 6, 12-14]. Sclerotinia in **bulbs/corms** *(off-label - spray and dip)* [2]. Sooty moulds in **winter wheat** [2, 3, 5, 8, 12, 13]. Sooty moulds in **spring wheat** [5]. Sooty moulds in **wheat** [6, 14]. Stagonospora in **bulbs/corms** *(off-label)* [2]. Storage rots in **pears** *(off-label - dip)* [2]. Storage rots in **apples** [2, 6, 8, 15]. Storage rots in **pears** [6]. Trichoderma in **mushrooms** *(off-label - spawn treatment)* [2]. Wet bubble in **mushrooms** [4]. Wormcast formation in **managed amenity turf** [18]. Wormcast formation in **turf** [9, 10, 17].

Efficacy

- Products vary in the diseases listed as controlled for several crops. Labels must be consulted for full details and for rates and timings
- Mostly applied as spray or drench. Spray treatments normally applied at first sign of disease and repeated after 1 mth if required
- Allow 4 mth intervals where used for worm cast control. Rain or irrigation after treatment may improve control [10, 17, 18]
- On turf apply as preventative treatment in spring or autumn during periods of high disease risk [10, 17, 18]
- Apply as a drench to control soil-borne diseases in cucumbers and tomatoes and as a pre-planting dip treatment for bulbs
- On apples the addition of a non-ionic wetter aids penetration and improves disease control [14]
- Apply by incorporation into casing to control mushroom diseases [2, 4]
- To delay appearance of resistant strains alternate treatment with non-MBC fungicide. Eyespot in cereals and *Botrytis cinerea* in many crops is now widely resistant
- Not compatible with alkaline products such as lime sulphur

Crop safety/Restrictions
- Maximum number of treatments (including applications of any product containing benomyl, carbendazim or thiophanate-methyl) varies with crop treated and product used - see labels for details
- Do not treat crops or turf suffering from drought or other physical or chemical stress
- Do not use on strawberry runner beds
- Apply as drench rather than spray where red spider mite predators are being used
- Consult processors before using on crops for processing

Special precautions/Environmental safety
- Harmful to fish or other aquatic life. Do not contaminate surface waters or ditches with chemical or used container [1-3, 5, 7, 9-14, 16-19]
- After use dipping suspension must not be discharged directly into ditches or drains. Preferred disposal method is via a soakaway [14]

Personal protective equipment/Label precautions
- A, C, H [1-19]; M [1-15, 17-19]; B, K [2-6, 8, 12-15]; D [2, 6, 8, 9, 14]
- U20c [2, 3, 5, 7-10, 16-19]; U20a [11]; U20b [1, 4, 6, 12-14]; C03 [1, 11]; E29, E31c [3]; E13c [1-3, 5, 7-14, 16-19]; E30a [1-5, 7-13, 16-19]; E01 [1, 11, 15]; E34 [15, 19]; E31a [10, 15, 18]; E32a [1, 2, 6, 8, 9, 11, 14]; E26 [5, 7, 12, 13, 17-19]; E31b [4, 5, 7, 12, 13, 16, 17, 19]; E15 [4, 6]; E24 [17, 18]

Latest application/Harvest interval
- Varies with crop and product used. See labels for details
- HI 2 d for strawberries; 7-14 d (depends on dose) for apples, pears, mushrooms; 21 d for oilseed rape.

Approval
- Approved for aerial application on winter wheat and oilseed rape [1, 4, 11, 15]. See notes in Section 1
- Following implementation of Directive 98/82/EC, approval for use of carbendazim on numerous crops was revoked in 1999. UK approvals for use of carbendazim on strawberries likely to be revoked in 2001 as a result of implementation of the MRL Directives
- Off-label approval unlimited for use on mushrooms (spawn treatment) (OLA 1144/95)[2]; unlimited for use on protected cucumbers on inert media (OLA1476/95)[2]; unlimited for use on protected and outdoor crops of lettuce (OLA 1751/96)[2]; unlimited for use on outdoor and protected ornamentals, bulbs, corms, freesias, chrysanthemums, pot plants, bedding plants (OLA 0009/99)[2]; unstipulated for use on protected celery (OLA 2078/99)[2]; unstipulated for use on protected tomatoes grown in soil/peat bags, inert substrates, hydroponic solution (OLA 2079/99)[2]; unlimited for use on pears (OLA 2256/99)[2]
- Approval expiry 30 Sep 2001 [19]

Maximum Residue Level (mg residue/kg food)
- peaches, nectarines, grapes 10; citrus fruits, strawberries, raspberries, currants, tomatoes, lettuce 5; potatoes 3; pome fruits, plums, onions, celery 2; bananas, cultivated mushrooms 1; aubergines, cucumbers, melons, squashes, Brussels sprouts 0.5; soya beans 0.2; tree nuts, bilberries, cranberries, wild berries, miscellaneous fruits (except bananas), beetroot, horseradish, Jerusalem artichokes, parsnips, parsley root, radishes, sweet potatoes, yams, garlic, shallots, spring onions, courgettes, watermelons, sweet corn, Chinese cabbage, kale, kohlrabi, spinach, beet leaves, watercress, chervil, chives, parsley, celery leaves, stem vegetables (except celery), wild mushrooms, lentils, oilseeds (except soya beans), hops, cereals, animal products 0.1

---

69   carbendazim + chlorothalonil

A systemic and protectant fungicide mixture

Products

Greenshield	Scotts	100:450 g/l	SC	07988

Uses

Anthracnose in *turf*. Dollar spot in *turf*. Fusarium patch in *turf*. Red thread in *turf*. Wormcast formation in *turf*.

Efficacy

- Apply at any time of year at the first sign of turf disease. Further treatments may be necessary if conditions remain favourable
- Established or severe infections of Anthracnose will not be controlled

Crop safety/Restrictions

- Do not mow or water within 24 h of treatment
- Do not mix with other pesticides, surfactants or fertilisers

Special precautions/Environmental safety

- Irritating to eyes, skin and respiratory system
- Dangerous to fish or other aquatic life. Do not contaminate surface waters or ditches with chemical or used container
- LERAP Category B
- Do not apply by hand-held equipment

Personal protective equipment/Label precautions

- A, C, H, M
- R04a, R04b, R04c, R04d, U05a, U09a, U19, U20a, C03, E01, E13b, E16a, E26, E30a, E31a

Approval

- Approved for aerial application on winter wheat. See notes in Section 1
- Following implementation of Directive 98/82/EC, approval for use of carbendazim and chlorothalonil on numerous crops was revoked in 1999

Maximum Residue Level (mg residue/kg food)

- see carbendazim and chlorothalonil entries

70	**carbendazim + chlorothalonil + maneb**

A broad-spectrum protectant and eradicant fungicide for cereals

Products

Tripart Victor	Tripart	80:150:200 g/l	SC	04359

Uses

Botrytis in *spring barley, spring wheat, winter barley, winter wheat*. Brown foot rot and ear blight in *spring wheat, winter wheat*. Brown rust in *spring barley, spring wheat, winter barley, winter wheat*. Eyespot in *spring barley, spring wheat, winter barley, winter wheat*. Net blotch in *spring barley, winter barley*. Powdery mildew in *spring wheat, winter wheat*. Rhynchosporium in *spring barley, winter barley*. Septoria diseases in *spring wheat, winter wheat*. Sooty moulds in *spring wheat, winter wheat*. Yellow rust in *spring barley, spring wheat, winter barley, winter wheat*.

Efficacy

- Correct timing crucial for optimum results. First treatment should be made between when flag leaf just visible and start of ear emergence
- Highly effective against late diseases on flag leaf and ear of winter cereals

Crop safety/Restrictions

- Maximum number of treatments (including application of any product containing benomyl, carbendazim or thiophanate-methyl) 2 per crop.

Special precautions/Environmental safety

- Irritating to eyes, skin and respiratory system
- Dangerous to fish or other aquatic life. Do not contaminate surface waters or ditches with chemical or used container
- LERAP Category B

Personal protective equipment/Label precautions

- A, C
- R04a, R04b, R04c, U05a, U08, U19, U20a, C03, E01, E13b, E16a, E16b, E30a, E32a

Latest application/Harvest interval
- Ear emergence complete (GS 59) for wheat; up to and including grain watery-ripe (GS 71) for barley

Maximum Residue Level (mg residue/kg food)
- see carbendazim, chlorothalonil and maneb entries

---

71   carbendazim + cymoxanil + oxadixyl + thiram

A broad spectrum fungicide seed treatment for peas

Products

Apron Elite	Novartis	16.7:6.7:16.7:33.4% w/w	WG	08770

Uses

Ascochyta in **peas** *(seed treatment)*. Damping off in **peas** *(seed treatment)*. Downy mildew in **peas** *(seed treatment)*.

Efficacy
- Apply through continuous flow seed treaters which should be calibrated before use

Crop safety/Restrictions
- Ensure moisture content of treated seed satisfactory and store in a dry place
- Check calibration of seed drill with treated seed before drilling and sow as soon as possible after treatment
- Consult before using on crops for processing

Special precautions/Environmental safety
- May cause sensitization by skin contact
- Dangerous to fish or other aquatic life. Do not contaminate surface waters or ditches with chemical or used container
- Do not use treated seed as food or feed

Personal protective equipment/Label precautions
- A, C, D, H, M
- M03, R04e, U05a, U08, U20b, C03, E01, E13b, E26, E30a, E31a, E34, S02, S04b, S05, S07

Maximum Residue Level (mg residue/kg food)
- see carbendazim entry

---

72   carbendazim + epoxiconazole

A systemic fungicide mixture for managed amenity turf

Products

Capricorn	Aventis Environ.	125:125 g/l	SC	10035

Uses

Fusarium patch in **managed amenity turf**.

Efficacy
- Treat at the first sign of disease and repeat necessary
- Apply after cutting and delay further mowing for at least 48 h to allow systemic distribution within the plant
- Only moderate control of Fusarium patch claimed

Crop safety/Restrictions
- Maximum number of treatments 2 per yr
- Do not apply during drought conditions or to frozen turf
- Use only on established turf

Special precautions/Environmental safety
- Irritant. May cause sensitization by skin contact
- Harmful to fish or other aquatic life. Do not contaminate surface waters or ditches with chemical or used container
- LERAP Category B

Personal protective equipment/Label precautions
- A, C, H, M
- R04, R04e, U04a, U05a, U09b, U20c, C03, E01, E13c, E16a, E16b, E26, E30a, E31b, E34

73	carbendazim + flusilazole

A broad-spectrum systemic and protectant fungicide for cereals

Products

1	Contrast	DuPont	125:250 g/l	SC	06150
2	Landgold Flusilazole MBC	Landgold	125:250 g/l	SC	08528
3	Punch C	DuPont	125:250 g/l	SC	06801
4	Standon Flusilazole Plus	Standon	125:250 g/l	SC	07403

Uses

Brown rust in **spring wheat, winter wheat** [1-4]. Brown rust in **spring barley, winter barley** [1, 3]. Canker in **spring oilseed rape** *(reduction)*, **winter oilseed rape** *(reduction)* [1, 3]. Eyespot in **spring wheat, winter wheat** [1-4]. Eyespot in **spring barley, winter barley** [1, 3]. Light leaf spot in **spring oilseed rape, winter oilseed rape** [1, 3]. Light leaf spot in **oilseed rape** [2, 4]. Phoma leaf spot in **spring oilseed rape, winter oilseed rape** [1, 3]. Powdery mildew in **spring wheat, winter wheat** [1-4]. Powdery mildew in **spring barley, sugar beet, winter barley** [1, 3]. Rust in **sugar beet** [1, 3]. Septoria in **spring wheat, winter wheat** [1-4]. Yellow rust in **spring wheat, winter wheat** [1-4]. Yellow rust in **spring barley, winter barley** [1, 3].

Efficacy
- Apply at early stage of disease development or in routine preventive programme
- Most effective timing of treatment on cereals varies with disease. See label for details
- Higher rate active against both MBC-sensitive and MBC-resistant eyespot
- Rain occurring within 2-3 h of spraying may reduce effectiveness
- To prevent build-up of resistant strains of cereal mildew tank mix with approved morpholine fungicide [1, 3]
- Treat oilseed rape in autumn when leaf lesions first appear and spring from the start of stem extension when disease appears

Crop safety/Restrictions
- Maximum number of treatments (including applications of any product containing flusilazole) 3 for winter wheat, 2 for spring wheat, oilseed rape
- Do not apply to crops under stress or during frosty weather

Special precautions/Environmental safety
- Harmful if swallowed
- Irritating to eyes
- Dangerous to fish or other aquatic life. Do not contaminate surface waters or ditches with chemical or used container

Personal protective equipment/Label precautions
- A, C [1-4]; H, M [1, 2, 4]
- M03, R03c, R04a, U05a, U11, U19, U20b, C03 [1-4]; E26 [1, 3]; E01, E13b, E30a, E31b, E34 [1-4]

Latest application/Harvest interval
- Normally up to and including grain watery ripe stage (GS 71) for wheat, ear emergence complete stage (GS 59) for barley, before first flower opened stage for oilseed rape. However, timings vary with crop and dose - see labels for details
- HI 7 wk for sugar beet

Maximum Residue Level (mg residue/kg food)
- see carbendazim entry

74   carbendazim + iprodione

A systemic and contact fungicide mixture

Products

1 Calidan	RP Agric.	87.5:175 g/l	SC	06536
2 Calidan	Aventis	87.5:175 g/l	SC	09980
3 Vitesse	Aventis Environ.	87.5:175 g/l	SC	10042
4 Vitesse	RP Amenity	87.5:175 g/l	SC	06537

Uses

Alternaria in *oilseed rape* [1, 2]. Anthracnose in *turf* [3, 4]. Botrytis in *oilseed rape* [1, 2]. Fusarium patch in *turf* [3, 4]. Light leaf spot in *oilseed rape* [1, 2]. Pink patch in *turf* [3, 4]. Red thread in *turf* [3, 4]. Sclerotinia stem rot in *oilseed rape* [1, 2]. Timothy leaf spot in *turf* [3, 4].

Efficacy
- Best results obtained by application at first signs of disease, repeated monthly as necessary [4]
- Maximum efficacy against Anthracnose achieved by treatment at early disease development stage. Curative treatments for well-established Anthracnose are not recommended [4]
- May be used all year round, but is best suited for spring or late summer/early autumn application [4]
- Timing for oilseed rape varies according to disease. See label [1]
- Complete control of Botrytis and Alternaria in rape may require 2 treatments separated by not less than 3 wk [1]

Crop safety/Restrictions
- Maximum number of treatments 2 per crop for oilseed rape [1]. (NB Oilseed rape may be treated with 3 applications of products containing benomyl, carbendazim or thiophanate-methyl per season)
- Where grass is being mown, apply after mowing. Delay further mowing for at least 48 h after treatment [4]

Special precautions/Environmental safety
- Harmful to fish or other aquatic life. Do not contaminate surface waters or ditches with chemical or used container

Personal protective equipment/Label precautions
- A, C, H, M
- U20c, E13c, E26, E30a, E31b

Latest application/Harvest interval
- HI 3 wk for oilseed rape [1]

Maximum Residue Level (mg residue/kg food)
- see carbendazim and iprodione entries

75   carbendazim + maneb

A broad-spectrum systemic and protectant fungicide

Products

1 Headland Dual	Headland	62:400 g/l	SC	03782
2 MC Flowable	United Phosphorus	62:400 g/l	SC	08198
3 Multi-W FL	PBI	50:320 g/l	SC	09447

Uses

Brown rust in *spring wheat* [1, 2]. Brown rust in *winter wheat* [1-3]. Eyespot in *winter wheat* [1, 2]. Powdery mildew in *spring wheat* [1]. Powdery mildew in *winter wheat* [1, 3]. Septoria diseases in *spring wheat* [1]. Septoria diseases in *winter wheat* [1, 3]. Sooty moulds in *spring wheat* [1]. Sooty moulds in *winter wheat* [1-3]. Sooty moulds in *winter barley* [2]. Yellow rust in *winter barley* [1]. Yellow rust in *spring wheat* [1, 2]. Yellow rust in *winter wheat* [1-3].

Efficacy
- Apply to wheat from flag leaf just visible (GS 37) until first ears visible (GS 51), best when flag leaf ligules visible (GS 39) or, if not possible to spray earlier, from ear emergence complete (GS 59) until grain watery ripe (GS 71)
- Treatment gives useful control of late attacks of rust or mildew but for early attacks or established infection tank-mix with specific mildew or rust fungicide
- Do not spray if frost or rain expected

Crop safety/Restrictions
- Maximum number of treatments (including applications of any product containing benomyl, carbendazim or thiophanate-methyl) 2 per crop

Special precautions/Environmental safety
- Harmful if swallowed [1, 2]
- Irritating to eyes and skin [2, 3]
- Irritating to respiratory system [2]
- Harmful to fish or other aquatic life. Do not contaminate surface waters or ditches with chemical or used container

Personal protective equipment/Label precautions
- A, C [1-3]; H, M [1, 3]
- M03, R03c [1, 2]; R04c [2]; R04a, R04b [2, 3]; U04a [1, 2]; U02, U19, U20b [2]; U05a, U08 [2, 3]; U20a [3]; C02 [1] (7 d); C03 [2, 3]; E31b, E34 [1, 2]; E01, E13c, E26, E30a [2, 3]; E32a [3]

Latest application/Harvest interval
- Up to and including grain watery-ripe stage (GS 71)
- HI 7 d

Maximum Residue Level (mg residue/kg food)
- see carbendazim and maneb entries

---

76   carbendazim + metalaxyl

A protectant fungicide for use in fruit and cabbage storage

Products

Ridomil mbc 60 WP	Novartis	50:10% w/w	WP	08437

Uses

Crown rot in **water lily** *(off-label)*. Phytophthora in **stored cabbages**. Storage rots in **apples, pears**.

Efficacy
- Apply as post-harvest drench or dip to prevent spread of storage diseases
- Produce must be treated immediately after harvest and methods must achieve thorough coverage of each fruit/cabbage
- Treated produce should be allowed to drain thoroughly without any further rinsing or cleaning prior to storage
- Little curative activity on crop infected at or before harvest
- Best results achieved when crop stored in a controlled environment
- May be used with calcium chloride for bitter pit control in susceptible apple cultivars
- Other mixtures recommended for broader spectrum control - see label

Crop safety/Restrictions
- Maximum number of treatments (including applications of any product containing benomyl, carbendazim or thiophanate-methyl) 1 per batch
- No variety restrictions yet found necessary for use as a fruit or cabbage dip or drench

Special precautions/Environmental safety
- Irritating to eyes and skin
- Harmful to fish or other aquatic life. Do not contaminate surface waters or ditches with chemical or used container (remove fish before treating water-lily)

Personal protective equipment/Label precautions
- A, B, C, D, H, K, M
- R04a, R04b, U05a, U08, U20a, C03, E01, E13c, E29, E30a, E32a

Latest application/Harvest interval
- Treated fruit must not be processed or sold for at least 4 wk, cabbage for 7 wk

Maximum Residue Level (mg residue/kg food)
- see carbendazim and metalaxyl entries

---

77   carbendazim + prochloraz

A broad-spectrum systemic and contact fungicide for cereals and oilseed rape

Products

1 Novak	Aventis	100:267 g/l	SE	08020
2 Sportak Alpha HF	Aventis	100:267 g/l	SE	07225

Uses

Botrytis in *oilseed rape* [1, 2]. Canker in *oilseed rape* [1, 2]. Dark leaf spot in *oilseed rape* [1, 2]. Disease control/foliar feed in *protected ornamentals* *(off-label)*, *sugar beet seed crops* *(off-label)* [2]. Eyespot in *winter wheat* [1, 2]. Glume blotch in *winter wheat* [1, 2]. Leaf spot in *winter wheat* [1, 2]. Light leaf spot in *oilseed rape* [1, 2]. Powdery mildew in *winter wheat* [1, 2]. Sclerotinia stem rot in *oilseed rape* [1, 2]. White leaf spot in *oilseed rape* [1, 2].

Efficacy
- To protect against eyespot in high risk situations apply to winter barley from when leaf sheaths begin to become erect to first node detectable (GS 30-31). May also be used to control eyespot already in crop (up to 10% of tillers affected)
- Applied for eyespot control, product will also control Rhynchosporium and mildew and protect against new infections of net blotch, mildew and Septoria. Where MBC resistance is not a problem product can be used against eyespot and leaf spot but prochloraz alone is preferable where these diseases are resistant
- Tank-mix with tridemorph, fenpropimorph or fenpropidin to control established mildew
- May be applied to winter barley up to full ear emergence (GS 59), but if used earlier in season, prochloraz alone or plus a non-MBC fungicide is preferred treatment
- Apply in oilseed rape at first signs of disease and repeat if necessary. Timing varies with disease - see label for details.
- A period of at least 3 h without rain should follow spraying

Crop safety/Restrictions
- Maximum number of treatments (including applications of any product containing benomyl, carbendazim or thiophanate-methyl) 2 per crop for cereals; 2 at normal rate or 2 at split plus 1 at normal rate for oilseed rape

Special precautions/Environmental safety
- Harmful in contact with skin [1]
- Irritating to eyes [1, 2]
- Irritating to skin [1]
- May cause sensitization by skin contact [2]
- Flammable [1]
- Dangerous to fish or other aquatic life. Do not contaminate surface waters or ditches with chemical or used container

Personal protective equipment/Label precautions
- A, C [1, 2]; H, M [1]
- R04a, R04e, U02, U05a, U08, U20b, C03 [1, 2]; E26 [2]; E01, E13b, E30a, E31b [1, 2]

Latest application/Harvest interval
- Before flowering (GS 60) for wheat
- HI 6 wk for oilseed rape

Approval
- Off-label approval unstipulated for use on ornamental plant production under protection (OLA 2217/99)[2]; unstipulated for use on sugar beet seed crops (OLA 2218/99)[2]

Maximum Residue Level (mg residue/kg food)
- see carbendazim entry

---

78   carbendazim + propiconazole

A contact and systemic fungicide for winter cereals

Products

1	Hispor 45 WP	Novartis	20:25% w/w	WP	08418
2	Sparkle 45 WP	Novartis	20:25% w/w	WP	08450

Uses

Brown rust in *winter wheat*. Eyespot in *winter wheat*. Powdery mildew in *winter wheat*. Septoria in *winter wheat*. Yellow rust in *winter wheat*.

Efficacy

- Major benefit obtained from spring treatment. Control provided for about 30 d
- Apply in spring from stage when leaf sheath begins to lengthen until first node detectable (GS 30-31). Further sprays may be applied up to fully emerged ear (GS 59)

Crop safety/Restrictions

- Maximum number of treatments (including applications of any product containing benomyl, carbendazim or thiophanate-methyl) 2 per crop

Special precautions/Environmental safety

- Irritating to eyes and skin
- Dangerous to fish or other aquatic life. Do not contaminate surface waters or ditches with chemical or used container
- Harmful to bees. Do not apply to crops in flower or to those in which bees are actively foraging. Do not apply when flowering weeds are present

Personal protective equipment/Label precautions

- A, C
- R04a, R04b, U05a, U09a, U19, U20a, C02, C03, E01, E12e, E13b, E30a

Latest application/Harvest interval

- Up to and including grain watery-ripe stage (GS 71)
- HI 35 d

Maximum Residue Level (mg residue/kg food)

- see carbendazim and propiconazole entries

79   carbendazim + tebuconazole

A systemic fungicide mixture for oilseed rape

Products

1	Bayer UK 413	Bayer	133:167 g/l	SC	08277
2	Tricur	Bayer	133:167 g/l	SC	10281

Uses

Alternaria in *oilseed rape*. Canker in *oilseed rape*. Light leaf spot in *oilseed rape*. Phoma leaf spot in *oilseed rape*. Ring spot in *oilseed rape* (reduction). Sclerotinia stem rot in *oilseed rape*.

Efficacy

- Best results against light leaf spot and Phoma achieved from two spray programme starting in early autumn/winter
- Treat Alternaria in oilseed rape at onset of disease
- Treat Sclerotinia in oilseed rape at early flower to first petal fall

Crop safety/Restrictions

- Maximum number of treatments (including applications of any product containing benomyl, carbendazim or thiophanate-methyl) 3 per crop for oilseed rape

Special precautions/Environmental safety

- Harmful to fish or other aquatic life. Do not contaminate surface waters or ditches with chemical or used container

Personal protective equipment/Label precautions

- A, C, H, M
- U20b, C03, E01, E13c, E15, E30a, E31b, E34

Latest application/Harvest interval
- Most seeds green stage (GS 6,3) for oilseed rape

Maximum Residue Level (mg residue/kg food)
- see carbendazim entry

80   carbendazim + tecnazene

A protectant fungicide and sprout suppressant for stored potatoes

Products

1 Hortag Tecnacarb	Hortag	1.8:6% w/w	DS	02929
2 Tripart Arena Plus	Tripart	2:6% w/w	DP	05602

Uses

Dry rot in **seed potatoes**. Gangrene in **seed potatoes**. Silver scurf in **seed potatoes**. Skin spot in **seed potatoes**.

Efficacy
- Potatoes should be dormant, have a mature skin and be dry and free from dirt
- Treat tubers as they go into store using a dusting machine
- Ensure even cover of tubers. Cover clamps with straw etc to aid vapour-phase transmission of a.i. Pack boxes as tightly together as possible
- Effectiveness of treatment is reduced if ventilation in store is inadequate or excessive
- Treatment can give protection for 3-4 mth but does not cure blemishes already present on tubers. It will not control sprouting if tubers have already broken dormancy

Crop safety/Restrictions
- Maximum number of treatments (including applications of any product containing benomyl, carbendazim or thiophanate-methyl) 1 per batch
- Treated tubers must not be removed for sale or processing, including washing, for at least 6 wk after application
- Air seed potatoes for 6 wk and ensure that chitting has commenced before planting out. Treatment may delay emergence and possibly slightly reduce ware yield

Special precautions/Environmental safety
- Dangerous to fish or other aquatic life. Do not contaminate surface waters or ditches with chemical or used container [2]

Personal protective equipment/Label precautions
- A, C, D, H, M [2]
- U19, U20c [1, 2]; E15, E25 [1]; E32a [1, 2]; E13b, E30a [2]

Latest application/Harvest interval
- 6 wk before sale or planting

Maximum Residue Level (mg residue/kg food)
- see carbendazim and tecnazene entries

81   carbendazim + vinclozolin

A protectant and systemic fungicide for oilseed rape

Products

Konker	BASF	165:250 g/l	SC	03988

Uses

Alternaria in **oilseed rape** *(reduction)*. Botrytis in **oilseed rape** *(reduction)*. Light leaf spot in **oilseed rape** *(reduction)*. Sclerotinia stem rot in **oilseed rape**.

Efficacy
- Best results achieved if applied before any disease becomes well established
- Apply from start of rapid spring growth up to and including full flower
- Timing depends on diseases present and weather conditions. See label for details

Crop safety/Restrictions
- Maximum total dose equivalent to two full dose treatments
- Do not apply if rain or frost is expected or when the crop is wet

Special precautions/Environmental safety
- Irritant. May cause sensitisation by skin contact
- Harmful to fish or other aquatic life. Do not contaminate surface waters or ditches with chemical or used container
- Must be applied only by vehicle mounted or trailed hydraulic sprayers
- Vehicles must be fitted with a cab and a forced air filtration unit with a pesticide filter complying with HSE Guidance Note PM74, or equivalent

Personal protective equipment/Label precautions
- A, C, H, K, M
- R04, R04e, U05a, U08, U19, U20b, C03, E01, E13c, E30a, E31c

Latest application/Harvest interval
- HI 7 wk for oilseed rape

Maximum Residue Level (mg residue/kg food)
- see carbendazim and vinclozolin entries

82   carbetamide

A residual pre- and post-emergence carbamate herbicide available only in mixtures

83   carbetamide + diflufenican + oxadiazon

A residual herbicide mixture for amenity vegetation

Products

Helmsman	Aventis Environ.	1:0.1:2% w/w	GR	09934

Uses

Annual dicotyledons in **amenity trees and shrubs, amenity vegetation**. Annual meadow grass in **amenity trees and shrubs, amenity vegetation**. Perennial dicotyledons in **amenity trees and shrubs, amenity vegetation**.

Efficacy
- Apply to soil surface in late winter or early spring using hand shaker or gravity fed applicator. Uniform application is essential for satisfactory weed control
- Chemical absorbed by emerging shoots of susceptible weeds as they grow through the treated soil
- Adequate moisture is essential. In dry conditions irrigate after application. Effectiveness may be reduced if drought conditions follow application
- Do not disturb treated soil by hoeing otherwise weed control may be impaired.

Crop safety/Restrictions
- Maximum number of treatments 1 per yr
- Do not apply to plants grown in containers
- Do not apply to areas under-planted with bulbs or annual plants
- In some cases heavy rain shortly after application may splash granules onto the lower leaves of shrubs which may result in transient damage

Special precautions/Environmental safety
- Irritating to eyes
- Not to be used on food crops
- Extremely dangerous to fish or other aquatic life. Do not contaminate surface waters or ditches with chemical or used container
- Do not apply on slopes where heavy rain soon after application could cause surface run-off

Personal protective equipment/Label precautions
- A
- R04a, U05a, U19, U20b, C01, C03, E01, E13a, E30a, E32a

84	carbofuran

A systemic carbamate insecticide and nematicide for soil treatment

Products

Tripart Nex	Tripart	5% w/w	GR	05165

Uses

Aphids in *carrots, parsnips, swedes, turnips*. Beet leaf miner in *fodder beet, mangels, sugar beet*. Cabbage root fly in *broccoli, brussels sprouts, cabbages, calabrese, cauliflowers, chinese cabbage, collards, kale, swedes, turnips, winter oilseed rape*. Cabbage stem flea beetle in *winter oilseed rape*. Cabbage stem weevil in *broccoli, brussels sprouts, cabbages, calabrese, cauliflowers, chinese cabbage, collards, kale, swedes, turnips*. Carrot fly in *carrots, parsnips*. Docking disorder vectors in *fodder beet, mangels, sugar beet*. Flea beetles in *broccoli, brussels sprouts, cabbages, calabrese, cauliflowers, chinese cabbage, collards, fodder beet, kale, mangels, red beet, sugar beet, swedes, turnips*. Free-living nematodes in *carrots, fodder beet, mangels, parsnips, sugar beet*. Frit fly in *maize, sweetcorn*. Millipedes in *fodder beet, mangels, sugar beet*. Onion fly in *bulb onions*. Potato cyst nematode in *potatoes*. Pygmy beetle in *fodder beet, mangels, sugar beet*. Rape winter stem weevil in *winter oilseed rape*. Springtails in *fodder beet, mangels, sugar beet*. Stem nematodes in *bulb onions*. Tortrix moths in *sugar beet*. Turnip root fly in *swedes, turnips*. Vine weevil in *hardy ornamental nursery stock, pot plants, strawberries*. Wireworms in *fodder beet, mangels, sugar beet*.

Efficacy

- Apply through a suitably calibrated granule applicator and incorporate into the soil
- Method of application, rate and timing vary with pest and crop. See label for details
- Performance is reduced in dry soil conditions
- Do not apply to potatoes in very wet or water-logged soils
- Controls only first generation carrot fly, use a follow-up application of a suitable specific insecticide for later attacks
- May be applied in mixture with granular fertilizers

Crop safety/Restrictions

- Maximum number of treatments 1 per crop or yr
- When applied at drilling do not allow granules to come into contact with crop seed
- Use on carrots is limited to crops growing in mineral soils
- Use only on onions intended for harvest as mature bulbs
- Apply on ornamentals as surface treatment to moist soil after planting or potting up and follow immediately by thorough watering. Do not treat plants grown under cover, see label for details of species which have been treated safely
- On strawberries only apply after last harvest of year

Special precautions/Environmental safety

- This product contains an anticholinesterase carbamate compound. Do not use if under medical advice not to work with such compounds
- Harmful if swallowed
- Dangerous to game, wild birds and animals. Bury or remove spillages
- Keep in original container, tightly closed, in a safe place, under lock and key
- Dangerous to fish or other aquatic life. Do not contaminate surface waters or ditches with chemical or used container
- To minimise risk of enhanced biodegradation do not apply more than once every two yr to any cropping area

Personal protective equipment/Label precautions

- A, B, H, K, M; C (or D+E)
- M02, M03, R03c, U02, U04a, U05a, U09a, U13, U19, U20a, C03, E01, E06a, E10a, E13b, E30b, E32a, E34

FOR FULL CONDITIONS OF USE ALWAYS READ THE PRODUCT LABEL

Latest application/Harvest interval
- At drilling when broadcast or used as a band treatment; 6 wk before harvest when applied post-emergence; at least 4 wk before release for sale or supply for ornamentals and pot plants

Approval
- Approval expiry 31 Dec 2001

Maximum Residue Level (mg residue/kg food)
- hops 10; radishes 0.5; carrots, parsnips, garlic, onions, shallots 0.3; broccoli, cauliflowers, kohlrabi, tea 0.2; tree nuts (except hazelnuts), grapes, blackberries, dewberries, loganberries, raspberries, bilberries, cranberries, currants, gooseberries, wild berries, miscellaneous fruits, beetroot, horseradish, Jerusalem artichokes, parsley root, salsify, sweet potatoes, yams, spring onions, fruiting vegetables, leafy vegetables and herbs, peas (with and without pods), asparagus, cardoons, fennel, globe artichokes, rhubarb, fungi, lentils, peas, poppy seed, sesame seed, mustard seed, wheat, rye, barley, triticale, maize, animal products 0.1

85   carbon dioxide (commodity substance)

A gas for the control of trapped rodents and other vertebrates

Products
carbon dioxide	various	99.9%	GA

Uses
Birds in **traps**. Mice in **traps**. Rats in **traps**.

Efficacy
- Use to destroy trapped rodent pests
- Use to control birds covered by general licences issued by the Agriculture and Environment Departments under Section 16(1) of the Wildlife and Countryside Act (1981) for the control of opportunistic bird species, where birds have been trapped or stupefied with alphachloralose/seconal

Special precautions/Environmental safety
- Operators must wear self-contained breathing apparatus when carbon dioxide levels are greater than 0.5% v/v
- Operators must be suitably trained and competent
- Unprotected persons and non-target animals must be excluded from the treatment enclosures and surrounding areas unless the carbon dioxide levels are below 0.5% v/v

Personal protective equipment/Label precautions
- G

Approval
- Only to be used where a licence has been issued in accordance with Section 16(1) of the Wildlife and Countryside Act 1981

86   carbosulfan

A systemic carbamate insecticide for control of soil pests

Products
1	Landgold Carbosulfan 10G	Landgold	10% w/w	GR	09019
2	Marshal 10G	FMC	10% w/w	GR	09804
3	Marshal/suSCon	Fargro	10% w/w	CG	06978
4	Posse 10G	Bayer	10% w/w	GR	09805

Uses
Aphids in **carrots, parsnips, sugar beet** [1, 2, 4]. Aphids in **fodder beet, mangels** [2, 4]. Cabbage root fly in **broccoli, brussels sprouts, cabbages, calabrese, cauliflowers, swedes, turnips** [1, 2, 4]. Cabbage root fly in **collards** [2, 4]. Cabbage stem weevil in **broccoli, brussels sprouts, cabbages, calabrese, cauliflowers, swedes, turnips** [1, 2, 4]. Cabbage stem weevil in **collards** [2, 4]. Carrot fly in **carrots, parsnips** [1]. Flea beetles in **broccoli,**

brussels sprouts, cabbages, calabrese, cauliflowers, sugar beet, swedes, turnips [1, 2, 4]. Flea beetles in *collards, fodder beet, mangels* [2, 4]. Free-living nematodes in *carrots, parsnips, sugar beet* [1, 2, 4]. Free-living nematodes in *fodder beet, mangels* [2, 4]. Large pine weevil in *forest nurseries (off-label - growing medium treatment), forestry* [3]. Mangold fly in *sugar beet* [1, 2, 4]. Mangold fly in *fodder beet, mangels* [2, 4]. Millipedes in *sugar beet* [1, 2, 4]. Millipedes in *fodder beet, mangels* [2, 4]. Pygmy mangold beetle in *sugar beet* [1, 2, 4]. Pygmy mangold beetle in *fodder beet, mangels* [2, 4]. Springtails in *sugar beet* [1, 2, 4]. Springtails in *fodder beet, mangels* [2, 4]. Symphylids in *sugar beet* [1, 2, 4]. Symphylids in *fodder beet, mangels* [2, 4]. Wireworms in *sugar beet* [1, 2, 4]. Wireworms in *fodder beet, mangels* [2, 4].

Efficacy
- At recommended rates seed is not damaged by contact with product
- Apply with suitable granule applicator feeding directly into seed furrow or immediately behind drill coulter (behind seed drill boot for brassicas) or use bow-wave technique
- For transplanted brassicas apply sub-surface with `Leeds' coulter
- See label for details of suitable applicators and settings. Correct calibration is essential
- Where used in forestry, granules should be placed in the planting hole by hand or metered applicator before or after placing the tree [3]
- Where applied annually in intensive brassica growing areas enhanced biodegradation by soil organisms may lead to unsatisfactory control

Crop safety/Restrictions
- Maximum number of treatments 1 per crop or tree
- Do not mix with compost for blockmaking or use in Hassy trays for brassicas
- Safe to Douglas Fir and Sitka Spruce. Other conifers may vary in their sensitivity [3]
- Forest trees may take 10-15 d to achieve full protection. An additional pre-planting insecticide dip or spray is advised [3]

Special precautions/Environmental safety
- This product contains an anticholinesterase carbamate compound. Do not use if under medical advice not to work with such compounds
- Harmful if swallowed [1, 2]
- Dangerous to game, wild birds and animals. Bury or remove spillages
- Dangerous to fish or other aquatic life. Do not contaminate surface waters or ditches with chemical or used container
- Keep in original container, tightly closed, in a safe place, under lock and key

Personal protective equipment/Label precautions
- A [3]; D, E [1, 3]; B, K, M [1-4]; C [1-4] (or D+E); H [1, 2, 4]
- M02 [1-4]; M03, R03c [1, 2, 4]; U10 [3]; U02, U04a, U05a, U13, U20a [1-4]; U09a, U19 [1, 2, 4]; C03 [1-4]; E30a [3]; E01, E10a, E13b [1-4]; E30b, E32a, E34 [1, 2, 4]

Latest application/Harvest interval
- HI brassicas 56 d [1, 2]; sugar beet, fodder beet, mangels, swedes, turnips, carrots, parsnips 100 d

Approval
- Off-label approval unstipulated for treatment of growing medium on module-raised trees in forest nurseries (OLA 2383/99)[3]
- Approval expiry 20 Apr 2001 [1]

Maximum Residue Level (mg residue/kg food)
- carrots, parsnips, tea 0.1; tree nuts, berries and small fruit, miscellaneous fruits, beetroot, celeriac, horseradish, Jerusalem artichokes, parsley root, radishes, salsify, sweet potatoes, yams, garlic, shallots, spring onions, tomatoes, peppers, aubergines, cucumbers, gherkins, courgettes, sweet corn, leaf vegetables and herbs, legume vegetables, asparagus, cardoons, fennel, globe artichokes, rhubarb, mushrooms, pulses, oilseeds (except sunflower seed, cotton seed), potatoes, cereals, animal products 0.05

87   carboxin

An carboxamide fungicide available only in mixtures

88   carboxin + imazalil + thiabendazole

A fungicide seed dressing for barley and oats

Products

Vitaflo Extra	Uniroyal	315:23:29 g/l	FS	07048

Uses

Covered smut in **spring barley, winter barley**. Leaf spot in **spring oats, winter oats**. Leaf stripe in **spring barley, winter barley**. Loose smut in **spring barley** *(partial control)*, **spring oats, winter barley** *(partial control)*, **winter oats**. Net blotch in **spring barley** *(moderate control)*, **winter barley** *(moderate control)*. Seedling blight and foot rot in **spring barley** *(partial control)*, **winter barley** *(partial control)*.

Efficacy

- Apply through suitable liquid flowable seed treating equipment of the batch treatment or continuous flow type where a secondary mixing auger is fitted
- Preferably treat seed several days before sowing and store treated seed in a cool, dry, well ventilated place. Disease control may be reduced following prolonged storage
- Stocks of treated seed carried over to following season should be tested for germination
- Product should not be used to control loose smut on barley to be grown for seed multiplication or retrieval purposes
- Where resistant strains of loose smut occur on winter barley, product will not be effective

Crop safety/Restrictions

- Maximum number of treatments 1 per batch of seed
- Do not treat seed with moisture content above 16%
- Do not apply to seed already treated with a fungicide
- Do not apply to cracked, split or sprouted seed
- Do not store treated seed for more than 3 mth
- Product available in large (1000 l) returnable containers. They should be re-circulated for 1 h before use and operated as directed on the label

Special precautions/Environmental safety

- Harmful to fish or other aquatic life. Do not contaminate surface waters or ditches with chemical or used container
- Do not handle seed unnecessarily
- Do not use treated seed as food or feed
- Harmful to game and wild life. Bury spillages
- Do not reuse sacks or containers that have been used for treated seed for food or feed
- Do not apply treated seed from the air

Personal protective equipment/Label precautions

- A, C, D, H, M
- U05a, U07, U09a, U20b, C03, E01, E13c, E26, E30a, E33, E34, E36, S01, S02, S04b, S05, S06, S07

Latest application/Harvest interval

- Before drilling

Approval

- Imazalil included in Annex I under EC Directive 91/414

Maximum Residue Level (mg residue/kg food)

- see imazalil and thiabendazole entries

89   carboxin + thiram

A fungicide seed dressing for cereals

Products

Anchor	Uniroyal	200:200 g/l	FS	08684

Uses

Bunt in **spring wheat** *(seed treatment)*, **winter wheat** *(seed treatment)*. Fusarium foot rot and seedling blight in **spring barley** *(seed treatment)*, **spring oats** *(seed treatment)*, **spring rye** *(seed treatment)*, **spring wheat** *(seed treatment)*, **triticale** *(seed treatment)*, **winter barley** *(seed treatment)*, **winter oats** *(seed treatment)*, **winter rye** *(seed treatment)*, **winter wheat** *(seed treatment)*. Leaf stripe in **spring barley** *(seed treatment - reduction)*, **winter barley** *(seed treatment - reduction)*. Loose smut in **spring barley** *(seed treatment - reduction)*, **winter barley** *(seed treatment - reduction)*. Septoria seedling blight in **spring wheat** *(seed treatment)*, **winter wheat** *(seed treatment)*.

Efficacy

- Apply through suitable liquid flowable seed treating equipment of the batch treatment or continuous flow type where a secondary mixing auger is fitted
- Do not store treated seed from one season to the next
- Drill flow may be affected by treatment. Always re-calibrate seed drill before use

Crop safety/Restrictions

- Maximum number of treatments 1 per batch of seed
- Do not treat seed with moisture content above 16%
- Do not apply to cracked, split or sprouted seed

Special precautions/Environmental safety

- Harmful to fish or other aquatic life. Do not contaminate surface waters or ditches with chemical or used container
- Do not use treated seed as food or feed
- Treated seed harmful to game and wildlife

Personal protective equipment/Label precautions

- A, D, H
- U09a, U20c, E13c, E26, E27, E30a, E31a, E34, S01, S02, S04b, S05, S06, S07

Approval

- Accepted by BLRA for use as a seed treatment on malting barley

---

90   carfentrazone-ethyl

A triazolinone foliar applied herbicide for cereals

Products

Aurora 50 WG	FMC	50% w/w	WG	09807

Uses

Cleavers in **durum wheat, spring barley, spring wheat, triticale, winter barley, winter wheat**. Ivy-leaved speedwell in **durum wheat, spring barley, spring wheat, triticale, winter barley, winter wheat**.

Efficacy

- Best results achieved from good spray cover applied to small actively growing weeds
- Recommended for use as two spray programme with one application in autumn and one in spring

Crop safety/Restrictions

- Maximum number of treatments 2 per crop
- Treat from 2 leaf stage of crop
- Do not treat crops under stress from drought, waterlogging, cold, pests, diseases, nutrient or lime deficiency or any factors reducing plant growth
- Do not treat cereals undersown with clover or other legumes
- In the event of crop failure cereals, ryegrass, maize, oilseed rape, peas, sunflowers, Phacelia, vetches, carrots or onions may be planted within 1 mth of treatment. Any crop may be sown 3 mth after treatment
- Follow label instructions for sprayer cleaning

Special precautions/Environmental safety

- Irritant

FOR FULL CONDITIONS OF USE ALWAYS READ THE PRODUCT LABEL

- Extremely dangerous to fish or other aquatic life. Do not contaminate surface waters or ditches with chemical or used container
- Avoid drift on to broad leaved plants outside the target area

Personal protective equipment/Label precautions
- A, H
- R04, U05a, U08, U13, U14, C03, E01, E13a, E30a, E32a, E34

Latest application/Harvest interval
- Before 3rd node detectable (GS 33)

91   carfentrazone-ethyl + flupyrsulfuron-methyl

A foliar and residual acting herbicide for cereals

Products

Lexus Class WSB	DuPont	33.3:16.7% w/w	WB	08637

Uses

Annual dicotyledons in **triticale, winter oats, winter rye, winter wheat**. Blackgrass in **triticale, winter oats, winter rye, winter wheat**.

Efficacy
- Best results obtained when applied to small actively growing weeds
- Good spray cover of weeds must be obtained
- Increased degradation of active ingredient in high soil temperatures reduces residual activity
- Weed control may be reduced in dry soil conditions but susceptible weeds germinating soon after treatment will be controlled if adequate soil moisture present
- Blackgrass should be treated from 1 leaf but before first node. Strains of blackgrass resistant to other herbicides may not be controlled

Crop safety/Restrictions
- Maximum number of treatments 1 per crop
- Apply to recommended crops from 2 leaf stage (GS 12) up to 31 Dec on oats, rye, triticale, or before first node detectable (GS 31) on wheat
- Do not use on barley, on crops undersown with grasses or legumes, or any other broad-leaved crop
- Do not apply within 7 d of rolling
- Do not treat any crop suffering from drought, waterlogging, pest or disease attack, nutrient deficiency, or any other stress factors
- Only cereals, oilseed rape, field beans, clover or grass may be sown in the yr of harvest of a treated crop
- In the event of crop failure only winter wheat may be sown 1-3 mth after treatment. Land should be ploughed and cultivated to 15 cm minimum before resowing
- Slight chlorosis and stunting may occur in certain conditions. Recovery is rapid and yield not affected

Special precautions/Environmental safety
- Irritant. May cause sensitization by skin contact
- Dangerous to fish or other aquatic life. Do not contaminate surface waters or ditches with chemical or used container
- Take extreme care to avoid damage by drift outside the target area or onto surface waters or ditches
- Spraying equipment should not be drained or flushed onto land planted, or to be planted, with trees or crops other than cereals and should be thoroughly cleansed after use - see label for instructions

Personal protective equipment/Label precautions
- A
- R04, R04e, U05a, U08, U14, U19, U20b, U22, C03, E01, E13b, E15, E30a

Latest application/Harvest interval
- Before 31 Dec in yr of sowing for winter oats, winter rye, triticale; before 1st node detectable (GS 31) for winter wheat

92   carfentrazone-ethyl + isoproturon

A foliar and residual acting herbicide for cereals

Products

Affinity	FMC	0.75:50% w/w	WG	09745

Uses

Annual dicotyledons in **winter barley, winter wheat**. Annual meadow grass in **winter barley, winter wheat**. Chickweed in **winter barley, winter wheat**. Cleavers in **winter barley, winter wheat**. Loose silky bent in **winter barley, winter wheat**. Mayweeds in **winter barley, winter wheat**. Rough meadow grass in **winter barley, winter wheat**.

Efficacy
- Good weed control depends on burying and dispersing any straw to 15 cm before or during seedbed preparation
- Best results obtained by treatment post weed emergence in crops growing in firm seedbeds with clods no greater than fist size
- Residual activity reduced on soils with more than 10% organic matter. Treat crops on such soils in spring
- Weed control may be reduced if heavy rain falls within a few hours of treatment or if prolonged dry weather follows spring application
- Do not harrow at any time after treatment

Crop safety/Restrictions
- Maximum number of treatments 1 per crop
- Treat from 2 leaf stage of crop. Do not apply pre-emergence to wheat or barley
- Crops should be drilled to 25 mm and the seeds well covered with soil. Ensure good root system is established on broadcast crops
- Crops drilled in Sep may be damaged if a period of rapid growth follows treatment
- Do not treat undersown crops or those to be undersown
- Transient necrotic spotting may occur after treatment particularly on crops subject to wide diurnal temperature fluctuations. Symptoms disappear within 3-4 wk and yield is not affected
- Do not roll in spring 1 wk before or after application
- In the event of crop failure only wheat, barley, maize, durum wheat, oats or flax may be sown 1 mth after treatment in Dec/Jan. Only maize, wheat or barley may be sown 1 mth after treatment in Feb/Mar. Any crop may be sown 3 mth after treatment

Special precautions/Environmental safety
- Irritant. May cause sensitization by skin contact
- Dangerous to fish or other aquatic life. Do not contaminate surface waters or ditches with chemical or used container
- Do not apply where soils are cracked, to avoid run-off through drains
- Take extreme care to avoid drift onto broad-leaved plants outside the target area or onto ponds waterways or ditches, or onto land intended for cropping

Personal protective equipment/Label precautions
- A, C, D, H
- R04, R04e, U05a, U08, U13, U14, C03, E01, E13b, E26, E30a, E31b, E34

Latest application/Harvest interval
- Before pseudostem erect stage (GS 30)

Approval
- Accepted by BLRA for use on malting barley

93   carfentrazone-ethyl + mecoprop-P

A foliar applied herbicide for cereals

Products

Platform S	FMC	1.5:60% w/w	WG	09706

Uses

Charlock in **spring barley, spring oats, spring wheat, winter barley, winter oats, winter wheat**. Chickweed in **spring barley, spring oats, spring wheat, winter barley, winter oats, winter wheat**. Cleavers in **spring barley, spring oats, spring wheat, winter barley, winter oats, winter wheat**. Red dead-nettle in **spring barley, spring oats, spring wheat, winter barley, winter oats, winter wheat**. Speedwells in **spring barley, spring oats, spring wheat, winter barley, winter oats, winter wheat**.

Efficacy

- Best results obtained when weeds have germinated and growing vigorously in warm moist conditions
- Treatment of large weeds and poor spray coverage may result in reduced weed control

Crop safety/Restrictions

- Maximum number of treatments 1 per crop. The total amount of mecoprop-P applied in a single yr must not exceed the maximum total dose approved for any single product for the crop/situation
- Can be used on all varieties of wheat and barley in autumn or spring from the beginning of tillering
- Do not treat crops suffering from stress from any cause
- Early sown crops may be prone to damage if treated after period of rapid growth in autumn
- Do not treat crops undersown or to be undersown
- In the event of crop failure, any cereal, maize, oilseed rape, peas, vetches or sunflowers may be sown 1 mth after a spring treatment. Any crop may be planted 3 mth after treatment

Special precautions/Environmental safety

- Irritant. Risk of serious damage to eyes
- May cause sensitization by skin contact
- Dangerous to fish or other aquatic life. Do not contaminate surface waters or ditches with chemical or used container

Personal protective equipment/Label precautions

- A, C, H, M
- R04, R04d, R04e, U05a, U08, U11, U13, U14, U20b, C03, E01, E13b, E30a, E31b; E07 (2 wk)

Latest application/Harvest interval

- Before 3rd node detectable (GS 33)

Approval

- Accepted by BLRA for use on malting barley

---

94   carfentrazone-ethyl + metsulfuron-methyl

A foliar applied herbicide mixture for cereals

Products

1	Ally Express	DuPont	40:10% w/w	WG	08640
2	Standon Carfentrazone MM	Standon	40:10% w/w	WG	09159

Uses

Annual dicotyledons in **durum wheat, triticale** [1]. Annual dicotyledons in **spring barley, spring oats, spring wheat, winter barley, winter oats, winter wheat** [1, 2].

Efficacy

- Best results achieved from applications made in good growing conditions
- Good spray cover of weeds must be obtained
- Growth of weeds is inhibited within hours of treatment but the time taken for visible colour changes to appear will vary according to species and weather
- Product has short residual life in soil. Under normal moisture conditions susceptible weeds germinating soon after treatment will be controlled
- Can be used on all soil types
- Apply in the spring from the 3-leaf stage on all crops

Crop safety/Restrictions
- Maximum number of treatments 1 per crop
- Slight necrotic spotting of crops can occur under certain crop and soil conditions. Recovery is rapid and there is no effect on grain yield or quality
- Do not use on any crop suffering stress from drought, waterlogging, cold, pest or disease attack, nutrient deficiency
- Do not use on crops undersown with grass or legumes or on any broad-leaved crop
- Do not apply within 7 d of rolling
- Do not apply to a crop already treated with any sulfonylurea product except those containing flupyrsulfuron alone or in mixture with carfentrazone-ethyl or thifensulfuron-methyl
- Only cereals, oilseed rape, field beans or grass may be sown as a following crop in the same calendar yr. Any crop may follow in the next spring. In the event of failure of a treated crop, only winter wheat may be sown 1-3 mth later after ploughing and cultivating to at least 15 cm

Special precautions/Environmental safety
- Irritant. May cause sensitization by skin contact
- Extremely dangerous to fish or other aquatic life. Do not contaminate surface waters or ditches with chemical or used container
- LERAP Category B
- Take extreme care to avoid drift onto broad-leaved plants outside the target area or onto ponds waterways or ditches, or onto land intended for cropping
- Spraying equipment should not be drained or flushed onto land planted, or to be planted, with trees or crops other than cereals and should be thoroughly cleansed after use - see label for instructions

Personal protective equipment/Label precautions
- A, H
- R04, R04e, U05a, U08, U14, U19, U20b, C03, E01, E13a, E16a, E16b, E30a, E32a

Latest application/Harvest interval
- Before 3rd node detectable (GS 33)

Approval
- Metsulfuron-methyl included in Annex I under EC Directive 91/414
- Accepted by BLRA for use on malting barley

---

95   carfentrazone-ethyl + thifensulfuron-methyl

A foliar applied herbicide mixture for cereals

Products

Harmony Express	DuPont	25:25% w/w	WG	09467

Uses

Annual dicotyledons in *spring barley, spring oats, spring wheat, winter barley, winter oats, winter wheat*. Cleavers in *spring barley, spring oats, spring wheat, winter barley, winter oats, winter wheat*. Knotgrass in *spring barley, spring oats, spring wheat, winter barley, winter oats, winter wheat*. Mayweeds in *spring barley, spring oats, spring wheat, winter barley, winter oats, winter wheat*. Speedwells in *spring barley, spring oats, spring wheat, winter barley, winter oats, winter wheat*.

Efficacy
- Best results achieved from applications made in good growing conditions
- Good spray cover of weeds must be obtained
- Growth of weeds is inhibited within hours of treatment but the time taken for visible colour changes to appear will vary according to species and weather
- Can be used on all soil types

Crop safety/Restrictions
- Maximum number of treatments 1 per crop
- Do not treat winter barley before 1 Feb in yr of harvest. Treat other crops in the autumn or spring from the 3-leaf stage

- Slight necrotic spotting of crops can occur under certain crop and soil conditions. Recovery is rapid and there is no effect on grain yield or quality
- Do not use on any crop suffering stress from drought, waterlogging, cold, pest or disease attack, nutrient deficiency
- Do not use on crops undersown with grass or legumes or on any broad-leaved crop
- Do not apply within 7 d of rolling
- Do not apply in tank mix with any sulfonylurea or in sequence to a crop already treated with any sulfonylurea product except those containing flupyrsulfuron or metsulfuron-methyl alone or in mixture with carfentrazone-ethyl
- Only cereals, oilseed rape, field beans or grass may be sown as a following crop in the harvest yr. Any crop may follow in the next spring. In the event of failure of a treated crop, only winter wheat may be sown before the normal harvest date, after ploughing and cultivating to at least 15 cm

Special precautions/Environmental safety
- Extremely dangerous to fish or other aquatic life. Do not contaminate surface waters or ditches with chemical or used container
- LERAP Category B

Personal protective equipment/Label precautions
- U05a, U20b, C03, E01, E13a, E16a, E16b, E30a

Latest application/Harvest interval
- Before 3rd node detectable (GS 33)

96   chlordane

A persistent organochlorine earthworm killer, all approvals for which were revoked in 1992. Chlordane is an environmental hazard and is banned under the EC Directive 70/117/EEC

97   chlorfenvinphos

A contact and ingested soil-applied organophosphorus insecticide

Products

Birlane 24	Cyanamid	240 g/l	EC	07002

Uses

Cabbage root fly in **broccoli, brussels sprouts, cabbages, cauliflowers, kohlrabi** (off-label), **radishes** (off-label), **swedes** (off-label), **turnips** (off-label). Carrot fly in **carrots, celeriac** (off-label), **parsnips**. Colorado beetle in **potatoes** (off-label).

Efficacy
- Soil incorporation recommended for most treatments. Application method, timing, rate and frequency vary with pest, crop and soil type. See labels for details
- Overall pre-planting sprays are less effective than band treatment and should not be used in areas of heavy cabbage root fly infestation
- Cabbage and cauliflower grown in peat blocks can be protected from cabbage root fly by incorporation into the peat used for blocking
- Efficacy reduced on highly organic or very dry soils unless crop irrigated

Crop safety/Restrictions
- Maximum number of treatments 1 per crop for most crops, but varies with dose rate and product. See label for details
- On carrots the maximum total dose applied per crop must not exceed the equivalent of 3 (on mineral soils) or 4 (on organic soils) full dose applications
- Do not apply in conjunction with seed treatments containing gamma-HCH

Special precautions/Environmental safety
- Chlorfenvinphos is subject to the Poisons Rules 1982 and the Poisons Act 1972. See notes in Section 1
- This product contains an anticholinesterase organophosphorus compound. Do not use if under medical advice not to work with such compounds
- Keep in original container, tightly closed, in a safe place, under lock and key

- Toxic in contact with skin and if swallowed
- Irritating to eyes and skin
- Flammable
- Dangerous to bees. Do not apply to crops in flower or to those in which bees are actively foraging except as directed. Do not apply when flowering weeds are present
- Extremely dangerous to fish or other aquatic life. Do not contaminate surface waters or ditches with chemical or used container
- LERAP category A
- Must not be applied by hand-held sprayers

Personal protective equipment/Label precautions

- A, C, H
- M01, M04, R02a, R02c, R04a, R04b, R07d, U02, U04a, U05a, U10, U13, U19, U20b, C03; C02 (3 wk); E01, E12d, E13a, E16c, E16d, E30b, E31b, E34

Latest application/Harvest interval

- HI brassicas, kohlrabi, carrots, celeriac, potatoes, parsnips 3 wk

Approval

- Off-label approval unlimited for use on kohlrabi, celeriac (OLA 1747/96); unstipulated for use on radishes (OLA 2057/99); to Dec 2001 for use on swedes, turnips (OLA 0866/00)
- Approval expiry 31 Dec 2001

Maximum Residue Level (mg residue/kg food)

- citrus fruits 1; carrots, horseradish, parsnips, parsley root, salsify, swedes, turnips, garlic, onions, shallots, celery, rhubarb, potatoes 0.5; meat 0.2; tomatoes, peppers, aubergines, cucumbers, gherkins, courgettes, cauliflowers, Brussels sprouts, head cabbage, lettuce, beans, peas, leeks 0.1; pome fruits, apricots, peaches, nectarines, plums, grapes, strawberries, cane fruits, bilberries, cranberries, currants, gooseberries, bananas, mushrooms 0.05; milk 0.008

---

98   chloridazon

A residual pyridazinone herbicide for beet crops

Products

1	Better DF	Sipcam	65% w/w	SG	06250
2	Better Flowable	Sipcam	430 g/l	SC	04924
3	Burex 430 SC	Agricola	430 g/l	SC	09494
4	Gladiator DF	Tripart	65% w/w	SG	06342
5	Portman Weedmaster	Portman	430 g/l	SC	06018
6	Pyramin DF	BASF	65% w/w	SG	03438
7	Sculptor	Sipcam	430 g/l	SC	08836
8	Starter Flowable	Truchem	430 g/l	SC	03421
9	Takron	BASF	430 g/l	SC	06237
10	Tripart Gladiator	Tripart	430 g/l	SC	00986

Uses

Annual dicotyledons in *fodder beet, mangels, sugar beet* [1-10]. Annual dicotyledons in *bulb onions* (off-label), *leeks* (off-label), *salad onions* (off-label) [6]. Annual meadow grass in *fodder beet, mangels, sugar beet* [1, 2, 4-10]. Annual meadow grass in *bulb onions* (off-label), *leeks* (off-label), *salad onions* (off-label) [6].

Efficacy

- Absorbed by roots of germinating weeds and best results achieved pre-emergence of weeds or crop when soil moist and adequate rain falls after application
- Where used pre-emergence spray as soon as possible after drilling in mid-Mar to mid-Apr on fine, firm, clod-free seedbed
- Where crop drilled after mid-Apr or soil dry apply pre-drilling and incorporate to 2.5 cm immediately afterwards
- Application rate depends on soil type. See label for details
- Various tank mixes recommended on sugar beet for pre- and post-emergence use and as repeated low dose treatments. See label for details

Crop safety/Restrictions
- Maximum number of treatments 1 per crop for fodder beet and mangels; 1 per crop [4, 5], 1 pre-emergence + 1 post-emergence [1], 1 pre-emergence + 3 post-emergence for sugar beet [1, 6, 9], for sugar beet, fodder beet, mangels [3, 7]
- Maximum dose 1.4 kg product/ha for onions and leeks [6]
- Do not use on Coarse Sands, Sands or Fine Sands or where organic matter exceeds 5%
- Crop vigour may be reduced by treatment of crops growing under unfavourable conditions including poor tilth, drilling at incorrect depth, soil capping, physical damage, pest or disease damage, excess seed dressing, trace-element deficiency or a sudden rise in temperature after a cold spell
- In the event of crop failure only sugar or fodder beet, mangels or maize may be re-drilled on treated land after cultivation
- Winter cereals may be sown in autumn after ploughing. Any spring crop may follow treated beet crops harvested normally

Special precautions/Environmental safety
- Irritating to eyes and skin [5, 7, 8, 10]
- Irritating to respiratory system [5, 10]
- May cause sensitization by skin contact [3, 9]
- Harmful to fish or other aquatic life. Do not contaminate surface waters or ditches with chemical or used container

Personal protective equipment/Label precautions
- A [3, 5, 7-10]; C [5]
- M03 [9]; R04a, R04b [5, 7, 8, 10]; R04c [5, 10]; R04 [9]; R04e [3, 9]; U08 [1-10]; U05a [3, 5, 7-10]; U20a [2-5, 10]; U20b [1, 6-9]; U19 [5, 7-9]; U14, U15 [7, 8]; C03 [3, 5, 7-10]; E13c, E30a [1-10]; E01, E31b [3, 5, 7-10]; E26 [3, 5, 7-9]; E32a [1, 2, 4, 6]; E34 [5]

Latest application/Harvest interval
- Pre-emergence for fodder beet and mangels; pre-emergence [4], 8-true leaf stage [1, 2, 4], before leaves of crop meet in row [6-9] for sugar beet; up to and including second true leaf stage for onions and leeks [6]

Approval
- Off-label approval unlimited for use on salad and bulb onions and leeks (OLA 0732/97)[6]

---

99   chloridazon + ethofumesate

A contact and residual herbicide for beet crops

Products

1 Gremlin	Sipcam	285:176 g/l	SC	09468
2 Magnum	BASF	275:170 g/l	SC	08635
3 Spectron	Aventis	211:200 g/l	SC	07284
4 Stefes Spectron	Stefes	211:200 g/l	SC	09337

Uses

Annual dicotyledons in **sugar beet** [1, 2, 4]. Annual dicotyledons in **fodder beet** [2, 4]. Annual dicotyledons in **mangels** [4]. Annual meadow grass in **sugar beet** [1, 2, 4]. Annual meadow grass in **fodder beet** [2, 4]. Annual meadow grass in **mangels** [4]. Blackgrass in **fodder beet, mangels, sugar beet** [4].

Efficacy
- Apply at or immediately after drilling but pre-emergence of crop or weeds [3, 4]; apply pre- or post-crop emergence up to cotyledon stage of weeds [2]
- Mixtures with triallate must be applied pre-drilling and incorporated
- Best results achieved on a fine firm clod-free seedbed when soil moist and adequate rain falls after spraying. Efficacy may be reduced if heavy rain falls just after incorporation
- Effectiveness may be reduced under conditions of low pH
- May be used on soil classes Loamy Sand - Silty Clay Loam [2]; may be used on all soils except Sands and those with more than 5% organic matter [3, 4]. Additional restrictions apply for some tank mixtures
- May be applied by conventional or repeat low dose method. See label for details
- Recommended for tank mixing with triallate on sugar beet incorporated pre-drilling and with various other beet herbicides early post-emergence. See labels for details

Crop safety/Restrictions
- Maximum number of treatments 1 pre-emergence for fodder beet, 1 pre-emergence plus 3 post-emergence for sugar beet [2]; 1 pre-emergence or 3 post-emergence on sugar beet [1]
- Crop vigour may be reduced by treatment of crops growing under unfavourable conditions including poor tilth, drilling at incorrect depth, soil capping, physical damage, excess nitrogen, excess seed dressing, trace element deficiency or a sudden rise in temperature after a cold spell. Frost after pre-emergence treatment may check crop growth
- In the event of crop failure only sugar beet, fodder beet or mangels may be re-drilled
- Any crop may be sown 3 mth after spraying following ploughing to 15 cm. Where a tank mixture with triallate has been used, grass and oat crops should not be sown until the following spring

Special precautions/Environmental safety
- Harmful if swallowed [1]
- Harmful to fish or other aquatic life. Do not contaminate surface waters or ditches with chemical or used container

Personal protective equipment/Label precautions
- A, H [1, 2]; C [1]
- R04, R04e [2]; R03c [1]; U08, U19, U20a [1-4]; U05a [2]; C03 [1, 2]; E26 [3]; E13c, E30a [1-4]; E31b [1, 3, 4]; E31c [2]; E01 [1, 2]; E34 [1]

Latest application/Harvest interval
- Pre-crop emergence [3, 4]; pre-emergence for fodder beet and before crop leaves meet between rows for sugar beet [1, 2]

---

100   chloridazon + propachlor

A residual pre-emergence herbicide for use in onions and leeks

Products

Ashlade CP	Nufarm Whyte	86:400 g/l	SC	06481

Uses

Annual dicotyledons in **bulb onions, bulb onions** *(off-label - post-emergence)*, **leeks, leeks** *(off-label - post-emergence)*, **salad onions, salad onions** *(off-label - post-emergence)*. Annual grasses in **bulb onions, bulb onions** *(off-label - post-emergence)*, **leeks, leeks** *(off-label - post-emergence)*, **salad onions, salad onions** *(off-label - post-emergence)*.

Efficacy
- Best results achieved from application to firm, moist, weed-free seedbed when adequate rain falls afterwards
- Apply pre-emergence of sown crops, preferably soon after drilling, before weeds emerge. Loose or fluffy seedbeds must be consolidated before application
- Apply to transplanted crops when soil has settled after planting
- Do not use on soils with more than 10% organic matter

Crop safety/Restrictions
- Maximum number of treatments 1 per crop
- Ensure crops are drilled to 20 mm depth
- Crops stressed by nutrient deficiency, pests or diseases, poor growing conditions or pesticide damage may be checked by treatment, especially on sandy or gravelly soils
- In the event of crop failure only onions, leeks or maize should be planted
- Any crop can follow a treated onion or leek crop harvested normally as long as the ground is cultivated thoroughly before drilling

Special precautions/Environmental safety
- Irritating to eyes and skin
- Harmful to fish or other aquatic life. Do not contaminate surface waters or ditches with chemical or used container

Personal protective equipment/Label precautions
- A, C
- M03, R04a, R04b, U02, U04a, U05a, U08, U13, U19, U20a, C03, E01, E13c, E30a, E31b

Latest application/Harvest interval
- Pre-emergence of crop; before 2 true leaf stage for onions and leeks (off-label).
- HI 12 wk for chives

FOR FULL CONDITIONS OF USE ALWAYS READ THE PRODUCT LABEL

Approval
- Off-label Approval unlimited for use post-emergence on bulb onions, leeks (OLA 1362/98)

101	chlormequat

A plant-growth regulator for reducing stem growth and lodging

Products

1	Adjust	Mandops	620 g/l	SL	05589
2	AgriGuard Chlormequat 700	AgriGuard	700 g/l	SL	09282
3	Atlas 3C:645 Chlormequat	Nufarm Whyte	645 g/l	SL	07700
4	Atlas Chlormequat 46	Nufarm Whyte	460 g/l	SL	07704
5	Atlas Chlormequat 700	Nufarm Whyte	700 g/l	SL	07708
6	Atlas Terbine	Nufarm Whyte	730 g/l	SL	07709
7	Atlas Tricol	Nufarm Whyte	670 g/l	SL	07707
8	Barclay Holdup	Barclay	700 g/l	SL	06799
9	Barclay Holdup 600	Barclay	600 g/l	SL	08794
10	Barclay Holdup 640	Barclay	640 g/l	SL	08795
11	Barleyquat B	Mandops	620 g/l	SL	07051
12	BASF 3C Chlormequat 720	BASF	720 g/l	SL	06514
13	Bettaquat B	Mandops	620 g/l	SL	07050
14	Fargro Chlormequat	Fargro	460 g/l	SL	02600
15	Greencrop Carna	Greencrop	600 g/l	SL	09403
16	Greencrop Coolfin	Greencrop	700 g/l	SL	09449
17	Landgold CCC 720	Landgold	720 g/l	SL	08527
18	Mandops Chlormequat 700	Mandops	700 g/l	SL	06002
19	Manipulator	Mandops	620 g/l	SL	05871
20	Portman Chlormequat 700	Portman	700 g/l	SL	03465
21	Quadrangle Chlormequat 700	Quadrangle	700 g/l	SL	03401
22	Renown	Stefes	640 g/l	SL	09058
23	Sigma PCT	Nufarm Whyte	460 g/l	SL	08663
24	Stabilan 700	Nufarm Whyte	700 g/l	SL	08006
25	Stefes CCC 640	Stefes	640 g/l	SL	06993
26	Stefes CCC 700	Stefes	700 g/l	SL	07116
27	Stefes CCC 720	Stefes	720 g/l	SL	05834
28	Tripart Brevis	Tripart	700 g/l	SL	03754
29	Uplift	United Phosphorus	700 g/l	SL	07527
30	Whyte Chlormequat 700	Nufarm Whyte	700 g/l	SL	09641

Uses

Increasing yield in **spring barley** [1, 11, 18]. Increasing yield in **winter barley** [1, 2, 8-11, 15, 16, 18-20, 22, 24, 28]. Lodging control in **spring barley** [1, 11]. Lodging control in **winter barley** [1-8, 10-12, 16, 17, 19, 20, 22-24, 27-30]. Lodging control in **spring wheat, winter wheat** [1-10, 12, 13, 15-28, 30]. Lodging control in **rye, triticale** [12]. Lodging control in **winter rye** [2, 16, 28]. Lodging control in **spring oats, winter oats** [2-10, 12, 15-18, 20-22, 24-28, 30]. Lodging control in **oats, wheat** [29]. Stem shortening in **pot plants** [14]. Stem shortening in **pelargoniums, poinsettias** [4, 5, 12, 14, 30]. Stem shortening in **bedding plants, camellias, hibiscus trionum, lilies, ornamentals** [4, 5, 14, 30].

Efficacy
- Most effective results on cereals normally achieved from application from Apr onwards, on wheat and rye from leaf sheath erect to first node detectable (GS 30-31), on oats at second node detectable (GS 32), on winter barley from mid-tillering to leaf sheath erect (GS 25-30). However, recommendations vary with product. See label for details
- Results on barley can be variable
- In tank mixes with other pesticides optimum timing for herbicide action may differ from that for growth reduction. See label for details of tank mix recommendations
- Addition of approved non-ionic wetter recommended for products approved on oats

- At least 6 h, preferably 24 h, required before rain for maximum effectiveness. Do not apply to wet crops
- May be used on cereals undersown with grass or clovers

Crop safety/Restrictions

- Maximum number of treatments 1 per crop for cereals (2 per crop for split applications on winter wheat); varies with species for ornamentals [4, 5, 12]
- Maximum total dose equivalent to one full dose trreatment [1, 11, 13]
- Do not use on very late sown spring wheat or oats or on crops under stress
- Mixtures with liquid nitrogen fertilizers may cause scorch and are specifically excluded on some labels
- Do not use in tank mix with cyanazine [5, 24]
- Ornamentals to be treated must be well established and growing vigorously. Do not treat in strong sunlight or when temperatures are likely to fall below 10°C
- Temporary yellow spotting may occur on poinsettias. It can be minimised by use of a non-ionic wetting agent - see label

Special precautions/Environmental safety

- Harmful in contact with skin [2, 11, 13, 14, 16, 19, 20, 24, 26, 28, 29]
- Harmful if swallowed
- Wash equipment thoroughly with water and wetting agent immediately after use and spray out. Spray out again before storing or using for another product. Traces can cause harm to susceptible crops sprayed later

Personal protective equipment/Label precautions

- A [1-30]; C [20, 26]; H [26]
- M03 [1-30]; R03c [1-28, 30]; R03a [2, 14, 16, 19, 20, 24, 26, 28, 29]; U08 [1-30]; U05a [1-4, 8-30]; U19 [1-3, 5-30]; U13 [8, 14, 18, 20, 21]; U20b [1, 2, 6, 9-17, 19, 22-30]; U20a [3-5, 7, 8, 18, 20, 21]; U02, U04a [17, 25-27]; U05b [5-7]; C03 [1-30]; C01 [8, 14, 18, 20, 21]; E01, E30a [1-30]; E15 [1, 3-30]; E34 [1-8, 10-18, 20-30]; E26 [2, 8-10, 14-18, 20-22, 24-30]; E31b [1-7, 9-17, 19, 22-27, 29, 30]; E08, E31a [8, 18, 20, 21, 28]

Latest application/Harvest interval

- Varies with product. See label for details

Approval

- Approved for aerial application on wheat and oats [2-7, 16, 28, 30]; on wheat, oats and barley [8-10, 12, 15, 18, 29]; wheat and barley [23]. See notes in Section 1
- Accepted by BLRA for use on malting barley
- Approval expiry 31 Oct 2001 [24]; replaced by MAPP 09302

Maximum Residue Level (mg residue/kg food)

- oats 5; pears 3; wheat, rye, barley, triticale 2; grapes 1; tree nuts, olives, oilseeds, tea, hops 0.1; citrus, quinces, stone fruit, berries and small fruit, miscellaneous fruit (except olives), root and tuber vegetables, bulb vegetables, fruiting vegetables, brassica vegetables, leaf vegetables, fresh herbs, legume vegetables, stem vegetables, fungi, pulses, sorghum, buckwheat, millet, rice 0.05

102   chlormequat + 2-chloroethylphosphonic acid

A plant growth regulator for use in cereals

Products

1	Barclay Banshee XL	Barclay	305:155 g/l	SC	09201
2	Greencrop Tycoon	Greencrop	305:155 g/l	SL	09571
3	Strate	Aventis	360:180 g/l	SL	10020
4	Sypex	BASF	305:155 g/l	SL	04650
5	Upgrade	RP Agric.	360:180 g/l	SL	06177
6	Upgrade	Aventis	360:180 g/l	SL	10029

Uses

Lodging control in **spring barley, winter barley, winter wheat**.

Efficacy

- Apply before lodging has started. Best results obtained when crops growing vigorously

- Only crops growing under conditions of high fertility should be treated
- Recommended dose varies with growth stage. See labels for details and recommendations for use of sequential treatments
- Do not spray when crop wet or rain imminent
- Product must always be used with specified wetters - see label [1, 2]

Crop safety/Restrictions
- Maximum number of treatments 1 per crop; maximum total dose equivalent to one full dose treatment
- Do not spray during cold weather or periods of night frost, when soil is very dry, when crop diseased or suffering pest damage, nutrient deficiency or herbicide stress
- If used on seed crops grown for certification inform seed merchant beforehand
- Do not use on wheat variety Moulin or on any winter varieties sown in spring [1, 2]
- Do not use on spring barley Triumph [2]
- Do not treat barley on soils with more than 10% organic matter [1, 2]
- Do not use straw from treated cereals as a horticultural growth medium or as a mulch [1, 2]
- Do not use in programme with any other product containing 2-chloroethylphosphonic acid [1, 2]

Special precautions/Environmental safety
- Harmful in contact with skin [3, 5, 6]
- Harmful if swallowed
- Irritating to eyes [3, 5, 6]
- Harmful to fish or other aquatic life. Do not contaminate surface waters or ditches with chemical or used container [3, 5, 6]

Personal protective equipment/Label precautions
- A [1-6]; C [3, 5, 6]
- M03, R03c [1-6]; R03a, R04a [3, 5, 6]; U05a, U08, U19 [1-6]; U20b [2-6]; U13 [3, 5, 6]; U20a [1]; C03, E01, E26, E30a, E34 [1-6]; E31c [1, 4]; E15 [1, 2, 4]; E31b [2, 3, 5, 6]; E13c [3, 5, 6]

Latest application/Harvest interval
- Before flag leaf ligule/collar just visible (GS 39) or 1st spikelet visible (GS 51) for wheat or barley at top dose; or before flag leaf sheath opening (GS 47) for winter wheat at reduced dose [1]
- Before flag leaf ligule/collar just visible (GS 39) on barley and up to before flag leaf sheath opening (GS 47) on winter wheat [5]

Approval
- Accepted by BLRA for use on malting barley

Maximum Residue Level (mg residue/kg food)
- see chlormequat and 2-chloroethylphosphonic acid entries

---

103   chlormequat + 2-chloroethylphosphonic acid + imazaquin

A plant growth regulator mixture for cereals

Products

1 Satellite	BASF	216:480:0.8 g/l	KL	BASF
2 Satellite	Cyanamid	216:480:0.8 g/l	KL	08969

Uses

Lodging control in **spring barley, winter barley, winter wheat**.

Efficacy
- Best results obtained when crops growing vigorously
- Only crops growing under conditions of high fertility should be treated
- Do not spray when crop wet or rain imminent

Crop safety/Restrictions
- Maximum total dose equivalent to one full dose treatment
- All varieties of winter wheat, winter barley and spring barley (except Triumph) may be treated. Do not treat spring varieties sown in winter
- Do not spray during cold weather or periods of night frost, when soil is very dry, when crop diseased or suffering pest damage, nutrient deficiency or herbicide stress

- Do not use on crops being grown for seed
- Do not treat undersown crops
- Do not use within 10 d of any herbicide or liquid fertiliser treatment

Special precautions/Environmental safety
- Harmful if swallowed
- Irritating to skin
- Risk of serious damage to eyes
- Harmful to fish or other aquatic life. Do not contaminate surface waters or ditches with chemical or used container

Personal protective equipment/Label precautions
- A, C, H
- M03, R03c, R04b, R04d, U05a, U08, U13, U19, U20b, C03, E01, E13c, E30a, E31b, E34

Latest application/Harvest interval
- Before flag leaf ligule/collar just visible (GS 39)

Approval
- Accepted by BLRA for use on malting barley
- Product registration number not available at time of printing [1]

Maximum Residue Level (mg residue/kg food)
- see chlormequat and 2-chloroethylphosphonic acid entries

---

104 chlormequat + 2-chloroethylphosphonic acid + mepiquat chloride

A plant growth regulator for reducing lodging in cereals

Products

Cyclade	BASF	230:155:75 g/l	SL	08958

Uses

Lodging control in *spring barley, winter barley, winter wheat*.

Efficacy
- Best results achieved in a vigorous, actively growing crop with adequate fertility and moisture
- Optimum timing on all crops is from second node detectable stage (GS 32)
- Recommended for use as part of an intensive growing system which includes provision for optimum fertiliser treatment and disease control
- May be applied to crops undersown with grasses or clovers
- Must be used with a non-ionic wetting agent

Crop safety/Restrictions
- Maximum number of treatments 1 per crop
- Do not apply to stressed crops or those on soils of low fertility unless receiving adequate dressings of fertiliser
- Do not apply in temperatures above 21°C or if crop is wet or if rain expected
- Do not treat variety Moulin nor any winter varieties sown in spring
- Do not use in a programme with any other product containing 2-chloroethylphosphonic acid
- Do not apply to barley on soils with more than 10% organic matter (winter wheat may be treated)
- Treatment may cause some delay in ear emergence

Special precautions/Environmental safety
- Harmful: If swallowed
- Do not use straw from treated crops as a horticultural growth medium
- Notify seed merchant in advance if use on a seed crop is proposed

Personal protective equipment/Label precautions
- A
- M03, R03c, U05a, U08, U19, U20b, C03, E01, E15, E30a, E31c, E34

Latest application/Harvest interval
- Before first spikelet of ear visible (GS 51) using reduced dose on barley; before flag leaf sheath opening (GS 47) using reduced dose on winter wheat

Approval
- Accepted by BLRA for use on malting barley

Maximum Residue Level (mg residue/kg food)
- see chlormequat and 2-chloroethylphosphonic acid entries

105   chlormequat + choline chloride

A plant growth regulator for use in cereals and certain ornamentals

Products

1	Atlas Quintacel	Nufarm Whyte	640:64 g/l	SL	07706
2	Barclay Take 5	Barclay	645:- g/l	SL	08524
3	Cropsafe 5C Chlormequat	Hortichem	645:- g/l	SL	07897
4	New 5C Cycocel	BASF	645:- g/l	SL	01482
5	Portman Supaquat	Portman	640:- g/l	SL	03466

Uses

Increasing yield in **winter barley** [4, 5]. Lodging control in **winter barley** [1, 2]. Lodging control in **spring oats, spring wheat, winter oats, winter wheat** [1, 2, 4, 5]. Lodging control in **triticale, winter rye** [2, 4]. Lodging control in **autumn sown spring wheat, spring rye** [4]. Stem shortening in **bedding plants, camellias, hibiscus trionum, lilies, ornamentals** [1]. Stem shortening in **pelargoniums, poinsettias** [1, 3, 4].

Efficacy
- Influence on growth varies with crop and growth stage. Risk of lodging reduced by application at early stem extension. Root development and yield can be improved by earlier treatment
- Most effective results normally achieved from spring application. On winter barley an autumn treatment may also be useful. Timing of spray is critical and recommendations vary with product. See label for details
- Often used in tank-mixes with pesticides. Recommendations for mixtures and sequential treatments vary with product. See label for details
- Add authorised non-ionic wetter when spraying oats
- At least 6 h required before rain for maximum effectiveness. Do not apply to wet crops
- May be used on cereals undersown with grass or clovers

Crop safety/Restrictions
- Maximum number of treatments 1 per crop for spring wheat, oats, rye, triticale; 1 or 2 per crop for winter wheat and winter barley (depending on dose); 1-3 per crop for ornamentals. See label for details
- Do not spray very late sown spring crops, crops on soils of low fertility, crops under stress from any cause or if frost expected
- Do not use on spring barley
- Do not use in tank mix with cyanazine [2]
- Mixtures with liquid nitrogen fertilizers may cause scorch

Special precautions/Environmental safety
- Harmful in contact with skin [5]
- Harmful if swallowed [1, 2, 4, 5]
- Wash equipment thoroughly with water and wetting agent immediately after use and spray out. Traces can cause harm to susceptible crops sprayed later [1]

Personal protective equipment/Label precautions
- A [1-5]; C [5]
- M03, R03c [1, 2, 4, 5]; R03a [5]; U05a, U08, U19 [1, 2, 4, 5]; U20b [2, 4]; U20a [1, 5]; C03 [1, 2, 4, 5]; E31c [4]; E01, E15, E30a, E34 [1, 2, 4, 5]; E31b [1, 2, 5]

Latest application/Harvest interval
- Before 3rd node detectable (GS 33) for oats, winter wheat, autumn drilled spring wheat; before 2nd node detectable (GS 32) for rye; before 1st node detectable (GS 31) for spring wheat, winter barley, triticale

Approval
- Approved for aerial application on wheat, oats [1]; wheat, oats, barley, rye, triticale [2, 4, 5]. See notes in Section 1
- Accepted by BLRA for use on malting barley

Maximum Residue Level (mg residue/kg food)
- see chlormequat entry

106   chlormequat + choline chloride + imazaquin

A plant growth regulator mixture for winter wheat

Products

1 Meteor	Cyanamid	368:28:0.8 g/l	SL	06505
2 Meteor	BASF	368:28:0.8 g/l	SL	BASF
3 Standon Imazaquin 5C	Standon	368:28:0.8 g/l	SL	08813

Uses

Increasing yield in *winter wheat*. Lodging control in *winter wheat*.

Efficacy
- Apply as single dose from leaf sheath lengthening up to and including 1st node detectable or as split dose, the first from tillers formed to leaf sheath lengthening, the second from leaf sheath erect up to and including 1st node detectable
- Apply to crops during good growing conditions or to those at risk from lodging
- On soils of low fertility, best results obtained where adequate nitrogen fertiliser used
- Do not apply when crop wet or rain imminent

Crop safety/Restrictions
- Maximum number of applications 1 per crop (2 per crop at split dose)
- Do not treat durum wheat
- Do not apply to undersown crops

Special precautions/Environmental safety
- Harmful if swallowed. Irritating to eyes

Personal protective equipment/Label precautions
- A, C
- M03, R03c, R04a, U05a, U08, U13, U19, U20a, C03, E01, E15, E30a, E31b, E34

Latest application/Harvest interval
- Before second node detectable (GS 31)

Approval
- Product registration number not available at time of printing [2]

Maximum Residue Level (mg residue/kg food)
- see chlormequat entry

107   chlormequat + mepiquat chloride

A plant growth regulator for reducing lodging in cereals

Products

Stronghold	BASF	345:115 g/l	SL	09134

Uses

Lodging control in *winter wheat*.

Efficacy
- Apply during good growing conditions at the correct timings - see label
- Optimum timing is when leaf sheaths erect (GS 30)
- May be applied to crops undersown with grasses or clovers

Crop safety/Restrictions
- Maximum total dose equivalent to one full dose treatment
- Do not apply to stressed crops or those on soils of low fertility unless receiving adequate dressings of fertiliser
- Do not apply in temperatures above 21°C or if crop is wet or if rain expected
- Do not treat any winter varieties sown in spring
- Treatment may cause some delay in ear emergence
- Mixtures with liquid fertilisers may cause scorching in some circumstances

FOR FULL CONDITIONS OF USE ALWAYS READ THE PRODUCT LABEL

Special precautions/Environmental safety
- Harmful: If swallowed
- Do not use straw from treated crops as a horticultural growth medium
- Notify seed merchant in advance if use on a seed crop is proposed

Personal protective equipment/Label precautions
- A
- M03, R03c, U05a, U08, U19, U20b, C03, E01, E15, E30a, E31c, E34

Latest application/Harvest interval
- Before 3rd node detectable (GS 33)

Approval
- Accepted by BLRA for use on malting barley

Maximum Residue Level (mg residue/kg food)
- see chlormequat entry

108   chlormequat with di-1-p-menthene

A plant-growth regulator for reducing stem growth and lodging

Products

Podquat	Mandops	470 g/l	SL	03003

Uses

Increasing yield in *broad beans, field beans, oilseed rape, peas*. Lodging control in *broad beans, field beans, oilseed rape, peas*.

Efficacy
- On winter oilseed rape and beans either apply as soon as possible after 3-leaf stage (GS 1,3) until growth ceases followed by spring treatment in mid-Mar to early Apr or use a single spring spray. See label for details
- May be used at temperatures down to 1°C provided spray dries on leaves before rain, frost or snow occurs

Crop safety/Restrictions
- Do not apply to plants covered by frost

Special precautions/Environmental safety
- Harmful if swallowed

Personal protective equipment/Label precautions
- A
- M03, R03c, U05a, U08, U19, U20a, C03, E01, E15, E30a, E31a, E34

Approval
- UK approvals for use of chlormequat on oilseed rape likely to be revoked in 2001 as a result of implementation of the MRL Directives

Maximum Residue Level (mg residue/kg food)
- see chlormequat entry

109   2-chloroethylphosphonic acid

A plant growth regulator for cereals and various horticultural crops

Products

1	Aventis Cerone	Aventis	480 g/l	SL	09972
2	Barclay Coolmore	Barclay	480 g/l	SL	07917
3	Cerone	RP Agric.	480 g/l	SL	06185
4	Cerone	Aventis	480 g/l	SL	09985
5	Charger	Aventis	360 g/l	SL	09986
6	Ethrel C	Hortichem	480 g/l	SL	06995

Uses

Basal bud stimulation in *protected roses* [6]. Fruit ripening in *apples (off-label)*, *tomatoes* [6]. Increasing branching in *pelargoniums* [6]. Inducing flowering in *bromeliads* [6]. Lodging control in **spring barley, triticale, winter barley, winter rye, winter wheat** [1-5]. Stem shortening in **narcissi** [6].

Efficacy
- Best results achieved on crops growing vigorously under conditions of high fertility
- Optimum timing varies between crops and products. See labels for details
- Do not spray crops when wet or if rain imminent
- Best results on horticultural crops when temperature does not fall below 10°C [6]
- Use on tomatoes 17 d before planned pulling date [6]
- Apply as drench to daffodils when stems average 15 cm [6]
- Apply to glasshouse roses when new growth started after pruning [6]

Crop safety/Restrictions
- Maximum number of treatments 1 per crop or yr
- Do not spray crops suffering from stress caused by any factor, during cold weather or period of night frost nor when soil very dry
- Do not apply to cereals within 10 d of herbicide or liquid fertilizer application
- Do not spray wheat or triticale where the leaf sheaths have split and the ear is visible

Special precautions/Environmental safety
- Irritating to eyes and skin
- Harmful to fish or other aquatic life. Do not contaminate surface waters or ditches with chemical or used container
- Avoid accidental deposits on painted objects such as cars, trucks, aircraft

Personal protective equipment/Label precautions
- A, C [1-6]
- R04a, R04b, U05a, U08, U13 [1-6]; U20b [1, 3-5]; U20a [6]; C03, E01, E13c, E30a, E31b [1-6]; E34 [6]; S06 [2]

Latest application/Harvest interval
- Before 1st spikelet visible (GS 51) for spring barley, winter barley, winter rye; before flag leaf sheath opening (GS 47) for triticale, winter wheat
- HI tomatoes 5 d

Approval
- Approved for aerial application on winter barley [1-5]. See notes in Section 1
- Off-label approval unstipulated for use on apples for cider making (OLA 1297/99)[6]
- Accepted by BLRA for use on malting barley

Maximum Residue Level (mg residue/kg food)
- currants 5; pome fruits, cherries, tomatoes, peppers 3; rye, barley 0.5; wheat, triticale 0.2; tree nuts, tea, hops 0.1; apricots, peaches, plums, strawberries, blackberries, dewberries, loganberries, raspberries, bilberries, cranberries, gooseberries, wild berries, avocados, bananas, dates, kiwi fruit, kumquats, litchis, mangoes, passion fruit, pomegranates, root and tuber vegetables, garlic, shallots, spring onions, aubergines, cucumbers, gherkins, courgettes, melons, squashes, water melons, brassica vegetables, leaf vegetables and herbs, legume vegetables, stem vegetables, fungi, pulses, oilseeds, potatoes, oats, rice, animal products 0.05

110   2-chloroethylphosphonic acid + mepiquat chloride

A plant growth regulator for reducing lodging in cereals

Products

1 Barclay Banshee	Barclay	155:305 g/l	SL	08175
2 Standon Mepiquat Plus	Standon	155:305 g/l	SL	09373
3 Terpal	BASF	155:305 g/l	SL	02103
4 Terpitz	Me2	155:305 g/l	SL	09634

Uses
Increasing yield in **winter barley** *(low lodging situations)* [1, 3, 4]. Lodging control in **spring barley, triticale, winter barley, winter rye, winter wheat** [1-4].

Efficacy
- Best results achieved on crops growing vigorously under conditions of high fertility

- Recommended dose and timing vary with crop, cultivar, growing conditions, previous treatment and desired degree of lodging control. See label for details
- Add an authorised non-ionic wetter to spray solution
- May be applied to crops undersown with grass or clovers
- Do not apply to crops if wet or rain expected

Crop safety/Restrictions
- Maximum number of treatments varies with dose and timing - see labels
- Do not treat crops damaged by herbicides or stressed by drought, waterlogging etc
- Do not treat crops on soils of low fertility unless adequately fertilized
- Do not use in a programme with any other product containing 2-chloroethylphosphonic acid
- Late tillering may be increased with crops subject to moisture stress and may reduce quality of malting barley
- Do not apply to winter cultivars sown in spring or treat winter barley, triticale or winter rye on soils with more than 10% organic matter (winter wheat may be treated)
- Do not apply at temperatures above 21°C

Special precautions/Environmental safety
- Do not contaminate surface waters or ditches with chemical or used container
- Do not use straw from treated cereals as a mulch or growing medium

Personal protective equipment/Label precautions
- U20b [1-4]; E31c [3]; E15, E30a [1-4]; E31a [1]; E26 [2]; E31b [2, 4]

Latest application/Harvest interval
- Before ear visible (GS 49) for winter barley, spring barley, winter wheat and triticale; flag leaf just visible (GS 37) for winter rye

Approval
- Accepted by BLRA for use on malting barley

Maximum Residue Level (mg residue/kg food)
- see 2-chloroethylphosphonic acid entry

---

111   chlorophacinone

An anticoagulant rodenticide

Products

1	Drat	RP Amenity	2.5 g/l	CB	05238
2	Drat Rat Bait	B H & B	0.005% w/w	RB	H6743
3	Endorats	Irish Drugs	0.005% w/w	RB	H6744
4	Karate Ready to Use Rat Bait	DiverseyLever	0.006% w/w	RB	H6745
5	Karate Ready to Use Rodenticide Sachets	DiverseyLever	0.006%w/w	RB	H6746
6	Karate Ready-to-Use Rat and Mouse Bait	DiverseyLever	0.006% w/w	RB	08658
7	Karate Ready-to-Use Rodenticide Sachets	DiverseyLever	0.006% w/w	RB	08659
8	Rat & Mouse Bait	B H & B	0.005% w/w	RB	00764
9	Ruby Rat	Heatherington	0.005% w/w	RB	06059

Uses

Mice in *farm buildings* [1, 6-9]. Rats in *farm buildings* [1-9]. Voles in *farm buildings* [1, 2, 6-9].

Efficacy
- Chemical formulated with oil, thus improving weather resistance of bait
- Mix concentrate with any convenient bait, such as grain, apple, carrot or potato [1]
- Use in baiting programme
- Bait stations should be sited where rats active, by rat holes, along runs or in harbourages. Place bait in suitable containers
- Replenish baits every few days and remove unused bait when take ceases or after 7-10 d

Crop safety/Restrictions
- Resistance status of target population should be taken into account when considering choice of rodenticide
- Bait stations may be sited conveniently but bait should be inaccessible to non-target animals and protected from prevailing weather

Special precautions/Environmental safety
- For use only by professional operators
- Harmful in contact with skin and if swallowed
- Prevent access to baits by children, domestic animals and birds; see label for other precautions required
- Harmful to game, wild birds and animals
- Store unused sachets in a safe place. Do not store half-used sachets [5, 7]

Personal protective equipment/Label precautions
- A, C [1]
- M03 [1, 4, 5]; R03a, R03c [1]; U13 [1-8]; U20b [2, 4, 8]; U02, U04a, U05a, U08, U14, U15 [1]; U20a [1, 3, 6, 7]; U20c [5]; C03 [1]; E31a [1, 2, 8]; E30a [2-4, 6-8]; E01, E10b, E15, E30b [1]; E32a [3-7]; V01a, V03a, V04a [1-3, 6-8]; V02 [1-8]; V05 [1]; V01b, V03b, V04b [4, 5]

---

112   chloropicrin

A highly toxic horticultural soil fumigant

Products

1 Chloropicrin Fumigant	Dewco-Lloyd	99.5% w/w	FU	04216
2 K & S Chlorofume	K Fumigation	99.3% w/w	FU	08722

Uses

Crown rot in **strawberries** *(soil fumigation)*. Nematodes in **strawberries** *(soil fumigation)*. Red core in **strawberries** *(soil fumigation)*. Replant disease in **hardy ornamentals** *(soil fumigation)*, **top fruit** *(soil fumigation)*. Verticillium wilt in **strawberries** *(soil fumigation)*.

Efficacy
- Treat pre-planting
- Apply with specialised injection equipment
- For treating small areas or re-planting a single tree, a hand-operated injector may be used. Mark the area to be treated and inject to 22 cm at intervals of 22 cm
- Double roll within 1 h of treatment and leave undisturbed for at least 10 d

Crop safety/Restrictions
- Polythene sheeting (150 gauge) should be progressively laid over soil as treatment proceeds. The margin of the sheeting around the treated area must be embedded or covered with treated soil. Remove progressively after at least 4 d provided good air movement conditions prevail. Aerate soil for 15 d before planting
- Carry out a cress test before replanting treated soil

Special precautions/Environmental safety
- Chloropicrin is subject to the Poisons Rules 1982 and the Poisons Act 1972. See notes in Section 1
- Very toxic by inhalation, in contact with skin and if swallowed
- Irritating to eyes and skin [1]
- Irritating to repiratory system [2]
- Causes burns [2]
- Before use, consult the code of practice for the fumigation of soil with chloropicrin. 2 fumigators must be present
- Avoid treatment or vapour release when persistent still air conditions prevail
- Remove contaminated gloves, boots or other clothing immediately and ventilate them in the open air until all odour is eliminated
- Dangerous to livestock
- Dangerous to game, wild birds and animals [1]
- Dangerous to bees [1]

- Dangerous to fish. Do not contaminate surface waters or ditches with chemical or used container [1]

Personal protective equipment/Label precautions
- A, G, H, K, M [1, 2]
- M04, R04a, R04b [1]; R01a, R01b, R01c, R04c [1, 2]; R05b [2]; U13, U14, U15 [1]; U04a, U05a, U10, U19, U20a, C03 [1, 2]; E01, E10a, E12c, E13b, E32a [1]; E15, E30b, E34 [1, 2]; E02, E06a [1, 2] (until advised); E33, E36 [2]

Withholding period
- Dangerous to livestock. Keep all livestock out of treated areas until advised otherwise by fumigator

Latest application/Harvest interval
- At least 20 d before planting

---

113   chlorothalonil

A protectant chlorophenyl fungicide for use in many crops and turf

See also carbendazim + chlorothalonil
     carbendazim + chlorothalonil + maneb

Products

1	AgriGuard Chlorothalonil	AgriGuard	500 g/l	SC	09390
2	Atlas Cropguard	Nufarm Whyte	500 g/l	SC	09123
3	Barclay Corrib 500	Barclay	500 g/l	SC	08981
4	Bombardier FL	Unicrop	500 g/l	SC	07910
5	Bravo 500	BASF	500 g/l	SC	05637
6	Bravo 500	Zeneca	500 g/l	SC	09059
7	Clortosip 500	Sipcam	500 g/l	SC	09320
8	Daconil Turf	Scotts	500 g/l	SC	09265
9	Flute	Sipcam	720 g/l	SC	08953
10	Fusonil Turf	Rigby Taylor	500 g/l	SC	09695
11	Greencrop Orchid	Greencrop	500 g/l	SC	09566
12	Jupital DG	Zeneca	75% w/w	WG	09181
13	Mainstay	Quadrangle	500 g/l	SC	05625
14	Mycoguard	Chiltern	500 g/l	SC	08115
15	Repulse	Hortichem	500 g/l	SC	07641
16	Rover	Sipcam	500 g/l	SC	09848
17	Sipcam Echo 75	Sipcam	75% w/w	WG	08302
18	Standon Chlorothalonil 500	Standon	500 g/l	SC	08597
19	Tripart Faber	Tripart	500 g/l	SC	04549
20	Tripart Ultrafaber	Tripart	720 g/l	SC	05627

Uses

Alternaria in *flowerhead brassicas, leaf brassicas* [19]. Anthracnose in *turf* [8, 10]. Ascochyta in *peas* [1]. Ascochyta in *combining peas* [3-7, 9, 11-13, 15-18, 20]. Blight in *potatoes* [1-7, 9, 11-20]. Blight in *protected tomatoes* [4, 5, 7, 13, 15, 16]. Botrytis in *peas* [1]. Botrytis in *spring oilseed rape* [1, 17]. Botrytis in *protected tomatoes* [1, 4, 13, 15, 17]. Botrytis in *brussels sprouts* [1, 4-7, 12, 13, 15-17]. Botrytis in *cabbages, cauliflowers* [1, 4-7, 12, 15-17]. Botrytis in *protected cucumbers* [1, 4, 5, 7, 12, 13, 15-17]. Botrytis in *protected ornamentals* [1, 4, 5, 7, 12, 15-17]. Botrytis in *onions* [1, 4, 5, 7, 15, 16]. Botrytis in *broccoli* [1, 4, 5, 7, 16, 17]. Botrytis in *cane fruit* [1, 5, 6, 12, 13, 15, 20]. Botrytis in *winter oilseed rape* [1, 5, 6, 12, 13, 17, 20]. Botrytis in *gooseberries* [1, 7, 16]. Botrytis in *grapevines, ornamentals, peppers* [13]. Botrytis in *calabrese* [17]. Botrytis in *flowerhead brassicas* [19]. Botrytis in *leaf brassicas* [3, 9, 13, 18-20]. Botrytis in *blackberries, loganberries, raspberries* [4-7, 16, 17]. Botrytis in *strawberries* [5, 6, 12, 13, 17, 20]. Botrytis in *combining peas* [5, 7, 12, 13, 15-17]. Botrytis in *chinese cabbage, collards, kale, strawberries* (off-label) [6]. Botrytis in *oilseed rape, redcurrants, strawberries* (outdoor crops only) [7, 16]. Botrytis in *blackcurrants* [7, 16, 18]. Cane spot in *cane fruit* [6, 12, 13, 20]. Celery leaf spot in *celeriac* (off-label) [19]. Celery leaf spot in *celery* (qualified minor use) [4, 7, 16, 17]. Celery leaf spot in *celery* [6, 12, 13, 19]. Chocolate spot in *field beans* [1-6, 9, 11-14, 18-20]. Chocolate spot in *spring field beans, winter field beans* [7, 16, 17]. Currant leaf spot in *blackcurrants* [1, 4-7,

9, 12, 13, 15-20]. Currant leaf spot in *gooseberries, redcurrants* [1, 7, 16]. Currant leaf spot in *redcurrants (qualified minor use)* [4, 6, 12, 13, 15, 17, 20]. Dark leaf spot in *flowerhead brassicas, leaf brassicas* [11]. Dark leaf spot in *broccoli, brussels sprouts, cabbages, calabrese, cauliflowers* [17]. Dollar spot in *turf* [8, 10]. Downy mildew in *hops* [1, 4-7, 12, 13, 15-17, 20]. Downy mildew in *brussels sprouts* [1, 4, 5, 7, 13, 15, 16]. Downy mildew in *cabbages, cauliflowers* [1, 4, 5, 7, 15, 16]. Downy mildew in *broccoli* [1, 4, 5, 7, 16]. Downy mildew in *winter oilseed rape* [1, 5-7, 12, 13, 16, 17, 20]. Downy mildew in *spring oilseed rape* [1, 7, 16, 17]. Downy mildew in *brassicas* [12, 13]. Downy mildew in *grapevines* [13]. Downy mildew in *flowerhead brassicas* [19]. Downy mildew in *leaf brassicas* [3, 6, 9, 12, 13, 17-20]. Downy mildew in *brassica seed beds* [6]. Dry bubble in *mushrooms* [6, 20]. Fusarium patch in *turf* [8, 10]. Glume blotch in *winter wheat* [1-7, 9, 11, 14, 16-20]. Glume blotch in *spring wheat* [2, 3, 7, 9, 11, 14, 16-19]. Ink disease in *irises* [12, 13, 15]. Ink disease in *irises (qualified minor use)* [4, 7, 16, 17]. Late ear diseases in *winter wheat* [1, 5]. Leaf blotch in *spring barley, winter barley* [14]. Leaf blotch in *onions* [19]. Leaf mould in *protected tomatoes* [1, 4, 5, 7, 13, 15-17]. Leaf rot in *onions* [1, 4-7, 12, 13, 15-17, 20]. Leaf spot in *winter wheat* [1-3, 5-7, 9, 11-14, 16-20]. Leaf spot in *gooseberries* [1, 5-7, 13, 16, 17]. Leaf spot in *celery* [15]. Leaf spot in *spring wheat* [2, 3, 7, 9, 11, 14, 16-19]. Leaf spot in *gooseberries (qualified minor use)* [4, 6, 12, 13, 15, 20]. Mycosphaerella in *peas* [1, 7, 16]. Mycosphaerella in *combining peas* [4-6, 12-15, 17, 20]. Neck rot in *onions* [1, 4-7, 12, 13, 15-17, 20]. Powdery mildew in *gooseberries* [1, 7, 16]. Powdery mildew in *raspberries* [12, 13, 17]. Powdery mildew in *protected cucumbers* [6, 12, 17]. Powdery mildew in *cane fruit* [6, 20]. Powdery mildew in *redcurrants* [7, 16]. Powdery mildew in *blackcurrants* [7, 16, 18]. Red thread in *turf* [8, 10]. Rhizoctonia in *flowerhead brassicas, leaf brassicas* [19]. Rhynchosporium in *barley* [19]. Rhynchosporium in *spring barley, winter barley* [2]. Ring spot in *brussels sprouts* [13, 15, 17]. Ring spot in *broccoli, cabbages, cauliflowers* [15, 17]. Ring spot in *calabrese* [17]. Ring spot in *flowerhead brassicas* [19]. Ring spot in *leaf brassicas* [3, 6, 9, 12, 13, 18-20]. Rust in *leeks* [13]. Septoria leaf spot in *celery (qualified minor use)* [1]. Septoria leaf spot in *winter wheat* [4]. Septoria leaf spot in *celeriac (off-label)* [6, 20]. Wet bubble in *mushrooms* [6, 20].

Efficacy

- For some crops products differ in diseases listed as controlled. See label for details and for application rates, timing and number of sprays
- Apply as protective spray or as soon as disease appears and repeat as directed
- Activity against Septoria may be reduced where serious mildew or rust present. In such conditions mix with suitable mildew or rust fungicide
- May be used at reduced rate on combining peas in tank mix with Ronilan FL [5]
- May be used for preventive and curative treatment of turf [8, 10]
- For Botrytis control in strawberries important to start spraying early in flowering period and repeat at least 3 times at 10 d intervals

Crop safety/Restrictions

- Varies with crop and product - see labels for details
- On strawberries some scorching of calyx may occur with protected crops
- Do not mow or water turf for 24 h after treatment. Do not add surfactant or mix with liquid fertilizer [8, 10]

Special precautions/Environmental safety

- Irritating to eyes [2, 7, 8, 11, 13, 14, 16, 19, 20]
- Irritating to skin [1-8, 11, 13-16, 18-20]
- May cause sensitization by skin contact [7, 9, 10, 16]
- Irritating to respiratory system [1, 3-8, 11-13, 15-20]
- Risk of serious damage to eyes [1-12, 14-18]
- Do not harvest for human or animal consumption for specified interval after last application [2, 14, 19] (field beans 7d; celery 7d; onions 7d; blackcurrants 3d)
- Dangerous to fish or other aquatic life. Do not contaminate surface waters or ditches with chemical or used container
- LERAP Category B
- Do not allow direct spray from broadcast air-assisted sprayers to fall within 18 m of surface waters or ditches

FOR FULL CONDITIONS OF USE ALWAYS READ THE PRODUCT LABEL

- Do not spray from the air within 250 m horizontal distance of surface waters or ditches
- Operators must use a vehicle fitted with a cab and a forced air filtration unit with a pesticide filter complying with HSE Guidance Note PM74 or to an equivalent or higher standard when making broadcast or air-assisted applications
- Must only be applied by a pedestrian controlled sprayer or vehicle mounted/drawn equipment [8, 10]

Personal protective equipment/Label precautions
- A, C [1-20]; H [1-12, 14-18]; M [1, 3-12, 15-18]; J [1, 3-5, 7, 9, 11, 16, 18]; D [12, 17]
- R04b [1-8, 11, 13-16, 18-20]; R04c [1, 3-8, 11-13, 15-20]; R04a [2, 7, 8, 11, 13, 14, 16, 19, 20]; R04d [1-12, 14-18]; R04e [7, 9, 10, 16]; U05a, U09a, U19, U20a [1-20]; U13, U14, U15 [2-4, 7, 9-11, 13, 14, 16-19]; U02 [2-4, 7, 9-11, 14, 16-19]; C03 [1-20]; C02 [2, 14, 19] (field beans 7d; celery 7d; onions 7d; blackcurrants 3d); E01, E13b, E16a, E30a [1-20]; E16b [1-7, 9-20]; E31b [1-3, 5, 7, 9, 11, 13, 14, 16-19]; E26 [1-4, 9-11, 14, 18, 19]; E34 [1, 2, 5, 8, 13, 14]; E31a [6, 8, 15, 20]; E17 [1, 3-7, 11, 16-18] (18 m); E18 [1, 3-7, 9, 11, 16, 17]; E31c [4, 10]; E32a [12]

Latest application/Harvest interval
- Before anthesis for spring barley, spring wheat, winter barley, winter wheat [2, 14]; before grain watery ripe (GS 71) for spring wheat [1, 9], winter wheat [1, 5, 9]; ear emergence just complete (GS 59) for spring barley, spring wheat, winter barley, winter wheat [3]; before flowering for winter oilseed rape; end Aug in yr of harvest for blackcurrants, gooseberries, redcurrants, blueberries, loganberries, raspberries
- HI 8 wk for field beans; 6 wk for combining peas; 28 d or before 31 Aug for post-harvest treatment for blackcurrants, gooseberries, redcurrants; 14 d for onions, strawberries; 10 d for hops; 7 d for broccoli, Brussels sprouts, cabbages, cauliflowers, celery, onions, potatoes; 3 d for hops, cane fruit; 48 h for protected cucumbers, protected tomatoes

Approval
- Approval for aerial spraying on potatoes [3-7, 9, 11, 13, 15, 16, 19, 20]; on wheat, field beans, oilseed rape, brassicas, onions, peas [3]. See notes in Section 1
- Following implementation of Directive 98/82/EC, approval for use of chlorothalonil on numerous crops was revoked in 1999, although uses on blackcurrants, redcurrants and gooseberries were subsequently reinstated for some products
- Off-label approval unstipulated for use on outdoor strawberries (OLA2452/99)[6]
- Accepted by BLRA for use on malting barley and hops

Maximum Residue Level (mg residue/kg food)
- cranberries, tomatoes, peppers, aubergines, peas (with pods) 2; grapes (table), cucumbers 1; garlic, onions, shallots, Brussels sprouts 0.5; tea, wheat, rye, barley, oats, triticale 0.1; lettuce, maize, rice, animal products 0.01

---

114   chlorothalonil + cyproconazole

A systemic protectant and curative fungicide mixture for cereals

Products

1 Alto Elite	Novartis	375:40 g/l	SC	08467
2 Octolan	Novartis	375:40 g/l	SC	08480

Uses

Brown rust in *spring barley, winter barley, winter wheat*. Chocolate spot in *field beans*. Eyespot in *winter barley, winter wheat*. Grey mould in *peas*. Net blotch in *spring barley, winter barley*. Powdery mildew in *spring barley, winter barley, winter wheat*. Rhynchosporium in *spring barley, winter barley*. Rust in *field beans*. Septoria diseases in *winter wheat*. Yellow rust in *spring barley, winter barley, winter wheat*.

Efficacy
- Apply at first signs of infection or as soon as disease becomes active
- A repeat application may be made if re-infection occurs
- For established mildew tank-mix with approved formulation of tridemorph
- When applied prior to third node detectable (GS 33) a useful reduction of eyespot will be obtained
- If infection of eyespot anticipated tank-mix with prochloraz

Crop safety/Restrictions
- Maximum number of treatments 2 per crop
- If applied to winter wheat in spring at GS 30-33 straw shortening may occur but yield is not reduced

Special precautions/Environmental safety
- Irritant. Risk of serious damage to eyes
- Do not apply at concentrations higher than recommended
- Dangerous to fish or other aquatic life. Do not contaminate surface waters or ditches with chemical or used container
- LERAP category B

Personal protective equipment/Label precautions
- A, C, H, M
- R04, R04d, U05a, U11, U19, U20a, C03, E01, E13b, E16a, E16b, E26, E30a, E31b, E34

Latest application/Harvest interval
- Up to and including anthesis complete (GS 69) for winter wheat; up to and including emergence of ear complete (GS 59) for barley

Approval
- Accepted by BLRA for use on malting barley

Maximum Residue Level (mg residue/kg food)
- see chlorothalonil entry

---

115   chlorothalonil + fluquinconazole

A contact and systemic fungicide mixture for wheat

Products

Vista CT	Aventis	400:83 g/l	SC	09368

Uses

Brown rust in *spring wheat, winter wheat*. Glume blotch in *spring wheat, winter wheat*. Powdery mildew in *spring wheat* (reduction), *winter wheat* (reduction). Septoria leaf spot in *spring wheat, winter wheat*.

Efficacy
- Best results achieved from applications when disease first becomes active in crop
- Apply to give good foliar cover and increase volume in dense crops to improve spray penetration
- Adequate season-long protection of autumn sown wheat will usually require a programme of at least two fungicide treatments.

Crop safety/Restrictions
- Maximum total dose equivalent to two full dose treatments
- Product must not be applied after end of flowering (GS 69)

Special precautions/Environmental safety
- Harmful if swallowed
- Risk of serious damage to eyes
- May cause sensitization by skin contact
- LERAP Category B

Personal protective equipment/Label precautions
- A, C, H
- M03, R03c, R04d, R04e, U05a, U11, U14, U15, C03, E01, E16a, E16b, E22a, E26, E31b, E34

Latest application/Harvest interval
- Before grain watery ripe (GS 71)

Maximum Residue Level (mg residue/kg food)
- See chlorothalonil entry

116   chlorothalonil + mancozeb

A protectant fungicide mixture for potatoes

Products

Adagio	Interfarm	201:274 g/l	SC	10057

Uses

Blight in **potatoes**.

Efficacy

- Start spray treatments immediately after a blight warning or just before the haulm meets in the row
- It is essential to start the programme before blight appears in the crop
- Repeat treatments at 7, 10 or 14 d intervals depending on blight risk and continue until haulm is to be destroyed

Crop safety/Restrictions

- Maximum number of treatments 5 per crop
- Irrigated crops should be sprayed immediately after irrigation

Special precautions/Environmental safety

- Irritating to eyes, skin and respiratory system
- May cause sensitization by skin contact
- Dangerous to fish or other aquatic life. Do not contaminate surface waters or ditches with chemical or used container
- LERAP Category B
- Broadcast air assisted applications must only be made by equipment fitted with a cab with a forced air filtration unit plus a pesticide filter complying with HSE Guidance Note PM 74 or an equivalent or higher standard

Personal protective equipment/Label precautions

- A, C, H, M
- U02, U05a, U08, U13, U14, U15, U19, U20a, C03, E01, E13b, E16a, E16b, E30a, E31b

Latest application/Harvest interval

- HI 7 d

Maximum Residue Level (mg residue/kg food)

- see chlorothalonil and mancozeb entries

117   chlorothalonil + metalaxyl

A systemic and protectant fungicide for various field crops

Products

Folio	Novartis	500:75 g/l	SC	08547

Uses

Alternaria in **brussels sprouts** *(moderate control)*, **calabrese** *(moderate control)*, **cauliflowers** *(moderate control)*. Blight in **early potatoes, maincrop potatoes**. Downy mildew in **broad beans, brussels sprouts, bulb onions** *(qualified minor use)*, **calabrese, cauliflowers, field beans, salad onions** *(qualified minor use)*, **winter oilseed rape**. Ring spot in **brussels sprouts** *(reduction)*, **calabrese** *(reduction)*, **cauliflowers** *(reduction)*. White blister in **brussels sprouts, calabrese, cauliflowers**. White tip in **leeks** *(qualified minor use)*.

Efficacy

- Apply at first signs of disease (oilseed rape, beans), at first signs of disease or when weather conditions favourable to disease (Brussels sprouts, onions, leeks)
- Repeat treatment at 14-21 d (14 d for oilseed rape and beans) intervals if necessary
- Oilseed rape crops most likely to benefit from treatment are infected crops between cotyledon and 3-leaf stage (GS 1,0-1,3)
- Apply as protectant spray on potatoes as first part of spray programme to mid-Aug. Later use a protectant fungicide (preferably tin-based) up to complete haulm destruction or harvest

Crop safety/Restrictions
- Maximum total dose equivalent to 2 full doses on oilseed rape, beans, cauliflowers, calabrese; 3 full doses on Brussels sprouts, leeks onions, early potatoes; 5 full doses on maincrop potaties
- Do not treat potato crops showing active blight infection or a second potato crop in the same field in the same season

Special precautions/Environmental safety
- Irritating to skin and eyes
- Risk of serious damage to eyes
- Dangerous to fish or other aquatic life. Do not contaminate surface waters or ditches with chemical or used container
- LERAP Category B

Personal protective equipment/Label precautions
- A, C, P
- R04a, R04b, R04d, U05a, U09a, U20a, C03; C02 (early potatoes 7d; all other crops 14 d); E01, E13b, E16a, E16b, E26, E29, E30a, E31a, E34

Latest application/Harvest interval
- HI field beans, broad beans, Brussels sprouts, cauliflowers, calabrese, onions, leeks 14 d

Maximum Residue Level (mg residue/kg food)
- see chlorothalonil and metalaxyl entries

---

118   chlorothalonil + propamocarb hydrochloride

A contact and systemic fungicide mixture for blight control in potatoes

Products

1	Aventis Merlin	Aventis	375:375 g/l	SC	09719
2	Merlin	Aventis	375:375 g/l	SC	07943

Uses

Late blight in **potatoes**.

Efficacy
- Commence treatment early in the season as soon as there is risk of infection
- In the absence of a blight warning treatment should start just before potatoes meet along the row
- Use only as a protectant. Stop use when blight readily visible (1% leaf area destroyed)
- Repeat sprays at 10-14 d intervals depending on blight infection risk. See label for details
- Complete blight spray programme after end Aug up to haulm destruction with protectant fungicides preferably fentin based

Crop safety/Restrictions
- Maximum number of treatments 5 per crop
- Apply to dry foliage. Do not apply if rainfall or irrigation imminent

Special precautions/Environmental safety
- Irritating to eyes
- Risk of serious damage to eyes
- May cause sensitization by skin contact
- Dangerous to fish or other aquatic life. Do not contaminate surface waters or ditches with chemical or used container
- LERAP Category B

Personal protective equipment/Label precautions
- A, C, H
- R04a, R04d, R04e, U05a, U08, U11, U19, U20a, C03, E01, E13b, E16a, E16b, E30a, E31b, E34

Latest application/Harvest interval
- HI 7 d

Maximum Residue Level (mg residue/kg food)
- see chlorothalonil entry

119 chlorothalonil + tetraconazole

A systemic protectant and curative fungicide mixture for cereals

Products

Voodoo	Sipcam	250:62.5 g/l	SE	09414

Uses

Brown rust in *autumn sown spring barley, autumn sown spring wheat, spring barley, spring wheat, winter barley, winter wheat*. Crown rust in *spring oats, winter oats*. Net blotch in *autumn sown spring barley, spring barley, winter barley*. Powdery mildew in *autumn sown spring barley, autumn sown spring wheat, spring barley, spring wheat, winter barley, winter wheat*. Rhynchosporium in *autumn sown spring barley, spring barley, winter barley*. Septoria diseases in *autumn sown spring wheat, spring wheat, winter wheat*. Sooty moulds in *spring wheat, winter wheat*. Yellow rust in *autumn sown spring barley, autumn sown spring wheat, spring barley, spring wheat, winter barley, winter wheat*.

Efficacy

- For best results apply before onset of disease attack, or at an early stage of disease development. Further treatments may be necessary if disease pressure remains high
- May be used in a programme in combination with a number of other fungicides to improve overall control and reduce potential for development of resistance

Crop safety/Restrictions

- Maximum total dose equivalent to three full dose treatments on wheat; two full dose treatments on barley

Special precautions/Environmental safety

- Irritating to eyes and skin
- Harmful to fish or other aquatic life. Do not contaminate surface waters or ditches with chemical or used container
- LERAP Category B

Personal protective equipment/Label precautions

- A, C
- M03, R04a, R04b, U05a, U19, U20a, C03, E01, E13c, E16a, E16b, E30a, E31b, E34

Latest application/Harvest interval

- Before early flowering (GS 63) in wheat; before end of ear emergence (GS 59) in barley

Maximum Residue Level (mg residue/kg food)

- see chlorothalonil entry

120 chlorotoluron

A contact and residual urea herbicide for cereals

Products

1	Alpha Chlortoluron 500	Makhteshim	500 g/l	SC	04848
2	Atol	Nufarm Whyte	700 g/l	SC	10235
3	Lentipur CL 500	Nufarm Whyte	500 g/l	SC	08743
4	Luxan Chlorotoluron 500 Flowable	Luxan	500 g/l	SC	09165
5	MSS Chlorotoluron 500	Nufarm Whyte	500 g/l	SC	07871
6	NWA CTU 500	Nufarm Whyte	500 g/l	SC	10173
7	Portman Chlortoluron	Portman	500 g/l	SC	03068
8	Tolugan 700	Makhteshim	700 g/l	SC	07874
9	Tripart Ludorum	Tripart	500 g/l	SC	03059

Uses

Annual dicotyledons in *winter barley, winter wheat* [1-9]. Annual dicotyledons in *triticale* [1-6, 8, 9]. Annual dicotyledons in *durum wheat* [3, 4, 6, 9]. Annual grasses in *winter barley, winter wheat* [1-9]. Annual grasses in *triticale* [1-6, 8, 9]. Annual grasses in *durum wheat* [3, 4, 6, 9]. Blackgrass in *winter barley, winter wheat* [1-9]. Blackgrass in *triticale* [1-6, 8, 9].

Blackgrass in **durum wheat** [3, 4, 6, 9]. Rough meadow grass in **winter barley, winter wheat** [1-9]. Rough meadow grass in **triticale** [1-6, 8, 9]. Rough meadow grass in **durum wheat** [3, 4, 6, 9]. Wild oats in **winter barley, winter wheat** [1-9]. Wild oats in **triticale** [1-6, 8, 9]. Wild oats in **durum wheat** [3, 4, 6, 9].

Efficacy
- Best results achieved by application soon after drilling. Application in autumn controls most weeds germinating in early spring
- For wild oat control apply within 1 wk of drilling, not after 2-leaf stage. Blackgrass and meadow grasses controlled to 5 leaf, ryegrasses to 3 leaf stage
- Any trash or burnt straw should be buried and dispersed during seedbed preparation
- Do not use on soils with more than 10% organic matter
- Control may be reduced if prolonged dry conditions follow application
- Harrowing after treatment may reduce weed control

Crop safety/Restrictions
- Maximum number of treatments 1 per crop
- Use only on listed crop varieties. See label. Ensure seed well covered at drilling
- Apply only as pre-emergence spray in durum wheat, pre- or post-emergence in wheat, barley or triticale
- Do not apply pre-emergence to crops sown after 30 Nov
- Do not apply to crops severely checked by waterlogging, pests, frost or other factors
- Do not use on undersown crops or those due to be undersown
- Do not apply post-emergence in mixture with liquid fertilizers
- Do not roll for 7 d before or after application to an emerged crop
- Crops on stony or gravelly soils may be damaged, especially after heavy rain

Special precautions/Environmental safety
- Harmful if swallowed [1, 5, 7, 8]
- Irritating to eyes and skin [1, 2, 5, 7-9]
- Harmful to fish or other aquatic life. Do not contaminate surface waters or ditches with chemical or used container

Personal protective equipment/Label precautions
- A, C [1-3, 5, 7, 8]
- M03 [1, 2, 5, 8]; R04a, R04b [1, 2, 5, 7-9]; R03c [1, 5, 7, 8]; U08, U19 [1-9]; U05a [1, 2, 5, 7-9]; U20a [2, 5, 7-9]; U20b [1, 3, 4, 6]; C03 [1, 2, 5, 7-9]; E15 [9]; E31a [7, 9]; E30a, E31b [1-9]; E34 [1, 2, 4, 5, 7-9]; E01 [1, 2, 5, 7-9]; E26 [4, 7, 9]; E13c [1-8]

Latest application/Harvest interval
- Pre-emergence for durum wheat. Post emergence timings on other crops vary - see labels

Approval
- May be applied through CDA equipment. See labels for details [2, 3, 9]
- Approved for aerial application on wheat, barley, triticale [3, 9]; on what and barley [4]. See notes in Section 1
- Accepted by BLRA for use on malting barley

121   chlorotoluron + isoproturon

A urea herbicide mixture for cereals

Products
Tolugan Extra	Makhteshim	300:300 g/l	SC	09393

Uses
Annual dicotyledons in **winter barley, winter wheat**. Annual grasses in **winter barley, winter wheat**. Blackgrass in **winter barley, winter wheat**. Rough meadow grass in **winter barley, winter wheat**. Wild oats in **winter barley, winter wheat**.

Efficacy
- Best results achieved by application soon after drilling. Application in autumn controls most weeds germinating in early spring
- For wild oat control apply within 1 wk of drilling, not after 2-leaf stage. Blackgrass and meadow grasses controlled to 5 leaf, ryegrasses to 3 leaf stage
- Any trash or burnt straw should be buried and dispersed during seedbed preparation
- Do not use on soils with more than 10% organic matter
- Control may be reduced if prolonged dry conditions follow application
- Harrowing after treatment may reduce weed control
- Where strains of herbicide-resistant blackgrass occur control may be unsatisfactory

Crop safety/Restrictions
- Maximum number of treatments 1 per crop
- Use only on listed crop varieties. See label. Ensure seed well covered at drilling
- Apply only as post-emergence treatment
- Do not apply to crops severely checked by waterlogging, pests, frost or other factors
- Do not use on undersown crops or those due to be undersown
- Do not apply post-emergence in mixture with liquid fertilizers
- Do not roll for 7 d before or after application to an emerged crop
- Crops on stony or gravelly soils may be damaged, especially after heavy rain

Special precautions/Environmental safety
- Irritant. May cause sensitization by skin contact
- Dangerous to fish or other aquatic life. Do not contaminate surface waters or ditches with chemical or used container

Personal protective equipment/Label precautions
- A, C, H
- R04, R04e, U05a, U08, U19, U20b, C03, E01, E13b, E26, E30a, E31b, E34

Latest application/Harvest interval
- Before end of tillering (GS 29)

Approval
- Accepted by BLRA for use on malting barley

---

122   chlorpropham

A residual carbamate herbicide and potato sprout suppressant

Products

1	Atlas CIPC 40	Nufarm Whyte	400 g/l	EC	07710
2	BL 500	Wheatley	500 g/l	HN	00279
3	Comrade	United Phosphorus	400 g/l	EC	10181
4	Luxan Gro-Stop 300 EC	Luxan	300 g/l	EC	08602
5	Luxan Gro-Stop Basis	Luxan	300 g/l	EC	08601
6	Luxan Gro-Stop Fog	Luxan	300 g/l	HN	09388
7	Luxan Gro-Stop HN	Luxan	300 g/l	HN	07689
8	MSS CIPC 40 EC	Nufarm Whyte	400 g/l	EC	01403
9	MSS CIPC 5 G	Nufarm Whyte	5% w/w	GR	01402
10	MSS CIPC 50 LF	Nufarm Whyte	500 g/l	EC	03285
11	MSS CIPC 50 M	Nufarm Whyte	500 g/l	EC	01404
12	MTM CIPC 40	MTM Agrochem.	400 g/l	EC	05895
13	Standon CIPC 300 HN	Standon	300 g/l	HN	09187
14	Warefog 25	Nufarm Whyte	600 g/l	HN	06776

Uses

Annual dicotyledons in **carrots, flower bulbs, onions** [1, 3, 8, 12]. Annual dicotyledons in **leeks** [1, 8]. Annual dicotyledons in **annual flowers, blackcurrants, celery, chrysanthemums, gooseberries, lucerne, strawberries** [3, 12]. Annual dicotyledons in **lettuce, parsley** [3, 8, 12]. Annual grasses in **carrots, flower bulbs, onions** [1, 3, 8, 12]. Annual grasses in **leeks** [1, 8]. Annual grasses in **annual flowers, blackcurrants, celery, chrysanthemums, gooseberries, lucerne, strawberries** [3, 12]. Annual grasses in **lettuce, parsley** [3, 8, 12]. Chickweed in **carrots** [1, 3, 8, 12]. Chickweed in **flower bulbs, leeks, onions** [1, 8]. Chickweed in **annual flowers, blackcurrants, celery, chrysanthemums, gooseberries, lucerne, strawberries** [3, 12]. Chickweed in **lettuce, parsley** [3, 8, 12]. Polygonums in **carrots** [1, 3, 8, 12]. Polygonums in **flower bulbs, leeks, onions** [1, 8]. Polygonums in **annual flowers, blackcurrants, celery, chrysanthemums, gooseberries, lucerne, strawberries** [3, 12]. Polygonums in **lettuce, parsley** [3, 8, 12]. Sprout suppression in **potatoes** [2, 9-11, 14]. Sprout suppression in **ware potatoes** [4, 5, 7, 13]. Sprout suppression in **ware potatoes** *(fog)* [6]. Volunteer ryegrass in **grass seed crops** *(off-label)* [1, 8, 12].

Efficacy
- Apply weed control sprays to freshly cultivated soil. Adequate rainfall must occur after spraying. Activity is greater in cold, wet than warm, dry conditions

- For sprout suppression apply with suitable fogging or rotary atomiser equipment or sprinkle granules over dry tubers before sprouting commences. Repeat applications may be needed. See labels for details
- Effectiveness of fogging reduced if air spaces between tubers are blocked. best results obtained at 5-10°C and 75-80% humidity
- Cure potatoes according to label instructions before treatment and allow 3 wk between completion of loading into store and first treatment

Crop safety/Restrictions

- Maximum number of treatments 1 per batch (3 per batch [10]) for fogging potatoes; 1 per yr for grass seed crops [1, 3, 8, 12]; 1 per crop for carrots, onions, leeks [1]; 2 per yr for flower bulbs [1]
- Maximum total dose 2.55 l per 20 tonnes [11], 3 l per 50 tonnes [7], of potatoes; 60 ml per 1000 kg batch of potatoes [13]
- Apply weed control sprays to seeded crops pre-emergence of crop or weeds, to onions as soon as first crop seedlings visible, to planted crops a few days before planting, to bulbs immediately after planting, to fruit crops in late autumn-early winter. See label for further details
- Not to be used on grass seed crops if grass to be grazed or cut for fodder before 31 May following treatment [1, 3, 8, 12]
- Excess rainfall after application may result in crop damage
- Do not use on Sands, Very Light soils or soils low in organic matter
- Poor conditions at drilling or planting, soil compaction, surface capping, waterlogging or attack by pests may result in crop damage
- On crops under glass high temperatures and poor ventilation may cause crop damage
- Only clean, mature, disease-free potatoes should be treated for sprout suppression
- Do not use on potatoes for seed. Do not handle, dry or store seed potatoes or any other seed or bulbs in boxes or buildings in which potatoes are being or have been treated
- Do not fog potatoes with a high level of skin spot [7]
- Do not remove potatoes for sale or processing for at least 21 d (4 wk [7, 13]) after application
- A minimum interval of 45 d (4 wk [7]) must elapse between applications [10, 11]

Special precautions/Environmental safety

- Harmful in contact with skin [1, 3, 8, 12]
- Harmful by inhalation [2, 6, 7, 11, 13]
- Harmful if swallowed [1-3, 6-8, 10-13]
- Irritating to eyes [14] and skin [1, 2, 6-8, 10, 11, 13]
- Irritating to respiratory system [1, 2, 6-8, 11, 13, 14]
- May cause sensitization by skin contact and by inhalation [6, 7, 13]
- Highly flammable [2, 11]
- Flammable [10]
- Harmful to fish. Do not contaminate ponds, waterways or ditches with chemical or used container/Harmful to fish or other aquatic life. Do not contaminate surface waters or ditches with chemical or used container [3-7, 12, 13]
- Keep unprotected persons out of treated areas for at least 24 h after application [7, 11]

Personal protective equipment/Label precautions

- A [1-8, 10-14]; C [1-5, 8, 10-12, 14]; J [2, 10, 11]; D [2, 6, 7, 10, 11, 13, 14]; H [2, 4-7, 10, 11, 13]; M [2, 6, 7, 10, 11, 13]; E [2, 6, 7, 11, 13, 14]; G [10]
- M03 [1-3, 6-13]; R03c [1-3, 6-8, 10-13]; R04a [1, 2, 6-8, 10, 11, 13, 14]; R04b [1, 2, 6-8, 10, 11, 13]; R04c [1, 2, 6-8, 11, 13, 14]; R07c [2, 11]; R03b [2, 6, 7, 11, 13]; R03a [1, 3, 8, 12]; R07d [10]; R04e, R04f [6, 7, 13]; U05a [1-14]; U20b [2, 3, 8-12]; U19 [1, 2, 6-8, 10, 11, 13, 14]; U09a [2, 3, 6, 8, 12]; U08 [1, 3-5, 7, 10-14]; U20c [14]; U20a [1, 4-7, 13]; U02, U04a, U13, U14, U17 [6, 7, 13]; C03, E01 [1-14]; E30a [1-5, 8-12, 14]; E34 [1-13]; E15 [1, 2, 8-11, 14]; E02 [2, 7, 11, 13] (24 h); E31a [1, 2, 10, 14]; E26 [3, 9, 12]; E32a [4-7, 9, 13]; E31b [3, 8, 11, 12]; E13c [3-7, 12, 13]

Latest application/Harvest interval

- 14 d after planting for leeks, onions [1]; 21 d before removal for sale or processing for potatoes [2]; 8 wk (4 wk [7]) before marketing or processing for ware potatoes [4, 5]; 48 h before drilling for carrots [1]; before 4 leaf stage (Silt soils only) for onions [1]; before leaves unfurl (tulip); up to 5 cm high (others) for flower bulbs [1]

Approval
- Some products are formulated for application by thermal fogging. See labels for details [2, 7, 13, 14]
- Off-label approval to May 2002 for use on grass seed crops (OLA 0240/98)[8], (OLA 0241/98)[12], OLA 0242/98)[1]

123　chlorpropham with cetrimide

A soil-acting herbicide for lettuce under cold glass

Products

Croptex Pewter	Hortichem	80:80 g/l	SC	02507

Uses

Annual dicotyledons in *protected lettuce*. Annual grasses in *protected lettuce*. Chickweed in *protected lettuce*. Polygonums in *protected lettuce*.

Efficacy
- Apply to drilled lettuce under cold glass within 24 h post-drilling, to transplanted crops pre-planting
- Adequate irrigation must be applied before or after treatment
- Best results achieved on firm soil of fine tilth, free from clods and weeds

Crop safety/Restrictions
- Do not apply to crop foliage or use where seed has chitted
- Excess irrigation may cause temporary check to crop under certain circumstances
- Do not apply where tomatoes or brassicas are growing in the same house
- In the event of crop failure only lettuce should be grown within 2 mth

Special precautions/Environmental safety
- Harmful in contact with skin and if swallowed
- Irritating to eyes, skin and respiratory system
- Highly flammable
- Dangerous to fish or other aquatic life. Do not contaminate surface waters or ditches with chemical or used container

Personal protective equipment/Label precautions
- A, C
- R03a, R03c, R04a, R04b, R04c, R07c, U05a, U08, U19, U20a, C03, E01, E13b, E26, E30a, E31a

Latest application/Harvest interval
- Before crop emergence or pre-planting

124　chlorpropham + fenuron

A residual herbicide for vegetables and ornamentals

Products

Croptex Chrome	Hortichem	80:15 g/l	EC	02415

Uses

Annual dicotyledons in *broad beans, field beans, flowers, leeks, maiden strawberries, onions, peas, protected bulbs, runner beans* (off-label), *spinach*. Annual grasses in *broad beans, field beans, flowers, leeks, maiden strawberries, onions, peas, protected bulbs, runner beans* (off-label), *spinach*. Chickweed in *broad beans, field beans, flowers, leeks, maiden strawberries, onions, peas, protected bulbs, runner beans* (off-label), *spinach*.

Efficacy
- Apply to soil freshly cultivated and free of established weeds. Adequate rainfall must occur after spraying. Activity is greater in cold, wet than warm, dry conditions

Crop safety/Restrictions
- Apply to seeded crops pre-emergence of crop or weeds, to transplanted onions and leeks 10-14 d post-planting, to transplanted flowers and strawberries 5 d pre-transplanting

- Do not use on Sands, Very Light soils or soils low in organic matter
- Do not treat protected bulb or flower crops if other crops are being grown in same block of houses
- Poor conditions at drilling or planting, soil compaction, surface capping, waterlogging or attack by pests may result in crop damage
- In the event of crop failure only recommended crops should be replanted in treated soil
- Plough or cultivate to 15 cm after harvest to dissipate any residues

Special precautions/Environmental safety
- Harmful if swallowed and in contact with skin
- Irritating to eyes, skin and respiratory system
- Flammable
- Dangerous to fish or other aquatic life. Do not contaminate surface waters or ditches with chemical or used container

Personal protective equipment/Label precautions
- A, C
- M03, R03a, R03c, R04a, R04b, R04c, R07d, U05a, U08, U13, U19, U20a, C03, E01, E13b, E26, E30a, E32a, E34

Approval
- Off-label approval unlimited for use on outdoor and covered crops of runner beans (OLA0622/96)

---

125   chlorpropham + linuron

A residual and contact herbicide for use in bulb crops

Products

Profalon	Hortichem	200:100 g/l	EC	08900

Uses

Annual dicotyledons in *daffodils, narcissi, tulips*. Annual grasses in *daffodils, narcissi, tulips*.

Efficacy
- Weeds controlled by combined residual action, requiring adequate soil moisture, and contact effect on young seedlings, requiring dry leaf surfaces for good control
- Do not spray during or immediately prior to rainfall
- Do not cultivate after spraying unless necessary

Crop safety/Restrictions
- Maximum number of treatments 1 per crop
- Spray daffodils and narcissi pre-emergence or post-emergence before flower buds show. Spray tulips pre-emergence only
- Do not apply on Very Light soils or Silts low in humus or clay content

Special precautions/Environmental safety
- Irritating to skin, eyes and respiratory system
- Flammable
- Dangerous to fish or other aquatic life. Do not contaminate surface waters or ditches with chemical or used container
- LERAP Category B
- Do not apply by hand-held sprayers

Personal protective equipment/Label precautions
- A, C, H
- R04a, R04b, R04c, R07d, U04a, U05a, U08, U19, U20a, C03, E01, E13b, E16a, E16b, E26, E30a, E31b, E34

Latest application/Harvest interval
- Pre-emergence for tulips

126 chlorpropham + pentanochlor

A contact and residual herbicide for horticultural crops

Products

| Atlas Brown | Nufarm Whyte | 150:300 g/l | EC | 07703 |

Uses

Annual dicotyledons in *carrots, celeriac, celery, chrysanthemums, fennel, narcissi, outdoor leaf herbs* (off-label), *parsley, parsnips, tulips*. Annual meadow grass in *carrots, celeriac, celery, chrysanthemums, fennel, narcissi, outdoor leaf herbs* (off-label), *parsley, parsnips, tulips*.

Efficacy
- Apply as pre- or post-weed emergence spray
- Best results by application to weeds up to 2-leaf stage on fine, firm, moist seedbed
- Greatest contact action achieved under warm, moist conditions, the short residual action greatest in earlier part of year

Crop safety/Restrictions
- Maximum number of treatments 1 per yr for narcissi, tulips; 2 per yr for carrots, celeriac, celery, fennel, parsnips, chrysanthemums, outdoor leaf herbs
- Apply to carrots and related crops pre- or post-emergence after fully expanded cotyledon stage, to narcissi and tulips at any time before emergence
- Apply to chrysanthemums either pre-planting or after planting as carefully directed spray, avoiding foliage
- Any crop may be sown or planted after 4 wk following ploughing and cultivation

Special precautions/Environmental safety
- Irritating to eyes

Personal protective equipment/Label precautions
- C
- R04a, U05a, U19, U20a, C03, E01, E15, E30a, E31b

Latest application/Harvest interval
- Pre-emergence for narcissi, tulips
- HI 28 d for carrots, parsnips

Approval
- Off-label approval unlimited for use on outdoor leaf herbs (OLA 0249/93, 0602/93)

127 chlorpropham + tar acids + fenuron

A residual herbicide for vegetables and ornamentals

Products

| Atlas Red | Nufarm Whyte | 200:-:50 g/l | EC | 07724 |

Uses

Annual dicotyledons in *corms, flower bulbs, hardy ornamental nursery stock, leeks, onions, paths and drives, rhubarb, standing ground*. Annual grasses in *corms, flower bulbs, hardy ornamental nursery stock, leeks, onions, paths and drives, rhubarb, standing ground*. Chickweed in *corms, flower bulbs, hardy ornamental nursery stock, leeks, onions, paths and drives, rhubarb, standing ground*.

Efficacy
- Apply to soil freshly cultivated and free of established weeds. Adequate rainfall must occur after spraying. Activity is greater in cold, wet than warm, dry conditions
- Weeds emerging during dry weather may not be controlled even if rain falls shortly after emergence

Crop safety/Restrictions
- Apply to bulbs immediately after planting and/or just before emergence

- Apply to seeded onions and leeks immediately after sowing and repeat after crook stage when crop fully waxed up
- Apply to rhubarb and nursery stock in dormant season, avoiding foliage of conifers or evergreens
- Do not use on Sands, Very Light soils or soils low in organic matter
- Poor conditions at drilling or planting, soil compaction, surface capping, waterlogging or pest attack may result in crop damage
- In the event of crop failure only recommended crops should be planted in treated soil
- Plough or cultivate to 150 mm after harvest to dissipate any residues

Special precautions/Environmental safety
- Harmful if swallowed and in contact with skin
- Irritating to eyes, skin and respiratory system
- Flammable
- Wash equipment thoroughly with water and wetting agent immediately after use and spray out. Spray out again before storing or using another product

Personal protective equipment/Label precautions
- A, C
- M03, R03a, R03c, R04a, R04b, R04c, R07d, U05a, U13, U19, U20b, C03, E01, E15, E30a, E31a, E34

Latest application/Harvest interval
- Feb for rhubarb; 2 mth before planting standing ground; not specified for other crops or situations

---

128   chlorpyrifos

A contact and ingested organophosphorus insecticide and acaricide

Products

1	Alpha Chlorpyrifos 48 EC	Makhteshim	480 g/l	EC	04821
2	Ballad	Headland	480 g/l	EC	09775
3	Barclay Clinch II	Barclay	480 g/l	EC	08596
4	Choir	Nufarm Whyte	480 g/l	EC	09778
5	Crossfire 480	Aventis Environ.	480 g/l	EC	09929
6	Crossfire 480	RP Amenity	480 g/l	EC	08142
7	Dursban 4	Dow	480 g/l	EC	07815
8	Greencrop Pontoon	Greencrop	480 g/l	EC	09667
9	Lorsban T	Rigby Taylor	480 g/l	EC	07813
10	Lorsban WG	Dow	75% w/w	WG	10139
11	Maraud	Scotts	480 g/l	EC	09274
12	Pyrinex 48EC	Makhteshim	480 g/l	EC	08644
13	Spannit	PBI	480 g/l	EC	08744
14	Spannit Granules	PBI	6% w/w	GR	08984
15	Standon Chlorpyrifos 48	Standon	480 g/l	EC	08286
16	SuSCon Green Soil Insecticide	Fargro	10% w/w	GR	06312
17	suSCon Indigo	Fargro	10% w/w	GR	09902

Uses

Ambrosia beetle in *cut logs* [1, 3, 4, 8, 9, 12]. Ambrosia beetle in *cut timber* [3]. Aphids in *blackcurrants* [1, 10, 12, 15]. Aphids in *spring barley, spring oats, spring wheat, winter barley, winter oats, winter wheat* [1, 12]. Aphids in *apples, gooseberries, pears, plums, raspberries, strawberries* [1, 3, 4, 7, 8, 10, 12, 13, 15]. Aphids in *brassicas* [1, 3, 4, 8, 12, 13]. Aphids in *redcurrants, whitecurrants* [15]. Aphids in *currants* [3, 4, 7, 8, 13]. Aphids in *oats, wheat* [3, 4, 7, 8, 13, 15]. Aphids in *fennel* (off-label) [7]. Aphids in *broccoli, cabbages, calabrese, cauliflowers, chinese cabbage* [7, 15]. Apple blossom weevil in *apples* [1, 3, 4, 7, 8, 10, 12, 13, 15]. Apple blossom weevil in *pears* [3, 13]. Apple sucker in *apples* [15]. Black pine beetle in *forestry transplants* [1, 3, 4, 7, 9, 12]. Black pine beetle in *forestry transplant lines* [8]. Cabbage root fly in *brassicas* [1, 3, 4, 8, 12, 13]. Cabbage root fly in *brussels sprouts, mooli* (off-label), *radishes* (off-label) [7]. Cabbage root fly in *broccoli, cabbages,*

---

cauliflowers [7, 14, 15]. Cabbage root fly in **calabrese, chinese cabbage** [7, 15]. Capsids in **apples, gooseberries, pears** [1, 3, 4, 7, 8, 10, 12, 13, 15]. Capsids in **currants** [1, 3, 4, 7, 8, 12, 13]. Capsids in **blackcurrants, redcurrants, whitecurrants** [10, 15]. Caterpillars in **pears** [1, 3, 4, 7, 8, 10, 12, 13]. Caterpillars in **gooseberries** [1, 3, 4, 7, 8, 10, 12, 13, 15]. Caterpillars in **apples, currants** [1, 3, 4, 7, 8, 12, 13]. Caterpillars in **brassicas** [1, 3, 4, 8, 12, 13]. Caterpillars in **blackcurrants, redcurrants, whitecurrants** [10, 15]. Caterpillars in **broccoli, cabbages, calabrese, cauliflowers, chinese cabbage** [7, 15]. Codling moth in **apples, pears** [1, 3, 4, 7, 8, 10, 12, 13, 15]. Colorado beetle in **potatoes** *(off-label)* [7]. Cutworms in **carrots, potatoes** [1, 3, 4, 7, 8, 12, 13, 15]. Cutworms in **onions** [1, 3, 4, 7, 8, 12, 15]. Cutworms in **leaf brassicas** [1, 8, 12]. Cutworms in **brassicas** [3, 4, 13]. Cutworms in **fennel** *(off-label)* [7]. Cutworms in **broccoli, cabbages, calabrese, cauliflowers, chinese cabbage** [7, 15]. Damson-hop aphid in **plums** [1, 8, 10, 12, 15]. Elm bark beetle in **cut logs** [4, 9]. Frit fly in **spring oats, spring wheat, winter oats, winter wheat** [1, 12]. Frit fly in **managed amenity turf** [1, 2, 5, 6, 12]. Frit fly in **grassland, maize** [1, 3, 4, 7, 8, 12, 13, 15]. Frit fly in **barley** [13]. Frit fly in **amenity grass** [2-6, 9, 11, 15]. Frit fly in **oats, wheat** [3, 4, 7, 8, 13, 15]. Frit fly in **golf courses** [3, 4, 9, 11, 13]. Frit fly in **amenity turf** [8, 11, 15]. Insect pests in **edible podded peas** *(off-label)*, **flower bulbs** *(off-label)*, **peas** *(off-label)*, **protected brassica seedlings** *(off-label)*, **protected onions/leeks/garlic** *(off-label)* [13]. Insect pests in **green beans** *(off-label)* [7]. Larch shoot beetle in **cut logs** [1, 3, 4, 7-9, 12]. Larch shoot beetle in **cut timber** [3]. Large pine weevil in **forestry transplants** [4]. Leatherjackets in **spring barley, spring oats, spring wheat, winter barley, winter oats, winter wheat** [1, 12]. Leatherjackets in **managed amenity turf** [1, 2, 5, 6, 12]. Leatherjackets in **grassland, sugar beet** [1, 3, 4, 7, 8, 12, 13, 15]. Leatherjackets in **amenity grass** [2-6, 9, 11, 15]. Leatherjackets in **brassicas** [3, 4, 13]. Leatherjackets in **oats, wheat** [3, 4, 7, 8, 13, 15]. Leatherjackets in **golf courses** [3, 4, 9, 11, 13]. Leatherjackets in **broccoli, cabbages, calabrese, cauliflowers, chinese cabbage** [7, 15]. Leatherjackets in **amenity turf** [8, 11, 15]. Mealy aphids in **plums** [1, 8, 10, 12]. Pear sucker in **pears** [15]. Pine shoot beetle in **cut logs** [1, 3, 4, 7-9, 12]. Pine shoot beetle in **cut timber** [3]. Pine weevil in **forestry transplants** [1, 3, 4, 7, 9, 12]. Pine weevil in **forest nurseries** [1, 3, 9, 12]. Pine weevil in **forestry transplant lines** [8]. Pygmy mangold beetle in **sugar beet** [3, 4, 7, 13]. Raspberry beetle in **raspberries** [1, 3, 4, 7, 8, 10, 12, 13, 15]. Raspberry cane midge in **raspberries** [1, 3, 4, 7, 8, 10, 12, 13, 15]. Red spider mites in **apples, gooseberries, pears, plums, raspberries, strawberries** [1, 3, 4, 7, 8, 10, 12, 13, 15]. Red spider mites in **currants** [1, 3, 4, 7, 8, 12, 13]. Red spider mites in **blackcurrants, redcurrants, whitecurrants** [10, 15]. Sawflies in **apples** [1, 3, 4, 7, 8, 10, 12, 13, 15]. Sawflies in **pears** [3, 7, 13]. Sciarid flies in **ornamental plant production** [17]. Strawberry blossom weevil in **strawberries** [7, 10]. Suckers in **apples, pears** [1, 3, 4, 7, 8, 10, 12, 13]. Summer-fruit tortrix moth in **apples** [15]. Thrips in **oats, wheat** [3, 4, 13]. Tortrix moths in **apples, pears, plums, strawberries** [1, 3, 4, 7, 8, 10, 12, 13, 15]. Vine weevil in **strawberries** [1, 3, 4, 7, 8, 10, 12, 15]. Vine weevil in **container-grown ornamentals, ornamental plant production** [16]. Vine weevil in **conifers** [8]. Vine weevil in **forestry transplant lines** [9]. Wheat bulb fly in **spring oats, spring wheat, winter oats, winter wheat** [1, 12]. Wheat bulb fly in **oats, wheat** [3, 4, 7, 8, 13, 15]. Wheat-blossom midges in **spring oats, spring wheat, winter oats, winter wheat** [1, 12]. Wheat-blossom midges in **oats, wheat** [3, 4, 7, 8, 13, 15]. Whitefly in **brassicas** [4]. Whitefly in **broccoli, cabbages, calabrese, cauliflowers, chinese cabbage** [7]. Winter moth in **apples, pears, plums** [1, 3, 4, 7, 8, 10, 12, 13, 15]. Woolly aphid in **apples** [1, 3, 4, 7, 8, 10, 12, 13, 15]. Woolly aphid in **pears** [3, 4, 7, 13].

Efficacy

- Brassicas raised in plant-raising beds may require retreatment at transplanting
- Activity may be reduced when soil temperature below 5°C or on organic soils
- In dry conditions the effect of granules applied as a surface band may be reduced [14]
- For vine weevil control in hardy ornamental nursery stock incorporate in growing medium when plants first potted from rooted cutting stage. Treat the fresh growing medium when plants are potted on into larger containers [16]
- For turf pests apply from Nov where high larval populations detected or damage seen [2, 5, 6, 9, 11]

- For sciarid control in ornamental plant production it is essential to incorporate thoroughly into growing media by hand or machine shortly before potting. Product controls larvae for several months from a single application [17]
- Control of sciarids in ornamental plant production not guaranteed if cuttings struck into untreated media are potted into treated media [17]
- Where pear suckers resistant to chlorpyrifos occur control is unlikely to be satisfactory

Crop safety/Restrictions

- Maximum number of treatments and timing vary with crop, product and pest. See label for details
- On carrots the maximum total dose applied per crop must not exceed the equivalent of 3 (on mineral soils) or 4 (on organic soils) full dose applications
- Do not treat potatoes under severe drought stress. The variety Desiree is particularly susceptible
- On lettuce apply only to strong well developed plants when damage first seen, or on professional advice
- Do not apply to sugar beet under stress or within 4 d of applying a herbicide
- Do not mix with highly alkaline materials. Not compatible with zineb [7]
- In apples use pre-blossom up to pink/white bud and post-blossom after petal fall
- Cuttings from treated grass should not be used as a mulch for at least 12 mth
- Test tolerance of glasshouse ornamentals before widescale use for propagating unrooted cuttings, or potting unusual plants and new species, and when using any media with a high content of non-peat ingredients [17]

Special precautions/Environmental safety

- Products contain an anticholinesterase organophosphorus compound. Do not use if under medical advice not to work with such compounds
- Harmful in contact with skin [3, 5-9, 11, 13, 15]
- Harmful if swallowed [1-9, 11-13, 15]
- May cause lung damage if swallowed [7]
- Irritating to eyes [2, 4, 13, 17]
- Risk of serious damage to eyes [1, 12]
- Irritating to skin [1-9, 11-13, 15]
- May cause sensitization by skin contact [1, 12]
- Flammable [1, 3, 5-9, 11-13, 15]
- Not to be used on food crops [16, 17]
- Dangerous (high risk [2, 4]) to bees. Do not apply to crops in flower or to those in which bees are actively foraging except as directed on crop. Do not apply when flowering weeds are present [1, 3, 5-13, 15]
- Dangerous (extremely dangerous [2, 4, 8, 10, 11, 16, 17]) to fish or other aquatic life. Do not contaminate surface waters or ditches with chemical or used container [1, 3, 5-7, 9, 12-15]
- LERAP Category A (except 14, 16, 17)
- Do not allow direct spray from broadcast air-assisted sprayers to fall within 18 m of surface waters or ditches; direct spray away from water [4, 8, 10, 11]
- Do not use product in growing media used for aquatic or semi-aquatic plants, and not for propagation of any edible crop [17]

Personal protective equipment/Label precautions

- A [1-17]; H [1, 3, 4, 7-9, 12, 13, 15]; C [1, 2, 4, 12, 13]
- M01 [1, 2, 4-17]; M03 [1-13, 15]; M05 [2, 4, 7]; R03c, R04b [1-9, 11-13, 15]; R07d [1, 3, 5-9, 11-13, 15]; R04d, R04e [1, 12]; R03a [3, 5-9, 11, 13, 15]; R04g [7]; R04a [2, 4, 13, 17]; U20b [1-12, 15-17]; U02 [1, 12-14, 16, 17]; U08 [1-9, 11-13, 15]; U05a [1, 3-9, 11-13, 15]; U19 [1, 3-13, 15]; U20a [13, 14]; U09a [10]; C03 [1-9, 11-13, 15]; C01 [16, 17]; C02 [13, 14]; E01 [1-9, 11-17]; E30a [1-17]; E13b [1, 3, 5-7, 9, 12-15]; E16c, E16d, E34 [1-13, 15]; E31b [1, 3-13, 15]; E12c [1, 7, 9-13]; E07 [1-4, 7, 8, 10, 12, 15] (14 d); E26 [15-17]; E32a [13, 14, 16, 17]; E13a [2, 4, 8, 10, 11, 16, 17]; E12d [3, 5, 6, 8, 15]; E17 [4, 8, 10, 11] (18 m); E12a [2, 4]

Withholding period

- Keep livestock out of treated areas for at least 14 d after treatment [1-4, 7, 8, 10, 12, 15]

FOR FULL CONDITIONS OF USE ALWAYS READ THE PRODUCT LABEL

Latest application/Harvest interval
- Varies with product. See individual labels

Approval
- Following implementation of Directive 98/82/EC, approval for use of chlorpyrifos on numerous crops was revoked in 1999
- Off-label approval unlimited for aerial application to control Colorado beetle in potatoes as required by Statutory Notice (OLA 0796/91) [7]; unlimited for use on a wide range of seedling brassicas and vegetables but after 1999 excluding Brussels sprouts, kale, collards, spring greens (OLA 1199/96)[13]; unlimited for use on ornamental bulbs (OLA 1498/96)[13]; unlimited for use on phaseolus green beans (kidney beans) harvested dry (OLA 1121/98)[7]; unstipulated for use on fennel (OLA 2822/99)[7]; unstipulated for use on radish and mooli (OLA 2935/99)[7]
- Approval expiry 31 Jul 2001 [7]

Maximum Residue Level (mg residue/kg food)
- kiwi fruit 2; pome fruits, grapes, tomatoes, peppers, aubergines 0.5; citrus fruits 0.3; plums, strawberries 0.2; carrots, tea 0.1; tree nuts, apricots, bilberries, cranberries, currants, gooseberries, wild berries, avocados, dates, figs, kumquats, litchis, mangoes, olives, passion fruit, pineapples, pomegranates, beetroot, celeriac, parsnips, horseradish, Jerusalem artichokes, parsley root, radishes, swedes, turnips, salsify, sweet potatoes, yams, garlic, shallots, spring onions, cucurbits, sweetcorn, kohlrabi, cress, spinach, beet leaves, watercress, witloof, mushrooms, pulses, oilseeds, potatoes, cereals, poultry 0.05; hops 0.1; milk, eggs 0.01

129   chlorpyrifos + disulfoton

A systemic and contact organophosphorus insecticide for brassicas

Products

Twinspan	PBI	4:6% w/w	GR	08983

Uses

Aphids in **broccoli, cabbages, cauliflowers**. Cabbage root fly in **broccoli, cabbages, cauliflowers**.

Efficacy
- Apply with suitable band applicator as bow-wave treatment at drilling followed if necessary by surface band treatment 2 d after singling, or as sub-surface band treatment at transplanting. See label for details
- Can be used on all mineral and organic soils
- Crops treated in plant-raising beds must be treated again at transplanting
- Effect of surface band application may be reduced in dry weather

Crop safety/Restrictions
- Do not treat mini-cauliflowers

Special precautions/Environmental safety
- This product contains an anticholinesterase organophosphorus compound. Do not use if under medical advice not to work with such compounds
- Harmful if swallowed and by inhalation
- Keep in original container, tightly closed, in a safe place, under lock and key
- Dangerous to game, wild birds and animals
- Dangerous to fish or other aquatic life. Do not contaminate surface waters or ditches with chemical or used container

Personal protective equipment/Label precautions
- A, B, H, K, M; C (or D+E)
- M01, M03, R03b, R03c, U02, U04a, U05a, U09a, U13, U14, U15, U19, U20a, C03; C02 (6 wk); E01, E10a, E13b, E30b, E32a, E34

Latest application/Harvest interval
- HI 6 wk

Approval
- Approval expiry 31 Dec 2001

Maximum Residue Level (mg residue/kg food)
- see chlorpyrifos and disulfoton entries

130 chlorpyrifos-methyl

An organophosphorus insecticide and acaricide for grain store use

Products

Reldan 22	Dow	225 g/l	EC	08191

Uses

Grain storage pests in *stored cereals*. Pre-harvest hygiene in *grain stores*.

Efficacy
- May be applied pre-harvest to surfaces of empty store and grain handling machinery and as admixture with grain
- Apply to grain after drying to moisture content below 14%, cooling and cleaning
- Insecticide may become depleted at grain surface if grain is being cooled by continuous extraction of air from the base leading to reduced control of grain store pests especially mites
- Resistance to organophosphorus compounds sometimes occurs in insect and mite pests of stored products

Crop safety/Restrictions
- Maximum number of treatments 1 per batch or 1 per store, prior to storage
- Min 90 d must elapse between treatment of grain and removal from store for consumption or processing

Special precautions/Environmental safety
- This product contains an anticholinesterase organophosphorus compound. Do not use if under medical advice not to work with such compounds
- May cause lung damage if swallowed
- Risk of serious damage to eyes
- Extremely dangerous to fish or other aquatic life. Do not contaminate surface waters or ditches with chemical or used container

Personal protective equipment/Label precautions
- A, C, H, M
- M01, M05, R04, R04d, R04g, U05a, U08, U19, U20b, C03, E01, E13a, E26, E30a, E31b, E34

Latest application/Harvest interval
- 8 wk before malting barley for stored grain; before use of storage facility for grain stores

Approval
- Following implementation of Directive 98/82/EC, approval for use of chlorpyrifos-methyl on cereal grain was revoked in 1999

Maximum Residue Level (mg residue/kg food)
- pome fruits, strawberries, tomatoes, peppers, aubergines 0.5; grapes 0.2; tea, hops 0.1; grapefruit, limes, pomelos, tree nuts, apricots, cherries, plums, cane fruits, bilberries, cranberries, currants, gooseberries, wild berries, miscellaneous fruits, root and tuber vegetables, bulb vegetables, cucurbits, brassicas, leaf vegetables, legumes, stem vegetables, mushrooms, pulses, oilseeds, potatoes, rice, meat 0.05; milk, eggs 0.01

131 chlorsulfuron

A sulfonylurea herbicide formulations of which were voluntarily withdrawn in the UK in 1988

132 chlorthal-dimethyl

A residual benzoic acid herbicide for use in horticulture

Products

Dacthal W-75	Hortichem	75% w/w	WP	05500

Uses

Annual dicotyledons in **blackcurrants, broccoli, brussels sprouts, cabbages, calabrese, cauliflowers, courgettes** *(off-label)*, **fodder rape, gooseberries, kale, leeks, marrows** *(off-label)*, **mustard, oilseed rape, onions, ornamentals, raspberries, runner beans, sage, strawberries, swedes, turnips**. Slender speedwell in **turf**.

Efficacy

- Best results on fine firm weed-free soil when adequate rain or irrigation follows
- Recommended alone on roses, runner beans, various ornamentals, strawberries and turf, in tank-mix with propachlor on brassicas, onions, leeks, sage, ornamentals, established strawberries and newly planted soft fruit
- Apply after drilling or planting prior to weed emergence. Rates and timing vary with crop and soil type. See label for details
- Do not use on organic soils
- For control of slender speedwell in turf apply when weeds growing actively. Do not mow for at least 3 d after treatment

Crop safety/Restrictions

- Maximum number of treatments 1 per crop for brassicas, oilseed rape, mustard, leeks, onions, sage, newly planted bush and cane fruit; 1 per season for established strawberries
- Apply to brassicas pre-emergence or after 3-4 true leaf stage
- Do not apply mixture with propachlor to newly planted strawberries after rolling or application of other herbicides
- Do not use on strawberries between flowering and harvest
- Many types of ornamental have been treated successfully. See label for details. For species of unknown susceptibility treat a small number of plants first
- Do not use on turf where bent grasses form a major constituent of sward
- Do not plant lettuce within 6 mth of application, seeded turf within 2 mth, other crops within 3 mth. In the event of crop failure deep plough before re-drilling or planting
- Do not use on dwarf French beans

Personal protective equipment/Label precautions

- U20a, E15, E30a, E32a

Latest application/Harvest interval

- Before crop emergence for sown crops

Approval

- Off-label approval to Jun 2000 for use on outdoor courgettes and marrows (OLA 1108/97)

---

133 choline chloride

A plant growth regulator available only in mixtures

See also chlormequat + choline chloride
chlormequat + choline chloride + imazaquin

---

134 cinidon-ethyl

A novel phthalimide contact herbicide for cereals

Products

| Lotus | BASF | 200 g/l | EC | 09231 |

Uses

Annual dicotyledons in **durum wheat, spring barley, winter barley, winter rye, winter wheat**. Chickweed in **durum wheat, spring barley, winter barley, winter rye, winter wheat**. Cleavers in **durum wheat, spring barley, winter barley, winter rye, winter wheat**. Speedwells in **durum wheat, spring barley, winter barley, winter rye, winter wheat**.

Efficacy

- Controls sensitive species that have emerged at time of treatment
- Ensure good even spray cover for best results
- Weed spectrum can be widened by mixture with mecoprop-P (see label)

Crop safety/Restrictions
- Maximum total dose equivalent to one full dose application
- Do not treat crops suffering from herbicide damage or physical stress
- Some necrotic spotting may occur after treatment but has no effect on final yield. Effect can be greater when mixture with mecoprop-P used, especially in spring, but yield not normally affected
- Do not roll or harrow for 7d before or after treatment
- In the event of crop failure only winter and spring wheat and barley, oilseed rape, maize or sunflowers may be planted

Special precautions/Environmental safety
- Irritating to skin. May cause sensitization by skin contact
- Risk of serious damage to eyes
- Dangerous to fish or other aquatic life. Do not contaminate surface waters or ditches with chemical or used container
- Treated cereals must not be used as animal feed or bedding for at least 35 d after last application

Personal protective equipment/Label precautions
- A, C, H
- M03, R04b, R04d, R04e, U05a, U08, U20b, C03, E01, E13b, E30a, E31c, E34

Withholding period
- Do not allow animals to graze for 35 d after treatment

Latest application/Harvest interval
- HI 35 d for cutting for stock feed or grazing
- Before first node detectable (GS 31) for spring barley and winter rye; before third node detectable (GS 33) for winter wheat, durum wheat and winter barley

Approval
- Accepted by BLRA for use on malting barley

135	**clodinafop-propargyl**

A contact acting herbicide for annual grass weed control in cereals

Products

1	Greencrop Boulevard	Greencrop	240 g/l	EC	09960
2	Landgold Clodinafop	Landgold	240 g/l	EC	09017
3	Marathon	Me2	240 g/l	EC	09959
4	Standon Clodinafop 240	Standon	240 g/l	EC	09258
5	Topik	Novartis	240 g/l	EC	08461

Uses

Blackgrass in *durum wheat, spring rye, spring wheat, triticale, winter rye, winter wheat*. Rough meadow grass in *durum wheat, spring rye, spring wheat, triticale, winter rye, winter wheat*. Wild oats in *durum wheat, spring rye, spring wheat, triticale, winter rye, winter wheat*.

Efficacy
- Spray in autumn, winter or spring from 1 true leaf stage (GS 11) to before second node detectable (GS 32)
- Products contain a herbicide safener (cloquintocet-mexyl) that improves crop tolerance to clodinafop-propargyl
- Spray when majority of weeds have germinated but before competition reduces yield
- Optimum control achieved when all grass weeds emerged. Wait for delayed germination on dry or cloddy seedbed
- A mineral oil additive is recommended to give more consistent control of very high blackgrass populations or for late season treatments. See label for details
- Weed control not affected by soil type, organic matter or straw residues
- Control may be reduced if rain falls within 1 h of treatment

Crop safety/Restrictions
- Maximum total dose 0.25 l/ha/crop
- Do not treat crops under stress or suffering from waterlogging, pest attack, disease or frost

- Do not treat crops undersown with grass mixtures
- Do not mix with products containing MCPA, mecoprop, 2,4-D or 2,4-DB
- MCPA, mecoprop, 2,4-D or 2,4-DB should not be applied within 21 d before, or 7 d after, treatment
- Only a broad leaved crop may be sown after failure of a treated crop. After normal harvest any broad leaved crop or wheat, durum wheat, rye, triticale or barley should be sown

Special precautions/Environmental safety
- Irritating to skin and eyes
- May cause sensitization by skin contact [1, 3, 4]
- Flammable [2, 5]
- Harmful (dangerous [1, 3, 4]) to fish or other aquatic life. Do not contaminate surface waters or ditches with chemical or used container

Personal protective equipment/Label precautions
- A, C, H, K [1-5]
- R07d [2, 5]; R04a, R04b [1-5]; R04e [1, 3, 4]; U20a [2, 5]; U02, U05a, U09a [1-5]; U20b [1, 3, 4]; C03 [1-5]; E13c [2, 5]; E30a, E32a [1-5]; E01, E13b, E26 [1, 3, 4]

Latest application/Harvest interval
- Before second node detectable stage (GS 32)

---

136   clodinafop-propargyl + diflufenican

A contact and residual herbicide mixture for broad spectrum weed control in cereals

Products

1	Amazon	Novartis	30:50 g/l	EC	08128
2	Amazon	Aventis	30:50 g/l	EC	10266
3	Lucifer	RP Agric.	30:50 g/l	EC	08129

Uses

Annual dicotyledons in *durum wheat, triticale, winter rye, winter wheat*. Blackgrass in *durum wheat, triticale, winter rye, winter wheat*. Rough meadow grass in *durum wheat, triticale, winter rye, winter wheat*. Wild oats in *durum wheat, triticale, winter rye, winter wheat*.

Efficacy
- Optimum weed control obtained when all grass weeds emerged and broad leaved weeds at susceptible stage before they compete with the crop
- Products contain a herbicide safener (cloquintocet-mexyl) that improves crop tolerance to clodinafop-propargyl
- Spray in autumn, winter or spring from first leaf unfolded stage (GS 11) to end Feb. Soil should be moist at time of treatment
- Delay treatment if dry or cloddy seedbeds favour late weed germination
- Grass weed control unaffected by seedbed condition but control of broad leaved weeds requires a fine, firm seedbed cleared of trash and straw
- Always use a recommended adjuvant. See label
- Speed of action depends on temperature and growing conditions and may appear slow in dry or cold weather

Crop safety/Restrictions
- Maximum number of treatments 1 per crop
- Do not treat crops under stress or suffering from waterlogging, pest attack, disease or frost
- Do not treat broadcast crops or those undersown with grass mixtures
- Do not mix with products containing MCPA, mecoprop, 2,4-D or 2,4-DB
- MCPA, mecoprop, 2,4-D or 2,4-DB should not be applied within 21 d before, or 7 d after, treatment
- Do not use on Sands, stony or gravelly soils or soil with over 10% organic matter
- Do not roll autumn treated crops until spring. Do not harrow at any time after treatment
- Only listed crops may be sown after failure of a treated crop - see label

- After normal harvest cereals, field beans, oilseed rape, onions, leaf brassicas or sugar beet seed crops may be sown in the autumn. Land must be ploughed or cultivated to 15 cm except for cereals
- Successive treatments of any products containing diflufenican can lead to soil build up and inversion ploughing must precede sowing any non-cereal crop. Even where ploughing occurs some crops may be damaged - see label

Special precautions/Environmental safety
- Irritating to skin and eyes
- Harmful to fish or other aquatic life. Do not contaminate surface waters or ditches with chemical or used container
- LERAP Category B

Personal protective equipment/Label precautions
- A, C, H, K
- R04a, R04b, U02, U05a, U09a, U20b, C03, E13c, E16a, E16b, E26, E30a, E32a

Latest application/Harvest interval
- Before end of Feb in year of harvest

137 clodinafop-propargyl + trifluralin

A contact and residual herbicide for winter wheat

Products

Hawk	Novartis	12:383 g/l	EC	08417

Uses
Annual dicotyledons in **winter wheat**. Blackgrass in **winter wheat**. Rough meadow grass in **winter wheat**. Wild oats in **winter wheat**.

Efficacy
- Product must be applied with specified adjuvant - see label
- Product contains a herbicide safener (cloquintocet-mexyl) that improves crop tolerance to clodinafop-propargyl
- Optimum weed control obtained when most grass weeds emerged and broad leaved weeds at susceptible stage before they compete with the crop. Pre-emergence control of annual meadow-grass and broad leaved weeds may not be satisfactory on medium and heavy soils
- Spray in autumn, winter or spring from first to third leaf unfolded stage (GS 11-13) but before ear at 1 cm stage (GS 30)
- Delay treatment if dry or cloddy seedbeds favour late weed germination
- Optimum weed control obtained in crops growing in a fine, firm seedbed cleared of trash and straw
- Wild oats and other weeds germinating from beneath treated zone not controlled
- Always use a recommended adjuvant/wetter. See label
- Speed of action depends on temperature and growing conditions and may appear slow in dry or cold weather
- Rain within 1 h of application may reduce grass weed control

Crop safety/Restrictions
- Maximum number of treatments 1 per crop
- Do not treat crops under stress or suffering from waterlogging, pest attack, disease or frost
- Do not treat crops undersown with grass mixtures
- Do not mix with products containing MCPA, mecoprop, 2,4-D or 2,4-DB
- MCPA, mecoprop, 2,4-D or 2,4-DB should not be applied within 21 d before, or 7 d after, treatment
- Do not use on Sands, stony or gravelly soils or soils with over 10% organic matter
- After normal harvest any broad leaved crop (except sugar beet), barley, wheat, durum wheat, rye or triticale may be sown. Before drilling or planting subsequent crops soil must be mouldboard ploughed to 15 cm

- See label for details of crops that may be safely drilled or planted within 5 mth of treatment. If a treated crop should fail after 5 mth but before normal harvest, sow only a broad-leaved crop (not sugar beet)
- 12 mth must elapse after treatment before sugar beet is drilled

Special precautions/Environmental safety
- Irritating to eyes and skin
- May cause sensitization by skin contact
- Harmful to fish or other aquatic life. Do not contaminate surface waters or ditches with chemical or used container
- Crops must not be treated if there is risk of run-off (eg from frozen foliage)

Personal protective equipment/Label precautions
- A, C, H, K
- R04a, R04b, R04e, U02, U05a, U09a, U20a, C03, E13c, E26, E30a, E32a

Latest application/Harvest interval
- Before ear at 1 cm stage (GS 30)

---

138   clofentezine

A selective ovicidal tetrazine acaricide for use in top fruit

Products

Apollo 50 SC	Aventis	500 g/l	SC	07242

Uses

Red spider mites in *apples, blackberries* (off-label - protected crops), *cherries, pears, plums, raspberries* (off-label - protected crops), *raspberries* (off-label), *strawberries* (off-label - protected crops).

Efficacy
- Acts on eggs and early motile stages of mites. For effective control total cover of plants is essential, particular care being needed to cover undersides of leaves
- For red spider mite control spray apples and pears between bud burst and pink bud, plums and cherries between white bud and first flower. Rust mite is also suppressed
- On established infestations apply in conjunction with an adult acaricide

Crop safety/Restrictions
- Maximum number of treatments 1 per yr
- Product safe on predatory mites, bees and other predatory insects

Personal protective equipment/Label precautions
- U08, U20b, E15, E30a, E32a

Latest application/Harvest interval
- 28 d for apple and pear, 8 wk for plum and cherry

Approval
- Off-label approval unlimited for use on blackcurrants, strawberries, raspberries (OLA 1250/95); unlimited for use on protected raspberries and blackberries (OLA 1645/98); to Jul 2003 for use on protected strawberries (OLA 1646/98)

---

139   clopyralid

A foliar translocated picolinic herbicide for a wide range of crops

See also benazolin + clopyralid
bromoxynil + clopyralid

Products

Dow Shield	Dow	200 g/l	SL	05578

Uses

Annual dicotyledons in *broad-leaved trees* (off-label), *broccoli, brussels sprouts, cabbages, calabrese, cauliflowers, cereals, conifers* (off-label), *established grassland, fodder beet, fodder rape, grass seed crops* (off-label), *kale, kohlrabi* (off-label), *linseed, maize, mangels, oilseed rape, onions, outdoor herbs (off-label), red beet, sage (off-label), spinach beet (off-label), spinach (off-label), strawberries, sugar beet, swedes, sweetcorn, turnips, woody ornamentals.* Corn marigold in *broccoli, brussels sprouts, cabbages, calabrese, cauliflowers, cereals, established grassland, fodder beet, fodder rape, kale, linseed, maize, mangels, oilseed rape, onions, red beet, strawberries, sugar beet, swedes, sweetcorn, turnips, woody ornamentals.* Creeping thistle in *asparagus (off-label), broccoli, brussels sprouts, cabbages, calabrese, cauliflowers, cereals, established grassland, established grassland (off-label - weed wiper), fodder beet, fodder rape, honesty (off-label), kale, linseed, maize, mangels, oilseed rape, onions, red beet, strawberries, sugar beet, swedes, sweetcorn, turnips, woody ornamentals.* Mayweeds in *broccoli, brussels sprouts, cabbages, calabrese, cauliflowers, cereals, established grassland, fodder beet, fodder rape, honesty (off-label), kale, linseed, maize, mangels, oilseed rape, onions, red beet, strawberries, sugar beet, swedes, sweetcorn, turnips, woody ornamentals.* Perennial dicotyledons in *broad-leaved trees (off-label), broccoli, brussels sprouts, cabbages, calabrese, cauliflowers, cereals, conifers (off-label), established grassland, fodder beet, fodder rape, grass seed crops (off-label), kale, linseed, maize, mangels, oilseed rape, onions, red beet, sage (off-label), strawberries, sugar beet, swedes, sweetcorn, turnips, woody ornamentals.*

Efficacy
- Best results achieved by application to young actively growing weed seedlings. Treat creeping thistle at rosette stage and repeat 3-4 wk later as directed
- High activity on weeds of Compositae family. For most crops recommended for use in tank mixes. See label for details
- Do not apply when crop damp or when rain expected within 6 h

Crop safety/Restrictions
- Maximum total dose 2.0 l/ha/crop for winter oilseed rape; 1.5 l/ha/crop for beet crops, vegetable brassicas, fodder rape, swedes, turnips, onions, maize, sweetcorn, spring oilseed rape, strawberries; 1.0 l/ha/yr for established grassland, ornamental plant production; 0.5 l/ha/crop for linseed; 0.35 l/ha/crop for cereals
- Timing of application varies with weed problem, crop and other ingredients of tank mixes. See label for details
- Do not apply to cereals later than the second node detectable stage (GS 32)
- Do not use straw from treated cereals in compost or any other form for glasshouse crops. Straw may be used for strawing down strawberries
- Straw from treated grass seed crops or linseed should be baled and carted away. If incorporated do not plant winter beans in same year
- Do not use on onions at temperatures above 20°C or when under stress
- Do not treat maiden strawberries or runner bed or apply to early leaf growth during blossom period or within 4 wk of picking. Aug or early Sep sprays may reduce yield
- Apply as directed spray in woody ornamentals, avoiding leaves, buds and green stems. Do not apply in root zone of families Compositae or Papilionaceae
- Do not plant susceptible autumn-sown crops in same year as treatment. Do not apply later than Jul where susceptible crops are to be planted in spring. See label for details

Special precautions/Environmental safety
- Wash spray equipment thoroughly with water and detergent immediately after use. Traces of product can damage susceptible plants sprayed later

Personal protective equipment/Label precautions
- A, C
- U08, U19, U20b, E15, E26, E30a, E31b; E07 (7 d)

Withholding period
- Keep livestock out of treated areas for at least 7 d and until foliage of any poisonous weeds such as ragwort has died and become unpalatable

Latest application/Harvest interval
- Before 3rd node detectable (GS 33) for cereals; before flower buds visible from above for oilseed rape and linseed.

- HI grassland 7 d; strawberries 4 wk; maize, sweetcorn, onions, Brussels sprouts, broccoli, cabbage, cauliflowers, calabrese, kale, fodder rape, oilseed rape, swedes, turnips, sugar beet, red beet, fodder beet, mangels, sage, honesty 6 wk

Approval
- Off-label approval unlimited for use on established grassland (wiper application), grass seed crops (OLA 0662/92), honesty (OLA 0480/93); unlimited for use on coniferous and broadleaved trees in forestry (OLA 0757/92); to Jul 2002 for use on outdoor herbs (see OLA notice for list) (OLA 1359/97); to Mar 2001 for use on outdoor kohlrabi (OLA 0495/96); to Oct 2001 for use on spinach and spinach beet (OLA 2360/96); unstipulated for use on asparagus (OLA0520/00)
- Accepted by BLRA for use on malting barley

140   clopyralid + 2,4-D + MCPA

A translocated herbicide mixture for grassland

Products

Lonpar	Dow	35:150:175 g/l	SL	08686

Uses

Creeping thistle in **established grassland**.

Efficacy
- Treatment must be made when weeds and grass are actively growing
- Important to ensure sufficient leaf area for uptake especially on established thistles with extensive root system. Treat at rosette stage
- On large well established thistles and where there is a large soil seed reservoir further treatment in the following year may be needed
- To allow maximum translocation do not cut grass for 28 d after treatment

Crop safety/Restrictions
- Maximum number of treatments 1 or 2 per yr depending on dose used - see label
- Do not spray in drought, very hot or very cold weather or if rain expected within 6 h
- Do not treat grass less than 1 yr old or sports or amenity turf
- Product kills or severely checks clover and should not be used where clover is an important constituent of the sward
- Very occasionally some yellowing of sward may occur after treatment which is quickly outgrown
- Do not drill grass, grass mixtures, kale, swedes or turnips into the sward within 6 wk of spraying

Special precautions/Environmental safety
- Harmful if swallowed
- Risk of serious damage to eyes
- Harmful to fish or other aquatic life. Do not contaminate surface waters or ditches with chemical or used container
- Wash spray equipment thoroughly with water and detergent immediately after use. Traces of product can damage susceptible plants sprayed later
- Susceptible crops (see label) must not be planted in the same calendar yr as treatment. Ensure that remains of treated crop have completely decayed before planting a subsequent susceptible crop. Where a susceptible crop is to be planted in the spring, product must not be sprayed later than end Jul in previous yr

Personal protective equipment/Label precautions
- A, C, H, M
- R03c, R04d, U05a, U15, C03, E01, E13c, E26, E34; E07 (2 wk)

141 clopyralid + diflufenican + MCPA

A selective herbicide for use in established turf

Products

1 Spearhead	Aventis Environ.	20:15:300 g/l		SL	09941
2 Spearhead	RP Amenity	20:15:300 g/l		SL	07342

Uses

Annual dicotyledons in *turf*. Perennial dicotyledons in *turf*.

Efficacy

- Best results achieved by application when grass and weeds are actively growing
- Treatment during early part of the season is recommended, but not during drought

Crop safety/Restrictions

- Maximum number of treatments 1 per yr
- Only use when sward is satisfactorily established and regular mowing has begun
- Turf sown in spring or early summer may be ready for treatment after 2 mth. Later sown turf should not be sprayed until growth is resumed in the following spring
- Avoid mowing within 3-4 d before or after treatment
- Do not use cuttings from treated area as a mulch for any crop
- Avoid drift. Small amounts of spray can cause serious injury to herbaceous plants, vegetables, fruit and glasshouse crops

Special precautions/Environmental safety

- Irritating to eyes and skin
- Harmful to fish or other aquatic life. Do not contaminate surface waters or ditches with chemical or used container
- LERAP Category B

Personal protective equipment/Label precautions

- A, C
- R04a, R04b, U02, U05a, U08, U19, U20b, C03, E01, E07, E13c, E16a, E16b, E26, E30a, E31b

Withholding period

- Keep livestock out of treated areas

142 clopyralid + fluroxypyr + MCPA

A translocated herbicide mixture for use in sports and amenity turf

Products

Greenor	Rigby Taylor	20:40:200 g/l		ME	07848

Uses

Annual dicotyledons in *managed amenity turf*.

Efficacy

- Best results achieved when weeds actively growing and turf grass competitive
- Treatment should normally be between Apr-Sep when the soil is moist
- Do not apply during drought unless irrigation is applied
- Allow 3 d before or after mowing established turf to ensure sufficient weed leaf surface present to allow uptake and movement

Crop safety/Restrictions

- Maximum number of treatments 2 per yr
- Treat young turf only in spring when at least 2 mth have elapsed since sowing
- Allow 5 d after mowing young turf before treatment
- Do not treat grass under stress from frost, drought, waterlogging, trace element deficiency, disease or pest attack
- Do not treat if night temperatures are low, when frost is imminent or during prolonged cold weather
- Product selective on a number of turf grass species (see label) but consultation or testing recommended before treatment of any cultivar

Special precautions/Environmental safety
- Irritating to eyes and skin
- Harmful to fish or other aquatic life. Do not contaminate surface waters or ditches with chemical or used container
- Wash spray equipment thoroughly with water and detergent immediately after use. Traces of product can damage susceptible plants sprayed later

Personal protective equipment/Label precautions
- A, C
- R04a, R04b, U05a, U08, U19, U20a, C03, E01, E13c, E26, E30a, E31b

Approval
- Fluroxypyr included in Annex I under EC Directive 91/414

---

143   clopyralid + fluroxypyr + triclopyr

A foliar acting herbicide mixture for grassland

Products

Pastor	Dow	50:75:100 g/l	EC	07440

Uses

Docks in **established grassland**. Stinging nettle in **established grassland**. Thistles in **established grassland**.

Efficacy
- May be applied in spring or autumn depending on weeds present
- Treatment must be made when weeds and grass are actively growing
- Important to ensure sufficient leaf area for uptake especially on established docks and thistles
- On large well established docks and where there is a large soil seed reservoir further treatment in the following year may be needed
- To allow maximum translocation do not cut grass for 4 wk after treatment

Crop safety/Restrictions
- Maximum number of treatments 1 per yr at full dose or 2 per yr at half dose
- Application during active growth ensures minimal check to grass
- Product may be used in established grassland which is under non-rotational setaside arrangements
- Do not spray in drought, very hot or very cold weather
- Do not treat grass less than 1 yr old or sports or amenity turf
- Product kills or severely checks clover and should not be used where clover is an important constituent of the sward
- Very occasionally some yellowing of sward may occur after treatment which is quickly outgrown
- Do not roll or harrow 10 d before or 7 d after treatment
- Residues in incompletely decayed plant tissue may affect succeeding susceptible crops such as peas, beans and other legumes, carrots and related crops, potatoes, tomatoes, lettuce
- Do not plant susceptible autumn-sown crops in the same yr as treatment with product. Spring sown crops may follow if treatment was before end Jul in the previous yr
- Do not allow spray or drift to reach other crops, amenity plantings, gardens, ponds, lakes or water courses

Special precautions/Environmental safety
- Irritating to eyes and skin
- May cause lung damage if swallowed
- May cause sensitization by skin contact
- Harmful to fish or other aquatic life. Do not contaminate surface waters or ditches with chemical or used container
- Wash spray equipment thoroughly with water and detergent immediately after use. Traces of product can damage susceptible plants sprayed later

Personal protective equipment/Label precautions
- A, C
- R04a, R04b, R04e, R04g, U02, U05a, U08, U19, U20b, C01, C03, E01, E13c, E26, E30a, E31b, E34; E07 (7 d)

Withholding period
- Keep livestock out of treated areas for at least 7 d following treatment and until foliage of poisonous weeds such as ragwort has died and become unpalatable

Latest application/Harvest interval
- HI 7 d

Approval
- Fluroxypyr included in Annex I under EC Directive 91/414

---

144   clopyralid + propyzamide

A post-emergence herbicide for winter oilseed rape

Products

1	Matrikerb	Rohm & Haas	4.3:43% w/w	WP	02443
2	Matrikerb	PBI	4.3:43% w/w	WP	09604

Uses

Annual dicotyledons in *winter oilseed rape*. Annual grasses in *winter oilseed rape*. Barren brome in *winter oilseed rape*. Mayweeds in *winter oilseed rape*.

Efficacy
- Apply from Oct to end Jan. Mayweed and groundsel may be controlled after emergence but before crop large enough to shield seedlings
- Best results achieved on fine, firm, moist soils when weeds germinating or small
- Do not use on soils with more than 10% organic matter
- Effectiveness reduced by surface organic debris, burnt straw or ash
- Do not apply if rainfall imminent or frost present on foliage

Crop safety/Restrictions
- Maximum number of treatments 1 per crop
- Apply to crop as soon as possible after 3-true leaf stage (GS 1,3)
- Minimum period between spraying and drilling a following crop varies from 10 to 40 wk. See label for details

Personal protective equipment/Label precautions
- A, C
- U09a, U19, U20a, C02, E01, E15, E30a, E32a, E34

Latest application/Harvest interval
- HI oilseed rape 6 wk; evening primrose 14 wk

Maximum Residue Level (mg residue/kg food)
- see propyzamide entry

---

145   clopyralid + triclopyr

A perennial and woody weed herbicide for use in grassland

Products

Grazon 90	Dow	60:240 g/l	EC	05456

Uses

Brambles in *amenity grass, established grassland*. Broom in *amenity grass, established grassland*. Docks in *amenity grass, established grassland*. Gorse in *amenity grass, established grassland*. Perennial dicotyledons in *amenity grass, established grassland, established grassland (off-label - weed wiper)*. Stinging nettle in *amenity grass, established grassland*. Thistles in *amenity grass, established grassland*.

Efficacy
- For good results must be applied to actively growing weeds
- Spray stinging nettle before flowering, docks in rosette stage in spring, creeping thistle before flower stems 15 cm high, brambles, broom and gorse in Jun-Aug
- Allow 2-3 wk regrowth after grazing or mowing before spraying perennial weeds
- Do not cut grass for 21 d before or 28 d after spraying

Crop safety/Restrictions
- Maximum number of treatments 1 per yr
- Only use on permanent pasture or leys established for at least 1 yr
- Do not apply overall where clover is an important constituent of sward
- Do not roll or harrow within 7 d before or after spraying. Do not direct drill kale, swedes, turnips, grass or grass mixtures within 6 wk of spraying. Do not plant susceptible autumn-sown crops (eg winter beans) in same year as treatment. Do not spray after end Jul where susceptible crops to be planted next spring
- Do not allow drift onto other crops, amenity plantings or gardens

Special precautions/Environmental safety
- Harmful if swallowed
- Irritating to skin. May cause sensitization by skin contact
- Risk of serious damage to eyes
- Not to be used on food crops
- Dangerous to fish or other aquatic life. Do not contaminate surface waters or ditches with chemical or used container
- Do not apply by hand-held rotary atomiser equipment

Personal protective equipment/Label precautions
- A, C, H, M
- R03c, R04b, R04d, R04e, U02, U05a, U08, U19, U20b, C01, C03, E01, E13b, E23, E26, E30a, E31b, E34; E07 (7 d)

Withholding period
- Keep livestock out of treated areas for at least 7 d after spraying and until foliage of any poisonous weeds such as ragwort or buttercup has died down and become unpalatable

Latest application/Harvest interval
- 7 d before grazing or harvest

Approval
- Off-label approval unlimited for use on established grassland via a tractor mounted/drawn weed wiper (OLA 0692/95)

---

146   copper ammonium carbonate

A protectant copper fungicide

Products

Croptex Fungex	Hortichem	8% w/w (copper)	SL	02888

Uses

Blight in **outdoor tomatoes**. Cane spot in **loganberries, raspberries**. Celery leaf spot in **celery**. Currant leaf spot in **blackcurrants**. Damping off in **seedlings of ornamentals**. Leaf curl in **peaches**. Leaf mould in **outdoor tomatoes, protected tomatoes**. Powdery mildew in **chrysanthemums, cucumbers**.

Efficacy
- Apply spray to both sides of foliage
- With protected crops keep foliage dry before and after spraying

Crop safety/Restrictions
- Maximum number of treatments 5 per crop for celery; 3 per yr for blackcurrants and loganberries; 2 per yr for peaches and raspberries
- Do not spray plants which are dry at the roots
- Ventilate glasshouse immediately after spraying

Special precautions/Environmental safety
- Harmful if swallowed. Risk of serious damage to eyes

- Harmful to fish or other aquatic life. Do not contaminate surface waters or ditches with chemical or used container
- Harmful to livestock

Personal protective equipment/Label precautions
- M03, R03c, R04d, U05a, U20a, C03, E01, E06b, E13c, E30a, E31a, E34

Withholding period
- Keep all livestock out of treated areas for at least 3 wk. Bury or remove spillages

---

147   copper hydroxide

An inorganic root pruning and root development agent

Products

Spin Out	Fargro	71 g/l	SL	07610

Uses

Root control in **container-grown stock**.

Efficacy
- Product should be painted on inner surfaces of containers prior to planting
- Used containers should be free from any loose soil or dirt
- Apply using conventional painting techniques such as brush, sponge, pad or airless paint sprayer
- Thorough coverage with a single coating to a minimum thickness of 0.075 mm is sufficient for root control

Crop safety/Restrictions
- Can be used in production of container grown forestry seedlings, hardy ornamental and herbaceous plant species
- Use at any stage of plant development from seedlings to large container grown trees
- Tested on a wide range of ornamental species (see label).   Efficacy and tolerance should be tested on a small scale for species not listed
- Performance characteristics may vary under some conditions such as when substrates with a pH of less than 5.0 are used
- Product must not be diluted

Special precautions/Environmental safety
- Harmful to fish or other aquatic life. Do not contaminate surface waters or ditches with chemical or used container
- Knapsack sprayer or similar equipment is not suitable

Personal protective equipment/Label precautions
- A, D
- U11, U12, U20b, E13c, E26, E30a, E32a

Latest application/Harvest interval
- Before planting

---

148   copper oxychloride

A protectant copper fungicide and bactericide

Products

1 Cuprokylt	Unicrop	50% w/w (copper)	WP	00604
2 Cuprokylt FL	Unicrop	270 g/l (copper)	SC	08299
3 Cuprosana H	Unicrop	6% w/w (copper)	DP	00605
4 Headland Inorganic Liquid Copper	Headland	256 g/l (copper)	SC	07799

Uses

Bacterial blight in **cob nuts** *(off-label)*, **hazel nuts** *(off-label)*, **walnuts** *(off-label)* [2]. Bacterial canker in **plums** [1, 2]. Bacterial canker in **cherries** [1, 2, 4]. Bacterial canker in **cob nuts** *(off-label)*, **hazel nuts** *(off-label)*, **walnuts** *(off-label)* [2]. Bacterial rot in **bulb onions** *(off-label)*, **garlic** *(off-label)*, **leeks** *(off-label)*, **salad onions** *(off-label)*, **shallots** *(off-label)* [1]. Blight in *outdoor tomatoes, potatoes* [1, 2, 4]. Buck-eye rot in *tomatoes* [1, 2, 4]. Cane spot in *loganberries, raspberries* [1, 2]. Canker in *apples, pears* [1, 2, 4]. Celery leaf spot in *celery* [1, 2, 4]. Damping off in *tomatoes* [1, 2]. Downy mildew in *hops* [1-4]. Downy mildew in *grapevines* [4]. Foot rot in *tomatoes* [1, 2]. Leaf curl in *peaches* [1, 2]. Purple blotch in *blackberries* [4]. Pythium in *watercress (off-label)* [2]. Rhizoctonia in *watercress (off-label)* [2]. Rust in *blackcurrants* [1, 2, 4]. Spear rot in *calabrese (off-label)* [1].

Efficacy

- Spray crops at high volume when foliage dry but avoid run off. Do not spray if rain expected soon
- Spray interval commonly 10-14 d but varies with crop, see label for details
- If buck-eye rot occurs, spray soil surface and lower parts of tomato plants to protect unaffected fruit [1, 2]
- A follow-up spray in the following spring should be made to top fruit severely infected with bacterial canker

Crop safety/Restrictions

- Maximum number of treatments 3 per crop for apples, blackberries, grapevines, pears [4]. Not specified for other products
- Some peach cultivars are sensitive to copper. Treat non-sensitive varieties only [1, 2]
- Slight damage may occur to leaves of cherries and plums [1, 2]

Special precautions/Environmental safety

- Harmful to livestock.
- Harmful to fish or other aquatic life. Do not contaminate surface waters or ditches with chemical or used container

Personal protective equipment/Label precautions

- A [4]
- U20a [1, 4]; U19 [3]; U20c [2, 3]; E31a [1]; E13c, E30a [1-4]; E06b [1-4] (3 wk); E32a [3]; E26, E31c [2]

Withholding period

- Keep all livestock out of treated areas for at least 3 wk

Latest application/Harvest interval

- Before bud burst for apples and pears.
- HI calabrese 3 d [1]

Approval

- Approved for aerial application on potatoes [1]. See notes in Section 1
- Off-label approval unlimited for use on calabrese (OLA 0993/92) [1]; unstipulated for use on cob nuts, hazel nuts, walnuts (OLA 0385/99)[2]; unstipulated for use on onions, shallots, garlic, leeks (OLA 1127/99)[1]; unstipulated for use on watercress during propagation (OLA 1538/00)[2]
- Accepted by BLRA for use on hops

---

149   copper oxychloride + maneb + sulphur

A protectant fungicide and yield stimulant for wheat and barley

Products

Tripart Senator Flowable	Tripart	10:160:640 g/l	SC	04561

Uses

Glume blotch in *wheat*. Leaf blotch in *barley*. Leaf spot in *wheat*. Mildew in *barley, wheat*.

Efficacy

- Use in a protectant programme covering period from first node detectable to beginning of ear emergence (GS 31-51), see label for details
- Treatment has little effect on established disease
- In addition to fungicidal effects treatment can also give nutritional benefits
- Do not apply when foliage is wet or rain imminent

Crop safety/Restrictions

- Maximum number of treatments 3 per crop
- Do not apply to crops suffering stress from any cause

- Do not roll or harrow within 7 d of treatment
- Do not apply with any trace elements, liquid fertilizers or wetting agent

Special precautions/Environmental safety
- Harmful by inhalation and if swallowed
- Irritating to eyes and skin
- Harmful to fish or other aquatic life. Do not contaminate surface waters or ditches with chemical or used container

Personal protective equipment/Label precautions
- A, C
- M03, R03b, R03c, R04a, R04b, U05a, U08, U19, U20a, C03, E01, E13c, E30a, E32a, E34

Latest application/Harvest interval
- Before ear fully emerged (GS 59)

Approval
- Accepted by BLRA for use on malting barley

Maximum Residue Level (mg residue/kg food)
- see maneb entry

---

150   copper oxychloride + metalaxyl

A systemic and protectant fungicide mixture

Products

Ridomil Plus	Novartis	35:15% w/w	WP	08353

Uses

Collar rot in **apples**. Downy mildew in **brussels sprouts** *(off-label)*, **cabbages** *(off-label)*, **calabrese** *(off-label)*, **cauliflowers** *(off-label)*, **grapevines** *(off-label)*, **hops, protected courgettes** *(off-label)*, **protected cucumbers** *(off-label)*, **protected gherkins** *(off-label)*, **spinach beet** *(off-label)*, **spinach** *(off-label)*. Red core in **strawberries**.

Efficacy
- Drench soil at base of apples in Sep to Dec or Mar in first 2 yr after planting only
- Apply drench to maiden strawberries immediately after planting and to established plants immediately new growth begins in autumn
- Spray hops to run-off. Increase dose and volume with hop growth

Crop safety/Restrictions
- Maximum number of treatments 3 per crop for brassicas; 1 per yr for strawberries; 8 per yr for hops, cucumbers
- Do not use on fruiting apple trees
- Use only on hops when well established

Special precautions/Environmental safety
- Irritating to eyes and skin
- Harmful to fish or other aquatic life. Do not contaminate surface waters or ditches with chemical or used container
- Harmful to livestock

Personal protective equipment/Label precautions
- A, C
- R04a, R04b, U05a, U09a, U19, U20a, C02, C03, E01, E06b, E13c, E30a, E32a, E34

Withholding period
- Keep all livestock out of treated areas for at least 3 wk. Bury or remove spillages

Latest application/Harvest interval
- At planting for maiden strawberries; after harvest but before 30 Nov for established strawberries.
- HI 14 d

Approval
- Off-label Approval unlimited for use in Brussels sprouts, cabbages, calabrese, cauliflowers (OLA 1383/97); unlimited for use on outdoor grapevines (OLA 1362/97); unlimited for use on

outdoor and protected spinach, spinach beet (OLA 1344/98); to Jun 2003 for use on protected cucumbers, gherkins, courgettes (OLA 1344/98)
- Accepted by BLRA for use on hops

Maximum Residue Level (mg residue/kg food)
- see metalaxyl entry

151   copper sulphate

See Bordeaux Mixture

152   copper sulphate + sulphur

A contact fungicide for mildew and blight control

Products

Top-Cop	Stoller	6:670 g/l	SC	04553

Uses

Blight in **potatoes, tomatoes**. Downy mildew in **hops**. Mildew in **grapevines, leaf brassicas, swedes, turnips**. Powdery mildew in **hops, sugar beet**.

Efficacy
- Apply at first blight warning or when signs of disease appear and repeat every 7-10 d as necessary
- Timing varies with crop and disease, see label for details

Crop safety/Restrictions
- Do not apply to hops at or after the burr stage

Special precautions/Environmental safety
- Irritating to skin, eyes and respiratory system
- Dangerous to fish or other aquatic life. Do not contaminate surface waters or ditches with chemical or used container
- Harmful to livestock

Personal protective equipment/Label precautions
- R04a, R04b, R04c, U05a, U08, U13, U14, U15, U20a, C03, E01, E06b, E13b, E30a, E31a

Withholding period
- Keep livestock out of treated areas for at least 3 wk

Approval
- Accepted by BLRA for use on hops

153   coumatetralyl

An anticoagulant coumarin rodenticide

Products

1	Racumin Bait	Bayer	0.0375% w/w	RB	H6754
2	Racumin Contact Powder	Bayer	0.75% w/w	TP	H6755

Uses

Rats in **farm buildings**.

Efficacy
- Chemical is available as a bait and tracking powder and it is recommended that both formulations are used together
- Place bait in runs in sheltered positions near rat holes. Use at least 250 g per baiting point and examine at least every 2 d. Replenish as long as bait being eaten
- When no signs of activity seen for about 10 d remove unused bait
- Apply tracking powder in a 2.5-5 cm wide layer inside and across entrance to holes or blow into holes with dusting machine. Lay a 3 mm thick layer along runs in 30 cm long patches. Only use on exposed runs if well away from buildings

Special precautions/Environmental safety
- Prevent access to bait or dust by children, birds and non-target animals, particularly dogs, cats, pigs

- Do not lay baits or dust where food, feed or water could become contaminated
- Remove all remains of bait and bait containers or exposed dust after treatment (except where used in sewers) and dispose of safely (e.g. burn/bury). Do not dispose of in refuse sacks or on open rubbish tips
- Search for rodent bodies (except where used in sewers) and dispose of safely (e.g. burn/bury). Do not dispose of in refuse sacks or on open rubbish tips

Personal protective equipment/Label precautions
- U20b [1]; U13 [1, 2]; U20c [2]; E30a, E32a, V01b, V02, V03b, V04b [1, 2]

154   cyanazine

A contact and residual triazine herbicide

Products

1	Fortrol	Cyanamid	500 g/l	SC	07009
2	I T Cyanazine	I T Agro	500 g/l	SC	10179
3	Standon Cyanazine 50	Standon	500 g/l	SA	09230

Uses

Annual dicotyledons in **broad beans** *(Scotland only)*, **broccoli** *(off-label)*, **cabbages** *(off-label)*, **calabrese** *(off-label)*, **cauliflowers** *(off-label)*, **collards** *(off-label)*, **farm forestry** *(off-label)*, **forestry transplants** *(off-label)*, **kale** *(off-label)* [1]. Annual dicotyledons in **narcissi, onions, peas, sweetcorn, tulips, winter barley, winter oilseed rape, winter wheat** [1-3]. Annual dicotyledons in **broad beans, maize, spring field beans, ware potatoes** [2]. Annual grasses in **broad beans** *(Scotland only)*, **broccoli** *(off-label)*, **cabbages** *(off-label)*, **calabrese** *(off-label)*, **cauliflowers** *(off-label)*, **collards** *(off-label)*, **farm forestry** *(off-label)*, **forestry transplants** *(off-label)*, **kale** *(off-label)* [1]. Annual grasses in **narcissi, onions, peas, sweetcorn, tulips, winter barley, winter oilseed rape, winter wheat** [1-3]. Annual grasses in **broad beans, maize, spring field beans, ware potatoes** [2]. Charlock in **honesty** *(off-label)* [1].

Efficacy
- Weeds controlled before emergence or at young seedling stage. Chemical must be carried into soil by rainfall
- Best results achieved when applied during mild, bright weather. Avoid applications in dull, cold or wet conditions
- Numerous tank mixtures recommended to broaden weed spectrum. See label for details of tank mix partners and timings
- Residual activity normally persists for about 2 mth but on cereals all previous crop and weed residues must be removed or buried at least 15 cm deep
- Do not use as pre-emergence treatment on soils with more than 10% organic matter
- Some strains of blackgrass have developed resistance to many blackgrass herbicides which may lead to poor control

Crop safety/Restrictions
- Maximum number of treatments 1 per crop (2 per crop for collards), unspecified for narcissi and tulips
- Apply pre- or post-emergence to winter cereals which must be drilled at least 25 mm deep. Do not spray within 7 d of rolling or harrowing
- Variety restrictions apply in peas, field beans and broad beans. See label for details of tolerant varieties
- Apply to oilseed rape after 1 Nov from 5-leaf stage when winter hardened. Do not apply after 31 Jan
- Apply pre-emergence to potatoes as soon as possible after planting and at least 7 d before any shoots emerge
- On onions only apply post-emergence to crops on fen soils with more than 10% organic matter after 2-true leaf stage
- On maize, sweetcorn, broad beans and field beans use as pre-emergence spray or as a pre-drilling incorporated treatment in maize or sweetcorn
- On flower bulbs treat pre- or early post-emergence
- Do not use on Sands, Very Light or stony soils
- Do not treat crops stressed by frost, waterlogging, drought, chemical damage etc. Heavy rain shortly after treatment may lead to damage, especially on lighter soils

FOR FULL CONDITIONS OF USE ALWAYS READ THE PRODUCT LABEL

Special precautions/Environmental safety
- Harmful if swallowed and in contact with skin
- Harmful to fish or other aquatic life. Do not contaminate surface waters or ditches with chemical or used container

Personal protective equipment/Label precautions
- A, C [1-3]
- M03, R03a, R03c [1-3]; U20a [1]; U05a, U08, U19 [1-3]; U20b [2, 3]; C03 [1-3]; E31a [1]; E01, E13c, E30a, E34 [1-3]; E26, E31b [2, 3]

Latest application/Harvest interval
- Pre-emergence for broad beans and sweetcorn; before 31 Jan for winter oilseed rape; pre-emergence or before flower buds visible for peas; pre-emergence before end Oct or post-emergence before 2nd node detectable (GS 32) for cereals; before 5 true leaf stage for collards, before flower buds appear above the developing leaves for honesty.
- HI onions, leeks 8 wk, calabrese 11 wk

Approval
- Off-label Approval unlimited for use in farm forestry (OLA 0602/94)[1]; to Feb 2000 for use on broccoli, cabbage, calabrese, cauliflower, collards, kale (OLA 0332/96)[1]; unlimited for use in forest tree establishment (OLA 0317/97)[1]
- Accepted by BLRA for use on malting barley

155 cyanazine + pendimethalin

A contact and residual herbicide mixture

Products

1 Activus	Cyanamid	150:264 g/l	SC	09174
2 Bullet	Cyanamid	150:264 g/l	SC	08049

Uses

Annual dicotyledons in *winter barley, winter wheat* [1]. Annual dicotyledons in *combining peas, fodder maize, spring field beans, ware potatoes* [2]. Annual meadow grass in *winter barley, winter wheat* [1]. Annual meadow grass in *combining peas, fodder maize, spring field beans, ware potatoes* [2]. Chickweed in *winter barley, winter wheat* [1]. Field pansy in *winter barley, winter wheat* [1]. Mayweeds in *winter barley, winter wheat* [1]. Rough meadow grass in *winter barley, winter wheat* [1]. Rough meadow grass in *combining peas, fodder maize, spring field beans, ware potatoes* [2].

Efficacy
- Spray as soon as possible after planting [2]. Treat winter wheat and barley early post-emergence [1]
- Best results achieved in crops growing on firm, fine seedbeds, where light rain falls soon after treatment. Do not disturb soil after application. Weed control may be reduced if prolonged dry conditions follow application
- Weeds present before crop emergence should be controlled with a contact herbicide as a tank-mix or in sequence [2]
- Weed control may be reduced on soils with a high Kd factor, where OM exceeds 6% or ash content is high. Surface organic crop residues should be dispersed
- Weeds germinating more than 2 mth after spraying may not be controlled
- Coarse irrigation of potatoes that washes soil particles to the base of the ridges may reduce weed control [2]

Crop safety/Restrictions
- Maximum number of treatments 1 per crop
- Do not use on Sands, (Very Light [1]), very stony or gravelly soils or those with more than 10% organic matter
- Do not apply to cereals within 24 h of frost. Frost occurring within 5 d of a post-emergence spray may cause crop damage [1]
- Wheat and barley seed should be covered with minimum 32 mm soil [1]; peas and spring beans with minimum 25 mm, and fodder maize with minimum 50 mm [2]
- Do not use on crops grown under protection such as polythene sheeting [2]

- Do not treat potato crops grown for certified seed or variety Russet Burbank. Slight distortion and discolouration of intial potato shoots may occur on Very Light soils if heavy rain rain falls between application and crop emergence [2]
- Do not treat pea variety Vedette [2]
- Do not treat after crop emergence or later than 7 d before emergence of most advanced potato shoots or after growing points of peas or beans are within 13 mm of soil surface [2]
- After a dry season land must be ploughed to 150 mm before drilling ryegrass
- Before drilling a winter crop plough or cultivate to 150 mm
- In the event of crop failure land must be ploughed or cultivated to 150 mm and then an interval of 8 wk must elapse before drilling any crop

Special precautions/Environmental safety
- Harmful if swallowed [2]
- Dangerous to fish or other aquatic life. Do not contaminate surface waters or ditches with chemical or used container

Personal protective equipment/Label precautions
- A [1, 2]
- M03 [1]; R03c [2]; U05a, U08, U13, U19, U20b, C03, E01, E13b, E30a, E31b, E34 [1, 2]

Latest application/Harvest interval
- Pre-emergence of broad-leaved crops [2]; before leaf sheath erect (GS 30) for winter barley; before tillering for winter wheat [1]

Approval
- Accepted by BLRA for use on malting barley

156 cyanazine + terbuthylazine

A foliar and soil acting herbicide for cereals

Products

Angle	Novartis	306:261 g/l	SC	08385

Uses

Annual dicotyledons in **winter barley, winter wheat**. Annual meadow grass in **winter barley, winter wheat**.

Efficacy
- Best results obtained from application made early post-emergence of weeds
- Weeds germinating after application are controlled by root uptake, those present at application by leaf and root uptake
- Weed control may be reduced if heavy rain falls shortly after application or if there is prolonged dry weather
- Weeds germinating more than 2 mth after application may not be controlled
- Reduced weed control may result from presence of more than 10% organic matter or of surface ash and trash

Crop safety/Restrictions
- Maximum number of treatments 1 per crop
- May be applied from 1 leaf unfolded on main shoot up to the 2 tiller stage (GS 11-22)
- Plant crop at normal depth of 25 mm; it is important that crop is well covered
- Do not apply to undersown crops or those due to be undersown
- Only spray healthy crops. Do not use on crops under stress or suffering from waterlogging, pest or disease attack, frost or when frost imminent
- Do not use on Sands or Very Light soils. There is a risk of crop damage on stony or gravelly soils, especially if heavy rain follows application
- Early crops (Sep drilled) may be prone to damage if spraying precedes or coincides with period of rapid growth in autumn

Special precautions/Environmental safety
- Harmful in contact with skin and if swallowed
- Irritating to skin and eyes
- May cause sensitization by skin contact
- Harmful to fish or other aquatic life. Do not contaminate surface waters or ditches with chemical or used container

Personal protective equipment/Label precautions
- A, C
- R03a, R03c, R04a, R04b, R04e, U08, U20a, E13c, E30a, E31b, E34

Latest application/Harvest interval
- Before main shoot and 2-tiller stage (GS 22)

Approval
- Accepted by BLRA for use on malting barley
- Approval expiry 31 Oct 2001

---

157   cycloxydim

A translocated post-emergence oxime herbicide for grass weed control

Products

1	Greencrop Valentia	Greencrop	200 g/l	EC	10197
2	Landgold Cycloxydim	Landgold	200 g/l	EC	06269
3	Laser	BASF	200 g/l	EC	05251
4	Standon Cycloxydim	Standon	200 g/l	EC	08830

Uses

Annual grasses in **brussels sprouts, cabbages, cauliflowers, field beans, fodder beet, mangels, peas, potatoes, sugar beet, swedes, winter oilseed rape** [1-4]. Annual grasses in **calabrese** [1, 3]. Annual grasses in **bulb onions, carrots, dwarf beans, farm forestry, flower bulbs, forestry, leeks, linseed, parsnips, salad onions, spring oilseed rape, strawberries** [1, 3, 4]. Annual grasses in **amenity vegetation** *(off-label)*, **soya beans** *(off-label)* [3]. Annual grasses in **broad beans** [4]. Black bent in **brussels sprouts, cabbages, cauliflowers, field beans, fodder beet, mangels, peas, potatoes, sugar beet, swedes, winter oilseed rape** [1-4]. Black bent in **calabrese** [1, 3]. Black bent in **bulb onions, carrots, dwarf beans, flower bulbs, leeks, linseed, parsnips, salad onions, spring oilseed rape, strawberries** [1, 3, 4]. Black bent in **broad beans** [4]. Blackgrass in **brussels sprouts, cabbages, cauliflowers, field beans, fodder beet, mangels, peas, potatoes, sugar beet, swedes, winter oilseed rape** [1-4]. Blackgrass in **calabrese** [1, 3]. Blackgrass in **bulb onions, carrots, dwarf beans, flower bulbs, leeks, linseed, parsnips, salad onions, spring oilseed rape, strawberries** [1, 3, 4]. Blackgrass in **broad beans** [4]. Couch in **brussels sprouts, cabbages, cauliflowers, field beans, fodder beet, mangels, peas, potatoes, sugar beet, swedes, winter oilseed rape** [1-4]. Couch in **calabrese** [1, 3]. Couch in **bulb onions, carrots, dwarf beans, flower bulbs, leeks, linseed, parsnips, salad onions, spring oilseed rape, strawberries** [1, 3, 4]. Couch in **broad beans** [4]. Creeping bent in **brussels sprouts, cabbages, cauliflowers, field beans, fodder beet, mangels, peas, potatoes, sugar beet, swedes, winter oilseed rape** [1-4]. Creeping bent in **calabrese** [1, 3]. Creeping bent in **bulb onions, carrots, dwarf beans, flower bulbs, leeks, linseed, parsnips, salad onions, spring oilseed rape, strawberries** [1, 3, 4]. Creeping bent in **broad beans** [4]. Green cover in **land temporarily removed from production** [1, 3]. Onion couch in **calabrese** [1, 3]. Onion couch in **brussels sprouts, bulb onions, cabbages, carrots, cauliflowers, dwarf beans, field beans, flower bulbs, fodder beet, leeks, linseed, mangels, parsnips, peas, potatoes, salad onions, spring oilseed rape, strawberries, sugar beet, swedes, winter oilseed rape** [1, 3, 4]. Onion couch in **broad beans** [4]. Perennial grasses in **farm forestry, forestry** [1, 3, 4]. Perennial grasses in **amenity vegetation** *(off-label)*, **soya beans** *(off-label)* [3]. Volunteer cereals in **brussels sprouts, cabbages, cauliflowers, field beans, fodder beet, mangels, peas, potatoes, sugar beet, swedes, winter oilseed rape** [1-4]. Volunteer cereals in **calabrese** [1, 3]. Volunteer cereals in **bulb onions, carrots, dwarf beans, flower bulbs, leeks, linseed, parsnips, salad onions, spring oilseed rape, strawberries** [1, 3, 4]. Volunteer cereals in **broad beans** [4]. Wild oats in **brussels sprouts, cabbages, cauliflowers, field beans, fodder beet, mangels, peas, potatoes, sugar beet, swedes, winter oilseed rape** [1-4]. Wild oats in **calabrese** [1, 3]. Wild oats in **bulb onions, carrots, dwarf beans, flower bulbs, leeks, linseed, parsnips, salad onions, spring oilseed rape, strawberries** [1, 3, 4]. Wild oats in **broad beans** [4].

Efficacy
- Best results achieved when weeds small and have not begun to compete with crop. Effectiveness reduced by drought, cool conditions or stress. Weeds emerging after application are not controlled
- Foliage death usually complete after 3-4 wk but longer under cool conditions, especially late treatments to winter oilseed rape
- Perennial grasses should have sufficient foliage to absorb spray and should not be cultivated for at least 14 d after treatment
- On established couch pre-planting cultivation recommended to fragment rhizomes and encourage uniform emergence
- Split applications to volunteer wheat and barley at GS 12-14 will often give adequate control in winter oilseed rape. See label for details
- Apply to dry foliage when rain not expected for at least 2 h
- Must be used with Actipron (see label)

Crop safety/Restrictions
- Maximum number of treatments 1 per crop for spring oilseed rape, early potatoes; 2 per crop (the second at reduced dose) for other crops. See label for details
- Recommended time of application varies with crop. See label for details
- On peas a crystal violet wax test should be done if leaf wax likely to have been affected by weather conditions or other chemical treatment. The wax test is essential if other products are to be sprayed before or after treatment
- May be used on ornamental bulbs when crop 5-10 cm tall. Product has been used on tulips, narcissi, hyacinths and irises but some subjects may be more sensitive and growers advised to check tolerance on small number of plants before treating the rest of the crop
- May be applied to setaside where the green cover is made up predominantly of tolerant crops listed on label. Use on industrial crops of linseed and oilseed rape on setaside also permitted.
- Do not apply to crops damaged or stressed by adverse weather, pest or disease attack or other pesticide treatment
- Prevent drift onto other crops, especially cereals and grass
- Guideline intervals for sowing succeeding crops after failed treated crop: field beans, peas, sugar beet, rape, kale, swedes, radish, white clover, lucerne 1 wk; dwarf French beans 4 wk; wheat, barley, maize 8 wk
- Oats should not be sown after failure of a treated crop

Special precautions/Environmental safety
- Irritating to skin and eyes
- Harmful to fish or other aquatic life. Do not contaminate surface waters or ditches with chemical or used container [1, 3, 4]

Personal protective equipment/Label precautions
- A, C [1-4]
- R04a, R04b, U05a, U08 [1-4]; U20b [1, 3, 4]; U20a [2]; C03, E01, E30a [1-4]; E31c [3, 4]; E13c [1, 3, 4]; E15 [2]; E31b [1, 2]; E26 [1]

Latest application/Harvest interval
- Before canopy prevents adequate spray penetration.
- HI cabbage, cauliflower, salad onions 4 wk; peas, dwarf French beans 5 wk; bulb onions, carrots, parsnips, early potatoes, strawberries 6 wk; sugar and fodder beet, leeks, mangels, early potatoes, maincrop potatoes, field beans, swedes, Brussels sprouts 8 wk; winter and spring oilseed rape, linseed 12 wk

Approval
- Off-label approval to Mar 2001 for use on broad beans (OLA 0792/96)[3]; unlimited for use in forestry, amenity vegetation (OLA 2585/96)[3]; to Mar 2002 for use in outdoor soya beans (OLA 0610/97)[3]

---

158   cyhexatin

An organotin acaricide approvals for use of which were revoked in 1988 because of evidence of teratogenicity

---

159   cymoxanil

A urea fungicide available only in mixtures

See also carbendazim + cymoxanil + oxadixyl + thiram

160   cymoxanil + mancozeb

A protectant and systemic fungicide for potato blight control

Products

1	Ashlade Solace	Nufarm Whyte	4.5:68% w/w	WP	08087
2	Besiege WSB	DuPont	4.5:68% w/w	WB	08075
3	Curzate M68	DuPont	4.5:68% w/w	WP	08072
4	Curzate M68 WSB	DuPont	4.5:68% w/w	WB	08073
5	Me2 Cymoxeb	Me2	4.5:68% w/w	WP	09486
6	Standon Cymoxanil Extra	Standon	4.5:68% w/w	WP	09442
7	Systol M	DuPont	4.5:68% w/w	WP	08085

Uses

Blight in **potatoes**.

Efficacy

- Apply immediately after blight warning or as soon as local conditions dictate and repeat at 7-14 d intervals until haulm dies down or is burnt off
- Spray interval should not be more than 10 d in irrigated crops

Crop safety/Restrictions

- Maximum number of treatments not specified
- At least 7 d (10 d [6]) must elapse between treatments

Special precautions/Environmental safety

- Irritating to eyes
- May cause sensitization by skin contact
- Harmful to fish or other aquatic life. Do not contaminate surface waters or ditches with chemical or used container
- Do not allow packs to become wet during storage [2]
- Keep product away from fire or sparks [2]

Personal protective equipment/Label precautions

- A [1-7]; C [1, 3, 5-7]
- R04a, R04e, U05a, U08, U19, U20b [1-7]; U22 [2, 4]; C03 [1-7]; E32a [1-4, 6, 7]; E01, E13c, E30a [1-7]; E31a [5]

Latest application/Harvest interval

- HI zero

Approval

- Approved for aerial application on potatoes [6]. See notes in Section 1

Maximum Residue Level (mg residue/kg food)

- see mancozeb entry

161   cymoxanil + mancozeb + oxadixyl

A systemic and contact protective fungicide for potatoes

Products

1	Ripost Pepite	Novartis	3.2:56.8% w/w	SG	08485
2	Trustan WDG	DuPont	3.2:56:8% w/w	WG	05050

Uses

Blight in **potatoes**.

Efficacy
- Apply first spray as soon as risk of blight infection or official warning. In absence of warning, spray just before crop meets within row
- Repeat spray every 10-14 d according to blight risk
- Do not treat crops already showing blight infection
- Do not apply within 2-3 h of rainfall or irrigation and only apply to dry foliage
- Use fentin based fungicide for final spray in blight control programme

Crop safety/Restrictions
- Maximum number of treatments 5 (including other phenylamide-based fungicides) per season

Special precautions/Environmental safety
- Irritating to eyes, skin and respiratory system
- May cause sensitization by skin contact
- Harmful to fish or other aquatic life. Do not contaminate surface waters or ditches with chemical or used container

Personal protective equipment/Label precautions
- A, C [1, 2]
- R04a, R04b, R04c, R04e [1, 2]; U20a [2]; U05a, U10, U11, U19 [1, 2]; U20b [1]; C03, E01, E13c, E30a, E32a [1, 2]

Latest application/Harvest interval
- Onset of senescence
- HI 7 d

Maximum Residue Level (mg residue/kg food)
- see mancozeb entry

---

162   cypermethrin

A contact and stomach acting pyrethroid insecticide

Products

1	Afrisect 10	Stefes	100 g/l	EC	09114
2	Cyperkill 10	Chiltern	100 g/l	EC	04119
3	Cyperkill 5	Mitchell Cotts	50 g/l	EC	00625
4	Permasect C	Nufarm Whyte	100 g/l	EC	09200
5	Toppel 10	United Phosphorus	100 g/l	EC	08772

Uses

Aphids in *autumn sown spring barley, autumn sown spring wheat* [1]. Aphids in *outdoor lettuce* [1-4]. Aphids in *apples* [1-3, 5]. Aphids in *winter barley, winter wheat* [1, 3]. Aphids in *barley (summer), wheat (summer)* [2]. Aphids in *vining peas, winter cereals* [2, 5]. Aphids in *peas* [3]. Aphids in *broccoli, brussels sprouts, cabbages, calabrese, cauliflowers, cherries, hops, lettuce, ornamentals, pears, plums* [5]. Apple sucker in *apples* [1, 2]. Asparagus beetle in *asparagus (off-label)* [5]. Barley yellow dwarf vectors in *rye, triticale* [4]. Barley yellow dwarf virus vectors in *winter barley, winter wheat* [1-4]. Barley yellow dwarf virus vectors in *autumn sown spring barley, autumn sown spring wheat* [1, 2, 4]. Barley yellow dwarf virus vectors in *winter cereals* [5]. Cabbage stem flea beetle in *winter oilseed rape* [1-4]. Cabbage stem flea beetle in *brussels sprouts, cabbages, oilseed rape* [5]. Capsids in *pears* [1-3, 5]. Capsids in *apples, ornamentals* [5]. Caterpillars in *fodder beet, mangels, potatoes, red beet, sugar beet* [1-3]. Caterpillars in *outdoor lettuce* [1-4]. Caterpillars in *apples* [1-3, 5]. Caterpillars in *kale* [1, 3, 4]. Caterpillars in *brussels sprouts, cabbages, cauliflowers* [1, 3-5]. Caterpillars in *brassicas* [2]. Caterpillars in *broccoli, calabrese* [3-5]. Caterpillars in *celery, cherries, lettuce, ornamentals, pears, plums* [5]. Codling moth in *apples* [1-3, 5]. Cutworms in *red beet* [1-3]. Cutworms in *fodder beet, mangels, outdoor lettuce* [1-4]. Cutworms in *potatoes, sugar beet* [1-5]. Cutworms in *ornamentals* [5]. Frit fly in *grass re-seeds* [1-5]. Leaf miners in *ornamentals (off-label), protected chicory (off-label), protected courgettes (off-label), protected cucumbers (off-label), protected endives (off-label), protected gherkins (off-label), protected lettuce (off-label)* [3]. Pea and bean weevils in *peas* [1]. Pea and bean weevils in *spring field beans, winter field beans* [1, 2]. Pea and bean weevils in *vining peas* [2-5]. Pea and bean weevils in *field beans* [3-5]. Pea moth in *peas* [1]. Pea moth in *vining peas* [2-5]. Pod midge in *spring oilseed rape, winter oilseed rape* [1-4]. Pod midge in *oilseed rape* [5]. Pollen beetles in *spring oilseed rape, winter oilseed rape* [1-4]. Pollen beetles in *oilseed rape* [5]. Rape winter stem weevil in *oilseed rape* [5]. Sawflies in *apples* [5]. Seed weevil in *spring oilseed rape, winter oilseed rape* [1-4]. Seed weevil in *oilseed rape* [5]. Suckers in *pears* [5]. Thrips in *ornamentals* [5]. Tortrix moths in *pears* [1-3]. Tortrix moths in *apples* [1-3, 5]. Whitefly in *ornamentals, protected celery, protected cucumbers, protected lettuce, protected ornamentals* [5]. Winter moth in *apples, pears* [1, 2]. Yellow cereal fly in *winter barley, winter wheat* [1-4]. Yellow cereal fly in *autumn sown spring barley, autumn sown spring wheat* [1, 2, 4]. Yellow cereal fly in *spring barley, spring wheat* [3]. Yellow cereal fly in *rye, triticale* [4]. Yellow cereal fly in *winter cereals* [5].

Efficacy
- Products combine rapid action, good persistence, and high activity on Lepidoptera.
- As effect is mainly via contact good coverage is essential for effective action. Spray volume should be increased on dense crops
- A repeat spray after 10-14 d is needed for some pests of outdoor crops, several sprays at shorter intervals for whitefly and other glasshouse pests
- Rates and timing of sprays vary with crop and pest. See label for details
- Add a non-ionic wetter to improve results on leaf brassicas. In Brussels sprouts use of a drop-leg sprayer may be beneficial
- Where aphids in hops, pear suckers or glasshouse whitefly resistant to cypermethrin occur control is unlikely to be satisfactory

Crop safety/Restrictions
- Maximum number of treatments varies with crop and product. See label or approval notice for details
- Test spray sample of new or unusual ornamentals before committing whole batches

Special precautions/Environmental safety
- Harmful if swallowed [1-4]
- May cause lung damage if swallowed [2-4]
- Irritating to eyes and skin [1-5]
- May cause sensitization by skin contact [1-4]
- Flammable
- Dangerous to bees. Do not apply to crops in flower or to those in which bees are actively foraging except as directed on peas. Do not apply when flowering weeds are present [5]
- Extremely dangerous to fish or other aquatic life. Do not contaminate surface waters or ditches with chemical or used container
- LERAP Category A
- Do not spray cereals after 31 Mar within 6 m of the edge of the growing crop

Personal protective equipment/Label precautions
- A, C [1-5]
- M03 [1-4]; R04a, R04b, R07d [1-5]; R03c, R04e [1-4]; R04g [2-4]; U05a, U10, U19, U20b [1-5]; U02, U04a, U11 [1-4]; C03 [1-5]; C02 [2] (7 d); E01, E13a, E16c, E16d, E30a, E31b [1-5]; E26 [2-5]; E34 [1-4]; E12c [5]

Latest application/Harvest interval
- Varies with product and crop. See labels for details

Approval
- Following implementation of Directive 98/82/EC, approval for use of cypermethrin on numerous crops was revoked in 1999
- Off-label approval unstipulated for use on asparagus (OLA3134/98)[5]
- Accepted by BLRA for use on malting barley and hops

Maximum Residue Level (mg residue/kg food)
- hops 30; citrus fruits, apricots, peaches, nectarines, wild berries, lettuces, herbs 2; Chinese cabbage, kale, wild mushrooms 1; grapes, cane fruits, tomatoes, peppers, aubergines, broccoli, cauliflowers, spinach, beet leaves, beans (with pods), leeks 0.5; cucumbers, gherkins, courgettes, kohlrabi, poppy seed, sesame seed, sunflower seed, rape seed, cotton seed, meat (except poultry) 0.2; garlic, onions, shallots 0.1; tree nuts, cranberries, bilberries, currants, gooseberries, miscellaneous fruits, root and tuber vegetables (except radishes, swedes, turnips), spring onions, sweetcorn, watercress, asparagus, celery, cardoons, rhubarb, cultivated mushrooms, beans, peanuts, soya beans, mustard seed, potatoes, triticale, maize, rice, poultry, eggs 0.05; milk 0.02

163   cyproconazole

A contact and systemic conazole fungicide for cereals and other field crops

See also chlorothalonil + cyproconazole

Products

1	AgriGuard Cyproconazole	AgriGuard	100 g/l	SC	09406
2	Alto 240 EC	Novartis	240 g/l	EC	08354
3	Aplan 240 EC	Novartis	240 g/l	EC	08355
4	Barclay Shandon	Barclay	100 g/l	SL	06464
5	Greencrop Gentian	Greencrop	100 g/l	SL	09380
6	Landgold Cyproconazole 100	Landgold	100 g/l	SL	06463
7	Standon Cyproconazole	Standon	100 g/l	SL	07751

Uses

Brown rust in **spring barley, winter barley, winter wheat** [1-7]. Brown rust in **rye** [1-3, 5]. Chocolate spot in **field beans** *(with chlorothalonil)* [1-3, 5]. Crown rust in **spring oats, winter oats** [1-3, 5]. Eyespot in **winter barley, winter wheat** [1-7]. Eyespot in **spring barley** [4]. Glume blotch in **winter wheat** [6]. Late ear diseases in **winter wheat** [2, 3]. Light leaf spot in **winter oilseed rape** [1-3, 5]. Net blotch in **spring barley, winter barley** [1-7]. Phoma leaf spot in **winter oilseed rape** [1-3, 5]. Powdery mildew in **spring barley, winter barley, winter wheat** [1-7]. Powdery mildew in **rye, spring oats, sugar beet, winter oats** [1-3, 5]. Ramularia leaf spots in **sugar beet** [1-3, 5]. Rhynchosporium in **spring barley, winter barley** [1-7]. Rust in **field beans, sugar beet** [1-3, 5]. Rust in **leeks** [2, 3]. Septoria diseases in **winter wheat** [1-5, 7]. Septoria leaf spot in **winter wheat** [6]. Yellow rust in **spring barley, winter barley, winter wheat** [1-7].

Efficacy

- Apply at start of disease development or as preventive treatment and repeat as necessary
- Most effective time of treatment varies with disease and use of tank mixes may be desirable. See label for details
- Product alone gives useful reduction of cereal eyespot. Where high infections probable a tank mix with prochloraz recommended
- On oilseed rape a two spray autumn/spring programme recommended for high risk situations and on susceptible varieties

Crop safety/Restrictions

- Maximum number of treatments 3 per crop
- Application to winter wheat in spring between the start of stem elongation and the third node detectable stage (GS 30-33) may cause straw shortening, but does not cause loss of yield

Special precautions/Environmental safety

- Irritating to eyes and skin [2, 3]
- Harmful (dangerous [2, 3]) to fish or other aquatic life. Do not contaminate surface waters or ditches with chemical or used container [1, 4-7]

Personal protective equipment/Label precautions

- A [1-7]; C, H [2, 3]
- R04a, R04b, U05a, U08, U20b, C03 [2, 3]; E30a, E31b [1-7]; E13c [1, 4-7]; E26 [1-5, 7]; E01, E13b [2, 3]

Latest application/Harvest interval

- HI 6 wk for field beans; 14 d for sugar beet
- Before grain watery ripe (GS 71) for wheat, rye [1, 5]; up to and including beginning of anthesis (GS 61) for wheat [4, 6, 7]; up to and including emergence of ear complete (GS 59) for barley, oats
- Before lowest pods more than 2 cm long (GS 5,1) for winter oilseed rape [1, 5]

Approval

- Accepted by BLRA for use on malting barley

164   cyproconazole + cyprodinil

A broad spectrum fungicide mixture for cereals

Products

Radius	Novartis	5.33:40 % w/w	WG	09387

Uses

Brown rust in **winter wheat**. Eyespot in **spring barley, winter barley, winter wheat**. Net blotch in **spring barley, winter barley**. Powdery mildew in **spring barley, winter barley, winter wheat**. Rhynchosporium in **spring barley, winter barley**. Septoria diseases in **winter wheat**. Yellow rust in **winter wheat**.

Efficacy
* Best results obtained from treatment at early stages of development. Later treatments may be required if disease pressure remains high
* Eyespot is fully controlled by treatment in spring during stem extension. Control may be reduced when very dry conditions follow application

Crop safety/Restrictions
* Maximum total dose equivalent to two full dose treatments

Special precautions/Environmental safety
* Irritating to skin. May cause sensitization by skin contact
* Dangerous to fish or other aquatic life. Do not contaminate surface waters or ditches with chemical or used container
* LERAP Category B

Personal protective equipment/Label precautions
* A, H
* R04b, R04e, U05a, U20a, C03, E01, E13b, E16a, E16b, E26, E29, E30a

Latest application/Harvest interval
* Before first spikelet of inflorescence visible (GS 51) for barley; before caryopsis watery ripe (GS 71) for winter wheat

Approval
* Accepted by BLRA for use on malting barley

165   cyproconazole + prochloraz

A broad spectrum protective and curative fungicide for cereals

Products

1	Profile	Aventis	48:320 g/l	EC	08134
2	Sportak Delta 460 HF	Aventis	48:320 g/l	EC	07431

Uses

Brown rust in **spring barley, spring wheat, winter barley, winter wheat**. Eyespot in **spring barley, spring wheat, winter barley, winter wheat**. Glume blotch in **spring wheat, winter wheat**. Leaf spot in **spring wheat, winter wheat**. Net blotch in **spring barley, winter barley**. Powdery mildew in **spring barley, spring wheat, winter barley, winter wheat**. Rhynchosporium in **spring barley, winter barley**. Yellow rust in **spring barley, spring wheat, winter barley, winter wheat**.

Efficacy
* To control foliar diseases apply before infection starts spreading to younger leaves and repeat if necessary
* For established mildew use in tank mixture with an approved morpholine fungicide
* For eyespot control apply in spring from when leaf sheaths start to erect up to and including 3rd node detectable stage (GS 30-33). Effective against eyespot and *Septoria tritici* resistant to MBC fungicides
* Eyespot already in crop on up to 10% tillers also controlled

- When applied to control eyespot may give some control of sharp eyespot and Fusarium if present
- See label for details of timing for foliar disease control. Where infection from Septoria may occur following a rain-splash event treat as soon as possible afterwards to protect second leaf and flag leaf
- Good spray cover of stem bases and leaves is essential
- A period of at least 3 h without rain should follow spraying for full effectiveness

Crop safety/Restrictions
- Maximum number of treatments 2 per crop
- Application to winter wheat in spring at GS 30-33 may cause straw shortening but does not cause loss of yield

Special precautions/Environmental safety
- Irritating to skin
- Harmful to fish or other aquatic life. Do not contaminate surface waters or ditches with chemical or used container

Personal protective equipment/Label precautions
- A [1, 2]
- R04b, U05a, U09a, U20b, C03 [1, 2]; E26 [2]; E01, E13c, E30a, E31b [1, 2]

Latest application/Harvest interval
- Up to and including emergence of ear complete (GS 59) on barley; before beginning of anthesis (GS 60) on wheat, rye

Approval
- Accepted by BLRA for use on malting barley

---

166   cyproconazole + propiconazole

A broad-spectrum mixture of conazole fungicides for wheat and barley

Products
Menara	Novartis	160:250 g/l	EC	09321

Uses
Brown rust in **spring barley, winter barley, winter wheat**. Late ear diseases in **winter wheat**. Net blotch in **spring barley, winter barley**. Powdery mildew in **spring barley, winter barley, winter wheat**. Rhynchosporium in **spring barley, winter barley**. Septoria diseases in **winter wheat**. Yellow rust in **winter wheat**.

Efficacy
- Treatment should be made at first sign of disease
- Useful reduction of eyespot when applied in spring during stem extension (GS 30-32)

Crop safety/Restrictions
- Maximum total dose equivalent to one full dose treatment
- Application to winter wheat in spring may cause straw shortening but does not cause loss of yield

Special precautions/Environmental safety
- Irritating to eyes and skin
- Dangerous to fish or other aquatic life. Do not contaminate surface waters or ditches with chemical or used container

Personal protective equipment/Label precautions
- A, C, H
- R04a, R04b, U05a, U08, U20b, C03, E01, E13b, E26, E29, E30a, E31b

Latest application/Harvest interval
- Before beginning of anthesis (GS 60) for barley; before grain watery ripe (GS 71) for wheat

Approval
- Accepted by BLRA for use on malting barley

Maximum Residue Level (mg residue/kg food)
- See propiconazole entry

167   cyproconazole + quinoxyfen

A contact and systemic fungicide mixture for cereals

Products

Divora	Novartis	80:75 g/l	SC	08960

Uses

Brown rust in **winter wheat**. Eyespot in **winter wheat** *(reduction)*. Powdery mildew in **winter wheat**. Septoria diseases in **winter wheat**. Yellow rust in **winter wheat**.

Efficacy
- Best results are achieved when disease is first detected and seen to be active. A second treatment may be needed under prolonged disease pressure
- Treatment for Septoria control should be made as soon as possible after weather that favours infection to prevent spread to the second and flag leaves
- Useful reduction of eyespot when applied between GS 30-34 to control other diseases

Crop safety/Restrictions
- Maximum number of treatments equivalent to a total of two full doses
- Application to winter wheat in spring may cause straw shortening but does not cause loss of yield

Special precautions/Environmental safety
- May cause sensitization by skin contact
- Dangerous to fish or other aquatic life. Do not contaminate surface waters or ditches with chemical or used container
- LERAP Category B

Personal protective equipment/Label precautions
- A, C, H
- R04e, U05a, U20b, C03, E01, E13b, E16a, E16b, E26, E30a, E31b

Latest application/Harvest interval
- Flag leaf sheath open (GS 49) for winter wheat

Approval
- Accepted by BLRA for use on malting barley

168   cyproconazole + tridemorph

A broad spectrum contact and systemic fungicide for cereals

Products

1 Alto Major	Novartis	80:350 g/l	EC	08468
2 Moot	Novartis	80:350 g/l	EC	08479

Uses

Brown rust in **barley, winter wheat**. Eyespot in **winter barley, winter wheat**. Net blotch in **barley**. Powdery mildew in **barley, winter wheat**. Rhynchosporium in **barley**. Septoria diseases in **winter wheat**. Sooty moulds in **winter wheat**. Yellow rust in **barley, winter wheat**.

Efficacy
- Apply when disease is first active
- Most effective time of treatment varies with disease. Tank mixtures and/or repeat treatments may be desirable. See label for details
- When applied between GS 30-34 for control of other diseases, useful reduction of eyespot is achieved

Crop safety/Restrictions
- Maximum number of treatments 2 per crop in the spring/summer (winter wheat, spring barley), 1 in autumn and 2 in spring/summer (winter barley)
- Maximum individual dose of tridemorph on cereals 375 g/ha
- Application to winter wheat in spring at GS 30-33 may cause straw shortening but does not cause loss of yield

Special precautions/Environmental safety
- Harmful if swallowed
- Irritating to eyes and skin
- Flammable
- Harmful to fish or other aquatic life. Do not contaminate surface waters or ditches with chemical or used container
- Tridemorph may cause harm to the unborn child. Women of child-bearing age should not come into contact with products containing tridemorph
- A closed transfer system meeting BS 6356 Part 9 or equivalent must be used when transferring product from container to sprayer and application equipment must have closed cabs

Personal protective equipment/Label precautions
- A, C
- M03, R03c, R04a, R04b, R07d, U02, U04a, U05a, U09a, U20b, C03, E01, E13c, E26, E30a, E31b, E34; E07 (14 d)

Withholding period
- Keep livestock out of treated areas for at least 14 d following treatment

Latest application/Harvest interval
- Up to and including emergence of ear complete (GS 59) (winter and spring barley), up to and including anthesis complete (GS 69) (winter wheat)

Approval
- All remaining approvals for tridemorph in UK were finally revoked in February 2000 with a 2-year use-up period for existing stocks
- Accepted by BLRA for use on malting barley
- Approval expiry 30 Sep 2001 [1]
- Approval expiry 30 Sep 2001 [2]

---

169   cyprodinil

An anilinopyrimidine systemic broad spectrum fungicide for cereals

See also cyproconazole + cyprodinil

Products

1	Barclay Amtrak	Barclay	75% w/w	WG	09562
2	Standon Cyprodinil	Standon	75% w/w	WG	09345
3	Unix	Novartis	75% w/w	WG	08764

Uses

Eyespot in **winter barley, winter wheat**. Net blotch in **spring barley, winter barley**. Powdery mildew in **spring barley, spring wheat, winter barley, winter wheat**. Rhynchosporium in **spring barley, winter barley**.

Efficacy
- Best results obtained from treatment at early stages of disease development
- For best control of eyespot spray before or during the period of stem extension in spring. Control may be reduced if very dry conditions follow treatment

Crop safety/Restrictions
- Maximum number of treatments equal to one and two thirds full dose on wheat and twice full dose on barley

Special precautions/Environmental safety
- Dangerous to fish or other aquatic life. Do not contaminate surface waters or ditches with chemical or used container
- LERAP Category B

Personal protective equipment/Label precautions
- A, H [1-3]
- U20a [2, 3]; U05a [1-3]; U14 [2]; U20b [1]; C03 [1-3]; E26, E29 [3]; E01, E13b, E16a, E16b, E30a [1-3]; E32a [1, 2]

Latest application/Harvest interval
- Up to and including first awns visible (GS 49) for spring barley, winter barley; up to and including grain watery ripe (GS 71) for spring wheat, winter wheat

Approval
- Accepted by BLRA for use on malting barley

170   cyromazine

A triazine insect larvicide

Products

Neporex 2SG	Novartis A H	2% w/w	SG	08589

Uses

Flies in *livestock houses, manure heaps*.

Efficacy

- Apply to surface of manure
- Granules can be scattered dry or applied in water
- Commence treatment when houseflies begin to breed
- Ensure that all breeding sites are treated
- Repeat after 2-3 wk for heavy populations
- Product affects housefly larvae moulting and has no effect on adults. Use an adulticide if large numbers of adult flies present

Crop safety/Restrictions

- Do not use continuously in intensive or controlled environment animal units. If treatment in subsequent stocking cycles is required use different a.i. and different control method
- Do not feed treated poultry manure to stock
- Spread treated manure only onto grassland or arable land prior to cultivation
- Allow at least 4 wk between spreading and grazing or cropping
- Do not use on manure intended for production of mushroom compost

Special precautions/Environmental safety

- Harmful to fish. Do not contaminate surface waters or ditches with chemical or used container

Personal protective equipment/Label precautions

- A, B, C, H, M
- U05a, U20a, C03, E01, E13c, E30a, E31a

171   2,4-D

A translocated phenoxy herbicide for cereals, grass and amenity use

See also amitrole + 2,4-D + diuron
*   clopyralid + 2,4-D + MCPA*

Products

1	Agricorn D II	FCC	500 g/l	SL	09415
2	Barclay Haybob II	Barclay	490 g/l	SL	08532
3	Depitox	Nufarm Whyte	490 g/l	SL	08000
4	Dicotox Extra	RP Amenity	400 g/l	EC	05330
5	Dicotox Extra	Aventis Environ.	400 g/l	EC	09930
6	Dioweed 50	United Phosphorus	500 g/l	SL	08050
7	Dormone	RP Amenity	465 g/l	SL	05412
8	Dormone	Aventis Environ.	465 g/l	SL	09932
9	Easel	Nufarm Whyte	500 g/l	SL	09878
10	Headland Staff	Headland	470 g/l	SL	07189
11	Herboxone	Headland	500 g/l	SL	10032
12	HY-D	Agrichem	470 g/l	SL	06278
13	Luxan 2,4-D	Luxan	490 g/l	SL	09379
14	MSS 2,4-D Amine	Nufarm Whyte	500 g/l	SL	01391
15	MSS 2,4-D Ester	Nufarm Whyte	500 g/l	EC	01393
16	Syford	Vitax	500 g/l	SL	02062

Uses

Annual dicotyledons in *barley, wheat* [1-3, 6, 10, 12, 14]. Annual dicotyledons in *established grassland* [1-3, 6, 9-13, 15, 16]. Annual dicotyledons in *amenity turf* [1, 4-9, 14-16]. Annual dicotyledons in *winter rye* [1, 6, 10-14]. Annual dicotyledons in *sports turf* [1, 6-8, 14]. Annual dicotyledons in *spring rye* [1, 6, 9, 13-15]. Annual dicotyledons in *spring oats* [12]. Annual dicotyledons in *grassland, undersown cereals* [14]. Annual dicotyledons in *grass seed crops*

[16]. Annual dicotyledons in *rye* [2, 3]. Annual dicotyledons in **apple orchards, pear orchards** [2, 3, 13]. Annual dicotyledons in **conifer plantations, forestry** [4, 5, 9, 15]. Annual dicotyledons in **winter oats** [9-12, 14, 15]. Annual dicotyledons in **spring barley, spring wheat, winter barley, winter wheat** [9, 11, 13, 15]. Aquatic weeds in **aquatic situations** [14]. Aquatic weeds in **water or waterside areas** [7, 8]. Heather in **conifer plantations, forestry** [4, 5, 9, 15]. Perennial dicotyledons in **barley, wheat** [1-3, 6, 10, 12, 14]. Perennial dicotyledons in **established grassland** [1-3, 6, 9-13, 15, 16]. Perennial dicotyledons in **amenity turf** [1, 4-9, 14-16]. Perennial dicotyledons in **winter rye** [1, 6, 10-14]. Perennial dicotyledons in **sports turf** [1, 6-8, 14]. Perennial dicotyledons in **spring rye** [1, 6, 9, 13-15]. Perennial dicotyledons in **spring oats** [12]. Perennial dicotyledons in **grassland, undersown cereals** [14]. Perennial dicotyledons in **grass seed crops** [16]. Perennial dicotyledons in **rye** [2, 3]. Perennial dicotyledons in **apple orchards, pear orchards** [2, 3, 13]. Perennial dicotyledons in **conifer plantations, forestry** [4, 5, 9, 15]. Perennial dicotyledons in **water or waterside areas** [7, 8]. Perennial dicotyledons in **winter oats** [9-12, 14, 15]. Perennial dicotyledons in **spring barley, spring wheat, winter barley, winter wheat** [9, 11, 13, 15]. Willows in **conifer plantations, forestry** [4, 5, 9, 15]. Woody weeds in **conifer plantations, forestry** [4, 5, 9, 15].

Efficacy
- Best results achieved by spraying weeds in seedling to young plant stage when growing actively in a strongly competing crop
- Most effective stage for spraying perennials varies with species. See label for details
- Spray aquatic weeds when in active growth between May and Sep [7, 8]
- Do not spray if rain falling or imminent
- Do not cut grass or graze for at least 7 d after spraying

Crop safety/Restrictions
- Maximum number of treatments normally 1 per crop and in forestry or 2 per yr in grassland. Check individual labels
- Spray winter cereals in spring when leaf-sheath erect but before first node detectable (GS 31), spring cereals from 5-leaf stage to before first node detectable (GS 15-31)
- Do not use on newly sown leys containing clover
- Do not spray grass seed crops after ear emergence
- Do not spray within 6 mth of laying turf or sowing fine grass
- Selective treatment of resistant conifers can be made in Aug when growth ceased and plants hardened off, spray must be directed if applied earlier. See label for details
- Do not plant conifers until at least 1 mth after treatment
- Do not spray crops stressed by cold weather or drought or if frost expected
- Do not use shortly before or after sowing any crop
- Do not direct drill brassicas or grass/clover mixtures within 3 wk of application
- Do not roll or harrow within 7 d before or after spraying

Special precautions/Environmental safety
- Harmful if swallowed [2, 3, 11, 13] and in contact with skin [1, 4-10, 12, 14-16]
- Irritating to eyes [4-6]
- Irritating to skin [2, 4, 5, 13]
- Risk of serious damage to eyes [2, 11, 13]
- May cause sensitization by skin contact [4, 5]
- Harmful to fish or other aquatic life. Do not contaminate surface waters or ditches with chemical or used container
- May be used to control aquatic weeds in presence of fish if used in strict accordance with directions for waterweed control and precautions needed for aquatic use [7, 8, 14]
- Water containing the herbicide must not be used for irrigation purposes within 3 wk of treatment or until the concentration in water is below 0.05 ppm [7, 8, 14]
- Do not dump surplus herbicide in water or ditch bottoms [7, 8, 12]
- Do not use treated water for irrigation purposes within 3 wk of treatment [7, 8, 12]

Personal protective equipment/Label precautions
- A, C [1-16]; H, M [1, 2, 4-16]; D [1, 2, 6, 12, 13]
- M03, R03c [1-16]; R03a [1, 4-10, 12, 14-16]; R04a [4-6]; R04b [2, 4, 5, 13]; R04e [4, 5]; R04d [2, 11, 13]; U05a, U08 [1-16]; U20b [2-5, 7-11, 13-16]; U20a [1, 6, 12]; C03, E01, E13c, E30a, E34 [1-16]; E07 [1-16] (2 wk); E31b [1, 3-6, 9, 12-16]; E26 [2, 3, 9-11, 13-15]; E19 [7, 8, 12]; E21 [7, 8, 12] (3 wk); E31a [1, 2, 6, 10-12]; E14b [2, 11]

Withholding period
- Keep livestock out of treated areas for at least 2 wk and until foliage of any poisonous weeds such as ragwort has died and become unpalatable

Latest application/Harvest interval
- Before first node detectable stage (GS 31) in cereals. Refer to labels for other crops

Approval
- May be applied through CDA equipment (see label for details) [16]. See notes in Section 1 on ULV application
- Approved for aquatic weed control [7, 8]. See notes in Section 1 on use of herbicides in or near water
- Accepted by BLRA for use on malting barley

172   2,4-D + dicamba

A translocated herbicide for use on turf

Products

New Estermone	Vitax	200:35 g/l	EC	06336

Uses

Annual dicotyledons in *amenity turf, lawns*. Perennial dicotyledons in *amenity turf, lawns*.

Efficacy
- Best results achieved by application when weeds growing actively in spring or early summer (later with irrigation and feeding)
- More resistant weeds may need repeat treatment after 3 wk
- Do not use during drought conditions or mow for 3 d before or after treatment

Crop safety/Restrictions
- Maximum number of treatments 3 per yr
- Do not treat newly sown or turfed areas
- Avoid spray drift onto cultivated crops or ornamentals
- Do not re-seed for 6 wk after application

Special precautions/Environmental safety
- Harmful to fish or other aquatic life. Do not contaminate surface waters or ditches with chemical or used container

Personal protective equipment/Label precautions
- A, C, H, M
- U08, U20b, E13c, E30a, E31a; E07 (2 wk)

Withholding period
- Keep livestock out of treated areas for at least 2 wk and until foliage of any poisonous weeds such as ragwort has died and become unpalatable

173   2,4-D + dicamba + triclopyr

A translocated herbicide for perennial and woody weed control

Products

1 Broadsword	United Phosphorus	200:85:65 g/l	EC	09140
2 Nufarm Nu-Shot	Nufarm Whyte	200:85:65 g/l	EC	09139

Uses

Annual dicotyledons in *established grassland, forestry, non-crop areas*. Brambles in *established grassland, forestry, non-crop areas*. Docks in *established grassland, forestry, non-crop areas*. Gorse in *established grassland, forestry, non-crop areas*. Japanese knotweed in *established grassland, forestry, non-crop areas*. Perennial dicotyledons in *established grassland, forestry, non-crop areas*. Rhododendrons in *established grassland, forestry, non-crop areas*. Stinging nettle in *established grassland, forestry, non-crop areas*. Thistles in *established grassland, forestry, non-crop areas*. Woody weeds in *established grassland, forestry, non-crop areas*.

Efficacy
- Apply as foliar spray to herbaceous or woody weeds. Timing and growth stage for best results vary with species. See label for details
- Dilute with water for stump treatment and apply after felling up to the start of regrowth. Treat any regrowth with a spray to the growing foliage
- May be applied at 1/3 dilution in weed wipers or 1/8 dilution with ropewick applicators
- Do not roll or harrow grassland for 7 d before or after spraying

Crop safety/Restrictions
- Do not use on pasture established less than 1 yr or on grass grown for seed
- Where clover a valued constituent of sward only use as a spot treatment
- Do not graze for 7 d or mow for 14 d after treatment
- Do not direct drill grass, clover or brassicas for at least 6 wk after grassland treatment
- Sprays may be applied in pines, spruce and fir providing drift is avoided. Optimum time is mid-autumn when tree growth ceased but weeds not yet senescent
- Do not plant trees for 1-3 mth after spraying depending on dose applied. See label
- Avoid spray drift into greenhouses or onto crops or ornamentals. Vapour drift may occur in hot conditions

Special precautions/Environmental safety
- Harmful if swallowed
- Irritating to eyes and skin
- Flammable
- Dangerous to fish or other aquatic life. Do not contaminate surface waters or ditches with chemical or used container
- Docks and other weeds may become increasingly palatable after treatment and may be preferentially grazed. Where ragwort or other poisonous weeds are present keep livestock out until the weeds have died and become unpalatable

Personal protective equipment/Label precautions
- A, C, D, H, M
- M03, R03c, R04a, R04b, R07d, U05a, U08, U19, U20b, C03, E01, E13b, E26, E30a, E31b, E34; E07 (7 d)

Withholding period
- Keep livestock out of treated area for at least 2 wk or until foliage of any poisonous weeds such as ragwort or buttercup has died and become unpalatable

---

174   2,4-D + dichlorprop + MCPA + mecoprop-P

A translocated herbicide for use in apple and pear orchards

Products

UPL Camppex	United Phosphorus	34.5:133:52.8:82.1 g/l	SL	09121

Uses

Annual dicotyledons in **apples, pears**. Chickweed in **apples, pears**. Cleavers in **apples, pears**. Perennial dicotyledons in **apples, pears**.

Efficacy
- Use on emerged weeds in established apple and pear orchards (from 1 yr after planting) as directed application
- Spray when weeds in active growth and at growth stage recommended on label
- Effectiveness may be reduced by rain within 12 h
- Do not roll, harrow or cut grass crops on orchard floor within at least 3 d before or after spraying

Crop safety/Restrictions
- Applications must be made around, and not directly to, trees. Do not allow drift onto trees
- Do not spray during blossom period. Do not spray to run-off
- Avoid spray drift onto neighbouring crops

Special precautions/Environmental safety
- Harmful in contact with skin and if swallowed

- Irritating to eyes
- Harmful to fish or other aquatic life. Do not contaminate surface waters or ditches with chemical or used container

Personal protective equipment/Label precautions
- A, C, H, M
- M03, R04a, U05a, U08, U20b, C03, E01, E13c, E30a, E31b, E34; E07 (2 wk)

Withholding period
- Keep livestock out of treated areas for at least 2 wk and until foliage of any poisonous weeds such as ragwort has died and become unpalatable

Latest application/Harvest interval
- Before blossom period

Maximum Residue Level (mg residue/kg food)
- see dichlorprop entry

175   2,4-D + mecoprop-P

A translocated herbicide for use in amenity turf

Products

1	Mascot Selective-P	Rigby Taylor	78.5:100 g/l	SL	06105
2	Supertox 30	RP Amenity	93.5:95 g/l	SL	09102
3	Supertox 30	Aventis Environ.	93.5:95 g/l	SL	09946
4	Sydex	Vitax	125:125 g/l	SL	06412

Uses

Annual dicotyledons in *golf courses, lawns* [1]. Annual dicotyledons in *amenity turf, sports turf* [1, 4]. Annual dicotyledons in *managed amenity turf* [2, 3]. Annual dicotyledons in *established grassland, grass seed crops* [4]. Perennial dicotyledons in *golf courses, lawns* [1]. Perennial dicotyledons in *amenity turf, sports turf* [1, 4]. Perennial dicotyledons in *managed amenity turf* [2, 3]. Perennial dicotyledons in *established grassland, grass seed crops* [4].

Efficacy
- May be applied from Apr to Sep, best results in May-Jun when weeds in active growth. Less susceptible weeds may require second treatment not less than 6 wk later
- For best results apply fertiliser 1-2 wk before treatment
- Do not spray during drought conditions or when rain imminent
- Do not close mow infrequently mown turf for 3-4 d before or after treatment. On fine turf do not mow for 24 h before or after treatment

Crop safety/Restrictions
- Maximum number of treatments 2 per yr. The total amount of mecoprop-P applied in a single yr must not exceed the maximum total dose approved for any single product for the crop/ situation
- Do not use first 4 mowings for mulching unless composted for at least 6 mth (do not use first 2 mowings for composting on some labels)
- Newly laid turf or newly sown grass should not be treated for at least 6 mth
- Grass cuttings should not be used for mulching but may be composted and used 6 mth later
- Do not apply during frosty weather
- Do not roll or harrow within 7 d before or after treatment

Special precautions/Environmental safety
- Harmful in contact with skin and if swallowed [1, 4]
- Irritating to eyes [2-4] and skin [1]
- Harmful (dangerous [4]) to fish or other aquatic life. Do not contaminate surface waters or ditches with chemical or used container
- Avoid drift onto broad-leaved plants outside the target area

Personal protective equipment/Label precautions
- A, C, H, M [1-4]

- M03 [1, 4]; R04b [1]; R03a, R03c [1, 4]; R04a [1-4]; U13 [1]; U05a, U08, U20b [1-4]; U11, U15 [2, 3]; C03 [1-4]; E32a [1]; E34 [1, 4]; E01, E30a [1-4]; E07 [1-4] (2 wk); E13c [1-3]; E13b [4]; E26, E31b [2-4]; E23 [2, 3]

Withholding period
- Keep livestock out of treated areas for at least 2 wk and until foliage of any poisonous weeds such as ragwort has died and become unpalatable

176   2,4-D + picloram

A persistent translocated herbicide for non-crop land

Products

Atladox HI	Nomix-Chipman	240:65 g/l	SL	05559

Uses

Annual dicotyledons in **amenity grass, non-crop grass, road verges**. Brambles in **amenity grass, non-crop grass, road verges**. Creeping thistle in **amenity grass, non-crop grass, road verges**. Docks in **amenity grass, non-crop grass, road verges**. Japanese knotweed in **amenity grass, non-crop grass, road verges**. Perennial dicotyledons in **amenity grass, non-crop grass, road verges**. Ragwort in **amenity grass, non-crop grass, road verges**. Scrub clearance in **amenity grass, non-crop grass, road verges**. Woody weeds in **amenity grass, non-crop grass, road verges**.

Efficacy
- Apply as overall foliar spray during period of active growth when foliage well developed

Crop safety/Restrictions
- Maximum number of treatments 1 per yr
- Do not apply around desirable trees or shrubs where roots may absorb chemical
- Prevent leaching into areas where desirable plants are present
- Avoid drift of spray onto desirable plants
- Do not use cuttings from treated grass for mulching or composting

Special precautions/Environmental safety
- Irritating to eyes
- Flammable
- Harmful to fish or other aquatic life. Do not contaminate surface waters or ditches with chemical or used container

Personal protective equipment/Label precautions
- A, C
- R04a, R07d, U05a, U08, U20b, C03, E01, E07, E13c, E26, E30a, E31a

Withholding period
- Keep livestock out of treated areas for at least 2 wk and until foliage of any poisonous weeds such as ragwort has died and become unpalatable

177   dalapon

A translocated chloroalkanoic acid grass herbicide no longer marketed alone but still approved for use

178   daminozide

A hydrazide plant growth regulator for use in certain ornamentals

Products

1 B-Nine	Hortichem	85% w/w	SP	07844
2 Dazide	Fine	85% w/w	SP	02691

Uses

Internode reduction in **ornamentals, poinsettias** [1]. Internode reduction in **azaleas, bedding plants, chrysanthemums, hydrangeas** [1, 2]. Internode reduction in **pot plants** [2].

Efficacy
- Best results obtained by application in late afternoon when glasshouse has cooled down
- Spray when foliage dry
- A reduced rate tank mix with chlormequat + choline chloride is recommended for use on poinsettias. See label for details [1]

Crop safety/Restrictions
- Maximum number of treatments 2 per crop
- Apply only to turgid, well watered plants. Do not water for 24 h after spraying
- Do not use on chrysanthemum Fandango
- Do not mix with other spray chemicals except as recommended above

Special precautions/Environmental safety
- Irritating to eyes [2]
- Harmful to fish or other aquatic life. Do not contaminate surface waters or ditches with chemical or used container [2]

Personal protective equipment/Label precautions
- H [2]; A, C [1, 2]
- R04a [1, 2]; U09a, U20a [2]; U05a, U19, C03 [1, 2]; E13c, E31a [2]; E01, E30a [1, 2]; E15 [1]

Approval
- Sales of daminozide for use on food crops were halted worldwide by the manufacturer in Oct 1989. Sales for use on flower crops were not affected

Maximum Residue Level (mg residue/kg food)
- tea, hops 0.1; tree nuts, oilseeds, animal products 0.05; citrus fruits, pome fruits, stone fruits, berries and small fruit, miscellaneous fruits, vegetables, pulses, potatoes, cereals 0.02

---

179   dazomet

A methyl isothiocyanate releasing soil fumigant

Products

Basamid	Hortichem	97% w/w	GR	07204

Uses

Nematodes in **field crops, protected crops, vegetables**. Soil insects in **field crops, protected crops, vegetables**. Soil-borne diseases in **field crops, protected crops, vegetables**. Weed seeds in **field crops, protected crops, vegetables**.

Efficacy
- Dazomet acts by releasing methyl isothiocyanate in contact with moist soil
- Soil sterilization is carried out after harvesting one crop and before planting the next
- The soil must be of fine tilth, free of clods and evenly moist to the depth of sterilization
- Soil moisture must not be less than 50% of water-holding capacity or oversaturated. If too dry, water at least 7-14 d before treatment
- Do not treat ground where water table may rise into treated layer
- In order to obtain short treatment times it is recommended to treat soils when soil temperature is above 7°C. Treatment should be used outdoors before winter rains make soil too wet to cultivate - usually early Nov
- For club root control treat only in summer when soil temperature above 10°C
- Where onion white rot a problem unlikely to give effective control where inoculum level high or crop under stress
- Apply granules with suitable applicators, mix into soil immediately to desired depth and seal surface with polythene sheeting, by flooding or heavy rolling. See label for suitable application and incorporation machinery
- With 'planting through' technique polythene seal is left in place to form mulch into which new crop can be planted

Crop safety/Restrictions
- Maximum number of treatments 1 per crop or batch of soil
- With 'planting through' technique no gas release cultivations are made and safety test with cress is particularly important. Conduct cress test on soil samples from centre as well as

edges of bed. Observe minimum of 30 d from application to cress test in soils at 10°C or above, at least 50 d in soils below 10°C. See label for details
- In all other situations 14-28 d after treatment cultivate lightly to allow gas to disperse and conduct cress test after a further 14-28 d (timing depends on soil type and temperature). Do not treat structures containing live plants or any ground within 1 m of live plants

Special precautions/Environmental safety
- Harmful if swallowed
- Do not contaminate surface waters or ditches with chemical or used container

Personal protective equipment/Label precautions
- A, M
- M03, R03c, U04a, U05a, U09a, U19, U20a, C03, E01, E15, E30a, E32a, E34

Latest application/Harvest interval
- Pre-planting

---

180   2,4-DB

A translocated phenoxy herbicide for use in lucerne

Products

DB Straight	United Phosphorus	300 g/l	SL	07523

Uses

Annual dicotyledons in *cereals undersown lucerne, lucerne*. Thistles in *cereals undersown lucerne, lucerne*.

Efficacy
- Best results achieved on young seedling weeds under good growing conditions. Treatment less effective in cold weather and dry soil conditions
- Rain within 12 h may reduce effectiveness

Crop safety/Restrictions
- In direct sown lucerne spray when seedlings have reached first trifoliate leaf stage. Optimum time 3-4 trifoliate leaves
- In spring barley and spring oats undersown with lucerne spray from when cereal has 1 leaf unfolded and lucerne has first trifoliate leaf
- In spring wheat undersown with lucerne spray from when cereal has 3 leaves unfolded and lucerne has first trifoliate leaf
- Do not treat any lucerne after fourth trifoliate leaf
- Do not allow spray drift onto neighbouring crops

Special precautions/Environmental safety
- Harmful to fish or other aquatic life. Do not contaminate surface waters or ditches with chemical or used container

Personal protective equipment/Label precautions
- U08, U20a, E13c, E26, E30a, E31b; E07 (2 wk)

Withholding period
- Keep livestock out of treated areas until foliage of any poisonous weeds such as ragwort has died and become unpalatable

Approval
- Accepted by BLRA for use on malting barley

---

181   2,4-DB + linuron + MCPA

A translocated herbicide for undersown cereals and grass

Products

Alistell	Zeneca	220:30:30 g/l	EC	06515

Uses

Annual dicotyledons in *cereals undersown clovers, seedling grassland*.

Efficacy
- Best results achieved on young seedling weeds growing actively in warm, moist weather
- May be applied at any time of year provided crop at correct stage and weather suitable
- Avoid spraying if rain falling or imminent

Crop safety/Restrictions
- Maximum number of treatments 1 per crop or yr
- Spray winter cereals when fully tillered but before first node detectable (GS 29-30)
- Spray spring wheat from 5-fully expanded leaf stage (GS 15), barley and oats from 2-fully expanded leaves (GS 12)
- Do not spray cereals undersown with lucerne, peas or beans
- Apply to clovers after 1-trifoliate leaf, to grasses after 2-fully expanded leaf stage
- Do not spray in conditions of drought, waterlogging or extremes of temperature
- In frosty weather clover leaf scorch may occur but damage normally outgrown
- Do not use on sand or soils with more than 10% organic matter
- Do not roll or harrow within 7 d before or after spraying
- Avoid drift of spray or vapour onto susceptible crops

Special precautions/Environmental safety
- Harmful in contact with skin and if swallowed
- Irritating to eyes and skin
- Dangerous to fish or other aquatic life. Do not contaminate surface waters or ditches with chemical or used container
- LERAP Category B
- Do not apply by hand-held sprayers

Personal protective equipment/Label precautions
- A, C, H
- M03, R03a, R03c, R04a, R04b, U05a, U08, U13, U19, U20b, C03, E01, E13b, E16a, E30a, E31b, E34; E07 (2 wk)

Withholding period
- Keep livestock out of treated areas for at least 2 wk and until foliage of any poisonous weeds such as ragwort has died and become unpalatable

Latest application/Harvest interval
- Before first node detectable (GS 31) for undersown cereals; 2 wk before grazing for grassland

Approval
- Accepted by BLRA for use on malting barley

---

182   2,4-DB + MCPA

A translocated herbicide for cereals, clovers and leys

Products

1 Agrichem DB Plus	Agrichem	243:40 g/l	SL	00044
2 Headland Cedar	Headland	240:40 g/l	SL	09948
3 MSS 2,4-DB + MCPA	Nufarm Whyte	280 g/l (total)	SL	01392
4 Redlegor	United Phosphorus	244:44 g/l	SL	07519

Uses

Annual dicotyledons in *leys* [1]. Annual dicotyledons in *cereals* [1, 3]. Annual dicotyledons in *undersown cereals* [1, 3, 4]. Annual dicotyledons in *grass re-seeds, spring barley, spring oats, spring wheat, undersown barley* (red or white clover), *undersown oats* (red or white clover), *undersown wheat* (red or white clover), *winter barley, winter oats, winter wheat* [2]. Annual dicotyledons in *clovers, grassland* [3]. Annual dicotyledons in *seedling leys* [3, 4]. Perennial dicotyledons in *leys* [1]. Perennial dicotyledons in *cereals* [1, 3]. Perennial dicotyledons in *undersown cereals* [1, 3, 4]. Perennial dicotyledons in *grass re-seeds, spring barley, spring oats, spring wheat, undersown barley* (red or white clover), *undersown oats* (red or white clover), *undersown wheat* (red or white clover), *winter barley, winter oats, winter wheat* [2]. Perennial dicotyledons in *clovers, grassland* [3]. Perennial dicotyledons in *seedling leys* [3, 4]. Polygonums in *cereals, clovers* [3]. Polygonums in *seedling leys, undersown cereals* [3, 4].

Efficacy
- Best results achieved on young seedling weeds under good growing conditions
- Spray thistles and other perennials when 10-20 cm high provided clover at correct stage
- Effectiveness may be reduced by rain within 12 h, by very cold conditions or drought

Crop safety/Restrictions
- Apply in spring to winter cereals from leaf sheath erect stage, to spring barley or oats from 2-leaf stage (GS 12), to spring wheat from 5-leaf stage (GS 15)
- Spray clovers as soon as possible after first trifoliate leaf, grasses after 2-3 leaf stage
- Red clover may suffer temporary distortion after treatment
- Do not spray established clover crops or lucerne
- Do not roll or harrow within 7 d before or after spraying
- Do not spray immediately before or after sowing any crop
- Avoid drift onto neighbouring sensitive crops

Special precautions/Environmental safety
- Harmful if swallowed [1, 2] and in contact with skin [3]
- Irritating to skin [1, 2]
- Risk of serious damage to eyes [1-3]
- Harmful to fish or other aquatic life. Do not contaminate surface waters or ditches with chemical or used container

Personal protective equipment/Label precautions
- A, C [1-3]
- M03 [3]; R03c, R04d [1-3]; R04b [1, 2]; R03a [3]; U08, U20b [1-4]; U05a [2, 3]; C03 [3]; E31c [1]; E13c, E26 [1-4]; E07 [1-4] (2 wk); E30a [1-3]; E31b [3, 4]; E01, E34 [2, 3]; E31a [2]

Withholding period
- Keep livestock out of treated areas until foliage of any poisonous weeds such as ragwort has died and become unpalatable

Latest application/Harvest interval
- Before first node detectable (GS 31) for cereals

Approval
- Accepted by BLRA for use on malting barley

---

183   DE 570

A sulfonyl urea herbicide for cereals

Products

Boxer	Dow	100 g/l	SC	09819

Uses

Annual dicotyledons in *spring barley, spring oats, spring wheat, winter barley, winter oats, winter wheat*. Charlock in *spring barley, spring oats, spring wheat, winter barley, winter oats, winter wheat*. Chickweed in *spring barley, spring oats, spring wheat, winter barley, winter oats, winter wheat*. Cleavers in *spring barley, spring oats, spring wheat, winter barley, winter oats, winter wheat*. Volunteer oilseed rape in *spring barley, spring oats, spring wheat, winter barley, winter oats, winter wheat*.

Efficacy
- Best results obtained from application in good growing conditions
- Apply after 1 Feb once crop has 3 leaves
- Product is mainly absorbed by leaves of weeds and is effective on all soil types

Crop safety/Restrictions
- Maximum total dose on any crop equivalent to one full dose treatment
- Do not roll or harrow within 7 d before or after application
- Do not spray when crops under stress from cold, drought, pest damage, nutrient deficiency or any other cause
- Only cereals, oilseed rape, field beans or grass may be sown as a following crop in the same calendar yr as treatment. Oilseed rape may show some temporary reduction of vigour after a dry summer, but yields are not affected

- Cereals, oilseed rape, field beans, grass, linseed, peas, sugar beet, potatoes, maize or clover (for use in grass/clover mixtures) may be sown in the calendar yr following treatment
- In the event of failure of a treated crop in spring only spring wheat, spring barley, spring oats, maize or ryegrass may be sown
- Do not spray in tank mixture or sequence with a product containing any other ALS-inhibitor (e.g.sulfonylurea herbicides) except those containing metsulfuron-methyl, thifensulfuron-methyl or tribenuron-methyl

Special precautions/Environmental safety

- Extremely dangerous to fish or other aquatic life. Do not contaminate surface waters or ditches with chemical or used container
- LERAP Category B
- See label for detailed instructions on tank cleaning

Personal protective equipment/Label precautions

- U05a, C03, E01, E13a, E16a, E16b, E26, E34

Latest application/Harvest interval

- Before flag leaf sheath extending stage (GS 41) for all crops

---

184   deltamethrin

A pyrethroid insecticide with contact and residual activity

Products

1	Aventis Decis	Aventis	25 g/l	EC	09710
2	Crackdown	AgrEvo Environ.	10 g/l	SC	H5097
3	Decis	Aventis	25 g/l	EC	07172
4	Landgold Deltaland	Landgold	25 g/l	EC	09906
5	Pearl Micro	Aventis	6.25% w/w	GR	08620
6	Thripstick	Aquaspersions	0.125 g/l	AL	02134

Uses

American serpentine leaf miner in *aubergines* (off-label - fog), *courgettes* (off-label - fog), *cucumbers* (off-label - fog), *gherkins* (off-label - fog), *onions* (off-label - fog), *peppers* (off-label - fog), *tomatoes* (off-label - fog) [3]. Aphids in *peppers* [1, 3]. Aphids in *apples* [1, 3-5]. Aphids in *cereals, hops, ornamentals, peas, plums, protected cucumbers, protected ornamentals, protected pot plants, protected tomatoes* [1, 3, 5]. Aphids in *chinese cabbage* (off-label), *grass seed crops* (off-label), *lettuce* (off-label), *radicchio* (off-label) [3]. Aphids in *spring barley, spring oats, spring wheat, winter barley, winter oats, winter wheat* [4]. Aphids in *protected peppers* [5]. Apple sucker in *apples* [4]. Barley yellow dwarf vectors in *winter barley, winter wheat* [4]. Barley yellow dwarf virus vectors in *cereals* [1, 3, 5]. Beet virus yellows vectors in *winter oilseed rape* [1, 3-5]. Brassica pod midge in *mustard, oilseed rape* [1, 3, 5]. Cabbage seed weevil in *mustard* [1, 3-5]. Cabbage seed weevil in *oilseed rape* [1, 3, 5]. Cabbage seed weevil in *spring oilseed rape, winter oilseed rape* [4]. Cabbage stem flea beetle in *winter oilseed rape* [1, 3-5]. Cabbage stem flea beetle in *winter oilseed rape (in store)* [2]. Cabbage stem weevil in *mustard* [1, 3-5]. Cabbage stem weevil in *oilseed rape* [1, 3, 5]. Cabbage stem weevil in *spring oilseed rape, winter oilseed rape* [4]. Capsids in *apples* [1, 3-5]. Capsids in *flowers, nursery stock, woody ornamentals* [1, 3, 5]. Caterpillars in *peppers* [1, 3]. Caterpillars in *outdoor lettuce, protected cucumbers, protected ornamentals, protected pot plants, protected tomatoes* [1, 3, 5]. Caterpillars in *chinese cabbage* (off-label), *lettuce* (off-label), *radicchio* (off-label) [3]. Caterpillars in *apples, broccoli, brussels sprouts, cabbages, cauliflowers, kale, plums, swedes, turnips* [4]. Caterpillars in *protected peppers* [5]. Caterpillars in *tomato houses* (off-label) [6]. Codling moth in *apples* [1, 3-5]. Colorado beetle in *potatoes* (off-label - statutory notice) [3]. Damson-hop aphid in *hops, plums* [4]. Flea beetles in *sugar beet* [1, 3-5]. Flea beetles in *leaf brassicas, root brassicas* [1, 3, 5]. Flea beetles in *evening primrose* (off-label), *herbs* (off-label), *protected lettuce* (off-label), *radicchio* (off-label) [3]. Flour beetles in *grain stores* [2]. Fruit tree tortrix moth in *plums* [4]. Grain beetles in *grain stores* [2]. Grain moths in *grain stores* [2]. Grain storage pests in *grain stores* [2]. Insect pests in *carrots* (off-label), *celery* (off-label), *fennel* (off-label), *garlic* (off-label), *leeks* (off-label), *onions* (off-label), *salad onions (off-label)* [3]. Leaf miners in **tomato houses** *(off-label)* [6]. Leafhoppers in **herbs** *(off-label)*, **protected lettuce** *(off-label)*, **radicchio** *(off-label)* [3]. Mealy bugs in **peppers** [1, 3]. Mealy bugs in **protected cucumbers, protected ornamentals, protected pot plants, protected tomatoes** [1, 3, 5]. Mealy bugs in **protected peppers** [5]. Pea and bean weevils in **broad beans, peas** [1, 3-5]. Pea and bean weevils in **field beans** [1, 3, 5]. Pea and bean weevils in **spring field beans, winter field beans** [4]. Pea moth in **peas** [1, 3-5]. Pear sucker in **pears** [4]. Plum fruit moth in **plums** [1, 3, 5]. Pollen beetles in **mustard** [1, 3-5]. Pollen beetles in **oilseed rape** [1, 3, 5]. Pollen beetles in **evening primrose** *(off-label)* [3]. Pollen beetles in **spring oilseed rape, winter oilseed rape** [4]. Rape winter stem weevil in **winter oilseed rape** [1, 3, 5]. Rape winter stem weevil in **winter oilseed rape** *(in store)* [2]. Raspberry beetle in **raspberries** [1, 3-5]. Sawflies in **apples, plums** [1, 3-5]. Scale insects in **peppers** [1, 3]. Scale insects in **flowers, nursery stock, protected cucumbers, protected ornamentals, protected pot plants, protected tomatoes, woody ornamentals** [1, 3, 5]. Scale insects in **protected peppers** [5]. Suckers in **apples, pears** [1, 3, 5]. Thrips in **flowers, nursery stock, woody ornamentals** [1, 3, 5]. Thrips in **protected cucumbers** [6]. Tortrix moths in **apples** [1, 3-5]. Western flower thrips in **aubergines** *(off-label - fog)*, **courgettes** *(off-label - fog)*, **cucumbers** *(off-label - fog)*, **gherkins** *(off-label - fog)*, **onions** *(off-label - fog)*, **peppers** *(off-label - fog)*, **tomatoes** *(off-label - fog)* [3]. Whitefly in **peppers** [1, 3]. Whitefly in **protected cucumbers, protected ornamentals, protected pot plants, protected tomatoes** [1, 3, 5]. Whitefly in **protected peppers** [5]. Yellow cereal fly in **cereals** [1, 3, 5].

Efficacy
- A contact and stomach poison with 3-4 wk persistence, particularly effective on caterpillars and sucking insects
- Normally applied at first signs of damage with follow-up treatments where necessary at 10-14 d intervals. Rates, timing and recommended combinations with other pesticides vary with crop and pest. See label for details [3]
- Spray is rainfast within 1 h. May be applied in frosty weather provided foliage not covered in ice [3]
- Temperatures above 35°C may reduce effectiveness or persistence
- Spray Thripstick without dilution onto polythene covering on glasshouse floor. Spray before planting cucumbers where thrips are a known problem or when pest problem seen. Repeat every 8 wk to ensure complete control [6]
- Where aphids in hops, pear suckers or glasshouse whitefly resistant to deltamethrin occur control is unlikely to be satisfactory

Crop safety/Restrictions
- Maximum number of treatments varies with crop and pest, 4 per crop for wheat and barley, only 1 application between 1 Apr and 31 Aug. See label or off-label approval notice for other crops
- Do not apply more than 1 aphicide treatment to cereals in summer
- Do not spray crops suffering from drought or other physical stress
- Consult processer before treating crops for processing

Special precautions/Environmental safety
- Harmful in contact with skin [1-4]
- Harmful if swallowed [1-5]
- Irritating to eyes and skin [1-5]
- Flammable [1-4]
- Dangerous to bees. Do not apply to crops in flower or to those in which bees are actively foraging except as directed (see labels). Do not apply when flowering weeds are present [1-5]
- Extremely dangerous to fish or other aquatic life. Do not contaminate surface waters or ditches with chemical or used container
- Do not apply to a cereal crop if any product containing a pyrethroid insecticide or dimethoate has been applied to that crop after the start of ear emergence (GS 51)
- LERAP category A (except 2, 6)
- Do not spray cereals after 31 Mar in the year of harvest within 6 m of the outside edge of the crop
- Reduced volume spraying must not be used on cereals after 31 Mar in yr of harvest

Personal protective equipment/Label precautions
- A, C, H [1-6]

FOR FULL CONDITIONS OF USE ALWAYS READ THE PRODUCT LABEL

- M03, R03c, R04a, R04b [1-5]; R03a, R07d [1-4]; U02, U05b, U06, U09a [6]; U20b [1, 3-6]; U05a [1-6]; U19 [1-3, 5, 6]; U04a, U08 [1-5]; U20a [2]; C03 [1-6]; E31a [6]; E01, E13a, E26, E30a [1-6]; E27 [2, 6]; E12d, E16c, E16d [1, 3-5]; E34 [1-5]; E31b [1-4]; E32a [5]; E12c [2]

Latest application/Harvest interval
- Early dough (GS 83) for barley, oats, wheat [3, 5]; end of flowering for mustard, oilseed rape [3, 5]
- HI outdoor leeks 3 d

Approval
- Off-label approval unlimited for control of Colorado beetle in potatoes as required by Statutory Notice (OLA 1440/97)[3]; unlimited for use in tomato houses (OLA 0180/92)[6]; to Mar 2001 for use on outdoor fennel (OLA 0508/98)[3]unlimited for use on carrots (OLA 1845/00)[3]; unlimited for use on onions, garlic (OLA 1847/00)[3]; unlimited for use on lettuce (OLA 1691/ 96)[3]; unlimited for use on onions, tomatoes, gherkins, aubergines, peppers, cucumbers, courgettes (OLA 1839/00)[3]; unlimited for use on evening primrose, grass seed crops, sorrel, radicchio, marjoram (OLA 1841/00)[3]; unlimited for use on protected spring onions, celery, Chinese cabbage (OLA 1849/00)[3]; to Jul 2003 for use on protected lettuce, raddichio, leafy herbs (OLA 1857/00)[3]; unstipulated for use on outdoor leeks (OLA 1851/00)[3]; unstipulated for use on salad leaf crops (see notice for details) (OLA 1853/00)[3]; unstipulated for use on potatoes (OLA 1855/00)[3]; unstipulated for use on outdoor lettuce, Chinese cabbage, cabbages (OLA 1843/00)[3]
- Accepted by BLRA for use in stores for malting barley (empty stores only) [2], and on malting barley crops and hops

Maximum Residue Level (mg residue/kg food)
- tea, hops 5; pulses, cereals 1; Chinese cabbage, kale, lettuces, spinach, beet leaves, herbs 0.5; currants, gooseberries, tomatoes, peppers, aubergines, beans (with pods) 0.2; pome fruits, stone fruits, grapes, garlic, onions, shallots, cucumbers, gherkins, courgettes, broccoli, cauliflowers, Brussels sprouts, head cabbages, peas (with pods), rape seed 0.1; miscellaneous fruit 0.05; citrus fruits, tree nuts, strawberries, bilberries, cranberries, wild berries, root and tuber vegetables, melons, squashes, watermelons, kohlrabi, watercress, peas (without pods), stem vegetables (except globe artichokes), mushrooms, oilseeds (except rape seed), early potatoes, poultry, eggs 0.01

---

185   deltamethrin + pirimicarb

An insecticide mixture combining systemic, contact and stomach activity

Products

1 Evidence	Aventis	7.5:100 g/l	EC	06934
2 Patriot EC	Aventis	5:100 g/l	EC	08990

Uses

Aphids in *barley (on ears)*, *oats (on ears)*, *wheat (on ears)* [2]. Blackfly in *sugar beet* [1, 2]. Caterpillars in *broccoli, brussels sprouts, cabbages, calabrese, cauliflowers, kale, swedes, turnips* [1, 2]. Green aphid in *broccoli, brussels sprouts, cabbages, calabrese, cauliflowers, kale, potatoes, sugar beet, swedes, turnips* [1, 2]. Green aphid in *oilseed rape* [2]. Mealy aphids in *broccoli, brussels sprouts, cabbages, calabrese, cauliflowers, kale, swedes, turnips* [1, 2]. Mealy aphids in *oilseed rape* [2]. Pea aphid in *peas* [1, 2]. Virus vectors in *potatoes* [1]. Virus yellows vectors in *sugar beet* [1, 2].

Efficacy
- Best results obtained from treatment in warm, calm conditions, ideally in morning or evening when uptake by the plant will be at its highest
- Peas should be inspected from beginning of flowering and treated according to current thresholds
- Treat potatoes when first aphids found to control virus spread and preserve processing quality. Other ware crops should be treated from late Jun onwards before population build-up occurs and seed crops must be treated at 80% emergence
- Higher water volume recommended where crop canopy dense
- Treat other crops when pests first seen or when warnings are issued
- Treat cereals when at least two-thirds of heads infested and numbers increasing [2]
- Where aphids resistant to deltamethrin + pirimicarb occur repeat treatments are likely to result in lower levels of control

Crop safety/Restrictions
- Maximum number of treatments 4 per crop on potatoes, edible brassicas; 2 per crop on peas and sugar beet; 1 per crop on cereals. See label for restrictions on oilseed rape
- Do not spray any crop when wilting or in temperatures above 25°C
- Do not treat wet crops liable to run-off
- Sugar beet may be treated as a band spray provided the pressure and water volume requirements are met

Special precautions/Environmental safety
- Harmful if swallowed
- Risk of serious damage to eyes
- Irritating to respiratory system
- Flammable
- Dangerous to bees. Do not apply to crops in flower or to those in which bees are actively foraging except as directed on peas. Do not apply when flowering weeds are present
- Extremely dangerous to fish or other aquatic life. Do not contaminate surface waters or ditches with chemical or used container.
- LERAP Category A
- Harmful to livestock
- Must not be applied to cereals if a pyrethroid or dimethoate product has been used at ear emergence (GS 51)
- Product formulation can release traces of other pesticides left in the sprayer. Sprayer should be thoroughly washed out before spraying this product

Personal protective equipment/Label precautions
- A, D, E, H [1, 2]
- M02 [1]; M03 [1, 2]; M05 [2]; R03c, R04c, R04d, R07d [1, 2]; U08 [1]; U04a, U05a, U19, U20b [1, 2]; U11 [2]; C03, E01, E12d, E13a, E16c, E16d, E30a, E31b, E34 [1, 2]; E06b [1, 2] (7 d); E26 [2]

Withholding period
- Livestock should be kept out of treated areas for at least 7 d

Latest application/Harvest interval
- Before grain watery ripe (GS 71) for barley, oats, wheat; end of flowering for oilseed rape
- HI 3 d

Approval
- Accepted by BLRA for use on malting barley

Maximum Residue Level (mg residue/kg food)
- see deltamethrin entry

---

186   desmedipham

A contact carbamate herbicide available only in mixtures

---

187   desmedipham + ethofumesate + phenmedipham

A selective contact and residual herbicide for beet

Products

1 Aventis Betanal Progress OF	Aventis	25:151:76 g/l	EC	09722
2 Betanal Progress OF	Aventis	25:151:76 g/l	EC	07629

Uses

Annual dicotyledons in *sugar beet*. Annual meadow grass in *sugar beet*.

Efficacy
- Best results achieved if weeds are not larger than fully expanded cotyledon stage at spraying
- Where a pre-emergence band spray has been applied treatment must be timed according to size of the untreated weeds between the rows

- Repeat applications as each flush of weeds reaches cotyledon size normally necessary for season long control
- Sequential treatments should be applied when the previous one is still showing an effect on the weeds
- Various mixtures with other beet herbicides are recommended. See label for details

Crop safety/Restrictions
- Maximum total dose 4.5 l/ha product per crop
- Apply first treatment when majority of crop plants have reached the fully expanded cotyledon stage
- Do not spray crops stressed by nutrient deficiency, wind damage, pest or disease attack, or previous herbicide treatments
- If temperature likely to exceed 21°C spray after 5 pm
- Frost within 7 d of treatment may cause check from which the crop may not recover
- Crystallisation may occur if spray volume exceeds that recommended or spray mixture not used within 2 h, especially if the water temperature is below 5°C
- Before use, wash out sprayer to remove all traces of previous products, especially hormone and sulfonyl urea weedkillers

Special precautions/Environmental safety
- Harmful if swallowed
- Irritating to skin
- Risk of serious damage to eyes
- Harmful to fish or aquatic life. Do not contaminate surface waters or ditches with chemical or used container

Personal protective equipment/Label precautions
- A, C, H
- M03, R03c, R04b, R04d, U05a, U08, U20a, C03, E01, E13c, E30a, E31b, E34

188   desmedipham + phenmedipham

A mixture of contact herbicides for use in sugar beet

Products

Betanal Compact	Aventis	34:129 g/l	EC	07247

Uses

Annual dicotyledons in **sugar beet**.

Efficacy
- Best results obtained from treatment when earliest germinating weeds have reached cotyledon stage
- Further treatments must be applied as each flush of weeds reaches cotyledon stage but allowing a minimum of 7 d between each spray
- Where a pre-emergence band spray has been applied, the first treatment should be timed according to the size of the weeds in the untreated area between the rows
- Product should be applied overall as a fine spray to optimise weed cover and spray retention
- Various tank mixtures and sequences recommended to widen weed spectrum and add residual activity - see label for details

Crop safety/Restrictions
- Maximum number of treatments 3 per crop
- Product safe to use on all soil types
- Spray in evening if daytime temperatures above 21°C expected
- Avoid or delay treatment if frost expected within 7 d
- Avoid or delay treating crops under stress from wind damage, manganese or lime deficiency, pest or disease attack etc
- Check from which recovery may not be complete may occur if treatment made during conditions of sharp diurnal temperature fluctuation

Special precautions/Environmental safety
- Irritant. Risk of serious damage to eyes

- Dangerous to fish or other aquatic life. Do not contaminate surface waters or ditches with chemical or used container
- Keep interval between spray preparation and completion of treatment to 2 h or less to avoid possibility of crystallisation especially if water temperature below 5°C
- Product may cause non-reinforced PVC pipes and hoses to soften and swell. Wherever possible, use reinforced PVC or synthetic rubber hoses

Personal protective equipment/Label precautions
- A, H
- R04, R04d, U05a, U08, U19, U20a, C03, E01, E13b, E26, E30a, E31b, E34

Latest application/Harvest interval
- Before crop leaves meet between rows

---

189   dicamba

A translocated benzoic herbicide for control of bracken and perennial weeds

See also 2,4-D + dicamba

Products

Cadence	Barclay	70% w/w	SG	09578

Uses

Chickweed in *established leys, permanent pasture*. Docks in *established leys, permanent pasture*.

Efficacy
- For weed control in established leys and pasture apply when weeds actively growing before flowering shoots appear. Large well-established docks may require a second treatment in the following yr

Crop safety/Restrictions
- Maximum total dose equivalent to one full dose per yr
- Avoid drift onto susceptible crops. Tomatoes may be affected by vapour drift at a considerable distance
- Do not roll crops within 1 wk of spraying. Do not treat crops where clover forms an important part of the sward

Special precautions/Environmental safety
- Irritating to eyes

Personal protective equipment/Label precautions
- A, C
- R04a, U05a, U08, U20b, C03; C02 (21 d); E01, E15, E29, E30a, E31b, E34

Withholding period
- Keep livestock out of treated areas for at least 3 wk and until foliage of any poisonous weeds such as ragwort and buttercups has died and become unpalatable

---

190   dicamba + dichlorprop + ferrous sulphate + MCPA

A herbicide/fertilizer combination for moss and weed control in turf

Products

Renovator 2	Scotts	0.027:0.446:23.08:0.223 % w/w	GR	09272

Uses

Annual dicotyledons in *amenity turf, golf courses*. Mosses in *amenity turf, golf courses*. Perennial dicotyledons in *amenity turf, golf courses*.

Efficacy
- Apply from mid-Apr to mid-Aug when weeds are growing
- Apply with a suitable calibrated fertilizer distributor
- For best control of moss scarify vigorously after 2 wk to remove dead moss
- Where regrowth of moss or weeds occurs a repeat treatment may be made after 6 wk
- Do not cut grass for at least 3 d before and at least 4 d after treatment

---

FOR FULL CONDITIONS OF USE ALWAYS READ THE PRODUCT LABEL

- Avoid treatment of wet grass or during drought. If no rain falls within 48 h water in thoroughly
- Do not apply during freezing conditions or when rain imminent

Crop safety/Restrictions
- Do not treat new turf until established for 6-9 mth
- The first 4 mowings after treatment should not be used to mulch cultivated plants unless composted at least 6 mth
- Avoid walking on treated areas until it has rained or they have been watered
- Do not re-seed or turf within 8 wk of last treatment

Personal protective equipment/Label precautions
- U09a, U20a, E15, E30a, E32a

Maximum Residue Level (mg residue/kg food)
- see dichlorprop entry

---

191   dicamba + dichlorprop + MCPA

A translocated herbicide mixture for turf

Products

1 Cleanrun 2	Scotts	0.03:0.45:0.22% w/w	MG	09271
2 Intrepid	Scotts	20.8:333:166.5 g/l	SL	09266

Uses

Annual dicotyledons in **turf**. Perennial dicotyledons in **turf**.

Efficacy
- Apply as directed on label between Apr and Sep when weeds growing actively
- If no rain falls within 48 h of treatment irrigate thoroughly [1]
- Do not mow for at least 3 d before and 3-4 d after treatment so that there is sufficient leaf growth for spray uptake and sufficient time for translocation. Se label for details
- If re-growth occurs or new weeds germinate re-treatment recommended

Crop safety/Restrictions
- Do not use during drought unless irrigation is carried out before and after treatment
- Do not apply during freezing conditions or when heavy rain is imminent
- Do not treat new turf until established for about 6-9 mth after seeding or turfing
- The first 4 mowings after treatment should not be used to mulch cultivated plants unless composted for at least 6 mth
- Avoid walking where possible on treated areas until after rain or irrigation [1]
- Do not re-seed turf within 8 wk of last treatment

Special precautions/Environmental safety
- Harmful if swallowed [2]
- Irritating to skin [2]
- Risk of serious damage to eyes [2]
- May cause sensitization by skin contact [2]

Personal protective equipment/Label precautions
- A, C [2]
- M03, R03c, R04b, R04d, R04e, U05a, U08, U19, U20a [2]; U09a [1]; C03, E01, E31b, E34 [2]; E15, E30a [1, 2]; E32a, S06 [1]

Maximum Residue Level (mg residue/kg food)
- see dichlorprop entry

---

192   dicamba + maleic hydrazide + MCPA

A herbicide/plant growth regulator mixture for amenity grass

Products

Mazide Selective	Vitax	6:200:75 g/l	SL	05753

Uses

Annual dicotyledons in *amenity grass, road verges*. Growth retardation in *amenity grass, road verges*. Perennial dicotyledons in *amenity grass, road verges*.

Efficacy

- Best results achieved by application in Apr-May when grass and weeds growing actively but before weeds have started to flower
- May be used either as one annual spray or as spring spray repeated after 8-10 wk

Crop safety/Restrictions

- Maximum number of treatments 2 per yr
- Do not use on fine turf

Special precautions/Environmental safety

- Harmful to fish or other aquatic life. Do not contaminate surface waters or ditches with chemical or used container

Personal protective equipment/Label precautions

- U08, U20b, E07, E13c, E26, E30a, E31a

Withholding period

- Keep livestock out of treated areas until foliage of any poisonous weeds such as ragwort has died and become unpalatable

Maximum Residue Level (mg residue/kg food)

- see maleic hydrazide entry

---

193   dicamba + MCPA + mecoprop

A translocated herbicide for cereals, grassland, amenity grass and orchards

Products

1 Barclay Hat-Trick	Barclay	25:200:400 g/l	SL	07579
2 Quad-Ban	Quadrangle	19.5:245:86.5 g/l	SL	03114

Uses

Annual dicotyledons in *leys, permanent pasture* [1, 2]. Annual dicotyledons in *barley, grass seed crops, oats, rye, wheat* [2]. Chickweed in *leys, permanent pasture* [1, 2]. Chickweed in *barley, grass seed crops, oats, rye, wheat* [2]. Cleavers in *barley, grass seed crops, leys, oats, permanent pasture, rye, wheat* [2]. Docks in *leys, permanent pasture* [1, 2]. Docks in *barley, grass seed crops, oats, rye, wheat* [2]. Mayweeds in *barley, grass seed crops, leys, oats, permanent pasture, rye, wheat* [2]. Perennial dicotyledons in *leys, permanent pasture* [1, 2]. Perennial dicotyledons in *barley, grass seed crops, oats, rye, wheat* [2]. Polygonums in *barley, grass seed crops, leys, oats, permanent pasture, rye, wheat* [2].

Efficacy

- Best results achieved on young, actively growing weeds up to 15 cm high in a strongly competitive crop, perennials when well developed but before flowering. See label for details of susceptibility
- Spray grassland and turf from early spring to Oct when grasses growing actively
- Do not spray in rain, when rain imminent or in drought

Crop safety/Restrictions

- Maximum number of treatments 1 per crop on cereals or 2 per yr on other crops
- Spray winter cereals from main leaf sheath erect and at least 5 cm high but before first node detectable (GS 31), spring cereals from 5-leaf stage to before first node detectable (GS 15-31) (some formulations only recommend treatment in spring)
- Do not spray cereals undersown with clovers or legumes, to be undersown with grass or legumes or grassland where clovers or other legumes are important
- Spray newly sown grass or undersown crops when grass seedlings have 2-3 leaves
- Do not spray grass seed crops later than 5 wk before emergence of seed heads
- In orchards avoid drift of spray onto tree foliage. Do not spray during blossom period
- Do not roll, harrow, cut, graze for 3-7 d before or after treatment (products vary)

Special precautions/Environmental safety
- Harmful in contact with skin and if swallowed
- Irritating to eyes
- Irritating to skin [2]
- Harmful to fish or other aquatic life. Do not contaminate surface waters or ditches with chemical or used container

Personal protective equipment/Label precautions
- A, C [1, 2]
- M03 [1, 2]; R04b [2]; R03a, R03c, R04a, U05a, U08, U20a, C03, E01, E07, E13c, E30a, E31b, E34 [1, 2]

Withholding period
- Keep livestock out of treated areas for at least 2 wk and until foliage of any poisonous weeds such as ragwort has died and become unpalatable

Latest application/Harvest interval
- Before first node detectable (GS 31) for cereals; 5 wk before head emergence for grass seed crops
- HI normally 7 d before cutting or grazing grass - check labels

Approval
- Accepted by BLRA for use on malting barley

---

194   dicamba + MCPA + mecoprop-P

A translocated herbicide for cereals, grassland, amenity grass and orchards

Products

1	Banlene Super	Aventis	18:252:42 g/l	SL	10053
2	Field Marshal	United Phosphorus	18:360:80	SL	08956
3	Headland Relay P	Headland	25:200:200 g/l	SL	08580
4	Headland Relay Turf	Headland	25:200:200 g/l	SL	08935
5	Hycamba Plus	Agrichem	20.4:125.9:240.2 g/l	SL	10180
6	Hyprone-P	Agrichem	16:101:92 g/l	SL	09125
7	Hysward-P	Agrichem	16:101:92 g/l	SL	09052
8	Mascot Super Selective-P	Rigby Taylor	15:100:100 g/l	SL	06106
9	Mircam Plus	Nufarm Whyte	19.5:245:43.3 g/l	SL	09884
10	Pasturol Plus	FCC	25:200:200 g/l	SL	08581
11	Tribute	Nomix-Chipman	18:252:42 g/l	SL	06921
12	Tribute Plus	Nomix-Chipman	18:252:42 g/l	SL	09493
13	Tritox	Scotts	15:178:54 g/l	SL	07764
14	UPL Grassland Herbicide	United Phosphorus	25:200:200	SL	08934

Uses

Annual dicotyledons in *apples, pears, rye* [1]. Annual dicotyledons in *leys* [1-3, 10, 14]. Annual dicotyledons in *barley, oats, wheat* [1, 2, 6]. Annual dicotyledons in *permanent pasture* [1, 3, 10, 14]. Annual dicotyledons in *grass seed crops* [1, 6, 9]. Annual dicotyledons in *established grassland* [1, 7]. Annual dicotyledons in *cereals undersown (grass only)* [2]. Annual dicotyledons in *sports turf* [3, 4, 8, 10-13]. Annual dicotyledons in *amenity grass* [3, 5, 7, 10]. Annual dicotyledons in *amenity turf* [4, 7-9, 11-13]. Annual dicotyledons in *managed amenity turf* [5]. Annual dicotyledons in *newly sown grass, undersown barley, undersown oats, undersown wheat* [6]. Annual dicotyledons in *grassland, spring barley, spring oats, spring wheat, winter barley, winter oats, winter wheat* [9]. Chickweed in *established grassland, rye* [1]. Chickweed in *permanent pasture* [1, 14]. Chickweed in *leys* [1, 2, 14]. Chickweed in *barley, oats, wheat* [1, 2, 6]. Chickweed in *grass seed crops* [9]. Chickweed in *cereals undersown (grass only)* [2]. Chickweed in *newly sown grass, undersown barley, undersown oats, undersown wheat* [6]. Chickweed in *amenity turf, grassland, spring barley, spring oats, spring wheat, winter barley, winter oats, winter wheat* [9]. Cleavers in *rye* [1]. Cleavers in *barley, oats, wheat* [1, 2, 6]. Cleavers in *leys, permanent pasture* [14]. Cleavers in *newly sown grass, undersown barley, undersown oats, undersown wheat* [6]. Cleavers in *grass seed crops* [6, 9]. Cleavers in *amenity turf, grassland, spring barley, spring oats, spring wheat, winter barley, winter oats, winter wheat* [9]. Docks in *apples, pears* [1]. Docks in *leys, permanent pasture* [1, 14]. Docks in *grass seed crops* [1, 6]. Docks in *established grassland* [1, 7]. Docks in *amenity grass, amenity turf* [7]. Mayweeds in *established grassland, rye* [1]. Mayweeds in *permanent pasture* [1, 14]. Mayweeds in *leys* [1, 2, 14]. Mayweeds in *barley, oats, wheat* [1, 2, 6]. Mayweeds in *grass seed crops* [1, 6, 9]. Mayweeds in *cereals undersown (grass only)* [2]. Mayweeds in *newly sown grass, undersown barley, undersown oats, undersown wheat* [6]. Mayweeds in *amenity turf, grassland, spring barley, spring oats, spring wheat, winter barley, winter oats, winter wheat* [9]. Perennial dicotyledons in *apples, pears, rye* [1]. Perennial dicotyledons in *leys* [1-3, 10, 14]. Perennial dicotyledons in *barley, oats, wheat* [1, 2, 6]. Perennial dicotyledons in *permanent pasture* [1, 3, 10, 14]. Perennial dicotyledons in *grass seed crops* [1, 6, 9]. Perennial dicotyledons in *established grassland* [1, 7]. Perennial dicotyledons in *cereals undersown (grass only)* [2]. Perennial dicotyledons in *sports turf* [3, 4, 8, 10-13]. Perennial dicotyledons in *amenity grass* [3, 5, 7, 10]. Perennial dicotyledons in *amenity turf* [4, 7-9, 11-13]. Perennial dicotyledons in *managed amenity turf* [5]. Perennial dicotyledons in *newly sown grass, undersown barley, undersown oats, undersown wheat* [6]. Perennial dicotyledons in *grassland, spring barley, spring oats, spring wheat, winter barley, winter oats, winter wheat* [9]. Polygonums in *rye* [1]. Polygonums in *leys* [1, 2, 14]. Polygonums in *barley, oats, wheat* [1, 2, 6]. Polygonums in *permanent pasture* [14]. Polygonums in *cereals undersown (grass only)* [2]. Polygonums in *newly sown grass, undersown barley, undersown oats, undersown wheat* [6]. Polygonums in *grass seed crops* [6, 9]. Polygonums in *amenity turf, grassland, spring barley, spring oats, spring wheat, winter barley, winter oats, winter wheat* [9].

Efficacy
- Treatment should be made when weeds growing actively. Weeds hardened by winter weather may be less susceptible
- For best results apply in fine warm weather, preferably when soil is moist. Do not spray if rain expected within 6 h or in drought
- Application of fertilizer 1-2 wk before spraying aids weed control in turf
- Where a second treatment later in the season is needed in amenity situations and on grass allow 4-6 wk between applications to permit sufficient foliage regrowth for uptake

Crop safety/Restrictions
- Maximum number of treatments (including other mecoprop-P products) 1 per crop on cereals or 2 per yr on other crops and amenity grass. The total amount of mecoprop-P applied in a single yr must not exceed the maximum total dose approved for any single product for the crop/situation
- Apply to winter cereals from the leaf sheath erect stage (GS 30), and to spring cereals from the 5 expanded leaf stage (GS 15)
- Do not apply to cereals after the first node is detectable (GS 31), or to grass under stress from drought or cold weather
- Do not spray cereals undersown with clovers or legumes, to be undersown with grass or legumes or grassland where clovers or other legumes are important
- Do not spray leys established less than 18 mth or orchards established less than 3 yr
- Spray grass seed crops 4-6 wk before flower heads begin to emerge (timothy 6 wk)
- Do not roll or harrow within 7 d before or after treatment, or graze for at least 7 d afterwards (longer if poisonous weeds present)
- Do not use on turf in year of establishment. Allow 6-8 wk after treatment before seeding bare patches [3, 4, 7, 8, 12, 13]
- The first mowings after use should not be used for mulching unless composted for 6 mth [3-5, 7, 8, 12, 13]
- Turf should not be mown for 24 h before or after treatment (3-4 d for closely mown turf) [3-5, 7, 8, 12, 13]
- Turf containing bulbs may be treated once the foliage has died down completely [3, 4, 7, 8, 12, 13]
- Avoid drift onto all broad-leaved plants outside the target area

Special precautions/Environmental safety
- Harmful in contact with skin [11, 12] and if swallowed [2-4, 10, 14]
- Irritating to eyes

- Risk of serious damage to eyes [5]
- Irritating to skin [5-8]
- Harmful to fish or other aquatic life. Do not contaminate surface waters or ditches with chemical or used container

Personal protective equipment/Label precautions

- A, C, H, M [1-14]
- M03 [2-4, 9, 10, 14]; R04a [1-4, 6-14]; R04b [5-9]; R03a [2-4, 9-12, 14]; R03c [2-4, 9, 10, 14]; R04d [5]; U19 [8]; U08 [1-14]; U20b [1, 2, 4-9, 11-14]; U05a [1, 3-12, 14]; C03 [1-14]; E32a [8]; E01, E13c, E30a [1-14]; E34 [1-4, 8-14]; E07 [1-3, 5-14] (2 wk); E26 [1-7, 9-12, 14]; E31b [1-4, 9-12, 14]; E31a [13]; E31c [5-7]; S06 [3, 10]

Withholding period

- Keep livestock out of treated areas for at least 2 wk and until foliage of any poisonous weeds such as ragwort has died and become unpalatable

Latest application/Harvest interval

- Before first node detectable (GS 31) for cereals; 4-6 wk before head emergence for grass seed crops
- HI 7 d before cutting or grazing for leys, permanent pasture

Approval

- Accepted by BLRA for use on malting barley

---

195   dicamba + mecoprop-P

A translocated post-emergence herbicide for cereals and grassland

Products

1 Camber	Headland	42:319 g/l	SL	09901
2 Condox	Zeneca	112:133 g/l	SL	09429
3 Di-Farmon R	Headland	42:319 g/l	SL	08472
4 Dockmaster	Nufarm Whyte	18.7:150 g/l	SL	10033
5 Foundation	Novartis	84:600 g/l	SL	08475
6 Hyban-P	Agrichem	18.7:150 g/l	SL	09129
7 Hygrass-P	Agrichem	18.7:150 g/l	SL	09130
8 Mircam	Nufarm Whyte	18.7:150 g/l	SL	09888

Uses

Annual dicotyledons in *established grassland* [1, 3, 5]. Annual dicotyledons in *barley, oats, wheat* [1, 3, 5, 6]. Annual dicotyledons in *established leys* [2]. Annual dicotyledons in *permanent pasture* [2, 7]. Annual dicotyledons in *spring barley, spring oats, spring wheat, winter barley, winter oats, winter wheat* [4, 8]. Annual dicotyledons in *rye, triticale* [6]. Annual dicotyledons in *leys* [7]. Chickweed in *established grassland* [1, 3]. Chickweed in *barley, oats, wheat* [1, 3, 5, 6]. Chickweed in *spring barley, spring oats, spring wheat, winter barley, winter oats, winter wheat* [4, 8]. Chickweed in *rye, triticale* [6]. Cleavers in *established grassland* [1, 3]. Cleavers in *barley, oats, wheat* [1, 3, 5, 6]. Cleavers in *spring barley, spring oats, spring wheat, winter barley, winter oats, winter wheat* [4, 8]. Cleavers in *rye, triticale* [6]. Docks in *established leys* [2]. Docks in *permanent pasture* [2, 7]. Docks in *leys* [7]. Mayweeds in *established grassland* [1, 3]. Mayweeds in *barley, oats, wheat* [1, 3, 5, 6]. Mayweeds in *spring barley, spring oats, spring wheat, winter barley, winter oats, winter wheat* [4, 8]. Mayweeds in *rye, triticale* [6]. Perennial dicotyledons in *established leys* [2]. Perennial dicotyledons in *permanent pasture* [2, 7]. Perennial dicotyledons in *spring barley, spring oats, spring wheat, winter barley, winter oats, winter wheat* [4, 8]. Perennial dicotyledons in *established grassland* [5]. Perennial dicotyledons in *barley, oats, rye, triticale, wheat* [6]. Perennial dicotyledons in *leys* [7]. Polygonums in *established grassland* [1, 3]. Polygonums in *barley, oats, wheat* [1, 3, 5, 6]. Polygonums in *spring barley, spring oats, spring wheat, winter barley, winter oats, winter wheat* [4, 8]. Polygonums in *rye, triticale* [6]. Thistles in *established leys* [2]. Thistles in *permanent pasture* [2, 7]. Thistles in *leys* [7].

Efficacy

- Best results by application in warm, moist weather when weeds are actively growing
- Do not spray in cold or frosty conditions
- Do not spray if rain expected within 6 h

Crop safety/Restrictions
- Maximum number of treatments 1 per crop for cereals and 1 or 2 per yr on grass depending on label. The total amount of mecoprop-P applied in a single yr must not exceed the maximum total dose approved for any single product for the crop/situation
- Apply to winter sown crops from 5 expanded leaf stage (GS 15)
- Apply to spring sown cereals from 5 expanded leaf stage but before first node is detectable (GS 15-31)
- Treat grassland just before perennial weeds flower [3, 5]
- Transient crop prostration may occur after spraying but recovery is rapid
- Do not treat undersown grass until tillering begins
- Do not spray cereals undersown with clover or legume mixtures
- Do not roll, harrow within 7 d before or after spraying
- Do not treat crops suffering from stress from any cause
- Avoid treatment when drift may damage neighbouring susceptible crops

Special precautions/Environmental safety
- Harmful if swallowed and in contact with skin [2-5, 8]
- Irritating to eyes [5, 7] and skin [2-4, 6, 8]
- Harmful to fish or other aquatic life. Do not contaminate surface waters or ditches with chemical or used container

Personal protective equipment/Label precautions
- A, C [1-8]; H [1-4, 6-8]; M [1, 3, 4, 7, 8]; D [2]
- M03, R04a [1-8]; R03a, R03c [1-5, 8]; R04b [1-4, 6, 8]; U20a [3]; U05a [1-8]; U08 [1-5, 8]; U20b [1, 2, 4-8]; U09a [6, 7]; C03, E01, E13c, E30a [1-8]; E07 [1-8] (2 wk); E26 [1, 3-8]; E34 [1-6, 8]; E31b [1, 3-5, 8]; E31c [2, 6, 7]

Withholding period
- Keep livestock out of treated areas for at least 2 wk and until foliage of poisonous weeds such as ragwort has died and become unpalatable

Latest application/Harvest interval
- Before 1st node detectable for cereals; 7 d before cutting or 14 d before grazing grass

Approval
- Accepted by BLRA for use on malting barley

---

196   dichlobenil

A residual benzonitrile herbicide for woody crops and non-crop uses

Products

1 Casoron G	Zeneca	6.75% w/w	GR	08065
2 Casoron G	Rigby Taylor	6.75% w/w	GR	09326
3 Casoron G	Uniroyal	6.75% w/w	GR	09022
4 Casoron G	Nomix-Chipman	6.75% w/w	GR	09023
5 Casoron G4	Uniroyal	4% w/w	GR	09215
6 Luxan Dichlobenil Granules	Luxan	6.75% w/w	GR	09250
7 Sierraron G	Scotts	6.75% w/w	GR	09675
8 Standon Dichlobenil 6G	Standon	6.75% w/w	GR	08874

Uses

Annual weeds in *rhubarb (off-label)* [1]. Annual weeds in *ornamental trees* [2-5]. Annual weeds in *established woody ornamentals, roses* [2-8]. Annual weeds in *soft fruit, top fruit* [6, 8]. Aquatic weeds in *aquatic situations* [1-4, 6, 7]. Perennial dicotyledons in *ornamental trees* [2-5]. Perennial dicotyledons in *established woody ornamentals, roses* [2-8]. Perennial dicotyledons in *soft fruit, top fruit* [6, 8]. Perennial grasses in *rhubarb (off-label)* [1]. Perennial grasses in *ornamental trees* [2-5]. Perennial grasses in *established woody ornamentals, roses* [2-8]. Perennial grasses in *soft fruit, top fruit* [6, 8]. Total vegetation control in *non-crop areas* [1-4, 6-8]. Total vegetation control in *land not intended to bear vegetation* [1-4, 7]. Volunteer potatoes in *potato dumps (blight prevention)* [1, 3, 6, 8].

Efficacy
- Best results achieved by application in winter to moist soil during cool weather, particularly if rain follows soon after. Do not disturb treated soil by hoeing
- Lower rates control annuals, higher rates perennials, see label for details

---

- Residual activity lasts 3-6 mth with selective, up to 12 mth with non-selective rates
- Apply to potato dump sites before emergence of potatoes for blight prevention [5, 6, 8]
- For control of emergent, floating and submerged aquatics apply to water surface in early spring. Intended for use in still or sluggish flowing water [2-4, 6, 7]
- Do not use on fen peat or moss soils

Crop safety/Restrictions

- Maximum number of treatments 1 per yr
- Apply to crops in dormant period. See label for details of timing and rates
- Do not treat crops established for less than 2 yr. See label for lists of resistant and sensitive species. Do not treat stone fruit or Norway spruce (Christmas trees)
- Do not apply within 300 m of glasshouses or hops or near areas underplanted with bulbs, annuals or herbaceous stock
- Do not apply to frozen, snow-covered or waterlogged ground or when crop foliage wet
- Do not apply to sites less than 18 mth before replanting or sowing
- Store well away from corms, bulbs, tubers and seed

Special precautions/Environmental safety

- Irritant. May cause sensitization by skin contact [6]
- Harmful to fish or other aquatic life. Do not contaminate surface waters or ditches with chemical or used container
- Do not dump surplus herbicide in water or ditch bottoms [2-7]
- Use not permitted on reservoirs that form part of a water supply system [2-7]
- Do not use treated water for irrigation purposes within 2 wk of treatment or until concentration of dichlobenil falls below 0.3 ppm[2-7]

Personal protective equipment/Label precautions

- A, H [6]
- M03, R04, R04e [6]; U20c [1-6, 8]; U05a [6]; U20b [7]; C03 [6]; E13c, E30a, E32a [1-8]; E19 [1-4, 6, 7]; E25, E29 [5, 8]; E01, E34 [6]; E21 [7] (2 wk)

Latest application/Harvest interval

- Early Mar for shrubs and trees; early Apr for apples, pears, currants, gooseberries; before bud movement for cane fruit, roses; before growth starts for potato dumps

Approval

- Approved for aquatic weed control [2-4, 6, 7]. See notes in Section 1 on use of herbicides in or near water
- Off-label approval unstipulated for use on established rhubarb (OLA 3083/98)[1]

---

197   dichlofluanid

A protectant trihalomethylthio fungicide for horticultural crops

Products

Elvaron WG	Bayer	50% w/w	WG	04855

Uses

Black spot in **roses**. Botrytis in **aubergines** *(off-label)*, **cane fruit, currants, gooseberries, grapevines, non-edible ornamentals** *(off-label)*, **peppers** *(off-label)*, **protected tomatoes, strawberries**. Cane blight in **raspberries**. Cane spot in **raspberries**. Downy mildew in **protected brassica seedlings**. Fire in **tulips**. Leaf mould in **protected tomatoes**. Mildew in **raspberries, roses**.

Efficacy

- Apply as a protective programme of sprays. See label for recommended timings
- Treatment also gives reduction of mildew and spur blight on raspberries and useful control of mildew on strawberries
- Apply to brassica seedlings under cover as soon as first seedlings appear and repeat after 3, 5, 7 and 10 d using a wetting agent. Thereafter apply 3 sprays at weekly intervals

Crop safety/Restrictions

- Maximum number of treatments 8 per crop for brassicas under cover, tomatoes, peppers, aubergines, non-edible ornamentals under glass and outdoor grapes for wine-making; 6 per crop for currants; 4 per crop for gooseberries. For other fruit crops number depends on dose applied and restriction on total for season

- Do not use on strawberries under glass or polythene. Plants which have been under cover recently may still be susceptible to leaf scorch
- On raspberries up to 7 applications may be made, up to 6 on currants, up to 8 on grapes for winemaking

Special precautions/Environmental safety
- Harmful to fish or other aquatic life. Do not contaminate surface waters or ditches with chemical or used container

Personal protective equipment/Label precautions
- U09a, U19, U20a, E13c, E30a, E32a

Latest application/Harvest interval
- Before 2 true leaf stage for Brussels sprouts, calabrese; before 8 expanded leaves for planted out cabbage, cauliflower, broccoli.
- HI raspberries 1 wk; strawberries, currants 2 wk; broccoli, cabbage 3 wk; gooseberries, loganberries, blackberries, Rubus hybrids, grapes (fresh consumption) 3 wk; grapes (winemaking) 5 wk; protected tomatoes, peppers, aubergines, non-edible ornamentals 3 d

Approval
- Off-label approval unlimited on protected peppers, aubergines and non-edible ornamentals (OLA 0167/93)

---

198   dichlorophen

A chlorophenol moss-killer, fungicide, bactericide and algicide

Products

1	50/50 Liquid Mosskiller	Vitax	360 g/l	SL	07191
2	Enforcer	Scotts	360 g/l	SL	09288
3	Fungo	Dax	340 g/l	SL	H4768
4	Mascot Mosskiller	Rigby Taylor	360 g/l	SL	02439
5	Mossicide	PBI	360 g/l	SL	09606
6	Nomix-Chipman Mosskiller	Nomix-Chipman	340 g/l	SL	06271
7	Panacide M	Coalite	360 g/l	SL	05611
8	Panacide M21	Coalite	165 g/l	SL	H4923
9	Panacide TS	Coalite	480 g/l	SL	05612
10	Panaclean 736	Coalite	45 g/l	SL	H5075
11	Super Mosstox	Aventis Environ.	340 g/l	SL	09942
12	Super Mosstox	RP Amenity	340 g/l	SL	05339

Uses

Algae in **paths and drives** [3]. Algae in **hard surfaces** [3, 8, 10]. Fungus diseases in **paths and drives, turf** [1]. Fungus diseases in **hard surfaces** [1, 2, 6]. Mosses in **hard surfaces** [1-6, 8, 10-12]. Mosses in **turf** [1, 2, 4, 7, 9, 11, 12]. Mosses in **paths and drives** [1, 3, 4, 11, 12]. Mosses in **lawns** [5, 6]. Mosses in **amenity turf, sports turf** [6]. Red thread in **turf** [11, 12].

Efficacy
- For moss control in turf spray at any time when moss is growing. Supplement spraying with other measures to improve fertility and drainage
- Rake out dead moss 2-3 wk after treatment
- Treat hard surfaces when rain not expected and brush away dead material when treatment fully dried [6]

Crop safety/Restrictions
- Dose must not exceed 1 litre in 20 litres water on turf, or 1 litre in 60 litres on hard surfaces [2]

Special precautions/Environmental safety
- Irritating to eyes and skin
- Risk of serious damage to eyes [3]
- Keep unprotected persons and animals away while treatment in progress [1, 4, 7] and for 48 h after treatment or until paint is dry [6]
- Prevent surface run-off from entering storm drains [1-8, 10]
- Dangerous (harmful [11, 12]) to fish or other aquatic life. Do not contaminate surface waters or ditches with chemical or used container [1-4, 6-10]

Personal protective equipment/Label precautions
- M [1, 2, 4, 6, 7, 9, 11, 12]; C [1-4, 6, 7, 9, 11, 12]; A, H [1-4, 6-12]; E [10]
- M03 [6, 7, 9]; M05 [3]; R04a, R04b [1-12]; R04d [3]; U05a [1-12]; U04a [1, 2, 4-7, 9]; U02 [1, 4-10]; U15 [1, 4, 6]; U14 [1, 2, 4, 6]; U20b [1, 3, 4, 6]; U17 [2, 4, 6]; U08 [1, 4]; U09a [2, 3, 5-12]; U20c [11, 12]; U20a [2, 5, 7-10]; U19 [8, 10]; C03 [1-12]; C09 [1-10]; E01, E30a [1-12]; E31a [1, 4, 6, 7, 9, 11, 12]; E26 [1, 4, 11, 12]; E02 [1, 2, 4-10] (48 hr); E13b [1-4, 6-10]; E20 [1-8, 10]; E13c [11, 12]; E15 [3]; E32a [8, 10]

Withholding period
- Keep unprotected persons out of treated areas for 48 h or until surfaces are dry

199   dichlorophen + ferrous sulphate

A mosskiller/fertilizer mixture for use on turf

Products

SHL Lawn Sand Plus	Sinclair	0.3:10% w/w	GR	04439

Uses

Mosses in *amenity turf*.

Efficacy
- Apply to established turf from late spring to early autumn when soil moist but not when grass wet or damp with dew
- Heavy infestations may need a repeat treatment

Crop safety/Restrictions
- Maximum number of treatments 2 per yr
- Do not treat newly sown grass or freshly laid turf for the first year
- Do not apply during drought or freezing conditions or when rain imminent
- Avoid walking on treated areas until it has rained or turf has been watered
- Do not mow for 3-4 d before or after application

Special precautions/Environmental safety
- Harmful to fish or other aquatic life. Do not contaminate surface waters or ditches with chemical or used container

Personal protective equipment/Label precautions
- U09a, U20a, E13c, E30a

200   1,3-dichloropropene

A halogenated hydrocarbon soil nematicide

Products

Telone II	Dow	94% w/w	VP	05749

Uses

Free-living nematodes in *potatoes*. Nematodes in *hops, raspberries, strawberries*. Potato cyst nematode in *potatoes*. Stem and bulb nematodes in *narcissi*. Stem nematodes in *strawberries*. Virus vectors in *hops, raspberries, strawberries*.

Efficacy
- Before treatment soil should be in friable seed bed condition above 5°C, with adequate moisture and all crop remains decomposed. Tilling to 30 cm improves results
- Do not use on heavy clays or soils with many large stones
- Apply with sub-surface A-blade injector combined with soil-sealing roller or, in hops, with a hollow tined injector. See label for details of suitable machinery
- Leave soil undisturbed for 21 d after treatment. Wet or cold soils need longer exposure
- For control of potato-cyst nematode contact firm's representative, check nematode level and use in integrated control programme
- For use in hops apply in May or Jun following the yr of grubbing and leave as long as possible before replanting

- Do not use water to clean out apparatus nor use aluminium or magnesium alloy containers which may corrode

Crop safety/Restrictions
- Maximum number of treatments 2 per yr for hops, 1 per yr for other crops
- Do not drill or plant until odour of fumigant is eliminated
- Do not use fertilizer containing ammonium salts after treatment. Use only nitrate nitrogen fertilizers until crop well established and soil temperature above 18°C
- Allow at least 6 wk after treatment before planting raspberries or strawberries

Special precautions/Environmental safety
- Toxic if swallowed. Harmful in contact with skin and by inhalation
- May cause lung damage if swallowed
- Irritating to eyes, skin and respiratory system
- May cause sensitization by skin contact
- Flammable
- Harmful to fish or other aquatic life. Do not contaminate surface waters or ditches with chemical or used container

Personal protective equipment/Label precautions
- A, C, H, M
- M04, M05, R02c, R03a, R03b, R04a, R04b, R04c, R04e, R04g, R07d, U02, U04a, U05a, U09a, U14, U15, U19, U20a, C03, E01, E13c, E24, E28, E30b, E33, E34

Latest application/Harvest interval
- 6 wk before planting for raspberries, strawberries; pre-planting of crop for hops, narcissi, potatoes

Approval
- Accepted by BLRA for use on hops

201   dichlorprop

A translocated phenoxy herbicide for use in cereals

See also dicamba + dichlorprop + ferrous sulphate + MCPA
dicamba + dichlorprop + MCPA

Products

| MSS 2,4-DP | Nufarm Whyte | 500 g/l | SL | 01394 |

Uses

Annual dicotyledons in **barley, oats, undersown cereals, wheat**. Black bindweed in **barley, oats, undersown cereals, wheat**. Perennial dicotyledons in **barley, oats, undersown cereals, wheat**. Redshank in **barley, oats, undersown cereals, wheat**.

Efficacy
- Best results achieved by application to young seedling weeds in good growing conditions in a strongly competing crop
- Effectiveness may be reduced by rain within 6 h of spraying, by very cold weather or by drought conditions

Crop safety/Restrictions
- Spray winter crops in spring from fully tillered stage to before first node detectable (GS 29-30), spring crops from 1-leaf unfolded to before first node detectable (GS 11-30)
- Spray crops undersown with grass as soon as possible after grass begins to tiller
- Do not spray crops undersown with legumes, lucerne, peas or beans
- Do not roll or harrow within 7 d before or after treatment
- Avoid spray drifting onto nearby susceptible crops

Special precautions/Environmental safety
- Harmful in contact with skin and if swallowed

Personal protective equipment/Label precautions
- A, C, P
- M03, R03a, R03c, U05a, U08, U20b, C03, E01, E07, E13c, E26, E30a, E31b, E34

Withholding period
- Keep livestock out of treated areas until foliage of any poisonous weeds such as ragwort has died and become unpalatable

Approval
- Accepted by BLRA for use on malting barley

Maximum Residue Level (mg residue/kg food)
- tea, hops 0.1; all other products (except cereals and animal products) 0.05

---

202   dichlorprop + ferrous sulphate + MCPA

A herbicide/fertilizer combination for moss and weed control in turf

Products

SHL Granular Feed, Weed and Mosskiller	Sinclair	0.4:10.9:0.3% w/w	GR	09925

Uses

Annual dicotyledons in **managed amenity turf**. Buttercups in **managed amenity turf**. Mosses in **managed amenity turf**. Perennial dicotyledons in **managed amenity turf**.

Efficacy
- Apply from late spring to early autumn when grass in active growth and soil moist
- A repeat treatment may be needed after 4-6 wk to control perennial weeds or if moss regrows

Crop safety/Restrictions
- Do not treat newly sown grass or freshly laid turf for the first year
- Do not apply during drought or freezing conditions or when rain imminent
- Avoid walking on treated areas until it has rained or turf has been watered
- Do not mow for 3-4 d before or after application
- Do not use first 4 mowings after treatment for composting

Special precautions/Environmental safety
- Harmful to fish or other aquatic life. Do not contaminate surface waters or ditches with chemical or used container

Personal protective equipment/Label precautions
- U09a, U20a, E13c, E30a

Maximum Residue Level (mg residue/kg food)
- see dichlorprop entry

---

203   dichlorprop + MCPA

A translocated herbicide for use in cereals and turf

Products

1	Redipon Extra	United Phosphorus	350:150 g/l	SL	07518
2	SHL Granular Feed and Weed	Sinclair	0.4:0.3% w/w	GR	09926
3	SHL Turf Feed and Weed	Sinclair	0.474:0.366 g/l	DP	04437

Uses

Annual dicotyledons in **barley, oats, wheat** [1]. Annual dicotyledons in **managed amenity turf** [2]. Annual dicotyledons in **amenity grass** [3]. Black bindweed in **barley, oats, wheat** [1]. Buttercups in **managed amenity turf** [2]. Buttercups in **amenity grass** [3]. Clovers in **amenity grass** [3]. Daisies in **amenity grass** [3]. Dandelions in **amenity grass** [3]. Hemp-nettle in **barley, oats, wheat** [1]. Perennial dicotyledons in **barley, oats, wheat** [1]. Perennial dicotyledons in **managed amenity turf** [2]. Perennial dicotyledons in **amenity grass** [3]. Redshank in **barley, oats, wheat** [1].

Efficacy
- Best results achieved by application to young seedling weeds in active growth
- Do not spray during cold weather, if rain or frost expected, if crop wet or in drought
- Use on turf from Apr-Sep. Avoid close mowing for 3-4 d before or after treatment [3]

Crop safety/Restrictions
- Apply to winter cereals in spring from fully tillered, leaf-sheath erect stage but before first node detectable (GS 31), to spring barley when crop has 5 leaves unfolded but before first node detectable (GS 15-31)
- Do not use on undersown crops
- Do not spray crops suffering from herbicide damage or stress
- Do not roll or harrow for 7 d before or after spraying
- Avoid drift of spray onto nearby susceptible crops
- Do not use on turf in year of sowing [3]

Special precautions/Environmental safety
- Harmful if swallowed and in contact with skin

Personal protective equipment/Label precautions
- A, C [1]
- M03, R03a, R03c [1, 3]; U19 [3]; U05a, U08, U20b [1, 3]; U09a, U20a [2]; C03, E01, E15, E31b, E34 [1, 3]; E07 [1, 3] (2 wk); E30a [1-3]; E26 [1]; E13c [2]

Withholding period
- Keep livestock out of treated areas until foliage of any poisonous weeds such as ragwort has died and become unpalatable

Latest application/Harvest interval
- Before first node detectable (GS 31)

Approval
- Accepted by BLRA for use on malting barley

Maximum Residue Level (mg residue/kg food)
- see dichlorprop entry

---

204   dichlorprop-P + MCPA + mecoprop-P

A translocated herbicide mixture for winter and spring cereals

Products

Hymec Triple	Agrichem	310:160:130 g/l	SL	09949

Uses

Annual dicotyledons in *durum wheat, spring barley, spring wheat, winter barley, winter wheat*. Chickweed in *durum wheat, spring barley, spring wheat, winter barley, winter wheat*. Cleavers in *durum wheat, spring barley, spring wheat, winter barley, winter wheat*. Field pansy in *durum wheat, spring barley, spring wheat, winter barley, winter wheat*. Mayweeds in *durum wheat, spring barley, spring wheat, winter barley, winter wheat*. Poppies in *durum wheat, spring barley, spring wheat, winter barley, winter wheat*.

Efficacy
- Best results obtained if application is made while majority of weeds are at seedling stage but not if temperatures are too low
- Optimum results achieved by spraying when temperature is above 10°C. If temperatures are lower delay spraying until growth becomes more active

Crop safety/Restrictions
- Maximum number of treatments 1 per crop
- Do not spray in windy conditions where spray drift may cause damage to neighbouring crops, especially sugar beet, oilseed rape, peas, turnips and most horticultural crops including lettuce and tomatoes under glass

Special precautions/Environmental safety
- Harmful if swallowed
- Risk of serious damage to eyes
- Harmful to fish or other aquatic life. Do not contaminate surface waters or ditches with chemical or used container

Personal protective equipment/Label precautions
- A, C, H, M
- M03, R03c, R04d, U05a, U08, U20b, C03, E01, E13c, E26, E30a, E31a, E34

Latest application/Harvest interval
- Before second node detectable (GS 32) for all crops

Approval
- Accepted by BLRA for use on malting barley

---

205   dichlorvos

A contact and fumigant organophosphorus insecticide

Products

1	Luxan Dichlorvos 600	Luxan	600 g/l	EC	08297
2	Luxan Dichlorvos Aerosol 15	Luxan	120 g/l	AE	08298
3	Nuvan 500 EC	Novartis A H	500 g/l	EC	08590

Uses

Beetles in **poultry houses** [3]. Flies in **poultry houses** [3]. Mosquitoes in **poultry houses** [3]. Non-indigenous leaf miners in **protected brassica seedlings** (off-label), **protected herbs** (off-label) [3]. Poultry ectoparasites in **poultry houses** [3]. Western flower thrips in **bulb onions** (off-label), **garlic** (off-label), **protected ornamentals** (off-label), **protected vegetables** (off-label), **salad onions** (off-label), **shallots** (off-label), **spinach** (off-label) [1]. Western flower thrips in **protected cucumbers** [1, 2]. Western flower thrips in **protected brassica seedlings** (off-label), **protected herbs** (off-label) [3].

Efficacy
- For use in poultry houses apply as surface spray, cold fog or mist using suitable applicator. Direct away from poultry. See label for details [3]
- Product mainly active in vapour phase and is inactive on insect eggs. Best results obtained from programme of treatments on dry crops when temperatures 15-25°C [1, 2]

Crop safety/Restrictions
- Do not use on chrysanthemums in flower or on roses. Phytotoxicity may occur on cucumbers and other cucurbits
- Do not allow temperature to drop after treatment as condensation may lead to phytotoxicity. Inspect for phytotoxicity before retreatment [1, 2]
- Treatment of cucumbers at flowering may cause flower or fruit abortion [1, 2]
- Do not apply directly to growing cucumbers; apply between rows [1, 2]
- Do not apply other pesticides with product or one day before or after treatment, especially sulphur based formulations [1, 2]

Special precautions/Environmental safety
- Dichlorvos is subject to the Poisons Rules 1982 and the Poisons Act 1972. See notes in Section 1
- This product contains an anticholinesterase organophosphorus compound. Do not use if under medical advice not to work with such compounds
- To be used only by operators trained in the use of the chemical and familiar with the precautionary measures to be observed
- Toxic in contact with skin or if swallowed
- Toxic (very toxic [1, 2]) by inhalation [3]
- May cause sensitization by skin contact [1, 2]
- Flammable [1, 3]
- Corrosive [1, 2]
- Keep in original container, tightly closed, in a safe place, under lock and key
- Keep unprotected persons out of treated areas for at least 12 h
- Harmful to game, wild birds and animals
- Extremely dangerous to bees. Do not apply to crops in flower or to those in which bees are actively foraging. Do not apply when flowering weeds are present
- Extremely dangerous to fish or other aquatic life. Do not contaminate surface waters or ditches with chemical or used container
- Do not enter treated areas within 12 h and thoroughly ventilate before doing so [1, 2]

Personal protective equipment/Label precautions
- D, J, K, M [1, 2]; A, H [1-3]; C [1, 3]
- M01, M04 [1-3]; R01b, R04e, R05 [1, 2]; R02a, R02c [1-3]; R07d [1, 3]; R02b [3]; U10, U11 [1, 2]; U02, U04a, U05a, U08, U13, U19, U20a, C03 [1-3]; C02 [3]; E01, E10b, E12b, E13a, E30b, E34 [1-3]; E02 [1-3] (12 h); E31b [1, 3]

Latest application/Harvest interval
- HI Brussels sprouts, cabbage 14 wk; cauliflower 9 wk; broccoli, calabrese 7 wk; kohlrabi 1 wk; cucurbits 48 h; other edible crops 24 h

Approval
- Product approved for application as a spray, a low volume mist or as a thermal fog (see label for detailed instructions and protective clothing requirements) [1]
- Off-label Approval unstipulated for high volume application to protected non-edible ornamentals and a range of protected vegetables (see approval notice) for control of western flower thrips (OLA 0625/99)[1]; unlimited for use as a fog in protected non-edible ornamentals and a range of protected vegetables (see approval notice) for control of western flower thrips and non-indigenous leaf miners (OLA 0626/99)[1]

Maximum Residue Level (mg residue/kg food)
- lettuce 1; carrots, horseradish, parsnips, parsley root, salsify, swedes, turnips, garlic, onions, shallots, tomatoes, peppers, aubergines, cucumbers, gherkins, courgettes, cauliflowers, Brussels sprouts, head cabbages, beans (with pods), peas (with pods), celery, leeks, rhubarb, cultivated mushrooms, potatoes 0.5; citrus fruits, pome fruits, apricots, peaches, nectarines, plums, grapes, cane fruits, bilberries, cranberries, currants, gooseberries, bananas 0.1; meat, eggs 0.05; milk 0.02

---

206   diclofop-methyl

A translocated phenoxypropionic herbicide available only in mixtures

---

207   diclofop-methyl + fenoxaprop-P-ethyl

A foliar acting herbicide mixture for grass control in wheat and barley

Products

1 Aventis Tigress Ultra	Aventis	250:20 g/l	EW	09725
2 Corniche	Aventis	250:20 g/l	EW	08947
3 Tigress Ultra	Aventis	250:20 g/l	EW	08946

Uses

Blackgrass in **spring barley, winter barley, winter wheat**. Ryegrass in **spring barley, winter barley, winter wheat**. Wild oats in **spring barley, winter barley, winter wheat**.

Efficacy
- Apply from emergence of crop up to before second node detectable (GS 32)
- Wild oats and blackgrass controlled from 2 fully expanded leaves to end of tillering but before 1st node detectable
- Ryegrass from seed controlled from 2 fully expanded leaves up to 3 tillers
- Spray is rainfast from 1 h after spraying. Do not apply to wet or icy foliage
- Can be sprayed in frosty weather provided crop hardened off
- Any conditions resulting in moisture stress may reduce effectiveness especially with later applications to spring barley
- Performance not affected by soils with high OM content or high Kd factor

Crop safety/Restrictions
- Maximum number of treatments 1 per crop
- Do not apply to undersown crops or those to be undersown
- Broadcast crops should be sprayed post-emergence after root system well established
- Do not roll or harrow within 1 wk
- Do not spray crops under stress from drought, waterlogging etc

---

- Do not spray immediately before or after a sudden drop in temperature, a period of warm days/cold nights or when extremely low temperatures forecast
- Applications may be followed by transient leaf discoloration

Special precautions/Environmental safety
- Irritating to eyes
- Dangerous to fish or other aquatic life. Do not contaminate surface waters or ditches with chemical or used containers
- Interval of 7 d must elapse before or after application of any other product
- Must not be mixed with weedkillers containing hormones or bifenox

Personal protective equipment/Label precautions
- A, C, H
- R04a, U05a, U20b, C03; C02 (6 wk); E01, E13b, E26, E30a, E31b; E07 (7 d)

Withholding period
- Keep livestock out of treated areas for at least 7 d

Latest application/Harvest interval
- Before flag leaf sheath erect (GS 41).
- HI 6 wk

Approval
- Accepted by BLRA for use on malting barley

---

208   dicloran

A protectant nitroaniline fungicide used as a glasshouse fumigant

Products

Fumite Dicloran Smoke	Hortichem	40% w/w	FU	09291

Uses

Botrytis in **protected lettuce**. Rhizoctonia in **protected lettuce**.

Efficacy
- Treat at first sign of disease and at 14 d intervals if necessary
- For best results fumigate in late afternoon or evening when temperature is at least 16°C. Keep house closed overnight or for at least 4 h. Do not fumigate in bright sunshine, windy conditions or when temperature too high
- Water plants, paths and straw used as mulches several hours before fumigating. Ensure that foliage is dry before treatment

Crop safety/Restrictions
- Do not fumigate young seedlings or plants being hardened off
- Treat only crops showing strong growth. Do not treat crops which are dry at the roots

Special precautions/Environmental safety
- Irritating to eyes and respiratory system
- Ventilate glasshouse thoroughly before re-entering
- Harmful to fish or other aquatic life. Do not contaminate surface waters or ditches with chemical or used container

Personal protective equipment/Label precautions
- R04a, R04c, U05a, U19, U20a, C02, C03, E13c, E30a, E32a

Latest application/Harvest interval
- HI lettuce 14 d

---

209   dicofol

A non-systemic organochlorine acaricide for horticultural use

Products

Kelthane	Landseer	180 g/l	EC	09260

Uses

Red spider mites in **apples, hops, strawberries**. Tarsonemid mites in **strawberries**.

Efficacy
- Spray apples when winter eggs hatched and summer eggs laid and repeat 3 wk later if necessary. Spray other crops at any time before stated harvest interval
- Strawberries may be sprayed after picking where necessary
- Apply fumigant treatment at first sign of infestation and repeat every 14 d
- Do not fumigate in bright sunshine, when foliage wet or roots dry
- Resistance has developed in some areas, in which case sprays are unlikely to give satisfactory control, especially after second or subsequent application

Crop safety/Restrictions
- Maximum number of treatments 2 per yr for apples and hops, 2 per crop for strawberries, protected cucumbers and tomatoes
- Do not spray apples within 4 wk of petal fall
- Do not spray cucumber or tomato seedlings before mid-May or in bright sunshine

Special precautions/Environmental safety
- Harmful in contact with skin and if swallowed
- Irritating to eyes and skin
- May cause sensitization by skin contact
- Flammable

Personal protective equipment/Label precautions
- A, C, H
- M03, R03a, R03c, R04a, R04b, R04e, R07d, U02, U05a, U09a, U19, U20b, C03, E01, E15, E30a, E31a, E34

Latest application/Harvest interval
- HI strawberries 7 d; apples, hops 28 d

Approval
- Revocation of all approvals for products containing dicofol was published in May 2000 with a two-year use-up period
- Accepted by BLRA for use on hops

Maximum Residue Level (mg residue/kg food)
- hops 50; apricots, peaches, nectarines, plums, currants, garlic, cultivated mushrooms 5; citrus fruit, strawberries, bananas 2; pome fruit, grapes 1; tomatoes, peppers, cucurbits, legume vegetables 0.5; cotton seed 0.1; tree nuts, oilseeds (except cotton seed), eggs 0.05; cane fruit, bilberries, cranberries, gooseberries, wild berries, miscellaneous fruit (except bananas), root and tuber vegetables, bulb vegetables, aubergines, sweet corn, brassica vegetables, leaf vegetables and fresh herbs, peas (without pods), stem vegetables, fungi, pulses, potatoes, cereals 0.02

---

210 dieldrin

A persistent organochlorine insecticide, all approvals for which were revoked in 1989. Dieldrin is an environmental hazard and is banned under the EC Directive 70/117/EEC

---

211 difenacoum

An anticoagulant coumarin rodenticide

See also calciferol + difenacoum

Products

1	Deosan Rataway	Deosan	0.005% w/w	RB	08803
2	Deosan Rataway Bait Bags	Deosan	0.005% w/w	RB	08805
3	Killgerm Rat Rods	Killgerm	0.005% w/w	RB	05154
4	Killgerm Wax Bait	Killgerm	0.005% w/w	RB	04096
5	Neosorexa	Sorex	0.005% w/w	RB	07757
6	Neosorexa Ratpacks	Sorex	0.005% w/w	RB	04653
7	Ratak	Scotts	0.005% w/w	RB	06832
8	Ratak Wax Blocks	Scotts	0.005% w/w	RB	06829
9	Sorexa Gel	Sorex	0.005% w/w	PC	08315

Uses

Mice in **farm buildings, farmyards** [1-8]. Mice in **farm buildings/yards** [9]. Rats in **farm buildings, farmyards** [1-8].

Efficacy

- Difenacoum is a chronic poison and rodents need to feed several times before accumulating a lethal dose. Effective against rodents resistant to other commonly used anticoagulants
- Ready-to-use in baiting programme for indoor and outdoor use
- Lay small baits about 1 m apart throughout infested areas for mice, larger baits for rats near holes and along runs; place blocks 5-10 m apart depending on severity of infestation
- Cover baits by placing in bait boxes, drain pipes or under boards
- Inspect bait sites frequently and top up as long as there is evidence of feeding

Special precautions/Environmental safety

- Only for use by farmers, horticulturists and other professional users
- Cover bait to prevent access by children, animals or birds

Personal protective equipment/Label precautions

- A [9]
- M05 [7, 8]; U13 [1-9]; U20b [3-5, 7-9]; U20a [1, 2, 6]; E32a [1-4, 6-9]; E30a [1-6, 9]; E30b [7, 8]; V01a, V03a, V04a [1-9]; V02 [1-6, 9]

212   difenoconazole

A diphenyl-ether triazole protectant and curative fungicide

Products

1	Landgold Difenoconazole	Landgold	250 g/l	EC	09964
2	Plover	Novartis	250 g/l	EC	08429

Uses

Alternaria in **brussels sprouts, cabbages, calabrese, cauliflowers, spring oilseed rape, winter oilseed rape**. Brown rust in **winter wheat**. Disease control/foliar feed in **flowers** *(off-label)*, **protected flowers** *(off-label)*. Light leaf spot in **spring oilseed rape, winter oilseed rape**. Ring spot in **brussels sprouts, cabbages, calabrese, cauliflowers**. Septoria in **winter wheat**. Stem canker in **spring oilseed rape, winter oilseed rape**. Yellow rust in **winter wheat**.

Efficacy

- For most effective control of Septoria, apply as part of a programme of sprays which includes a suitable flag leaf treatment
- Adequate control of yellow rust may require an earlier appropriate treatment
- Treat oilseed rape in autumn from 4 expanded true leaf stage (GS 1,4). A repeat spray may be made in spring at the beginning of stem extension (GS 2,0) if visible symptoms develop
- Improved control of established infections on oilseed rape achieved by mixture with carbendazim. See label
- In cabbages a 3-spray programme should be used starting at the first sign of disease and repeated at 14-21 d intervals
- Product is fully rainfast 2 h after application

Crop safety/Restrictions

- Maximum number of treatments 3 per crop for cabbages; 2 per crop for oilseed rape; 1 per crop for wheat
- Apply to wheat any time from ear fully emerged stage but before early milk-ripe stage (GS 59-73)
- Recommended for use on all varieties of winter and spring sown oilseed rape and cabbage

Special precautions/Environmental safety

- Irritating to eyes
- May cause sensitization by skin contact
- Harmful to fish or other aquatic life. Do not contaminate surface waters or ditches with chemical or used container

Personal protective equipment/Label precautions
- A, C, H [1, 2]
- R04a, R04e [1, 2]; U07 [2]; U05a, U09a, U20b, C03 [1, 2]; E31a, E33, E34, E36 [2]; E01, E13c, E26, E29 [1, 2]; E30a, E31b [1]

Latest application/Harvest interval
- Before grain early milk-ripe stage (GS 73)

Approval
- Off-label approval unstipulated for use on outdoor asparagus (OLA 1539/98)[2]

---

213   difenzoquat

A post-emergence quaternary ammonium herbicide for wild oat control

Products

1 Avenge 2	BASF	150 g/l	SL	10210
2 Avenge 2	Cyanamid	150 g/l	SL	03241

Uses

Wild oats in **barley, durum wheat, maize, rye, ryegrass seed crops, triticale, wheat**.

Efficacy
- Autumn and winter spraying controls wild oats from 2-leaf stage to mid-tillering, spring treatment to end of tillering
- Dose rate varies with season and growth-stage of weed. See label for details
- Spring treatment on barley provides control of powdery mildew as well as wild oats
- Do not apply if rain expected within 6 h
- Recommended for tank mixture with a range of herbicides, growth regulators and other pesticides. See label for details. Other products can be applied after 24 h

Crop safety/Restrictions
- Maximum number of treatments 2 per crop for winter cereals, durum wheat and ryegrass seed crops; 1 per crop for spring cereals and maize
- Apply to named varieties of winter and spring wheat, durum and triticale (see label) and all varieties of other recommended crops
- Apply to crops from 2-leaf stage to 50% of plants with 3 nodes detectable (GS 12-33)
- Apply to crops undersown with ryegrass and clover either before emergence or after grass has reached 3-leaf stage
- Spray ryegrass seed crops after 3-leaf stage, maize as soon as weeds at susceptible stage
- Do not spray crops suffering stress from waterlogging, drought or other factors
- Temporary yellowing may follow application, especially under extremes of temperature, but is normally outgrown rapidly

Special precautions/Environmental safety
- Harmful if swallowed and in contact with skin
- Irritating to skin
- Risk of serious damage to eyes
- Harmful to fish or other aquatic life. Do not contaminate surface waters or ditches with chemical or used container
- Harmful to livestock

Personal protective equipment/Label precautions
- A, C
- M03, R03a, R03c, R04b, R04d, U05a, U08, U14, U15, U19, U20a, C03, E01, E06b, E13c, E30a, E31a, E34

Withholding period
- Keep all livestock out of treated areas for at least 6 wk

Latest application/Harvest interval
- Before flag leaf just visible (GS 39) for cereals and maize; 6 wk before grazing for ryegrass seed crops

Approval
- Accepted by BLRA for use on malting barley

---

214   diflubenzuron

A selective, persistent, contact and stomach acting insecticide

Products

1 Dimilin 25-WP	Hortichem	25% w/w	WP	08902
2 Dimilin Flo	Uniroyal	480 g/l	SC	08769
3 Dimilin Flo	Zeneca	480 g/l	SC	08985

Uses

Browntail moth in *amenity trees and shrubs, hedges, nursery stock, ornamentals* [1-3]. Bud moth in *apples, pears* [1-3]. Carnation tortrix moth in *amenity trees and shrubs, hedges, nursery stock, ornamentals* [2, 3]. Caterpillars in *broccoli, brussels sprouts, cabbages, calabrese, cauliflowers* [1-3]. Clouded drab moth in *apples, pears* [1-3]. Codling moth in *apples, pears* [1-3]. Fruit tree tortrix moth in *apples, pears* [1-3]. Houseflies in *livestock houses, manure heaps, refuse tips* [2]. Lackey moth in *amenity trees and shrubs, hedges, nursery stock, ornamentals* [1-3]. Oak leaf roller moth in *forestry* [1-3]. Pear sucker in *pears* [1-3]. Pine beauty moth in *forestry* [1-3]. Pine looper in *forestry* [1-3]. Plum fruit moth in *plums* [1-3]. Rust mite in *apples, pears, plums* [1-3]. Sciarid flies in *mushrooms* [2]. Small ermine moth in *amenity trees and shrubs, hedges, nursery stock, ornamentals* [1-3]. Tomato moth in *protected peppers* *(off-label)* [2]. Tortrix moths in *plums* [1-3]. Winter moth in *amenity trees and shrubs, apples, blackcurrants, forestry, hedges, nursery stock, ornamentals, pears, plums* [1-3].

Efficacy

- Most active on young caterpillars and most effective control achieved by spraying as eggs start to hatch
- Dose and timing of spray treatments vary with pest and crop. See label for details
- Addition of wetter recommended for use on brassicas and for pear sucker control in pears
- Apply as casing mixing treatment on mushrooms or as post-casing drench

Crop safety/Restrictions

- Maximum number of treatments 3 per yr for apples, pears; 2 per yr for plums, blackcurrants; 2 per crop for brassicas; 1 per spawning for mushrooms
- Before treating ornamentals check varietal tolerance on a small sample
- Do not use as a compost drench or incorporated treatment on ornamental crops
- Do not spray protected plants in flower or with flower buds showing colour
- For use only on the food crops specified on the label

Special precautions/Environmental safety

- Dangerous to fish or other aquatic life. Do not contaminate surface waters or ditches with chemical or used container [1]
- Do not apply directly to livestock
- Negligible effect on many beneficial insects (see label) and may therefore be used with biological control agents and in integrated control programmes
- LERAP Category B

Personal protective equipment/Label precautions

- A, C, H, M [1]
- U20c [2, 3]; U05a, U09a, U19, U20a, C03 [1]; E15, E18 [2]; E17 [2] (plums 20 m; blackcurrants, forestry, ornamentals over 50 cm 10 m); E16a, E16b, E30a [1-3]; E05, E32a [2, 3]; E01, E13b, E31a [1]

Withholding period

- Do not apply directly to livestock/poultry

Latest application/Harvest interval

- HI apples, pears, plums, blackcurrants, brassicas 14 d [1, 3]

Approval

- Approved for aerial application in forestry when average wind velocity does not exceed 18 knots and gusts do not exceed 20 knots [3]. See notes in Section 1
- Off-label approval unstipulated for use on protected sweet peppers (OLA 2433/99)[2]

Maximum Residue Level (mg residue/kg food)

- citrus fruits, pome fruits, plums, tomatoes, peppers, aubergines, Brussels sprouts, head cabbages 1; cultivated mushrooms 0.1; meat, milk, eggs 0.05

215   diflufenican

A shoot absorbed anilide herbicide available only in mixtures

See also bromoxynil + diflufenican + ioxynil
carbetamide + diflufenican + oxadiazon
clodinafop-propargyl + diflufenican
clopyralid + diflufenican + MCPA

216   diflufenican + flurtamone

A contact and residual herbicide mixture for cereals

Products

1 Bacara	Aventis	100:250 g/l	SC	09976
2 Bacara	RP Agric.	100:250 g/l	SC	08323

Uses

Annual dicotyledons in **winter barley, winter wheat**. Annual meadow grass in **winter barley, winter wheat**. Blackgrass in **winter barley, winter wheat**. Loose silky bent in **winter barley, winter wheat**. Volunteer oilseed rape in **winter barley, winter wheat**.

Efficacy

- Apply pre-emergence or from when crop has first leaf unfolded before susceptible weeds pass recommended size
- Best results obtained on firm, fine seedbeds with adequate soil moisture present at and after application
- Good weed control requires ash, trash and burnt straw to be buried during seed bed preparation
- Loose fluffy seedbeds should be rolled before application and the final seed bed should be fine and firm without large clods
- Do not use on soils with more than 10% organic matter
- Speed of control depends on weather conditions and activity can be slow under cool conditions

Crop safety/Restrictions

- Maximum number of treatments 1 per crop
- Crops should be drilled to a normal depth of 25 mm and the seed well covered. Do not treat broadcast crops
- Do not use on crops being grown for seed or crops undersown or to be undersown
- Do not treat frosted crops or when frost is imminent. Severe frost after application, or any other stress, may lead to transient discoloration or scorch
- Do not use on Sands or Very Light soils or those that are very stony or gravelly. Do not use on any soil that is waterlogged or likely to be waterlogged by heavy rain
- Do not harrow at any time after application and do not roll autumn treated crops until spring
- Take particular care to match spray swaths otherwise crop discoloration and biomass reduction may occur which may lead to yield reduction
- In the event of crop failure winter wheat mat be redrilled immediately after normal cutlivation, and winter barley may be sown after ploughing. Fields must be ploughed and 20 wk must elapse before sowing spring crops of wheat, barley, oilseed rape, peas, field beans or potatoes
- Do not broadcast or direct drill oilseed rape or other brassica crops as a following crop on treated land. See label for detailed advice on preparing land for subsequent autumn cropping in the normal rotation
- Successive treatments of any products containing diflufenican can lead to soil build-up and inversion ploughing must precede sowing any following non-cereal crop. Even where ploughing occurs some crops may be damaged

Special precautions/Environmental safety
- Dangerous to fish or other aquatic life. Do not contaminate surface waters or ditches with chemical or used container
- LERAP Category B

Personal protective equipment/Label precautions
- A, H
- U20c, E13b, E16a, E16b, E26, E30a, E31b

Latest application/Harvest interval
- Before 2nd node detectable (GS32)

217   diflufenican + flurtamone + isoproturon

A contact and residual herbicide for winter cereals

Products

1 Ingot	Aventis	27:67:400 g/l	SC	09997
2 Ingot	RP Agric.	27:67:400 g/l	SC	08610

Uses

Annual dicotyledons in *winter barley, winter wheat*. Annual meadow grass in *winter barley, winter wheat*. Blackgrass in *winter barley, winter wheat*. Loose silky bent in *winter barley, winter wheat*.

Efficacy
- Apply from when crop has first leaf unfolded before susceptible weeds pass recommended size
- Best results obtained on firm, fine seedbeds with adequate soil moisture
- Speed of control depends on weather conditions and activity will be reduced during prolonged dry periods especially in spring. Weed control, especially of grasses, may also be reduced in wet seasons or where heavy rain falls shortly after treatment
- Do not treat on soils with more than 10% organic matter

Crop safety/Restrictions
- Maximum number of treatments 1 per crop. Maximum total dose of isoproturon 2.5 kg a.i./ha per crop
- Crops should be drilled to a normal depth of 25 mm and the seed well covered. Do not treat broadcast crops
- Do not apply pre-emergence to wheat or barley
- Do not use on crops being grown for seed or crops undersown or to be undersown
- Do not treat frosted crops or when frost is imminent. Severe frost after application may cause transient discoloration or scorch
- Do not use on Sands or Very Light soils or those that are very stony or gravelly
- Do not harrow at any time after application
- Take particular care to match spray swaths otherwise crop discoloration and biomass reduction may occur which may lead to yield reduction
- Do not broadcast oilseed rape or other brassica crops as a following crop on treated land. See label for details of time and land preparation before sowing other crops in the event of failure of a treated crop
- Successive treatments of any products containing diflufenican can lead to soil build-up and inversion ploughing must precede sowing any following non-cereal crop. Even where ploughing occurs some crops may be damaged

Special precautions/Environmental safety
- Dangerous to fish or other aquatic life. Do not contaminate surface waters or ditches with chemical or used container
- LERAP Category B
- Do not spray where soils are cracked, to avoid run-off through drains

Personal protective equipment/Label precautions
- A, C, H
- U20c, E13b, E16a, E16b, E26, E30a, E31b

Latest application/Harvest interval
- Before 2nd node detectable (GS32)

Approval
- Accepted by BLRA for use on malting barley

218 diflufenican + isoproturon

A contact and residual herbicide for use in winter cereals

Products

1	DFF + IPU WDG	Aventis	9.74:78.5% w/w	WG	10204
2	Grenadier	RP Agric.	41.7:500 g/l	SC	08136
3	Javelin	Aventis	62.5:500 g/l	SC	09998
4	Javelin	RP Agric.	62.5:500 g/l	SC	06192
5	Javelin Gold	RP Agric.	20:500 g/l	SC	06200
6	Javelin Gold	Aventis	20:500 g/l	SC	09999
7	Landgold DFF 625	Landgold	62.5:500 g/l	SC	06274
8	Panther	Aventis	50:500 g/l	SC	10008
9	Panther	RP Agric.	50:500 g/l	SC	06491
10	Standon Diflufenican-IPU	Standon	50:500 g/l	SC	09175
11	Tolkan Turbo	RP Agric.	20:500 g/l	SC	06795

Uses

Annual dicotyledons in *winter barley, winter wheat* [1-11]. Annual dicotyledons in *triticale, winter rye* [2-6, 8-11]. Annual grasses in *winter barley, winter wheat* [1-11]. Annual grasses in *triticale, winter rye* [2-6, 8-11]. Blackgrass in *winter barley, winter wheat* [1-11]. Blackgrass in *triticale, winter rye* [2-6, 8-11]. Wild oats in *winter barley, winter wheat* [1-11]. Wild oats in *triticale, winter rye* [2-6, 8-11].

Efficacy
- May be applied in autumn or spring (but only post-emergence on wheat or barley). Best control normally achieved by early post-emergence treatment
- Best results by application to fine, firm seedbed moist at or after application
- Weeds controlled from before emergence to 6 true leaf stage
- Apply to moist, but not waterlogged, soils
- Any trash or ash should be buried during seedbed preparation

Crop safety/Restrictions
- Maximum number of treatments 1 per crop. Maximum total dose of isoproturon 2.5 kg a.i./ha per yr
- Spray winter wheat and barley post-emergence before end Feb [3-7], before crop reaches second node detectable stage (GS 32) [2, 8-11]
- On triticale and winter rye only treat named varieties and apply as pre-emergence spray
- Drill crop to normal depth (25 mm) and ensure seed well covered
- Do not use on other cereals, broadcast or undersown crops or crops to be undersown
- Do not use on Sands or Very Light soils, or those that are very stony or gravelly or on soils with more than 10% organic matter
- Do not spray when heavy rain is forecast or on crops suffering from stress, frost, deficiency, pest or disease attack
- Do not harrow after application nor roll autumn-treated crops until spring
- See label for details of time and land preparation needed before sowing other crops (12 wk for most crops). In the event of crop failure winter wheat can be redrilled without ploughing, other crops only after ploughing - see label for details
- Successive treatments of any products containing diflufenican can lead to soil build up and inversion ploughing must precede sowing any non-cereal crop. Even where ploughing occurs some crops may be damaged

Special precautions/Environmental safety
- Irritant. May cause sensitization by skin contact [2]
- Irritating to eyes [1]

- Harmful (dangerous [2, 4, 7, 10, 11]; extremely dangerous [1]) to fish or other aquatic life. Do not contaminate surface waters or ditches with chemical or used container [5, 9]
- Do not spray where soils are cracked, to avoid run-off through drains
- LERAP Category B

Personal protective equipment/Label precautions
- A, C, H [1-4, 7, 11]
- R04, R04e [2]; R04a [1]; U20c [1-4, 11]; U20b [5-10]; U08, U19 [10]; E26, E31b [2-11]; E16a, E16b, E30a [1-11]; E13b [2-4, 7, 10, 11]; E13c [5, 6, 8, 9]; E13a, E32a [1]

Latest application/Harvest interval
- Before second node detectable (GS 32) [2, 5, 9-11], before end Feb [1, 4, 7], for wheat and barley; pre-emergence of crop for triticale and rye

Approval
- Accepted by BLRA for use on malting barley

---

219   diflufenican + terbuthylazine

A foliar and residual acting herbicide

Products

Bolero	Novartis	200:400	SC	08392

Uses

Annual dicotyledons in **winter barley, winter wheat**. Annual meadow grass in **winter barley, winter wheat**.

Efficacy
- Apply to healthy crops from when crop has 1 leaf unfolded (GS 11) up to the end of tillering. Best results obtained from applications made early post-emergence of weeds
- Product is foliar and soil acting. Weeds germinating after treatment are controlled by root uptake
- Prolonged dry weather or heavy rain after treatment may reduce control
- Weed control may be reduced by presence of high levels of organic matter (over 10%) or surface trash

Crop safety/Restrictions
- Maximum number of treatments 1 per crop
- Ensure seed covered by 2.5 cm consolidated soil before spraying
- Do not use on crops under stress or suffering from water-logging, pest attack, disease or frost. Do not spray when frost imminent
- Do not apply to undersown crops or those to be undersown
- Sep drilled crops may be damaged if treatment precedes or coincides with a period of rapid growth
- Do not use on Sands or Very Light soils. On stony or gravelly soils crop damage may occur if heavy rain follows treatment
- After harvest treated soils should be inversion ploughed to 15 cm before drilling field beans, leaf brassicas, winter oilseed rape, winter onions or sugar beet seed crops. Autumn sown cereals can be drilled as normal
- Restrictions apply to crops that may be sown after failure of a treated crop. See label for details of crops, cultivations and intervals
- Successive treatments of any products containing diflufenican can lead to soil build up and inversion ploughing must precede sowing any non-cereal crop. Even where ploughing occurs some crops may be damaged - see label

Special precautions/Environmental safety
- Harmful in contact with skin and if swallowed
- Harmful to fish or other aquatic life. Do not contaminate surface waters or ditches with chemical or used container

Personal protective equipment/Label precautions
- A
- R03a, R03c, U08, U20a, E13c, E26, E30a, E31b, E34

Latest application/Harvest interval
- Before ear at 1 cm stage (GS 30)

Approval
- Accepted by BLRA for use on malting barley

220   diflufenican + trifluralin

A contact and residual herbicide for use in winter cereals

Products

1 Ardent	Aventis	40:400 g/l	SC	09968
2 Ardent	RP Agric.	40:400 g/l	SC	06203

Uses

Annual dicotyledons in *triticale, winter barley, winter rye, winter wheat*. Annual meadow grass in *winter barley, winter wheat*.

Efficacy
- Apply pre- or early post-emergence of crop up to maximum recommended weed size
- Weeds controlled pre-emergence up to 4 leaves (2 leaves for annual meadow grass)
- Best results achieved on firm, fine moist seedbeds
- Rolling after autumn treatment will reduce weed control. Roll in spring if necessary
- Speed of control depends on temperature and growing conditions; activity can be slow under cool conditions
- Do not use on soils with more than 10% organic matter; this may include newly ploughed grassland for a time

Crop safety/Restrictions
- Maximum number of treatments 1 per crop
- Ensure that crop is evenly drilled and seed well covered. Do not treat broadcast crops or those that are frosted or stressed
- Do not treat undersown crops or those to be undersown
- Do not harrow after application
- Do not use on oats or any spring sown cereals
- Do not use on Sands or Very Light soils, or those that are very stony or gravelly
- Crops occasionally show transient leaf discoloration after treatment. Symptoms are quickly outgrown and yield not affected
- See label for details of time and land preparation needed in the event of crop failure and for normal subsequent cropping
- Successive treatments of any products containing diflufenican can lead to soil build-up and inversion ploughing must precede sowing any following non-cereal crop. Even where ploughing occurs some crops may be damaged

Special precautions/Environmental safety
- Irritating to eyes
- Harmful to fish or other aquatic life. Do not contaminate surface waters or ditches with chemical or used container
- LERAP Category B

Personal protective equipment/Label precautions
- A, C
- R04a, U05a, U08, U13, U20b, C03, E01, E13c, E16a, E16b, E26, E30a, E31b

Latest application/Harvest interval
- Before 2nd tiller stage (GS 22), or end Nov, whichever is sooner

Approval
- Accepted by BLRA for use on malting barley

221   di-1-p-menthene

An antitranspirant and coating agent available only in mixtures

See also chlormequat with di-1-p-menthene

222 dimethoate

A contact and systemic organophosphorus insecticide and acaricide

Products

1	Barclay Dimethosect	Barclay	400 g/l	EC	08538
2	BASF Dimethoate 40	BASF	400 g/l	EC	00199
3	Greencrop Pelethon	Greencrop	400 g/l	EC	09477
4	P A Dimethoate 40	Portman	400 g/l	EC	01527
5	Rogor L40	Isagro	400 g/l	EC	07611

Uses

Aphids in *fodder beet (excluding myzus persicae)*, *mangels (excluding myzus persicae)*, *red beet (excluding myzus persicae)*, *sugar beet (excluding myzus persicae)* [1-3]. Aphids in *cereals* [1-4]. Aphids in *peas* [1-5]. Aphids in *potatoes (excluding myzus persicae)* [1, 2, 4, 5]. Aphids in *broad beans, carrots, french beans, runner beans* [1, 4]. Aphids in *grass seed crops, watercress (off-label)* [2]. Aphids in *sugar beet seed crops (excluding myzus persicae)* [2, 5]. Aphids in *ware potatoes (excluding myzus persicae)* [3]. Aphids in *apples, cane fruit, cherries, currants, flowers, gooseberries, ornamentals, pears, plums, protected carnations, roses, strawberries, tomatoes, woody ornamentals* [4]. Aphids in *beet crops (excluding myzus persicae)* [4, 5]. Aphids in *mangel seed crops (excluding myzus persicae)*, *spinach, wheat* [5]. Bryobia mites in *apples, pears* [4]. Cabbage aphid in *brassicas* [4]. Capsids in *apples, cane fruit, gooseberries, pears* [4]. Insect pests in *brassica seed beds (off-label)*, *bulb onions (off-label)*, *celeriac (off-label)*, *chinese cabbage (off-label)*, *garlic (off-label)*, *kale (off-label)*, *kohlrabi (off-label)*, *leeks (off-label)*, *salad onions (off-label)*, *shallots (off-label)* [2]. Leaf miners in *fodder beet, mangels, red beet, sugar beet* [1-3]. Leaf miners in *flowers, ornamentals, protected carnations, roses, woody ornamentals* [4]. Leaf miners in *beet crops, mangel seed crops, spinach, sugar beet seed crops* [5]. Mangold fly in *beet crops* [4]. Pea midge in *peas* [1-5]. Red spider mites in *apples, cane fruit, cherries, currants, gooseberries, pears, plums, strawberries* [4]. Sawflies in *apples, pears, plums* [4]. Suckers in *apples, pears* [4]. Thrips in *peas* [1-5]. Wheat bulb fly in *wheat* [1-5].

Efficacy

- Chemical has quick knock-down effect and systemic activity lasts for up to 14 d
- With some crops, products differ in range of pests listed as controlled. Uses section above provides summary. See labels for details
- For most pests apply when pest first seen and repeat 2-3 wk later or as necessary. Timing and number of sprays varies with crop and pest. See labels for details
- Best results achieved when crop growing vigorously. Systemic activity reduced when crops suffering from drought or other stress
- In hot weather apply in early morning or late evening
- Do not tank mix with alkaline materials. See label for recommended tank-mixes
- Where aphids or spider mites resistant to organophosphorus compounds occur control is unlikely to be satisfactory and repeat treatments may result in lower levels of control
- Consult processor before spraying crops grown for processing

Crop safety/Restrictions

- Maximum number of treatments 8 per crop for hops, tomatoes; 7 per crop for seed potatoes; 6 per crop for brassicas, peas, outdoor lettuce, strawberries, swedes, turnips; 4 per crop for cereals, grass seed crops, carrots, apples, pears, plums, cherries, gooseberries, cane fruit; 3 per crop for blackcurrants; 2 per crop for beans, ware potatoes; 1 per crop for beet crops (see also below), celery; 1 per crop for protected lettuce between Oct and Feb
- On beet crops only one treatment per crop may be made for control of leaf miners (max 84 g a.i./ha) and black bean aphid (max 420 g a.i./ha)
- On carrots the maximum total dose applied per crop must not exceed the equivalent of 3 (on mineral soils) or 4 (on organic soils) full dose applications
- In potatoes and beet crops resistant strains of peach-potato aphid (*Myzus persicae*) are common and dimethoate products must not be used to control this pest
- Test for varietal susceptibility on all unusual plants or new cultivars

Special precautions/Environmental safety
- This product contains an anticholinesterase organophosphorus compound. Do not use if under medical advice not to work with such compounds
- Harmful if swallowed [1, 3, 4] and in contact with skin [2, 5]
- Irritating to skin [1, 3] and eyes [4]
- May cause sensitization by skin contact
- Flammable
- Harmful to game, wild birds and animals. Bury all spillages
- Harmful to livestock
- Likely to cause adverse effects on beneficial arthropods
- Dangerous to bees. Do not apply to crops in flower, except as directed on peas, or to those in which bees are actively foraging. Do not apply when flowering weeds are present
- Surface residues may also cause bee mortality following spraying
- Dangerous to fish or other aquatic life. Do not contaminate surface waters or ditches with chemical or used container
- LERAP Category A
- Do not treat cereals after 1 Apr within 6 m of edge of crop
- Must not be applied to cereals if any product containing a pyrethroid insecticide or dimethoate has been sprayed after the start of ear emergence (GS 51)

Personal protective equipment/Label precautions
- J [1, 2, 4, 5]; A, C, H, M [1-5]
- M01 [1-5]; M03 [1, 3-5]; R03c, R04e, R07d [1-5]; R03a [2, 5]; R04a [1, 3, 4]; R04b [1, 3]; U10 [2]; U13 [2, 4, 5]; U02 [1, 2, 4, 5]; U05a, U19, U20a [1-5]; U04a [1-3, 5]; U08 [1, 3-5]; C03 [1-5]; C02 [2, 3] (7 d); E30b [2]; E01, E13b, E16c, E16d, E26, E31b, E34 [1-5]; E06b [1-5] (7 d); E10b, E12d [1-3]; E12c [4, 5]; E30a [1, 3-5]; S04b [4, 5]

Withholding period
- Keep all livestock out of treated areas for at least 7 d

Latest application/Harvest interval
- Before 30 Jun in yr of harvest for potatoes and beet crops; before 31 Mar in yr of harvest for aerial application to cereals
- HI apples, pears 35 d; protected lettuce, blackcurrants 28 d; field beans, plums, cherries, raspberries, gooseberries, strawberries 21 d; cereals, grass seed crops, carrots, peas, beans, hops 14 d; watercress 10 d; brassicas, outdoor lettuce, tomatoes, celery 7 d

Approval
- Approved for aerial application on cereals, peas, ware potatoes, sugar beet [1, 2, 5]. See notes in Section 1
- Off label approval unlimited for use on compost for propagation of protected and outdoor watercress (OLA 0394/94)[2]; unlimited for use on outdoor and protected vegetables (see approval notice for details) (OLA 0389/94)[2]; to Dec 2003 for use on watercress beds (OLA 0159/99)[2]; to July 2003 for use in outdoor and protected seedlings of salad onions, bulb onions, garlic, shallots, leeks, and mature outdoor crops of leeks, bulb onions, shallots, garlic (OLA 1511/00)[2]
- Accepted by BLRA for use pre-harvest on malting barley

Maximum Residue Level (mg residue/kg food)
- citrus fruits, apricots, peaches, nectarines, plums, bilberries, cranberries, currants, gooseberries, cucumbers, gherkins, courgettes, cauliflowers, Brussels sprouts, head cabbages, lettuce, beans (with pods) 2; pome fruits, grapes, cane fruits, bananas, carrots, horseradish, parsnips, parsley root, salsify, swedes, turnips, garlic, onions, shallots, tomatoes, peppers, aubergines, peas (with pods), celery, leeks, rhubarb, cultivated mushrooms 1; potatoes 0.05

---

223   dimethomorph

A cinnamic acid fungicide with translaminar activity available only in mixtures

---

224 dimethomorph + mancozeb

A systemic and protectant fungicide for potato blight control

Products

1 Invader	BASF	7.5:66.7% w/w	WG	BASF
2 Invader	Cyanamid	7.5:66.7% w/w	WG	06989

Uses

Blight in **potatoes**.

Efficacy

- Commence treatment as soon as there is a risk of blight infection
- In the absence of a warning treatment should start before the crop meets along the rows
- Repeat treatments every 10-14 d depending in the degree of infection risk
- Irrigated crops should be regarded as at high risk and treated every 10 d
- For best results good spray coverage of the foliage is essential

Crop safety/Restrictions

- Maximum number of treatments 8 per crop

Special precautions/Environmental safety

- Irritant. May cause sensitization by skin contact
- Oxidising agent. Contact with combustible material may cause fire
- Dangerous to fish or other aquatic life. Do not contaminate surface waters or ditches with chemical or used container
- LERAP Category B

Personal protective equipment/Label precautions

- A
- R04e, R08, U02, U04a, U05a, U08, U13, U19, U20a, C03, E01, E13b, E16a, E16b, E30a

Latest application/Harvest interval

- HI 7 d before harvest

Approval

- Product registration number not available at time of printing [1]

Maximum Residue Level (mg residue/kg food)

- see mancozeb entry

225 dinocap

A protectant dinitrophenyl fungicide for powdery mildew control

Products

Karathane Liquid	Landseer	350 g/l	EC	09262

Uses

Powdery mildew in **apples, chrysanthemums, grapevines** (off-label), **roses, strawberries**.

Efficacy

- Spray at 7-14 d intervals to maintain protective film
- Product must be applied to roses before disease becomes established
- Regular use on apples suppresses red spider mites and rust mites

Crop safety/Restrictions

- Maximum total dose on apples equivalent to 10 full dose treatments
- Maximum number of treatments on strawberries 5 per yr
- When applied to apples during blossom period may cause spotting but no adverse effect on pollination or fruit set. Do not apply to Golden Delicious during blossom period
- Do not apply when temperature is above 24°C. Do not apply with white oils
- Certain chrysanthemum cultivars may be susceptible
- May cause petal spotting on white roses

Special precautions/Environmental safety

- Harmful if swallowed
- Irritating to eyes and skin. May cause sensitization by skin contact

- Dangerous to fish or other aquatic life. Do not contaminate surface waters or ditches with chemical or used container
- LERAP Category B

Personal protective equipment/Label precautions
- A, C, H, J, M
- M03, R03c, R04a, R04b, R04e, U04a, U05a, U10, U11, U19, U20a, C03, E01, E13b, E16a, E16b, E26, E30a, E31a, E34

Latest application/Harvest interval
- HI apples 14 d, strawberries 7 d

Approval
- Off-label approval unstipulated for use on outdoor grapevines (HI 21 d) (OLA 1543/99)

226   dinoseb

A dinitrophenol herbicide and dessicant, all approvals for which were revoked in 1988 because of evidence of teratogenicity and the potential danger to operators. It is banned under the EC Directive 70/117/EEC

227   diphacinone

An anticoagulant rodenticide

Products

1 Tomcat Rat and Mouse Bait	Antec	0.005%w/w	RB	07171
2 Tomcat Rat and Mouse Blox	Antec	0.005% w/w	BB	07230

Uses

Mice in *agricultural premises*. Rats in *agricultural premises*.

Efficacy
- Products formulated using human food grade ingredients, flavour enhancers and paraffin
- Rodents must consume bait for 3-5 d to produce mortality after 6-15 d
- Maintain uninterrupted supply of fresh bait for 10-15 d or until signs of rodent activity cease
- Remove and replace stale, damp or mouldy bait
- Bait product [1] available loose or in sachets

Special precautions/Environmental safety
- Prevent access to the baits by children, birds and other animals, particularly dogs, cats and pigs
- Do not apply direct to livestock
- Remove exposed milk and collect eggs before application

Personal protective equipment/Label precautions
- M03, U13, U20a, E15, E30a, E32a, V01a, V03a, V04a

Approval
- Approval expiry 31 May 2001 [1]

228   diquat

A non-residual bipyridyl contact herbicide and crop desiccant

Products

1 Barclay Desiquat	Barclay	200 g/l	SL	09063
2 Greencrop Boomerang	Greencrop	200 g/l	SL	09563
3 Landgold Diquat	Landgold	200 g/l	SL	09020
4 Midstream	Scotts	100 g/l	PC	09267
5 Reglone	Zeneca	200 g/l	SL	09646
6 Standon Diquat	Standon	200 g/l	SL	05587

Uses

Annual dicotyledons in *flower bulbs, potatoes, row crops (between row treatment), sugar beet, vegetables* [2]. Annual dicotyledons in *row crops* [5]. Aquatic weeds in *areas of water* [4, 5]. Chemical stripping in *hops* [2, 5]. Pre-harvest desiccation in *laid barley and oats (stockfeed only), linseed, potatoes* [1-3, 5, 6]. Pre-harvest desiccation in *field beans (stock or pigeon feed only), oilseed rape* [1-3, 6]. Pre-harvest desiccation in *combining peas* [1, 3, 6]. Pre-harvest desiccation in *clover seed crops, peas* [2, 5]. Pre-harvest desiccation in *Echium plantaginium (off-label), spring field beans (stock or pigeon feed only), spring oilseed rape, winter field beans (stock or pigeon feed only), winter oilseed rape* [5].

Efficacy

- Acts rapidly on green parts of plants and rainfast in 15 min
- Best results for potato desiccation achieved by spraying in bright light and low humidity conditions [1, 3, 5, 6]
- Treatments for weed control normally require co-application with paraquat. Use of wetter for improved weed control essential for most uses except potato desiccation, treatment of hops and aquatic weed control (except *Lemna*)
- For pre-harvest desiccation, apply to potatoes when tubers the desired size and to other crops when mature or approaching maturity. See label for details of timing and of period to be left before harvesting potatoes and for timing of pea and bean desiccation sprays [1, 3, 5, 6]
- Spray linseed when seed matured evenly over whole field; direct combining can normally begin 10-20 d after spraying [1, 3, 5, 6]
- Apply as hop stripping treatment when shoots have reached top wire [5]
- Apply to floating and submerged aquatic weeds in still or slow moving water [5]
- Do not apply in muddy water or activity will be reduced [4, 5]
- Apply undiluted to submerged weeds in fast-flowing streams and rivers and in static or sluggishly flowing waterbodies. Heavy infestations may need successive treatments at 14-21 d intervals. Use special equipment (see label) [4, 5]

Crop safety/Restrictions

- Maximum number of treatments 3 per crop on hops [5], 1 per crop in most other situations - see labels for details
- Do not apply potato haulm destruction treatment when soil dry. Tubers may be damaged if spray applied during or shortly after dry periods. See label for details of maximum allowable soil-moisture deficit and varietal drought resistance scores
- Treated laid barley and oats may be used only for stock feed; treated peas must be harvested dry; treated field beans may be used for pigeon and animal feed only
- If potato tubers are to be stored leave 14 d after treatment before lifting
- Do not add wetters to desiccant sprays for potatoes or for water weed control except for *Lemna* control.
- Consult processor before adding Agral or other wetter on peas
- For weed control in row crops apply as overall spray before crop emergence or before transplanting

Special precautions/Environmental safety

- Harmful if swallowed [1-3, 5, 6]
- Harmful in contact with skin [4]
- Irritating to skin and eyes
- Harmful to livestock [1-3, 5, 6]
- Do not use treated straw or haulm as animal feed or bedding within 4 d of spraying [1-3, 5, 6]
- Do not use treated water for human consumption within 24 h or for overhead irrigation within 10 d of treatment [4, 5]
- Do not dump surplus herbicide in water or ditch bottoms
- Do not contaminate surface waters or ditches with chemical or used container
- Do not use on barley, oats and field beans intended for human consumption [1-3, 5, 6]

Personal protective equipment/Label precautions

- A, C [1-6]
- M03, R04a, R04b [1-6]; R03c [1-3, 5, 6]; R03a [4]; U20a [1, 3, 6]; U04b [1, 3, 4, 6]; U05a [1-6]; U19 [1-3, 5, 6]; U08 [1, 3, 5, 6]; U09b [4]; U02, U20b [2, 4, 5]; U09a, U13 [2]; U04a [2, 5]; C03, E01, E30a, E34 [1-6]; E19 [1, 3-6]; E21 [1, 3-6] (10 d); E15, E31a [1-3, 5, 6]; E06b [1-3, 5, 6] (24 h); E08 [1, 3, 5, 6] (4 d); E32a [4]; E26 [2]

Withholding period
- Keep all livestock out of treated areas and away from treated water for at least 24 h [1-3, 5, 6]

Latest application/Harvest interval
- Varies with crop and product to ensure best results - see labels
- HI zero when used as a crop desiccant but see label for advisory intervals

Approval
- Approved for aquatic weed control [4, 5]. See notes in Section 1 on use of herbicides in or near water
- Off-label Approval unstipulated for use on outdoor *Echium plantagineum* (OLA 1844/99) [5]
- Accepted by BLRA for use on hops

Maximum Residue Level (mg residue/kg food)
- milk 0.01; meat/vegetable products, potatoes, peas, oilseed rape 0.05

---

229   diquat + paraquat

A non-selective non-residual bipyridyl contact herbicide

Products

1 PDQ	Zeneca	80:120 g/l	SL	10241
2 Speedway 2	Scotts	2.5:2.5% w/w	WG	09273

Uses

Annual dicotyledons in *apples, blackcurrants, bush fruit, cane fruit, cherries, cultivated land/soil, damsons, flower bulbs, forestry, gooseberries, hardy ornamentals, hops, pears, plums, potatoes, raspberries, row crops, strawberries, stubbles* [1]. Annual dicotyledons in *non-crop areas* [1, 2]. Annual dicotyledons in *turf (sward destruction/turf renovation)* [2]. Annual grasses in *apples, blackcurrants, bush fruit, cane fruit, cherries, cultivated land/soil, damsons, flower bulbs, forestry, gooseberries, hardy ornamentals, hops, pears, plums, potatoes, raspberries, row crops, strawberries, stubbles* [1]. Annual grasses in *non-crop areas* [1, 2]. Annual grasses in *turf (sward destruction/turf renovation)* [2]. Chemical stripping in *hops* [1]. Green cover in *field margins, land temporarily removed from production* [1]. Perennial grasses in *non-crop areas, turf (sward destruction/turf renovation)* [2]. Perennial non-rhizomatous grasses in *apples, blackcurrants, bush fruit, cane fruit, cherries, cultivated land/soil, damsons, flower bulbs, forestry, gooseberries, hardy ornamentals, hops, non-crop areas, pears, plums, potatoes, raspberries, row crops, strawberries, stubbles* [1]. Sward destruction in *grassland* [1]. Volunteer cereals in *apples, blackcurrants, bush fruit, cane fruit, cherries, cultivated land/soil, damsons, forestry, gooseberries, hops, non-crop areas, pears, plums, potatoes, raspberries, row crops, strawberries, stubbles* [1].

Efficacy
- Rapid kill obtained under bright conditions but most effective results on difficult weeds obtained from slower action in winter
- Apply to young emerged weeds less than 15 cm high, annual grasses must have at least 2 leaves when sprayed [1]
- Addition of approved non-ionic wetter recommended for control of certain species and with low dose rates. See label for details [1]
- Interval between spraying and cultivation varies. See label for details [1]
- For chemical stripping apply in Jul or after hops have reached top wire. Do not use on hops under drought conditions [1]
- Chemical rapidly inactivated in moist soil, activity reduced in dirty or muddy water
- Spray is rainfast in 10 min

Crop safety/Restrictions
- Maximum number of treatments 2 per yr for grassland destruction
- In non-crop areas do not use around green bark [2]
- Apply up to just before sown crops emerge or just before planting [1]
- On sandy or immature peat soils and on forest nursery seedbeds allow 3 d between spraying and planting [1]
- Where trash or dying weeds are left on surface allow at least 3 d before planting

---

- In potatoes spray earlies up to 10% emergence, maincrop to 40% emergence, provided plants are less than 15 cm high. Do not use post-emergence on potatoes from diseased or small tubers or under very hot, dry conditions [1]
- Use guarded no-drift sprayers to kill inter-row weeds and strawberry runners [1]
- Apply to fruit crops as a directed spray, preferably in dormant season [1]
- If spraying bulbs at end of season ensure all crop foliage is detached from bulbs. Do not use on very sandy soils [1]

Special precautions/Environmental safety

- Paraquat is subject to the Poisons Rules 1982 and the Poisons Act 1972. See notes in Section 1
- Toxic
- Harmful in contact with skin and if swallowed
- Irritating to skin
- Risk of serious damage to eyes
- Harmful to livestock. Paraquat may be harmful to hares; stubbles must be sprayed early in the day
- Keep in original container, tightly closed, in a safe place, under lock and key
- Do not put in a food or drinks container

Personal protective equipment/Label precautions

- A, C [1, 2]; H, M [1]
- M04 [2]; M03, M05, R02, R03a, R03c, R04b, R04d [1]; U04b, U05a, U09a, U19, U20b [1, 2]; C03 [1]; E02, E31a [2]; E01, E15, E30b, E34 [1, 2]; E11, E31c [1]; E06b [1] (24 h)

Withholding period

- Keep all livestock out of treated areas for at least 24 h

Latest application/Harvest interval

- Varies according to situation - see label
- HI zero for potatoes

Approval

- Accepted by BLRA for use on hops

Maximum Residue Level (mg residue/kg food)

- see paraquat entry

---

230   disulfoton

A systemic organophosphorus aphicide and insecticide

See also chlorpyrifos + disulfoton

Products

| Disulfoton P 10 | United Phosphorus | 10% w/w | GR | 08023 |

Uses

Aphids in **broad beans, broccoli, brussels sprouts, cabbages, carrots, cauliflowers, field beans, fodder beet, mangels, parsnips, potatoes, sugar beet, sugar beet stecklings**. Carrot fly in **carrots, parsnips**. Mangold fly in **fodder beet, mangels, sugar beet**.

Efficacy

- For effective results granules applied at drilling or planting should be incorporated in soil. See label for details of suitable application machinery
- One application is normally sufficient, a split application may be used on carrots
- Persistence may be reduced in fen peat soils

Crop safety/Restrictions

- Maximum number of treatments 1 (soil) + 1 (foliage) for Brussels sprouts, cabbages, cauliflowers; 1 (soil) or 2 (foliage) 11 kg/ha (total) for beet crops; 1 per crop for broad beans, celery, field beans, French beans, marrows, parsley, potatoes, runner beans, strawberries; 1 per crop (soil or foliage) for carrots, parsnips
- Consult processor before treating crops grown for processing

Special precautions/Environmental safety
- This product contains an anticholinesterase organophosphorus compound. Do not use if under medical advice not to work with such compounds
- Toxic in contact with skin, by inhalation or if swallowed
- Keep in original container, tightly closed, in a safe place, under lock and key
- Dangerous to livestock
- Dangerous to game, wild birds and animals
- Applied correctly treatment will not harm bees
- Dangerous to fish or other aquatic life. Do not contaminate surface waters or ditches with chemical or used container

Personal protective equipment/Label precautions
- A, B, H, J, K, M; C (or D + E)
- M01, M04, R02a, R02b, R02c, U02, U04a, U05a, U11, U12, U13, U19, U20a, C03; C02 (6 wk); E01, E10a, E13b, E26, E30b, E32a, E34

Withholding period
- Keep all livestock out of treated areas for at least 6 wk. Bury or remove spillages

Latest application/Harvest interval
- HI 6 wk

Approval
- Approved for aerial application on brassicas, beans, carrots. See notes in Section 1
- Approval expiry 31 Dec 2001

Maximum Residue Level (mg residue/kg food)
- barley, sorghum 0.2; wheat 0.1; tea 0.05; all other products 0.02

---

231   dithianon

A protectant and eradicant dicarbonitrile fungicide for scab control

Products

1 Dithianon Flowable	BASF	750 g/l	SC	10219
2 Dithianon Flowable	Cyanamid	750 g/l	SC	07007

Uses

Scab in **apples, pears**.

Efficacy
- Apply at bud-burst and repeat every 10-14 d until danger of scab infection ceases
- Application at high rate within 48 h of a Mills period prevents new infection
- Spray programme also reduces summer infection with apple canker

Crop safety/Restrictions
- Maximum number of treatments 8 per crop
- Do not use on Golden Delicious apples after green cluster
- Do not mix with lime sulphur or highly alkaline products

Special precautions/Environmental safety
- Harmful if swallowed.
- Irritating to eyes and skin

Personal protective equipment/Label precautions
- M03, R03c, R04a, R04b, U05a, U08, U19, U20a, C03, E01, E15, E26, E30a, E31b, E34

Latest application/Harvest interval
- HI 4 wk

---

232   diuron

A residual urea herbicide for non-crop areas and woody crops

See also amitrole + 2,4-D + diuron
bromacil + diuron

Products

1 Atlas Diuron	Nufarm Whyte	500 g/l	SC	08214
2 Chipko Diuron 80	Nomix-Chipman	80% w/w	WP	00497

Products (Continued)

3	Chipman Diuron 80	Nomix-Chipman	80% w/w	WP	08054
4	Chipman Diuron Flowable	Nomix-Chipman	500 g/l	SC	05701
5	Diuron 80 WP	Aventis Environ.	80% w/w	WP	09931
6	Diuron 80 WP	RP Amenity	80% w/w	WP	05199
7	Freeway	RP Amenity	500 g/l	SC	06047
8	Freeway	Aventis Environ.	500 g/l	SC	09933
9	MSS Diuron 50 FL	Nufarm Whyte	500 g/l	SC	07160
10	MSS Diuron 500 FL	Nufarm Whyte	500 g/l	SC	08171
11	Unicrop Flowable Diuron	Unicrop	500 g/l	SC	02270

Uses

Annual dicotyledons in *trees and shrubs* [1, 10]. Annual dicotyledons in *land not intended to bear vegetation* [1, 3, 10]. Annual dicotyledons in *apples, asparagus (off-label), blackcurrants (off-label), pears* [11]. Annual dicotyledons in *established woody ornamentals, woody nursery stock* [2, 5-8]. Annual dicotyledons in *amenity trees and shrubs* [3, 7, 8]. Annual dicotyledons in *apple orchards, pear orchards* [7, 8]. Annual grasses in *land not intended to bear vegetation, trees and shrubs* [1, 10]. Annual grasses in *apples, asparagus (off-label), blackcurrants (off-label), pears* [11]. Annual grasses in *established woody ornamentals, woody nursery stock* [2, 5-8]. Annual grasses in *amenity trees and shrubs, apple orchards, pear orchards* [7, 8]. Annual meadow grass in *amenity trees and shrubs, land not intended to bear vegetation* [3]. Annual weeds in *non-crop areas* [2, 4-9]. Annual weeds in *farm buildings, industrial sites, paths, railway tracks, road verges* [4]. Annual weeds in *trees and shrubs* [4, 9]. Perennial weeds in *farm buildings, industrial sites, non-crop areas, paths, railway tracks, road verges* [4].

Efficacy

- Best results when applied to moist soil and rain falls soon afterwards
- Length of residual activity may be reduced on heavy or highly organic soils or those with ash substrates
- Selective rates must be applied to weed-free soil and activity persists for 2-3 mth
- Around trees and shrubs spray in late winter or early spring [7, 8]
- Treat hard surfaces between Feb and end May [2-4]
- Application for total vegetation control may be at any time of year, best results obtained in late winter to early spring
- Application to frozen ground not recommended

Crop safety/Restrictions

- Maximum number of treatments 1 per yr on land not intended for cropping (high rate), 2 per yr around trees and shrubs (low rate) [3, 5-8]; one per year for apples, pears [11]
- Apply to weed-free soil in apple and pear orchards established for at least 1 yr during Feb-Mar
- Do not treat trees and shrubs less than 5 cm tall or established less than 12 mth. See label for list of sensitive species [1, 3, 7, 8]
- Do not use on Sands or Very Light soils or those that are gravelly or where less than 1% organic matter
- Do not apply to areas intended for replanting during next 12 mth (2 yr for vegetables)
- Do not use on lawns, grass tennis courts or similar areas of turf [3]
- Do not apply non-selective rates on or near desirable plants where chemical may be washed into contact with roots
- Application for amenity use should only take place between the beginning of Feb and end of Apr (May in Scotland)

Special precautions/Environmental safety

- Irritating to eyes and skin
- Irritating to respiratory system [2, 3, 11]
- Harmful to fish or other aquatic life. Do not contaminate surface waters or ditches with chemical or used container
- Do not apply over drains or in drainage channels or gullies
- Avoid run-off when using on paved and similar surfaces

Personal protective equipment/Label precautions

- A [1, 3, 4, 7, 8]; C [1, 3, 4]; H, M [7, 8]

- R04a, R04b [1-11]; R04c [2, 3, 11]; U05a [1-11]; U19, U20b [1-9, 11]; U08 [1-10]; U09a [11]; U20a [10]; C03 [1, 3-11]; E13c, E30a [1-11]; E32a [2, 3, 5, 6]; E31c [11]; E01 [1, 3-11]; E26 [1, 4, 9-11]; E34 [1, 10, 11]; E31a [4]; E31b [1, 7-10]

Approval
- Off-label approval unlimited for use on established blackcurrants (OLA 1318/95)[11]; unlimited for use on outdoor asparagus (OLA 2352/98)[11]

---

233   diuron + glyphosate

A non-selective residual herbicide mixture for non-crop and amenity use

Products

1	Touché	Nomix-Chipman	217.6:145.3 g/l	RH	07913
2	Xanadu	Aventis Environ.	125:100 g/l	EW	09943
3	Xanadu	Monsanto	125:100 g/l	EW	09228

Uses

Annual dicotyledons in *amenity trees and shrubs, land not intended to bear vegetation*. Annual grasses in *amenity trees and shrubs, land not intended to bear vegetation*. Perennial dicotyledons in *amenity trees and shrubs, land not intended to bear vegetation*. Perennial grasses in *amenity trees and shrubs, land not intended to bear vegetation*.

Efficacy
- Best results achieved from treatment when weeds are green and actively growing. Symptoms may be slow to appear in poor growing conditions
- Perennial grasses are susceptible when tillering and making new rhizome growth, normally when plants have 4-5 new leaves
- Perennial dicotyledons most susceptible if treated at or near flowering but will be severely checked if treated at other times when growing actively
- Most species of germinating weeds are controlled. See label for more resistant species
- Weed control may be reduced on heavy or highly organic soils and when weeds are suffering stress in any situation
- Treat hard surfaces between Feb and end May
- At least 6 and preferably 24 h rain-free must follow spraying. Rain soon after treatment may reduce initial weed control
- Apply with Nomix System equipment [1]

Crop safety/Restrictions
- Maximum number of treatments 1 per yr
- Ornamental trees and shrubs should be established for at least 12 mth and be at least 50 mm tall
- Among ornamental trees and shrubs take care to avoid foliage or the stems of young plants
- Do not allow spray to contact desired plants or crops
- Do not use on tree and shrub nurseries or before planting
- Do not treat ornamental trees and shrubs on Sands or Very Light soils or those that are gravelly or with less than 1% organic matter [1]
- Do not decant, connect directly to Nomix applicator [1]

Special precautions/Environmental safety
- Not to be used on food crops [2, 3]
- Harmful to fish or other aquatic life. Do not contaminate surface waters or ditches with chemical or used container
- Do not apply over drains or in drainage channels, gullies or similar structures

Personal protective equipment/Label precautions
- H, M [1]
- U02 [1]; U09a, U19, U20b [1-3]; C01 [2, 3]; E26 [1]; E13c, E30a, E32a [1-3]

Approval
- Approved for use through ULV applicators [1]

Maximum Residue Level (mg residue/kg food)
- see glyphosate entry

234   diuron + paraquat

A total herbicide with contact and residual activity

Products

Dexuron	Nomix-Chipman	300:100 g/l	SC	07169

Uses

Annual dicotyledons in **established woody ornamentals, woody nursery stock**. Annual grasses in **established woody ornamentals, woody nursery stock**. Perennial dicotyledons in **established woody ornamentals, woody nursery stock**. Perennial grasses in **established woody ornamentals, woody nursery stock**. Total vegetation control in **non-crop areas, paths**.

Efficacy

* Apply to emerged weeds at any time of year. Best results achieved by application in spring or early summer
* Effectiveness not reduced by rain soon after treatment

Crop safety/Restrictions

* Maximum number of treatments 1 per yr
* Avoid contact of spray with green bark, buds or foliage of desirable trees or shrubs
* In nurseries use low dose rate as inter-row spray, not more than once per year

Special precautions/Environmental safety

* Paraquat is subject to the Poisons Rules 1982 and the Poisons Act 1972. See notes in Section 1
* Keep in original container, tightly closed, in a safe place, under lock and key
* Toxic if swallowed
* Harmful in contact with skin. Irritating to eyes and skin
* Harmful to fish or other aquatic life. Do not contaminate surface waters or ditches with chemical or used container

Personal protective equipment/Label precautions

* A, C
* M04, R02c, R03a, R04a, R04b, U02, U04a, U05a, U09a, U19, U20b, C03, E01, E13c, E26, E30b, E31a, E34; E06b (24 h)

Withholding period

* Keep livestock out of treated areas for at least 2 wk and until foliage of any poisonous weeds such as ragwort has died and become unpalatable

Maximum Residue Level (mg residue/kg food)

* see paraquat entry

235   DNOC

A dinitrophenyl insecticide all approvals for which were revoked in December 1989

236   dodemorph

A systemic morpholine fungicide for powdery mildew control

Products

F238	BASF	385 g/l	EC	00206

Uses

Powdery mildew in **roses**.

Efficacy

* Spray roses every 10-14 d during mildew period or every 7 d and at increased dose if cleaning up established infection or if disease pressure high
* Add Citowett when treating rose varieties which are difficult to wet
* Product has negligible effect on *Phytoseiulus* spp being used to control red spider mites

Crop safety/Restrictions
- Do not use on seedling roses
- Do not apply to roses under hot, sunny conditions, particularly under glass, but spray early in the morning or during the evening. Increase the humidity some hours before spraying
- Check tolerance of new varieties before treating rest of crop

Special precautions/Environmental safety
- Irritating to skin. Risk of serious damage to eyes
- Flammable
- Harmful to fish or other aquatic life. Do not contaminate surface waters or ditches with chemical or used container

Personal protective equipment/Label precautions
- A, C
- R04b, R04d, R07d, U04a, U05a, U09a, U14, U15, U19, U20c, C03, E01, E13c, E26, E30a, E31c

---

237 dodine

A protectant and eradicant guanidine fungicide

Products

Radspor FL	Truchem	450 g/l	SC	01685

Uses

Currant leaf spot in **blackcurrants, gooseberries**. Scab in **apples, pears**.

Efficacy
- Apply protective spray on apples and pears at bud-burst and at 10-14 d intervals until late Jun to early Jul
- Apply post-infection spray within 36 h of rain responsible for initiating infection. Where scab already present spray prevents production of spores
- Apply to blackcurrants immediately after flowering, repeat every 10-14 d to within 1 mth of harvest and once or twice post-harvest

Crop safety/Restrictions
- Do not apply in very cold weather (under 5°C) or under slow drying conditions to pears or dessert apples during bloom or immediately after petal fall
- Do not mix with lime sulphur or tetradifon

Special precautions/Environmental safety
- Harmful: In contact with skin
- Irritating to eyes and skin
- May cause sensitization by skin contact
- Harmful to fish or other aquatic life. Do not contaminate surface waters or ditches with chemical or used container

Personal protective equipment/Label precautions
- A, C
- M03, R03a, R03c, R04a, R04b, R04e, U05a, U08, U19, U20a, C03, E01, E13c, E30a, E31a, E34

Latest application/Harvest interval
- Early Jul for dessert apples; pre-blossom for culinary apples and pears
- HI blackcurrants 1 mth

---

238 endosulfan

A contact and ingested organochlorine insecticide and acaricide

Products

Thiodan 20 EC	Aventis	200 g/l	EC	07335

Uses

Big-bud mite in ***blackcurrants***. Blackberry mite in ***blackberries***. Bulb scale mite in ***narcissi***. Tarsonemid mites in ***strawberries***.

Efficacy

- Adjust spray volume to achieve total cover. See label for minimum dilutions
- On blackcurrants apply 3 sprays, at first flower, end of flowering and fruit set
- On blackberries apply 3 sprays at 14 d intervals before flowering
- On strawberries apply immediately after whole crop picked; where a second crop to be picked in autumn within 1 wk of mowing old foliage
- For effective control on strawberries it is essential that the spray penetrates the crowns of the plants
- Apply as drench to boxed narcissi a few days after bringing into greenhouse

Crop safety/Restrictions

- When timing applications on blackcurrants it is important to consider different cultivars separately. Applications should not be made over a whole plantation of mixed cultivars regardless of variations in maturity

Special precautions/Environmental safety

- Endosulfan is subject to the Poisons Rules 1982 and the Poisons Act 1972. See notes in Section 1
- Toxic if swallowed. Harmful in contact with skin
- Flammable
- Keep in original container, tightly closed, in a safe place, under lock and key
- Keep unprotected persons out of treated areas for at least 1 d
- Dangerous to livestock
- Harmful to bees. Do not apply at flowering stage except as directed on hops, oilseed rape and mustard. Keep down flowering weeds.
- Dangerous to fish or other aquatic life. Do not contaminate surface waters or ditches with chemical or used container

Personal protective equipment/Label precautions

- A, C, H, J, K, L, M
- M04, R02c, R03a, R07d, U02, U04a, U05a, U09a, U13, U19, U20a, C03; C02 (6 wk); E01, E12e, E13b, E30b, E31b, E34; E02 (1 d); E06a (3 wk)

Withholding period

- Keep all livestock out of treated areas for at least 3 wk

Latest application/Harvest interval

- HI blackcurrants, blackberries, strawberries 6 wk

Approval

- UK approvals for use of endosulfan on blackcurrants, blackberries, strawberries likely to be revoked in 2001 as a result of implementation of the MRL Directives
- Off-label approval unlimited for use on outdoor and protected ornamentals (OLA 0656/92)

Maximum Residue Level (mg residue/kg food)

- strawberries, blackberries, currants, gooseberries 2; citrus fruit, pome fruit, stone fruit, grapes, raspberries, kiwi fruit, olives, onions, fruiting vegetables (except sweet corn), brassica vegetables (except kohlrabi), lettuce, spinach, legume vegetables, cardoons, celery, globe artichokes, leeks, cultivated mushrooms 1; cotton seed 0.3; beetroot, carrots, celeriac, parsley root, swedes, turnips, potatoes, maize 0.2; tree nuts, peanuts, poppy seed, sesame seed, wheat, rye, barley, oats, triticale, meat 0.1; dewberries, loganberries, bilberries, cranberries, wild berries, avocados, dates, figs, kumquats, litchis, mangoes, passion fruit, pineapples, pomegranates, horseradish, Jerusalem artichokes, parsnips, radishes, salsify, sweet potatoes, yams, garlic, shallots, spring onions, sweet corn, kohlrabi, watercress, witloof, herbs, asparagus, fennel, rhubarb, wild mushrooms, pulses, sorghum, buckwheat, millet, rice 0.05; milk, dairy produce, eggs 0.004

239   epoxiconazole

A systemic, protectant and curative triazole fungicide for use in cereals

See also carbendazim + epoxiconazole

Products

1	AgriGuard Epoxiconazole	AgriGuard	125 g/l	SC	09407
2	Landgold Epoxiconazole	Landgold	125 g/l	SC	09821
3	Opus	BASF	125 g/l	SC	08319
4	Standon Epoxiconazole	Standon	125 g/l	SC	09517

Uses

Brown rust in **winter wheat** [1, 2, 4]. Brown rust in **barley** [1, 3]. Brown rust in **spring barley, winter barley** [2, 4]. Brown rust in **rye, triticale, wheat** [3]. Eyespot in **winter barley** *(reduction)*, **winter wheat** *(reduction)* [1-4]. Fusarium ear blight in **winter wheat** *(reduction)* [1, 2, 4]. Net blotch in **barley** [1, 3]. Net blotch in **spring barley, winter barley** [2, 4]. Powdery mildew in **winter wheat** [1, 2, 4]. Powdery mildew in **barley** [1, 3]. Powdery mildew in **spring barley, winter barley** [2, 4]. Powdery mildew in **oats, rye, triticale, wheat** [3]. Rhynchosporium in **barley** [1, 3]. Rhynchosporium in **spring barley, winter barley** [2, 4]. Rhynchosporium in **rye** [3]. Septoria in **winter wheat** [1, 2, 4]. Septoria in **triticale, wheat** [3]. Sooty moulds in **winter wheat** *(reduction)* [1, 2, 4]. Sooty moulds in **wheat** *(reduction)* [3]. Yellow rust in **winter wheat** [1, 2, 4]. Yellow rust in **barley** [1, 3]. Yellow rust in **spring barley, winter barley** [2, 4]. Yellow rust in **rye, triticale, wheat** [3].

Efficacy

- Apply at the start of foliar disease attack
- Optimum effect against eyespot achieved by spraying between leaf-sheath erect and second node detectable stages (GS 30-32)
- Best control of ear diseases of wheat obtained by treatment during ear emergence
- For Septoria spray after third node detectable stage (GS 33) when weather favouring disease development has occurred
- Mildew control improved by use of tank mixtures. See label for details

Crop safety/Restrictions

- Maximum total dose equivalent to two full dose treatments
- Product may cause damage to broad-leaved plant species
- Avoid spray drift onto neighbouring crops

Special precautions/Environmental safety

- Dangerous to fish or other aquatic life. Do not contaminate surface waters or ditches with chemical or used container
- LERAP Category B

Personal protective equipment/Label precautions

- A [1-4]
- U05a [1-4]; U20b [2-4]; U20a [1]; C03, E01, E13b, E16a, E30a [1-4]; E31c [2-4]; E26, E31b [1]; E16b [1, 2, 4]

Latest application/Harvest interval

- Up to and including flowering just complete (GS 69) in wheat, rye, triticale; up to and including emergence of ear just complete (GS 59) in barley, oats

Approval

- Accepted by BLRA for use on malting barley up to ear emergence

240   epoxiconazole + fenpropimorph

A systemic, protectant and curative fungicide mixture for cereals

Products

1	Barclay Riverdance	Barclay	84:250 g/l	SE	09658
2	Eclipse	BASF	84:250 g/l	SE	07361
3	Greencrop Galore	Greencrop	84:250 g/l	SE	09561

Products (Continued)

4 Landgold Epoxiconazole FM	Landgold	84:250 g/l	SE	08806
5 Opus Team	BASF	84:250 g/l	SE	07362
6 Standon Epoxifen	Standon	84:250 g/l	SE	08972

Uses

Brown rust in **barley** [1-3, 5]. Brown rust in **winter wheat** [1, 3, 4, 6]. Brown rust in **rye, triticale, wheat** [2, 5]. Brown rust in **spring barley, winter barley** [4, 6]. Eyespot in **winter barley** *(reduction)*, **winter wheat** *(reduction)* [1-6]. Fusarium ear blight in **winter wheat** *(reduction)* [1, 3, 4, 6]. Net blotch in **barley** [1-3, 5]. Net blotch in **spring barley, winter barley** [4, 6]. Powdery mildew in **barley** [1-3, 5]. Powdery mildew in **winter wheat** [1, 3, 4, 6]. Powdery mildew in **oats, rye, triticale, wheat** [2, 5]. Powdery mildew in **spring barley, winter barley** [4, 6]. Rhynchosporium in **barley** [1-3, 5]. Rhynchosporium in **rye** [2, 5]. Rhynchosporium in **spring barley, winter barley** [4, 6]. Septoria in **winter wheat** [1, 3, 4, 6]. Septoria in **triticale, wheat** [2, 5]. Sooty moulds in **wheat** *(reduction)* [2, 5]. Sooty moulds in **winter wheat** *(reduction)* [3]. Yellow rust in **barley** [1-3, 5]. Yellow rust in **winter wheat** [1, 3, 4, 6]. Yellow rust in **rye, triticale, wheat** [2, 5]. Yellow rust in **spring barley, winter barley** [4, 6].

Efficacy
- Apply at the start of foliar disease attack
- Optimum effect against eyespot achieved by spraying between leaf-sheath erect and second node detectable stages (GS 30-32)
- Best control of ear diseases obtained by treatment during ear emergence
- For Septoria spray after third node detectable stage (GS 33) when weather favouring disease development has occurred

Crop safety/Restrictions
- Maximum total dose equivalent to two full dose treatments
- Product may cause damage to broad-leaved plant species
- Avoid spray drift onto neighbouring crops

Special precautions/Environmental safety
- Irritating to skin
- Harmful (dangerous [4, 6]) to fish or other aquatic life. Do not contaminate surface waters or ditches with chemical or used container [1-3, 5]
- LERAP category B

Personal protective equipment/Label precautions
- A [1-6]
- R04b, U05a, U20b, C03 [1-6]; E31c [2, 5]; E01, E16a, E30a [1-6]; E13c [1-3, 5]; E13b [4, 6]; E16b, E26 [1, 3, 4, 6]; E31b [1, 4, 6]; E31a [3]

Latest application/Harvest interval
- Up to and including flowering just complete (GS 69) in wheat, rye, triticale; up to and including emergence of ear just complete (GS 59) in barley, oats

Approval
- Accepted by BLRA for use on malting barley up to ear emergence

241   epoxiconazole + fenpropimorph + kresoxim-methyl

A protectant, systemic and curative fungicide mixture for cereals

Products

1 Mantra	BASF	125:150:125 g/l	SE	08886
2 Standon Kresoxim Super	Standon	125:150:125 g/l	SE	09794

Uses

Brown rust in **rye, spring wheat, triticale** [1]. Brown rust in **spring barley, winter barley, winter wheat** [1, 2]. Eyespot in **rye** *(reduction)*, **triticale** *(reduction)*, **winter oats** *(reduction)* [1]. Eyespot in **winter barley** *(reduction)*, **winter wheat** *(reduction)* [1, 2]. Fusarium ear blight in **spring wheat** *(reduction)* [1]. Fusarium ear blight in **winter wheat** *(reduction)* [1, 2]. Net blotch in **spring barley, winter barley** [1, 2]. Powdery mildew in **oats, rye, spring wheat,** *triticale* [1]. Powdery mildew in *spring barley, winter barley, winter wheat* [1, 2]. Rhynchosporium in *rye* [1]. Rhynchosporium in *spring barley, winter barley* [1, 2]. Septoria diseases in *spring wheat, triticale* [1]. Septoria diseases in *winter wheat* [1, 2]. Sooty moulds in *winter wheat* *(reduction)* [1, 2]. Yellow rust in *rye, spring wheat, triticale* [1]. Yellow rust in *spring barley, winter barley, winter wheat* [1, 2].

Efficacy
- For best results spray at the start of foliar disease attack and repeat if infection conditions persist
- Optimum effect against eyespot obtained by treatment between leaf sheaths erect and first node detectable stages (GS 30-32)
- For protection against ear diseases apply during ear emergence

Crop safety/Restrictions
- Maximum total dose equivalent to two full dose treatments

Special precautions/Environmental safety
- Irritant. May cause sensitization by skin contact
- Dangerous to fish or other aquatic life. Do not contaminate surface waters or ditches with chemical or used container
- LERAP Category B
- Avoid spray drift onto neighbouring crops. Product may damage broad-leaved species

Personal protective equipment/Label precautions
- A [1, 2]
- R04, R04e, U05a, U14, U20b, C03 [1, 2]; E31c [1]; E01, E13b, E16a, E30a [1, 2]; E16b, E26, E29, E31b [2]

Latest application/Harvest interval
- Mid flowering (GS 65) for wheat, rye, triticale ; completion of ear emergence (GS 59) for barley, oats

Approval
- Kresoxim-methyl included in Annex I under EC Directive 91/414
- Accepted by BLRA for use on malting barley up to ear emergence

242   epoxiconazole + kresoxim-methyl

A protectant, systemic and curative fungicide mixture for cereals

Products

1 Barclay Avalon	Barclay	125:125 g/l	SC	09466
2 Landgold Strobilurin KE	Landgold	125:125 g/l	SC	09908
3 Landmark	BASF	125:125 g/l	SC	08889
4 Me2 KME	Me2	125:125 g/l	SC	09594
5 Standon Kresoxim- Epoxiconazole	Standon	125:125 g/l	SC	09281

Uses

Brown rust in *spring barley, winter barley, winter wheat* [1-5]. Brown rust in *rye, spring wheat, triticale* [3]. Eyespot in *winter barley* *(reduction)*, *winter wheat* *(reduction)* [1-5]. Eyespot in *rye* *(reduction)*, *triticale* *(reduction)*, *winter oats* *(reduction)* [3]. Fusarium ear blight in *winter wheat* *(reduction)* [1-5]. Fusarium ear blight in *spring wheat* *(reduction)* [3]. Net blotch in *spring barley, winter barley* [1-5]. Powdery mildew in *spring barley, winter barley, winter wheat* [1-5]. Powdery mildew in *oats, rye, spring wheat, triticale* [3]. Rhynchosporium in *spring barley, winter barley* [1-5]. Rhynchosporium in *rye* [3]. Septoria diseases in *winter wheat* [1-5]. Septoria diseases in *spring wheat, triticale* [3]. Sooty moulds in *winter wheat* *(reduction)* [1-5]. Yellow rust in *spring barley, winter barley, winter wheat* [1-5]. Yellow rust in *rye, spring wheat, triticale* [3].

Efficacy
- For best results spray at the start of foliar disease attack and repeat if infection conditions persist
- Optimum effect against eyespot obtained by treatment between leaf sheaths erect and first node detectable stages (GS 30-32)
- For protection against ear diseases apply during ear emergence

Crop safety/Restrictions
- Maximum total dose equivalent to two full dose treatments

Special precautions/Environmental safety
- Irritant. May cause sensitisation by skin contact
- Dangerous to fish or other aquatic life. Do not contaminate surface waters or ditches with chemical or used container
- LERAP Category B
- Avoid spray drift onto neighbouring crops. Product may damage broad-leaved species

Personal protective equipment/Label precautions
- A [1-5]
- R04, R04e, U05a, U14 [1-5]; U20b [1, 3, 4]; U20a [2, 5]; C03, E01, E13b, E16a, E26, E30a [1-5]; E31c [1, 3]; E16b [1, 2, 4, 5]; E31b [2, 5]; E31a [4]

Latest application/Harvest interval
- Mid flowering (GS 65) for wheat, rye, triticale; completion of ear emergence (GS 59) for barley, oats

Approval
- Kresoxim-methyl included in Annex I under EC Directive 91/414
- Accepted by BLRA for use on malting barley up to ear emergence

---

243   esfenvalerate

A contact and ingested pyrethroid insecticide

Products

1	Sumi-Alpha	BASF	25 g/l	EC	BASF
2	Sumi-Alpha	Cyanamid	25 g/l	EC	07207

Uses

Aphids in **winter barley, winter wheat**.

Efficacy
- Crops at high risk (e.g. after grass or in areas with history of BYDV) should be treated when aphids first seen or by mid-Oct. Otherwise treat in late Oct-early Nov
- High risk crops will need a second treatment
- Product also recommended between onset of flowering and milky ripe stages (GS 61-73) for control of summer cereal aphids

Crop safety/Restrictions
- Do not use if another pyrethroid or dimethoate has been applied to crop after start of ear emergence (GS 51)

Special precautions/Environmental safety
- Harmful if swallowed, irritating to skin and eyes
- Extremely dangerous to bees. Do not apply at flowering stage except as directed on oilseed rape and peas. Keep down flowering weeds
- Extremely dangerous to fish or other aquatic life. Do not contaminate surface waters or ditches with chemical or used container
- LERAP Category A
- Store product in dark away from direct sunlight

Personal protective equipment/Label precautions
- A, C, H
- M03, R03c, R04a, R04b, U04a, U05a, U08, U19, U20b, C03, E01, E12b, E13a, E16c, E16d, E22a, E26, E30a, E31b, E34

Latest application/Harvest interval
- HI 20 d

Approval
- Esfenvalerate included in Annex I under EC Directive 91/414
- Accepted by BLRA for use pre-harvest on malting barley
- Product registration number not available at time of printing [1]

244   ethofumesate

A benzofuran herbicide for grass weed control in various crops

See also bromoxynil + ethofumesate + ioxynil
chloridazon + ethofumesate

Products

1	AgriGuard Ethofumesate Flo	AgriGuard	500 g/l	SC	09478
2	Barclay Keeper	Barclay	200 g/l	EC	08835
3	Barclay Keeper 500 FL	Barclay	500 g/l	SC	09438
4	Landgold Ethofumesate 200	Landgold	200 g/l	EC	08980
5	MSS Thor	Nufarm Whyte	200 g/l	EC	08817
6	Nortron Flo	Aventis	500 g/l	SC	08154
7	Salute	United Phosphorus	200 g/l	EC	07660
8	Standon Ethofumesate 200	Standon	200 g/l	EC	09360
9	Stefes Fumat 2	Stefes	200 g/l	EC	07856

Uses

Annual dicotyledons in **fodder beet, mangels, sugar beet** [1-9]. Annual dicotyledons in **red beet** [1-4, 6-9]. Annual dicotyledons in **garlic** *(off-label)*, **horseradish** *(off-label)*, **onions** *(off-label)* [6]. Annual grasses in **grass seed crops, leys** [2, 3, 9]. Annual grasses in **amenity turf, established grassland** [3, 9]. Annual grasses in **garlic** *(off-label)*, **horseradish** *(off-label)*, **onions** *(off-label)*, **strawberries** *(off-label)* [6]. Annual meadow grass in **fodder beet, mangels, sugar beet** [1-9]. Annual meadow grass in **red beet** [1-4, 6-9]. Blackgrass in **fodder beet, mangels, sugar beet** [1-9]. Blackgrass in **red beet** [1-4, 6-9]. Blackgrass in **amenity turf, established grassland, grass seed crops, leys** [3, 7-9]. Chickweed in **grass seed crops, leys** [2, 3, 7-9]. Chickweed in **amenity turf, established grassland** [3, 7-9]. Cleavers in **grass seed crops, leys** [2, 3, 7-9]. Cleavers in **amenity turf, established grassland** [3, 7-9]. Clover in **strawberries** *(off-label)* [6]. Volunteer cereals in **amenity turf, established grassland, grass seed crops, leys** [3, 7-9].

Efficacy

- Most products may be applied pre- or post-emergence of crop or weeds but some restricted to pre-emergence or post-emergence use only. Check label
- Apply in beet crops in tank mixes with other pre- or post-emergence herbicides. Recommendations vary for different mixtures. See label for details
- In grass crops apply to moist soil as soon as possible after sowing or post-emergence when crop in active growth, normally mid-Oct to mid-Dec. See label for details
- Volunteer cereals not well controlled pre-emergence, weed grasses should be sprayed before fully tillered
- Grass crops may be sprayed during rain or when wet. Not recommended in very dry conditions or prolonged frost

Crop safety/Restrictions

- Maximum number of treatments 1 pre-emergence [5]; 1 pre- plus 1 post-emergence (2 at reduced dose) per crop or yr for beet crops and grassland
- Safe timing on beet crops varies with other ingredient of tank mix. See label for details
- Do not use on Sands or Heavy soils, Very Light soils containing a high percentage of stones, or soils with more than 5-10% organic matter (percentage varies according to label)
- May be used in Italian, hybrid and perennial ryegrass, timothy, cocksfoot, meadow fescue and tall fescue. Apply pre-emergence to autumn-sown leys, post-emergence after 2-3 leaf stage. See label for details
- Do not use on swards reseeded without ploughing
- Clovers will be killed or severely checked
- Any crop may be sown 3 mth after application of mixtures in beet crops following ploughing, 5 mth after application in grass crops
- Do not graze or cut grass for 14 d after, or roll less than 7 d before or after spraying

Special precautions/Environmental safety
- Flammable [2, 4, 5, 7-9]
- Harmful to fish or other aquatic life. Do not contaminate surface waters or ditches with chemical or used container

Personal protective equipment/Label precautions
- A [2]
- M05 [5]; M03 [2]; R07d [2, 4, 5, 7-9]; U08, U19 [1-9]; U20a [1, 2, 4-9]; U05a [2, 3]; U20b [3]; C03 [2, 3]; E13c, E30a, E31b [1-9]; E26 [1-8]; E01 [2, 3]; E34 [4]

Withholding period
- Do not graze grass for 14 d after treatment

Latest application/Harvest interval
- Pre-emergence of beet crops [5]; before crops meet across rows for beet crops and mangels; 14 d before cutting or grazing for grass leys; pre-emergence for horseradish

Approval
- Off-label approval unlimited for use on strawberries, onions, garlic, horseradish (OLA 2104/98)[6]

---

245   ethofumesate + metamitron + phenmedipham

A contact and residual herbicide mixture for sugar beet

Products

Betanal Trio WG	Aventis	6.5:28:6.5% w/w	WG	07537

Uses

Annual dicotyledons in **sugar beet**. Annual meadow grass in **sugar beet**.

Efficacy
- Best results obtained from a series of treatments applied as an overall fine spray commencing when earliest germinating weeds are no larger than fully expanded cotyledon
- Apply subsequent sprays as each new flush of weeds reaches early cotyledon and continue until weed emergence ceases (maximum 3 sprays)
- Product must be applied with Actipron
- Product may follow certain pre-emergence treatments and be used in conjunction with other post-emergence sprays - see label for details
- Where a pre-emergence band spray has been applied, the first treatment should be timed according to the size of the weeds in the untreated area between the rows

Crop safety/Restrictions
- Maximum number of treatments 3 per crop; maximum total dose 6 kg product per ha
- Product may be used on all soil types but residual activity may be reduced on those with more than 5% organic matter
- Crop tolerance may be reduced by stress caused by growing conditions, effects of pests, disease or other pesticides, nutrient deficiency etc
- Only beet crops should be sown within 4 mth of last treatment; winter cereals may be sown after this interval
- Any spring crop may be sown in the year following use
- Mould-board ploughing to 150 mm followed by thorough cultivation recommended before planting any crop

Special precautions/Environmental safety
- Irritating to eyes
- Harmful to fish or other aquatic life. Do not contaminate surface waters or ditches with chemical or used container

Personal protective equipment/Label precautions
- A, C
- R04a, U05a, U11, U20c, C03, E01, E13c, E30a, E32a

Latest application/Harvest interval
- Before crop meets between rows

246   ethofumesate + phenmedipham

A contact and residual herbicide for use in beet crops

Products

1	Barclay Goalpost	Barclay	100:80 g/l	EC	09497
2	Betanal Tandem	Aventis	100:80 g/l	EC	07254
3	Betosip Combi	Sipcam	100:80 g/l	EC	08630
4	Stefes Medimat 2	Stefes	100:80 g/l	EC	07577
5	Stefes Tandem	Stefes	100:80 g/l	EC	08906

Uses

Annual dicotyledons in *sugar beet* [1, 3-5]. Annual dicotyledons in *fodder beet, mangels* [1, 4, 5]. Annual dicotyledons in *red beet* [4, 5]. Annual meadow grass in *fodder beet, mangels, sugar beet* [1, 4]. Annual meadow grass in *red beet* [4]. Blackgrass in *fodder beet, mangels, sugar beet* [1, 4]. Blackgrass in *red beet* [4].

Efficacy

- Best results achieved by repeat applications to cotyledon stage weeds. Larger susceptible weeds not killed by first treatment usually checked and controlled by second application
- Apply on all soil types at 7-10 d intervals
- On soils with more than 5% organic matter residual activity may be reduced
- Do not spray wet foliage or if rain imminent
- Spray must be applied low volume. See label for details

Crop safety/Restrictions

- Maximum number of treatments normally 3 per crop - see labels for details
- Apply reduced dose from when the crop has fully expanded cotyledons or full dose from 2 fully expanded true leaf stage [1, 3, 4]
- Spray in evening if daytime temperatures above 21°C expected
- Avoid or delay treatment if frost expected within 7 d
- Avoid or delay treating crops under stress from wind damage, manganese or lime deficiency, pest or disease attack etc
- Check from which recovery may not be complete may occur if treatment made during conditions of sharp diurnal temperature fluctuation

Special precautions/Environmental safety

- Harmful in contact with skin [4, 5]
- Irritating to eyes [1, 3]
- Irritating to skin [1, 2, 4, 5]
- Irritating to respiratory system [2, 4, 5]
- Flammable [1, 2, 4, 5]
- May cause sensitisation by skin contact [3]
- Harmful to fish or aquatic life. Do not contaminate surface waters or ditches with chemical or used container
- Extra care necessary to avoid drift because product is recommended for use as a fine spray
- Spray volumes must not exceed those recommended [1, 3, 4]
- Interval between mixing spray and completion of spraying should not exceed 2 h to avoid crystallization [1, 3, 4]
- Product may cause non-reinforced PVC pipes and hoses to soften and swell. Wherever possible use reinforced PVC or synthetic rubber [1, 3, 4]

Personal protective equipment/Label precautions

- A, C [1-5]
- R03a, R04c [2, 4, 5]; R04b, R07d [1, 2, 4, 5]; R04e [3]; R04a [1, 3]; U08 [2-5]; U05a, U19 [1-5]; U20a [2, 4, 5]; U10, U11, U12, U20b [1]; C03, E01, E13c, E30a, E31b [1-5]; E34 [2, 4, 5]; E26 [1-3, 5]; E24 [2, 5]

Latest application/Harvest interval

- Before crop foliage meets in the rows

247　ethoprophos

An organophosphorus nematicide and insecticide

Products

1	Aventis Mocap 10G	Aventis	10% w/w	GR	09973
2	Mocap 10G	Aventis	10% w/w	GR	10003
3	Mocap 10G	RP Agric.	10% w/w	GR	06773

Uses

Potato cyst nematode in **potatoes**. Wireworms in **potatoes**.

Efficacy

- Broadcast shortly before or during final soil preparation with suitable fertilizer spreader and incorporate immediately to 10-15 cm. See label for details
- Treatment can be applied on all soil types. Control of pests reduced on organic soils
- Effectiveness dependent on soil moisture. Drought after application may reduce control

Crop safety/Restrictions

- Maximum number of treatments 1 per crop

Special precautions/Environmental safety

- This product contains an anticholinesterase organophosphorous compound. Do not use if under medical advice not to work with such compounds
- Harmful by inhalation, in contact with skin and if swallowed
- May cause sensitization by skin contact
- Dangerous to game, wild birds and animals
- Dangerous to fish or other aquatic life. Do not contaminate surface waters or ditches with chemical or used container
- Do not harvest crops for human or animal consumption for at least 8 wk after application

Personal protective equipment/Label precautions

- A, B, H, K, M; C (or D + E)
- M01, M03, R03a, R03b, R03c, R04e, U02, U04a, U05a, U08, U13, U19, U20a, C03; C02 (8 wk); E01, E10a, E13b, E30a, E32a, E34

Latest application/Harvest interval

- Pre-planting of crop
- HI 8 wk

248　etridiazole

A protective thiadiazole fungicide for soil or compost incorporation

Products

1	Aaterra WP	Zeneca	35% w/w	WP	06625
2	Standon Etridiazole 35	Standon	35% w/w	WP	08778

Uses

Damping off in **cabbages** *(seedlings and transplants)*, **cauliflowers** *(seedlings and transplants)*, **celery** *(seedlings and transplants)*, **cucumbers** *(seedlings and transplants)*, **mustard and cress** *(seedlings and transplants)*, **nft tomatoes, tomatoes** *(seedlings and transplants)*. Phytophthora in **container-grown stock, hardy ornamental nursery stock, nft tomatoes, tulips, watercress** *(off-label)*. Pythium in **tulips, watercress** *(off-label)*. Root diseases in **rockwool tomatoes** *(off-label)*.

Efficacy

- Best results obtained when incorporated thoroughly into soil or compost. Drench application also recommended
- Do not apply to wet soil
- Treat compost as soon as possible before use

Crop safety/Restrictions

- Maximum number of treatments depends on application method (see label)
- After drenching wash spray residue from crop foliage

- Do not use on Escallonia, Pyracantha, Gloxinia spp., pansies or lettuces
- Germination of lettuce in previously treated soil may be impaired
- Do not drench seedlings until well established
- When treating compost for blocking reduce dose by 50%
- Test on small numbers of plants in advance when treating subjects of unknown susceptibility or using compost with more than 20% inert material

Special precautions/Environmental safety
- Irritating to eyes and skin

Personal protective equipment/Label precautions
- A, C [1, 2]
- R04a, R04b [1, 2]; U19 [1]; U02, U04a, U05a, U08, U20b, C03 [1, 2]; E34 [1]; E01, E15, E30a, E32a [1, 2]

Latest application/Harvest interval
- 24 h after seeding watercress.
- HI tomatoes, cucumbers, mustard and cress 3 d

Approval
- Off-label approval unlimited for use on tomatoes grown on rock wool (OLA 0600/94)[1]; unlimited for use on watercress propagation beds (OLA 1213/96)[1]

---

249   etrimfos

A contact organophosphorus insecticide for stored grain crops

Products

1	Satisfar	Nickerson Seeds	525 g/l	EC	04180
2	Satisfar Dust	Nickerson Seeds	2% w/w	DP	04085

Uses

Grain beetles in *grain stores* [1]. Grain beetles in *stored cereals, stored oilseed rape* [1, 2]. Grain storage mites in *grain stores* [1]. Grain storage mites in *stored cereals, stored oilseed rape* [1, 2]. Grain storage pests in *grain stores* [1]. Grain storage pests in *stored cereals, stored oilseed rape* [1, 2]. Grain weevils in *grain stores* [1]. Grain weevils in *stored cereals, stored oilseed rape* [1, 2].

Efficacy
- Spray internal surfaces of clean store with knapsack or motorized sprayer
- Allow sufficient time for pests to emerge from hiding places and make contact with chemical before grain is stored
- Apply as admixture treatment to grain as it enters store using a suitable stored grain sprayer or automatic seed treater if using dust
- Controls malathion-resistant beetles and gamma-HCH-resistant mites

Crop safety/Restrictions
- Maximum number of treatments 1 per batch
- Cool grain to below 15°C before treatment and storage. Moisture content of grain should not exceed 16%

Special precautions/Environmental safety
- This product contains an anticholinesterase organophosphorus compound. Do not use if under medical advice not to work with such compounds
- Harmful to game, wild birds and animals
- Dangerous to fish or other aquatic life. Do not contaminate surface waters or ditches with chemical or used container

Personal protective equipment/Label precautions
- A, C [1, 2]
- M03, R04a, R07d [1]; U09a, U20a [2]; U19 [1, 2]; U05a, U08, U20b, C03 [1]; E10b, E13b, E30a, E32a [1, 2]; E01 [1]

Latest application/Harvest interval
- On entry into store [2]; on removal from store [1]

Approval
- Accepted by BLRA for use in stores for malting barley
- Approval expiry 31 Dec 2001 [1]
- Approval expiry 31 Dec 2001 [2]

Maximum Residue Level (mg residue/kg food)
- cereals (except rice) 5

250   fatty acids

A soap concentrate insecticide and acaricide

Products

Savona	Koppert	49% w/w	SL	06057

Uses

Aphids in **broad beans, brussels sprouts, cabbages, cucumbers, fruit trees, lettuce, peas, peppers, pumpkins, runner beans, tomatoes, watercress** (off-label), **woody ornamentals**. Mealy bugs in **broad beans, brussels sprouts, cabbages, cucumbers, fruit trees, lettuce, peas, peppers, pumpkins, runner beans, tomatoes, woody ornamentals**. Red spider mites in **broad beans, brussels sprouts, cabbages, cucumbers, fruit trees, lettuce, peas, peppers, pumpkins, runner beans, tomatoes, woody ornamentals**. Scale insects in **broad beans, brussels sprouts, cabbages, cucumbers, fruit trees, lettuce, peas, peppers, pumpkins, runner beans, tomatoes, woody ornamentals**. Whitefly in **broad beans, brussels sprouts, cabbages, cucumbers, fruit trees, lettuce, peas, peppers, pumpkins, runner beans, tomatoes, woody ornamentals**.

Efficacy
- Use only soft or rain water for diluting spray
- For glasshouse use apply when insects first seen and repeat as necessary
- To control whitefly spray when required and use biological control after 12 h

Crop safety/Restrictions
- Do not use on new transplants, newly rooted cuttings or plants under stress
- Do not use on specified susceptible shrubs. See label for details

Special precautions/Environmental safety
- Harmful to fish or other aquatic life. Do not contaminate surface waters or ditches with chemical or used container

Personal protective equipment/Label precautions
- U20c, E13c, E26, E30a, E31a

Latest application/Harvest interval
- HI zero

Approval
- Off-label approval to Apr 1999 for use on watercress grown outdoors (OLA 0735/94)

251   fenarimol

A systemic curative and protective pyrimidine fungicide

Products

1	Rimidin	Rigby Taylor	120 g/l	SC	05907
2	Rubigan	Dow	120 g/l	SC	05489

Uses

Dollar spot in **managed amenity turf, sports turf** [1]. Fusarium patch in **managed amenity turf, sports turf** [1]. Powdery mildew in **apples, blackcurrants, gooseberries, marrows** (off-label), **protected peppers** (off-label), **protected tomatoes** (off-label), **pumpkins** (off-label), **raspberries, roses, squashes** (off-label), **strawberries** [2]. Red thread in **managed amenity turf, sports turf** [1]. Scab in **apples** [2].

Efficacy
- Recommended spray interval varies from 7-14 d depending on crop and climatic conditions. See label for timing details [2]
- Efficient coverage and short spray intervals essential, especially for scab control [2]
- Spray strawberry runner beds regularly throughout season [2]
- For turf disease control spray as preventive treatment and repeat as necessary. See label for details [1]
- Do not mow within 24 h after treatment [1]

Crop safety/Restrictions
- Maximum number of treatments 15 per yr for apples; 3 per yr for raspberries
- Do not use on trees that are under stress from drought, severe pest damage, mineral deficiency or poor soil conditions

Special precautions/Environmental safety
- Dangerous to fish or other aquatic life. Do not contaminate surface waters or ditches with chemical or used container

Personal protective equipment/Label precautions
- A, C
- U08, U19; C02 (14 d); E13b, E26, E30a, E31b

Latest application/Harvest interval
- HI tomatoes, peppers 2 d; other crops 14 d

Approval
- Off-label approval unlimited for use on protected tomatoes, protected peppers (OLA 0645/94) [2]; unstipulated for use on marrows, squashes, pumpkins (OLA 2651/99)[2]

Maximum Residue Level (mg residue/kg food)
- hops 5; currants, gooseberries 1; pome fruits, grapes 0.3; tea 0.05; citrus fruits, tree nuts, blackberries, dewberries, loganberries, bilberries, cranberries, wild berries, miscellaneous fruits, root and tuber vegetables, bulb vegetables, sweet corn, brassica vegetables, leaf vegetables and herbs, beans (with and without pods), stem vegetables (except globe artichokes), fungi, pulses, oilseeds, potatoes, rye, oats, triticale, maize, rice, animal products (except liver and kidney) 0.02

---

252 fenazaquin

A mitochondrial electron transport inhibitor

Products

Matador 200 SC	Dow	200 g/l	SC	07960

Uses

Red spider mites in **apples**. Two-spotted spider mite in **ornamentals**.

Efficacy
- Acts by contact to give rapid knockdown
- Treat apples after petal fall when most overwintered eggs have hatched but before damage is seen
- Product should be used as part of a pest control programme. A further acaricide treatment may be necessary after application
- Control may be reduced where water volumes are reduced or in orchards where water volumes above 750 l/ha are required for good crop cover

Crop safety/Restrictions
- Maximum number of treatments 1 per yr
- Other mitochondrial electron transport inhibitor (METI) acaricides should not be applied to the same crop in the same calendar yr either separately or in mixture
- Do not apply to roses. Do not treat new ornamental species or varieties without first testing a few plants on a small scale
- Do not use on edible crops other than apples
- Do not treat ornamentals when in blossom or under stress

Special precautions/Environmental safety
- Harmful if swallowed
- Extremely dangerous to fish or other aquatic life. Do not contaminate surface waters or ditches with chemical or used container
- LERAP Category B

Personal protective equipment/Label precautions
- A
- M03, R03c, U05a, U19, C03; C02 (30 d); E01, E13a, E16a, E16b, E20, E30a, E31a, E34; E17 (15 m)

Latest application/Harvest interval
- HI 30 d for apples

253   fenbuconazole

A systemic protectant and curative triazole fungicide for top fruit

Products

1	Indar 5EW	Landseer	50 g/l	EW	09518
2	Kruga 5EC	Interfarm	50 g/l	EC	09863
3	Reward 5EC	Interfarm	50 g/l	EC	09862
4	Surpass 5EC	Interfarm	50 g/l	EC	09861

Uses

Brown rust in *spring barley, winter barley, winter wheat* [2-4]. Glume blotch in *winter wheat* [2-4]. Leaf spot in *winter wheat* [2-4]. Powdery mildew in *apples (reduction), pears (reduction)* [1]. Powdery mildew in *winter wheat (reduction)* [2-4]. Rhynchosporium in *spring barley, winter barley* [2-4]. Scab in *apples, pears* [1]. Yellow rust in *winter wheat* [2-4].

Efficacy
- Most effective when used as part of a routine preventative programme from bud burst to onset of petal fall [1]
- After petal fall, tank mix with other protectant fungicides to enhance scab control [1]
- Safe to use on all main commercial varieties of apples and pears in UK [1]
- See label for recommended spray intervals. In periods of rapid growth or high disease pressure, a 7 d interval should be used [1]

Crop safety/Restrictions
- Maximum total dose on top fruit equivalent to ten full doses per yr [1]
- Consult processors before using on pears for processing [1]

Special precautions/Environmental safety
- Irritant. May cause serious damage to eyes
- Do not harvest for human or animal consumption for at least 4 wk after last application
- Harmful to fish or other aquatic life. Do not contaminate surface waters or ditches with chemical or used container

Personal protective equipment/Label precautions
- A, C [1-4]
- R04 [1]; R04d [1-4]; R04b [2-4]; U20a [1]; U05a, U08 [1-4]; U19, U20b [2-4]; C02 [1] (4 wk); C03 [1-4]; E13c, E32a [1]; E01, E26 [1-4]; E13b, E30a, E31b [2-4]

Latest application/Harvest interval
- HI 28 d for top fruit

254   fenbuconazole + propiconazole

A broad spectrum systemic fungicide mixture with protectant, curative and eradicant properties for use in wheat

Products

Graphic	Novartis	37.5:47 g/l	EC	08415

Uses

Brown rust in *winter wheat*. Powdery mildew in *winter wheat* *(moderate control)*. Septoria in *winter wheat*. Yellow rust in *winter wheat*.

Efficacy

- Spray at start of disease attack or as part of disease control programme. Apply a second treatment 2-5 wk later if disease pressure remains high or if wet weather continues
- Treat in spring from ear at 1 cm stage (GS 30)

Crop safety/Restrictions

- Maximum number of treatments 2 per crop
- Use only in spring or summer
- If used in a programme with other products containing propiconazole do not apply more than 500 g a.i./ha to the crop in any one season

Special precautions/Environmental safety

- Irritating to skin
- Risk of serious damage to eyes
- Dangerous to fish or other aquatic life. Do not contaminate surface waters or ditches with chemical or used container
- Do not harvest crops for human or animal consumption for at least 5 wk after application

Personal protective equipment/Label precautions

- A, C
- R04b, R04d, U02, U05a, U09a, U20a, C03; C02 (5 wk); E01, E13b

Latest application/Harvest interval

- Before start of flowering (GS 61)

Approval

- Accepted by BLRA for use on malting barley

Maximum Residue Level (mg residue/kg food)

- see propiconazole entry

255　fenbutatin oxide

A selective contact and ingested organotin acaricide

Products

Torq	Fargro	50% w/w	WP	08370

Uses

Two-spotted spider mite in *protected cucumbers, protected ornamentals, protected peppers (off-label), protected tomatoes, tunnel grown strawberries*.

Efficacy

- Active on larvae and adult mites. Spray may take 7-10 d to effect complete kill but mites cease feeding and crop damage stops almost immediately
- Apply as soon as mites first appear and repeat as necessary
- On tunnel-grown strawberries apply when mites first appear, usually before flowering starts, and repeat 10-14 d later. A post-harvest spray is also recommended

Crop safety/Restrictions

- Maximum number of treatments 2 per crop pre-harvest + 1 post harvest for tunnel grown strawberries
- Allow at least 10 d between spray applications and do not apply within 10 d of a previous spray
- Do not add wetters or mix with anything other than water
- Do not use white petroleum oil within 28 d of treatment, or any other pesticide within 7 d
- Do not apply to crops which are under stress for any reason
- On subjects of unknown susceptibility test treat on a small number of plants in advance

Special precautions/Environmental safety

- Irritating to eyes, skin and respiratory system
- Extremely dangerous to fish or other aquatic life. Do not contaminate surface waters or ditches with chemical or used container
- Suitable for use in IPM programmes with *Encarsia* or *Phytoseiulus* being used for biological control

---

Personal protective equipment/Label precautions
- A, C, D, H, M
- M03, R04a, R04b, R04c, U05a, U08, U19, U20a, C03, E01, E13a, E30a, E32a, E34

Latest application/Harvest interval
- HI glasshouse cucumbers, tomatoes 3 d; tunnel-grown strawberries 7 d

Approval
- Off-label approval unlimited for use on protected peppers (OLA 0857/97)
- UK approvals for use of fenbutatin oxide on peppers likely to be revoked in 2001 as a result of implementation of the MRL Directives

Maximum Residue Level (mg residue/kg food)
- citrus 5; bananas 3; pome fruit, grapes 2; strawberries, tomatoes, aubergines 1; cucumbers, courgettes 0.5; tea, hops 0.1; tree nuts, stone fruits, cane fruit, other small fruit, wild berries, other miscellaneous fruit, root and tuber vegetables, bulb vegetables, other fruiting vegetables, cucurbits (inedible peel), sweet corn, brassica vegetables, leaf vegetables and fresh herbs, legume vegetables, stem vegetables, fungi, pulses, oilseeds, potatoes 0.05

---

256   fenhexamid

A protectant fungicide for soft fruit

Products

Teldor	Bayer	51% w/w	WG	08955

Uses

Botrytis in *blackberries, blackcurrants, gooseberries, grapevines, loganberries, raspberries, redcurrants, rubus hybrids, strawberries, whitecurrants*.

Efficacy
- Use as part of a programme of sprays throughout the flowering period to achieve effective control of Botrytis
- To minimise possibility of development of resistance, no more than two sprays of the product may be applied consecutively. Other fungicides from a different chemical group should then be used for at least two consecutive sprays. If only two applications are made on grapevines, only one may include fenhexamid
- Complete spray cover of all flowers and fruitlets throughout the blossom period is essential for successful control of Botrytis
- Spray programmes should normally start at the start of flowering

Crop safety/Restrictions
- Maximum number of treatments 2 per yr on grapevines; 4 per yr on other listed crops but no more than 2 sprays may be applied consecutively

Special precautions/Environmental safety
- Harmful to fish or other aquatic life. Do not contaminate surface waters or ditches with chemical or used container

Personal protective equipment/Label precautions
- U08, U19, U20b, E13c, E30a, E32a

Latest application/Harvest interval
- HI 7 d for raspberries, loganberries, blackberries, blackcurrants, redcurrants, whitecurrants, gooseberries; 1 d for strawberries

---

257   fenitrothion

A broad spectrum, contact organophosphorus insecticide

Products

1	Dicofen	PBI	500 g/l	EC	09598
2	EC-Kill	Antec	500 g/l	EC	H5468
3	Killgerm Fenitrothion 40 WP	Killgerm	38.69% w/w	WP	H4858

Products (Continued)

4	Killgerm Fenitrothion 50 EC	Killgerm	500 g/l	EC	H4722
5	Micromite	Killgerm	500 g/l	EC	H4480
6	Sectacide 50 EC	Killgerm	500 g/l	EC	H4939

Uses

Ants in *refuse tips* [2]. Ants in *food storage areas, livestock houses* [2, 5]. Aphids in *apples, pears, peas, plums* [1]. Apple blossom weevil in *apples* [1]. Bedbugs in *refuse tips* [2]. Bedbugs in *food storage areas, livestock houses* [2, 5]. Beetles in *refuse tips* [2]. Beetles in *food storage areas, livestock houses* [2, 5]. Capsids in *apples, blackcurrants, gooseberries* [1]. Caterpillars in *pears, plums* [1]. Cockroaches in *refuse tips* [2]. Cockroaches in *food storage areas, livestock houses* [2, 5]. Codling moth in *apples* [1]. Crickets in *refuse tips* [2]. Crickets in *food storage areas, livestock houses* [2, 5]. Earwigs in *refuse tips* [2]. Earwigs in *food storage areas, livestock houses* [2, 5]. Fleas in *refuse tips* [2]. Fleas in *food storage areas, livestock houses* [2, 5]. Flies in *refuse tips* [2-4, 6]. Flies in *livestock houses* [2, 4-6]. Flies in *food storage areas* [2, 5]. Flies in *poultry houses* [3]. Flour beetles in *grain stores* [1]. Frit fly in *maize, sweetcorn* [1]. Grain beetles in *grain stores* [1]. Grain weevils in *grain stores* [1]. Leatherjackets in *barley, durum wheat, oats, rye, triticale, wheat* [1]. Mites in *refuse tips* [2]. Mites in *food storage areas, livestock houses* [2, 5]. Moths in *refuse tips* [2]. Moths in *food storage areas, livestock houses* [2, 5]. Pea and bean weevils in *peas* [1]. Pea midge in *peas* [1]. Pea moth in *peas* [1]. Poultry house pests in *poultry houses* [2, 5]. Raspberry beetle in *raspberries* [1]. Raspberry cane midge in *raspberries* [1]. Saddle gall midge in *barley, durum wheat, oats, rye, triticale, wheat* [1]. Sawflies in *apples, blackcurrants, gooseberries* [1]. Silverfish in *refuse tips* [2]. Silverfish in *food storage areas, livestock houses* [2, 5]. Suckers in *apples* [1]. Thrips in *barley, peas, wheat* [1]. Tortrix moths in *apples, strawberries* [1]. Wheat-blossom midges in *wheat* [1]. Winter moth in *apples* [1].

Efficacy

- Number and timing of sprays vary with disease and crop. See label for details
- Apply as bran bait for leatherjacket control in cereals
- For cane midge add suitable spreader and apply to lower 60 cm of young canes
- For control of grain store pests apply to all surfaces of empty stores before filling with grain. Remove dust and debris before applying
- Where pear suckers resistant to fenitrothion occur control is unlikely to be satisfactory [1]

Crop safety/Restrictions

- Maximum number of treatments 1 per crop for cereals, maize, sweetcorn; 2 per crop or yr for peas, plums, blackcurrants, gooseberries, strawberries, grain stores; 3 per crop for apples, pears; 4 per crop for raspberries
- Do not mix with magnesium sulphate or highly alkaline materials
- On raspberries may cause slight yellowing of leaves of some varieties which closely resembles that caused by certain viruses

Special precautions/Environmental safety

- This product contains an anticholinesterase organophosphorus compound. Do not use if under medical advice not to work with such compounds
- Products for use in livestock houses etc must be used by professional operators
- Harmful in contact with skin or if swallowed
- May cause lung damage if swallowed [4, 6]
- Irritating to eyes and skin [1]
- Causes severe burns [1]
- Flammable
- Harmful to livestock [1]
- Harmful to game, wild birds and animals [1]
- Harmful to fish or other aquatic life. Do not contaminate surface waters or ditches with chemical or used container
- Dangerous to bees [1, 3]
- Do not apply directly to livestock and poultry. Remove exposed milk and collect eggs before application. Protect milk machinery and milk containers from contamination

FOR FULL CONDITIONS OF USE ALWAYS READ THE PRODUCT LABEL

Personal protective equipment/Label precautions
- C [1, 5]; A [1-6]; D [5]; H [2-6]
- M01, M03 [1-6]; M05 [4, 6]; R04a, R04b, R05a [1]; R03a, R03c, R07d [1, 2, 4-6]; R04g [4, 6]; U08, U14, U15 [1]; U19 [1-6]; U05a [1, 2, 4-6]; U13 [1, 2, 5]; U02 [1, 3, 4, 6]; U20a [1, 2]; U16, U17 [5]; U20b [3-6]; U09a [2-6]; U04a [4, 6]; C02 [1]; C03 [1-6]; C04, C05, C08 [5]; C09 [3, 4, 6]; C07 [3]; E10b [1]; E13c, E30a [1-6]; E01 [1, 2, 4-6]; E31a, E34 [1, 2, 5]; E12c [1, 3]; E05 [3-6]; E02 [2, 4-6] (until surfaces dry); E04, E15, E32a [3, 4, 6]

Withholding period
- Keep unprotected persons and animals out of treated areas for 48 h or until surfaces are dry

Latest application/Harvest interval
- HI raspberries 7 d; other crops 2 wk

Approval
- Approved for aerial application on cereals, peas [1]. See notes in Section 1
- Accepted by BLRA for use in stores for malting barley (empty stores only) [1], and on malting barley crops
- Approval expiry 20 Apr 2001 [1]

Maximum Residue Level (mg residue/kg food)
- cereals 5; citrus fruits 2; pome fruits, apricots, peaches, nectarines, plums, grapes, cane fruits, bilberries, cranberries, currants, gooseberries, bananas, carrots, horseradish, parsnips, parsley root, salsify, swedes, turnips, garlic, onions, shallots, tomatoes, peppers, aubergines, cucumbers, gherkins, courgettes, cauliflowers, Brussels sprouts, head cabbages, lettuce, beans (with pods), peas (with pods), celery, leeks, rhubarb, cultivated mushrooms 0.5; potatoes 0.05

---

258   fenitrothion + permethrin + resmethrin

An organophosphate/pyrethroid insecticide mixture for grain stores

Products

Turbair Grain Store Insecticide	Graincare	10:20:2 g/l	UL	02238

Uses

Flour beetles in **grain stores**. Grain beetles in **grain stores**. Grain moths in **grain stores**. Grain weevils in **grain stores**.

Efficacy
- Apply as a surface spray in empty stores using a suitable fan-assisted ULV sprayer
- Clean store thoroughly before applying
- Combines knock-down effect with up to 5 mth residual activity
- Should not be mixed with other sprays

Special precautions/Environmental safety
- This product contains an anticholinesterase organophosphorus compound. Do not use if under medical advice not to work with such compounds
- Irritating to eyes, skin and respiratory system
- Highly flammable
- Extremely dangerous to fish or other aquatic life. Do not contaminate surface waters or ditches with chemical or used container

Personal protective equipment/Label precautions
- A, C
- M01, R04a, R04b, R04c, R07c, U05a, U08, U19, U20a, C03, E01, E13a, E30a, E31a, E34

Approval
- Product formulated for application as a ULV spray. See label for details
- Accepted by BLRA for use in stores for malting barley
- Approval expiry 20 Apr 2001

Maximum Residue Level (mg residue/kg food)
- see fenitrothion and permethrin entries

259   fenoxaprop-P-ethyl

A phenoxypropionic acid herbicide for use in wheat

See also diclofop-methyl + fenoxaprop-P-ethyl

Products

1	Aventis Cheetah Super	Aventis	55 g/l	EW	09730
2	Cheetah Super	Aventis	55 g/l	EW	08723
3	Triumph	Aventis	120 g/l	EC	08740

Uses

Blackgrass in **spring wheat, winter wheat**. Canary grass in **spring wheat, winter wheat**. Loose silky bent in **spring wheat, winter wheat**. Rough meadow grass in **spring wheat, winter wheat**. Wild oats in **spring wheat, winter wheat**.

Efficacy

- Treat weeds from 2 fully expanded leaves up to flag leaf ligule just visible; for awned canary-grass from 2 leaves to the end of tillering
- A second application may be made in spring where susceptible weeds emerge after an autumn application
- Spray is rainfast 1 h after application
- Product may be sprayed in frosty weather provided crop hardened off but do not spray wet foliage or leaves covered with ice
- Dry conditions resulting in moisture stress may reduce effectiveness

Crop safety/Restrictions

- Maximum total dose equivalent to two full dose treatments
- Treat from crop emergence to flag leaf fully emerged (GS 41).
- Do not apply to barley, durum wheat, undersown crops or crops to be undersown
- Do not roll or harrow within 1 wk of spraying
- Do not spray crops under stress, suffering from drought, waterlogging or nutrient deficiency or those grazed or if soil compacted
- Broadcast crops should be sprayed post-emergence after plants have developed well-established root system
- Avoid spraying immediately before or after a sudden drop in temperature or a period of warm days/cold nights
- Do not mix with hormone weedkillers

Special precautions/Environmental safety

- Irritating to eyes and skin
- Dangerous to fish or other aquatic life. Do not contaminate surface waters or ditches with chemical or used container

Personal protective equipment/Label precautions

- A, C, H [1-3]
- R04b [1-3]; R04a [3]; U05a, U20b [1-3]; U09a [1, 2]; U08 [3]; C03, E01, E13c, E30a, E31b [1-3]; E26 [3]

Latest application/Harvest interval

- Before flag leaf sheath extending (GS 41)

260   fenoxaprop-P-ethyl + isoproturon

A foliar and root-acting grass killer for use in wheat

Products

Puma X	Aventis	14:300 g/l	SE	08779

Uses

Annual meadow grass in **spring wheat, winter wheat**. Blackgrass in **spring wheat, winter wheat**. Canary grass in **spring wheat, winter wheat**. Chickweed in **spring wheat, winter wheat**. Corn marigold in **spring wheat, winter wheat**. Hemp-nettle in **spring wheat, winter** *wheat*. Mayweeds in *spring wheat, winter wheat*. Wild oats in *spring wheat, winter wheat*.

Efficacy
- Annual grass weeds listed are controlled up to and including end of tillering (GS 29) (annual meadow grass up to and including 1 tiller)
- Chickweed and mayweeds controlled up to 6 leaf stage, hemp-nettle and corn marigold up to 2 leaves
- Emerged weeds controlled on all soil types and foliar activity on grass weeds unaffected by high organic matter. Residual activity reduced on soils with more than 10% organic matter
- If prolonged dry weather follows application the speed of action is slower and weed control may be reduced, especially in spring
- In seasons of above average rainfall or where heavy rain falls shortly after application weed control may be reduced
- Foliar activity on grass weeds is unaffected by rainfall from 1 h after application

Crop safety/Restrictions
- Maximum number of treatments 1 per crop. Maximum total dose of isoproturon 2.5 kg a.i./ha per crop
- Apply from emergence of crop up to and including 1st node stage (GS 31)
- Only treat autumn sown spring wheat after 1 Jan
- Broadcast crops should be sprayed post-emergence after plants have a well-established root system
- Do not apply to barley, durum wheat, undersown crops or those due to be undersown
- Do not spray if frost imminent or after onset of frosty weather
- On free draining, stony or gravelly soils there is a risk of crop damage especially if heavy rain falls soon after application
- Do not spray crops lacking nutrient or under stress from drought, pest or disease attack, waterlogging or soil compaction
- Early sown crops (Sep) may be damaged if spraying precedes or coincides with a period of rapid growth in autumn
- Do not roll within 1 wk before or after or harrow within 1 wk before or at any time after application
- Do not mix with hormone herbicides or bifenox products. See label for permitted mixtures

Special precautions/Environmental safety
- Irritating to skin
- Dangerous to fish or aquatic life. Do not contaminate surface waters or ditches with chemical or used container

Personal protective equipment/Label precautions
- A, C
- R04b, U04a, U05a, U08, U20a, C03, E01, E13b, E26, E30a, E31b

Latest application/Harvest interval
- Before second node detectable (GS 32)

---

261   fenoxycarb

An insect specific growth regulator for top fruit

Products

1 Insegar	Novartis	250 g/l	WP	08558
2 Insegar WG	Novartis	25% w/w	WG	09789

Uses

Summer-fruit tortrix moth in *apples, pears*.

Efficacy
- Best results from application at 5th instar stage before pupation. Product prevents transformation from larva to pupa
- Correct timing best identified from pest warnings

- Because of mode of action rapid knock-down of pest is not achieved and larvae continue to feed for a period after treatment
- Adequate water volume necessary to ensure complete coverage of leaves

Crop safety/Restrictions
- Maximum number of treatments 2 per crop
- Use on all varieties of apples and pears
- Consult processors before use

Special precautions/Environmental safety
- High risk to bees. Do not apply to crops in flower or to those in which bees are actively foraging. Do not apply when flowering weeds are present
- Dangerous to fish or other aquatic life. Do not contaminate surface waters or ditches with chemical or used container
- Do not allow direct spray from broadcast air-assisted sprayers to fall within 8 m of surface waters or ditches. Direct spray away from water
- Apply to minimise off-target drift to reduce effects on non-target organisms. Some margin of safety to beneficial arthropods is indicated.

Personal protective equipment/Label precautions
- D [1]; A, H [1, 2]; C [2]
- U05a, U20a [1, 2]; U23 [2]; C03 [1, 2]; E31b, E34 [1]; E01, E12a, E13b, E22b, E29, E30a [1, 2]; E17 [1, 2] (8 m); E32a [2]

Latest application/Harvest interval
- HI 42 d

---

262   fenpropathrin

A contact and ingested pyrethroid acaricide and insecticide

Products

1	Meothrin	BASF	100 g/l	EC	BASF
2	Meothrin	Cyanamid	100 g/l	EC	07206

Uses

Caterpillars in *apples, blackcurrants, hops*. Damson-hop aphid in *hops*. Red spider mites in *apples, blackcurrants, hops*. Two-spotted spider mite in *hops, roses, strawberries*.

Efficacy
- Acts on motile stages of mites and gives rapid kill
- Apply on apples as post-blossom spray and repeat 3-4 wk later if necessary
- Apply on hops when hatch of spring eggs complete in May and repeat if necessary
- Apply on roses when pest first seen and repeat as necessary
- Apply on strawberries as a pre-flowering spray and repeat after harvest if necessary
- Where aphids resistant to fenpropathrin occur in hops control is unlikely to be satisfactory and repeat treatments may result in lower levels of control

Crop safety/Restrictions
- Maximum number of treatments 2 per crop for apples and hops; 2 per yr, 1 pre-flowering and 1 post-harvest for strawberries; 3 per yr for blackcurrants

Special precautions/Environmental safety
- Toxic in contact with skin or if swallowed. Irritating to eyes and skin
- Flammable
- Extremely dangerous to bees. Do not apply to crops in flower or to those in which bees are actively foraging except as directed in blackcurrants. Do not apply when flowering weeds are present
- Extremely dangerous to fish or other aquatic life. Do not contaminate surface waters or ditches with chemical or used container
- Do not operate air-assisted sprayers within 18 m of surface water or ditches. Direct spray away from water
- LERAP Category A

Personal protective equipment/Label precautions
- A, C, H, J
- M04, R02a, R02c, R04a, R04b, R07d, U02, U04a, U05a, U08, U19, U20a, C02, C03, E01, E12b, E13a, E16c, E16d, E30a, E31a, E34

Latest application/Harvest interval
- Pre-flowering for strawberries; 14 d after end of flowering for blackcurrants.
- HI apples and hops 7 d

Approval
- Accepted by BLRA for use on hops
- Product registration number not available at time of printing [1]

---

263   fenpropidin

A systemic, curative and protective piperidine (morpholine) fungicide

Products

1	Landgold Fenpropidin 750	Landgold	750 g/l	EC	08973
2	Mallard	Novartis	750 g/l	EC	08662
3	Tern	Novartis	750 g/l	EC	08660

Uses

Brown rust in **barley, wheat**. Glume blotch in **wheat**. Leaf blotch in **barley**. Leaf spot in **wheat**. Powdery mildew in **barley, wheat**. Yellow rust in **barley, wheat**.

Efficacy
- Best results obtained when applied at early stage of disease development. See label for details of recommended timing alone and in mixtures
- Disease control enhanced by vapour-phase activity. Control can persist for 4-6 wk
- Alternate with triazole fungicides to discourage build-up of resistance

Crop safety/Restrictions
- Maximum number of treatments 3 per crop (up to 2 in yr of harvest) for winter crops; 2 per crop for spring crops

Special precautions/Environmental safety
- Harmful in contact with skin and if swallowed
- Irritating to skin
- Risk of serious damage to eyes
- May cause sensitization by skin contact
- Dangerous to fish or other aquatic life. Do not contaminate surface waters or ditches with chemical or used container

Personal protective equipment/Label precautions
- A, C, H
- M03, R03a, R03c, R04b, R04d, R04e, U02, U04a, U05a, U05b, U10, U20a, C03; C02 (5 wk); E01, E13b, E29, E30a, E31c, E34

Latest application/Harvest interval
- Up to and including ear emergence complete (GS 59).
- HI 5 wk

Approval
- Accepted by BLRA for use on malting barley

---

264   fenpropidin + fenpropimorph

A contact and systemic fungicide mixture for use in wheat and barley

Products

1	Agrys	Novartis	480:270 g/l	EC	08382
2	Boscor	Novartis	188:562 g/l	EC	08682

Uses

Brown rust in *spring barley, spring wheat, winter barley, winter wheat* [1, 2]. Powdery mildew in *spring barley, spring wheat, winter barley, winter wheat* [1, 2]. Rhynchosporium in *spring barley (moderate control)*, *winter barley (moderate control)* [1]. Yellow rust in *spring barley, spring wheat, winter barley, winter wheat* [1].

Efficacy

- Spray at start of disease attack or as part of a disease control programme
- Repeat treatment may be needed 3-4 wk later if disease pressure high
- Best results obtained when application made at early stage of disease development. Treatment of established infections will be less effective

Crop safety/Restrictions

- Maximum total dose equivalent to two full dose treatments

Special precautions/Environmental safety

- Irritating to skin
- Risk of serious damage to eyes
- Flammable [2]
- Harmful to fish or other aquatic life. Do not contaminate surface waters or ditches with chemical or used container
- Do not allow direct spray from ground-based vehicle-mounted/drawn sprayers to fall within 6 m of surface waters or ditches. Direct spray away from water [2]

Personal protective equipment/Label precautions

- A, C, H [1, 2]
- R04b, R04d [1, 2]; R07d [2]; U09a, U19, U20a [1]; U02, U05a [1, 2]; U04a, U10 [2]; C03 [1, 2]; E16 [1]; E01, E13c, E26, E30a, E31b [1, 2]; E29, E34 [2]

Latest application/Harvest interval

- Before start of flowering (GS 60)

Approval

- Accepted by BLRA for use on malting barley

---

265   fenpropidin + prochloraz

A contact and systemic fungicide mixture for cereals

Products

Sponsor	Aventis	250:250 g/l	EC	08674

Uses

Brown rust in *barley, wheat*. Eyespot in *barley, wheat*. Glume blotch in *wheat*. Leaf blotch in *barley*. Leaf spot in *wheat*. Net blotch in *barley*. Powdery mildew in *barley, wheat*. Yellow rust in *barley, wheat*.

Efficacy

- For eyespot apply in spring from when leaf sheaths erect up to and including third node detectable stage (GS 33)
- Product effective against strains of eyespot resistant to benzimidazole fungicides
- Protection of winter cereals throughout the season usually requires at least 2 treatments. See label for details of timing
- For foliar diseases treat in spring or early summer at first sign of infection on new growth
- Best protection against ear diseases only achieved by spraying at full ear emergence

Crop safety/Restrictions

- Maximum number of treatments 2 per crop

Special precautions/Environmental safety

- Harmful if swallowed
- Harmful to fish or other aquatic life. Do not contaminate surface waters or ditches with chemical or used container
- Do not allow direct spray from vehicle mounted/drawn hydraulic sprayers to fall within 6 m, or from hand-held sprayers to fall within 2 m, of surface waters or ditches. Direct spray away from water
- Do not harvest crops for human or animal consumption for at least 5 wk after last application

Personal protective equipment/Label precautions
- A, C, H
- M03, R03c, R04e, U02, U04a, U05a, U09a, U20b, C03; C02 (5 wk); E01, E13c, E26, E30a, E31b, E34

Latest application/Harvest interval
- Up to and including ear emergence complete (GS 59)

Approval
- Accepted by BLRA for use on malting barley

266 fenpropidin + propiconazole

A systemic curative and protective fungicide for cereals

Products

1 Prophet	Novartis	375:125 g/l	EC	08433
2 Sheen	Novartis	450:100 g/l	EC	08442
3 Zulu	Novartis	450:100 g/l	EC	08464

Uses

Brown rust in *spring barley, spring wheat, winter barley, winter wheat* [1-3]. Leaf blotch in *spring barley, winter barley* [2, 3]. Net blotch in *spring barley, winter barley* [1-3]. Powdery mildew in *spring barley, spring wheat, winter barley, winter wheat* [1-3]. Rhynchosporium in *spring barley, winter barley* [1]. Septoria diseases in *spring wheat, winter wheat* [1-3]. Yellow rust in *spring barley, spring wheat, winter barley, winter wheat* [1-3].

Efficacy
- Best results obtained when applied at early stage of disease development or as part of a disease control programme. Repeat treatment may be necessary if disease pressure remains high

Crop safety/Restrictions
- Maximum number of treatments (including any products containing propiconazole) 3 per crop (2 in yr of harvest) for winter crops; 2 per crop for spring crops [2, 3]; 2 per crop for all uses [1]
- Transient crop scorch can result on some wheat varieties, particularly those with erect flag leaves. Avoid spraying crops under stress

Special precautions/Environmental safety
- Harmful in contact with skin and if swallowed [2, 3]
- Irritating to skin
- Risk of serious damage to eyes
- May cause sensitization by skin contact
- Risk of serious damage to eyes
- Dangerous to fish or other aquatic life. Do not contaminate surface waters or ditches with chemical or used container
- Do not allow direct spray from ground-based vehicle-mounted/drawn sprayers to fall within 6 m, or from hand-held sprayers to within 2 m, of surface waters or ditches. Direct spray away from water

Personal protective equipment/Label precautions
- H [1]; A, C [1-3]
- M03 [2, 3]; R04b, R04d, R04e [1-3]; R03a, R03c [2, 3]; U11 [1]; U02, U04a, U05a, U10, U20a, C03 [1-3]; C02 [1-3] (5 wk); E26, E29 [1]; E01, E13b, E16 [1-3]; E30a, E31b, E34 [2, 3]

Latest application/Harvest interval
- Up to and including grain watery ripe (GS 71) for wheat; up to and including emergence of ear just complete (GS 59) for barley.
- HI 5 wk

Approval
- Accepted by BLRA for use on malting barley

Maximum Residue Level (mg residue/kg food)
- see propiconazole entry

267   fenpropidin + propiconazole + tebuconazole

A protectant, curative and eradicant fungicide mixture

Products

Gladio	Novartis	375:125:125 g/l	EC	08413

Uses

Brown rust in **spring barley, spring wheat, winter barley, winter wheat**. Net blotch in **spring barley, winter barley**. Powdery mildew in **spring barley, spring wheat, winter barley, winter wheat**. Rhynchosporium in **spring barley, winter barley**. Septoria diseases in **spring wheat, winter wheat**. Yellow rust in **spring barley, spring wheat, winter barley, winter wheat**.

Efficacy
- Spray at first signs of disease and before infection spreads to new growth
- Control of established infections may be less good
- Repeat treatment 3-5 wk later if re-infection occurs or disease pressure remains high

Crop safety/Restrictions
- Maximum number of treatments equivalent to two full dose treatments

Special precautions/Environmental safety
- Irritating to skin. Risk of serious damage to eyes
- May cause sensitization by skin contact
- Harmful to bees. Do not apply to crops in flower or to those in which bees are actively foraging. Do not apply when flowering weeds are present
- Extremely dangerous to fish or other aquatic life. Do not contaminate surface waters or ditches with chemical or used container

Personal protective equipment/Label precautions
- A, C, H
- R04b, R04d, R04e, U05a, U09b, U20a, C03, E01, E12e, E13a, E26, E30a, E31b

Latest application/Harvest interval
- Before beginning of anthesis (GS 61) for wheat; up to and including ear emergence (GS 59) for barley

Approval
- Accepted by BLRA for use on malting barley

Maximum Residue Level (mg residue/kg food)
- see propiconazole entry

268   fenpropidin + tebuconazole

A broad spectrum fungicide mixture for cereals

Products

Monicle	Bayer	300:200 g/l	EC	07375

Uses

Botrytis in **wheat**. Brown rust in **barley, wheat**. Fusarium ear blight in **wheat**. Glume blotch in **wheat**. Net blotch in **barley**. Powdery mildew in **barley, wheat**. Rhynchosporium in **barley**. Septoria leaf spot in **wheat**. Sooty moulds in **wheat**. Yellow rust in **barley, wheat**.

Efficacy
- Best disease control and yield benefit obtained when applied at early stage of disease development before infection spreads to new growth
- To protect the flag leaf and ear from Septoria diseases apply from flag leaf emergence to ear fully emerged (GS 37-59)
- Applications once foliar symptoms of *Septoria tritici* are already present on upper leaves will be less effective

Crop safety/Restrictions
- Maximum total dose 2.5 l/ha per crop

- Occasional slight temporary leaf speckling may occur on wheat but this has not been shown to reduce yield response or disease control
- Do not treat durum wheat

Special precautions/Environmental safety
- Harmful if swallowed and in contact with skin
- Irritating to skin
- Risk of serious damage to eyes
- Extremely dangerous to fish or other aquatic life. Do not contaminate surface waters or ditches with chemical or used container

Personal protective equipment/Label precautions
- A, C, H
- M03, R03a, R03c, R04b, R04d, U02, U04a, U05a, U11, U20a, C03; C02 (5 wk); E01, E13a, E30a, E31b, E34

Latest application/Harvest interval
- Before grain milky ripe stage (GS 73) for winter wheat; up to and including ear emergence just complete (GS 59) for barley and spring wheat
- HI 5 wk for crops for human or animal consumption

Approval
- Accepted by BLRA for use on malting barley

269　fenpropimorph

A contact and systemic morpholine fungicide

See also azoxystrobin + fenpropimorph
epoxiconazole + fenpropimorph
epoxiconazole + fenpropimorph + kresoxim-methyl
fenpropidin + fenpropimorph

Products

1	Aura	Novartis	750 g/l	EC	08388
2	Corbel	BASF	750 g/l	EC	00578
3	Mistral	Novartis	750 g/l	EC	08425
4	Standon Fenpropimorph 750	Standon	750 g/l	EC	08965

Uses

Brown rust in **barley, triticale, wheat** [1-4]. Brown rust in **field beans** [3]. Crown rot in **carrots** *(off-label)*, **horseradish** *(off-label)*, **parsley root** *(off-label)*, **parsnips** *(off-label)*, **salsify** *(off-label)* [2]. Powdery mildew in **barley, oats, rye, wheat** [1-4]. Powdery mildew in **hops** *(off-label)*, **soft fruit** *(off-label)* [2]. Rhynchosporium in **barley** [1-4]. Rust in **leeks** [1-4]. Rust in **field beans** [1, 2, 4]. Rust in **red beet** *(off-label)*, **sugar beet seed crops** *(off-label)* [2]. Yellow rust in **barley, triticale, wheat** [1-4].

Efficacy
- On cereals spray at start of disease attack. See labels for recommended tank mixes. Follow-up treatments may be needed if disease pressure remains high
- On field beans a second application may be needed after 2-3 wk
- Product rainfast after 2 h

Crop safety/Restrictions
- Maximum number of treatments 2 per crop for spring cereals and field beans; 3 per crop at 1.0 l/ha or 4 at 0.75 l/ha (with 1 treatment applied in autumn) for winter cereals; 6 per crop for leeks
- Scorch may occur if applied during frosty weather or in high temperatures
- Consult processors before using on crops for processing
- An interval of at least 10 d on field beans and 14 d on leeks must elapse between applications

Special precautions/Environmental safety
- Harmful by inhalation [1]
- Irritating to eyes and skin
- Irritating to respiratory system [1]

- Dangerous to fish or other aquatic life. Do not contaminate surface waters or ditches with chemical or used container

Personal protective equipment/Label precautions
- A, C [1-4]
- M03 [1, 3]; R04a, R04b [1-4]; R03b, R04c [1]; U05a, U08, U14, U15, U19, U20a, C03 [1-4]; E31c [2]; E01, E13b, E30a [1-4]; E26 [2-4]; E34 [1, 3]; E31b [1, 3, 4]; E29 [3]

Latest application/Harvest interval
- HI leeks 3 wk; cereals, field beans 5 wk

Approval
- Off-label approval unlimited for use in red beet (OLA 1246/94)[2]; unlimited for use on cane and soft fruit (OLA 0787/95)[2]; unlimited for use in sugar beet seed crops (OLA 1807/96)[2]; to 2001 for use in hops (OLA 2078/96)[2]; unlimited for use on outdoor carrot, parsnip, parsley root, salsify, horseradish (OLA 2483/96)[2]
- Accepted by BLRA for use on malting barley and hops

270 fenpropimorph + flusilazole

A broad-spectrum eradicant and protectant fungicide mixture for cereals

Products

Colstar	DuPont	375:160 g/l	EC	06783

Uses

Brown rust in **barley, winter wheat**. Net blotch in **barley**. Powdery mildew in **barley, winter wheat**. Rhynchosporium in **barley**. Septoria diseases in **winter wheat**. Yellow rust in **barley, winter wheat**.

Efficacy
- Disease control is more effective if treatment made at an early stage of disease development
- Treat winter cereals in spring or early summer before diseases spread to new growth
- Spring barley should be treated when diseases are first evident
- Treatment may be repeated after 3-4 wk if necessary

Crop safety/Restrictions
- Maximum number of treatments (including other products containing flusilazole) 3 per crop (winter wheat), 2 per crop (winter or spring barley)
- Do not apply to crops under stress
- Do not apply during frosty weather

Special precautions/Environmental safety
- Irritating to eyes
- Dangerous to fish or other aquatic life. Do not contaminate surface waters or ditches with chemical or used container

Personal protective equipment/Label precautions
- A, C
- R04a, U05a, U11, U19, U20b, C03, E01, E13b, E26, E30a, E31b

Latest application/Harvest interval
- Before beginning of anthesis (GS 60) for winter wheat; up to and including completion of ear emergence (GS 59) for barley

Approval
- Accepted by BLRA for use on malting barley

271 fenpropimorph + kresoxim-methyl

A protectant and systemic fungicide mixture for cereals

Products

1	Ensign	BASF	300:150 g/l	SE	08362
2	Greencrop Monsoon	Greencrop	300:150 g/l	SE	09573
3	Landgold Strobilurin KF	Landgold	300:150 g/l	SE	09196
4	Standon Kresoxim FM	Standon	300:150 g/l	SE	08922

Uses

Powdery mildew in *spring barley, spring oats, spring rye, spring wheat, triticale, winter barley, winter oats, winter rye, winter wheat*. Rhynchosporium in *spring barley, spring rye, triticale, winter barley, winter rye*. Septoria in *spring rye* (reduction), *triticale* (reduction), *winter rye* (reduction), *winter wheat* (reduction).

Efficacy

- For best results spray at the start of foliar disease attack and repeat if infection conditions persist

Crop safety/Restrictions

- Maximum total dose for all crops 1.4 litres product/ha

Special precautions/Environmental safety

- Irritant. May cause sensitization by skin contact
- Dangerous to fish or other aquatic life. Do not contaminate surface waters or ditches with chemical or used container

Personal protective equipment/Label precautions

- A [1-4]
- R04, R04e, U05a, U14, U20b, C03 [1-4]; E22b [1]; E31c [1, 3, 4]; E01, E13b, E26, E30a [1-4]; E31b [2]

Latest application/Harvest interval

- Completion of ear emergence (GS 59) for barley and oats; completion of flowering (GS 69) for wheat, rye and triticale

Approval

- Kresoxim-methyl included in Annex I under EC Directive 91/414
- Accepted by BLRA for use on malting barley

---

272 fenpropimorph + prochloraz

A fungicide mixture for late season disease control in cereals

Products

Sprint HF	AgrEvo	375:225 g/l	EC	07292

Uses

Brown rust in *barley, spring wheat, winter rye, winter wheat*. Eyespot in *barley, spring wheat, winter rye, winter wheat*. Net blotch in *barley*. Powdery mildew in *barley, spring wheat, winter rye, winter wheat*. Rhynchosporium in *barley, winter rye*. Septoria in *spring wheat, winter rye, winter wheat*. Yellow rust in *barley, spring wheat, winter rye, winter wheat*.

Efficacy

- In wheat, if disease present, spray as soon as ligule of flag leaf visible (GS 39). If no disease, delay until first signs appear or use as protectant treatment before flowering
- In barley spray when disease appears on new growth or as protective spray when flag leaf ligule just visible (GS 39)

Crop safety/Restrictions

- Maximum number of treatments 2 per crop
- Do not apply to crops suffering from stress

Special precautions/Environmental safety

- Harmful in contact with skin
- Dangerous to fish or other aquatic life. Do not contaminate surface waters or ditches with chemical or used container

Personal protective equipment/Label precautions

- A, C
- M03, R03a, U05a, U08, U20b, C03; C02 (5 wk); E01, E13b, E26, E30a, E31b, E34

Latest application/Harvest interval

- Up to and including ear emergence just complete (GS 59).
- HI 5 wk

Approval

- Accepted by BLRA for use on malting barley

273 fenpropimorph + propiconazole

A fungicide mixture for control of leaf diseases of cereals

Products

1 Belvedere	Makhteshim	375:125 g/l	EC	08084
2 Decade	Novartis	375:125 g/l	EC	08402
3 Glint	Novartis	375:125 g/l	EC	08414
4 Mantle	Novartis	375:125 g/l	EC	08424

Uses

Brown rust in *winter barley, winter wheat*. Mildew in *spring barley, winter barley, winter wheat*. Net blotch in *spring barley, winter barley*. Rhynchosporium in *spring barley, winter barley*. Septoria in *winter wheat*. Sooty moulds in *winter wheat*. Yellow rust in *winter barley, winter wheat*.

Efficacy
- Apply at start of disease attack
- Spray winter wheat between GS 30-32 against early Septoria, mildew and rusts, at flag leaf against late foliar and ear disease, and repeat 28 d later if necessary [2, 3]
- Spray spring barley immediately disease appears in crop
- Spray winter barley at risk from foliar disease at GS 30-31, and repeat at flag leaf (GS 39) if necessary, or spray at awn emergence (GS 49) [2, 3]. See label for details

Crop safety/Restrictions
- Maximum number of treatments 2 per crop [2, 3]; 2 per crop for spring wheat and barley, 3 per crop for winter crops (1 in autumn at least 3 mth before 2 in spring/summer) [4]
- If used in programme with other products containing fenpropimorph observe maximum total doses per season permitted for wheat and barley

Special precautions/Environmental safety
- Harmful by inhalation. Irritating to eyes, skin and respiratory system
- Dangerous to fish or other aquatic life. Do not contaminate surface waters or ditches with chemical or used container

Personal protective equipment/Label precautions
- A, C
- M03, R03b, R04a, R04b, R04c, U05a, U08, U14, U15, U19, U20a, C03, E01, E13b, E30a, E31b, E34

Latest application/Harvest interval
- HI 5 wk

Approval
- Accepted by BLRA for use on malting barley

Maximum Residue Level (mg residue/kg food)
- see propiconazole entry

274 fenpropimorph + quinoxyfen

A systemic fungicide mixture for cereals

Products

Orka	Dow	250:66.7 g/l	EW	08879

Uses

Powdery mildew in *durum wheat, rye, spring barley, spring oats, spring wheat, triticale, winter barley, winter oats, winter wheat*.

Efficacy
- For best results treat at early stage of disease development before infection spreads to new crop growth. Further treatment may be necessary if disease pressure remains high
- For control of established infections and broad spectrum disease control use in tank mixtures. See label
- Product rainfast after 1 h
- Systemic activity may be reduced in severe drought

Crop safety/Restrictions
- Maximum total dose 3.0 l product per ha
- Apply only in the spring from mid-tillering stage (GS 25)
- Crop scorch may occur when treatment made in high temperatures

Special precautions/Environmental safety
- Irritant. May cause sensitization by skin contact
- Dangerous to fish or other aquatic life. Do not contaminate surface waters or ditches with chemical or used container
- LERAP Category B

Personal protective equipment/Label precautions
- A, C, H
- R04, R04e, U05a, U14, C03, E01, E13b, E16a, E16b, E34

Latest application/Harvest interval
- First awns visible (GS 49)

Approval
- Accepted by BLRA for use on malting barley

---

275   fenpyroximate

A mitochondrial electron transport inhibitor acaricide for apples

Products

Sequel	Hortichem	51.3 g/l	SC	09886

Uses

Fruit tree red spider mite in *apples*.

Efficacy
- Kills motile stages of fruit tree red spider mite. Best results achieved if applied in warm weather
- Total spray cover of trees essential. Use higher volumes for large trees
- Apply when majority of winter eggs have hatched

Crop safety/Restrictions
- Maximum number of treatments 1 per yr
- Other mitochondrial electron transport inhibitor (METI) acaricides should not be applied to the same crop in the same calendar yr either separately or in mixture
- Do not apply when apple crops or pollinator are in flower
- Consult processor before use on crops for processing

Special precautions/Environmental safety
- Irritant. Risk of serious damage to eyes
- Dangerous to fish or other aquatic life. Do not contaminate surface waters or ditches with chemical or used container
- Do not allow direct spray from air assisted sprayers to fall within 38 m of surface waters or ditches. Direct spray away from water

Personal protective equipment/Label precautions
- A, C, H
- R04, R04d, R04e, U05a, U08, U14, U20b, C03, E01, E13b, E22b, E26, E30a, E31c; E17 (38 m)

Latest application/Harvest interval
- HI 2 wk

---

276   fentin acetate

An organotin fungicide available only in mixtures

277   fentin acetate + maneb

A curative and protectant fungicide for use in potatoes

Products

| Brestan 60 SP | Aventis | 54:16% w/w | WB | 07305 |

Uses

Blight in **potatoes**.

Efficacy

- Commence spraying before infection occurs, before haulm meets across the rows or at first blight warning, whichever occurs first
- Repeat at 7-14 d intervals until growth ceases or haulm burnt off
- Do not spray if rain imminent

Crop safety/Restrictions

- Maximum number of treatments 6 per crop

Special precautions/Environmental safety

- Fentin acetate is subject to the Poisons Rules 1982 and the Poisons Act 1972. See notes in Section 1
- Harmful if swallowed. Irritating to eyes, skin and respiratory system
- Keep in original container, tightly closed, in a safe place, under lock and key
- Harmful to livestock
- Harmful to fish or other aquatic life. Do not contaminate surface waters or ditches with chemical or used container

Personal protective equipment/Label precautions

- A, C, H
- M03, R03c, R04a, R04b, R04c, U02, U04a, U05a, U08, U13, U19, U20b, C03, E01, E06b, E13c, E26, E30a, E32a, E34

Withholding period

- Keep all livestock out of treated areas for at least 1 wk

Latest application/Harvest interval

- HI 7 d

Maximum Residue Level (mg residue/kg food)

- see fentin hydroxide and maneb entries

278   fentin hydroxide

A curative and protectant organotin fungicide

Products

1	Barclay Fentin Flow	Barclay	480 g/l	SC	07914
2	Barclay Fentin Flow 532	Barclay	532 g/l	SC	09434
3	Farmatin 560	Aventis	532 g/l	SC	09877
4	Farmatin 560	AgrEvo	532 g/l	SC	07320
5	Greencrop Rosette	Greencrop	480 g/l	SC	09648
6	Keytin	Chiltern	532 g/l	SC	08894
7	MSS Flotin 480	Nufarm Whyte	480 g/l	SC	07616
8	Super-Tin 80 WP	PBI	80% w/w	WB	07605

Uses

Blight in **potatoes**.

Efficacy

- Apply to potatoes before haulm meets across rows or on receipt of blight warning and repeat at 7-14 d intervals throughout season. Late sprays protect against tuber blight
- Best results achieved by using for 2 final sprays of blight control programme. Allow at least 14 d after complete death of haulm before lifting unless for immediate sale

Crop safety/Restrictions

- Maximum number of treatments - see product labels

- Drought stress [2, 4, 6] or mixtures with emulsifiable concentrates [7] may cause localised leaf spotting
- Product may sometimes harden the foliage of young potato plants [1]
- A minimum interval of 7 d (10 d [4]) between applications must be observed

Special precautions/Environmental safety
- Fentin hydroxide is subject to the Poisons Rules 1982 and the Poisons Act 1972. See notes in Section 1
- Very toxic by inhalation [8]
- Harmful if swallowed [1, 5, 7] and in contact with skin [2-4, 6, 8]
- Irritating to skin [8] and eyes [1-4, 6, 7]
- Risk of serious damage to eyes [8]
- Irritating to respiratory system [1, 5, 7, 8]
- Keep in original container, tightly closed, in a safe place, under lock and key [1, 2, 7, 8]
- Harmful to livestock
- Harmful (dangerous [8]) to fish or other aquatic life. Do not contaminate surface waters or ditches with chemical or used container

Personal protective equipment/Label precautions
- A, C, H [1-8]; M [1, 5]; K [2, 5, 6]
- M03 [1-7]; M04 [8]; R03c, R04b [1-8]; R03a [2-4, 6, 8]; R04a [1-7]; R01b, R04d [8]; R04c [1, 5, 7, 8]; U02, U04a, U05a, U08, U19, U20a [1-8]; U13, U14 [2-8]; U15 [2-4, 6]; U22 [8]; C03 [1-8]; C02 [6] (7 d); E01 [1-8]; E06b [1-8] (7 d); E13c, E34 [1-7]; E30a [3, 4, 6]; E32a [2-4]; E13b [8]; E31b [1, 6-8]; E30b [1, 2, 5, 7, 8]; E26 [1, 2, 5, 7]; E31c [5]

Withholding period
- Keep all livestock out of treated areas for at least 1 wk

Latest application/Harvest interval
- HI potatoes 0 d

Maximum Residue Level (mg residue/kg food)
- hops 0.5; potatoes, tea 0.1; all other produce 0.05

---

279   fenuron

A urea herbicide available only in mixtures

See also chlorpropham + fenuron

---

280   ferbam

A dithiocarbamate fungicide available only in mixture

---

281   ferbam + maneb + zineb

A protectant dithiocarbamate complex fungicide

Products

Trimanzone WP	Intracrop	10:65:10% w/w	WP	05860

Uses

Blight in **potatoes**. Fungus diseases in **leeks** *(off-label)*.

Efficacy
- For blight control apply before haulm meets across the rows or as soon as blight warning received, whichever occurs first. Repeat every 10-14 d

Crop safety/Restrictions
- Observe minimum period of 10 d between applications

Special precautions/Environmental safety
- Irritating to skin, eyes and respiratory system

Personal protective equipment/Label precautions
- A, C
- R04a, R04b, R04c, U05a, U08, U20a, C02, C03, E01, E15, E30a, E32a

Latest application/Harvest interval
- HI potatoes, leeks 7 d

Approval
- Off-label approval unlimited for use on leeks (OLA 0599/92)

Maximum Residue Level (mg residue/kg food)
- see maneb entry

---

282   ferrous sulphate

A herbicide/fertilizer combination for moss control in turf

See also dicamba + dichlorprop + ferrous sulphate + MCPA
dichlorophen + ferrous sulphate

Products

1	Elliott's Lawn Sand	Elliott	17% w/w	SA	04860
2	Elliott's Mosskiller	Elliott	45% w/w	GR	04909
3	Greenmaster Autumn	Scotts	18.1% w/w	GR	07508
4	Greenmaster Mosskiller	Scotts	26.8% w/w	GR	07509
5	Maxicrop Moss Killer & Conditioner	Maxicrop	16.4% w/w	SL	04635
6	SHL Lawn Sand	Sinclair	10% w/w	SA	05254
7	Taylors Lawn Sand	Rigby Taylor	7.35% w/w	SA	04451
8	Vitax Microgran 2	Vitax	20% w/w	MG	04541
9	Vitax Turf Tonic	Vitax	15% w/w	SA	04354

Uses

Mosses in **turf**.

Efficacy
- Apply autumn treatment from Sep onward but not when heavy rain expected or in frosty weather [3]
- Apply from Mar to Sep except during drought or when soil frozen [4]
- For best results apply when light showers expected, mow 3 d before treatment and do not mow for 3-4 d afterwards
- Do not walk on treated areas until well watered and water after 2 d if no rain
- Rake out dead moss thoroughly 7-14 d after treatment. See label
- Fertilizer component of autumn treatment encourages root growth, that of mosskiller formulations promotes tillering

Crop safety/Restrictions
- Maximum number of treatments - see labels
- If spilt on paving, concrete, clothes etc brush off immediately to avoid discolouration
- Observe label restrictions for interval before cutting after treatment

Special precautions/Environmental safety
- Harmful to fish or other aquatic life. Do not contaminate surface waters or ditches with chemical or used container [3, 4]

Personal protective equipment/Label precautions
- U20b [3, 4, 8, 9]; U20c [3-5, 7]; U20a [1, 2, 6]; E01 [9]; E15 [1, 2, 5, 7-9]; E30a [1-9]; E32a [1-4, 6-9]; E26, E31a [5]; E13c [3, 4]

---

283   flamprop-M-isopropyl

A translocated, post-emergence arylalanine wild-oat herbicide

Products

1	Commando	BASF	200 g/l	EC	10215
2	Commando	Cyanamid	200 g/l	EC	07005

Uses

Wild oats in *durum wheat, spring barley, spring rye, spring wheat, triticale, winter barley, winter rye, winter wheat*.

Efficacy

- Must be applied when crop and weeds growing actively under conditions of warm, moist days and warm nights
- Timing determined mainly by crop growth stage. Best control at later stages of wild oats but not after weeds visible above crop
- Best results achieved before 3rd node detectable stage (GS 33) in barley, or 4th node detectable (GS 34) in wheat
- Good spray coverage and retention essential. Spray rainfast after 2 h
- Addition of Swirl adjuvant oil recommended for use on winter wheat and winter barley
- Do not treat thin, open crops
- See label for recommended tank mixes and sequential treatments

Crop safety/Restrictions

- Maximum number of treatments 1 per crop
- May be used on crops undersown with ryegrass and clover
- Do not use on crops under stress or during periods of high temperature
- Do not roll or harrow within 7 d of spraying
- Do not mix with any herbicide other than fluroxypyr

Special precautions/Environmental safety

- Irritating to eyes and skin
- Flammable
- Do not use straw from barley treated after stage GS 33 or wheat treated after stage GS 34 as feed or bedding for animals
- Harmful to fish or other aquatic life. Do not contaminate surface waters or ditches with chemical or used container

Personal protective equipment/Label precautions

- A, C
- R04a, R04b, R07d, U02, U04a, U05a, U08, U19, U20a, C03, E01, E13c, E26, E30a, E31b, E34

Latest application/Harvest interval

- Before flag leaf sheath opening (GS 47) for wheat; before first awns visible (GS 49) for barley; before 3rd node detectable (GS 33) for rye and triticale; before 4th node detectable (GS 34) for durum wheat

Approval

- Approved for aerial application to cereals. See notes in Section 1
- Accepted by BLRA for use on malting barley

284　fluazifop-P-butyl

A phenoxypropionic acid grass herbicide for broadleaved crops

Products

Fusilade 250 EW	Zeneca	250 g/l	EW	06531

Uses

Annual grasses in *blackcurrants, broad beans* (off-label), *carrots, chinese cabbage* (off-label), *collards* (off-label), *combining peas, farm forestry, field beans, field margins, flax, flowers* (off-label), *fodder beet, garlic* (off-label), *gooseberries, hops, kale* (off-label), *kale* (stockfeed only), *linseed, linseed/flax for industrial use, lucerne* (off-label), *navy beans* (off-label), *oilseed rape for industrial use, onions, ornamentals* (off-label), *parsnips* (off-label), *raspberries, red beet* (off-label), *spring oilseed rape, strawberries, sugar beet, swedes* (off-label), *swedes* (stockfeed only), *turnips* (off-label), *turnips* (stockfeed only), *vining peas, winter oilseed rape.* Barley cover crops in *row crops.* Barren brome in *field margins.* Blackgrass in *field beans, hops.* Green cover in *land temporarily removed from production, setaside.* Perennial grasses in *blackcurrants, broad beans* (off-label), *carrots, chinese cabbage* (off-label), *collards* (off-label), *combining peas, farm forestry, field beans, flax, flowers* (off-label), *fodder beet, garlic* (off-label), *gooseberries, hops, kale* (off-label), *kale (stockfeed only)*, *linseed, linseed/flax for industrial use, navy beans (off-label)*, *oilseed rape for industrial use, onions, ornamentals (off-label)*, *parsnips (off-label)*, *raspberries, red beet (off-label)*, *spring oilseed rape, strawberries, sugar beet, swedes (off-label)*, *swedes (stockfeed only)*, *turnips (off-label)*, *turnips (stockfeed only)*, *vining peas, winter oilseed rape*. Volunteer cereals in *blackcurrants, carrots, combining peas, farm forestry, field beans, field margins, flax, fodder beet, gooseberries, hops, kale (stockfeed only)*, *linseed, linseed/flax for industrial use, oilseed rape for industrial use, onions, raspberries, spring oilseed rape, strawberries, sugar beet, swedes (stockfeed only)*, *turnips (stockfeed only)*, *vining peas, winter oilseed rape*. Wild oats in *blackcurrants, carrots, combining peas, farm forestry, field beans, field margins, flax, fodder beet, gooseberries, hops, kale (stockfeed only)*, *linseed, linseed/flax for industrial use, oilseed rape for industrial use, onions, raspberries, spring oilseed rape, strawberries, sugar beet, swedes (stockfeed only)*, *turnips (stockfeed only)*, *vining peas, winter oilseed rape*.

Efficacy
- Best results achieved by application when weed growth active under warm conditions with adequate soil moisture. Agral or other specified adjuvant must always be added to spray. See label
- Spray weeds from 2-expanded leaf stage to fully tillered, couch from 4 leaves when majority of shoots have emerged, with a second application if necessary
- Control may be reduced under dry conditions. Do not cultivate for 2 wk after spraying couch
- Annual meadow grass is not controlled
- May also be used to remove grass cover crops

Crop safety/Restrictions
- Maximum number of treatments normally 2 per crop for sugar and fodder beet, winter oilseed rape (including crops for industrial use); 2 per yr for strawberries and farm forestry; 1 per crop or yr for spring oilseed rape (including crops for industrial use) and other crops
- Apply to sugar and fodder beet from 1-true leaf to 50% ground cover
- Apply to winter oilseed rape from 1-true leaf to established plant stage
- Apply to spring oilseed rape from 1-true leaf but before 5-true leaves
- Apply in fruit crops after harvest. See label for timing details on other crops
- Before using on onions or peas use crystal violet test to check that leaf wax is sufficient
- Do not sow cereals for at least 8 wk after application of high rate or 2 wk after low rate
- Do not apply through CDA sprayer, with hand-held equipment or from air
- Avoid treatment before spring growth has hardened or when buds opening

Special precautions/Environmental safety
- Irritating to skin
- Harmful to fish or other aquatic life. Do not contaminate surface waters or ditches with chemical or used container
- Do not apply by hand-held equipment
- Do not treat bush and cane fruit or hops between flowering and harvest
- Oilseed rape, linseed and flax for industrial use must not be harvested for human or animal consumption nor grazed
- Do not use for forestry establishment on land not previously under arable cultivation or improved grassland
- Treated vegetation in field margins, setaside etc, must not be grazed or harvested for human or animal consumption and unprotected persons must be kept out of treated areas for at least 24 h

Personal protective equipment/Label precautions
- A, C, H
- M03, R04b, U05a, U08, U20a, C03, E01, E13c, E30a, E31b, E34

Latest application/Harvest interval
- Before 5 leaf stage for spring oilseed rape; before flowering for blackcurrants, gooseberries, hops, raspberries, strawberries; before flower buds visible for field beans, peas, linseed, flax, oilseed rape; 2 wk before sowing cereals or grass for field margins, land temporarily removed from production, spring oilseed rape for industrial use, grass crops for farm forestry
- HI beet crops, carrots, kale, 8 wk; onions 4 wk

Approval
• Off-label approval unlimited for use on ornamentals, flowers, garlic, broad beans, dry harvested beans, parsnips, red beet, kale, Chinese cabbage, spring greens, swedes, turnips (OLA 2768/96); unlimited for use on lucerne (OLA 2699/98)
• Accepted by BLRA for use on hops

285   fluazinam

A pyridinamine fungicide for use in potatoes

Products

1	Barclay Cobbler	Barclay	500 g/l	SC	08349
2	Landgold Fluazinam	Landgold	500 g/l	SC	08060
3	Shirlan	Zeneca	500 g/l	SC	07091
4	Shirlan Programme	Zeneca	500 g/l	SC	08761
5	Standon Fluazinam 500	Standon	500 g/l	SC	08670

Uses

Blight in **potatoes**.

Efficacy
• Commence treatment at the first blight risk warning (before blight enters the crop). Products are rainfast within 1 h
• In the absence of a warning, treatment should start before foliage of adjacent plants meets in the rows
• Spray at 7-14 d intervals depending on severity of risk (see label)
• Ensure complete coverage of the foliage and stems, increasing volume as haulm growth progresses, in dense crops and if blight risk increases

Crop safety/Restrictions
• Maximum number of treatments 10 per crop

Special precautions/Environmental safety
• Irritant. May cause sensitization by skin contact
• Dangerous to fish or other aquatic life. Do not contaminate surface waters or ditches with chemical or used container
• LERAP Category B
• Do not use with hand-held sprayers

Personal protective equipment/Label precautions
• A, C, H [1-5]
• M03 [3, 4]; R04, R04e, U05a [1-5]; U20a [2-4]; U02, U04a, U08 [1, 3-5]; U20b [1, 5]; C03, E01, E13b, E16a, E26, E30a [1-5]; E31c, E34 [3, 4]; E31b [1, 2, 5]

Latest application/Harvest interval
• HI 7 d before harvest

286   fludioxonil

A cyanopyrrole fungicide seed treatment for wheat and barley

Products

Beret Gold	Novartis	25 g/l	FS	08390

Uses

Bunt in **winter wheat** *(seed treatment)*. Covered smut in **spring barley** *(seed treatment)*, **winter barley** *(seed treatment)*. Foot rot in **spring barley** *(seed treatment)*, **winter barley** *(seed treatment)*. Fusarium foot rot and seedling blight in **spring barley** *(seed treatment)*, **winter barley** *(seed treatment)*, **winter wheat** *(seed treatment)*. Leaf stripe in **spring barley** *(seed treatment - reduction)*, **winter barley** *(seed treatment - reduction)*. Snow mould in **spring barley** *(seed treatment)*, **winter barley** *(seed treatment)*, **winter wheat** *(seed treatment)*.

Efficacy
- Apply direct to seed using conventional seed treatment equipment. Continuous flow treaters should be calibrated using product before use
- Effective against benzimidazole-resistant strains of *Fusarium nivale*
- Product may reduce flow rate of seed through drill. Recalibrate with treated seed before drilling

Crop safety/Restrictions
- Maximum number of treatments 1 per seed batch
- Do not apply to cracked, split or sprouted seed
- Sow treated seed within 6 mth

Special precautions/Environmental safety
- Harmful to fish or other aquatic life. Do not contaminate surface waters or ditches with chemical or used container
- Do not use treated seed as food or feed
- Treated seed harmful to game and wildlife

Personal protective equipment/Label precautions
- A, H
- U20a, E13c, E26, E32a, S01, S02, S04b, S05, S06, S07

Latest application/Harvest interval
- Before drilling

Approval
- Accepted by BLRA for use as a seed treatment on malting barley

---

287   flupyrsulfuron-methyl

A sulfonylurea herbicide for winter wheat

See also carfentrazone-ethyl + flupyrsulfuron-methyl

Products

Lexus 50 DF	DuPont	50% w/w	WG	09026

Uses

Annual dicotyledons in **winter wheat**. Blackgrass in **winter wheat**.

Efficacy
- Best results achieved from applications made in good growing conditions
- Good spray cover of weeds must be obtained
- Growth of weeds is inhibited within hours of treatment but visible symptoms may not be apparent for up to 4 wk
- Product has moderate residual life in soil. Under normal moisture conditions susceptible weeds germinating soon after treatment will be controlled
- Product may be used on all soil types but residual activity and weed control is reduced on highly alkaline soils
- Apply in the spring from the 3-leaf stage on all crops

Crop safety/Restrictions
- Maximum number of treatments 1 per crop
- Treat from 2 leaf stage of crop
- Do not use on wheat undersown with grasses or legumes, or any other broad-leaved crop
- Do not apply within 7 d of rolling
- Do not treat any crop suffering from drought, waterlogging, pest or disease attack, nutrient deficiency, or any other stress factors
- Slight chlorosis and stunting may occur in certain conditions. Recovery is rapid and yield not affected
- After treatment other 'ALS-inhibiting' herbicides should not be used on the same crop except metsulfuron-methyl, thifensulfuron-methyl or tribenuron-methyl (alone or in mixtures)
- Only cereals, oilseed rape, field beans, clover or grass may be sown in the yr of harvest of a treated crop
- In the event of crop failure only winter or spring wheat may be sown within 3 mth of treatment. Land should be ploughed and cultivated to 15 cm minimum before resowing

---

Special precautions/Environmental safety
- Extremely dangerous to fish or other aquatic life. Do not contaminate surface waters or ditches with chemical or used container
- Take extreme care to avoid drift onto broad-leaved plants outside the target area or onto ponds waterways or ditches, or onto land intended for cropping
- Spraying equipment should not be drained or flushed onto land planted, or to be planted, with trees or crops other than cereals and should be thoroughly cleansed after use - see label for instructions

Personal protective equipment/Label precautions
- U08, U19, U20b, E13a, E30a, E32a

Latest application/Harvest interval
- Before 1st node detectable (GS 31)

288   flupyrsulfuron-methyl + metsulfuron-methyl

A sulfonylurea herbicide mixture for winter wheat

Products

1 Lexus XPE WSB	DuPont	33.3:16.7% w/w	WB	08542
2 Standon Flupyrsulfuron MM	Standon	33.3:16.7% w/w	WG	09098

Uses

Annual dicotyledons in *winter wheat*. Blackgrass in *winter wheat*.

Efficacy
- Best results obtained when applied to small actively growing weeds
- Good spray cover of weeds must be obtained
- Increased degradation of active ingredient in high soil temperatures reduces residual activity
- Product has moderate residual life in soil. Under normal moisture conditions susceptible weeds germinating soon after treatment will be controlled
- Product may be used on all soil types but residual activity and weed control is reduced on highly alkaline soils
- Blackgrass should be treated from 1 leaf up to mid-tillering. Strains of blackgrass resistant to other herbicides may not be controlled

Crop safety/Restrictions
- Maximum number of treatments 1 per crop
- Must only be applied in spring after 1 Feb, from 2 leaf stage of crop
- Do not use on wheat undersown with grasses or legumes, or any other broad-leaved crop
- Do not apply within 7 d of rolling
- Do not treat any crop suffering from drought, waterlogging, pest or disease attack, nutrient deficiency, or any other stress factors
- Slight chlorosis and stunting may occur in certain conditions. Recovery is rapid and yield not affected
- Do not apply in sequence, or in tank mixture, with any product containing any other sulfonyl urea
- Only cereals, oilseed rape, field beans, clover or grass may be sown in the yr of harvest of a treated crop
- In the event of crop failure only winter wheat may be sown within 3 mth of treatment. Land should be ploughed and cultivated to 15 cm minimum before resowing

Special precautions/Environmental safety
- Extremely dangerous to fish or other aquatic life. Do not contaminate surface waters or ditches with chemical or used container
- LERAP Category B
- Take extreme care to avoid drift onto broad-leaved plants outside the target area or onto ponds waterways or ditches, or onto land intended for cropping
- Spraying equipment should not be drained or flushed onto land planted, or to be planted, with trees or crops other than cereals and should be thoroughly cleansed after use - see label for instructions

Personal protective equipment/Label precautions
- U08, U19, U20b, E13a, E16a, E16b, E30a, E32a

Latest application/Harvest interval
- Beginning of stem extension (GS 30)

Approval
- Metsulfuron-methyl included in Annex I under EC Directive 91/414

---

289 flupyrsulfuron-methyl + thifensulfuron-methyl

A sulfonylurea herbicide mixture for winter wheat

Products

Lexus Millenium	DuPont	10:40% w/w	WG	09206

Uses

Annual dicotyledons in *winter wheat*. Blackgrass in *winter wheat*.

Efficacy
- Best results obtained when applied to small actively growing weeds
- Good spray cover of weeds must be obtained
- Increased degradation of active ingredient in high soil temperatures reduces residual activity
- Product has moderate residual life in soil. Under normal moisture conditions susceptible weeds germinating soon after treatment will be controlled
- Product may be used on all soil types but residual activity and weed control is reduced on highly alkaline soils
- Blackgrass should be treated from 1 leaf up to mid-tillering. Strains of blackgrass resistant to other herbicides may not be controlled

Crop safety/Restrictions
- Maximum number of treatments 1 per crop
- Do not treat winter wheat at or after 1st node detectable stage (GS 31)
- Do not use on wheat undersown with grasses or legumes, or any other broad-leaved crop
- Do not apply within 7 d of rolling
- Do not treat any crop suffering from drought, waterlogging, pest or disease attack, nutrient deficiency, or any other stress factors
- Slight chlorosis and stunting may occur in certain conditions. Recovery is rapid and yield not affected
- Do not apply to a crop already treated with any other product containing thifensulfuron-methyl. Only specified products containing metsulfuron-methyl or tribenuron-methyl may follow treatment. See label
- Only cereals, oilseed rape, field beans or grass may be sown in the yr of harvest of a treated crop. Any crop may be sown the following spring
- In the event of crop failure only winter wheat may be sown before normal harvest date. Land should be ploughed and cultivated to 15 cm minimum before resowing

Special precautions/Environmental safety
- May cause sensitization by skin contact
- Extremely dangerous to fish or other aquatic life. Do not contaminate surface waters or ditches with chemical or used container
- LERAP Category B

Personal protective equipment/Label precautions
- A, H
- R04e, U05a, U08, U20b, C03, E01, E13a, E16a, E16b, E26, E30a, E32a

Latest application/Harvest interval
- Before 1st node detectable (GS 31)

290   fluquinconazole

A protectant, eradicant and systematic triazole fungicide for winter wheat

See also chlorothalonil + fluquinconazole

Products

1 Flamenco	Aventis	100 g/l	SC	09913
2 Flamenco	AgrEvo	100 g/l	SC	09385

Uses

Brown rust in **winter wheat**. Powdery mildew in **winter wheat**. Septoria diseases in **winter wheat**.

Efficacy
- Best results obtained from treatment when disease first becomes active in crop but before infection spreads to younger leaves
- Adequate disease protection throughout the season will usually require a programme of at least two fungicide treatments
- May also be applied as a protectant at end of ear emergence (but no later) if crop still disease free

Crop safety/Restrictions
- Maximum total dose equivalent to two full doses
- Ensure good spray coverage and increase volume in dense crops
- Do not apply after the start of anthesis (GS 59)

Special precautions/Environmental safety
- Harmful: If swallowed
- Irritating to eyes
- May cause sensitization by skin contact
- Dangerous to fish or other aquatic life. Do not contaminate surface waters or ditches with chemical or used container

Personal protective equipment/Label precautions
- A, C, H, M
- M03, R03c, R04a, R04e, U05a, U11, U14, U15, C03, E01, E13b, E22a, E26, E31c, E34

Latest application/Harvest interval
- Before beginning of anthesis (GS 59)

291   fluquinconazole + prochloraz

A broad-spectrum triazole mixture for use as a spray and seed treatment in winter wheat

Products

1 Aventis Foil	Aventis	54:174 g/l	SE	09709
2 Foil	AgrEvo	54:174 g/l	SE	09363
3 Jockey	Aventis	167:31.2 g/l	FS	10076

Uses

Brown rust in **winter wheat** [1, 2]. Brown rust in **winter barley** [2]. Bunt in **winter wheat** *(seed treatment)* [3]. Fusarium root rot in **winter wheat** *(seed treatment)* [3]. Leaf spot in **winter wheat** *(seed treatment)* [3]. Powdery mildew in **winter wheat** *(moderate control)* [1, 2]. Powdery mildew in **winter barley** [2]. Rhynchosporium in **winter barley** [2]. Septoria leaf spot in **winter wheat** [1, 2]. Take-all in **winter wheat** *(reduction (seed treatment))* [3]. Yellow rust in **winter wheat** [1, 2].

Efficacy
- Best results obtained from treatment as disease first becomes active in crop [1, 2]
- Treat before infection spreads to younger leaves [1, 2]
- Ensure good even coverage of seed to obtain reliable disease control [3]
- Seed treatment provides early control of leaf spot but may require foliar treatment for later infection [3]

Crop safety/Restrictions
- Maximum total dose equivalent to two full doses (spray); 1 per seed batch (seed dressing)
- Ensure good spray coverage and increase volume in dense crops [1, 2]
- Do not apply after the start of anthesis (GS 59) [1, 2]
- Do not treat cracked, split or sprouted seed [3]

Special precautions/Environmental safety
- Harmful if swallowed
- Irritating to eyes [2]
- May cause lung damage if swallowed [1, 2]
- May cause sensitization by skin contact [1, 2]
- Dangerous to fish or other aquatic life. Do not contaminate surface waters or ditches with chemical or used container
- LERAP Category B [1]
- Do not handle treated seed unnecessarily [3]
- Do not use treated seed as food or feed [3]

Personal protective equipment/Label precautions
- C [1, 2]; A, H [1-3]; D [3]
- M05 [1, 2]; M03 [1-3]; R04a [2]; R04e, R04g [1, 2]; R03c, U05a [1-3]; U07, U20b [3]; C03 [1-3]; E31c [1, 2]; E01, E13b, E26, E34 [1-3]; E30a [2, 3]; E16a, E16b [1]; E29, E33, E36, S01, S02, S03, S04b, S05, S06, S07, S08 [3]

Latest application/Harvest interval
- Before beginning of anthesis (GS 59) (spray); before drilling (seed treatment)

292 fluroxypyr

A post-emergence aryloxyalkanoic acid herbicide

See also clopyralid + fluroxypyr + MCPA
clopyralid + fluroxypyr + triclopyr

Products

1	AgriGuard Fluroxypyr 200	AgriGuard	200 g/l	EC	09298
2	Barclay Hurler	Barclay	200 g/l	EC	08791
3	Barclay Hurler 200	Barclay	200 g/l	EC	09917
4	Greencrop Reaper	Greencrop	200 g/l	EC	09359
5	Standon Fluroxypyr	Standon	200 g/l	EC	08293
6	Starane 2	Dow	200 g/l	EC	05496
7	Tomahawk	Makhteshim	200 g/l	EC	09249

Uses

Annual dicotyledons in *barley, durum wheat, established grassland, oats, rye, seedling leys, triticale, wheat* [1-7]. Annual dicotyledons in *forage maize* [2-5, 7]. Annual dicotyledons in *bulb onions (off-label)*, *maize* [6]. Black bindweed in *barley, durum wheat, established grassland, oats, rye, seedling leys, triticale, wheat* [1-7]. Black bindweed in *forage maize* [2-5, 7]. Black bindweed in *maize* [6]. Chickweed in *barley, durum wheat, established grassland, oats, rye, seedling leys, triticale, wheat* [1-7]. Chickweed in *forage maize* [2-5, 7]. Chickweed in *maize* [6]. Cleavers in *barley, durum wheat, established grassland, oats, rye, seedling leys, triticale, wheat* [1-7]. Cleavers in *forage maize* [2-5, 7]. Cleavers in *maize* [6]. Docks in *barley, durum wheat, established grassland, oats, rye, seedling leys, triticale, wheat* [1-7]. Docks in *forage maize* [2-5, 7]. Docks in *maize* [6]. Forget-me-not in *barley, durum wheat, established grassland, oats, rye, seedling leys, triticale, wheat* [1-7]. Forget-me-not in *forage maize* [2-5, 7]. Forget-me-not in *maize* [6]. Hemp-nettle in *barley, durum wheat, established grassland, oats, rye, seedling leys, triticale, wheat* [1-7]. Hemp-nettle in *forage maize* [2-5, 7]. Hemp-nettle in *maize* [6]. Volunteer potatoes in *barley, durum wheat, established grassland, oats, rye, seedling leys, triticale, wheat* [1-7]. Volunteer potatoes in *forage maize* [2-5, 7]. Volunteer potatoes in *bulb onions* (off-label), *maize* [6].

Efficacy
- Best results achieved under good growing conditions in a strongly competing crop
- A number of tank mixtures with HBN and other herbicides are recommended for use in autumn and spring to extend range of species controlled. See label for details
- Spray is rainfast in 1 h
- Do not spray if frost imminent

Crop safety/Restrictions
- Maximum number of treatments 1 per crop or yr
- Apply to new leys from 3 expanded leaf stage
- Timing varies in tank mixtures. See label for details
- Do not use on crops undersown with clovers or other legumes
- Crops undersown with grass may be sprayed provided grasses are tillering
- Do not treat crops suffering stress caused by any factor
- Do not roll or harrow for 7 d before or after treatment

Special precautions/Environmental safety
- Flammable
- May cause lung damage if swallowed [3, 6]
- Harmful to fish or other aquatic life. Do not contaminate surface waters or ditches with chemical or used container
- Wash spray equipment thoroughly with water and detergent immediately after use. Traces of product can damage susceptible plants sprayed later

Personal protective equipment/Label precautions
- M05 [3, 6]; R07d [1-7]; R03, R04g [3, 6]; U08, U20b [1-7]; U19 [2-7]; E13c, E26, E30a, E31b [1-7]; E07 [1-7] (3 d); E34 [1]

Withholding period
- Keep livestock out of treated areas for at least 3 d and until foliage of poisonous weeds such as ragwort has died and become unpalatable

Latest application/Harvest interval
- Before flag leaf sheath opening (GS 47) for winter wheat and barley; before flag leaf sheath extending (GS 41) for spring wheat and barley; before second node detectable (GS 32) for oats, rye, triticale and durum wheat; before 7 leaves unfolded and before buttress roots appear for maize

Approval
- Fluroxypyr included in Annex I under EC Directive 91/414
- Off-label approval to May 2002 for use in bulb onions (OLA 0943/97)[6]
- Accepted by BLRA for use on malting barley

293   fluroxypyr + mecoprop-P

A post-emergence herbicide for broadleaved weeds in amenity turf

Products
| Bastion T | Rigby Taylor | 72:300 g/l | EC | 06011 |

Uses
Annual dicotyledons in **amenity turf, managed amenity turf**. Perennial dicotyledons in **amenity turf, managed amenity turf**. Slender speedwell in **amenity turf, managed amenity turf**.

Efficacy
- Best results achieved when soil moist and weeds in active growth, normally Apr-Sep
- Do not apply in drought period unless irrigation applied
- Avoid mowing for 3 d before or after spraying (5 d before for young turf)
- Do not spray if turf wet

Crop safety/Restrictions
- Maximum number of treatments 2 per yr. The total amount of mecoprop-P applied in a single yr must not exceed the maximum total dose approved for any single product for the crop/situation

- Young turf must only be treated in spring provided that at least 2 mth have elapsed between sowing and application
- After mowing young turf allow 5 d before treatment
- Do not treat turf under stress from any cause, if night temperatures are low, if ground frost imminent or during prolonged cold weather
- Avoid drift onto all broad-leaved plants outside the target area

Special precautions/Environmental safety
- Harmful in contact with skin or if swallowed, irritating to eyes
- Harmful to fish or other aquatic life. Do not contaminate surface waters or ditches with chemical or used container
- Wash spray equipment thoroughly with water and detergent immediately after use. Traces of product can damage susceptible plants sprayed later

Personal protective equipment/Label precautions
- A, C
- M03, R03a, R03c, R04a, U05a, U08, U19, U20b, C03, E01, E13c, E26, E30a, E31b, E34

Approval
- Fluroxypyr included in Annex I under EC Directive 91/414

---

294   fluroxypyr + triclopyr

A foliar acting herbicide for docks in grassland

Products

1 Doxstar	Dow	100:100 g/l	EC	06050
2 Evade	Dow	20:60 g/l	ME	08071

Uses

Annual dicotyledons in *amenity grass* [2]. Docks in *established grassland* [1]. Perennial dicotyledons in *amenity grass* [2]. Woody weeds in *amenity grass* [2].

Efficacy
- For dock control apply in spring or autumn or, at lower dose, in spring and autumn [1]
- Treat docks in rosette stage, up to 200 mm high or across [1]
- Allow 2-3 wk growth after cutting or grazing to allow sufficient regrowth to occur before spraying [1]
- For woody weed control apply between May and end Aug when weeds are actively growing [2]
- Do not spray in drought, very hot or very cold weather
- Control may be reduced if rain falls within 2 h of application
- Do not cut grass for 28 d after spraying [1]

Crop safety/Restrictions
- Maximum number of treatments 2 per yr
- Do not spray grass less than 1 yr old
- Clover will be killed or severely checked by treatment
- Do not roll or harrow for 10 d before or 7 d after spraying
- Avoid contact with foliage, stems or branches of desirable vegetation either directly or by drift. Use of a tree guard recommended when applying near trees and shrubs [2]
- Do not sow kale, turnips, swedes or grass mixtures containing clover by direct drilling or minimum cultivation techniques within 6 wk of application [1]
- Do not plant winter beans or legumes in same yr as treatment. Restrictions apply to many other crops planted in spring following treatment. See label for details of cultivations and intervals required [2]

Special precautions/Environmental safety
- Irritating to skin
- Irritating to eyes [2]
- May cause lung damage if swallowed [1]
- Flammable [1]
- Harmful to fish or other aquatic life. Do not contaminate surface waters or ditches with chemical or used container
- LERAP Category B [2]

FOR FULL CONDITIONS OF USE ALWAYS READ THE PRODUCT LABEL

- Do not allow drift to come into contact with crops, amenity plantings, gardens, ponds, lakes or watercourses
- Wash spray equipment thoroughly with water and detergent immediately after use. Traces of product can damage susceptible plants sprayed later

Personal protective equipment/Label precautions
- A, C [1, 2]; H, M [2]
- R04g, R07d [1]; R04b [1, 2]; R04a [2]; U02, U19 [1]; U05a, U08, U20b [1, 2]; C01 [1]; C03 [1, 2]; E07 [1] (7 d); E01, E13c, E26, E30a, E31b, E34 [1, 2]; E16a, E16b [2]

Withholding period
- Keep livestock out of treated areas for at least 7 d and until foliage of any poisonous weeds such as ragwort has died and become unpalatable [1]

Latest application/Harvest interval
- 7 d before grazing or harvest of grass

Approval
- Fluroxypyr included in Annex I under EC Directive 91/414

295   flurtamone

A carotenoid synthesis inhibitor available only in mixtures

See also diflufenican + flurtamone
diflufenican + flurtamone + isoproturon

296   flusilazole

A systemic, protective and curative conazole fungicide for cereals and oilseed rape

See also carbendazim + flusilazole
fenpropimorph + flusilazole

Products

1	DUK 747	DuPont	400 g/l	EC	08239
2	Genie	DuPont	400 g/l	EC	08238
3	Lyric	DuPont	250 g/l	EW	08252
4	Sanction	DuPont	400 g/l	EC	08237

Uses

Brown rust in *spring barley, winter barley, winter wheat*. Eyespot in *spring barley, winter barley, winter wheat*. Light leaf spot in *oilseed rape*. Net blotch in *spring barley, winter barley*. Powdery mildew in *spring barley, winter barley, winter wheat*. Rhynchosporium in *spring barley, winter barley*. Septoria in *spring barley, winter barley, winter wheat*. Yellow rust in *spring barley, winter barley, winter wheat*.

Efficacy
- Use as a routine preventative spray or when disease first develops
- Best control of eyespot achieved by spraying between leaf-sheath erect and second node detectable stages (GS 30-32)
- Product active against both MBC-sensitive and MBC-resistant strains of eyespot
- Treat oilseed rape in autumn and/or spring at stem extension stage
- Rain occurring within 2 h after application may reduce effectiveness
- See label for recommended tank-mixes

Crop safety/Restrictions
- Maximum number of treatments (including other products containing flusilazole) on cereals depends on dose and timing - see labels for details. High rate must not be used more than once in any crop
- Do not apply to crops under stress or during frosty weather

Special precautions/Environmental safety
- Harmful if swallowed
- Dangerous to fish or other aquatic life. Do not contaminate surface waters or ditches with chemical or used container

Personal protective equipment/Label precautions
- A, C
- M03, R03c, U05a, U11, U19, U20a, C03, E01, E13b, E26, E30a, E31b, E34

Latest application/Harvest interval
- High dose before 3rd node detectable (GS 33) plus reduced dose before early milk stage (GS 72) on winter wheat and barley; before first flowers open (GS 4,0) on oilseed rape

Approval
- Accepted by BLRA for use on malting barley

297   fomesafen

A contact and residual diphenyl ether herbicide for use in leguminous crops

Products

Flex	Novartis	250 g/l	SL	08885

Uses

Annual dicotyledons in *green beans*. Volunteer oilseed rape in *green beans*.

Efficacy
- Best results obtained when weeds actively growing in warm moist conditions with adequate soil moisture
- Weeds should be treated at seedling stage. Larger weeds, and those hardened off by adverse conditions, may be less well controlled
- Residual activity gives control of later germinating weed seedlings
- Product may be used on all soil types but residual control of weeds germinating after treatment may be reduced on soils with high organic matter
- Weed spectrum may be widened by use in a programme with other herbicides. See label for details

Crop safety/Restrictions
- Maximum total dose 0.9 l/ha
- Apply to healthy crops only. Crops growing under stress from drought, waterlogging etc should not be treated
- Some crop damage, such as leaf crinkling, may occur. Effect is transient and should not affect yield

Special precautions/Environmental safety
- Irritating to eyes
- May cause sensitization by skin contact

Personal protective equipment/Label precautions
- A, C, H
- M03, R04a, R04e, U04a, U05a, U11, U14, U19, U20b, C03, E01, E15, E26, E30a, E31a

Latest application/Harvest interval
- HI 4 wk

Approval
- Off-label approval to Jul 2003 for use on outdoor soya beans (OLA 1365/97)

298   fomesafen + terbutryn

A residual herbicide for use in leguminous crops

Products

Reflex T	Novartis	80:400 g/l	SC	08884

Uses

Annual dicotyledons in *broad beans, peas* (spring sown), *spring field beans*.

Efficacy
- Apply after planting, but before the crop emerges
- Best results achieved when applied to moist, fine, firm tilth
- Rain soon after application is essential for optimum weed control
- Results may be unsatisfactory in dry, or excessively wet, soil conditions
- Weed control may be reduced on soils with more than 10% organic matter

Crop safety/Restrictions
- Maximum number of treatments 1 per crop
- Use only once every 5 yr
- All varieties of spring sown peas may be treated but forage varieties may be damaged from which recovery may not be complete. All varieties of spring sown field and broad beans may be treated
- Emerged crop leaves may be severely scorched or killed
- Do not use on Sands
- Plough or cultivate to at least 150 mm before drilling or planting another crop. Only cereals should be planted in the calendar year of use with a minimum interval of 4 mth after treatment

Special precautions/Environmental safety
- Irritant. May cause sensitization by skin contact
- Harmful to fish or other aquatic life. Do not contaminate surface waters or ditches with chemical or used container

Personal protective equipment/Label precautions
- A, C, H
- R04, R04e, U04a, U05a, U08, U19, U20b, C03, E01, E13c, E26, E30a, E31a, E34

Latest application/Harvest interval
- Pre-emergence of crop

299 formaldehyde (commodity substance)

An agricultural/horticultural and animal husbandry fungicide

Products

1 formaldehyde	various	38-40%	SL
2 paraformaldehyde	various	-	

Uses

Fungus diseases in **flower bulbs** *(dip)*, **glasshouses** *(spray, dip or fumigant)*, **mushroom houses** *(spray or fumigant)* [1]. Fungus diseases in **livestock houses** [2]. Soil-borne diseases in **soil and compost** *(drench)* [1].

Efficacy
- Use as a dip to sterilize flower bulbs [1]
- Use as drench to sterilize soil and compost, indoors and outdoors [1]
- Use as spray or fumigant in mushroom houses [1]
- Use as spray, dip or fumigant for glasshouse hygiene [1]
- Use as fumigant to sterilize animal houses [2]

Crop safety/Restrictions
- Observe maximum permitted concentrations. See PR 12, 1990, pp 9-10

Special precautions/Environmental safety
- Formaldehyde is subject to the Poisons Rules 1982 and the Poisons Act 1972. See notes in Section 1
- Operators must observe Occupational Exposure Standard as set out in HSE Guidance Note EH40/90 and ACOP 30. Control of Substances Hazardous to Health in Fumigation
- Operators must be supplied with a Section 6 (HSW) Safety Data Sheet before commencing work

Personal protective equipment/Label precautions
- A, D, P [1]

300 fosamine-ammonium

A contact phosphonic acid herbicide for woody weed control

Products

Krenite	DuPont	480 g/l	SL	01165

Uses

Woody weeds in *conifer plantations (off-label)*, *forestry, non-crop areas, waterside areas*.

Efficacy

- Apply as overall spray to foliage during Aug-Oct when growth ceased but before leaves have started to change colour. Effects develop in following spring
- Thorough coverage of leaves and stems needed for effective control
- Addition of non-ionic wetter or Actipron recommended. Rain within 24 h may reduce effectiveness
- Most deciduous species controlled or suppressed but evergreens resistant
- Little effect on underlying herbaceous vegetation

Crop safety/Restrictions

- Do not use in conifer plantations

Personal protective equipment/Label precautions

- U08, U20a, E15, E26, E30a, E31a

Approval

- May be used alongside river, canal and reservoir banks but should not be sprayed into water. See notes in Section 1 on use of herbicides in or near water
- Off-label approval unlimited for use in conifer plantations (OLA 1437/96)

---

301   fosetyl-aluminium

A systemic phosphonic acid fungicide for various horticultural crops

Products

| Aliette 80 WG | Hortichem | 80% w/w | WG | 09156 |

Uses

Collar rot in *apples*. Crown rot in *apples, strawberries (off-label)*. Damping off in *brassicas (off-label)*. Downy mildew in *brassicas (off-label)*, *broad beans, combining peas (off-label)*, *herbs (off-label)*, *hops, lettuce (off-label)*, *protected lettuce, spinach beet (off-label)*, *spinach (off-label)*, *vining peas (off-label)*, *watercress (off-label)*. Phytophthora in *chicory (off-label)*, *watercress (off-label)*. Phytophthora root rot in *capillary benches, protected pot plants*. Phytophthora stem rot in *capillary benches, protected pot plants*. Phytophthora wilt in *hardy ornamental nursery stock*. Pythium in *watercress (off-label)*. Red core in *strawberries, strawberries (off-label - spring treatment)*. Root rot in *capillary benches, protected pot plants*.

Efficacy

- Spray young orchards for crown rot protection after blossom when first leaves fully open and repeat after 4-6 wk. Apply as paste to bark of apples to control collar rot
- Apply to broad beans when infection appears (usually at flowering) and 14 d later. Consult before treating crops to be processed
- Spray autumn-planted strawberry runners 2-3 wk after planting or use dip treatment at planting. Spray established crops in late summer/early autumn after picking and repeat annually
- Apply to hops as early season basal spray or as foliar spray every 10-14 d from when training is completed
- May be used by wet incorporation in blocking compost for protected lettuce
- Apply as drench to rooted cuttings of hardy nursery stock after first potting and repeat mthly. Up to 6 applications may be needed
- See label for details of application to capillary benches and trays of young plants.

Crop safety/Restrictions

- Maximum number of treatments 1 per batch of compost for lettuce; 1 per yr for strawberries (root dip or foliar spray); 1 per yr for apples (bark paste), 2 per yr for apples (foliar spray); 2 per crop for broad beans, peas; 2 per yr for hops (basal spray); 6 per yr for hops (foliar spray); 1 per crop for brassicas; 1 per crop during propagation, 2 per crop after planting out for lettuce
- Use on strawberries only between harvest and 31 Dec

- Use in compost for protected lettuce only from Sep to Apr and follow all directions carefully to avoid severe crop injury. Crop maturity may be delayed by a few days
- Check tolerance of ornamental species before large-scale treatment

Personal protective equipment/Label precautions
- A, C, H
- U20c, E15, E30a, E32a

Latest application/Harvest interval
- Pre-planting for dipping autumn planted strawberry runners; up to 31 Dec for spraying autumn planted strawberry runners; pre-sowing for protected lettuce; pre-emergence for brassicas
- HI apples 5 mth (bark paste) or 4 wk (spray); broad beans, peas 17 d; hops 14 d

Approval
- Off-label approval to Jul 2003 for use on chicory in forcing sheds (OLA 0052/99); to Apr 2002 for use on watercress during propagation (OLA 0054/99); to Feb 2002 for use as a compost drench on on outdoor and protected crops of cauliflowers, calabrese, broccoli, Brussels sprouts, kale, cabbages, Chinese cabbage, collards (OLA 0056/99); to Feb 2002 for use on outdoor and protected lettuce varieties (OLA 0057/99; unlimited for use as a seed treatment on vining and combining peas (OLA 0058/99); to Mar 2003 for use on outdoor and protected strawberries (OLA 0566/99); unstipulated for use on outdoor grapevines (OLA 2907/99); unstipulated for use on outdoor spinach and spinach beet (OLA 1220/00); to Jul 2003 for use on cauliflowers, calabrese, broccoli, Brussels sprouts, cabbages, Chinese cabbage, collards, kale (OLA 2076/00)
- Accepted by BLRA for use on hops

302   fosthiazate

A contact nematicide for potatoes

Products

Nemathorin 10G	Zeneca	10% w/w	FG	08915

Uses

Potato cyst nematode in **potatoes**.

Efficacy
- Apply and incorporate granules in one operation to a uniform depth of 10-15 cm. Deeper incorporation will reduce control
- Application best achieved using equipment such as Horstine Farmery Microband Applicator, Matco or Stocks Micrometer applicators together with a rear mounted powered rotary cultivator
- Granules must not become wet or damp before use

Crop safety/Restrictions
- Maximum number of treatments 1 per crop
- Product must only be applied using tractor-mounted/drawn direct placement machinery. Do not use air assisted broadcast machinery
- Do not apply more than once every four years on the same area of land
- Consult before using on crops intended for processing

Special precautions/Environmental safety
- Harmful if swallowed
- Irritating to eyes and skin
- Dangerous to game, wild birds and animals
- Risk of explosion if dust in the air
- Harmful to fish or other aquatic life. Do not contaminate surface waters or ditches with chemical or used container
- Incorporation to 10-15 cm and ridging up of treated soil must be carried out immediately after application
- Failure completely to bury granules immediately after application is hazardous to wildlife

Personal protective equipment/Label precautions
- A, E, G, H, K, M
- M01, M03, R03c, R04a, R04b, U02, U04a, U05a, U09a, U13, U19, U20a, C03, E01, E10a, E13c, E26, E29, E32a, E34; E06a (13 wk)

Withholding period
- Dangerous to livestock. Keep all livestock out of treated areas for at least 13 wk

Latest application/Harvest interval
- At planting

---

303   fuberidazole

A benzimidazole (MBC) fungicide available only in mixtures

See also bitertanol + fuberidazole
bitertanol + fuberidazole + imidacloprid

---

304   fuberidazole + imidacloprid + triadimenol

A broad spectrum systemic fungicide and insecticide seed treatment for winter cereals

Products

Baytan Secur	Bayer	15:117:125 g/l	FS	09510

Uses

Blue mould in *winter wheat* *(seed treatment)*. Brown foot rot in *winter barley* *(seed treatment)*. Brown rust in *winter barley* *(seed treatment)*, *winter wheat* *(seed treatment)*. Bunt in *winter wheat* *(seed treatment)*. Covered smut in *winter barley* *(seed treatment)*. Fusarium foot rot and seedling blight in *winter barley* *(seed treatment - reduction)*, *winter oats* *(seed treatment - reduction)*, *winter wheat* *(seed treatment - reduction)*. Lodging control in *winter wheat* *(seed treatment - reduction)*. Loose smut in *winter barley* *(seed treatment)*, *winter oats* *(seed treatment)*, *winter wheat* *(seed treatment)*. Net blotch in *winter barley* *(seed treatment - seed-borne only)*. Powdery mildew in *winter barley* *(seed treatment)*, *winter oats* *(seed treatment)*, *winter wheat* *(seed treatment)*. Pyrenophora leaf spot in *winter oats* *(seed treatment)*. Septoria leaf spot in *winter wheat* *(seed treatment)*. Septoria seedling blight in *winter wheat* *(seed treatment - reduction)*. Snow rot in *winter barley* *(seed treatment - reduction)*. Virus vectors in *winter barley* *(seed treatment)*, *winter oats* *(seed treatment)*, *winter wheat* *(seed treatment)*. Wireworms in *winter barley* *(reduction of damage)*, *winter oats* *(reduction of damage)*, *winter wheat* *(reduction of damage)*. Yellow rust in *winter barley* *(seed treatment)*, *winter wheat* *(seed treatment)*.

Efficacy
- Apply through recommended seed treatment machinery
- Evenness of seed cover improved by simultaneous application of equal volumes of product and water
- Calibrate drill for treated seed and drill at 2.5-4 cm into firm, well prepared seedbed
- Use minimum 125 kg treated seed per ha and increase seed rate as drilling season progresses. Lower drilling rates and/or early drilling affect duration of BYDV protection needed and may require follow-up aphicide treatment
- When aphid activity unusually late, or is heavy and prolonged in areas of high risk, and mild weather predominates, follow-up treatment may be required
- In addition to seed-borne diseases early attacks of various foliar, air-borne diseases are controlled or suppressed. See label for details

Crop safety/Restrictions
- Maximum number of treatments 1 per batch of seed
- Do not use on naked oats
- Do not drill treated winter wheat seed after end of Nov
- Germination tests should be done on all batches of seed to be treated to ensure seed viability and suitability for treatment
- Do not use on seed with moisture content above 16%, on sprouted, cracked or skinned seed or on seed already treated with another seed treatment
- Store treated seed in cool, dry, well-ventilated store and drill as soon as possible, preferably in season of purchase
- Treatment may accentuate effects of adverse seedbed conditions on crop emergence

---

Special precautions/Environmental safety
- Harmful if swallowed
- May cause sensitisation by skin contact
- Dangerous to fish or other aquatic life. Do not contaminate surface waters or ditches with chemical or used container
- Do not handle seed unnecessarily
- Do not use treated seed as food or feed
- Dangerous to birds, game and other wildlife. Treated seed should not be left on the soil surface. Bury spillages
- Treated seed should not be broadcast, but drilled to a depth of 4 cm in a well prepared seedbed
- If seed is left on the soil surface the field should be harrowed and rolled to ensure good incorporation

Personal protective equipment/Label precautions
- A, H
- M03, R03c, R04e, U04a, U05a, U13, U20b, C03, E01, E03, E13b, E26, E30a, E32a, E34, S01, S02, S03, S04c, S05, S06, S07

Latest application/Harvest interval
- Before drilling

---

305 fuberidazole + triadimenol

A broad spectrum systemic fungicide seed treatment for cereals

Products

Baytan Flowable	Bayer	22.5:187.5 g/l	FS	02593

Uses

Blue mould in *triticale* (seed treatment), *wheat* (seed treatment). Brown foot rot in *barley* (seed treatment), *oats* (seed treatment), *rye* (seed treatment), *triticale* (seed treatment), *wheat* (seed treatment). Brown rust in *barley* (seed treatment), *rye* (seed treatment), *wheat* (seed treatment). Bunt in *wheat* (seed treatment). Covered smut in *barley* (seed treatment). Crown rust in *oats* (seed treatment). Leaf stripe in *spring barley* (seed treatment). Loose smut in *barley* (seed treatment), *oats* (seed treatment), *wheat* (seed treatment). Mildew in *barley* (seed treatment), *wheat* (seed treatment). Pyrenophora leaf spot in *oats* (seed treatment). Septoria in *wheat* (seed treatment). Yellow rust in *barley* (seed treatment), *wheat* (seed treatment).

Efficacy
- Apply through recommended seed treatment machinery
- Calibrate drill for treated seed and drill at 2.5-4 cm into firm, well prepared seedbed
- In addition to seed-borne diseases early attacks of various foliar, air-borne diseases are controlled or suppressed. See label for details

Crop safety/Restrictions
- Maximum number of treatments 1 per batch of seed
- Do not use on naked oats
- Do not drill treated winter wheat or rye seed after end of Nov. Seed rate should be increased as drilling season progresses
- Treated spring wheat may be drilled in autumn up to end of Nov or from Feb onward
- Germination tests should be done on all batches of seed to be treated to ensure seed viability and suitability for treatment
- Do not use on seed with moisture content above 16%, on sprouted, cracked or skinned seed or on seed already treated with another seed treatment
- Store treated seed in cool, dry, well-ventilated store and drill as soon as possible, preferably in season of purchase
- Treatment may accentuate effects of adverse seedbed conditions on crop emergence

Special precautions/Environmental safety
- Dangerous to fish or other aquatic life. Do not contaminate surface waters or ditches with chemical or used container
- Do not use treated seed as food or feed

Personal protective equipment/Label precautions
- A
- U20b, E03, E13b, E26, E30a, E32a, E34, S01, S02, S03, S04a, S05, S06, S07

Latest application/Harvest interval
- Before drilling

Approval
- Accepted by BLRA for use as a seed treatment on malting barley

306   furalaxyl

A protective and curative phenylamide (acylalanine) fungicide for ornamentals

Products

Fongarid	Novartis	25% w/w	WP	08407

Uses

Damping off in **bedding plants**. Phytophthora in **hardy ornamental nursery stock, pot plants**. Pythium in **hardy ornamental nursery stock, pot plants**.

Efficacy
- Apply by incorporation into compost or as a drench to obtain at least 12 wk protection
- Apply drench within 3 d of seeding or planting out as a protective treatment or as soon as first signs of root disease appear as curative treatment
- Do not apply to field or border soil

Crop safety/Restrictions
- Manufacturer's literature lists genera which have been treated successfully. With all subjects of unknown susceptibility treat a few plants before committing whole batches

Special precautions/Environmental safety
- Irritating to eyes and skin
- Harmful to fish or other aquatic life. Do not contaminate surface waters or ditches with chemical or used container

Personal protective equipment/Label precautions
- A, C
- R04a, R04b, U02, U05a, U09a, U19, U20a, C03, E01, E13c, E30a, E31b, E34

307   gamma-HCH

A contact, ingested and fumigant organochlorine insecticide

Products

1	Atlas Steward	Nufarm Whyte	560 g/l	SC	07728
2	Lindane Flowable	PBI	800 g/l	SC	09113

Uses

Chafer grubs in **strawberries** [1]. Cutworms in **cereals, grassland, maize, sugar beet** [1, 2]. Flea beetles in **sugar beet** [1]. Leatherjackets in **strawberries** [1]. Leatherjackets in **cereals, grassland, maize, sugar beet** [1, 2]. Millipedes in **sugar beet** [1, 2]. Pygmy mangold beetle in **sugar beet** [1, 2]. Springtails in **sugar beet** [1, 2]. Symphylids in **sugar beet** [1, 2]. Wireworms in **strawberries** [1]. Wireworms in **cereals, grassland, maize, sugar beet** [1, 2].

Efficacy
- Method of application, dose, timing and number of applications vary with formulation, crop and pest. See label for details
- Where glasshouse whitefly resistant to gamma-HCH occur control is unlikely to be satisfactory

Crop safety/Restrictions
- Maximum number of treatments 1 per crop for cereals, maize, sugar beet, strawberries; 1 per yr for grassland
- Do not use if potatoes or carrots are to be planted within 18 mth

Special precautions/Environmental safety
- Harmful if swallowed [2] and in contact with skin [1]
- Extremely dangerous to fish or aquatic life. Do not contaminate surface waters or ditches with chemical or used container [1, 2]

- LERAP Category B [1, 2]
- Dangerous to bees. Do not apply to crops in flower or to those in which bees are actively foraging. Do not apply when flowering weeds are present [2]
- Harmful to livestock [1]
- Must be incorporated into soil following application [1, 2]

Personal protective equipment/Label precautions
- A, C, H, K [1, 2]; M [2]
- M04 [1]; M03 [2]; R03a [1]; R03c, U05a, U08, U20b [1, 2]; U02, U19 [2]; C02 [1] (14 d); C03, E01, E13a, E16a, E16b, E26, E30a, E31a, E34 [1, 2]; E12c [2]

Withholding period
- Keep all livestock out of treated areas for at least 14 d [2]

Approval
- Approvals for marketing and use as a seed treatment were withdrawn in July 1999. Approvals for remaining agricultural and garden uses are to be withdrawn in 2001 followed by a 12-month use-up period

Maximum Residue Level (mg residue/kg food)
- strawberries, blackberries, loganberries, raspberries, bilberries, cranberries, currants, gooseberries 3; tomatoes, peppers, aubergines, cauliflowers, Brussels sprouts, head cabbages, lettuce, meat (sheep meat) 2; citrus fruits, pome fruits, apricots, peaches, nectarines, plums, bananas, swedes, turnips, garlic, onions, shallots, cucumbers, gherkins, courgettes, beans (with pods), celery, leeks, rhubarb, cultivated mushrooms, meat (except sheep meat) 1; grapes 0.5; carrots, horseradish, parsnips, parsley root, salsify, tea 0.2; peas (with pods), cereals, eggs 0.1; potatoes 0.05

---

308   gamma-HCH + thiophanate-methyl

An insecticide and lumbricide for use on turf

Products

Castaway Plus	RP Amenity	60:500 g/l	SC	05327

Uses

Earthworms in *turf*. Leatherjackets in *turf*.

Efficacy
- Apply by spraying in spring or autumn. Drenching is not required
- Do not mow for 2 d after spraying. If mown beforehand leave clippings
- Do not collect first clippings after treatment
- Do not spray during drought or if ground frozen
- Effectiveness is not impaired by rain or irrigation immediately after application
- Do not mix with any other product

Special precautions/Environmental safety
- Irritating to eyes and skin
- Harmful to bees. Do not apply to crops in flower or to those in which bees are actively foraging. Do not apply when flowering weeds are present
- Harmful to fish or other aquatic life. Do not contaminate surface waters or ditches with chemical or used container
- LERAP Category B

Personal protective equipment/Label precautions
- A, C, H, M
- M03, R04a, R04b, U02, U05a, U08, U19, U20c, C03, E01, E13c, E16a, E16b, E26, E30a, E34

Withholding period
- Keep livestock out of treated areas for at least 14 d

Approval
- Approvals for remaining agricultural and garden uses are to be withdrawn in 2001 followed by a 12-month use-up period
- Approval expiry 31 Jul 2001

Maximum Residue Level (mg residue/kg food)
- see gamma-HCH entry

309   gibberellins

A plant growth regulator for use in apples, pears etc

Products

1 Berelex	Whyte Chemicals	1 g/tablet	TB	06637
2 Novagib	Fine	10 g/l	SL	08954

Uses

Increasing fruit set in *pears* [1]. Increasing yield in *celery, rhubarb* [1]. Reducing fruit russeting in *apples* [2].

Efficacy

- Apply to apples at completion of petal fall and repeat 3 or 4 times at 7-10 d intervals. Number of sprays and spray interval depend on dose (see labels) [2]
- Useful in pears when blossom sparse, setting conditions poor or where frost has killed many flowers. Apply as single or split application. See label for details. Resulting fruit may be seedless [1]
- May be used on pears if 80% of flowers frosted but not effective on very severely frosted blossom. Conference pear responds well, Beurre Hardy only in some seasons. Young trees generally less responsive [1]
- Apply to celery 3 wk before harvest to increase head size, to rhubarb crowns on transfer to forcing shed or as drench at first signs of growth in field [1]

Crop safety/Restrictions

- Maximum number of treatments 1 per crop or yr on celery and rhubarb; 2 per yr on pears [1]
- Maximum total dose 2.0 l/ha per yr for apples
- Good results achieved on Cox's Orange Pippin, Discovery, Golden Delicious and Karmijn apples. For other cultivars test on a small number of trees
- Return bloom may be reduced in yr following treatment [2]
- Do not apply to pears after petal fall [1]
- Consult before treating crops grown for processing [2]

Personal protective equipment/Label precautions

- U20a [1]; U20b [2]; E27, E31a, E34 [1]; E15, E26, E30a, E31b [1, 2]; E01, E24, E29 [2]

Latest application/Harvest interval

- Before petal fall for pears [1]
- HI 3 wk for celery [1]; zero for apples [2]

Approval

- Approval expiry 31 Jan 2001 [1]; replaced by MAFF 08903

310   glufosinate-ammonium

A non-selective, non-residual phosphinic acid contact herbicide

Products

1 Aventis Challenge	Aventis	150 g/l	SL	09721
2 Aventis Harvest	Aventis	150 g/l	SL	09810
3 Challenge	Aventis	150 g/l	SL	07306
4 Challenge 60	Fargro	60 g/l	SL	08236
5 Dash	Nomix-Chipman	120 g/l	SL	05177
6 Harvest	Aventis	150 g/l	SL	07321
7 Nomix Touchweed	Nomix-Chipman	60 g/l	SL	09596

Uses

Annual dicotyledons in *cane fruit, cultivated land/soil* [1-4, 6, 7]. Annual dicotyledons in *apples, cherries, currants, damsons, grapevines, non-crop farm areas, pears, plums, potatoes, strawberries, stubbles, sugar beet, tree nuts, vegetables* [1-3, 6]. Annual dicotyledons in *bush fruit, fruit trees, ornamental trees, shrubs* [4, 7]. Annual dicotyledons in *nursery stock, woody ornamentals* [5]. Annual grasses in *cane fruit, cultivated land/soil* [1-4, 6, 7]. Annual grasses in *apples, cherries, currants, damsons, grapevines, non-crop farm areas, pears, plums, potatoes, strawberries, stubbles, sugar beet, tree nuts, vegetables* [1-3, 6]. Annual grasses in *bush fruit, fruit trees, ornamental trees, shrubs* [4, 7]. Annual grasses in *nursery stock, woody ornamentals* [5]. Green cover in *land temporarily removed from production* [1-3, 6]. Harvest management/desiccation in *barley (not seed crops), combining peas (not seed crops), field beans, linseed, oilseed rape, potatoes (not seed crops), wheat (not seed crops)* [1-3, 6]. Perennial dicotyledons in *cane fruit* [1-4, 6, 7]. Perennial dicotyledons in *apples, cherries, currants, damsons, grapevines, land temporarily removed from production, non-crop farm areas, pears, plums, strawberries, tree nuts* [1-3, 6]. Perennial dicotyledons in *bush fruit, cultivated land/soil, fruit trees, ornamental trees, shrubs* [4, 7]. Perennial dicotyledons in *nursery stock, woody ornamentals* [5]. Perennial grasses in *cane fruit* [1-4, 6, 7]. Perennial grasses in *apples, cherries, currants, damsons, grapevines, land temporarily removed from production, non-crop farm areas, pears, plums, strawberries, tree nuts* [1-3, 6]. Perennial grasses in *bush fruit, cultivated land/soil, fruit trees, ornamental trees, shrubs* [4, 7]. Perennial grasses in *nursery stock, woody ornamentals* [5]. Sward destruction in *grassland* [1-3, 6].

Efficacy

- Activity quickest under warm, moist conditions. Light rainfall 3-4 h after application will not affect activity. Do not spray wet foliage or if rain likely within 6 h
- For weed control uses treat when weeds growing actively. Deep rooted weeds may require second treatment [4, 7]
- Ploughing or other cultivations can follow 4 h after spraying
- On uncropped headlands apply in May/Jun to prevent weeds invading field
- Crops can normally be sown/planted immediately after spraying or sprayed post-drilling. On sand, very light or immature peat soils allow at least 3 d before sowing/planting or expected emergence
- For weed control in potatoes apply pre-emergence or up to 10% emergence on earlies and seed crops, up to 40% on maincrop, on plants up to 15 cm high
- In sugar beet and vegetables apply just before crop emergence, using stale seedbed technique
- In top and soft fruit, grapevines, woody ornamentals, nursery stock and forestry apply up to 3 treatments (2 treatments [4, 7]) between 1 Mar and 30 Sep as directed sprays
- For grass destruction apply before winter dormancy occurs. Heavily grazed fields should show active regrowth. Plough from the day after spraying
- Apply pre-harvest desiccation treatments 10-21 d before harvest (14-21 d for oilseed rape). See label for timing details on individual crops
- For potato haulm desiccation apply to listed varieties (not seed crops) at onset of senescence, 14-21 d before harvest

Crop safety/Restrictions

- Maximum number of treatments 4 per crop for potatoes (including 2 desiccant uses); 2 per crop (including 1 desiccant use) for oilseed rape, dried peas, field beans, linseed, wheat and barley; 3 (2 [4, 7]) per yr for fruit and forestry; 2 per yr for strawberries, ornamental trees and shrubs, non-crop land and setaside; 1 per crop for sugar beet, vegetables and other crops; 1 per yr for grassland destruction, on cultivated land prior to planting edible crops
- Pre-harvest desiccation sprays should not be used on seed crops of wheat, barley, peas or potatoes but may be used on seed crops of oilseed rape, field beans and linseed
- Do not desiccate potatoes in exceptionally wet weather or in saturated soil. See label for details
- Do not spray potatoes after emergence if grown from small or diseased seed or under very dry conditions
- For weed control uses application must be between 1 Mar and 30 Sep
- Do not allow spray to contact dormant or green buds, suckers, damaged or green bark and foliage of wanted plants [4, 7]

Special precautions/Environmental safety

- Harmful if swallowed and in contact with skin [1-3, 6]
- Irritating to eyes [1-3, 5, 6]
- Harmful to fish or other aquatic life. Do not contaminate surface waters or ditches with chemical or used container
- Treated pea haulm may be fed to livestock from 7 d after spraying, treated grain from 14 d [3, 6]
- Do not use straw from treated crops as animal feed or bedding [3, 6]
- Do not spray hedge bottoms [3, 6]
- Product supplied in small volume returnable container - see label for filling and mixing instructions [6]

Personal protective equipment/Label precautions
- A, C, H, M [1-7]
- M03 [1-3, 6]; R04a [1-3, 5, 6]; R03a, R03c [1-3, 6]; U20b [5]; U08 [1-7]; U05a [1-3, 5, 6]; U20a [1-4, 6, 7]; U04a, U13, U14, U15, U19 [1-3, 6]; U02 [4, 7]; C03 [1-3, 5, 6]; E26 [5]; E07, E13c, E30a [1-7]; E01 [1-3, 5, 6]; E09, E34 [1-3, 6]; E31a [4, 7]

Withholding period
- Keep livestock out of treated areas until foliage of any poisonous weeds such as ragwort has died and become unpalatable

Latest application/Harvest interval
- 30 Sep for use on non-crop land, top fruit, forestry, pre-drilling, pre-planting, pre-emergence in sugar beet, vegetables and other crops; before winter dormancy for grassland destruction.
- HI potatoes, oilseed rape, drying peas, field beans, linseed 7 d; wheat, barley 14 d

Approval
- Accepted by BLRA for use on malting barley

311   glyphosate (agricultural uses)

A translocated non-residual phosphonic acid herbicide

See also diuron + glyphosate

Products

1	AgriGuard Glyphosate 360	AgriGuard	360 g/l	SL	09184
2	Apache	Zeneca	220 g/l	SL	06748
3	Azural	Monsanto	360 g/l	SL	09582
4	Barbarian	Barclay	360 g/l	SL	07980
5	Barclay Barbarian	Barclay	360 g/l	SL	09865
6	Barclay Cleanup	Barclay	180 g/l	SL	09179
7	Barclay Dart	Barclay	180 g/l	SL	05129
8	Barclay Gallup	Barclay	360 g/l	SL	05161
9	Barclay Gallup 360	Barclay	360 g/l	SL	09127
10	Barclay Gallup Biograde 360	Barclay	360 g/l	SL	09840
11	Barclay Garryowen	Barclay	360 g/l	SL	08599
12	Buggy SG	Sipcam	36% w/w	WG	08573
13	Cardel Egret	Cardel	360 g/l	SL	09703
14	Clinic	Nufarm Whyte	360 g/l	SL	09378
15	Glyper	PBI	360 g/l	SL	07968
16	Greencrop Gypsy	Greencrop	360 g/l	SL	09432
17	Helosate	Helm	360 g/l	SL	06499
18	I T Glyphosate	I T Agro	360 g/l	SL	07212
19	Landgold Glyphosate 360	Landgold	360 g/l	SL	05929
20	Portman Glider	Portman	480 g/l	SL	04695
21	Portman Glyphosate	Portman	360 g/l	SL	05891
22	Portman Glyphosate 360	Portman	360 g/l	SL	04699
23	Portman Glyphosate 480	Portman	480 g/l	SL	07194
24	Roundup	Monsanto	360 g/l	SL	01828
25	Roundup Biactive	Monsanto	360 g/l	SL	06941
26	Roundup Biactive Dry	Monsanto	42% w/w	WG	06942
27	Roundup GT	Monsanto	400 g/l	SL	08068
28	Roundup Rapide	Monsanto	400 g/l	SL	08067
29	Stacato	Sipcam	360 g/l	SL	05892
30	Standon Glyphosate 360	Standon	360 g/l	SL	05582
31	Stefes Glyphosate	Stefes	360 g/l	SL	05819
32	Stefes Kickdown	Stefes	360 g/l	SL	06329
33	Sting ECO	Monsanto	120 g/l	SL	08291
34	Stride	Zeneca	440 g/l	SL	06750
35	Touchdown	Zeneca	330 g/l	SL	06326

Uses

Annual and perennial weeds in **stubbles** [1-32, 34, 35]. Annual and perennial weeds in **non-crop farm areas** [1-5, 8-11, 13, 14, 16, 21, 22, 24-26, 29, 31, 34]. Annual and perennial weeds in **grassland** [1, 3, 8, 11-15, 22, 24-26, 31]. Annual and perennial weeds in **cultivated land/ soil** [1, 3, 8, 11, 13, 14, 16, 21, 31, 32]. Annual and perennial weeds in **field crops** *(wiper application)* [1, 3, 8, 11, 14, 16]. Annual and perennial weeds in **land temporarily removed from production** [2, 12, 25, 26, 34]. Annual and perennial weeds in **land not intended to bear vegetation** [35]. Annual dicotyledons in **non-crop farm areas, stubbles** [33]. Annual dicotyledons in **cultivated land/soil** [4, 5, 9, 10, 33]. Annual grasses in **non-crop farm areas, stubbles** [33]. Annual grasses in **cultivated land/soil** [4, 5, 9, 10, 33]. Annual weeds in **barley, durum wheat, field beans, linseed, mustard, oats, oilseed rape, wheat** [25, 26]. Annual weeds in **cultivated land/soil** [6, 7]. Annual weeds in **grassland** [6, 7, 17, 23-29, 31-33]. Annual weeds in **peas, sugar beet, swedes, turnips** [6, 7, 25, 26]. Black bent in **oilseed rape for industrial use** [2]. Black bent in **oilseed rape** [2, 4, 5, 9, 10]. Black bent in **barley, field beans, stubbles, wheat** [4, 5, 9, 10]. Black bent in **combining peas** [4, 5, 9, 10, 34]. Black bent in **durum wheat, spring oats, winter oats** [5, 9, 10]. Bolters in **sugar beet** *(wiper application)* [1, 3-5, 8-11, 14, 16, 24-26, 31, 32]. Bracken in **grassland** [2, 34, 35]. Couch in **stubbles** [1-32, 34, 35]. Couch in **barley, wheat** [1-5, 8-17, 19-32, 34]. Couch in **oilseed rape** [1-5, 8-17, 19, 21-30, 32, 34, 35]. Couch in **combining peas** [1-5, 8-17, 19, 22-28, 30-32, 34]. Couch in **field beans** [1, 3-5, 8-17, 19, 22-28, 30-32]. Couch in **oats** [1, 3, 8, 11-17, 19-30]. Couch in **linseed** [2, 12, 15, 17, 23-26, 34]. Couch in **mustard** [2, 12, 15, 17, 24-26, 34]. Couch in **oilseed rape for industrial use** [2, 34, 35]. Couch in **land not intended to bear vegetation** [35]. Couch in **spring oats, winter oats** [5, 9, 10]. Couch in **durum wheat** [5, 9, 10, 15, 23, 25, 26]. Couch in **grassland** [6-8, 11, 12, 15]. Creeping bent in **oilseed rape for industrial use** [2]. Creeping bent in **oilseed rape** [2, 4, 5, 9, 10]. Creeping bent in **barley, field beans, stubbles, wheat** [4, 5, 9, 10]. Creeping bent in **combining peas** [4, 5, 9, 10, 34]. Creeping bent in **durum wheat, spring oats, winter oats** [5, 9, 10]. Destruction of short term leys in **grassland** [28]. Green cover in **land temporarily removed from production** [2, 10, 34, 35]. Green cover in **non-crop areas** [2, 34]. Harvest management/desiccation in **linseed** [1-3, 10, 12-15, 17, 23-26, 34]. Harvest management/desiccation in **oilseed rape** [1-5, 8-19, 19, 21-32, 34, 35]. Harvest management/desiccation in **wheat** [1-3, 8, 11-18, 18-32, 34]. Harvest management/desiccation in **combining peas** [1-3, 8, 11-17, 19, 22-28, 30-32, 34]. Harvest management/desiccation in **barley** [1, 2, 8, 11, 12, 15-32, 34]. Harvest management/desiccation in **oats** [1, 3, 8, 11-30]. Harvest management/desiccation in **field beans** [1, 3, 8, 11-17, 19, 22-28, 30-32]. Harvest management/desiccation in **durum wheat** [1, 3, 8, 11, 13-16, 23, 25, 26]. Harvest management/desiccation in **mustard** [2, 3, 12-15, 17, 18, 23-28, 34]. Harvest management/desiccation in **oilseed rape for industrial use, spring field beans, winter field beans** [2, 34, 35]. Harvest management/desiccation in **spring barley, spring wheat, winter barley, winter wheat** [35]. Japanese knotweed in **land not intended to bear vegetation** [35]. Perennial dicotyledons in **oilseed rape, oilseed rape for industrial use** [2]. Perennial dicotyledons in **grassland** *(wiper application)* [24-26, 31]. Perennial weeds in **non-crop farm areas** [4, 5, 9, 10]. Rushes in **grassland** [2, 34, 35]. Sward destruction in **grassland** [1-19, 21-27, 29-32, 34, 35]. Volunteer cereals in **stubbles** [1-22, 24-26, 29-35]. Volunteer cereals in **barley, durum wheat, field beans, linseed, mustard, oats, oilseed rape, wheat** [25, 26]. Volunteer cereals in **cultivated land/soil** [4-7, 9, 10, 33]. Volunteer cereals in **peas, sugar beet, swedes, turnips** [6, 7, 25, 26]. Volunteer oilseed rape in **stubbles** [35]. Volunteer potatoes in **stubbles** [1-32, 34, 35, 35]. Weed beet in **sugar beet** *(wiper application)* [1, 3, 8, 11, 14, 16]. Wild oats in **cereals** *(wiper application)* [1, 3, 8, 11, 14, 24-26].

Efficacy

- For best results apply to actively growing weeds with enough leaf to absorb chemical
- Products are formulated as either isopropylamine [1, 4, 6-8, 11, 14-33], ammonium [3, 5, 9, 10, 12, 13], or trimesium [2, 34, 35] salts of glyphosate and may vary in the details of efficacy claims. See individual product labels
- Annual weed grasses should have at least 5 cm of leaf and annual broad-leaved weeds at least 2 expanded true leaves

- Perennial grass weeds should have 4-5 new leaves and be at least 10 cm long when treated. Perennial broad-leaved weeds should be treated at or near flowering but before onset of senescence
- Volunteer potatoes and polygonums are not controlled by harvest-aid rates
- In order to allow translocation, do not cultivate before treating perennials and do not apply other pesticides, lime, fertilizer or farmyard manure within 5 d of treatment
- Recommended intervals after treatment and before cultivation vary. See labels
- A rainfree period of at least 6 h (preferably 24 h) should follow spraying
- Do not tank-mix with other pesticides or fertilizers as such mixtures may lead to reduced control. Adjuvants are obligatory for some products and recommended for some uses with others. See labels
- Do not spray weeds affected by drought, waterlogging, frost or high temperatures

Crop safety/Restrictions
- Maximum number of applications normally 1 per crop (pre-harvest), 1 per yr (grassland destruction, non-crop land, green cover treatment) and 1 in other situations but check label for details
- Do not treat cereals grown for seed or undersown crops
- Disperse decaying matter by thorough cultivation before sowing or planting a following crop
- Consult grain merchant before treating crops grown on contract or intended for malting
- With wiper application weeds should be at least 10 cm taller than crop

Special precautions/Environmental safety
- Harmful if swallowed [2, 6, 7, 32, 34, 35]
- Irritating to skin [4, 7, 8, 11, 15, 17-24, 29-32] and eyes [1-3, 13, 14, 16, 22, 34, 35]
- Risk of serious damage to eyes [5-7, 9, 12, 32]
- Harmful (dangerous [6, 7]) to fish or other aquatic life. Do not contaminate surface waters or ditches with chemical or used container [1, 2, 4, 5, 8, 9, 11, 12, 14-24, 27-35]
- Do not mix, store or apply in galvanized or unlined mild steel containers or spray tanks
- Do not leave spray in spray tanks for long period and make sure tanks are well vented
- Take extreme care to avoid drift
- Treated poisonous plants must be removed before grazing or conserving [8, 11, 15, 24-26, 31]
- Do not use in covered areas such as greenhouses or under polythene
- For field edge treatment direct spray away from hedge bottoms
- Do not use on grassland if crop to be used for animal feed or bedding [2, 34, 35]

Personal protective equipment/Label precautions
- A [1-26, 29-32, 34, 35]; C [1-9, 11-26, 29-32, 34, 35]; M [1-3, 12-17, 21, 22, 24-26, 34, 35]; H [2, 3, 10, 12, 14-17, 22, 24-26, 34, 35]; N [1, 3, 13-17, 21, 24]; F [3, 14, 15, 17, 24]
- M03 [2, 6, 7, 32, 34, 35]; R04a [1-4, 7, 8, 11, 13-24, 29-32, 34, 35]; R04b [1, 3, 4, 7, 8, 11, 13-24, 29-32]; R05a [7]; R03c [2, 6, 7, 32, 34, 35]; R04d [5-7, 9, 12, 32]; U05a [1-9, 11-24, 29-32, 34, 35]; U08 [3-5, 7-24, 29-32]; U19 [1, 3, 4, 6-8, 10-24, 29-32]; U02 [1, 3, 5-24, 29-32]; U20b [1, 3, 5, 6, 9, 11-14, 16, 17, 24, 26-29]; U20a [2, 4, 7, 8, 10, 15, 18-23, 25, 30-32, 34, 35]; U20c [33]; U09a [1, 6]; C03 [1-26, 29-32, 34, 35]; E30a [1-35]; E01 [1-26, 29-32, 34, 35]; E26 [1, 2, 6-8, 10, 11, 15-17, 19, 20, 22, 24, 27, 28, 30-33]; E13c [1-5, 8-24, 27-35]; E19 [12, 14, 15, 17, 20, 22, 24]; E31b [1, 3-6, 9, 10, 12-18, 21, 23-29, 33]; E34 [1, 2, 6-8, 16, 19, 20, 22, 30-35]; E32a [2, 34, 35]; E31c [34, 35]; E15 [25, 26]; E07 [4]; E31a [11]; E13b [6]

Withholding period
- Keep livestock out of treated areas until foliage of any poisonous weeds such as ragwort has died and become unpalatable [4]
- Exclude livestock from treated fields. Livestock may not graze or be fed the treated forage nor may it be used for hay, silage or bedding [2, 34, 35]

Latest application/Harvest interval
- 24 h-5 d before cultivating stubbles depending on product and dose (see labels for details); pre-emergence for autumn crops and spring cereals; 72 h post-drilling for sugar beet, peas, swedes, turnips, onions and leeks; post-leaf fall but before white bud or green cluster for top fruit.
- HI grass 5 d; wheat, barley, oats, field beans, combining peas, linseed 7 d; mustard 8 d; oilseed rape 14 d

FOR FULL CONDITIONS OF USE ALWAYS READ THE PRODUCT LABEL

Approval
- May be applied through CDA equipment. See label for details. When applying through rotary atomisers the spray droplet spectra must have a minimum Volume Median Diameter (VMD) of 200 microns
- Accepted by BLRA for use on malting barley
- Approval expiry 31 Aug 2001 [23]; replaced by MAFF 09182

Maximum Residue Level (mg residue/kg food)
- wild mushrooms 50; barley, oats, sorghum, soya 20; linseed, mustard seed, rape seed 10; wheat, rye, triticale 5; meat (cattle, goat, sheep kidney), beans, peas 2; meat (pig kidney) 0.5; all other products (except beans, peas) 0.1

---

312   glyphosate (top fruit, horticulture, forestry, amenity etc.)

A translocated non-residual phosphonic acid herbicide

See also diuron + glyphosate

Products

1	AgriGuard Glyphosate 360	AgriGuard	360 g/l	SL	09184
2	Azural	Monsanto	360 g/l	SL	09582
3	Barbarian	Barclay	360 g/l	SL	07980
4	Barclay Barbarian	Barclay	360 g/l	SL	09865
5	Barclay Cleanup	Barclay	180 g/l	SL	09179
6	Barclay Dart	Barclay	180 g/l	SL	05129
7	Barclay Gallup	Barclay	360 g/l	SL	05161
8	Barclay Gallup 360	Barclay	360 g/l	SL	09127
9	Barclay Gallup Amenity	Barclay	360 g/l	SL	06753
10	Barclay Gallup Biograde 360	Barclay	360 g/l	SL	09840
11	Barclay Gallup Biograde Amenity	Barclay	360 g/l	SL	10203
12	Barclay Garryowen	Barclay	360 g/l	SL	08599
13	Buggy SG	Sipcam	36% w/w	WG	08573
14	Cardel Egret	Cardel	360 g/l	SL	09703
15	CDA Vanquish	RP Amenity	120 g/l	RH	08577
16	CDA Vanquish	Aventis Environ.	120 g/l	RH	09927
17	Clinic	Nufarm Whyte	360 g/l	SL	09378
18	GLY-480	Powaspray	480 g/l	SL	09837
19	Glyfos ProActive	Nomix-Chipman	360 g/l	SL	07800
20	Glyper	PBI	360 g/l	SL	07968
21	Greencrop Gypsy	Greencrop	360 g/l	SL	09432
22	Helosate	Helm	360 g/l	SL	06499
23	Hilite	Nomix-Chipman	144 g/l	RH	06261
24	MSS Glyfield	Nufarm Whyte	360 g/l	SL	08009
25	Portman Glyphosate	Portman	360 g/l	SL	05891
26	Portman Glyphosate 360	Portman	360 g/l	SL	04699
27	Rival	Monsanto	360 g/l	SL	09220
28	Roundup	Monsanto	360 g/l	SL	01828
29	Roundup Biactive	Monsanto	360 g/l	SL	06941
30	Roundup Biactive Dry	Monsanto	42% w/w	WG	06942
31	Roundup Pro Biactive	Monsanto	360 g/l	SL	06954
32	Spasor	RP Amenity	360 g/l	SL	07211
33	Spasor	Aventis Environ.	360 g/l	SL	09945
34	Spasor Biactive	RP Amenity	360 g/l	SL	07651
35	Spasor Biactive	Aventis Environ.	360 g/l	SL	09940
36	Stefes Glyphosate	Stefes	360 g/l	SL	05819
37	Stefes Kickdown	Stefes	360 g/l	SL	06329
38	Stirrup	Nomix-Chipman	144 g/l	RH	06132
39	Touchdown LA	Scotts	330 g/l	SL	09270

Uses

Annual and perennial weeds in *cherries, plums* [1-4, 7, 8, 10, 12-14, 17, 20-22, 26, 28-30, 36, 37]. Annual and perennial weeds in *damsons* [1-4, 7, 8, 10, 12-14, 17, 20-22, 28-30, 36, 37]. Annual and perennial weeds in *apples, pears* [1-4, 7, 8, 10, 12, 14, 17, 20-22, 26, 28-30, 36,

37]. Annual and perennial weeds in *forestry* [1, 3, 4, 7-13, 17, 19-27, 31, 36-38]. Annual and perennial weeds in *amenity vegetation* [15, 16, 27, 31]. Annual and perennial weeds in *non-crop areas* [18, 20, 22, 27, 31, 34, 35, 39]. Annual and perennial weeds in *land not intended to bear vegetation* [2, 13-16, 19, 24, 27, 31, 37]. Annual and perennial weeds in *woody ornamentals* [23, 31-35, 38]. Annual and perennial weeds in *hedges, top fruit* [23, 38]. Annual and perennial weeds in *conifers* [27, 31]. Annual and perennial weeds in *amenity grass (wiper application), amenity trees and shrubs (wiper application)* [27, 31-33]. Annual and perennial weeds in *hard surfaces, walls* [27, 31, 34, 35]. Annual and perennial weeds in *fencelines* [27, 31, 34, 35, 39]. Annual and perennial weeds in *farm buildings/yards, land clearance* [27, 31, 39]. Annual and perennial weeds in *asparagus (off-label), blackcurrants (off-label), grapevines (off-label), tree nuts (off-label)* [28]. Annual and perennial weeds in *amenity grass* [31-33]. Annual and perennial weeds in *industrial sites* [9, 11, 19, 27, 31]. Annual and perennial weeds in *road verges* [9, 11, 19, 27, 31, 34, 35]. Annual and perennial weeds in *amenity areas (wiper application)* [9, 11, 27, 31]. Annual and perennial weeds in *paths and drives* [9, 11, 27, 31, 34, 35, 39]. Annual and perennial weeds in *amenity areas* [9, 11, 31, 34, 35]. Annual and perennial weeds in *amenity trees and shrubs* [9, 23, 27, 31-33, 38, 39]. Annual and perennial weeds in *broad-leaved trees* [9, 27, 31]. Annual weeds in *leeks, onions* [5, 6, 29]. Bracken in *forestry* [1, 3, 4, 7-13, 17, 19-22, 24, 27, 31, 36, 37]. Bracken in *tolerant conifers* [27, 31]. Bracken in *amenity trees and shrubs, fencelines, land clearance, non-crop areas, paths and drives* [27, 31, 39]. Brambles in *tolerant conifers* [27, 31]. Chemical thinning in *forestry* [3, 4, 7-12, 17, 20, 22, 27, 31]. Couch in *forestry* [1, 3, 4, 7-12, 17, 20-22, 27, 31, 36, 37]. Heather in *forestry* [9, 11, 13, 19, 24]. Perennial dicotyledons in *forestry (wiper application)* [27, 31, 36]. Perennial dicotyledons in *orchards (wiper application)* [30]. Perennial grasses in *tolerant conifers* [13, 19, 24, 27, 31]. Perennial grasses in *aquatic situations* [3, 4, 8-11, 13, 17, 19, 20, 22, 24, 27, 30-35]. Reeds in *aquatic situations* [3, 4, 8-11, 13, 17, 19, 20, 22, 24, 27, 30-35]. Rhododendrons in *forestry* [9, 11, 13, 19, 24, 27, 31]. Rushes in *forestry* [3, 4, 8, 10, 20, 22, 27, 31, 36, 37]. Rushes in *aquatic situations* [3, 4, 8-11, 13, 17, 19, 20, 22, 24, 27, 30-35]. Sedges in *aquatic situations* [3, 4, 8-11, 13, 17, 19, 20, 22, 24, 27, 30-35]. Sucker control in *apples, cherries, damsons, pears, plums* [28-30]. Sward destruction in *amenity grass* [23, 27, 31, 38]. Total vegetation control in *amenity vegetation, land not intended to bear vegetation* [15, 16, 19, 27, 31]. Total vegetation control in *industrial sites* [19, 23, 27, 31-35, 38]. Total vegetation control in *hard surfaces* [19, 23, 27, 31, 38]. Total vegetation control in *fencelines, road verges* [19, 27, 31-33]. Total vegetation control in *amenity areas* [23, 27, 31-33, 36, 38]. Total vegetation control in *non-crop areas, paths and drives, walls* [27]. Volunteer cereals in *leeks, onions* [5, 6, 29]. Waterlilies in *aquatic situations* [3, 4, 8-11, 13, 17, 19, 20, 22, 24, 27, 30-35]. Woody weeds in *forestry* [1, 3, 4, 7-13, 17, 19-25, 27, 31, 36-38]. Woody weeds in *tolerant conifers* [27, 31].

Efficacy

- For best results apply to actively growing weeds with enough leaf to absorb chemical
- Products are formulated as either isopropylamine [1-12, 14-38], ammonium [13], or trimesium [39] salts of glyphosate and may vary in the details of efficacy claims. See individual product labels
- Annual weed grasses should have at least 5 cm of leaf and annual broad-leaved weeds at least 2 expanded true leaves
- Perennial grass weeds should have 4-5 new leaves and be at least 10 cm long when treated. Perennial broad-leaved weeds should be treated at or near flowering but before onset of senescence
- Bracken must be treated at full frond expansion
- In order to allow translocation, do not cultivate before treating perennials and do not apply other pesticides, lime, fertilizer or farmyard manure within 5 d of treatment
- Recommended intervals after treatment and before cultivation vary. See labels
- A rainfree period of at least 6 h (preferably 24 h) should follow spraying
- Most products should not be tank-mixed with other pesticides or fertilizers as such mixtures may lead to reduced control, but some [39] are compatible with listed diuron-containing products. See labels for details

- Adjuvants are obligatory for some products and recommended for some uses with others. See labels
- Do not spray weeds affected by drought, waterlogging, frost or high temperatures
- Fruit tree suckers best treated in late spring
- Chemical thinning treatment can be applied as stump spray or stem injection
- For rhododendron control apply to stumps or regrowth. Addition of adjuvant Mixture B recommended for application to foliage by knapsack sprayer [28]
- Product formulated ready for use without dilution. Apply only through specified applicator (see label for details) [15, 16, 23, 38]

Crop safety/Restrictions

- Maximum number of treatments normally 1 per crop or season on field and edible crops and unrestricted for non-crop uses. Check labels for details
- Decaying remains of plants killed by spraying must be dispersed before direct drilling
- Do not use treated straw as a mulch or growing medium for horticultural crops
- For use in orchards, grapevines and tree nuts care must be taken to avoid contact with the trees. Do not use in orchards established less than 2 yr and keep off low-lying branches
- Do not spray root suckers in orchards in late summer or autumn
- Do not use under glass or polythene as damage to crops may result
- With wiper application weeds should be at least 10 cm taller than crop
- Do not use wiper techniques in soft fruit crops
- Certain conifers may be sprayed overall in dormant season. See label for details
- Use a tree guard when spraying in established forestry plantations

Special precautions/Environmental safety

- Harmful if swallowed [5, 6, 37, 39]
- Irritating to eyes [37, 38] and skin [1-3, 6, 7, 9, 12, 14, 17, 18, 20-22, 25, 27, 28, 32, 33]
- Risk of serious damage to eyes [4-6, 8, 13]
- Causes severe burns [6]
- Harmful (dangerous [5]) to fish or other aquatic life. Do not contaminate surface waters or ditches with chemical or used container [1-4, 7-14, 17-25, 27, 28, 32, 33, 36-39]
- Maximum permitted concentration in treated water 0.2 ppm [4, 8-11, 19, 22, 24, 27, 28]
- The Environment Agency or Local River Purification Authority must be consulted before use in or near water
- Do not mix, store or apply in galvanised or unlined mild steel containers or spray tanks
- Do not leave diluted chemical in spray tanks for long period and make sure that tanks are well vented
- Take extreme care to avoid drift onto other crops
- For field edge treatment direct spray away from hedge bottoms
- Do not dump surplus herbicide in water or ditch bottoms [4, 8, 10, 11, 13, 17, 19, 20, 22, 24, 32, 33]

Personal protective equipment/Label precautions

- A [1-25, 27-39]; C [1-14, 17-25, 27-33, 36-39]; M [1, 2, 4, 8, 10, 11, 13-17, 19-25, 27-35, 38, 39]; N [1, 2, 4, 8, 14, 17, 20-23, 25, 27, 28, 32, 33, 38]; H [10, 11, 13, 15-17, 19-22, 24, 27-31, 34, 35, 39]; F [17, 20, 22, 27, 28]; D [19, 24]
- M03 [5, 6, 37, 39]; R04a [1-3, 6, 7, 9, 12, 14, 17, 18, 20-22, 25, 27, 28, 32, 33, 37, 38]; R04b [1-3, 6, 7, 9, 12, 14, 17, 18, 20-22, 25, 27, 28, 32, 33]; R05a [6]; R03c [5, 6, 37, 39]; R04d [4-6, 8, 13]; U19 [1-3, 5-7, 9-25, 27, 28, 32, 33, 36-38]; U02 [1, 2, 4-14, 17-25, 27, 28, 32, 33, 36-38]; U05a [1-9, 12-14, 17, 18, 20-23, 25, 27, 28, 32, 33, 36-39]; U08 [2-4, 6-14, 17, 18, 20-23, 25, 27, 28, 32, 33, 36-38]; U20a [3, 6, 7, 9-11, 19, 20, 22, 24, 25, 28-30, 34-37, 39]; U20b [1, 2, 4, 5, 8, 12-18, 21, 23, 27, 31-33, 38]; U09a [1, 5]; C03 [1-18, 20-23, 25, 27-39]; E01, E30a [1-25, 27-39]; E13c [1-4, 7-14, 17-25, 27, 28, 32, 33, 36-39]; E31b [1-5, 8, 10, 11, 13-22, 24, 25, 27-31, 34, 35]; E26 [1, 5-7, 9-12, 15, 16, 19-24, 36-38]; E34 [1, 5-7, 9, 21, 22, 36, 37, 39]; E32a [23, 38, 39]; E19 [4, 8, 10, 11, 13, 17, 19, 20, 22, 24, 32, 33]; E15 [15, 16, 29-31, 34, 35]; E07 [3]; E31a [12]; E13b [5]

Latest application/Harvest interval

- 5 d before drilling for grassland; stubbles 24 h-5 d depending on product and dose (see labels for details); post-leaf fall but before white bud or green cluster for top fruit

Approval
- Approved for aquatic weed control. See notes in Section 1 on use of herbicides in or near water [4, 8-11, 17, 19, 22, 24, 28-35]
- May be applied through CDA equipment. See label for details. When applying through rotary atomisers the spray droplet spectra must have a minimum Volume Median Diameter (VMD) of 200 microns
- Off-label approval unlimited for use on grapevines and tree nuts (OLA 0337/92)[28]; unlimited for use on outdoor asparagus (OLA 0584/94)[28]; unlimited for use on blackcurrants (OLA 1407/96)[28]

Maximum Residue Level (mg residue/kg food)
- see glyphosate (agriculture) entry

---

313   glyphosate + oxadiazon

A non-selective herbicide for total vegetation control

Products

1 Zapper	Aventis Environ.	10.8:30% w/w	WB	09944
2 Zapper	RP Amenity	10.8:30% w/w	WB	08605

Uses

Annual and perennial weeds in *land not intended to bear vegetation* (soil surface treatment). Total vegetation control in *land not intended to bear vegetation* (soil surface treatment).

Efficacy
- Apply post-emergence of weeds at any time when they are actively growing from Mar to end Sep. Control may be reduced if weeds are under stress when treated
- Best results on perennials including docks, perennial sowthistle and willowherb obtained from treatment just before flowering or seed set
- A rainfree period of at least 6 h (preferably 24 h) should follow spraying
- Perennials such as dandelions or docks emerging after application from established rootstocks will not be controlled
- Pre-emergence activity is reduced on soils with more than 10% organic matter or where leaves have collected or a mat of organic matter has built up
- Use of wetting agents or adjuvants may reduce activity

Crop safety/Restrictions
- When used on sites that are to be cleared or grubbed 2 yr should elapse before sowing or planting anything. Soil should be ploughed or dug after clearance
- Take care to prevent heavy rain after application washing product onto sensitive areas such as newly sown grass or areas about to be planted

Special precautions/Environmental safety
- Dangerous to fish or other aquatic life. Do not contaminate surface waters or ditches with chemical or used container
- LERAP Category B

Personal protective equipment/Label precautions
- A, C
- U20c, U22, E13b, E16a, E16b, E30a, E32a

Maximum Residue Level (mg residue/kg food)
- see glyphosate (agricultural) entry

---

314   guazatine

A guanidine fungicide seed dressing for cereals

Products

1 Panoctine	RP Agric.	300 g/l	LS	06207
2 Ravine	Aventis	300 g/l	LS	10095

Uses

Fusarium foot rot and seedling blight in **barley** *(seed treatment - reduction)*, **oats** *(seed treatment - reduction)*, **wheat** *(seed treatment - reduction)*. Septoria seedling blight in **wheat** *(seed treatment)*.

Efficacy

- Apply with conventional seed treatment machinery
- After treating, bag seed immediately and keep in dry, draught-free store

Crop safety/Restrictions

- Maximum number of treatments 1 per batch
- Do not treat grain with moisture content above 16% and do not allow moisture content of treated seed to exceed 16%
- Do not apply to cracked, split or sprouted seed
- Treatment may lower germination capacity, particularly if seed grown, harvested or stored under adverse conditions

Special precautions/Environmental safety

- Harmful if swallowed and in contact with skin. Risk of serious damage to eyes
- Dangerous to fish or other aquatic life. Do not contaminate surface waters or ditches with chemical or used container
- Do not use treated seed as food or feed

Personal protective equipment/Label precautions

- A, C, D, H
- M03, R03a, R03c, R04d, U05a, U20b, C03, E01, E13b, E26, E30a, E31a, E34, S01, S02, S04b, S05, S06, S07

Latest application/Harvest interval

- Pre-drilling

Approval

- Accepted by BLRA for use as a seed treatment on malting barley

---

315   guazatine + imazalil

A fungicide seed treatment for barley and oats

Products

Panoctine Plus	RP Agric.	300:25 g/l	LS	06208

Uses

Brown foot rot in **barley** *(seed treatment - reduction)*, **oats** *(seed treatment - reduction)*. Foot rot in **barley** *(seed treatment)*. Fusarium foot rot and seedling blight in **barley** *(seed treatment - reduction)*, **oats** *(seed treatment - reduction)*. Leaf stripe in **barley** *(seed treatment)*. Net blotch in **barley** *(seed treatment)*. Pyrenophora leaf spot in **oats** *(seed treatment)*.

Efficacy

- Apply with conventional seed treatment machinery
- After treating, bag seed immediately and keep in dry, draught-free store

Crop safety/Restrictions

- Maximum number of treatments 1 per batch
- Do not treat grain with moisture content above 16% and do not allow moisture content of treated seed to exceed 16%
- Do not apply to cracked, split or sprouted seed
- Do not store treated seed for more than 6 mth
- Treatment may lower germination capacity, particularly if seed grown, harvested or stored under adverse conditions

Special precautions/Environmental safety

- Harmful if swallowed and in contact with skin. Risk of serious damage to eyes
- Dangerous to fish or other aquatic life. Do not contaminate surface waters or ditches with chemical or used container
- Do not use treated seed as food or feed

Personal protective equipment/Label precautions
- A, C, D, H
- M03, R03a, R03c, R04d, U05a, U09a, U20b, C03, E01, E13b, E26, E30a, E31a, E34, S01, S02, S04b, S05, S06, S07

Latest application/Harvest interval
- Pre-drilling

Approval
- Imazalil included in Annex I under EC Directive 91/414
- Accepted by BLRA for use on malting barley

Maximum Residue Level (mg residue/kg food)
- see imazalil entry

---

316   hymexazol

A systemic isoxazole fungicide for pelleting sugar beet seed

Products

Tachigaren 70 WP	Sumitomo	70% w/w	WP	02649

Uses

Aphanomyces cochlioides in **beetroot** *(off-label)*. Black leg in **sugar beet**. Damping off in **sugar beet**.

Efficacy
- Incorporate into pelleted seed using suitable seed pelleting machinery

Crop safety/Restrictions
- Maximum number of treatments 1 per batch of seed

Special precautions/Environmental safety
- Irritating to eyes, skin and respiratory system
- Harmful to fish of aquatic life. Do not contaminate surface waters or ditches with chemical or used container
- Do not use treated seed as food or feed
- Treated seed harmful to game and wildlife

Personal protective equipment/Label precautions
- A, C, F
- R04a, R04b, R04c, U05a, U20a, C03, E01, E13c, E30a, E32a, S01, S02, S05, S06

Approval
- Off-label approval unlimited for use on pelleted beetroot seed to be sown outdoors (OLA 0565/98)

---

317   imazalil

A systemic and protectant conazole fungicide

See also azaconazole + imazalil
carboxin + imazalil + thiabendazole
guazatine + imazalil

Products

1	Fungaflor	Hortichem	200 g/l	EC	05967
2	Fungaflor Smoke	Hortichem	15% w/w	FU	05969
3	Fungazil 100 SL	RP Agric.	100 g/l	LS	06202
4	Sphinx	Aventis	50 g/l	LS	07607
5	Stryper	Uniroyal	50 g/l	LS	08310

Uses

Brown foot rot in **spring barley** *(seed treatment)*, **winter barley** *(seed treatment)* [4, 5]. Dry rot in **seed potatoes, ware potatoes** [3]. Gangrene in **seed potatoes, ware potatoes** [3]. Leaf stripe in **spring barley** *(seed treatment)*, **winter barley** *(seed treatment)* [4, 5]. Net blotch in **spring barley** *(seed treatment)*, **winter barley** *(seed treatment)* [4, 5]. Powdery

---

mildew in **courgettes** *(off-label)*, **gherkins** *(off-label)* [1]. Powdery mildew in **protected cucumbers, protected ornamentals, protected roses** [1, 2]. Silver scurf in **seed potatoes, ware potatoes** [3]. Skin spot in **seed potatoes, ware potatoes** [3].

Efficacy
- Treat cucurbits before or as soon as disease appears and repeat every 10-14 d or every 7 d if infection pressure great or with susceptible cultivars [1, 2]
- Apply to clean soil-free potatoes post-harvest before putting into store or at first grading. A further treatment may be applied in early spring before planting [3]
- For best control of skin and wound diseases of ware potatoes treat as soon as possible after harvest, preferably within 7-10 d, before any wounds have healed [3]
- Apply through canopied hydraulic or spinning disc equipment preferably diluted with up to 2 l water per tonne of potatoes to obtain maximum skin cover and penetration [3]
- Apply seed treatment via conventional machine and sow treated seed as soon as possible [4, 5]

Crop safety/Restrictions
- Maximum number of treatments 1 for ware potatoes and 2 per batch of seed potatoes [3]; 1 per batch of barley seed [4, 5]
- Where possible apply to ware potatoes at least 6 wk before washing for sale or use. Consult processor before treating potatoes for processing [3]
- Do not spray cucurbits or ornamentals in full, bright sunshine. When spraying in the evening the spray should dry before nightfall. May cause damage if open flowers are sprayed. Do not use on rose cultivars Dr A.J. Verhage and Jack Frost [1]
- With ornamentals of unknown tolerance test on a few plants in first instance [1]
- Do not treat grain over 16% moisture content and keep seed dry and draught free after treatment [4, 5]
- Do not apply to cracked, split or sprouted seed [4, 5]

Special precautions/Environmental safety
- Harmful if swallowed [1, 3]
- Irritating to eyes and skin [1, 3]; irritating to eyes [2, 4, 5]
- Causes burns [1]
- Flammable [1]; highly flammable [3]
- Keep unprotected persons out of treated areas for at least 2 h [2]
- Harmful to bees. Do not apply to crops in flower or to those in which bees are actively foraging. Do not apply when flowering weeds are present [1]
- Harmful to fish or other aquatic life. Do not contaminate surface waters or ditches with chemical or used container [1, 3-5]
- Do not use treated seed as food or feed [3-5]

Personal protective equipment/Label precautions
- A, C [1, 3-5]; H [1, 4, 5]; D [3]
- M03 [3-5]; R05b, R07d [1]; R04a [1-5]; R03c, R04b [1, 3]; R07c [2]; U08 [1]; U19, U20a [1, 3]; U05a [2-5]; U04a [3]; U09a, U20b [4, 5]; C02 [1, 2] (1 d); C03 [2-5]; E12e [1]; E32a [1, 2]; E30a [1-5]; E34 [1, 3]; E13c [1, 3-5]; E02 [2]; E01 [2-5]; E26, E31a [3-5]; E03 [4, 5]; S01, S02, S05, S06 [3-5]; S03, S04a, S07 [4, 5]

Latest application/Harvest interval
- During storage and/or prior to chitting for potatoes [3].
- HI cucumbers 1 d [1, 2]

Approval
- Imazalil included in Annex I under EC Directive 91/414
- Off-label approval unstipulated for use on protected courgettes and gherkins (OLA 1491/99)[1]; to Jun 2001 for use on outdoor courgettes and gherkins (OLA 1492/99)[1]
- Accepted by BLRA for use as a seed treatment on malting barley

Maximum Residue Level (mg residue/kg food)
- citrus fruits; pome fruits, ware potatoes 5; bananas 2; cucumbers, gherkins, courgettes 0.2; tea, hops 0.1; all other products 0.02

318 imazalil + pencycuron

A fungicide mixture for treatment of seed potatoes

Products

1 Monceren IM	Bayer	0.6:12.5% w/w	DS	06259
2 Monceren IM Flowable	Bayer	20:250 g/l	FS	06731

Uses

Black scurf in **potatoes** *(tuber treatment)*. Silver scurf in **potatoes** *(tuber treatment - reduction)*. Stem canker in **potatoes** *(tuber treatment - reduction)*.

Efficacy
- Apply to clean seed tubers during planting (see label for suitable method) or sprinkle over tubers in chitting trays before loading into planter. It is essential to obtain an even distribution over tubers [1]
- Apply to clean seed tubers into or out of store (follow agronomic guidelines in manufacturer's literature) [2]
- If seed tubers become damp from light rain distribution of product should not be affected. Tubers in the hopper should be covered if a shower interrupts planting
- Use suitable misting equipment. See label for details [2]
- Treatment usually most conveniently carried out over roller table at the end of grading out [2]

Crop safety/Restrictions
- Maximum number of treatments 1 per batch
- May be used on seed tubers previously treated with a liquid fungicide but not before 8 wk have elapsed if this contained imazalil
- Do not use on tubers which have previously been treated with a dry powder seed treatment or hot water
- Do not tank mix with other potato fungicides or storage products [2]
- Treated tubers may only be used as seed. They must not be used for human or animal consumption

Special precautions/Environmental safety
- Operators must wear suitable respiratory equipment and gloves when handling product and when riding on planter. Wear gloves when handling treated tubers
- Irritating to eyes [2]
- Harmful to fish or other aquatic life. Do not contaminate surface waters or ditches with chemical or used container
- Do not use treated seed as food or feed
- Treated seed harmful to game and wildlife

Personal protective equipment/Label precautions
- D [1]; A [1, 2]; C [2]
- R04a [2]; U20a, E03, E13c, E30a, E32a, S01, S02, S03, S04b, S05, S06 [1, 2]

Latest application/Harvest interval
- Immediately before planting

Approval
- Imazalil included in Annex I under EC Directive 91/414

Maximum Residue Level (mg residue/kg food)
- see imazalil entry

319 imazalil + thiabendazole

A fungicide mixture for treatment of seed potatoes

Products

Extratect Flowable	Banks	100:300 g/l	FS	08704

Uses

Dry rot in **seed potatoes** *(tuber treatment - reduction)*. Gangrene in **seed potatoes** *(tuber treatment - reduction)*. Silver scurf in **potatoes** *(tuber treatment - reduction)*, **seed potatoes** *(tuber treatment - reduction)*. Skin spot in **potatoes** *(tuber treatment - reduction)*, **seed potatoes** *(tuber treatment - reduction)*. Stem canker in **potatoes** *(tuber treatment - reduction)*.

Efficacy

- Apply immediately after lifting with suitable spinning disc or low volume hydraulic applicator for disease reduction during storage
- Apply within 2 mth of planting for maximum disease reduction in the progeny crop

Crop safety/Restrictions

- Maximum number of treatments 1 per batch
- Do not mix with any other product
- May not be used on seed tubers previously treated with a product containing imazalil
- Delayed emergence noted under certain conditions but with no final effect on yield

Special precautions/Environmental safety

- Irritating to respiratory system. Risk of serious damage to eyes
- Harmful to fish or other aquatic life. Do not contaminate surface waters or ditches with chemical or used container
- Do not use treated seed as food or feed

Personal protective equipment/Label precautions

- A, C, D, H
- R04c, R04d, U05a, U09a, U11, U19, U20c, C03, E01, E13c, E30a, E31a, S01, S02, S05, S06

Latest application/Harvest interval

- Pre-planting

Approval

- Imazalil included in Annex I under EC Directive 91/414
- Approved for use through ULV equipment

Maximum Residue Level (mg residue/kg food)

- see imazalil and thiabendazole entries

---

320   imazalil + triticonazole

A conazole fungicide mixture for seed treatment of barley

Products

Robust	Aventis	12.5:12.5 g/l	LS	10176

Uses

Covered smut in **spring barley, winter barley**. Leaf stripe in **spring barley, winter barley**. Loose smut in **spring barley, winter barley**. Net blotch in **spring barley** *(moderate control)*, **winter barley** *(moderate control)*. Seedling blight and foot rot in **spring barley** *(moderate control)*, **winter barley** *(moderate control)*.

Efficacy

- Apply undiluted through conventional seed treatment machine
- Treated seed may be stored for up to 18 mth but germination should be checked befoe use and sowing rate adjusted if necessary

Crop safety/Restrictions

- Maximum number of treatments 1 per batch of seed
- Do not treat seed with moisture content above 16%
- Do not apply to cracked, split or sprouted seed
- Keep treated seed in a dry, draught free store
- Calibrate seed drill before sowing
- Product available in large (1000 l) returnable containers. They should be re-circulated for 30 min before use and operated as directed on the label

Special precautions/Environmental safety

- Irritant. May cause sensitization by skin contact
- Harmful to fish or other aquatic life. Do not contaminate surface waters or ditches with chemical or used container

- Do not handle treated seed unnecessarily
- Do not use treated seed as food or feed

Personal protective equipment/Label precautions
- A, C, H
- M03, R04, R04e, U05a, U09a, U20b, C03, E01, E13c, E26, E30a, E34, S01, S02, S03, S04b, S05, S06, S07

Latest application/Harvest interval
- Pre-drilling

Approval
- Imazalil included in Annex I under EC Directive 91/414

321   imazamethabenz-methyl

A post-emergence imidazolinone grass weed herbicide for use in winter cereals

Products

1 Dagger	BASF	300 g/l	SC	10218
2 Dagger	Cyanamid	300 g/l	SC	03737

Uses

Blackgrass in *winter barley, winter wheat*. Charlock in *winter barley, winter wheat*. Loose silky bent in *winter barley, winter wheat*. Onion couch in *winter barley, winter wheat*. Volunteer oilseed rape in *winter barley, winter wheat*. Wild oats in *winter barley, winter wheat*.

Efficacy
- Has contact and residual activity. Wild oats controlled from pre-emergence to 4 leaves + 3 tillers stage, charlock and volunteer oilseed rape to 10 cm. Good activity on onion couch up to 7.5 cm and useful suppression of cleavers
- When used alone add Agral for improved contact activity
- Best results achieved when applied to fine, firm, clod-free seedbed when soil moist
- Do not use on soils with more than 10% organic matter
- Effects of autumn/winter treatment normally persist to control spring flushes of weeds
- Tank mixture with isoproturon recommended for control of tillered blackgrass, with pendimethalin for general grass and dicotyledon control

Crop safety/Restrictions
- Maximum number of treatments 2 per crop (as split dose treatment)
- Apply from 2-fully expanded leaf stage of crop to before second node detectable (GS 12-32)
- Do not use on durum wheat
- Do not use on soils where surface water is likely to accumulate
- See label for guidance on following crops after normal harvesting
- In case of crop failure land must be ploughed to 15 cm and re-drilled in spring but only with specified crops - see label for details

Special precautions/Environmental safety
- Irritating to eyes

Personal protective equipment/Label precautions
- A, C
- M03, R04a, U05a, U08, U20a, C02, E01, E15, E26, E30a, E31b, E34

Latest application/Harvest interval
- Before 2nd node detectable (GS 32)

Approval
- Accepted by BLRA for use on malting barley

322 imazapyr

A non-selective translocated and residual imidazolinone herbicide for use on non-crop land and in forestry

Products

1 Arsenal	BASF	250 g/l	SL	10208
2 Arsenal	Nomix-Chipman	250 g/l	SL	05537
3 Arsenal	Cyanamid	250 g/l	SL	04064
4 Arsenal 50	Nomix-Chipman	50 g/l	SL	05567
5 Arsenal 50	BASF	50 g/l	SL	BASF
6 Arsenal 50	Cyanamid	50 g/l	SL	04070

Uses

Bracken in **non-crop areas** [1-6]. Bracken in **farm buildings/yards, fencelines, forestry** (site preparation), **industrial sites, railway tracks** [4-6]. Total vegetation control in **non-crop areas** [1-6]. Total vegetation control in **farm buildings/yards, fencelines, forestry** (site preparation), **industrial sites, railway tracks** [4-6].

Efficacy

- Chemical is absorbed through roots and foliage, kills underground storage organs and gives long term residual control. Complete kill may take several wk
- May be applied before weed emergence but gives best results from application at any time of year when weeds are growing actively
- Apply from beginning Jul to end Oct as a conifer site preparation treatment [4, 6]

Crop safety/Restrictions

- Maximum number of treatments 1 per yr
- Avoid drift onto desirable plants
- Do not apply to soil which may later be used to grow desirable plants
- Do not apply on or near desirable trees or plants or on areas where their roots may extend or in locations where the chemical may be washed or move into contact with their roots
- At least 5 mth must elapse between application and planting of named conifers only (Sitka spruce, Lodgepole pine, Corsican pine)
- Must not be used as site preparation treatment for broad-leaved tree species

Special precautions/Environmental safety

- Irritating to eyes
- Not to be used on food crops
- Dangerous to fish or other aquatic life. Do not contaminate surface waters or ditches with chemical or used container [4-6]
- LERAP Category B

Personal protective equipment/Label precautions

- A, C [1-6]; M [1-3]
- R04a, U02, U05a, U08, U09a, U20b, C03, E01, E16a, E16b, E26, E30a, E31b [1-6]; E15 [1-3]; E13b [4-6]

Approval

- May be applied through CDA equipment. See label for details [1-4, 6]
- Product registration number not available at time of printing [5]

323 imazaquin

An imidazolinone herbicide and plant growth regulator available only in mixtures

See also chlormequat + 2-chloroethylphosphonic acid + imazaquin
chlormequat + choline chloride + imazaquin

324 imidacloprid

A nitroimidazolidinimine insecticide for seed, soil, peat or foliar treatment

See also bitertanol + fuberidazole + imidacloprid
fuberidazole + imidacloprid + triadimenol

Products

1	Admire	Bayer	70% w/w	WG	07481
2	Gaucho	Bayer	70% w/w	WS	06590
3	Intercept 5GR	Scotts	5% w/w	GR	08126
4	Intercept 70WG	Scotts	70% w/w	WG	08585

Uses

Aphids in **lettuce** *(off-label - seed treatment)* [2]. Aphids in **bedding plants, hardy ornamental nursery stock, pot plants** [3, 4]. Beet virus yellows vectors in **sugar beet** *(seed treatment)* [2]. Damson-hop aphid in **hops** [1]. Flea beetles in **sugar beet** *(seed treatment)* [2]. Mangold fly in **sugar beet** *(seed treatment)* [2]. Millipedes in **sugar beet** *(seed treatment)* [2]. Pygmy beetle in **sugar beet** *(seed treatment)* [2]. Sciarid flies in **bedding plants, hardy ornamental nursery stock, pot plants** [3, 4]. Springtails in **sugar beet** *(seed treatment)* [2]. Symphylids in **sugar beet** *(seed treatment)* [2]. Vine weevil in **bedding plants, hardy ornamental nursery stock, pot plants** [3, 4]. Whitefly in **bedding plants, hardy ornamental nursery stock, pot plants** [3, 4].

Efficacy

- Apply to sugar beet seed as part of the normal commercial pelleting process using special treatment machinery [2]
- Treated seed should be drilled within the season of purchase [2]
- Apply as a directed stem base spray before most bines reach a height of 2 m. If necessary treat both sides of the crown at half the normal concentration [1]
- Base of hop plants should be free of weeds and debris at application [1]
- Bines emerging away from the main stock or adjacent to poles may require a separate application [1]
- Uptake and movement within hops requires soil moisture and good growing conditions [1]
- Control may be impaired in plantations greater than 3640 plants/ha [1]
- When applied as drench or incorporated as granules in moist compost compound is readily absorbed and translocated to aerial parts of plant [3, 4]

Crop safety/Restrictions

- Maximum number of treatments 1 per batch of seed [2] or growing medium [3, 4]; 1 per yr [1]
- To minimise likelihood of resistance do not treat all the hop crop in any one yr [1] and adopt a planned programme to alternate with pesticides of different types or use other measures when using in compost [3, 4]
- Product must not be used in compost that has already been treated with an imidacloprid-containing product [4]
- Product formulated for use only as a compost incorporation treatment into peat-based growing media [3]
- For use only on container grown ornamentals [3, 4]
- The safety of seeds sown into treated compost should not be assumed. Test treat before full-scale use [3]
- Product must not be used on crops for human or animal consumption and treated compost must not be re-used for this purpose [3, 4]

Special precautions/Environmental safety

- Irritant. May cause sensitization by skin contact [1, 2]
- High risk to bees. Do not apply to crops in flower or to those in which bees are actively foraging. Do not apply when flowering weeds are present [1, 4]
- Avoid spillage or other environmental contamination when incorporating into compost [3]
- Do not use treated seed as food or feed [2]
- Treated seed harmful to game and wildlife [2]

Personal protective equipment/Label precautions
- C, D, H [2]; A [1-3]
- R04e [1, 2]; R04 [1]; U20a [1, 2]; U05a [1, 4]; U20b [3, 4]; C03 [1, 4]; E15, E30a, E32a [1-4]; E01, E12a, E34 [1, 4]; S02, S04b, S05, S06, S07 [2]

Latest application/Harvest interval
- Before bines reach 2 m or before end 1st wk Jun for hops [1]; before drilling for sugar beet [2]; before sowing or planting for bedding plants, hardy ornamental nursery stock, pot plants [3]

Approval
- Off-label approval to May 2001 for use on lettuce seed (OLA 1041/96)[2]
- Accepted by BLRA for use on hops [1]

325   imidacloprid + tebuconazole + triazoxide

A broad spectrum fungicide and insecticide seed treatment for winter barley

Products

Raxil Secur	Bayer	233:20:20 g/l	LS	08966

Uses

Leaf stripe in **winter barley** *(seed treatment)*. Loose smut in **winter barley** *(seed treatment)*. Net blotch in **winter barley**. Virus vectors in **winter barley** *(seed treatment)*. Wireworms in **winter barley** *(reduction of damage)*.

Efficacy
- Best applied through recommended seed treatment machines
- Evenness of seed cover improved by simultaneous application of equal volumes of product and water or dilution of product with an equal volume of water
- Drill treated seed in the same season. Use minimum 125 kg treated seed per ha. Lower drilling rates and/or early drilling affect duration of BYDV protection needed and may require follow-up aphicide treatment
- In high risk areas where aphid activity is heavy and prolonged a follow-up aphicide treatment may be required
- Protection against foliar air-borne and splash-borne diseases later in the season will require appropriate fungicide follow-up sprays

Crop safety/Restrictions
- Maximum number of treatments 1 per batch of seed
- Slightly delayed and reduced emergence may occur but this is normally outgrown
- Any delay in field emergence, for whatever reason, may be accentuated by treatment
- Do not use on seed with more than 16% moisture content, or on sprouted, cracked or skinned seed

Special precautions/Environmental safety
- Harmful if swallowed
- May cause sensitisation by skin contact
- Dangerous to fish or other aquatic life. Do not contaminate surface waters or ditches with chemical or used container
- Do not use treated seed as food or feed
- Dangerous to birds, game and other wildlife. Treated seed should not be left on the soil surface. Bury spillages
- Treated seed should not be broadcast, but drilled to a depth of 4 cm in a well prepared seedbed
- If seed is left on the soil surface the field should be harrowed and rolled to ensure good incorporation

Personal protective equipment/Label precautions
- A, H
- M03, R03c, R04e, U04a, U05a, U13, U20b, C03, E01, E03, E13c, E26, E30a, E32a, E34, S01, S02, S03, S04c, S05, S06, S07

Latest application/Harvest interval
- Before drilling

Approval
- Accepted by BLRA as a seed treatment on malting barley

326   indol-3-ylacetic acid

A plant growth regulator for promoting rooting of cuttings

Products

1	Rhizopon A Powder	Fargro	1% w/w	DP	07131
2	Rhizopon A Tablets	Fargro	50 mg a.i.	TB	07132

Uses

Rooting of cuttings in *ornamentals*.

Efficacy

- Apply by dipping end of prepared cuttings into powder or dissolved tablet solution
- Shake off excess powder and make planting holes to prevent powder stripping off [1]
- Consult manufacturer for details of application by spray or total immersion

Crop safety/Restrictions

- Maximum number of treatments 1 per cutting
- Store product in a cool, dark and dry place
- Use solutions once only. Discard after use [2]
- Use plastic, not metallic, container for solutions [2]

Personal protective equipment/Label precautions

- U19 [1]; U20a, E15, E30a, E32a [1, 2]

Latest application/Harvest interval

- Before cutting insertion

327   4-indol-3-yl-butyric acid

A plant growth regulator promoting the rooting of cuttings

Products

1	Chryzoplus Grey	Fargro	0.8% w/w	DP	07984
2	Chryzopon Rose	Fargro	0.1% w/w	DP	07982
3	Chryzosan White	Fargro	0.6% w/w	DP	07983
4	Chryzotek Beige	Fargro	0.4% w/w	DP	07125
5	Chryzotop Green	Fargro	0.25% w/w	DP	07129
6	Rhizopon AA Powder (0.5%)	Fargro	0.5% w/w	DP	07126
7	Rhizopon AA Powder (1%)	Fargro	1% w/w	DP	07127
8	Rhizopon AA Powder (2%)	Fargro	2% w/w	DP	07128
9	Rhizopon AA Tablets	Fargro	50 mg a.i.	TB	07130
10	Seradix 1	Hortichem	0.1% w/w	DP	06191
11	Seradix 2	Hortichem	0.3% w/w	DP	06193
12	Seradix 3	Hortichem	0.8% w/w	DP	06194

Uses

Rooting of cuttings in *ornamentals*.

Efficacy

- Dip base of cuttings into powder immediately before planting
- Powders or solutions of different concentration are required for different types of cutting. Lowest concentration for softwood, intermediate for semi-ripe, highest for hardwood
- See label for details of concentration and timing recommended for different species
- Use of planting holes recommended for powder formulations to ensure product is not removed on insertion of cutting. Cuttings should be watered in if necessary

Crop safety/Restrictions

- Use of too strong a powder or solution may cause injury to cuttings
- No unused moistened powder should be returned to container

Personal protective equipment/Label precautions

- U19, U20a, E15, E30a, E32a

328   4-indol-3-yl-butyric acid + 2-(1-naphthyl)acetic acid with dichlorophen

A plant growth regulator for promoting rooting of cuttings

Products

Synergol	Hortichem	5.0:5.0 g/l	SL	07386

Uses

Rooting of cuttings in **ornamentals**.

Efficacy

- Dip base of cuttings into diluted concentrate immediately before planting
- Suitable for hardwood and softwood cuttings
- See label for details of concentration and timing for different species

Special precautions/Environmental safety

- Harmful if swallowed

Personal protective equipment/Label precautions

- A, C
- M03, R03c, U05a, U13, U20b, C03, E01, E15, E30a, E32a, E34

329   ioxynil

A contact acting HBN herbicide for use in turf and onions

See also benazolin + bromoxynil + ioxynil
bromoxynil + diflufenican + ioxynil
bromoxynil + ethofumesate + ioxynil
bromoxynil + ioxynil
bromoxynil + ioxynil + mecoprop-P
bromoxynil + ioxynil + triasulfuron

Products

1	Actrilawn 10	RP Amenity	100 g/l	SL	05247
2	Totril	Aventis	225 g/l	EC	10026
3	Totril	RP Agric.	225 g/l	EC	06116

Uses

Annual dicotyledons in **newly sown turf** [1]. Annual dicotyledons in **carrots** (off-label), **chives** (off-label), **garlic, leeks, onions, parsnips** (off-label), **shallots** [2, 3].

Efficacy

- Best results on seedling to 4-leaf stage weeds in active growth during mild weather
- In newly sown turf apply after first flush of weed seedlings, in spring normally 4 wk after sowing
- May be used on established turf from May to Sep under suitable conditions. Do not mow within 7 d of treatment

Crop safety/Restrictions

- Maximum number of treatments 1 per yr for turf; 1 per crop (more at split doses) for spring or autumn sown bulb onions, spring sown leeks and autumn sown salad onions; 1 per crop (more at split doses) for spring sown pickling and salad onions, transplanted onions, leeks, garlic and shallots. See label for details of split applications
- Apply to sown onion crops as soon as possible after plants have 3 true leaves or to transplanted crops when established
- Apply to newly sown turf after the 2-leaf stage of grasses
- Do not use on crested dogstail

Special precautions/Environmental safety

- Harmful if swallowed [1], if swallowed or in contact with skin [3]
- Irritating to eyes and skin [3]
- Do not apply by hand-held equipment or at concentrations higher than those recommended

- Dangerous to fish or other aquatic life. Do not contaminate surface waters or ditches with chemical or used container
- Harmful to bees. Do not apply to crops in flower or to those in which bees are actively foraging. Do not apply when flowering weeds are present

Personal protective equipment/Label precautions
- A, C [1-3]
- M03, R03c [1-3]; R03a, R04a [2, 3]; U05a, U08, U13, U19, U20b, C03 [1-3]; E13c [1]; E01, E26, E30a, E31b, E34 [1-3]; E07 [1-3] (6 wk); E12e, E13b [2, 3]

Withholding period
- Keep livestock out of treated areas for at least 6 wk after treatment and until foliage of any poisonous weeds such as ragwort has died and become unpalatable

Latest application/Harvest interval
- HI onions, shallots, garlic, leeks 14 d [3]

Approval
- Ioxynil was reviewed in 1995 and approvals for home garden use, and most hand held applications revoked. There are timing restrictions on grassland, leeks and onions and some other crops
- Off-label approval to Feb 2005 for use on outdoor carrots, parsnips (OLA 0326/00)[3]; to Jul 2003 for use on outdoor carrots, parsnips (OLA 1920/00)[2]; unstipulated for use on chives (OLA1918/00)[2]

Maximum Residue Level (mg residue/kg food)
- garlic, onions, shallots 0.1

---

330   iprodione

A protectant dicarboximide fungicide with some eradicant activity

See also carbendazim + iprodione

Products

1	Aventis Rovral Flo	Aventis	255 g/l	SC	09974
2	I T Iprodione	I T Agro	50% w/w	WP	08267
3	Landgold Iprodione 250	Landgold	255 g/l	SC	06465
4	Rovral Flo	RP Agric.	255 g/l	SC	06328
5	Rovral Flo	Aventis	255 g/l	SC	10013
6	Rovral Green	Aventis Environ.	250 g/l	SC	09938
7	Rovral Green	RP Amenity	250 g/l	SC	05702
8	Rovral Liquid FS	RP Agric.	500 g/l	FS	06366
9	Rovral Liquid FS	Aventis	500 g/l	FS	10014
10	Rovral WP	RP Agric.	50% w/w	WP	06091
11	Rovral WP	Aventis	50% w/w	WP	10015

Uses

Alternaria in *brassica seed crops* [1-5, 10, 11]. Alternaria in *leaf brassicas, mustard, oilseed rape, stubble turnips, winter wheat* [1, 3-5]. Alternaria in *brassicas* (seed treatment), *flower seeds* (seed treatment) [10, 11]. Alternaria in *fodder rape* (seed treatment), *leaf brassicas* (seed treatment), *mustard* (seed treatment), *ornamentals* (seed treatment), *stubble turnips* (seed treatment), *swedes* (seed treatment), *turnips* (seed treatment) [2]. Alternaria in *stored cabbages* [2, 10, 11]. Alternaria in *linseed* (seed treatment), *oilseed rape* (seed treatment) [2, 8-11]. Black scurf in *seed potatoes* (seed treatment) [2, 8, 9]. Black scurf and stem canker in *potatoes* (seed treatment) [10, 11]. Botrytis in *strawberries* [1-5, 10, 11]. Botrytis in *brassica seed crops, mustard, oilseed rape, winter wheat* [1, 3-5]. Botrytis in *chicory* (off-label), *courgettes* (off-label), *dwarf beans* (off-label), *endives* (off-label), *fennel* (off-label), *green beans* (off-label), *mange-tout peas* (off-label), *red beet* (off-label - dip), *runner beans* (off-label) [10, 11]. Botrytis in *rhubarb* (off-label) [11]. Botrytis in *cucumbers, outdoor lettuce, outdoor tomatoes, pot plants, protected lettuce, protected tomatoes, raspberries, stored cabbages* [2, 10, 11]. Botrytis in *cob nuts* (off-label), *grapevines* (off-label), *hazel nuts* (off-label), *outdoor lettuce* (off-label), *walnuts* (off-label) [4, 5]. Botrytis in *radicchio* (off-label) [4, 5, 10, 11]. Brown patch in *amenity grass, managed amenity turf* [6, 7]. Chocolate spot in *field beans* [1, 3-5]. Collar rot in *onions, salad onions* [1, 3-5]. Dollar spot in *amenity grass, managed amenity turf* [6, 7]. Fusarium patch in *amenity grass, managed amenity turf* [6, 7]. Glume blotch in *winter wheat* [1, 3-5]. Grey snow mould in *amenity grass, managed amenity turf* [6, 7]. Leaf rot in *onions, salad onions* [1, 3-5]. Melting out in *amenity grass, managed amenity turf* [6, 7]. Net blotch in *barley* [1, 3-5]. Phoma in *red beet* (off-label - dip) [10, 11]. Red thread in *amenity grass, managed amenity turf* [6, 7]. Sclerotinia in *chicory* (off-label), *courgettes* (off-label), *dwarf beans* (off-label), *endives* (off-label), *fennel* (off-label), *green beans* (off-label), *mange-tout peas* (off-label), *runner beans* (off-label) [10, 11]. Sclerotinia in *outdoor lettuce* (off-label) [4, 5]. Sclerotinia in *radicchio* (off-label) [4, 5, 10, 11]. Sclerotinia stem rot in *oilseed rape* [1, 3-5].

Efficacy
- Many diseases require a programme of 2 or more sprays at intervals of 2-4 wk. Recommendations vary with disease and crop - see label for details
- Use as a drench to control cabbage storage diseases. Spray ornamental pot plants and cucumbers to run-off [2, 10]
- Apply turf treatments after mowing and do not mow again for 24 h [4]
- Treatment harmless to *Encarsia* or *Phytoseiulus* being used for integrated pest control
- Apply to dry rapeseed and linseed prior to sowing [8]
- Best results on seed potatoes achieved using hydraulic sprayer with solid or hollow cone jets with air or liquid pressure atomisation [8]
- Apply to seed potatoes after harvest or after grading out and before traying out for chitting [8]

Crop safety/Restrictions
- Maximum number of treatments 1 per batch for seed treatments [8]; 1 per crop on cereals, vining peas, cabbage (as drench), chicory, borage; 2 per crop on field beans and stubble turnips; 3 per crop on brassicas (including seed crops), oilseed rape, protected winter lettuce (Oct-Feb); 4 per crop on strawberries, grapevines, salad onions, cucumbers; 5 per crop on raspberries; 6 per crop on bulb onions, tomatoes, curcurbits, peppers, aubergines and turf; 7 per crop on lettuce (Mar-Sep)
- Not to be used on protected lettuce or radicchio [4]
- A minimum of 3 wk must elapse between treatments on leaf brassicas [4]
- See label for pot plants showing good tolerance. Check other species before applying on a large scale [2, 10]
- Do not treat oilseed rape seed that is cracked or broken or of low viability [8]
- Do not excessively wet seed potato skins [8]
- Do not treat oats [4]

Special precautions/Environmental safety
- Irritating to eyes and skin [6, 7]
- Harmful to fish or other aquatic life. Do not contaminate surface waters or ditches with chemical or used container
- Treated brassica seed crops not to be used for human or animal consumption [2, 10]
- See label for guidance on disposal of spent drench liquor [2, 10]
- Do not use treated seed as food or feed [8-11]

Personal protective equipment/Label precautions
- A [1, 3-7]; C [6, 7]
- R04a, R04b [6, 7]; U20c [2, 6-11]; U08 [1, 3-7]; U05a [1, 4-7]; U04a [6, 7]; U19 [2, 10, 11]; U20b [1, 4, 5]; U20a [3]; C03 [1, 4-7]; E13c, E30a [1-11]; E34 [2, 6-11]; E31b [1, 3-7]; E01 [1, 4-7]; E26 [3, 6-9]; E32a [2, 8-11]; S01, S02, S05, S06 [8-11]; S07 [8, 9]

Latest application/Harvest interval
- Before grain watery ripe (GS 69) for wheat and barley; pre-planting for seed treatments.
- HI strawberries, protected tomatoes 1 d; outdoor tomatoes, cucurbits, peppers, aubergines 2 d; lettuce (Mar-Sep), onions, raspberries 7 d; grapevines 2 wk; brassicas, brassica seed crops, oilseed rape, beans, stubble turnips, borage 21 d; protected winter lettuce (Oct-Feb) 4 wk; cabbage (drench treatment) 2 mth

Approval
- Approved for aerial application on field beans, oilseed rape [3]. See notes in Section 1
- Following implementation of Directive 98/82/EC, approval for use of iprodione on kale was revoked in 1999

- Off-label Approval unlimited for use on outdoor crops of lettuce, cress, lamb's lettuce, scarole, radicchio (OLA 0715/95)[4], (OLA 1869/00)[5]; unlimited for use on grapevines (OLA 0478/93)[4], (OLA 1870/00)[5]; to Mar 2003 for use on pears as a post-harvest dip (OLA 0693/98)[10], (OLA 1914/00)[11]; unlimited for use on French, dwarf and runner beans, mange-tout peas (OLA 1565/98)[10], (OLA 1904/00)[11], and to Jul 2003 on the same crops with a shorter harvest interval (OLA 1501/95)[10]; unstipulated for use on chicory roots (OLA 1970/99)[10], (OLA 1906/00)[11]; unstipulated for pre-storage use on red beet (OLA2048/99)[10], (OLA 1910/00)[11]; to Oct 2003 for use on outdoor hazel nuts, cobnuts, walnuts (OLA 2583/99)[4]; to Jul 2003 for use on outdoor hazel nuts, cobnuts, walnuts (OLA 1871/00)[5]; unstipulated for use on outdoor endives, fennel, protected radicchio, protected courgettes (OLA1908/00)[11]; unstipulated for use on protected rhubarb (OLA1912/00)[11]
- Accepted by BLRA for use on malting barley

Maximum Residue Level (mg residue/kg food)
- pome fruits, bilberries, currants, gooseberries, lettuces, herbs 10; stone fruits, cane fruits, kiwi fruit, garlic, onions, shallots, tomatoes, peppers, aubergines, head cabbages 5; cucumbers, gherkins, courgettes 2; witloof 1; rape seed, wheat 0.5; pulses 0.2; kohlrabi, tea, hops 0.1; animal products 0.05; citrus fruits, tree nuts, cranberries, wild berries, miscellaneous fruit (except bananas and kiwi fruit), root and tuber vegetables (except carrots, radishes), squashes, water melons, sweetcorn, spinach, beet leaves, water cress, stem vegetables (except fennel), mushrooms, oilseeds (except linseed, rape seed, mustard seed), rye, oats, triticale, maize 0.02

331   iprodione + thiophanate-methyl

A protectant and systemic fungicide for oilseed rape

Products

1	Aventis Compass	Aventis	167:167 g/l	SC	10040
2	Compass	RP Agric.	167:167 g/l	SC	06190
3	Compass	Aventis	167:167 g/l	SC	10041
4	Snooker	Aventis	150:200 g/l	SC	10018
5	Snooker	RP Agric.	150:200 g/l	SC	07940

Uses

Alternaria in *oilseed rape* [1-3]. Alternaria in *carrots* (off-label), *horseradish* (off-label), *parsnips* (off-label) [2, 3]. Alternaria in *winter oilseed rape* [4, 5]. Botrytis in *Limnanthes alba (meadowfoam)* (off-label) [2, 3]. Crown rot in *carrots* (off-label), *horseradish* (off-label), *parsnips* (off-label) [2, 3]. Grey mould in *oilseed rape* [1-3]. Grey mould in *winter oilseed rape* [4, 5]. Light leaf spot in *oilseed rape* [1-3]. Light leaf spot in *winter oilseed rape* [4, 5]. Sclerotinia in *Echium plantaginium* (off-label) [2]. Sclerotinia stem rot in *oilseed rape* [1-3]. Sclerotinia stem rot in *winter oilseed rape* [4, 5]. Stem canker in *oilseed rape* [1-3].

Efficacy
- Timing of sprays on oilseed rape varies with disease, see label for details
- For season-long control product should be applied as part of a disease control programme

Crop safety/Restrictions
- Maximum number of treatments (refers to total sprays containing benomyl, carbendazim and thiophanate methyl) 2 per crop for winter oilseed rape
- Under conditions of crop stress, e.g. after prolonged dry conditions on light soils do not spray winter oilseed rape from aircraft in less than 100 l/ha in strong direct sunlight. Under these conditions aerial application should be made in early morning or evening [2]
- Treatment may extend duration of green leaf in winter oilseed rape

Special precautions/Environmental safety
- Irritating to eyes and skin. May cause sensitization by skin contact [4, 5]
- Treated haulm must not be fed to livestock [2]
- Harmful to fish or other aquatic life. Do not contaminate surface waters or ditches with chemical or used container

Personal protective equipment/Label precautions
- A, C, H, M [1-5]
- R04a, R04b, R04e [4, 5]; U20c [1-3]; U05a, U09a, U20b, C03 [4, 5]; E13c, E26, E30a, E31b [1-5]; E01 [4, 5]

Latest application/Harvest interval
- Before end of flowering for winter oilseed rape
- HI 3 wk

Approval
- Off-label approval unstipulated for use on *Limnanthes alba* (Meadowfoam) (OLA 0784/99)[2], (OLA 1934/00) [3]; to Mar 2004 for use on *Echium plantagineum* (OLA 0785/99)[2]; unstipulated for use on carrots, parsnips, horseradish (OLA 1863/99)[2], (OLA 1936/00) [3]

Maximum Residue Level (mg residue/kg food)
- see iprodione entry

332   isoproturon

A residual urea herbicide for use in cereals

See also carfentrazone-ethyl + isoproturon
chlorotoluron + isoproturon
diflufenican + flurtamone + isoproturon
diflufenican + isoproturon
fenoxaprop-P-ethyl + isoproturon

Products

1	Alpha Isoproturon 500	Makhteshim	500 g/l	SC	05882
2	Alpha Isoproturon 650	Makhteshim	650 g/l	SC	07034
3	Atlas Fieldgard	Nufarm Whyte	500 g/l	SC	08582
4	Atum WDG	RP Agric.	83% w/w	WG	07778
5	Barclay Guideline 500	Barclay	500 g/l	SC	09257
6	Bison 83 WG	Nufarm Whyte	83% w/w	WG	10063
7	DAPT Isoproturon 500 FL	DAPT	500 g/l	SC	08092
8	Isoguard	Chiltern	500 g/l	SC	08497
9	Landgold Isoproturon	Landgold	500 g/l	SC	06012
10	Luxan Isoproturon 500 Flowable	Luxan	500 g/l	SC	07437
11	Mysen	Portman	500 g/l	SC	07695
12	Portman Isotop SC	Portman	500 g/l	SC	07663
13	Standon IPU	Standon	500 g/l	SC	08671
14	Tolkan Liquid	Aventis	500 g/l	SC	10023
15	Tolkan Liquid	RP Agric.	500 g/l	SC	06172

Uses

Annual dicotyledons in **winter barley, winter wheat** [1-15]. Annual dicotyledons in **autumn sown spring wheat** [4, 14, 15]. Annual dicotyledons in **triticale** [4, 5, 11, 13-15]. Annual dicotyledons in **winter rye** [4, 5, 11, 14, 15]. Annual dicotyledons in **spring wheat** [5, 11, 13]. Annual grasses in **winter barley, winter wheat** [1-15]. Annual grasses in **autumn sown spring wheat** [4, 14, 15]. Annual grasses in **triticale** [4, 5, 11, 13-15]. Annual grasses in **winter rye** [4, 5, 11, 14, 15]. Annual grasses in **spring wheat** [5, 11, 13]. Annual meadow grass in **spring barley** *(off-label)* [14, 15]. Blackgrass in **winter barley, winter wheat** [1-5, 7-15]. Blackgrass in **autumn sown spring wheat** [4, 14, 15]. Blackgrass in **triticale** [4, 5, 11, 13-15]. Blackgrass in **winter rye** [4, 5, 11, 14, 15]. Blackgrass in **spring wheat** [5, 11, 13]. Rough meadow grass in **winter barley, winter wheat** [1-5, 7-15]. Rough meadow grass in **autumn sown spring wheat** [4, 14, 15]. Rough meadow grass in **triticale** [4, 5, 11, 13-15]. Rough meadow grass in **winter rye** [4, 5, 11, 14, 15]. Rough meadow grass in **spring wheat** [5, 11, 13]. Wild oats in **winter barley, winter wheat** [1-5, 7-15]. Wild oats in **autumn sown spring wheat** [4, 14, 15]. Wild oats in **triticale** [4, 5, 11, 13-15]. Wild oats in **winter rye** [4, 5, 11, 14, 15]. Wild oats in **spring wheat** [5, 11, 13].

Efficacy
- May be applied in autumn or spring (but only post-emergence on wheat or barley). Best control normally achieved by early post-emergence treatment
- See label for details of rates, timings and tank mixes for different weed problems
- Apply to moist soils. Effectiveness may be reduced in seasons of above average rainfall or by prolonged dry weather
- Residual activity reduced on soils with more than 10% organic matter. Only use on such soils in spring

Crop safety/Restrictions

- Maximum total dose 2.5 kg a.i./ha for any crop. The IPU stewardship programme guidelines recommend that where possible this should be reduced to 1.5 kg a.i./ha using mixtures or sequences with other herbicides
- Recommended timing of treatment varies depending on crop to be treated, method of sowing, season of application, weeds to be controlled and product being used. See label for details
- Do not apply pre-emergence to wheat or barley. On triticale and winter rye use pre-emergence only and on named varieties
- Do not use on durum wheats, undersown cereals or crops to be undersown
- Do not apply to very wet or waterlogged soils, or when heavy rainfall is forecast
- Do not use on very cloddy soils or if frost is imminent or after onset of frosty weather
- Crop damage may occur on free draining, stony or gravelly soils if heavy rain falls soon after spraying. Early sown crops may be damaged if spraying precedes or coincides with a period of rapid growth
- Do not roll for 1 wk before or after treatment or harrow for 1 wk before or any time after treatment

Special precautions/Environmental safety

- Irritant. May cause sensitization by skin contact [4]
- Irritating to eyes and skin [3, 8, 12]
- Dangerous (harmful [1, 2, 7, 10]) to fish or other aquatic life. Do not contaminate surface waters or ditches with chemical or used container [3-6, 8, 9, 12, 13]
- Do not spray where soils are cracked, to avoid run-off through drains

Personal protective equipment/Label precautions

- A, C [1-9, 12-14]; H [1-9, 13, 14]; D [4, 6]
- R04a, R04b [3, 8, 12]; R04, R04e [4]; U20a [1, 7, 9, 12]; U08 [1-4, 6-8, 11-15]; U19 [1-4, 7, 8, 12-15]; U05a [1-3, 7, 8, 11-13]; U20b [2-6, 8, 10, 11, 13-15]; U04a [13]; C03 [1-3, 7, 8]; E26, E30a [1-15]; E31b [1-3, 5, 7-15]; E13c [1, 2, 7, 10]; E01 [1-3, 7, 8]; E13b [3-6, 8, 9, 12, 13]; E15 [11, 14, 15]; E32a [4, 6]

Latest application/Harvest interval

- Not later than second node detectable stage (GS 32)

Approval

- May be applied through CDA equipment (see labels for details) [15]
- All approvals for aerial use of isoproturon were revoked in 1995
- Off-label approval unlimited for use on spring barley (OLA 0846/96)[15]; to Jul 2003 for use on spring barley (OLA 1938/00)[14]
- Accepted by BLRA for use on malting barley

---

333   isoproturon + pendimethalin

A contact and residual herbicide for use in winter cereals

Products

1	Encore	BASF	125:250 g/l	SC	BASF1
2	Encore	Cyanamid	125:250 g/l	SC	04737
3	Jolt	Cyanamid	125:250 g/l	SC	05488
4	Trump	Cyanamid	236:236 g/l	SC	03687
5	Trump	BASF	236:236 g/l	SC	BASF2

Uses

Annual dicotyledons in *triticale, winter barley, winter rye, winter wheat* [1-5]. Annual dicotyledons in *autumn sown spring wheat* [4, 5]. Annual grasses in *triticale, winter barley, winter rye, winter wheat* [1-5]. Annual grasses in *autumn sown spring wheat* [4, 5]. Blackgrass in *triticale, winter barley, winter rye, winter wheat* [1-5]. Blackgrass in *autumn sown spring wheat* [4, 5]. Wild oats in *triticale, winter barley, winter rye, winter wheat* [1-5]. Wild oats in *autumn sown spring wheat* [4, 5].

FOR FULL CONDITIONS OF USE ALWAYS READ THE PRODUCT LABEL

Efficacy
- May be applied in autumn or spring (but only post-emergence on wheat or barley)
- Annual grasses controlled from pre-emergence to 3-leaf stage, extended to 3-tiller stage with blackgrass by tank mixing with additional isoproturon. Best results on wild oats post-emergence in autumn
- Annual dicotyledons controlled pre-emergence and up to 4-leaf stage [1-3], up to 8-leaf stage [4, 5]. See label for details
- Apply to moist soils. Best results achieved by application to fine, firm seedbed. Trash, ash or straw should have been incorporated evenly
- Contact activity is reduced by rain within 6 h of application
- Activity may be reduced on soil with more than 6% organic matter or ash. Do not use on soils with more than 10% organic matter

Crop safety/Restrictions
- Maximum number of treatments 1 per crop. Maximum total dose of isoproturon 2.5 kg a.i./ha per crop
- Apply before first node detectable (GS 31) [1-3], to before leaf sheath erect (GS 30) [4, 5]
- Do not apply pre-emergence to wheat or barley. On winter rye and triticale use pre-emergence only on named cultivars
- Do not use on durum wheat or spring cereals
- May be used on autumn sown spring wheat [4, 5]
- Do not use pre-emergence on winter rye or triticale drilled after 30 Nov [4, 5]
- Do not use on crops suffering stress from disease, drought, waterlogging, poor seedbed conditions or other causes or apply post-emergence when frost imminent
- Do not apply to very wet or waterlogged soils, or when heavy rain is forecast
- Do not undersow treated crops
- Do not roll or harrow after application
- In the event of autumn crop failure spring wheat, spring barley, maize, potatoes, beans or peas may be grown following ploughing to at least 150 mm

Special precautions/Environmental safety
- Dangerous to fish or other aquatic life. Do not contaminate surface waters or ditches with chemical or used container
- Do not spray where soils are cracked, to avoid run-off through drains

Personal protective equipment/Label precautions
- A [1-5]; C [1-3]
- U08, U13, U19, U20a, E13b, E30a, E31b, E34 [1-5]

Latest application/Harvest interval
- Pre-emergence for winter rye and triticale; before leaf sheath erect (GS 30)[4, 5] or before 1st node detectable (GS 31)[1-3] for winter wheat and barley

Approval
- Accepted by BLRA for use on malting barley
- Product registration numbers not available at time of printing [1, 5]

---

334   isoproturon + simazine

A contact and residual herbicide for winter wheat and barley

Products

1 Alpha Protugan Plus	Makhteshim	375:60 g/l	SC	08799
2 Harlequin 500 SC	Makhteshim	450:50 g/l	SC	09779

Uses

Annual dicotyledons in *winter wheat* [1, 2]. Annual dicotyledons in *winter barley* [2]. Annual grasses in *winter wheat* [1, 2]. Annual grasses in *winter barley* [2]. Blackgrass in *winter wheat* [1, 2]. Blackgrass in *winter barley* [2].

Efficacy
- May be applied in autumn or spring (but only post-emergence of the crop)
- Annual grasses controlled up to early tillering, annual dicotyledons to 50 mm

- Apply to fine, firm, even seedbed. Any trash or burnt straw should be dispersed in preparing seedbed
- Apply to moist soils. Weed control may be reduced in excessively wet autumns or if prolonged dry weather follows application to dry soil

Crop safety/Restrictions
- Maximum number of treatments 1 per crop. Maximum total dose of isoproturon 2.5 kg a.i./ha per crop
- Apply from 2-leaf stage of crop to before leaf sheath erect (GS 12-30)
- Do not use on durum wheat, undersown crops or those due to be undersown
- With direct drilled crops soil surface should be broken by surface cultivation and seed covered by 12-25 mm soil
- Do not use on sand or on soils with more than 10% organic matter
- On stony or gravelly soils there is risk of crop damage, especially with heavy rain soon after application
- Do not apply when frost or heavy rain is forecast or to crops severely checked by frost, waterlogging, pest or disease attack.
- Early sown crops may be prone to damage if spraying precedes or coincides with period of rapid growth in autumn
- Do not harrow for 7 d before treatment or roll for 7 d before or after treatment in spring
- In the event of crop failure land should be inverted by mouldboard ploughing to at least 150 mm and harrowed before drilling/planting another crop
- Do not harrow after application

Special precautions/Environmental safety
- Irritating to skin [2]
- Dangerous to fish or other aquatic life. Do not contaminate surface waters or ditches with chemical or used container
- Do not spray where soils are cracked, to avoid run-off through drains
- LERAP Category B

Personal protective equipment/Label precautions
- A, C, H [1, 2]; D, M [2]
- R04b [2]; U05a, U08, U20b [1, 2]; U19 [2]; C03, E01, E13b, E16a, E16b, E26, E30a, E31b [1, 2]; E29 [2]

Latest application/Harvest interval
- End of tillering [1]; leaf sheath erect (GS 30) [2]

Approval
- Accepted by BLRA for use on malting barley

---

335   isoxaben

A soil-acting amide herbicide for use in grass and fruit

Products

1 Flexidor 125	Dow	125 g/l	SC	05104
2 Gallery 125	Rigby Taylor	125 g/l	SC	06889
3 Knot Out	Vitax	125 g/l	SC	05163

Uses

Annual dicotyledons in **apples, blackberries, blackcurrants, cherries, container-grown woody ornamentals, gooseberries, grapevines, hardy ornamental nursery stock, hops, outdoor carrots** *(off-label)*, **pears, plums, protected courgettes** *(off-label)*, **protected marrows** *(off-label)*, **protected ornamentals** *(off-label)*, **protected pumpkins** *(off-label)*, **protected squashes** *(off-label)*, **raspberries, strawberries** [1]. Annual dicotyledons in **forestry transplants** [1, 2]. Annual dicotyledons in **amenity trees and shrubs, managed amenity turf** [2]. Annual dicotyledons in **amenity grass** [2, 3]. Cleavers in **asparagus** *(off-label)* [1]. Fat-hen in **asparagus** *(off-label)* [1].

Efficacy
- When used alone apply pre-weed emergence
- Effectiveness is reduced in dry conditions. Weed seeds germinating at depth are not controlled

---

- Activity reduced on soils with more than 10% organic matter. Do not use on peaty soils
- Various tank mixtures are recommended for early post-weed emergence treatment (especially for grass weeds). See label for details
- Best results on turf achieved by applying to firm, moist seedbed within 2 d of sowing. Avoid disturbing soil surface after application [3]

Crop safety/Restrictions

- Maximum number of treatments 1 per crop, sowing or yr for amenity grass, ornamentals, strawberries, 2 per yr for hardy ornamental nursery stock and forestry transplants
- See label for details of crops which may be sown in the event of failure of a treated crop

Personal protective equipment/Label precautions

- A, C [1-3]; H [2]
- U20b, E15, E30a, E32a [1-3]; E07 [1-3] (50 d); E26, E31a, E34 [3]

Withholding period

- Keep all livestock out of treated areas for at least 50 d

Latest application/Harvest interval

- Before 1 Apr in yr of harvest for edible crops

Approval

- Off-label approval unlimited for use in protected ornamentals (OLA 0605/94)[1]; unlimited for use on outdoor carrots (OLA 2209/96)[1]; unlimited for use on outdoor asparagus (OLA 2318/97)[1]; unlimited for use on courgettes, marrows, squashes, pumpkins grown outdoors or under crop covers (OLA 2578/98)[1]
- Accepted by BLRA for use on hops

336   isoxaben + trifluralin

A residual herbicide mixture for amenity use

Products

1 Axit GR	Fargro	0.55:2.06% w/w	GR	08892
2 Premiere Granules	Dow	0.55:2.06% w/w	GR	07987

Uses

Annual dicotyledons in *container-grown ornamentals, hardy ornamental nursery stock* [1]. Annual dicotyledons in *amenity trees and shrubs, amenity vegetation* [2]. Annual meadow grass in *container-grown ornamentals, hardy ornamental nursery stock* [1]. Annual meadow grass in *amenity trees and shrubs, amenity vegetation* [2]. Fat-hen in *container-grown ornamentals, hardy ornamental nursery stock* [1]. Fat-hen in *amenity trees and shrubs, amenity vegetation* [2]. Groundsel in *amenity trees and shrubs, amenity vegetation* [2]. Sowthistle in *container-grown ornamentals, hardy ornamental nursery stock* [1]. Sowthistle in *amenity trees and shrubs, amenity vegetation* [2].

Efficacy

- Best results obtained when granules applied to firm, moist soil, free from clods. 20-30 mm irrigation or rainfall required within 3 d after treatment
- Weed control reduced under dry soil conditions
- Treatment should be made before mulching and before weed emergence
- Uniform cover with granules optimises results. Improved uniformity of cover may be achieved by spreading the granules twice at half dose at right angles
- Control may be reduced on soils with high organic matter or where organic manure has been applied

Crop safety/Restrictions

- Maximum number of treatments 1 per yr
- Ensure soil has settled and no cracks are present before application to transplanted trees and shrubs [2]
- May be used on Light, Medium and Heavy soils, but on Light soils early growth may be reduced under adverse conditions
- Do not use on soils with more than 10% organic matter (except in container grown hardy nursery stock)

- For newly planted containerised stock allow compost to settle 7-10 d before treatment [1]
- Do not use in the 12 mth prior to lifting field grown nursery stock if edible crops are to be grown as a following crop
- Land must be mouldboard ploughed to at least 20 cm before drilling or planting succeeding plants except well rooted forestry trees and ornamentals

Special precautions/Environmental safety
- Harmful to fish or other aquatic life. Do not contaminate surface waters or ditches with chemical or used container
- Do not allow direct applications of granules from ground based/vehicle mounted applicators to fall within 6 m, or from hand-held applicators to within 2 m, of surface waters or ditches. Direct applications away from water
- Product contains up to 1% crystalline silica. The MEL for respirable silica dust is 0.4 mg/m

Personal protective equipment/Label precautions
- A [1]
- U20b, E01, E13c, E16, E30a, E32a, E34 [1, 2]

337   kresoxim-methyl

A protectant strobilurin fungicide for apples

See also epoxiconazole + fenpropimorph + kresoxim-methyl
epoxiconazole + kresoxim-methyl
fenpropimorph + kresoxim-methyl

Products

Stroby WG	BASF	50% w/w	WG	08653

Uses

Powdery mildew in *apples* *(reduction)*. Scab in *apples*.

Efficacy
- Best results achieved from treatments starting at bud burst and repeated at 10-14 d intervals depending on severity of disease pressure
- Do not spray product more than three times consecutively. To minimise risk of development of resistance apply in a programme with products from different chemical groups.
- Product should not be used as final spray of the season
- Product may be applied in ultra low volumes (ULV) but disease control may be reduced

Crop safety/Restrictions
- Maximum number of treatments 3 per yr, and not more than three consecutively. Blocks of two or three applications must be separated by a minimum of two applications of a fungicide from a different cross-resistance group
- Consult before using on crops intended for processing

Special precautions/Environmental safety
- Dangerous to fish or other aquatic life. Do not contaminate surface waters or ditches with chemical or used container
- Do not allow direct spray from broadcast air-assisted sprayers to fall within 5 m of surface waters or ditches. Direct spray away from water
- Harmless to ladybirds and predatory mites
- Harmless to honey bees and may be applied during flowering. Nevertheless local beekeepers should be notified when treatment of orchards in flower is to occur

Personal protective equipment/Label precautions
- U20b, E13b, E30a, E31c

Latest application/Harvest interval
- HI 35 d

Approval
- Kresoxim-methyl included in Annex I under EC Directive 91/414
- Approved for use in ULV systems

338   lambda-cyhalothrin

A quick-acting contact and ingested pyrethroid insecticide

Products

1 Hallmark with Zeon Technology	Zeneca	100 g/l	CS	09809
2 Landgold Lambda-C	Landgold	50 g/l	EC	09205
3 Me2 Lambda	Me2	50 g/l	EC	09712
4 Standon Lambda-C	Standon	50 g/l	EC	08831

Uses

Aphids in *chestnuts (off-label)*, *cob nuts (off-label)*, *hazel nuts (off-label)*, *spring barley, spring oats, spring wheat, walnuts (off-label)*, *winter oats* [1]. Aphids in *durum wheat, potatoes, winter barley, winter wheat* [1-4]. Aphids in *winter oilseed rape* [2-4]. Barley yellow dwarf virus vectors in *spring barley, spring wheat, winter oats* [1]. Barley yellow dwarf virus vectors in *durum wheat, winter barley, winter wheat* [1-4]. Beet leaf miner in *sugar beet* [1]. Beet virus yellows vectors in *spring oilseed rape, winter oilseed rape* [1]. Bryobia mites in *gooseberries (off-label)* [1]. Cabbage seed weevil in *spring oilseed rape, winter oilseed rape* [4]. Cabbage stem flea beetle in *spring oilseed rape* [1]. Cabbage stem flea beetle in *winter oilseed rape* [1-4]. Capsids in *gooseberries (off-label)* [1]. Carrot fly in *carrots (off-label)*, *celery (off-label)*, *fennel (off-label)*, *parsnips (off-label)* [1]. Caterpillars in *chestnuts (off-label)*, *cob nuts (off-label)*, *gooseberries (off-label)*, *hazel nuts (off-label)*, *walnuts (off-label)* [1]. Caterpillars in *broccoli, brussels sprouts, cabbages, calabrese, cauliflowers* [1-4]. Cutworms in *carrots, chicory (off-label)*, *fennel (off-label)*, *lettuce* [1]. Cutworms in *sugar beet* [1-4]. Damson-hop aphid in *hops* [2-4]. Flea beetles in *spring oilseed rape, winter oilseed rape* [1]. Flea beetles in *sugar beet* [1-4]. Frit fly in *sweetcorn (off-label)* [1]. Insect pests in *soya bean (off-label)* [1]. Leaf miners in *red beet (off-label)* [1]. Leaf miners in *sugar beet* [2-4]. Pea and bean weevils in *peas, spring field beans, winter field beans* [1-4]. Pea aphid in *peas* [1]. Pea moth in *peas* [1-4]. Pear sucker in *pears* [1-4]. Pod midge in *spring oilseed rape, winter oilseed rape* [1-3]. Pollen beetles in *spring oilseed rape, winter oilseed rape* [1-4]. Red spider mites in *hops* [4]. Seed weevil in *spring oilseed rape, winter oilseed rape* [1-3]. Silver Y moth in *celery (off-label)*, *dwarf beans (off-label)*, *navy beans (off-label)*, *runner beans (off-label)* [1]. Two-spotted spider mite in *gooseberries (off-label)* [1]. Virus vectors in *winter oilseed rape* [2-4]. Whitefly in *broccoli, brussels sprouts, cabbages, calabrese, cauliflowers* [1]. Yellow cereal fly in *winter wheat* [1].

Efficacy

- Best results normally obtained from treatment when pest attack first seen. See label for detailed recommendations on each crop
- Repeat applications recommended in some crops where prolonged attack occurs, up to maximum total dose. See label for details
- If using mixtures, add lambda-cyhalothrin product to the tank last
- Where strains of aphids resistant to lambda-cyhalothrin occur control is unlikely to be satisfactory
- Addition of wetter recommended for control of certain pests in brassicas and oilseed rape [1, 2]
- Use of sufficient water volume to ensure thorough crop penetration recommended for optimum results
- Use of drop-legged sprayer gives improved results in crops such as Brussels sprouts

Crop safety/Restrictions

- Maximum number of applications per crop varies - see labels
- Maximum total dose equivalent to 2 maximum dose treatments on field beans, sugar beet, peas, lettuce, carrots; 3 maximum dose treatments on oilseed rape, pears; 4 maximum dose treatments on cereals, brassicas; 6 maximum dose treatments on hops; 8 maximum dose treatments on potatoes

Special precautions/Environmental safety

- Harmful in contact with skin [2-4]
- Harmful if swallowed [1-4]

- Irritating to eyes and skin [2-4]
- May cause sensitization by skin contact [1]
- Flammable [2-4]
- Keep in original container, tightly closed, in a safe place, under lock and key [4]
- Extremely dangerous to bees. Do not apply to crops in flower or to those in which bees are actively foraging, except as directed in peas and oilseed rape and hops. Do not apply when flowering weeds are present [2, 4]
- Do not apply to a cereal crop if any product containing a pyrethroid insecticide or dimethoate has been applied to the crop after the start of ear emergence (GS 51)
- Do not spray cereals in the spring/summer (ie after 1 Apr) within 6 m of edge of crop
- Extremely dangerous to fish or other aquatic life. Do not contaminate surface waters or ditches with chemical or used container
- LERAP Category A
- Do not allow spray from air-assisted applications to hops to fall within 50 m of surface waters or ditches

Personal protective equipment/Label precautions
- A, C, H, J, K, L, M [1-4]
- M03 [1-4]; M05 [1]; R03a, R04a, R04b, R07d [2-4]; R03c [1-4]; R04e [1]; U02, U05a, U08 [1-4]; U20b [1, 4]; U20a [2, 3]; C03 [1-4]; E12b, E26 [2-4]; E01, E13a, E16c, E16d, E30a, E34 [1-4]; E17 [1-4] (18 m); E31b [1, 4]; E31a [2, 3]

Latest application/Harvest interval
- Before late milk stage (GS 77) in cereals; before end flowering for oilseed rape
- HI 3 d for lettuce; 6 wk for spring oilseed rape; 8 wk for sugar beet; 9 wk for carrots (14 d off-label) [1]

Approval
- Accepted by BLRA for use on malting barley and hops
- Off-label approval unstipulated for use on carrots, parsnips (OLA 0283/00)[1]; unstipulated for use on dwarf beans, runner beans, navy beans (OLA 2455/00)[1]; unstipulated for use on red beet (OLA 0286/00)[1]; unstipulated for use on outdoor celery (OLA 0289/00)[1]; to Oct 2002 for use on outdoor gooseberries (OLA 0292/00)[1]; to Mar 2002 for use on outdoor soya beans (OLA 0304/00)[1]; unstipulated for use on hazlenuts, cobnuts, chestnuts, walnuts (OLA 0295/00)[1]; unstipulated for use on sweetcorn (OLA 0298/00)[1]; unstipulated for use on fennel (OLA 0301/00)[1]; to Jul 2003 for use on chicory for forcing (OLA 0801/00)[1]

Maximum Residue Level (mg residue/kg food)
- hops 10; tomatoes, aubergines, cress, lamb's lettuce, lettuce, scarole, chervil, chives, parsley, celery leaves, tea 1; meat (except poultry) 0.5; apricots, peaches, grapes, head cabbages, beans (with pods), peas (with pods), melons 0.2; pome fruits, cherries, plums, currants, gooseberries, cucumbers, gherkins, peppers, courgettes, flowerhead brassicas 0.1; tree nuts, Brussels sprouts, barley, milk and dairy produce 0.05; blackberries, dewberries, loganberries, raspberries, bilberries, cranberries, wild berries, miscellaneous fruits, root and tuber vegetables, garlic, onions, shallots, sweet corn, watercress, beans (without pods), peas (without pods), asparagus, wild mushrooms, pulses, oilseed, potatoes, wheat, rye, oats, triticale, maize, rice, poultry meat, eggs 0.02

---

339 lambda-cyhalothrin + pirimicarb

An insecticide mixture combining translaminar, contact, fumigant and stomach activity

Products

Dovetail	Zeneca	5:100 g/l	EC	07973

Uses

Aphids in **brassicas, carrots, cereals, lettuce, peas, potatoes, sugar beet**. Beet virus yellows vectors in **oilseed rape**. Cabbage stem flea beetle in **oilseed rape**. Caterpillars in **brassicas**. Cutworms in **carrots, lettuce, potatoes, sugar beet**. Flea beetles in **oilseed rape, sugar beet**. Leaf miners in **sugar beet**. Mealy aphids in **brassicas, oilseed rape**. Pea and bean weevils in **field beans, peas**. Pea midge in **peas**. Pea moth in **peas**. Pod midge in *oilseed rape.* Pollen beetles in *brassicas, oilseed rape.* Seed weevil in *oilseed rape.* Whitefly in *brassicas.*

Efficacy
- Best results obtained from treatment when pest attack first seen or after warning issued
- Repeat applications recommended in some crops where prolonged attacks occur, up to maximum total dose
- Addition of Agral recommended for certain uses in brassicas and oilseed rape. See label
- Use of drop-leg sprayers improves efficacy in crops such as Brussels sprouts
- Control unlikely to be satisfactory if aphids resistant to lambda-cyhalothrin or pirimicarb present

Crop safety/Restrictions
- Maximum total dose (product): peas, field beans, cereals, carrots, lettuce 3.0 l/ha; oilseed rape 4.5 l/ha; sugar beet, brassicas 6.0 l/ha; potatoes 12.0 l/ha

Special precautions/Environmental safety
- Harmful if swallowed. Irritating to skin. Risk of serious damage to eyes
- May cause sensitization by skin contact
- May cause lung damage if swallowed
- Flammable
- Harmful to fish or other aquatic life. Do not contaminate surface waters or ditches with chemical or used container
- LERAP category A
- Do not spray cereals after 1 Apr within 6 m of the edge of the crop
- Must not be applied to cereals if any product containing a pyrethroid insecticide or dimethoate has been sprayed after the start of ear emergence (GS 51)

Personal protective equipment/Label precautions
- A, C, H, J, K, M
- M03, R03c, R04b, R04d, R04e, R04g, R07d, U02, U05b, U08, U20a, C03, E01, E13c, E16c, E16d, E30a, E31c, E34; E06b (7 d)

Withholding period
- Harmful to livestock. Keep all livestock out of treated areas for at least 7 d following treatment

Latest application/Harvest interval
- Before late milky ripe (GS 77) for cereals; before end of flowering for oilseed rape
- HI carrots 9 wk; sugar beet 8 wk; brassicas, lettuce, potatoes, peas, field beans 3 d

Approval
- Accepted by BLRA for use pre-harvest on malting barley

Maximum Residue Level (mg residue/kg food)
- see lambda-cyhalothrin entry

---

340   Ienacil

A residual, soil-acting uracil herbicide for horticultural crops

Products

1 Agricola Lenacil FL	Agricola	440 g/l	SC	09481
2 Venzar Flowable	DuPont	440 g/l	SC	06907

Uses

Annual dicotyledons in *fodder beet, mangels, sugar beet* [1, 2]. Annual dicotyledons in *farm woodland (off-label)*, *red beet* [2]. Annual meadow grass in *fodder beet, mangels, sugar beet* [1, 2]. Annual meadow grass in *farm woodland (off-label)*, *red beet* [2].

Efficacy
- May be used pre or post emergence, alone or in mixture to broaden weed spectrum
- Mixtures with triallate for grass weed control should be applied and incorporated during final seed bed preparations
- Pre-emergence treatment alone or in mixture not recommended on all soil types. Consult label. Residual activity reduced on soils with high OM content

- Best results achieved on a firm moist tilth. Continuing presence of moisture from rain or irrigation gives improved residual control of later germinating weeds. Effectiveness may be reduced by dry conditions

Crop safety/Restrictions
- Maximum number of treatments in beet crops 1 pre-emergence + 3 post-emergence per crop
- Apply overall or as band spray to beet crops pre-drilling incorporated, pre- or post-emergence
- See label for soil type restrictions
- Heavy rain after application may cause damage to beet crops especially if followed by very hot weather
- Reduction in beet stand may occur where crop emergence or vigour is impaired by soil capping or pest attack
- Do not apply other residual herbicides within 3 mth before or after spraying
- Do not treat crops under stress from drought, low temperatures, nutrient deficiency, pest or disease attack, or waterlogging
- Succeeding crops should not be planted or sown for at least 4 mth after treatment following ploughing to at least 150 mm

Special precautions/Environmental safety
- Irritating to eyes, skin [1] and respiratory system [2]
- Harmful to fish or other aquatic life. Do not contaminate surface waters or ditches with chemical or used container

Personal protective equipment/Label precautions
- A, C [1, 2]
- R04c [2]; R04a, R04b, U05a, U08, U19, U20a, C03 [1, 2]; E31a, E32a [2]; E01, E13c, E30a [1, 2]; E26, E31b [1]

Latest application/Harvest interval
- Pre-emergence for red beet, fodder beet, spinach, spinach beet, mangels; before leaves meet over rows when used post-emergence

Approval
- Off-label approval unlimited for use in farm woodland (OLA 1282/97)[2]

341   lenacil + phenmedipham

A contact and residual herbicide for use in sugar beet

Products

1	Agricola Lens	Agricola	66.6:95 g/l	SC	10257
2	Stefes 880	Stefes	440:114 g/l	KL	08914

Uses

Annual dicotyledons in *fodder beet, mangels* [1]. Annual dicotyledons in *sugar beet* [1, 2].

Efficacy
- Apply at any stage of crop when weeds at cotyledon stage
- Germinating weeds controlled by root uptake for several weeks
- Best results achieved under warm, moist conditions on a fine, firm seedbed
- Treatment may be repeated up to a maximum of 3 applications on later weed flushes
- Residual activity may be reduced on highly organic soils or under very dry conditions
- Rain falling 1 h after application does not reduce activity

Crop safety/Restrictions
- Maximum number of treatments 3 per crop
- Do not apply when temperature above or likely to exceed 21°C on day of spraying or under conditions of high light intensity
- Do not spray any crop under stress from drought, waterlogging, cold, wind damage or any other cause
- Heavy rain after application may reduce stand of crop particularly in very hot weather. In severe cases yield may be reduced
- Do not sow or plant any crop within 4 mth of treatment. In case of crop failure only sow or plant beet crops, strawberries or other tolerant horticultural crop within 4 mth

Special precautions/Environmental safety
- Irritating to eyes, skin and respiratory system [2]
- Irritating to eyes and respiratory system [1]
- Harmful to fish or other aquatic life. Do not contaminate surface waters or ditches with chemical or used container

Personal protective equipment/Label precautions
- A, C [1, 2]
- M03 [1]; R04b [2]; R04a, R04c [1, 2]; U20a [2]; U05a, U08, U19 [1, 2]; U20b [1]; C03 [1, 2]; E26 [2]; E01, E13c, E30a, E31b [1, 2]

Latest application/Harvest interval
- Before crop leaves meet across rows

---

342   lindane

See gamma-HCH

---

343   linuron

A contact and residual urea herbicide for various field crops

See also 2,4-DB + linuron + MCPA
chlorpropham + linuron

Products

1 Afalon	Aventis	450 g/l	SC	08186
2 Alpha Linuron 50 SC	Makhteshim	500 g/l	SC	06967
3 Ashlade Linuron FL	Nufarm Whyte	480 g/l	SC	06221
4 DAPT Linuron 50 SC	DAPT	500 g/l	SC	08058
5 MSS Linuron 500	Nufarm Whyte	480 g/l	SC	08893
6 UPL Linuron 45% Flowable	United Phosphorus	450 g/l	SC	07435

Uses

Annual dicotyledons in *celeriac (off-label)*, *celery, garlic (off-label)*, *leeks (off-label)*, *onions (off-label)* [1]. Annual dicotyledons in *carrots, parsley, potatoes* [1-6]. Annual dicotyledons in *parsnips* [1, 2, 4, 6]. Annual dicotyledons in *spring oats* [2]. Annual dicotyledons in *spring barley, spring wheat* [3]. Annual dicotyledons in *spring cereals* [4-6]. Annual grasses in *celeriac (off-label)*, *garlic (off-label)*, *leeks (off-label)*, *onions (off-label)* [1]. Annual meadow grass in *celery* [1]. Annual meadow grass in *carrots, parsley, potatoes* [1-6]. Annual meadow grass in *parsnips* [1, 2, 4, 6]. Annual meadow grass in *spring barley, spring wheat* [2, 3]. Annual meadow grass in *spring cereals* [4-6]. Black bindweed in *celery* [1]. Black bindweed in *carrots, parsley, potatoes* [1-6]. Black bindweed in *parsnips* [1, 2, 4, 6]. Black bindweed in *spring oats* [2]. Black bindweed in *spring barley, spring wheat* [2, 3]. Black bindweed in *spring cereals* [4-6]. Chickweed in *celery* [1]. Chickweed in *carrots, parsley, potatoes* [1-6]. Chickweed in *parsnips* [1, 2, 4, 6]. Chickweed in *spring oats* [2]. Chickweed in *spring barley, spring wheat* [2, 3]. Chickweed in *spring cereals* [4-6]. Corn marigold in *spring oats* [2]. Corn marigold in *spring barley, spring wheat* [2, 3]. Corn marigold in *carrots, parsley, potatoes* [2-5]. Corn marigold in *parsnips* [2, 4]. Corn marigold in *spring cereals* [4, 5]. Fat-hen in *celery* [1]. Fat-hen in *carrots, parsley, potatoes* [1-6]. Fat-hen in *parsnips* [1, 2, 4, 6]. Fat-hen in *spring oats* [2]. Fat-hen in *spring barley, spring wheat* [2, 3]. Fat-hen in *spring cereals* [4-6]. Redshank in *celery* [1]. Redshank in *carrots, parsley, potatoes* [1-6]. Redshank in *parsnips* [1, 2, 4, 6]. Redshank in *spring oats* [2]. Redshank in *spring barley, spring wheat* [2, 3]. Redshank in *spring cereals* [4-6].

Efficacy
- Many weeds controlled pre-emergence or post-emergence to 2-3 leaf stage, some (annual meadow grass, mayweed) only susceptible pre-emergence. See label for details
- Best results achieved by application to firm, moist soil of fine tilth
- Little residual effect on soil with more than 10% organic matter

Crop safety/Restrictions

- Maximum number of treatments 1 per crop for celery, potatoes, cereals; 2 per crop for carrots, parsley, parsnips [1, 3-6]
- Maximum total dose equivalent to one full dose treatment [2]
- Apply only pre-crop emergence [6]
- Drill spring cereals at least 3 cm deep and apply pre-emergence of crop or weeds
- Do not use on undersown cereals or crops grown on Sands or Very Light soils or soils heavier than Sandy Clay Loam or with more than 10% organic matter
- Apply to potatoes well earthed up to a rounded ridge pre-crop emergence and do not cultivate after spraying. May be used in tank mix with glufosinate-ammonium [1]
- Apply to carrots or parsley at any time after drilling on organic soils, and within 4 d of drilling on other soils. Apply post-emergence as soon as weeds appear but after first rough leaf stage.
- Do not apply to emerged crops of carrots, parsnips or parsley under stress
- Recommendations for parsnips, parsley and celery vary. See label for details
- Potatoes, carrots and parsnips may be planted at any time after application. Lettuce should not be grown within 12 mth of treatment. Transplanted brassicas may be grown from 3 mth after treatment

Special precautions/Environmental safety

- Irritating to eyes and skin [3, 5, 6]
- Irritating to respiratory system [3, 5]
- Dangerous to fish or other aquatic life. Do not contaminate surface waters or ditches with chemical or used container [1-6]
- LERAP Category B
- Do not apply by hand-held sprayers

Personal protective equipment/Label precautions

- A, C, H [1-6]
- M03 [6]; R04a, R04b [3, 5, 6]; R04c [3, 5]; U08, U20b [1-6]; U19 [2-6]; U05a [3-6]; U09a [6]; C03 [3-6]; E13b, E16a, E30a [1-6]; E01 [3-6]; E31b [1, 3, 5, 6]; E07 [3, 5] (5 mth); E26 [1, 2, 6]; E32a [2, 4]; E34 [6]

Latest application/Harvest interval

- Pre-emergence for cereals and linseed; pre- or up to 40% emergence for potatoes (products differ, see label for details).
- HI onions, garlic, herbs 7 d, leeks 21 d

Approval

- Off-label approval to Jun 2001 for use on onions, leeks, garlic, celeriac (OLA 1660/98)[1]
- Accepted by BLRA for use on malting barley

344	magnesium phosphide

A phosphine generating compound used to control insect pests in stored commodities

Products

Degesch Plates	Rentokil	56% w/w	GE	07603

Uses

Insect pests in **stored cereals, stored plant material**.

Efficacy

- Product acts as fumigant by releasing poisonous hydrogen phosphide gas on contact with moisture in the air
- Place plates on the floor or wall of the building or on the surface of the commodity. Exposure time varies depending on temperature and pest. See label

Special precautions/Environmental safety

- Very toxic in contact with skin, by inhalation and if swallowed
- Highly flammable
- Only to be used by professional operators trained in the use of magnesium phosphide and familiar with the precautionary measures to be observed. See label for full precautions

- Keep in original container, tightly closed, in a safe place, under lock and key
- Do not allow plates or their spent residues to come into contact with food other than raw cereal grains
- Do not bulk spent plates and residues. Spontaneous ignition could result

Personal protective equipment/Label precautions
- A
- M04, R01a, R01b, R01c, R07c, U05b, U07, U13, U19, U20a, E02, E07, E15, E29, E30b, E32b

Withholding period
- Keep livestock out of treated areas

Approval
- Accepted by BLRA for use in stores for malting barley

---

345   malathion

A broad-spectrum contact organophosphorus insecticide and acaricide

Products

Malathion 60	United Phosphorus	600 g/l	EC	08018

Uses

Aphids in **carnations** (indoors), **chrysanthemums** (indoors), **cyclamens** (indoors), **dahlias** (indoors), **gladioli** (indoors), **lettuce** (indoors), **roses** (indoors), **strawberries, tomatoes** (indoors), **tulips** (indoors). Flea beetles in **watercress** (off-label). Leafhoppers in **carnations** (indoors), **chrysanthemums** (indoors), **cyclamens** (indoors), **dahlias** (indoors), **gladioli** (indoors), **roses** (indoors), **tomatoes** (indoors), **tulips** (indoors). Mealy bugs in **carnations** (indoors), **chrysanthemums** (indoors), **cyclamens** (indoors), **dahlias** (indoors), **gladioli** (indoors), **roses** (indoors), **tomatoes** (indoors), **tulips** (indoors). Midges in **watercress** (off-label). Mustard beetle in **watercress** (off-label). Scale insects in **roses** (indoors). Thrips in **carnations** (indoors), **chrysanthemums** (indoors), **cyclamens** (indoors), **dahlias** (indoors), **gladioli** (indoors), **roses** (indoors), **tomatoes** (indoors), **tulips** (indoors). Whitefly in **carnations** (indoors), **chrysanthemums** (indoors), **cyclamens** (indoors), **dahlias** (indoors), **gladioli** (indoors), **roses** (indoors), **tomatoes** (indoors), **tulips** (indoors).

Efficacy
- Spray when pest first seen and repeat as necessary, usually at 7-14 d intervals
- Number and timing of sprays vary with crop and pest. See label for details
- Repeat spray routinely for scale insect and whitefly control in glasshouses
- Where aphids resistant to malathion occur control is unlikely to be satisfactory

Crop safety/Restrictions
- Do not use on antirrhinums, crassula, ferns, fuchsias, gerberas, petunias, pileas, sweet peas or zinnias as damage may occur

Special precautions/Environmental safety
- This product contains an anticholinesterase organophosphorus compound. Do not use if under medical advice not to work with such compounds
- Dangerous to bees. Do not apply to crops in flower or to those in which bees are actively foraging. Do not apply when flowering weeds are present
- Extremely dangerous to fish or other aquatic life. Do not contaminate surface waters or ditches with chemical or used container
- Do not use by tractor mounted/drawn or air assisted sprayers

Personal protective equipment/Label precautions
- A, H, M
- M01, U08, U19, U20b, E12c, E13a, E22b, E26, E30a, E31a, E34

Latest application/Harvest interval
- HI 4 d; crops for processing 7 d

Approval
- Off-label approval unlimited for use on watercress beds (OLA 2719/97)

Maximum Residue Level (mg residue/kg food)
- cereals 8; garlic, onions, shallots, tomatoes, peppers, aubergines, cucumbers, gherkins, courgettes, cauliflowers, Brussels sprouts, head cabbages, lettuce, beans (with pods), peas (with pods), celery, leeks, rhubarb, cultivated mushrooms 3; citrus fruits 2; pome fruits, apricots, peaches, nectarines, plums, grapes, cane fruits, bilberries, cranberries, currants, gooseberries, bananas, carrots, horseradish, parsnips, parsley root, salsify, swedes, turnips, potatoes 0.5

346   maleic hydrazide

A pyridazinone plant growth regulator suppressing sprout and bud growth

See also dicamba + maleic hydrazide + MCPA
Products

1	Fazor	Dow	60% w/w	SG	05558
2	Mazide 25	Vitax	250 g/l	SL	02067
3	Regulox K	RP Amenity	250 g/l	SL	05405
4	Regulox K	Aventis Environ.	250 g/l	SL	09937
5	Royal MH 180	Nufarm Whyte	180 g/l	SL	07840
6	Source II	Chiltern	80% w/w	WB	08314

Uses

Growth retardation in **hedges** [2]. Growth suppression in **amenity grass** [2-5]. Growth suppression in **golf courses, road verges** [2, 5]. Growth suppression in **industrial sites, roadside grass, waste ground** [3, 4]. Growth suppression in **grass near water** [3-5]. Growth suppression in **areas around farm buildings** [5]. Sprout suppression in **onions** [1, 2, 5, 6]. Sprout suppression in **potatoes** [1, 6]. Sucker inhibition in **amenity trees and shrubs** [2-4]. Volunteer suppression in **potatoes** [1]. Volunteer suppression in **ware potatoes** [6].

Efficacy
- Apply to grass at any time of yr when growth active, best when growth starting in Apr-May and repeated when growth recommences
- Uniform coverage and 8 h dry weather necessary for effective results
- Mow 2-3 d before and 5-10 d after spraying for best results. Need for mowing reduced for up to 6 wk
- Apply to onions at 10% necking and not later than 50% necking stage when the tops are still green. Rain or irrigation within 24 h may reduce effectiveness
- Apply to second early or maincrop potatoes at least 3 wk before haulm destruction. Accurate timing essential but rain or irrigation within 24 h may reduce effectiveness. See label for details [1]
- To control suckers wet trunks thoroughly, especially pruned and basal bud areas [2]
- Spray hawthorn hedges in full leaf, privet 7 d after cutting, in Apr-May [2]

Crop safety/Restrictions
- Maximum number of treatments 2 per yr on amenity grass, land not intended for cropping and land adjacent to aquatic areas; 1 per crop on onions, potatoes and on or around tree trunks
- Do not apply in drought or when crops are suffering from pest, disease or herbicide damage. Do not treat fine turf or grass seeded less than 8 mth previously
- May be applied to grass along water courses but not to water surface [3-5]
- Do not treat onions more than 2 wk before maturing. Treated onions may be stored until Mar but must then be removed to avoid browning
- Only use on potatoes of good keeping quality; not on seed, first earlies or crops grown under polythene
- Do not treat potatoes within 3 wk of applying a haulm desiccant or if temperatures above 26°C
- Avoid drift onto nearby vegetables, flowers or other garden plants

Special precautions/Environmental safety
- Irritant. Risk of serious damage to eyes [6]
- Only apply to grass not to be used for grazing
- Do not use treated water for irrigation purposes within 3 wk of treatment or until concentration in water falls below 0.02 ppm [3-5]
- Maximum permitted concentration in water 2 ppm [3-5]

- Do not dump surplus product in water or ditch bottoms [3-5]
- Do not apply by knapsack sprayer [1]

Personal protective equipment/Label precautions
- H [1]; A [1, 6]; C [6]
- R04, R04d [6]; U20c [2-4]; U08 [1]; U20b [1, 5]; U02, U05a, U13, U14, U15, U19, U20a, U22, C03 [6]; E15, E30a [1-6]; E31a [2, 5]; E19, E21 [3-5]; E26 [3, 4, 6]; E31b [3, 4]; E32a [1]; E01 [6]

Latest application/Harvest interval
- 3 wk before haulm destruction for potatoes; before 50% necking for onions
- HI onions 4 d [2, 3, 5, 6], 7 d [1]; potatoes 3 wk

Approval
- Approved for use on grass near water [3, 5]. See notes in Section 1 on use of herbicides in or near water

Maximum Residue Level (mg residue/kg food)
- ware potatoes 50; garlic, onions, shallots 10; all other products 1

---

347 mancozeb

A protective dithiocarbamate fungicide for potatoes and other crops

See also benalaxyl + mancozeb
chlorothalonil + mancozeb
cymoxanil + mancozeb
cymoxanil + mancozeb + oxadixyl
dimethomorph + mancozeb

Products

1	Agrichem Mancozeb 80	Agrichem	80% w/w	WP	06354
2	Ashlade Mancozeb FL	Nufarm Whyte	412 g/l	SC	06226
3	Dithane 945	Interfarm	80% w/w	WP	09897
4	Dithane Dry Flowable	PBI	75% w/w	WG	04255
5	Helm 75 WG	PBI	75% w/w	WG	08309
6	Karamate Dry Flo	Landseer	75% w/w	WG	09259
7	Landgold Mancozeb 80 W	Landgold	80% w/w	WP	06507
8	Micene 80	Sipcam	80% w/w	WP	09112
9	Micene DF	Sipcam	77% w/w	WG	09957
10	Opie 80 WP	PBI	80% w/w	WP	08301
11	Penncozeb WDG	Nufarm Whyte	75% w/w	WG	07833
12	Tariff 75 WG	PBI	75% w/w	WG	08308

Uses

Black spot in *roses* [6]. Blight in *potatoes* [1-5, 7-12]. Brown rust in *spring wheat, winter wheat* [1, 3-5, 9-12]. Brown rust in *barley, wheat* [7]. Brown rust in *spring barley, winter barley* [8, 9, 11]. Currant leaf spot in *blackcurrants, gooseberries* [6]. Downy mildew in *winter oilseed rape* [1, 3-5, 10, 12]. Downy mildew in *bulb onions (off-label), garlic (off-label)* [3]. Downy mildew in *grapevines (off-label), lettuce, protected lettuce* [6]. Fire in *tulips* [6]. Net blotch in *winter barley* [5, 8-12]. Net blotch in *barley* [7]. Net blotch in *spring barley* [8, 9, 11]. Ray blight in *chrysanthemums* [6]. Rhynchosporium in *winter barley* [1, 3, 5, 9-12]. Rhynchosporium in *spring barley* [9, 11]. Rust in *pelargoniums, protected carnations, roses* [6]. Scab in *pears* [6]. Scab in *apples* [6, 8, 9, 11]. Septoria diseases in *spring wheat, winter wheat* [7, 9, 11]. Septoria leaf spot in *spring wheat, winter wheat* [1, 3-5, 8, 10, 12]. Sooty moulds in *spring wheat, winter wheat* [1, 3-5, 9-12]. Sooty moulds in *winter barley* [5, 9-12]. Sooty moulds in *barley, wheat* [7]. Sooty moulds in *spring barley* [9, 11]. Yellow rust in *winter wheat* [1, 3-5, 9-12]. Yellow rust in *barley, wheat* [7]. Yellow rust in *spring barley, spring wheat, winter barley* [9, 11].

Efficacy
- Mancozeb is a protectant fungicide and will give moderate control, suppression or reduction of the cereal diseases listed but in many cases mixture with carbendazim is essential to achieve satisfactory results. See labels for details
- Apply to cereals before any disease is well established. See label for recommended timing

- May be recommended for suppression or control of mildew in cereals depending on product and tank mix. See label for details
- Apply to potatoes before haulm meets across rows (usually mid-Jun) or at earlier blight warning, and repeat every 7-14 d depending on conditions and product used (see label)
- May be used on potatoes up to desiccation of haulm
- On oilseed rape apply as soon as disease develops between cotyledon and 5-leaf stage (GS 1,0-1,5)

Crop safety/Restrictions
- Maximum number of treatments varies with crop and product used - check labels for details
- Check labels for minimum interval that must elapse between treatments
- Apply to cereals from 4-leaf stage to before early milk stage (GS 71). Recommendations vary, see labels for details
- Treat winter oilseed rape before 6 true leaf stage (GS 1,6) and before 31 Dec
- On protected lettuce only 2 post-planting applications of mancozeb or of any combination of products containing EBDC fungicide (mancozeb, maneb, thiram, zineb) either as a spray or a dust are permitted within 2 wk of planting out and none thereafter.
- Avoid treating wet cereal crops or those suffering from drought or other stress

Special precautions/Environmental safety
- Irritating to eyes and skin [2, 7-9, 11]
- Irritating to respiratory system [1, 2, 4, 5, 7-12]
- May cause sensitization by skin contact [1, 4-6, 10, 12]
- Harmful to fish or other aquatic life. Do not contaminate surface waters or ditches with chemical or used container
- Keep away from fire and sparks [1, 10]
- Use contents of opened container as soon as possible and do not store opened containers until following season. If product becomes wet, effectiveness may be reduced and inflammable vapour produced [1, 10]

Personal protective equipment/Label precautions
- A [1-12]; D [1, 6-9]; H [7]; C [7-9]
- R04c [1, 2, 4, 5, 7-12]; R04e [1, 4-6, 10, 12]; R04a, R04b [2, 7-9, 11]; U05a [1-12]; U08 [2, 4, 5, 7-12]; U20a [2-5, 7, 8, 10, 12]; U19 [2, 8, 9, 11]; U20b [1, 6, 9, 11]; U13 [8, 9, 11]; C03, E01, E13c, E30a, E32a [1-12]; E34 [2]; E26 [2, 8, 9]

Latest application/Harvest interval
- Before early milk stage (GS 73) for cereals; before 6 true leaf stage and before 31 Dec for winter oilseed rape.
- HI potatoes 7 d; outdoor lettuce 14 d; protected lettuce 21 d; apples, pears, blackcurrants, gooseberries 4 wk

Approval
- Approved for aerial application on potatoes [1-5, 7, 8, 10-12]. See notes in Section 1
- Off-label approval unstipulated for use on grapevines (HI 63 d) (OLA 1666/99)[6]; to Jul 2003 for use on outdoor bulb onions (including those grown from sets) and garlic (OLA 1720/2000)[3]
- Accepted by BLRA for use on malting barley
- Approval expiry 31 Dec 2001 [4]

Maximum Residue Level (mg residue/kg food)
- hops 25; lettuces, herbs 5; oranges, apricots, peaches, nectarines, grapes, strawberries, peppers, aubergines 2; cherries, plums 1; garlic, onions, shallots, cucumbers 0.5; celeriac, witloof 0.2; tree nuts, oilseeds (except rape seed), tea 0.1; bilberries, cranberries, wild berries, blackberries, loganberries, miscellaneous fruit, root and tuber vegetables, horseradish, Jerusalem artichokes, parsley root, sweet potatoes, swedes, turnips, yams, spinach, beet leaves, watercress, asparagus, cardoons, fennel, globe artichokes, rhubarb, mushrooms, potatoes, maize, rice, animal products 0.05

348   mancozeb + metalaxyl

A systemic and protectant fungicide for potatoes and some other crops

Products

1	Fubol 75 WP	Novartis	67.5:7.5% w/w	WP	08409
2	Osprey 58 WP	Novartis	48:10% w/w	WP	08428

Uses

Blight in **potatoes**.

Efficacy

- Apply as protectant spray on potatoes immediately risk of blight in district or as crops begin to meet in (not across) rows and repeat every 10-21 d according to blight risk (every 10-14 d on irrigated crops)
- Use for first part of spray programme to mid-Aug. Later use a protectant fungicide (preferably tin-based) up to complete haulm destruction or harvest
- Recommended for potato blight control in Northern Ireland but specific Northern Ireland label must be consulted [2]
- Do not apply on potatoes for 2-3 h before rainfall or when raining

Crop safety/Restrictions

- Maximum number of treatments 5 per crop (including other phenylamide-based fungicides) for potatoes; 3 per crop on Brussels sprouts, cabbages, cauliflowers, broccoli, calabrese, onions, leeks; 2 per crop on oilseed rape, field and broad beans, apple orchards
- Do not treat potato crops showing active blight infection or a second potato crop in the same field in the same season

Special precautions/Environmental safety

- Irritating to skin, eyes and respiratory system
- Harmful to fish or other aquatic life. Do not contaminate surface waters or ditches with chemical or used container

Personal protective equipment/Label precautions

- A
- R04a, R04b, R04c, U05a, U08, U20a, C02, C03, E01, E13c, E30a, E32a

Latest application/Harvest interval

- Within 6 wk of drilling for carrots and parsnips; pre-planting for seedling brassicas.
- HI early and maincrop potatoes 1 d [1]; early potatoes 7 d, maincrop potatoes 14 d [2]; Brussels sprouts, onions 14 d; apples 4 wk; carrots 8 wk; raspberries 3 mth

Approval

- Approved for aerial application on potatoes. See notes in Section 1

Maximum Residue Level (mg residue/kg food)

- see mancozeb and metalaxyl entries

349   mancozeb + metalaxyl-M

A systemic and protectant fungicide mixture

Products

1	Fubol Gold	Novartis	64:4% w/w	WB	08812
2	Fubol Gold WG	Novartis	64:4% w/w	WG	10184

Uses

Blight in **potatoes** [1, 2]. Phytophthora fruit rot in **apple orchards** *(off-label)* [1]. White blister in **cabbages** *(off-label)*, **protected radishes** *(off-label)*, **radishes** *(off-label)*, **shallots** *(off-label)* [1].

Efficacy

- Commence blight programme before risk of infection occurs as crops begin to meet along the rows and repeat every 10-14 d according to blight risk. Do not exceed a 14 d interval between sprays

- If infection risk conditions occur earlier than the above growth stage commence spraying immediately
- Complete the blight programme using a protectant fungicide starting no later than 10 d after the last phenylamide spray. At least 2 such sprays should be applied

Crop safety/Restrictions

- Maximum number of treatments 5 per crop (including all other phenylamide fungicides)
- Monitor sprayed crops and stop using any phenylamide product if active blight is identified. Switch to a tin-based product programme within 7 d and continue up to haulm destruction or harvest
- After treating early potatoes destroy and remove any remaining haulm after harvest to minimise blight pressure on neighbouring maincrop potatoes

Special precautions/Environmental safety

- Irritant. May cause sensitization by skin contact [2]
- Dangerous to fish or other aquatic life. Do not contaminate surface waters or ditches with chemical or used container
- Do not harvest crops for human consumption for at least 7 d after final application

Personal protective equipment/Label precautions

- A [1, 2]
- R04, R04e [2]; U22 [1]; U05a, U08, U20b [1, 2]; C02 [1, 2] (7 d); E13b, E26, E30a, E32a, E34 [1, 2]; E29 [2]

Latest application/Harvest interval

- Before end of active haulm growth or before end Aug, whichever is earlier
- HI 7 d

Approval

- Off-label approval unstipulated for use on apple orchard floors, protected radishes (OLA 0936/99)[1]; unstipulated for use on outdoor radishes (OLA 0380/00)[1]; to Apr 2003 for use on shallots (OLA 0381/00)[1]; to Apr 2003 for use on cabbages (OLA 0382/00)[1]

Maximum Residue Level (mg residue/kg food)

- see mancozeb and metalaxyl entries

350   mancozeb + ofurace

A systemic and protectant fungicide mixture for potatoes

Products

Patafol	Nufarm Whyte	67:5.8% w/w	WP	07397

Uses

Blight in **potatoes, potatoes grown for seed** *(off-label)*.

Efficacy

- Apply to potatoes at first blight warning or just before foliage meets in the row and repeat at 10-14 d intervals
- Complete blight programme up to haulm destruction with organotin products
- Always apply before the crop is infected
- Do not apply if rain likely within 2-3 h of treatment
- After rain or irrigation wait until foliage is dry before spraying

Crop safety/Restrictions

- Maximum number of treatments 5 per crop
- Do not apply after the end of Aug

Special precautions/Environmental safety

- Irritating to eyes, skin and respiratory system
- Harmful to fish or other aquatic life. Do not contaminate surface waters or ditches with chemical or used container

Personal protective equipment/Label precautions

- A, C
- R04a, R04b, R04c, U05a, U08, U19, U20b, C03; C02 (7 d); E01, E13c, E29, E30a, E32a

Latest application/Harvest interval
- HI 7 d

Approval
- Off-label approval unlimited for use on protected crops of potatoes for seed (OLA 1599/96)

Maximum Residue Level (mg residue/kg food)
- see mancozeb entry

---

351   mancozeb + oxadixyl

A systemic and contact protectant fungicide for potato blight control

Products

Recoil	Novartis	56:10% w/w	WP	08483

Uses

Blight in **potatoes**. Root rot in **blackberries** *(off-label)*, **cane fruit** *(off-label)*, **dewberries** *(off-label)*, **loganberries** *(off-label)*, **raspberries** *(off-label)*.

Efficacy
- Commence treatment for potato blight early in season as soon as there is risk of infection
- In absence of a blight warning treatment should start before potatoes meet within row
- Repeat treatments every 10-14 d depending on degree of blight infection risk
- Use as a protectant treatment. Potatoes showing blight should not be treated
- Do not spray within 2-3 h of rainfall and only apply to dry foliage
- To complete blight spray programme after end of Aug make subsequent treatments up to haulm destruction with another fungicide, preferably fentin based

Crop safety/Restrictions
- Maximum number of treatments 5 per crop (including other phenylamide-based fungicides) for potatoes; 2 per yr for raspberries
- Do not treat raspberries on light soils with less than 2% organic matter

Special precautions/Environmental safety
- Irritating to skin, eyes and respiratory system. May cause sensitization by skin contact
- Do not spray within 2 m of surface water courses (raspberries, blackberries, dewberries, loganberries and cane fruit only)

Personal protective equipment/Label precautions
- A, C
- R04a, R04b, R04c, R04e, U02, U05a, U11, U12, U19, U20b, C03, E01, E15, E30a, E32a

Latest application/Harvest interval
- Before end of Aug.
- HI zero

Approval
- Approved for aerial application on potatoes. See notes in Section 1.
- Off-label approval unlimited for use as a soil drench on raspberries, blackberries, dewberries, loganberries, cane fruit (other than wild) (OLA 1711/97)

Maximum Residue Level (mg residue/kg food)
- see mancozeb entry

---

352   mancozeb + propamocarb hydrochloride

A systemic and contact protectant fungicide for potato blight control

Products

1	Aventis Tattoo	Aventis	301.6:248 g/l	SC	09811
2	Tattoo	Aventis	301.6:248 g/l	SC	07293

Uses

Blight in **potatoes**.

Efficacy
- Commence treatment as soon as there is risk of blight infection

- In the absence of a warning treatment should start before the crop meets within the row
- Repeat treatments every 10-14 d depending on degree of infection risk
- Irrigated crops should be regarded as at high risk and treated every 10 d
- Do not spray when rainfall is imminent and apply only to dry foliage
- To complete the spray programme after the end of Aug, make subsequent treatments up to haulm destruction with a protectant fungicide, preferably fentin based.

Crop safety/Restrictions
- Maximum number of treatments 5 per crop
- Do not use once blight has become readily visible

Special precautions/Environmental safety
- Irritant. May cause sensitization by skin contact
- Harmful to fish or other aquatic life. Do not contaminate surface waters or ditches with chemical or used container

Personal protective equipment/Label precautions
- A, H
- R04e, U05a, U08, U14, U19, U20a, C03, E01, E13c, E30a, E31b

Latest application/Harvest interval
- Before end of Aug.
- HI 14 d

Maximum Residue Level (mg residue/kg food)
- see mancozeb entry

---

353   maneb

A protectant dithiocarbamate fungicide

See also carbendazim + chlorothalonil + maneb
carbendazim + maneb
copper oxychloride + maneb + sulphur
fentin acetate + maneb
ferbam + maneb + zineb

Products

1	Agrichem Maneb 80	Agrichem	80% w/w	WP	05474
2	Maneb 80	Rohm & Haas	80% w/w	WP	01276
3	X-Spor SC	United Phosphorus	480 g/l	SC	08077

Uses

Blight in **potatoes** [1-3]. Brown rust in **winter wheat** [1-3]. Brown rust in **spring barley, winter barley** [2, 3]. Eyespot in **spring barley, winter barley, winter wheat** [3]. Net blotch in **spring barley, winter barley** [2]. Powdery mildew in **winter wheat** [1, 3]. Rhynchosporium in **spring barley, winter barley** [3]. Septoria in **winter wheat** [1, 2]. Septoria diseases in **winter wheat** [3]. Sooty moulds in **spring barley, winter barley** [1]. Sooty moulds in **winter wheat** [1, 3]. Yellow rust in **winter wheat** [1-3]. Yellow rust in **spring barley, winter barley** [2, 3].

Efficacy
- Maneb is a protectant fungicide and will give moderate control, suppression or reduction of the cereal diseases listed but in many cases mixture with carbendazim is essential to achieve satisfactory results - see labels for details
- Apply to potatoes before blight infection occurs, at blight warning or before haulms meet in row and repeat every 10-14 d
- Most effective on cereals as a preventative treatment before disease established, an early application at about first node stage (GS 31), a late application after flag leaf emergence (GS 37) and during ear emergence before watery ripe stage (GS 71)

Crop safety/Restrictions
- Maximum number of treatments normally 2 per crop in cereals (check labels)
- Do not apply if frost or rain expected, if crop wet or suffering from drought or physical or chemical stress
- A minimum of 10 d must elapse between applications to potatoes

Special precautions/Environmental safety
- Harmful if swallowed [3]
- Irritating to respiratory system [1, 2], eyes and skin [3]
- May cause sensitization by skin contact [1, 2]
- Dangerous (harmful [2]) to fish or other aquatic life. Do not contaminate surface waters or ditches with chemical or used container [1]
- Keep away from fire and sparks [1, 2]
- Use contents of opened container as soon as possible and do not store opened containers until following season. If product becomes wet, effectiveness may be reduced and inflammable vapour produced [1, 2]

Personal protective equipment/Label precautions
- A [1-3]; D [1]; C [3]
- M03 [3]; R04e [1, 2]; R04c [1-3]; R03c, R04a, R04b [3]; U20a [2]; U05a [1, 2]; U08 [1-3]; U20b [1, 3]; U04a [3]; C02 [2] (7 d); C03 [1-3]; E13c, E29, E34 [2]; E32a [1, 2]; E01, E30a [1-3]; E13b [1]; E15, E26, E31b [3]

Latest application/Harvest interval
- Before grain milky-ripe (GS 71) for winter wheat [3]; before flag leaf sheath opening (GS 45) for barley [3]; before early milk stage (GS 71) for cereals [1]
- HI potatoes, cereals 7 d

Approval
- Approved for aerial application on potatoes. See notes in Section 1
- Following implementation of Directive 98/82/EC, approval for use of maneb on several crops was revoked in 1999
- Accepted by BLRA for use on malting barley

Maximum Residue Level (mg residue/kg food)
- hops 25; lettuces, herbs 5; oranges, apricots, peaches, nectarines, grapes, strawberries, peppers, aubergines 2; cherries, plums 1; garlic, onions, shallots, cucumbers 0.5; celeriac, witloof 0.2; tree nuts, oilseeds (except rape seed), tea 0.1; bilberries, cranberries, wild berries, blackberries, loganberries, miscellaneous fruit, root and tuber vegetables, horseradish, Jerusalem artichokes, parsley root, sweet potatoes, swedes, turnips, yams, spinach, beet leaves, watercress, asparagus, cardoons, fennel, globe artichokes, rhubarb, mushrooms, potatoes, maize, rice, animal products 0.05

354 MCPA

A translocated phenoxyacetic herbicide for cereals and grassland

See also 2,4-DB + linuron + MCPA
2,4-DB + MCPA
benazolin + 2,4-DB + MCPA
bentazone + MCPA + MCPB
bifenox + MCPA + mecoprop-P
clopyralid + 2,4-D + MCPA
clopyralid + diflufenican + MCPA
clopyralid + fluroxypyr + MCPA
dicamba + dichlorprop + ferrous sulphate + MCPA
dicamba + dichlorprop + MCPA
dicamba + maleic hydrazide + MCPA
dicamba + MCPA + mecoprop-P
dichlorprop-P + MCPA + mecoprop-P

Products

1	Agricorn 500 II	FCC	500 g/l	SL	09155
2	Agritox 50	Nufarm Whyte	500 g/l	SL	07400
3	Agroxone	Headland	485 g/l	SL	09947
4	Barclay Meadowman II	Barclay	500 g/l	SL	08525
5	BASF MCPA Amine 50	BASF	500 g/l	SL	00209
6	Campbell's MCPA 50	MTM Agrochem.	500 g/l	SL	00381

Products (Continued)

7 Headland Spear	Headland	500 g/l	SL	07115
8 HY-MCPA	Agrichem	500 g/l	SL	06293
9 Luxan MCPA 500	Luxan	500 g/l	SL	07470
10 MCPA 25%	Nufarm Whyte	235 g/l	SL	07998
11 MSS MCPA 50	Nufarm Whyte	500 g/l	SL	01412
12 Quad MCPA 50%	Quadrangle	500 g/l	SL	01669

Uses

Annual dicotyledons in *established grassland* [1-12]. Annual dicotyledons in *barley, oats, wheat* [1, 2, 4-6, 11, 12]. Annual dicotyledons in *winter rye* [1, 5-7, 9, 12]. Annual dicotyledons in *lawns, road verges, turf, undersown cereals* [11]. Annual dicotyledons in *linseed* [2-4, 11]. Annual dicotyledons in *grass seed crops* [2-4, 9, 10, 12]. Annual dicotyledons in *rye* [2, 4]. Annual dicotyledons in *undersown barley (red clover or grass), undersown wheat (red clover or grass)* [3]. Annual dicotyledons in *spring barley, spring oats, spring wheat, winter barley, winter wheat* [3, 7-10]. Annual dicotyledons in *winter oats* [3, 7, 9, 10]. Annual dicotyledons in *newly sown grass* [3, 9, 10]. Annual dicotyledons in *undersown barley, undersown oats, undersown wheat* [5]. Annual dicotyledons in *amenity grass, managed amenity turf* [7]. Annual dicotyledons in *young leys* [8]. Buttercups in *established grassland* [1, 5-8, 11, 12]. Buttercups in *amenity grass, managed amenity turf* [7]. Charlock in *barley, oats, wheat* [1, 2, 4-6, 11, 12]. Charlock in *winter rye* [1, 5-7, 9, 12]. Charlock in *established grassland* [2-4, 9, 10]. Charlock in *rye* [2, 4]. Charlock in *undersown barley (red clover or grass), undersown wheat (red clover or grass)* [3]. Charlock in *linseed* [3, 11]. Charlock in *spring barley, spring oats, spring wheat, winter barley, winter wheat* [3, 7-10]. Charlock in *winter oats* [3, 7, 9, 10]. Charlock in *grass seed crops, newly sown grass* [3, 9, 10]. Charlock in *undersown barley, undersown oats, undersown wheat* [5]. Charlock in *young leys* [8]. Creeping thistle in *lawns, road verges, turf* [11]. Daisies in *lawns, road verges, turf* [11]. Dandelions in *established grassland* [1, 5, 6, 8, 11, 12]. Dandelions in *lawns, turf* [11]. Docks in *established grassland* [1, 5-8, 11, 12]. Docks in *road verges* [11]. Fat-hen in *barley, oats, wheat* [1, 2, 4-6, 11, 12]. Fat-hen in *winter rye* [1, 5-7, 9, 12]. Fat-hen in *established grassland* [2-4, 9, 10]. Fat-hen in *rye* [2, 4]. Fat-hen in *undersown barley (red clover or grass), undersown wheat (red clover or grass)* [3]. Fat-hen in *linseed* [3, 11]. Fat-hen in *spring barley, spring oats, spring wheat, winter barley, winter wheat* [3, 7-10]. Fat-hen in *winter oats* [3, 7, 9, 10]. Fat-hen in *grass seed crops, newly sown grass* [3, 9, 10]. Fat-hen in *undersown barley, undersown oats, undersown wheat* [5]. Fat-hen in *young leys* [8]. Hemp-nettle in *barley, oats, wheat* [1, 2, 4-6, 11, 12]. Hemp-nettle in *winter rye* [1, 5-7, 9, 12]. Hemp-nettle in *established grassland* [2-4, 9, 10]. Hemp-nettle in *rye* [2, 4]. Hemp-nettle in *undersown barley (red clover or grass), undersown wheat (red clover or grass)* [3]. Hemp-nettle in *linseed* [3, 11]. Hemp-nettle in *spring barley, spring oats, spring wheat, winter barley, winter wheat* [3, 7-10]. Hemp-nettle in *winter oats* [3, 7, 9, 10]. Hemp-nettle in *grass seed crops, newly sown grass* [3, 9, 10]. Hemp-nettle in *undersown barley, undersown oats, undersown wheat* [5]. Hemp-nettle in *young leys* [8]. Perennial dicotyledons in *established grassland* [1-12]. Perennial dicotyledons in *barley, wheat* [1, 2, 4-6, 11, 12]. Perennial dicotyledons in *winter rye* [1, 5-7, 9, 12]. Perennial dicotyledons in *lawns, road verges, turf* [11]. Perennial dicotyledons in *rye* [2, 4]. Perennial dicotyledons in *oats* [2, 4-6, 11, 12]. Perennial dicotyledons in *linseed, undersown barley (red clover or grass), undersown wheat (red clover or grass)* [3]. Perennial dicotyledons in *spring barley, spring oats, spring wheat, winter barley, winter wheat* [3, 7-10]. Perennial dicotyledons in *winter oats* [3, 7, 9, 10]. Perennial dicotyledons in *newly sown grass* [3, 9, 10]. Perennial dicotyledons in *grass seed crops* [3, 9, 10, 12]. Perennial dicotyledons in *amenity grass, managed amenity turf* [7]. Perennial dicotyledons in *young leys* [8]. Ragwort in *established grassland* [7]. Rushes in *established grassland* [7]. Stinging nettle in *industrial sites, land not intended to bear vegetation* [7]. Thistles in *established grassland* [5-8, 11, 12]. Thistles in *industrial sites, land not intended to bear vegetation* [7]. Wild radish in *barley, oats, wheat* [1, 2, 4-6, 11, 12]. Wild radish in *winter rye* [1, 5-7, 9, 12]. Wild radish in *established grassland* [2-4, 9, 10]. Wild radish in *rye* [2, 4]. Wild radish in *undersown barley (red clover or grass), undersown wheat (red clover or grass)* [3]. Wild radish in *linseed* [3, 11]. Wild radish in *spring barley, spring oats, spring wheat, winter barley, winter wheat* [3, 7-10]. Wild radish in *winter oats* [3, 7, 9, 10]. Wild radish in

grass seed crops, newly sown grass [3, 9, 10]. Wild radish in **undersown barley, undersown oats, undersown wheat** [5]. Wild radish in **young leys** [8].

Efficacy
- Best results achieved by application to weeds in seedling to young plant stage under good growing conditions when crop growing actively
- Spray perennial weeds in grassland before flowering. Most susceptible growth stage varies between species. See label for details
- Do not spray during cold weather, drought, if rain or frost expected or if crop wet

Crop safety/Restrictions
- Maximum number of treatments normally 1 per crop or yr except grass (2 per yr) for some products. See label
- Apply to winter cereals in spring from fully tillered, leaf sheath erect stage to before first node detectable (GS 31)
- Apply to spring barley and wheat from 5-leaves unfolded (GS 15), to oats from 1-leaf unfolded (GS 11) to before first node detectable (GS 31)
- Apply to cereals undersown with grass after grass has 2-3 leaves unfolded
- Do not use on cereals before undersowing
- Do not roll, harrow or graze for a few days before or after spraying; see label
- Recommendations for crops undersown with legumes vary. Red clover may withstand low doses after 2-trifoliate leaf stage, especially if shielded by taller weeds but white clover is more sensitive. See label for details
- Do not use on grassland where clovers are an important part of the sward
- Apply to grass seed crops from 2-3 leaf stage to 5 wk before head emergence
- Temporary wilting may occur on linseed but without long term effects
- Do not use on any crop suffering from stress or herbicide damage
- Do not direct drill brassicas or legumes within 6 wk of spraying grassland
- Avoid spray drift onto nearby susceptible crops

Special precautions/Environmental safety
- Harmful in contact with skin   [7, 9, 11]
- Harmful if swallowed
- Irritating to skin. Risk of serious damage to eyes [5, 7, 9, 10]
- Harmful to fish or other aquatic life. Do not contaminate surface waters or ditches with chemical or used container [1, 5]

Personal protective equipment/Label precautions
- A, C [1-7, 9-12]
- M03, R03c [1-12]; R04b, R04d [5, 7, 9, 10]; R03a [7, 9, 11]; U05a, U08 [1-12]; U20b [2-5, 7-11]; U20a [1, 6, 12]; U02 [9]; C03, E01, E26, E34 [1-12]; E07 [1-12] (2 wk); E30a [1-10, 12]; E31c [5, 8]; E13c [1, 5]; E15 [2-4, 6-12]; E31a [1-3, 6, 10, 12]; E31b [4, 7, 9, 11]

Withholding period
- Keep livestock out of treated areas until foliage of poisonous weeds such as ragwort has died and become unpalatable

Latest application/Harvest interval
- Before 1st node detectable (GS 31) for cereals; 4-6 wk before heading for grass seed crops; before crop 15 cm high for linseed

Approval
- Accepted by BLRA for use on malting barley

355   MCPA + MCPB

A translocated herbicide for undersown cereals and grassland

Products

1 Bellmac Plus	United Phosphorus	38:262 g/l	SL	07521
2 MSS MCPB + MCPA	Nufarm Whyte	25:275 g/l	SL	01413
3 Trifolex-Tra	Cyanamid	34:216 g/l	SL	07147
4 Trifolex-Tra	BASF	34:216 g/l	SL	BASF
5 Tropotox Plus	Unicrop	37.5:262.5 g/l	SL	06156

Uses

Annual dicotyledons in *leys* [1-4]. Annual dicotyledons in **undersown cereals** [1-5]. Annual dicotyledons in **sainfoin** [1, 2, 5]. Annual dicotyledons in **established grassland** [2-4]. Annual dicotyledons in **cereals** [2, 5]. Annual dicotyledons in **peas** [3, 4]. Annual dicotyledons in **clover seed crops, direct-sown seedling clovers, dredge corn containing peas, established leys, permanent pasture** [5]. Perennial dicotyledons in *leys* [1-4]. Perennial dicotyledons in **undersown cereals** [1-5]. Perennial dicotyledons in **sainfoin** [1, 2, 5]. Perennial dicotyledons in **established grassland** [2-4]. Perennial dicotyledons in **cereals** [2, 5]. Perennial dicotyledons in **peas** [3, 4]. Perennial dicotyledons in **clover seed crops, direct-sown seedling clovers, dredge corn containing peas, established leys, permanent pasture** [5].

Efficacy

- Best results achieved by application to weeds in seedling to young plant stage under good growing conditions when crop growing actively. Spray perennials when adequate leaf surface before flowering. Retreatment often needed in following year
- Spray leys and sainfoin before crop provides cover for weeds
- Rain, cold or drought may reduce effectiveness

Crop safety/Restrictions

- Maximum number of treatments 1 per crop or yr
- Apply to cereals from 2-expanded leaf stage to before jointing (GS 12-30) and, where undersown, after 1-trifoliate leaf stage of clover
- Apply to direct-sown seedling clover after 1-trifoliate leaf stage
- Apply to mature white clover for fodder at any stage. Do not spray red clover after flower stalk has begun to form
- Do not spray clovers for seed [2, 3]
- Apply to sainfoin after first compound leaf stage [1], second compound leaf [2]
- Do not spray peas [1, 2]. Only spray peas in tank mix with Fortrol from 4-leaf stage to before bud visible (GS 104-201) [3]
- Do not roll or harrow for a few days before or after spraying [1-3]
- Avoid spray drift onto nearby sensitive crops

Special precautions/Environmental safety

- Harmful to fish or other aquatic life. Do not contaminate surface waters or ditches with chemical or used container

Personal protective equipment/Label precautions

- U08 [1-5]; U05a [2-5]; U20a [2-4]; U20b [1, 5]; C03 [2-4]; E13c, E26, E30a, E31b [1-5]; E07 [1-5] (2 wk); E01 [2-5]; E34 [1]

Withholding period

- Keep livestock out of treated areas until foliage of any poisonous weeds such as ragwort has died and become unpalatable

Latest application/Harvest interval

- Before first node detectable (GS 31) for cereals

Approval

- Accepted by BLRA for use on malting barley
- Product registration number not available at time of printing [4]

---

356 MCPA + mecoprop-P

A translocated selective herbicide for amenity grass

Products

Greenmaster Extra	Scotts	0.49:0.29 % w/w	GR	07594

Uses

Annual dicotyledons in **amenity grass**. Perennial dicotyledons in **amenity grass**.

Efficacy

- Apply from Apr to Sep, when weeds growing actively and have large leaf area available for chemical absorption
- Granules contain NPK fertilizer to encourage grass growth

- Do not apply when heavy rain expected or during prolonged drought. Irrigate after 1-2 d unless rain has fallen
- Do not mow within 2-3 d of treatment

Crop safety/Restrictions
- Maximum total dose 105 g/sq m per yr. The total amount of mecoprop-P applied in a single yr must not exceed the maximum total dose approved for any single product for use on turf
- Avoid contact with cultivated plants
- Do not use first 4 mowings as compost or mulch unless composted for 6 mth
- Do not treat newly sown or turfed areas for at least 6 mth
- Do not reseed bare patches for 8 wk after treatment

Special precautions/Environmental safety
- Harmful to fish or other aquatic life. Do not contaminate surface waters or ditches with chemical or used container

Personal protective equipment/Label precautions
- A, C, H, M
- U20a, E13c, E15, E30a, E32a; E07 (2 wk)

Withholding period
- Keep livestock out of treated areas for at least 2 wk and until foliage of any poisonous weeds such as ragwort has died and become unpalatable

---

357 MCPB

A translocated phenoxybutyric herbicide

See also bentazone + MCPA + MCPB
bentazone + MCPB
MCPA + MCPB

Products

1	Bellmac Straight	United Phosphorus	400 g/l	SL	07522
2	Tropotox	Unicrop	400 g/l	SL	06179

Uses

Annual dicotyledons in *leys, undersown cereals* [1]. Annual dicotyledons in *blackcurrants, peas, white clover seed crops* [1, 2]. Perennial dicotyledons in *leys, undersown cereals* [1]. Perennial dicotyledons in *blackcurrants, peas, white clover seed crops* [1, 2].

Efficacy
- Best results achieved by spraying young seedling weeds in good growing conditions
- Best results on perennials by spraying before flowering
- Effectiveness may be reduced by rain within 12 h, by very cold or dry conditions

Crop safety/Restrictions
- Maximum number of treatments 1 per crop
- Apply to undersown cereals from 2-leaves unfolded to first node detectable (GS 12-31), and after first trifoliate leaf stage of clover
- Red clover seedlings may be temporarily damaged but later growth is normal
- Apply to white clover seed crops in Mar to early Apr, not after mid-May, and allow 3 wk before cutting and closing up for seed
- Apply to peas from 3-6 leaf stage but before flower bud detectable (GS 103-201). Consult PGRO (see Appendix 2) for information on susceptibility of cultivars
- Do not use on beans, vetches or established red clover
- Apply to blackcurrants in Aug after harvest and after shoot growth ceased but while bindweed still growing actively; direct spray onto weeds as far as possible
- Do not roll or harrow for 7 d before or after treatment
- Avoid drift onto nearby sensitive crops

Special precautions/Environmental safety
- Harmful in contact with skin and if swallowed
- Harmful to fish or other aquatic life. Do not contaminate surface waters or ditches with chemical or used container

Personal protective equipment/Label precautions
- A, C [1, 2]
- M03, R03a, R03c [1, 2]; U19 [2]; U05a, U08, U20b, C03, E01, E07, E13c, E26, E30a, E31b, E34 [1, 2]

Withholding period
- Keep livestock out of treated areas until foliage of any poisonous weeds such as ragwort has died and become unpalatable

Latest application/Harvest interval
- First node detectable stage (GS 31) for cereals; before flower buds appear in terminal leaf (GS 201) for peas

Approval
- Accepted by BLRA for use on malting barley

---

358 mecoprop-P

A translocated phenoxypropionic herbicide for cereals and grassland

See also 2,4-D + mecoprop-P
bifenox + MCPA + mecoprop-P
bromoxynil + ioxynil + mecoprop-P
carfentrazone-ethyl + mecoprop-P
dicamba + MCPA + mecoprop-P
dicamba + mecoprop-P
dichlorprop-P + MCPA + mecoprop-P
fluroxypyr + mecoprop-P
MCPA + mecoprop-P

Products

1	Clenecorn Super	FCC	600 g/l	SL	09382
2	Clovotox	RP Amenity	142.5 g/l	SL	05354
3	Clovotox	Aventis Environ.	142.5 g/l	SL	09928
4	Compitox Plus	Nufarm Whyte	600 g/l	SL	09558
5	Duplosan KV	BASF	600 g/l	SL	09431
6	Isomec	Nufarm Whyte	600 g/l	SL	09881
7	Landgold Mecoprop-P 600	Landgold	600 g/l	SL	09014
8	Optica	Headland	600 g/l	SL	09963

Uses

Annual dicotyledons in **grassland** [1, 4-6]. Annual dicotyledons in **grass seed crops, young leys** [1, 4-6, 8]. Annual dicotyledons in **barley, oats, wheat** [1, 5, 7]. Annual dicotyledons in **managed amenity turf** [2, 3, 8]. Annual dicotyledons in **spring barley, spring oats, spring wheat, winter barley, winter oats, winter wheat** [4, 6, 8]. Annual dicotyledons in **leys** [7]. Annual dicotyledons in **established grassland** [7, 8]. Chickweed in **grassland** [1, 4-6]. Chickweed in **grass seed crops, young leys** [1, 4-6, 8]. Chickweed in **barley, oats, wheat** [1, 5, 7]. Chickweed in **managed amenity turf** [2, 3, 8]. Chickweed in **spring oats, spring wheat, winter barley, winter oats, winter wheat** [4, 6, 8]. Chickweed in **leys** [7]. Chickweed in **established grassland** [7, 8]. Cleavers in **grassland** [1, 4-6]. Cleavers in **grass seed crops, young leys** [1, 4-6, 8]. Cleavers in **barley, oats, wheat** [1, 5, 7]. Cleavers in **spring barley, spring oats, spring wheat, winter barley, winter oats, winter wheat** [4, 6, 8]. Cleavers in **leys** [7]. Cleavers in **established grassland** [7, 8]. Cleavers in **managed amenity turf** [8]. Clovers in **managed amenity turf** [2, 3]. Perennial dicotyledons in **grass seed crops, grassland, young leys** [1, 4-6]. Perennial dicotyledons in **barley, oats, wheat** [1, 5, 7]. Perennial dicotyledons in **managed amenity turf** [2, 3]. Perennial dicotyledons in **spring barley, spring oats, spring wheat, winter barley, winter oats, winter wheat** [4, 6]. Perennial dicotyledons in **established grassland, leys** [7].

Efficacy
- Best results achieved by application to seedling weeds which have not been frost hardened, when soil warm and moist and expected to remain so for several days
- Do not spray during cold weather, periods of drought, if rain or frost expected or if crop wet

Crop safety/Restrictions
- Maximum number of treatments normally 1 per crop for spring cereals and newly sown grass; 2 per crop for winter cereals and established grass. Check labels for details
- The total amount of mecoprop-P applied in a single yr must not exceed the maximum total dose approved for any single product for the crop/situation
- Spray winter cereals from 1 leaf stage in autumn up to and including first node detectable in spring (GS 10-31) or up to second node detectable (GS 32) if necessary. Apply to spring cereals from first fully expanded leaf stage (GS 11) but before first node detectable (GS 31)
- Spray cereals undersown with grass after grass starts to tiller
- Do not spray cereals undersown with clovers or legumes or to be undersown with legumes or grasses
- Spray newly sown grass leys when grasses have at least 3 fully expanded leaves and have begun to tiller. Any clovers will be damaged
- Do not spray grass seed crops within 5 wk of seed head emergence
- Do not spray crops suffering from herbicide damage or physical stress
- Do not roll or harrow for 7 d before or after treatment
- Avoid drift onto beans, beet, brassicas, fruit and other sensitive crops

Special precautions/Environmental safety
- Harmful in contact with skin [1, 4, 6, 7]
- Harmful if swallowed [1, 4-8]
- Irritating to eyes [1-4, 7, 8]
- Irritating to skin [1, 4, 7, 8]
- Risk of serious damage to eyes [5, 6]
- Harmful to fish or other aquatic life. Do not contaminate surface waters or ditches with chemical or used container [1-3, 5, 6, 8]
- Do not contaminate surface waters or ditches with chemical or used container [4, 7]

Personal protective equipment/Label precautions
- A, C [1-8]; H [2, 3, 5, 8]; M [5, 8]
- M03 [1, 4-8]; R04a [1-4, 7, 8]; R03c [1, 4-8]; R03a [1, 4, 6, 7]; R04b [1, 4, 7, 8]; R04d [5, 6]; U05a, U08 [1-8]; U20b [1-6, 8]; U20a [7]; U13 [4]; C03 [1-8]; E30a [1-7]; E01, E26 [1-8]; E07 [1-8] (2 wk); E31b [1-3, 6-8]; E13c [1-3, 5, 6, 8]; E34 [1, 4-8]; E15 [4, 7]; E31c [5]; E31a [4]

Withholding period
- Keep livestock out of treated areas for at least 2 wk and until foliage of any poisonous weed, such as ragwort, has died and become unpalatable

Latest application/Harvest interval
- Generally before 1st node detectable (GS 31) for spring cereals and before 3rd node detectable (GS 33) for winter cereals, but individual labels vary.
- 5 wk before emergence of seed head for grass seed crops

Approval
- Accepted by BLRA for use on malting barley

359   mecoprop

A translocted phenoxypropionic herbicide only available in mixtures

See also dicamba + MCPA + mecoprop

360   mefluidide

A sulfonanilide plant growth regulator for suppression of grass

Products

Embark Lite	Intracrop	24 g/l	SL	08749

Uses

Growth suppression in *amenity grass, managed amenity turf*.

Efficacy
- Apply when grass growth active in Apr/May and repeat when growth re-commences
- Grass growth suppressed for up to 8 wk

- Mow (not below 2 cm) 2-3 d before or 5-10 d after spraying
- Only apply to dry foliage but do not use in drought
- A rainfree period of 8 h is needed after spraying
- May be tank-mixed with suitable 2,4-D or mecoprop product for control of dicotyledons
- Delay application for 3-4 wk after application of high-nitrogen fertilizer

Crop safety/Restrictions
- Maximum number of treatments 2 per yr
- Do not use on fine turf areas or on turf established for less than 4 mth
- Do not use on grass suffering from herbicide damage
- Treated turf may appear less dense and be temporarily discoloured but normal colour returns in 3-4 wk
- Do not reseed within 2 wk after application

Special precautions/Environmental safety
- Do not allow animals to graze treated areas

Personal protective equipment/Label precautions
- U05a, U20c, E01, E15, E26, E30a, E31b

361   mepiquat chloride

A quaternary ammonium plant growth regulator available only in mixtures

See also 2-chloroethylphosphonic acid + mepiquat chloride
chlormequat + 2-chloroethylphosphonic acid + mepiquat chloride
chlormequat + mepiquat chloride

362   mercuric oxide

An inorganic mercury fungicide, all approvals for which were revoked in 1992. Mercuric oxide is banned under the EC Directive 70/117/EEC

363   mercurous chloride

A soil applied inorganic mercury fungicide, all approvals for which were revoked in 1992. Mercurous chloride is banned under the EC Directive 70/117/EEC

364   metalaxyl

A phenylamide (acylalanine) fungicide available only in mixtures

See also carbendazim + metalaxyl
chlorothalonil + metalaxyl
copper oxychloride + metalaxyl
mancozeb + metalaxyl

365   metalaxyl + thiabendazole + thiram

A protective fungicide seed treatment for peas and beans

Products

Apron Combi FS	Novartis	233:120:100	FS	08386

Uses

Ascochyta in **peas - all types** *(seed treatment)*, **spring field beans** *(seed treatment)*, **winter field beans** *(seed treatment)*. Damping off in **peas - all types** *(seed treatment)*, **spring field beans** *(seed treatment)*, **winter field beans** *(seed treatment)*. Downy mildew in **peas - all types** *(seed treatment)*, **spring field beans** *(seed treatment)*, **winter field beans** *(seed treatment)*.

Efficacy
- Apply through continuous flow seed treaters which should be calibrated before use

Crop safety/Restrictions
- Ensure moisture content of treated seed satisfactory and store in a dry place
- Check calibration of seed drill with treated seed before drilling

Special precautions/Environmental safety
- Harmful if swallowed. Irritating to eyes
- Harmful to fish or other aquatic life. Do not contaminate surface waters or ditches with chemical or used container
- Do not use treated seed as food or feed
- Treated seed harmful to game and wildlife

Personal protective equipment/Label precautions
- A, C, D, H, M
- M03, R03c, R04a, U05a, U09a, U13, U20a, C03, E01, E13c, E30a, E31a, S01, S02, S04b, S05, S06

Maximum Residue Level (mg residue/kg food)
- see metalaxyl and thiabendazole entries

366   metalaxyl + thiram

A systemic and protectant fungicide for use in lettuce

Products

Favour 600 SC	Novartis	100:500 g/l	SC	08405

Uses

Downy mildew in *lettuce, protected lettuce*.

Efficacy
- On lettuce apply from 100% emergence and repeat at 14 d intervals
- Resistant strains of lettuce downy mildew may develop against which product may be ineffective. See label for details

Crop safety/Restrictions
- Maximum number of treatments; 2 post-planting applications of thiram or any combination of products containing thiram or other EBDC compound (mancozeb, maneb, zineb) either as a spray or a dust are permitted on protected lettuce within 2 wk of planting out and none thereafter. On winter lettuce, if only thiram based products are being used, 3 applications may be made within 3 wk of planting out and none thereafter
- Maximum number of treatments 5 per crop for outdoor lettuce; 3 per yr on outdoor parsley
- Do not use as a soil block treatment

Special precautions/Environmental safety
- Harmful if swallowed. Irritating to eyes, skin and respiratory system
- Harmful to fish or other aquatic life. Do not contaminate surface waters or ditches with chemical or used container

Personal protective equipment/Label precautions
- A, C
- M03, R03c, R04a, R04c, R04b, R04c, U05a, U08, U20b, C03; C02 (2 wk outdoor; 3 wk protected); E01, E13c, E30a, E31b, E34

Latest application/Harvest interval
- HI protected lettuce 3 wk; outdoor lettuce 2 wk

Maximum Residue Level (mg residue/kg food)
- see metalaxyl entry

367   metalaxyl-M

A phenylamide systemic fungicide

See also mancozeb + metalaxyl-M

Products

SL 567A	Novartis	480 g/l	EC	08811

Uses

Cavity spot in **carrots**. Phytophthora in **raspberries** *(off-label - soil drench)*. Pythium in **watercress** *(off-label)*. White blister in **horseradish** *(off-label)*.

Efficacy
- Best results achieved when applied to damp soil
- Efficacy may be reduced in prolonged dry weather. Irrigation immediately before or after application may be necessary
- Results may not be satisfactory on soils with high organic matter content
- Control of cavity spot on carrots overwintered in the ground or lifted in winter may be lower than expected

Crop safety/Restrictions
- Maximum number of treatments 1 per crop
- Do not use where carrots have been grown on the same site for either of the previous two yrs
- Consult before use on crops intended for processing

Special precautions/Environmental safety
- Harmful if swallowed
- Irritating to eyes
- May cause sensitization by skin contact
- Harmful to fish or other aquatic life. Do not contaminate surface waters or ditches with chemical or used container

Personal protective equipment/Label precautions
- A, C, H
- M03, R03c, R04a, R04e, U02, U04a, U05a, U07, U10, U20b, C03, E01, E13c, E26, E29, E30a, E31b, E33, E34, E36

Latest application/Harvest interval
- 6 wk after drilling for carrots

Approval
- Off-label extension of use (PPPR) to Jan 2001 on outdoor raspberry canes (OLA 1147/99); extension of use (PPPR) to Jan 2001 on watercress under propagation (OLA 1149/99); to Jan 2001 for use on horseradish (OLA 1397/99)

Maximum Residue Level (mg residue/kg food)
- see metalaxyl entry

368   metaldehyde

A molluscicide bait for controlling slugs and snails

Products

1	Chiltern Blues	Chiltern	6% w/w	PT	10071
2	Chiltern Hundreds	Chiltern	3% w/w	PT	10072
3	Devcol Morgan 6% Metaldehyde Slug Killer	Nehra	6% w/w	PT	07422
4	Dixie 6	Greencrop	6% w/w	PT	10052
5	Doff Horticultural Slug Killer Blue Mini Pellets	Doff Portland	3% w/w	PT	09666
6	Doff New Formula Metaldehyde Slug Killer Mini Pellets	Doff Portland	6% w/w	PT	09772
7	Escar-Go 6	Chiltern	6% w/w	PT	06076
8	Hardy	Chiltern	6% w/w	PT	06948

Products (Continued)

9	Luxan 9363	Luxan	6% w/w	PT	07359
10	Luxan Metaldehyde	Luxan	6% w/w	PT	06564
11	Metarex Green	De Sangosse	6% w/w	PT	08131
12	Metarex RG	De Sangosse	6% w/w	PT	06754
13	Mifaslug	FCC	6% w/w	PT	01349
14	Molotov	Chiltern	3% w/w	PT	08295
15	Optimol	PBI	4% w/w	PT	09061
16	PBI Slug Pellets	PBI	6% w/w	PT	09607
17	Regel	De Sangosse	6% w/w	PT	08155
18	Superflor 6% Metaldehyde Slug Killer Mini Pellets	CMI	6% w/w	PT	05453

Uses

Slugs in **all edible crops** [1-18]. Slugs in **all non-edible crops** [1-9, 11-18]. Slugs in **brassicas** *(seed admixture)*, **winter wheat** *(seed admixture)* [16]. Snails in **all edible crops** [1-18]. Snails in **all non-edible crops** [1-9, 11-18].

Efficacy

- Apply pellets by hand, fiddle drill, fertilizer distributor, by air (check label) or in admixture with seed. See labels for rates and timing.
- Best results achieved from an even spread of granules applied during mild, damp weather when slugs and snails most active. May be applied in standing crops
- Varieties of oilseed rape low in glucosinalates can be more acceptable to slugs than "single low" varieties and control may not be as good
- Do not apply when rain imminent or water glasshouse crops within 4 d of application
- To prevent slug build up apply at end of season to brassicas and other leafy crops
- To reduce tuber damage in potatoes apply twice in Jul and Aug

Special precautions/Environmental safety

- Dangerous to game, wild birds and animals
- Product contains proprietary cat and dog deterrent [4, 5, 5, 6, 6, 9, 11, 12, 15-17]
- Take care to avoid lodging of pellets in the foliage when making late applications to edible crops

Personal protective equipment/Label precautions

- A, H [2, 4, 6, 9-12, 14, 15, 17]; J [4, 6, 10, 15]
- U20a [3, 11-13, 15-17]; U20b [1, 7, 8, 18]; U20c [2, 4-6, 9, 10, 14]; E10a, E15, E32a [1-18]; E05a [1-18] (7 d); E30a [1-4, 6-18]; E01 [1-3, 6-18]; E31b, E34 [3, 11-13, 15-17]; E29 [1, 2, 4-8, 14]

Withholding period

- Keep poultry out of treated areas for at least 7 d

Approval

- Approved for aerial application on all crops [5, 6, 12, 13, 15, 18]. See notes in Section 1
- Accepted by BLRA for use on malting barley
- Approval expiry 31 Oct 2001 [3]; replaced by MAFF 09619

---

369 metamitron

A contact and residual triazinone herbicide for use in beet crops

See also ethofumesate + metamitron + phenmedipham

Products

1	Barclay Seismic	Barclay	70% w/w	WG	09323
2	Goltix 90	Bayer	90% w/w	WG	08654
3	Goltix Flowable	Bayer	700 g/l	SC	08986
4	Goltix WG	Bayer	70% w/w	WG	02430
5	Landgold Metamitron	Landgold	70% w/w	WG	06287
6	Marquise	Makhteshim	70% w/w	WG	08738
7	Standon Metamitron	Standon	70% w/w	WG	07885
8	Tripart Accendo	Tripart	70% w/w	WG	06110
9	Volcan	Sipcam	70% w/w	WG	09295

Uses

Annual dicotyledons in *fodder beet, mangels, sugar beet* [1-9]. Annual dicotyledons in *red beet* [1-5, 7-9]. Annual dicotyledons in *asparagus (off-label), outdoor leaf herbs (off-label)* [4]. Annual grasses in *fodder beet, mangels, sugar beet* [1-9]. Annual grasses in *red beet* [1-5, 7-9]. Annual meadow grass in *fodder beet, mangels, sugar beet* [1-9]. Annual meadow grass in *red beet* [1-5, 7-9]. Fat-hen in *fodder beet, mangels, sugar beet* [1-9]. Fat-hen in *red beet* [1-5, 7-9].

Efficacy

- May be used pre-emergence alone or post-emergence in tank mixture or with an authorised adjuvant oil
- Low dose programme (LDP). Apply a series of low-dose post-weed emergence sprays, including adjuvant oil, timing each treatment according to weed emergence and size. See label for details and for recommended tank mixes and sequential treatments. On mineral soils the LDP should be preceeded by pre-drilling or pre-emergence treatment
- Traditional application. Apply either pre-drilling before final cultivation with incorporation to 8-10 cm, or pre-crop emergence at or soon after drilling into firm, moist seedbed to emerged weeds from cotyledon to first true leaf stage
- On emerged weeds at or beyond 2-leaf stage addition of adjuvant oil advised
- Up to 3 post-emergence sprays may be used on soils with over 10% organic matter
- For control of wild oats and certain other weeds, tank mixes with other herbicides or sequential treatments are recommended. See label for details

Crop safety/Restrictions

- Maximum total dose equivalent to three full dose treatments for most products; 4 lower maximum dose treatments [6]
- Using traditional method post-crop emergence on mineral soils do not apply before first true leaves have reached 1 cm long
- Crop tolerance may be reduced by stress caused by growing conditions, effects of pests, disease or other pesticides, nutrient deficiency etc
- Only sugar beet, fodder beet or mangels may be drilled within 4 mth after treatment. Winter cereals may be sown in same season after ploughing, provided 16 wk passed since last treatment

Special precautions/Environmental safety

- Harmful if swallowed [3]
- Risk of serious damage to eyes [1]
- May cause sensitization by skin contact [1, 6]

Personal protective equipment/Label precautions

- A [1, 3]; C, H [1]
- M03 [3]; R04, R04e [1, 6]; R03c [3]; R04d [1]; U20a [2, 4-6, 8]; U20b [1, 3, 7]; U08, U14, U19 [6]; U05a [1, 6]; U20c [9]; C03 [1, 3, 6]; E15, E30a [1-9]; E34 [2-4, 8]; E32a [1, 3, 5-7, 9]; E01 [1, 3, 6]; E26 [3]

Latest application/Harvest interval

- Before crop foliage meets across rows

Approval

- Off-label approval unlimited for use on outdoor asparagus (OLA 2415/96)[4]; unstipulated for use on outdoor leafy herbs (OLA 1208)[4]

---

370   metam-sodium

A methyl isothiocyanate producing sterilant for glasshouse, nursery and outdoor soils

Products

1	Metham Sodium 400	United Phosphorus	400 g/l	SL	08051
2	Sistan	Unicrop	380 g/l	SL	01957

Uses

Dutch elm disease in *elm trees (off-label)* [2]. Nematodes in *chrysanthemums, ornamentals, outdoor tomatoes (Jersey only), protected carnations, protected crops, tomatoes* [1]. Nematodes in *glasshouse soils, nursery soils, outdoor soils, potting soils* [2]. Soil insects in *chrysanthemums, ornamentals, outdoor tomatoes* (Jersey only), *protected carnations, protected crops, tomatoes* [1]. Soil pests in *glasshouse soils, nursery soils, outdoor soils, potting soils* [2]. Soil-borne diseases in *chrysanthemums, ornamentals, outdoor tomatoes* (Jersey only), *protected carnations, protected crops, tomatoes* [1]. Soil-borne diseases in *glasshouse soils, nursery soils, outdoor soils, potting soils* [2]. Weed seeds in *chrysanthemums, ornamentals, outdoor tomatoes* (Jersey only), *protected carnations, protected crops, tomatoes* [1]. Weed seeds in *glasshouse soils, nursery soils, outdoor soils, potting soils* [2].

Efficacy
- Product acts by breaking down in contact with soil to release methyl isothiocyanate
- Apply to glasshouse soils as a drench, inject undiluted to 20 cm at 30 cm intervals and seal with water or apply to surface, rotavate and seal
- After treatment allow sufficient time (several weeks) for residues to dissipate and aerate soil by forking. Time varies with season. See label for details
- Apply to outdoor soils by drenching or sub-surface application
- Apply when soil temperatures exceed 7°C, preferably above 10°C, between 1 Apr and 31 Oct. Soil must be of fine tilth and adequate moisture content
- Product may also be used by mixing into potting soils
- Avoid using in equipment incorporating natural rubber parts
- Use in control programme against dutch elm disease to sever root grafts which can transmit disease from tree to tree. See OLA notice for details [2]

Crop safety/Restrictions
- No plants must be present during treatment. Do not treat glasshouses within 2 m of growing crops. Do not plant until soil is entirely free of fumes
- Cress test advised to check for absence of residues. See label for details

Special precautions/Environmental safety
- Harmful in contact with skin and if swallowed [2]
- Irritating to eyes, skin and respiratory system

Personal protective equipment/Label precautions
- A, C, H, M [1, 2]; D [1]
- M03, R03a, R03c [2]; R04a, R04b, R04c [1, 2]; U20b [2]; U02, U04a, U05a, U08, U19 [1, 2]; U20a [1]; C03 [1, 2]; E34 [2]; E02 [2] (24 h); E01, E15, E30a, E31a [1, 2]

Latest application/Harvest interval
- Pre-planting of crop

Approval
- Off-label approval unlimited for use around elm trees infected with Dutch elm disease (OLA 0592/93)[2]

---

371   metazachlor

A residual anilide herbicide for use in brassicas, nurseries and forestry

Products

1 Barclay Metaza	Barclay	500 g/l	SC	09169
2 Butisan S	BASF	500 g/l	SC	00357
3 Landgold Metazachlor 50	Landgold	500 g/l	SC	09726
4 Standon Metazachlor 50	Standon	500 g/l	SC	05581

Uses

Annual dicotyledons in *oilseed rape* [1, 2]. Annual dicotyledons in *broccoli, brussels sprouts, cabbages, calabrese, cauliflowers, swedes, turnips* [1-4]. Annual dicotyledons in *farm forestry* (off-label), *farm woodland, forestry, hardy ornamental nursery stock, honesty* (off-label), *kohlrabi* (off-label), *nursery fruit trees and bushes, ornamentals* [2]. Annual dicotyledons in *spring oilseed rape, winter oilseed rape* [2-4]. Annual grasses in *farm woodland, forestry, kohlrabi* (off-label) [2]. Annual meadow grass in *oilseed rape* [1, 2]. Annual meadow grass in *broccoli, brussels sprouts, cabbages, calabrese, cauliflowers, swedes, turnips* [1-4]. Annual meadow grass in *hardy ornamental nursery stock, nursery fruit trees and bushes, ornamentals* [2]. Annual meadow grass in *spring oilseed rape, winter oilseed rape* [2-4]. Blackgrass in *oilseed rape* [1, 2]. Blackgrass in *broccoli, brussels sprouts, cabbages, calabrese, cauliflowers, swedes, turnips* [1-4]. Blackgrass in *hardy ornamental nursery stock, nursery fruit trees and bushes, ornamentals* [2]. Blackgrass in *spring oilseed rape, winter oilseed rape* [2-4]. Triazine resistant groundsel in *broccoli, brussels sprouts, cabbages, calabrese, cauliflowers, hardy ornamental nursery stock, nursery fruit trees and bushes, ornamentals, spring oilseed rape, swedes, turnips, winter oilseed rape* [2].

Efficacy
- Activity is dependent on root uptake. For pre-emergence use apply to firm, moist, clod-free seedbed
- Some weeds (chickweed, mayweed, blackgrass etc) susceptible up to 2- or 4-leaf stage. Moderate control of cleavers achieved provided weeds not emerged and adequate soil moisture present
- Split pre- and post-emergence treatments recommended for certain weeds in winter oilseed rape on light and/or stony soils
- Effectiveness is reduced on soils with more than 10% organic matter
- Various tank-mixtures recommended. See label for details

Crop safety/Restrictions
- Maximum number of treatments 1 per crop for spring oilseed rape, swedes, turnips and brassicas; 2 per crop for winter oilseed rape and honesty (split dose treatment); 3 per yr for ornamentals, nursery stock, nursery fruit trees, forestry and farm forestry
- On winter oilseed rape may be applied pre-emergence from drilling until seed chits, post-emergence after fully expanded cotyledon stage (GS 1,0) or by split dose technique depending on soil and weeds. See label for details
- On spring oilseed rape, swedes, turnips recommended as a pre-emergence sequential treatment following trifluralin
- On spring oilseed rape may also be used pre-weed-emergence from cotyledon to 10-leaf stage of crop (GS 1,0-1,10)
- With pre-emergence treatment ensure seed covered by 15 mm of well consolidated soil. Harrow across slits of direct-drilled crops
- Ensure brassica transplants have roots well covered and are well established. Direct drilled brassicas should not be treated before 3 leaf stage
- In ornamentals and hardy nursery stock apply after plants established and hardened off as a directed spray or, on some subjects, as an overall spray. See label for list of tolerant subjects. Do not treat plants in containers
- Do not use on sand, very light or poorly drained soils
- Do not treat protected crops or spray overall on ornamentals with soft foliage
- Do not spray crops suffering from wilting, pest or disease
- Do not spray broadcast crops or if a period of heavy rain forecast
- Any crop can follow normally harvested treated oilseed rape. See label for details of crops which may be planted in event of crop failure

Special precautions/Environmental safety
- Harmful if swallowed [1-4]
- Irritating to skin [1-4]
- May cause sensitization by skin contact [1, 3]
- Harmful to fish or other aquatic life. Do not contaminate surface waters or ditches with chemical or used container [1, 3, 4]

Personal protective equipment/Label precautions
- A, C, H, M [1-4]
- M03, R03c, R04b [1-4]; R04e [1, 3]; U20a [2]; U05a, U08, U19 [1-4]; U20b [1, 3, 4]; C02 [2]; C03 [1-4]; E31c [2]; E01, E30a, E34 [1-4]; E13c, E26, E31b [1, 3, 4]; E07 [1, 3, 4] (2 wk or 5 wk for swedes, turnips)

Withholding period
- Keep livestock out of treated areas until foliage of any poisonous weeds such as ragwort has died and become unpalatable
- Keep livestock out of treated areas of swede and turnip for at least 5 wk following treatment

Latest application/Harvest interval
- Pre-emergence for swedes and turnips; before 10 leaf stage for spring oilseed rape; before end of Jan for winter oilseed rape
- HI brassicas 6 wk. Do not use on nursery fruit trees within 1 yr of fruit production

Approval
- Off-label approval unlimited for honesty grown outdoors (OLA 0551/93)[2]; unlimited for use in farm forestry (OLA 0277/94)[2]; to Mar 2001 for use on outdoor kohlrabi (OLA 0498/96)[2]

372   metazachlor + quinmerac

A residual herbicide mixture for oilseed rape

Products

1 Katamaran	BASF	375:125 g/l	SC	09049
2 Standon Metazachlor-Q	Standon	375:125 g/l	SC	09676

Uses

Annual dicotyledons in *winter oilseed rape*. Annual meadow grass in *winter oilseed rape*. Blackgrass in *winter oilseed rape*. Cleavers in *winter oilseed rape*. Poppies in *winter oilseed rape*.

Efficacy
- Activity is dependant on root uptake. Pre-emergence treatments should be applied to firm moist seedbeds. Applications to dry soil do not become effective until after rain has fallen
- Maximum activity achieved from treatment before weed emergence for some species
- Weed control may be reduced if excessive rain falls shortly after application especially on light soils

Crop safety/Restrictions
- Maximum total dose equivalent to one full dose treatment
- May be used on all soil types except Sands, Very Light Soils, and soils containing more than 10% organic matter. Crop vigour and/or plant stand may be reduced on brashy and stony soils
- To ensure crop safety it is essential that crop seed is well covered with soil to 15 mm. Loose or puffy seedbeds must be consolidated before treatment. Do not use on broadcast crops
- Crop vigour and possibly plant stand may be reduced if excessive rain falls shortly after treatment especially on light soils. The effects may be increased where TCA has been used
- Damage may occur in waterlogged conditions. Do not use on poorly drained soils
- Do not treat stressed crops. In frosty conditions transient scorch may occur
- In the event of crop failure after use, wheat or barley may be sown in the autumn after ploughing to 15 cm. Spring cereals or brassicas may be planted after ploughing in the spring

Special precautions/Environmental safety
- Irritant. May cause sensitization by skin contact
- Dangerous to fish or other aquatic life. Do not contaminate surface waters or ditches with chemical or used container
- LERAP Category B
- To reduce risk of movement to water do not apply to dry soil or if heavy rain is forecast. On clay soils create a fine consolidated seedbed

Personal protective equipment/Label precautions
- A [1, 2]
- R04, R04e [1, 2]; U19 [1]; U05a, U08, U14, U20b, C03 [1, 2]; E31c [1]; E01, E07, E13b, E16a, E16b, E30a [1, 2]; E26, E31b [2]

Withholding period
- Keep livestock out of treated areas until foliage of any poisonous weeds such as ragwort has died and become unpalatable

Latest application/Harvest interval
- End Jan in yr of harvest

373   metconazole

A conazole fungicide

Products

1	Caramba	BASF	60 g/l	SL	10213
2	Caramba	Cyanamid	60 g/l	SL	09864

Uses

Alternaria in **spring oilseed rape, winter oilseed rape**. Brown rust in **spring barley, winter barley, winter wheat**. Fusarium ear blight in **winter wheat** *(reduction)*. Leaf blotch in **winter wheat**. Net blotch in **spring barley** *(reduction)*, **winter barley** *(reduction)*. Powdery mildew in **spring barley** *(moderate control)*, **winter barley, winter wheat** *(moderate control)*. Rhynchosporium in **spring barley, winter barley**. Sclerotinia in **spring oilseed rape** *(reduction)*, **winter oilseed rape** *(reduction)*. Yellow rust in **winter wheat**.

Efficacy

- Best results from application when disease starts to develop in crop
- Treatment for leaf blotch should be made after GS 33 and when weather favouring development of the disease has occurred
- Treat mildew infections before 3% infection on any green leaf. A specific mildewicide will improve control of established infections
- Treat yellow rust before 1% infection on any leaf or as preventive treatment after GS 39
- For brown rust spray susceptible varieties before any of top 3 leaves has more than 2% infection
- Sclerotinia in oilseed rape should be treated at petal fall. A tank mixture with carbendazim may be needed for fully effective control

Crop safety/Restrictions

- Maximum total dose equivalent to two full dose treatments
- Only cereals. oilseed rape or sugar beet may be sown as following crops after treatment

Special precautions/Environmental safety

- Irritating to skin
- Risk of serious damage to eyes
- May cause sensitization by skin contact
- Flammable
- Dangerous to fish or other aquatic life. Do not contaminate surface waters or ditches with chemical or used container
- LERAP Category B but avoid treatment close to field boundary, even if permitted by LERAP assessment, to reduce effects on non-target insects or other arthropods
- Product available in returnable containers - see label for detailed operating procedures

Personal protective equipment/Label precautions

- A, C, H
- R04b, R04d, R04e, R07d, U05a, U11, U14, U20c, C03, E01, E13b, E16a, E16b, E22b, E26, E30a, E31a; E33, E34, E36 (returnable container)

Latest application/Harvest interval

- Up to and including milky ripe stage (GS 71) for cereals; 10% pods at final size for oilseed rape

Approval

- Accepted by BLRA for use on malting barley

374   methabenzthiazuron

A contact and residual urea herbicide for cereals

Products

Tribunil	Bayer	70% w/w	WP	02169

Uses

Annual dicotyledons in *durum wheat, spring barley, triticale, winter barley, winter oats, winter rye, winter wheat*. Annual meadow grass in *durum wheat, spring barley, triticale, winter barley, winter oats, winter rye, winter wheat*. Blackgrass in *winter barley, winter wheat*. Rough meadow grass in *durum wheat, spring barley, triticale, winter barley, winter oats, winter rye, winter wheat*.

Efficacy

- Apply pre-weed emergence on fine, firm seedbed or post-emergence to 1-2 true leaf stage of weeds, blackgrass (at higher rates) not beyond 2-leaf stage or after end Nov
- Do not use lower rates on soils with more than 10% organic matter
- Do not use higher rate for blackgrass control on soils with more than 4% organic matter or where seedbed preparation has not been preceded by ploughing

Crop safety/Restrictions

- Maximum number of treatments 1 per crop
- Apply in winter wheat or autumn sown spring wheat pre- or post-emergence up to 6 wk after drilling, provided weeds in correct stage
- Apply pre-emergence in winter barley, but not in crops drilled after mid-Oct on sand in E. Anglia
- Only use pre-emergence in spring barley drilled to at least 25 mm and not on sand
- Do not use on crops to be undersown with clover
- Use lower rate on direct drilled crops and roll before spraying where harrowed or disced after drilling

Special precautions/Environmental safety

- Harmful to fish or other aquatic life. Do not contaminate surface waters or ditches with chemical or used container

Personal protective equipment/Label precautions

- U08, U19, U20a, E07, E13c, E30a, E32a

Withholding period

- Keep livestock out of treated areas for at least 14 d after application

Latest application/Harvest interval

- Pre-emergence for spring barley, winter oats, winter rye, triticale and autumn sown spring wheat; post-emergence before 30 Oct for winter barley and durum wheat (pre-emergence before 30 Nov at higher rate); post-emergence before 1st node detectable (GS 31) for winter wheat (pre-emergence up to 6 wk after drilling but before 30 Nov at higher rate)

Approval

- Approved for aerial application on cereals, grass. See notes in Section 1
- Accepted by BLRA for use on malting barley

---

375 methiocarb

A stomach acting carbamate molluscicide and insecticide

Products

1	Decoy Wetex	Bayer	2% w/w	PT	09707
2	Draza	Bayer	4% w/w	PT	00765
3	Exit Wetex	Bayer	3% w/w	PT	10149
4	Huron	Bayer	3% w/w	PT	10148
5	Karan	Bayer	3% w/w	PT	09637
6	Lupus	Bayer	3% w/w	PT	09638
7	Rivet	Bayer	3% w/w	PT	09512

Uses

Cutworms in *sugar beet* (reduction) [1-7]. Leatherjackets in *cereals* (reduction), *potatoes* (reduction), *ryegrass leys* (reduction), *sugar beet* (reduction) [1-7]. Millipedes in *sugar beet* (reduction) [1-7]. Slugs in *brassicas, cereals, field crops, fruit crops, oilseed rape, ornamentals, peas, potatoes, ryegrass leys, strawberries, sugar beet* [1-7]. Slugs in *protected crops* (off-label) [2]. Snails in *blackcurrants, field crops, fruit crops, ornamentals, peas* [1-7]. Snails in *protected crops* (off-label) [2]. Strawberry seed beetle in *strawberries* [1-7].

Efficacy
- Use as a surface, overall application when pests active (normally mild, damp weather), pre-drilling or post-emergence. May also be used on cereals, ryegrass, oilseed rape or other brassicas in admixture with seed at time of drilling
- Also reduces populations of cutworms and millipedes
- Best on potatoes in late Jul to Aug, on peas before start of pods formation
- Apply to strawberries before strawing down, to blackcurrants at grape stage to prevent snails contaminating crop
- See label for details of suitable application equipment

Crop safety/Restrictions
- Do not allow pellets to lodge in edible crops

Special precautions/Environmental safety
- Harmful if swallowed
- This product contains an anticholinesterase carbamate compound. Do not use if under medical advice not to work with such compounds
- Dangerous to game, wild birds and animals
- Dangerous to fish or other aquatic life. Do not contaminate surface waters or ditches with chemical or used container
- Keep poultry out of treated areas for at least 7 d
- Avoid surface application within 6 m of field boundary to reduce effects on non-target species

Personal protective equipment/Label precautions
- A, H, J
- M02, M03, R03c, U05a, U20b, C03, E01, E05a, E10a, E13b, E22b, E30a, E32a, E34

Latest application/Harvest interval
- HI 7 d

Approval
- Approved for aerial application on all crops [1-7]. See notes in Section 1
- Off-label Approval unlimited for use on all protected edible and non-edible crops (OLA 1599/98)[2]
- Accepted by BLRA for use in malting barley

---

376   methomyl

An oxime carbamate insecticide available only in mixtures

---

377   methomyl + (Z)-9-tricosene

An insecticide bait for fly control in animal houses

Products

Golden Malrin Fly Bait	Novartis A H	1:0.025% w/w	GR	08821

Uses

Flies in *livestock houses*.

Efficacy
- Product contains a pheromone attractant and can be applied as a scatter bait.
- Alternatively sprinkle granules on to polystyrene ceiling tiles, old egg trays or similar coated with adhesive and hang in infested areas out of reach of animals
- Re-apply every 4-5 d, or when granules disappear, until fly population has been reduced to acceptable levels, then apply every 7 d

Crop safety/Restrictions
- Prevent access to bait by children and livestock
- Do not contaminate animal feed or use in areas where bait might come into contact with food

Special precautions/Environmental safety
- Product contains an anticholinesterase carbamate compound. Do not use if under medical advice not to work with such compounds
- Dangerous to fish or other aquatic life. Do not contaminate surface waters or ditches with chemical or used container

Personal protective equipment/Label precautions
- A, H
- M02, U13, U20a, C03, E13b, E30a, E31a, V01a, V02

Approval
- Approval expiry 20 Apr 2001

378   methoprene

A pheromone analogue insect growth regulator

Products

Apex 5E	Novartis A H	600 g/l	EC	08739

Uses

Sciarid flies in **mushroom houses**.

Efficacy
- Use as surface spray either on the compost or the casing
- Use pre-casing treatment only when early pest infestation occurs
- Product controls sciarids at pupal stage. Treatment will not control the generation of sciarids present at application

Crop safety/Restrictions
- Maximum number of treatments 1 per spawning on compost or casing
- Product prevents development of adults from eggs laid up to 14 d before or after treatment. Alternative treatments may be necessary if adult numbers increase subsequently
- Pre-casing treatment may not prevent development from larvae able to move into untreated casing prior to pupation

Special precautions/Environmental safety
- Dangerous to fish or other aquatic life. Do not contaminate surface waters or ditches with chemical or used container

Personal protective equipment/Label precautions
- A, C, H
- U05a, U19, U20c, C03, E01, E13b, E26, E30a, E32a

Latest application/Harvest interval
- Before casing on the compost or immediately after casing

379   2-methoxyethyl mercury acetate

An organomercury fungicide seed treatment all approvals for which were revoked with effect from 31 March 1992

380   methyl bromide

A highly toxic alkyl halide fumigant for food and non-food commodities

Products

1	Bromomethane 100%	Brian Jones	100% w/w	GA	09244
2	Sobrom BM 100	Brian Jones	100%	GA	04381

Uses

Food storage pests in **food storage areas** [1, 2]. Grain storage pests in **grain stores** [1]. Insect pests in **stored plant material** [1, 2].

Efficacy
- Chemical is released from containers as a highly toxic, colourless gas with only a slight odour
- Dose depends on temperature. At lower temperatures higher doses and longer exposures should be used
- See label for dosage and exposure required in different situations

Crop safety/Restrictions
- Maximum number of treatments 1 per infestation
- Air animal feed for at least 5 d following treatment
- Some types of produce, cut flowers and plants are susceptible to damage by methyl bromide fumigation. Check before treating
- Methyl bromide can cause taint to flours and foodstuffs containing reactive sulfur compounds. See label for guidance

Special precautions/Environmental safety
- Methyl bromide is subject to the Poisons Rules 1982 and the Poisons Act 1972. See notes in Section 1
- Not for retail sale
- Very toxic by inhalation. Fatal if swallowed.
- Very toxic in contact with skin [2]
- Wear suitable protective clothing (See HSE Guidance Note CS 12)
- Wear suitable respiratory equipment during fumigation and while removing sheets. Do *not* wear gloves or rubber boots
- Do not use application equipment incorporating natural rubber, PVC or aluminium, zinc, magnesium or their alloys
- Extinguish all naked flames, including pilot lights when applying fumigant
- Keep in original container, tightly closed in a safe place under lock and key. Do not re-use container and return to supplier when empty
- Keep unprotected persons out of treated area until it is shown by test to be clear of methyl bromide
- Harmful to fish or other aquatic life. Do not contaminate surface waters or ditches with chemical or used container [2]
- Special equipment is needed for application and the chemical may only be used by professional operators trained in its use and familiar with the precautionary measures to be observed
- All operations must be in accordance with the HSE guidance note CS 12. A minimum of 2 operators are required

Personal protective equipment/Label precautions
- D [1]
- M04 [1, 2]; R01a [2]; R01b, U04a, U05a, C03 [1, 2]; E27 [2]; E01, E13c, E30b, E34 [1, 2]; E02 [1, 2] (until clear); E07 [1] (until clear)

Withholding period
- Keep animals and livestock out of treated area until it is shown by test to be clear of methyl bromide

Latest application/Harvest interval
- 3 d before loading grains; 7 d before use of all other commodities and spaces

Maximum Residue Level (mg residue/kg food)
- oilseeds, cereals 0.1; all other products (except tree nuts, stone fruits, grapes, figs, pulses, animal products) 0.05

---

381   methyl bromide with amyl acetate

A highly toxic alkyl halide fumigant for soil and food storage areas

Products

Fumyl-O-Gas	Brian Jones	99.7%	GA	04833

Uses

Food storage pests in *food storage*. Nematodes in *field crops, protected crops*. Soil insects in *field crops, protected crops*. Soil-borne diseases in *field crops, protected crops*. Weed seeds in *field crops, protected crops*.

Efficacy
- Chemical is released from containers as a highly volatile, highly toxic gas and must be released under gas-proof sheeting
- Amyl acetate is added as a warning odourant

- Soil must be clear of trash and cultivated to at least 40 cm before treatment
- Gas must be confined in soil for 48-96 h

Crop safety/Restrictions
- Maximum number of treatments 1 per yr
- Crops may be planted 3-21 d after removal of gas-proof sheeting. For some crops a leaching irrigation is essential. See label for details

Special precautions/Environmental safety
- Methyl bromide is subject to the Poisons Rules 1982 and the Poisons Act 1972. See notes in Section 1
- Product not for retail sale
- Special equipment is needed for application and the chemical may only be used by professional operators trained in its use and familiar with the precautionary measures to be observed. Refer to HSE Guidance Note CS 12 before applying
- Very toxic by inhalation. Risk of serious damage to health by prolonged exposure
- Keep in original container, tightly closed, in a safe place, under lock and key
- Wear suitable respiratory equipment during fumigation and while removing sheeting
- Do not wear gloves or rubber boots
- Ventilate treated areas thoroughly when gas has cleared
- Keep unprotected persons out of treated areas for at least 24 h
- Dangerous to game, wild birds, animals and bees
- Harmful to fish or other aquatic life. Do not contaminate surface waters or ditches with chemical or used container

Personal protective equipment/Label precautions
- M04, R01a, R01b, U04a, U05a, C03, E01, E30b, E34; E02 (until clear)

Withholding period
- Keep livestock out of treated areas for at least 24 h

Latest application/Harvest interval
- Before sowing or planting

Maximum Residue Level (mg residue/kg food)
- see methyl bromide entry

---

382 methyl bromide with chloropicrin

A highly toxic alkyl halide fumigant for treatment of soil and stored products

Products

Sobrom BM 98	Brian Jones	98% w/w	GA	04189

Uses

Nematodes in *field crops, orchards, protected crops*. Soil insects in *field crops, orchards, protected crops*. Soil-borne diseases in *field crops, orchards, protected crops*. Weed seeds in *field crops, orchards, protected crops*.

Efficacy
- Chemical is released from containers as a highly volatile, highly toxic gas and for soil fumigation must be released under a plastic tarpaulin
- Chloropicrin is added as a warning odourant tear gas
- Soil must be clear of trash and cultivated to at least 40 cm before treatment
- Gas must be confined in soil for 48-96 h

Crop safety/Restrictions
- Maximum number of treatments 1 per yr
- Crops may be planted 3-21 d after removal of tarpaulin. For some crops a leaching irrigation is essential. See labels for details

Special precautions/Environmental safety
- Methyl bromide and chloropicrin are subject to the Poisons Rules 1982 and the Poisons Act 1972. See notes in Section 1
- Product not for retail sale

- Special equipment is needed for application and the chemical may only be used by professional operators trained in its use and familiar with the precautionary measures to be observed. Refer to HSE Guidance Note CS 12 before applying
- Very toxic by inhalation, in contact with skin and if swallowed
- Irritating to eyes, respiratory system and skin
- Keep in original container, tightly closed, in a safe place, under lock and key
- Wear suitable respiratory equipment during fumigation and while removing sheeting
- Do *not* wear gloves or rubber boots
- Ventilate treated areas thoroughly when gas has cleared
- Keep unprotected persons out of treated areas for at least 24 h
- Harmful to fish or other aquatic life. Do not contaminate surface waters or ditches with chemical or used container

Personal protective equipment/Label precautions
- M04, R01a, R01b, R01c, R04a, R04b, R04c, U04a, U05a, C03, E01, E13c, E30b, E34; E02, E07 (24 h)

Withholding period
- Keep livestock out of treated areas for at least 24 h

Latest application/Harvest interval
- Before sowing or planting

Maximum Residue Level (mg residue/kg food)
- see methyl bromide entry

---

383   metoxuron

A contact and residual urea herbicide for cereals and carrots

Products

Dosaflo	Novartis	500 g/l	SC	08473

Uses

Annual dicotyledons in *carrots, durum wheat, triticale, winter barley, winter wheat*. Annual grasses in *carrots, durum wheat, triticale, winter barley, winter wheat*. Barren brome in *durum wheat, triticale, winter barley, winter wheat*. Blackgrass in *durum wheat, triticale, winter barley, winter wheat*. Mayweeds in *carrots*.

Efficacy
- Best results achieved by application to weeds in seedling to young plant stage
- Best control of barren brome in autumn or winter from 1-3 leaf stage
- Effects on blackgrass reduced on soils of high organic matter or low pH

Crop safety/Restrictions
- Maximum number of treatments 2 per crop for cereals with minimum of 6 wk between applications
- Apply to named winter cereal varieties from 2-leaf unfolded stage to before first node detectable (GS 12-31). See label for lists of resistant and susceptible varieties
- Apply to carrots after 2-true leaf stage. Do not spray when soil very dry or wet
- Do not spray cereals on Sands or Very Light soils, or carrots on soils with more than 80% sand or less than 1% organic matter
- Do not apply during prolonged frosty weather, when frost or heavy rain imminent or ground waterlogged
- If treatment repeated in spring at least 6 wk must elapse after autumn/winter spray
- Do not roll for 7 d before or after spraying
- Do not cultivate after spraying or harrow or cultivate within 14 d beforehand
- Do not spray crops checked by pests, wind, frost or waterlogging until recovered
- No crop should be sown within 6 wk of treatment

Personal protective equipment/Label precautions
- U08, U19, U20b, E15, E26, E30a, E31b

Latest application/Harvest interval
- Before 1st node detectable (GS 31) for cereals

Approval
- Accepted by BLRA for use on malting barley

---

384   metribuzin

A contact and residual triazinone herbicide for use in potatoes

Products

1	Citation	United Phosphorus	70% w/w	WG	08685
2	Citation 70	United Phosphorus	70% w/w	WG	09370
3	Inter-Metribuzin WG	I T Agro	70% w/w	WG	09801
4	Lexone 70DF	DuPont	70% w/w	WG	04991
5	Sencorex WG	Bayer	70% w/w	WG	03755

Uses

Annual dicotyledons in *potatoes* [1-5]. Annual grasses in *potatoes* [1-5]. Annual weeds in *carrots* *(off-label)*, *parsnips* *(off-label)* [5]. Volunteer oilseed rape in *potatoes* [1-5].

Efficacy

- May be applied pre- or post-emergence of crop. Best results achieved on weeds at cotyledon to 1-leaf stage
- Apply to moist soil with well-rounded ridges and few clods
- Activity reduced by dry conditions and on soils with high organic matter content
- Do not cultivate after treatment
- On fen and moss soils pre-planting incorporation to 10-15 cm gives increased activity
- With named maincrop and second early varieties on soils with more than 10% organic matter shallow pre- or post-planting incorporation may be used. See label for details
- Effective control using a programme of reduced doses is made possible by using a spray of smaller droplets, thus improving retention. See label for details

Crop safety/Restrictions

- Maximum number of treatments 3 per crop subject to maximum permitted dose - see labels
- Apply pre-emergence only on named first earlies, pre- or post-emergence on named second earlies. On named maincrop varieties apply pre-emergence (except for certain varieties on Sands or Very Light soils) or post-emergence before longest shoots reach 15 cm. See label for details
- On stony or gravelly soils there is risk of crop damage, especially if heavy rain falls soon after application
- When days are hot and sunny delay spraying until evening
- Some varieties may be sensitive to post-emergence treatment if crop under stress
- Ryegrass, cereals or winter beans may be sown in same season provided at least 16 wk elapsed after treatment and ground ploughed to 15 cm and thoroughly cultivated soon after harvest
- Do not grow any vegetable brassicas on silt soils in Lincs, and lettuces or radishes anywhere in UK on land treated the previous yr. Other crops may be sown normally in spring of next yr

Special precautions/Environmental safety

- Dangerous to fish or other aquatic life. Do not contaminate surface waters or ditches with chemical or used container [5]
- LERAP Category B

Personal protective equipment/Label precautions

- U20a [4, 5]; U08, U13, U19 [1-5]; U20b [1-3]; E13b [5]; E15, E16a, E16b, E30a, E32a [1-5]; E29 [1, 3]

Latest application/Harvest interval

- Pre-emergence for earlies; before most advanced shoots reach 15 cm for maincrop

Approval

- Off-label approval to Aug 2001 for use on carrots, parsnips (OLA 1769/96)[5]
- Approval expiry 30 Nov 2001 [1]

385   metsulfuron-methyl

A contact and residual sulfonylurea herbicide for use in cereals and setaside

See also carfentrazone-ethyl + metsulfuron-methyl
*   flupyrsulfuron-methyl + metsulfuron-methyl*

Products

1	AgriGuard Metsulfuron	AgriGuard	20% w/w	WG	09569
2	Ally	DuPont	20% w/w	WG	02977
3	Ally WSB	DuPont	20% w/w	WB	06588
4	Jubilee (WSB)	DuPont	20% w/w	WB	06082
5	Jubilee 20 DF	DuPont	20% w/w	WG	06136
6	Landgold Metsulfuron	Landgold	20% w/w	WG	06280
7	Lorate 20 DF	DuPont	20% w/w	WG	06135
8	Standon Metsulfuron	Standon	20% w/w	WG	05670

Uses

Annual dicotyledons in **barley, oats, wheat** [1-8]. Annual dicotyledons in **triticale** [1-5, 7]. Annual dicotyledons in **linseed** [2-5, 7]. Chickweed in **barley, oats, wheat** [1-8]. Chickweed in **triticale** [1-5, 7]. Chickweed in **linseed** [2-5, 7]. Green cover in **land temporarily removed from production** [2]. Green cover in **setaside** [5, 7]. Mayweeds in **barley, oats, wheat** [1-8]. Mayweeds in **triticale** [1-5, 7]. Mayweeds in **linseed** [2-5, 7].

Efficacy

- Best results achieved on small, actively growing weeds up to 6-true leaf stage. Good spray cover is important
- Commonly used in tank-mixture on wheat and barley with other cereal herbicides to improve control of resistant dicotyledons (cleavers, fumitory, ivy-leaved speedwell), larger weeds and grasses. See label for recommended mixtures
- When tank mixing, always add metsulfuron-methyl product to tank first
- May be used on setaside where the green cover is made up predominantly of grass, wheat barley, oats or triticale [2, 5, 7]

Crop safety/Restrictions

- Maximum number of treatments 1 per crop or per yr on setaside
- Product must be applied only after 1 Feb
- Apply to wheat and oats from 2-leaf (GS 12), to barley, and triticale from 3-leaf stage (GS 13) until flag-leaf fully emerged (GS 39). Do not spray Igri barley before leaf sheath erect stage (GS 30)
- Apply to linseed from first pair of true leaves unfolded up to 30 cm high or before flower bud visible, whichever is the sooner
- Recommendations for oats, triticale and linseed apply to product alone
- Do not use on cereal crops undersown with grass or legumes
- Do not use on any crop suffering stress from drought, waterlogging, frost, deficiency, pest or disease attack or apply within 7 d of rolling
- Do not spray a cereal crop in tank mixture, or in sequence, with a product containing any other sulfonylurea except as directed on the label
- Only cereals, oilseed rape, field beans or grass may be sown in same calendar year after treating cereals with the product alone. Other restrictions apply to tank mixtures. See label for details. In the event of crop failure sow only wheat within 3 mth after treatment. Only cereals should be planted within 16 mth of applying to a linseed crop

Special precautions/Environmental safety

- Extremely dangerous to aquatic higher plants. Do not contaminate surface waters or ditches with chemical or used container
- LERAP category B
- Take extreme care to avoid damage by drift onto broad-leaved plants outside the target area, onto surface waters or ditches or onto land intended for cropping
- A range of broad leaved species will be fully or partially controlled when used in setaside, hence product may be suitable where wild flower borders or other forms of conservation headland are being developed [2, 5, 7]

FOR FULL CONDITIONS OF USE ALWAYS READ THE PRODUCT LABEL

- Before use on setaside as part of grant-aided scheme, ensure compliance with the management rules [2, 5, 7]
- Do not open the water soluble bag or touch it with wet hands or gloves [3, 4]
- Spraying equipment should not be drained or flushed onto land planted, or to be planted, with trees or crops other than cereals and should be thoroughly cleansed after use - see label for instructions

Personal protective equipment/Label precautions
- U05a, U20a [2, 4-8]; U08, U19 [1-8]; U22 [3, 4]; U20b [1]; C03, E01, E14a [2-8]; E16a, E16b, E30a, E32a [1-8]; E13a, E34 [1]

Latest application/Harvest interval
- Before flag leaf sheath extending stage for cereals (GS 41); before flower bud visible or up to 30 cm tall, whichever is earlier, for linseed; before 1 Aug in yr of treatment for setaside

Approval
- Metsulfuron-methyl included in Annex I under EC Directive 91/414
- Accepted by BLRA for use on malting barley

---

386   metsulfuron-methyl + thifensulfuron-methyl

A contact residual and translocated sulfonylurea herbicide mixture for use in cereals

Products

1	DP 928	DuPont	8.6:42.9% w/w	WB	09632
2	Harmony M	DuPont	7:68% w/w	WG	03990

Uses

Annual dicotyledons in *spring barley, spring wheat, winter wheat* [1, 2]. Chickweed in *spring barley, spring wheat, winter wheat* [1]. Cleavers in *spring barley, spring wheat, winter wheat* [2]. Field pansy in *spring barley, spring wheat, winter wheat* [2]. Mayweeds in *spring barley, spring wheat, winter wheat* [1]. Polygonums in *spring barley, spring wheat, winter wheat* [2]. Speedwells in *spring barley, spring wheat, winter wheat* [2].

Efficacy
- Best results by application to small, actively growing weeds up to 6-true leaf stage
- Effectiveness may be reduced if heavy rain occurs within 4 h of application or if soil conditions very dry
- Tank-mixture with mecoprop, mecoprop-P or reduced rate of fluroxypyr improves control of cleavers and other problem weeds
- When tank mixing, always add metsulfuron-methyl/thifensulfuron-methyl product to tank first and fully disperse before adding second product

Crop safety/Restrictions
- Maximum number of treatments 1 per crop
- Apply in spring to crops from 3-leaf stage to flag-leaf fully emerged (GS 13-39) [2]
- Product must only be used after 1 Feb and after crop has three leaves
- Do not use on durum wheat or winter barley or on any crop suffering stress from drought, waterlogging, frost, deficiency, pest or disease attack
- Do not spray in tank mixture, or in sequence, with a product containing any other sulfonylurea herbicide except as directed on the label
- Do not apply within 7 d of rolling
- Take extreme care to avoid drift onto broad-leaved crops or contamination of land planted or to be planted with any crops other than cereals
- Only cereals, oilseed rape, field beans or grass may be sown in same calendar year after treatment. In the event of crop failure sow only wheat within 3 mth after treatment

Special precautions/Environmental safety
- Extremely dangerous to aquatic higher plants. Do not contaminate surface waters or ditches with chemical or used container
- LERAP Category B
- Place water soluble bags whole, directly into spray tank [1]
- Spraying equipment should not be drained or flushed onto land planted, or to be planted, with trees or crops other than cereals and should be thoroughly cleansed after use - see label for instructions

Personal protective equipment/Label precautions
- U08, U20a [2]; U19 [1, 2]; U09a, U20b, U22 [1]; E14a, E32a [2]; E16a, E16b, E30a [1, 2]; E13a [1]

Latest application/Harvest interval
- Before flag leaf ligule/collar just visible (GS 39)

Approval
- Metsulfuron-methyl included in Annex I under EC Directive 91/414
- Accepted by BLRA for use on malting barley

---

387   metsulfuron-methyl + tribenuron-methyl

A sulfonylurea herbicide mixture for winter wheat

Products

DP 911 WSB	DuPont	13:26.1% w/w	WB	09867

Uses

Annual dicotyledons in **winter wheat**. Chickweed in **winter wheat**. Mayweeds in **winter wheat**.

Efficacy
- Best results obtained when applied to small actively growing weeds
- Product acts by foliar and root uptake. Good spray cover essential but performance may be reduced when soil conditions are very dry and residual effects may be reduced by heavy rain
- Weed growth inhibited within hours of treatment and many show marked colour changes as they die back

Crop safety/Restrictions
- Maximum number of treatments 1 per crop
- Product must only be used after 1 Feb and after crop has three leaves
- Do not apply to a crop suffering from drought, water-logging, low temperatures, pest or disease attack, nutrient deficiency, soil compaction or any other stress
- Do not use on crops undersown with grasses, clover or other legumes
- Do not apply within 7 d of rolling
- Do not apply in mixture, or in sequence, with a product containing any other sulfonylurea herbicide
- Only cereals may be sown to succeed a treated wheat crop. In the event of failure of a treated crop only winter wheat may be sown within 3 mth of application. Plough and cultivate to at least 150 mm beforehand

Special precautions/Environmental safety
- Irritant. May cause sensitization by skin contact
- Extremely dangerous to fish or other aquatic life. Do not contaminate surface waters or ditches with chemical or used container
- LERAP Category B
- Some non-target crops are highly sensitive. Take extreme care to avoid drift outside the target area, or onto ponds, waterways or ditches
- Place water soluble bags whole, directly into spray tank
- Spraying equipment should be thoroughly cleaned in accordance with manufacturer's instructions

Personal protective equipment/Label precautions
- A
- R04, R04e, U05a, U09a, U19, U20b, U22, C03, E01, E13a, E16a, E16b, E30a, E32a

Latest application/Harvest interval
- Before flag leaf sheath extending (GS 41)

Approval
- Metsulfuron-methyl included in Annex I under EC Directive 91/414

388 monolinuron

A residual urea herbicide for potatoes, leeks and French beans

Products

Arresin	Aventis	200 g/l	EC	07303

Uses

Annual dicotyledons in *french beans, leeks, potatoes*. Annual grasses in *french beans, leeks, potatoes*. Annual meadow grass in *french beans, leeks, potatoes*. Fat-hen in *french beans, leeks, potatoes*. General weed control in *horseradish* *(off-label)*. Polygonums in *french beans, leeks, potatoes*.

Efficacy

- Best results achieved by application on moist, firm, clod-free seedbed pre-weed emergence or on seedlings up to 2-3 leaf stage
- Light rain after application can improve control
- In potatoes or French beans on fen or peat soils apply post-weed emergence but before emergence of crop

Crop safety/Restrictions

- Maximum number of treatments 1 per crop
- May be used in tank-mixture with glufosinate-ammonium on potatoes up to 10% emergence on earlies, 40% emergence on maincrop. Do not apply alone post-emergence
- On early sown French beans apply at least 5 d pre-emergence
- Some bean cultivars may be sensitive on light soils, with application close to emergence with heavy rain after treatment or in unevenly drilled crop. See label for list of cultivars
- Apply to transplanted or direct-sown leeks after established and 180 mm high
- Only apply pre-emergence to crops grown under polythene
- Do not use on Very Light soils
- Do not sow or plant other crops within 2-3 mth, depending on rate applied. Lettuce must not be sown in same season

Special precautions/Environmental safety

- Irritating to skin, eyes and respiratory system
- Flammable
- Extremely dangerous to fish or other aquatic life. Do not contaminate surface waters or ditches with chemical or used container

Personal protective equipment/Label precautions

- A, C, H, J, M
- R04a, R04b, R04c, R07d, U04a, U05a, U08, U19, U20b, E01, E13a, E30a, E31b, E34

Latest application/Harvest interval

- Pre-emergence for French beans; 3 wk after transplanting for leeks; pre-emergence or up to 10% emergence in seed or early potatoes, pre-emergence or up to 40% emergence (only in mixture with glufosinate-ammonium) in maincrop potatoes

Approval

- Approvals for UK uses of monolinuron to be revoked with a maximum 18-month use-up period
- Approved for aerial application on potatoes and French beans. See notes in Section 1
- Off-label approval to Apr 2001 for use on outdoor horseradish (OLA 0920/96)
- Approval expiry 9 Sep 2001

389 myclobutanil

A systemic, protectant and curative conazole fungicide

Products

1	Systhane 20EW	Landseer	200 g/l	EW	09396
2	Systhane 6 W	PBI	6% w/w	WP	09611

Uses

American gooseberry mildew in *blackcurrants, gooseberries* [1]. Black spot in *roses* [2]. Blossom wilt in *cherries (off-label)*, *mirabelles (off-label)* [1]. Mildew in *roses* [2]. Powdery mildew in *apples, blackberries (off-label - protected crops)*, *hops (off-label)*, *pears, raspberries (off-label - protected crops)*, *strawberries* [1]. Rust in *plums (off-label)* [1]. Rust in *roses* [2]. Scab in *apples, pears* [1].

Efficacy

- Best results achieved when used as part of routine preventive spray programme from bud burst to end of flowering in apples and pears and from just before the signs of mildew infection in blackcurrants and gooseberries [1]
- In strawberries commence spraying at, or just prior to, first flower. Post-harvest sprays may be required on mildew-susceptible varieties where mildew is present and likely to be damaging [1]
- Spray at 7-14 d intervals depending on disease pressure and dose applied [1]
- For improved scab control in post-blossom period tank-mix with mancozeb or captan [1]
- Apply alone from mid-Jun for control of secondary mildew on apples and pears [1]
- On roses spray at first signs of disease and repeat every 2 wk. In high risk areas spray when leaves emerge in spring, repeat 1 wk later and then continue normal programme [2]

Crop safety/Restrictions

- Maximum total dose equivalent to ten full dose treatments in apples and pears, six full dose treatments in blackcurrants, gooseberries, strawberries [1]

Special precautions/Environmental safety

- Harmful to fish or other aquatic life. Do not contaminate surface waters or ditches with chemical or used container

Personal protective equipment/Label precautions

- A, H, J [1]
- U08 [1]; U20a [1, 2]; C03 [1]; C02 [1] (strawberries 3 d; other crops 14 d); E01, E26 [1]; E13c, E32a [1, 2]; E30a [2]

Latest application/Harvest interval

- HI 14 d (apples, pears, blackcurrants, gooseberries); 3 d (strawberries)

Approval

- Off-label approval unstipulated for use on cherries and mirabelles (HI 21 d) (OLA 1535/99)[1]; to Aug 2003 for use on plums (HI 3 d) (OLA1536/99)[1]; unstipulated for use on hops (HI14 d) (OLA1560/99)[1]; to Jul 2003 for use on protected raspberries, blackberries, Rubus hybrids (HI 3 d) (OLA1881/99)[1]
- Accepted by BLRA for use on hops

---

390   2-(1-naphthyl)acetic acid

A plant growth regulator and herbicide for sucker control

See also 4-indol-3-yl-butyric acid + 2-(1-naphthyl)acetic acid with dichlorophen

Products

1	Rhizopon B Powder (0.1%)	Fargro	0.1% w/w	SP	07133
2	Rhizopon B Powder (0.2%)	Fargro	0.2% w/w	SP	07134
3	Rhizopon B Tablets	Fargro	25 mg a.i.	TB	07135

Uses

Rooting of cuttings in *ornamentals*.

Efficacy

- Dip moistened base of cuttings into powder immediately before planting [1, 2]
- See label for details of concentrations recommended for promotion of rooting in cuttings of different species [3]
- Dip prepared cuttings in solution for 4-24 h depending on species [3]

Crop safety/Restrictions

- Maximum number of treatments 1 per cutting

Personal protective equipment/Label precautions
- U08, U19, U20a, E15, E30a, E32a, E34

Latest application/Harvest interval
- Before cutting insertion

391　(2-naphthyloxy)acetic acid

A plant growth regulator for setting tomato fruit

Products

Betapal Concentrate	Vitax	16 g/l	SL	00234

Uses

Increasing fruit set in **tomatoes**.

Efficacy
- Apply with any fine sprayer or syringe when first half dozen flowers are open
- Spray actual trusses in flower and repeat every 2 wk

Crop safety/Restrictions
- Do not spray growing head of tomato plants

392　napropamide

A soil applied amide herbicide for oilseed rape, fruit and woody ornamentals

Products

1	AgriGuard Napropamide 450 FL	AgriGuard	450 g/l	SC	09223
2	Devrinol	United Phosphorus	450 g/l	SC	09374
3	Devrinol	RP Agric.	450 g/l	SC	06195

Uses

Annual dicotyledons in **maiden strawberries** [1]. Annual dicotyledons in **blackcurrants, gooseberries, raspberries, strawberries** [1-3]. Annual dicotyledons in **container-grown woody ornamentals, newly planted woody ornamentals** [1, 3]. Annual dicotyledons in **apples, forest nurseries, ornamental plant production, pears, plums** [2]. Annual dicotyledons in **winter oilseed rape** [2, 3]. Annual dicotyledons in **established woody ornamentals, farm woodland** *(off-label)* [3]. Annual grasses in **maiden strawberries** [1]. Annual grasses in **blackcurrants, gooseberries, raspberries, strawberries** [1-3]. Annual grasses in **container-grown woody ornamentals, newly planted woody ornamentals** [1, 3]. Annual grasses in **apples, forest nurseries, ornamental plant production, pears, plums** [2]. Annual grasses in **winter oilseed rape** [2, 3]. Annual grasses in **established woody ornamentals, farm woodland** *(off-label)* [3]. Cleavers in **maiden strawberries** [1]. Cleavers in **blackcurrants, gooseberries, raspberries, strawberries** [1-3]. Cleavers in **container-grown woody ornamentals, newly planted woody ornamentals** [1, 3]. Cleavers in **apples, forest nurseries, ornamental plant production, pears, plums** [2]. Cleavers in **winter oilseed rape** [2, 3]. Cleavers in **established woody ornamentals** [3]. Groundsel in **maiden strawberries** [1]. Groundsel in **blackcurrants, gooseberries, raspberries, strawberries** [1-3]. Groundsel in **container-grown woody ornamentals, newly planted woody ornamentals** [1, 3]. Groundsel in **apples, forest nurseries, ornamental plant production, pears, plums** [2]. Groundsel in **winter oilseed rape** [2, 3]. Groundsel in **established woody ornamentals** [3].

Efficacy
- Best results obtained from treatment pre-emergence of weeds but product may be used in conjunction with contact herbicide such as paraquat for control of emerged weeds. Otherwise remove existing weeds before application
- Product broken down by sunlight, so application during summer not recommended. Most crops recommended for treatment between Nov and end-Feb

- Apply to winter oilseed rape as pre-drilling treatment in tank-mixture with trifluralin and incorporated within 30 min. Post-emergence use of specific grass weedkilller recommended where volunteer cereals serious
- Where minimal cultivation used to establish oilseed rape, tank-mixture may be applied directly to stubble and mixed into top 25 mm as part of surface cultivations
- Increase water volume to ensure adequate dampening of compost when treating containerised nursery stock with dense leaf canopy
- Do not use on soils with more than 10% organic matter

Crop safety/Restrictions
- Maximum number of treatments 1 per crop or yr
- Apply up to 14 d prior to drilling winter oilseed rape [2, 3]
- Do not use on Sands.
- Bush and cane fruit must be established for at least 10 mth before spraying and treated between 1 Nov and end Feb
- Apply to strawberries established for at least one season or to maiden crops as long as planted carefully and no roots exposed, between 1 Nov and end Feb. Do not treat runners of poor vigour or with shallow roots, or runner beds
- Do not treat ornamentals in containers of less than 1 litre
- Newly planted ornamentals should have no roots exposed. Do not treat stock of poor vigour or with shallow roots. Treatments made in Mar and Apr must be followed by 25 mm irrigation within 24 h
- Some phytotoxicity seen on yellow and golden varieties of conifers and container grown alpines. On any ornamental variety treat only a small number of plants in first season
- After use in fruit or ornamentals no crop can be drilled within 7 mth of treatment. Leaf, flowerhead, root and fodder brassica crops may be drilled after 7 mth; potatoes, maize, peas or dwarf beans after 9 mth; autumn sown wheat or grass after 18 mth; any crop after 2 yr
- After use in oilseed rape only oilseed rape, swedes, fodder turnips, brassicas or potatoes should be sown within 12 mth of application [2, 3]
- Soil should be mould-board ploughed to a depth of at least 200 mm before drilling or planting any following crop

Special precautions/Environmental safety
- Harmful to fish or other aquatic life. Do not contaminate surface waters or ditches with chemical or used container
- Do not apply by knapsack or hand-held sprayers

Personal protective equipment/Label precautions
- H [3]; A, C [1-3]
- U05a [3]; U09a, U20b [1-3]; C03, E01 [3]; E13c, E26, E30a, E31a, E34 [1-3]

Latest application/Harvest interval
- Pre-sowing for winter oilseed rape; before end Feb for strawberries, bush and cane fruit; before end of Apr for field and container grown ornamental trees and shrubs

Approval
- Off-label approval unlimited for use on farm woodland (OLA 1280/97)[3]
- Approval expiry 31 Dec 2001 [3]

---

393 nicotine

A general purpose, non-persistent, contact, alkaloid insecticide

Products

1	Nico Soap	United Phosphorus	75 g/l	LI	07517
2	Nicotine 40% Shreds	Dow	40% w/w	FU	05725
3	No-Fid	Hortichem	75 g/l	LI	07959
4	XL-All Insecticide	Vitax	70 g/l	LI	02369
5	XL-All Nicotine 95%	Vitax	950 g/l	LI	07402

Uses

Aphids in **protected ornamentals** [1-5]. Aphids in **apples, artichokes, asparagus, celeriac, chicory, chinese cabbage, chives, chrysanthemums, cucumbers, cucurbits, dwarf beans,** *edible podded peas, endives, fennel, field beans, flowerhead brassicas, garlic, kale, leeks, onions, peppers, potatoes, radicchio, radishes, shallots, strawberries, sweetcorn, tomatoes* [1, 3]. Aphids in *broad beans, carrots, leaf brassicas, lettuce, ornamentals, parsnips, peas, red beet, runner beans, spinach, swedes, turnips* [1, 3-5]. Aphids in *aubergines, protected herbs, protected potatoes, protected salad onions* [2]. Aphids in *protected courgettes, protected cucumbers, protected lettuce, protected peppers, protected tomatoes* [2, 4, 5]. Aphids in *celery, flower bulbs, flowers, navy beans, nursery stock, parsley, protected asparagus, protected celery, protected crops, protected flowers, protected marrows, protected melons, protected roses, soft fruit, top fruit, vegetables* [4, 5]. Capsids in *apples* [1, 3]. Capsids in *flower bulbs, flowers, protected crops, soft fruit, top fruit, vegetables* [4, 5]. Caterpillars in *apples, artichokes, asparagus, broad beans, brussels sprouts, cabbages, carrots, celeriac, celery, chicory, chives, cucurbits, dwarf beans, fennel, field beans, flowerhead brassicas, french beans, garlic, gooseberries, kale, leeks, onions, ornamentals, parsnips, peas, potatoes, protected ornamentals, red beet, runner beans, shallots, spinach, strawberries, swedes, sweetcorn, turnips* [1, 3]. General insect control in *french beans (off-label)* [4, 5]. Glasshouse whitefly in *aubergines, protected courgettes, protected cucumbers, protected herbs, protected lettuce, protected ornamentals, protected peppers, protected potatoes, protected salad onions, protected tomatoes* [2]. Green leafhopper in *aubergines, protected courgettes, protected cucumbers, protected herbs, protected lettuce, protected ornamentals, protected peppers, protected potatoes, protected salad onions, protected tomatoes* [2]. Leaf miners in *celery, chrysanthemums* [1, 3]. Leaf miners in *flower bulbs, flowers, leaf brassicas, protected crops* [4, 5]. Leafhoppers in *flower bulbs, flowers, protected crops, soft fruit, vegetables* [4, 5]. Potato virus vectors in *chitting potatoes* [2]. Sawflies in *gooseberries* [1, 3]. Sawflies in *flower bulbs, flowers, protected crops, soft fruit, top fruit, vegetables* [4, 5]. Thrips in *aubergines, protected courgettes, protected cucumbers, protected herbs, protected lettuce, protected ornamentals, protected peppers, protected potatoes, protected salad onions, protected tomatoes* [2]. Thrips in *flower bulbs, flowers, protected crops, vegetables* [4, 5]. Woolly aphid in *flowers, fruit crops* [4, 5].

Efficacy
- Apply as foliar spray, taking care to cover undersides of leaves and repeat as necessary or dip young plants, cuttings or strawberry runners before planting out
- Best results achieved by spraying at air temperatures above 16°C but do not treat in bright sunlight or windy weather [2]
- Fumigate glasshouse crops at temperatures of at least 16°C. See label for details of recommended fumigation procedure [2]
- Potatoes may be fumigated in chitting houses to control virus-spreading aphids [2]. Field crops of potatoes may also be treated [1, 3]
- May be used in integrated control systems to give partial control of organophosphorus-, organochlorine- and pyrethroid-resistant whitefly

Crop safety/Restrictions
- Maximum number of treatments normally not specified but 3 per crop for protected vegetables and hops [1]
- On plants of unknown sensitivity test first on a small scale

Special precautions/Environmental safety
- Nicotine is subject to the Poisons Rules 1982 and the Poisons Act 1972. See notes in Section 1 [2, 5]
- Toxic in contact with skin, by inhalation and if swallowed [5]
- Harmful in contact with skin, by inhalation and if swallowed [1, 2, 4, 5]
- Highly flammable [2]
- Keep unprotected persons out of treated glasshouses for at least 12 h
- Wear overall, hood, rubber gloves and respirator if entering glasshouse within 12 h of fumigating [2]
- Dangerous to livestock.
- Dangerous to game, wild birds and animals

- Harmful to bees. Do not apply to crops in flower or to those in which bees are actively foraging. Do not apply when flowering weeds are present
- Dangerous to fish or other aquatic life. Do not contaminate surface waters or ditches with chemical or used container
- Keep unprotected persons out of treated areas for at least 12 h following treatment and ensure adequate ventilation before re-entering

Personal protective equipment/Label precautions
- A [1-5]; D, J [2]; H [2, 5]; K, M [5]; C [1, 3, 5]
- M03 [1-4]; M04 [5]; R03a, R03b, R03c [1-5]; R07c [2]; R02a, R02b, R02c [5]; U19, U20a [1-5]; U08, U10, U15 [4, 5]; U02, U04a, U05a, U13 [1, 3-5]; U05b [2]; U09a [1, 3]; C02 [2, 4, 5] (24 h); C03 [1-5]; E32a, E34 [2, 4, 5]; E01, E10a, E12e, E13b [1-5]; E02 [1-5] (12 h); E06b, E31a [4, 5]; E30a, E31b [1, 3-5]; E30b [2, 5]; E06a [1-3] (12 h)

Withholding period
- Keep all livestock out of treated areas for at least 12 h

Latest application/Harvest interval
- HI 2 d for most crops - check labels

Approval
- Off-label Approval unlimited for use on French beans (OLA 0080/92)[4]

---

394   octhilinone

An isothiazolone fungicide for treating canker and tree wounds

Products

Pancil T	Landseer	1.0% w/w	PA	09261

Uses

Canker in *apples, cherries, pears, plums, woody ornamentals*. Pruning wounds in *apples, cherries, pears, plums, woody ornamentals*. Silver leaf in *apples, cherries, pears, plums, woody ornamentals*.

Efficacy
- Treat canker by removing infected shoots, cutting back to healthy tissue and painting evenly over wound
- Apply to pruning wounds immediately after cutting. Cover all cracks in bark and ensure that coating extends beyond edges of wound
- Best results achieved by application in dry conditions

Crop safety/Restrictions
- Only apply after harvest and before bud burst
- Do not apply to grafting cuts or bud. Do not apply under freezing conditions

Special precautions/Environmental safety
- Irritating to skin, risk of serious damage to eyes
- May cause sensitization by skin contact
- Dangerous to fish or other aquatic life. Do not contaminate surface waters or ditches with chemical or used container

Personal protective equipment/Label precautions
- A, C
- M03, R04b, R04d, R04e, U05a, U09a, U13, U19, U20b, C03, E01, E13b, E26, E30a, E31a, E34

Latest application/Harvest interval
- Before bud burst

---

395   ofurace

A phenylamide (acylalanine) fungicide available only in mixtures

See also mancozeb + ofurace

---

396 oxadiazon

A residual and contact oxadiazolone herbicide for fruit and ornamentals

See also carbetamide + diflufenican + oxadiazon
glyphosate + oxadiazon

Products

1	Ronstar 2G	Hortichem	2% w/w	GR	06492
2	Ronstar Liquid	Hortichem	250 g/l	EC	08974

Uses

Annual dicotyledons in **container-grown ornamentals** [1]. Annual dicotyledons in **apples, blackcurrants, gooseberries, grapevines, hops, pears, raspberries, woody ornamentals** [2]. Annual grasses in **container-grown ornamentals** [1]. Annual grasses in **apples, blackcurrants, gooseberries, grapevines, hops, pears, raspberries, woody ornamentals** [2]. Bindweeds in **apples, blackcurrants, gooseberries, grapevines, hops, pears, raspberries, woody ornamentals** [2]. Cleavers in **apples, blackcurrants, gooseberries, grapevines, hops, pears, raspberries, woody ornamentals** [2]. Knotgrass in **apples, blackcurrants, gooseberries, grapevines, hops, pears, raspberries, woody ornamentals** [2].

Efficacy

- Rain or overhead watering is needed soon after application for effective results
- Pre-emergence activity reduced on soils with more than 10% organic matter and, in these conditions, post-emergence treatment is more effective [2]
- Best results on bindweed when first shoots are 10-15 cm long
- Do not cultivate after treatment [2]

Crop safety/Restrictions

- See label for list of ornamental species which may be treated with granules. Treat small numbers of other species to check safety. Do not treat hydrangea, spiraea or genista
- Do not treat container stock under glass or use on plants rooted in media with high sand or non-organic content. Do not apply to plants with wet foliage
- Apply spray to apples and pears from Jan to Jul, avoiding young growth
- Treat bush fruit from Jan to bud-break, avoiding bushes, grapevines in Feb/Mar before start of new growth or in Jun/Jul avoiding foliage
- Treat hops cropped for at least 2 yr in Feb or in Jun/Jul after deleafing
- Treat woody ornamentals from Jan to Jun, avoiding young growth. Do not spray container stock overall
- Do not use more than 8 l/ha in any 12 mth period [2]

Special precautions/Environmental safety

- Irritating to skin [2] and eyes [1]
- Risk of serious damage to eyes [2]
- Flammable [2]
- Dangerous to fish or other aquatic life. Do not contaminate surface waters or ditches with chemical or used container

Personal protective equipment/Label precautions

- A, C [1, 2]
- R04a [1]; R04b [1, 2]; R04d, R07d [2]; U05a, U08, U14, U15, U20b, C03, E01, E13b, E30a, E32a, E34 [1, 2]

Latest application/Harvest interval

- Jul for apples, grapevines, hops, pears; Jun for raspberries, woody ornamentals

Approval

- Accepted by BLRA for use on hops

397   oxadixyl

A phenylamide (acylalanine) fungicide available only in mixtures

See also carbendazim + cymoxanil + oxadixyl + thiram
cymoxanil + mancozeb + oxadixyl
mancozeb + oxadixyl

398   oxamyl

A soil-applied, systemic oxime carbamate nematicide and insecticide

Products

Vydate 10G	DuPont	10% w/w	GR	02322

Uses

American serpentine leaf miner in **aubergines** *(off-label)*, **broad beans** *(off-label)*, **cucurbits** *(off-label)*, **garlic** *(off-label)*, **ornamentals** *(off-label)*, **peppers** *(off-label)*, **shallots** *(off-label)*, **soya bean** *(off-label)*, **tomatoes** *(off-label)*. Aphids in **potatoes, sugar beet**. Docking disorder vectors in **sugar beet**. Free-living nematodes in **potatoes**. Mangold fly in **sugar beet**. Millipedes in **sugar beet**. Non-indigenous leaf miners in **aubergines** *(off-label)*, **broad beans** *(off-label)*, **cucurbits** *(off-label)*, **garlic** *(off-label)*, **ornamentals** *(off-label)*, **peppers** *(off-label)*, **shallots** *(off-label)*, **soya bean** *(off-label)*, **tomatoes** *(off-label)*. Potato cyst nematode in **potatoes**. Pygmy beetle in **sugar beet**. Spraing vectors in **potatoes**. Stem nematodes in **garlic** *(off-label)*, **protected garlic** *(off-label)*, **protected onions** *(off-label)*.

Efficacy

- Apply granules with suitable applicator before drilling or planting. See label for details of recommended machines
- In potatoes incorporate thoroughly to 10 cm and plant within 3-4 d
- In sugar beet apply in seed furrow at drilling

Crop safety/Restrictions

- Maximum number of treatments 1 per crop or yr

Special precautions/Environmental safety

- This product contains an anticholinesterase carbamate compound. Do not use if under medical advice not to work with such compounds
- Harmful by inhalation or if swallowed
- Keep in original container, tightly closed, in a safe place, under lock and key
- Dangerous to fish or other aquatic life. Do not contaminate surface waters or ditches with chemical or used container
- Dangerous to game and wildlife. Cover granules completely. Bury spillages
- Wear protective gloves if handling treated compost or soil within 2 wk after treatment
- Allow at least 12 h, followed by at least 1 h ventilation, before entry of unprotected persons into treated glasshouses

Personal protective equipment/Label precautions

- A, B, H, K, M; C (or D)
- M02, M04, R03b, R03c, U02, U04a, U05a, U09a, U13, U19, U20a, C02, C03, E01, E10a, E13b, E30b, E32a, E34

Latest application/Harvest interval

- At drilling/planting for vegetables; before drilling/planting for potatoes and peas.
- HI tomatoes, ornamentals 2 wk

Approval

- Off-label Approval unlimited for use on outdoor and protected ornamentals, protected tomatoes, aubergines, peppers, soya beans and broad beans (OLA 0020/93); unlimited for use on protected onions and garlic (OLA 0925/94); unlimited for use on outdoor garlic for stem nematode control (OLA 0163/92)

399   oxycarboxin

A protectant, eradicant and systemic carboxamide fungicide

Products

| Plantvax 75 | Fargro | 75% w/w | WP | 01601 |

Uses

Rust in **carnations, chrysanthemums, ornamentals, pelargoniums, roses**.

Efficacy
- Apply as a routine to established carnation beds from Sep to Feb at 7-10 d intervals. If rust appears spray immediately and repeat several times
- On stock plants or cuttings of carnations or geraniums apply after striking and repeat 3 times at 10-14 d intervals. Remove and burn any infected cuttings
- Can be used as a drench on carnations grown in peat bags

Crop safety/Restrictions
- Spray product alone and preferably not within 2 d of other sprays
- Do not apply Spraying Oil for 14 d before or after treatment
- Can scorch leaves when taken up by roots. Avoid spraying when blooms open

Special precautions/Environmental safety
- Irritating to eyes
- Avoid working in spray
- Dangerous to fish or other aquatic life. Do not contaminate surface waters or ditches with chemical or used container

Personal protective equipment/Label precautions
- A, C
- R04a, U02, U05a, U09a, U19, U20b, C03, E01, E13b, E30a, E32a

400   paclobutrazol

A conazole plant growth regulator for ornamentals and fruit

Products

| 1 Bonzi | Zeneca | 4 g/l | SC | 06640 |
| 2 Cultar | Zeneca | 250 g/l | SC | 06649 |

Uses

Controlling vigour in **apples, cherries, pears, plums** [2]. Improving colour in **poinsettias** [1]. Increasing flowering in **azaleas, bedding plants, begonias, kalanchoes, lilies, roses, tulips** [1]. Increasing fruit set in **apples, cherries, pears, plums** [2]. Stem shortening in **azaleas, bedding plants, begonias, kalanchoes, lilies, roses, tulips** [1].

Efficacy
- Chemical is active via both foliage and root uptake
- Apply as spray to produce compact pot plants and to improve bract colour of poinsettias [1]
- Apply as drench to reduce flower stem length of potted tulips [1]
- Timing is critical and varies with species. See label for details [1]
- Apply to apple and pear trees under good growing conditions as pre-blossom spray (apples only) and post-blossom at 7-14 d intervals [2]
- Timing and dose of orchard treatment vary with species and cultivar. See label [2]
- Apply to plums and cherries as a trench drench in the spring [2]

Crop safety/Restrictions
- Maximum concentration 250 ml/10 l water [1]
- Maximum total dose 3.0 l/ha (pears), 4.0 l/ha (apples) [2]
- Maximum number of treatments 1 per yr for cherries, plums [2]
- Some varietal restrictions apply in top fruit (see label for details) [2]
- Do not use on trees of low vigour or under stress [2]
- Do not tank mix with or apply to apples and pears on same day as Thinsec [2]

- Do not use on trees from green cluster to 2 wk after full petal fall [2]
- Do not use in underplanted orchards. Consult label for guidance on following crops after grubbing [2]

Special precautions/Environmental safety
- Do not use on food crops [1]
- Harmful if swallowed [2]
- Harmful to fish or other aquatic life. Do not contaminate surface waters or ditches with chemical or used container

Personal protective equipment/Label precautions
- A, H, M [2]
- M03, R03c [2]; U08 [1]; U20a [1, 2]; U05a [2]; C01 [1]; C03 [2]; C02 [2] (2 wk); E31b [1]; E13c, E26, E30a, E34 [1, 2]; E01, E31c [2]; E07 [2] (2 yr)

Withholding period
- Keep livestock out of treated areas for at least 2 yr following treatment [2]

Latest application/Harvest interval
- Before end Apr for cherries; before beginning Apr for plums [2]
- HI apples, pears 14 d [2]

---

401   paraffin oil (commodity substance)

An agent for the control of birds by egg treatment

Products

Uses

Birds in *miscellaneous situations* (egg treatment).

Efficacy
- Egg treatment should be undertaken as soon as clutch is complete
- Eggs should be treated by complete immersion in liquid paraffin

Crop safety/Restrictions
- Treat eggs once only
- Use to control eggs of birds covered by licences issued by the Agriculture and Environment Departments under Section 16(1) of the Wildlife and Countryside Act (1981)

Personal protective equipment/Label precautions
- A, C

Approval
- Only to be used where a licence has been approved in accordance with Section 16(1) of the Wildlife and Countryside Act (1981)

---

402   paraquat

A non-selective, non-residual contact bipyridilium herbicide

See also diquat + paraquat
diuron + paraquat

Products

1	Barclay Total	Barclay	200 g/l	SL	08822
2	Dextrone X	Nomix-Chipman	200 g/l	SL	00687
3	Gramoxone 100	Zeneca	200 g/l	SL	06674

Uses

Annual dicotyledons in *orchards, redcurrants, row crops, whitecurrants* [1]. Annual dicotyledons in *forest nursery beds* (stale seedbed) [1, 2]. Annual dicotyledons in *flower bulbs, forestry, forestry transplant lines, non-crop areas* [1-3]. Annual dicotyledons in *blackcurrants, cultivated land/soil* (minimum cultivation), *gooseberries, hops, potatoes, raspberries, stubbles, sugar beet* [1, 3]. Annual dicotyledons in *woody ornamentals* [2]. Annual dicotyledons in *forest nursery beds, grapevines, grassland* (sward destruction/direct drilling), *hardy ornamentals, herbs* (off-label), *lettuce* (off-label), *lucerne* (off-label), *mint* (off-

---

label), **protected vegetables** *(off-label)*, **row crops** *(stale seedbed/inter-row)*, **top fruit** [3]. Annual grasses in **orchards, redcurrants, row crops, whitecurrants** [1]. Annual grasses in **flower bulbs, forest nursery beds** *(stale seedbed)*, **forestry, forestry transplant lines, non-crop areas** [1-3]. Annual grasses in **blackcurrants, cultivated land/soil** *(minimum cultivation)*, **gooseberries, grassland** *(sward destruction/direct drilling)*, **hops, potatoes, raspberries, stubbles, sugar beet** [1, 3]. Annual grasses in **woody ornamentals** [2]. Annual grasses in **grapevines, hardy ornamentals, herbs** *(off-label)*, **lettuce** *(off-label)*, **lucerne** *(off-label)*, **mint** *(off-label)*, **protected vegetables** *(off-label)*, **row crops** *(stale seedbed/inter-row)*, **top fruit** [3]. Barren brome in **field crops** *(stubble treatment)* [1]. Barren brome in **stubbles** [3]. Chemical stripping in **hops** [1, 3]. Creeping bent in **orchards, redcurrants, whitecurrants** [1]. Creeping bent in **flower bulbs, forestry, forestry transplant lines, non-crop areas** [1-3]. Creeping bent in **blackcurrants, cultivated land/soil** *(minimum cultivation)*, **gooseberries, grassland** *(sward destruction/direct drilling)*, **hops, raspberries, stubbles** [1, 3]. Creeping bent in **woody ornamentals** [2]. Creeping bent in **hardy ornamentals, top fruit** [3]. Firebreak desiccation in **forestry** [1-3]. Green cover in **land temporarily removed from production** [1, 3]. Green cover in **field margins** [3]. Perennial ryegrass in **orchards, redcurrants, whitecurrants** [1]. Perennial ryegrass in **non-crop areas** [1-3]. Perennial ryegrass in **blackcurrants, gooseberries, grassland** *(sward destruction/direct drilling)*, **hops, raspberries** [1, 3]. Perennial ryegrass in **top fruit** [3]. Rough meadow grass in **orchards, redcurrants, whitecurrants** [1]. Rough meadow grass in **non-crop areas** [1-3]. Rough meadow grass in **blackcurrants, gooseberries, grassland** *(sward destruction/direct drilling)*, **hops, raspberries** [1, 3]. Rough meadow grass in **top fruit** [3]. Runner desiccation in **strawberries** [1, 3]. Volunteer cereals in **row crops** [1]. Volunteer cereals in **cultivated land/soil** *(minimum cultivation)*, **potatoes, stubbles, sugar beet** [1, 3]. Volunteer cereals in **row crops** *(stale seedbed/inter-row)* [3]. Wild oats in **stubbles** [1, 3].

Efficacy

- Apply to green weeds preferably less than 15 cm high. Spray is rainfast after 10 min
- Addition of wetter recommended with lower application rates [1, 3]
- Only use clean water for mixing up spray
- Allow at least 4 h before cultivating, leave overnight if possible
- Spray in autumn to suppress couch when shoots have 2 leaves, repeat as necessary and plough after last treatment
- For direct-drilling land should be free of perennial weeds and protection against slugs should be provided. Allow 7-10 d before drilling into sprayed grass
- Best time for use in top fruit is Nov-Apr
- For forestry fire-break use apply in Jul-Aug and fire 7-10 d later

Crop safety/Restrictions

- Maximum number of treatments 1 per yr for lucerne, outdoor lettuce, protected vegetables and ornamentals; 2 per yr for mint and leafy herbs
- Chemical is inactivated on contact with soil. Crops can be sown or planted soon after spraying on most soils 24 h [2], 4 h [3], after 3 d on sandy or immature peat soils
- Do not use on straw or other artificial growing media
- Apply to lucerne in late Feb/early Mar when crop dormant
- Do not use on potatoes under hot, dry conditions or on hops under drought conditions
- For interrow use in row crops apply with guarded, no-drift sprayer
- Use as a carefully directed spray in blackcurrants, gooseberries, grapevines and other fruit crops, in raspberries only when dormant
- For stawberry runner control use guarded sprayer, not when flowers or fruit present
- Apply to bulbs pre-emergence (at least 3 d pre-emergence on sandy soils) or at end of season, provided no attached foliage and bulbs well covered (not on sandy soils)
- If using around glasshouses ensure vents and doors closed
- In forestry seedbed apply up to 3 d before seedling emergence

Special precautions/Environmental safety

- Paraquat is subject to the Poisons Rules 1982 and the Poisons Act 1972. See notes in Section 1
- Toxic if swallowed. Paraquat can kill if swallowed
- Harmful in contact with skin

- Irritating to eyes and skin
- Keep in original container, tightly closed, in a safe place, under lock and key
- Do not put in a food or drinks container
- Products in this profile are for professional use only
- Harmful to livestock. Paraquat may be harmful to hares; stubbles must be sprayed early in the day
- Do not contaminate surface waters or ditches with chemical or used container
- Paraquat may be harmful to hares. Where possible spray stubbles early in the day
- Observe label restrictions for maximum permitted concentration when spraying at low volume

Personal protective equipment/Label precautions
- A, C [1-3]
- M04, R02c, R03a, R04a, R04b [1-3]; U04b, U20b [2, 3]; U05a, U09a, U19 [1-3]; U02, U04a, U20a [1]; C03 [1-3]; E13c, E31a [2]; E01, E11, E26, E30b, E34 [1-3]; E06b [1-3] (24 h); E31c [3]; E15 [1, 3]; E31b [1]

Withholding period
- Keep all livestock out of treated areas for at least 24 h

Latest application/Harvest interval
- Before shoots 15 cm high and before 10% of shoots emerged for early potatoes, before 40% of shoots emerged for maincrop potatoes [1, 3]; before 31 Mar for lucerne; see labels for other crops
- HI mint, leafy herbs 8 wk [3]

Approval
- Off-label Approval to Apr 2001 for use on leafy herbs (OLA 0832/96)[3]; unlimited for use on dormant lucerne (OLA 1188/96)[3]; unlimited for use on a wide range of protected and outdoor vegetables (see OLA notice for details)(OLA 1497/96)[3]
- Accepted by BLRA for use in hops

Maximum Residue Level (mg residue/kg food)
- tea, hops 0.1; all other products (except cereals, animal products) 0.05

---

403   penconazole

A protectant conazole fungicide with antisporulant activity

See also captan + penconazole

Products

Topas	Novartis	100 g/l	EC	08458

Uses
Powdery mildew in **apples, blackcurrants, hops, ornamental trees**. Rust in **roses**. Scab in **ornamental trees**.

Efficacy
- Use as a protectant fungicide by treating at the earliest signs of disease
- Treat crops every 10-14 d (every 7-10 d in warm, humid weather) at first sign of infection or as a protective spray ensuring complete coverage. See label for details of timing
- Increase dose and volume with growth of hops but do not exceed 2000 l/ha. Little or no activity will be seen on established powdery mildew
- Antisporulant activity reduces development of secondary mildew in apples

Crop safety/Restrictions
- Maximum number of treatments 10 per yr for apples, 6 per yr for hops, 4 per yr for blackcurrants
- Check for varietal susceptibility in roses. Some defoliation may occur after repeat applications on Dearest

Special precautions/Environmental safety
- Irritating to eyes and skin
- Harmful to fish or other aquatic life. Do not contaminate surface waters or ditches with chemical or used container

---

Personal protective equipment/Label precautions
- A, C
- R04a, R04b, U02, U05a, U08, U20a, C02, C03, E01, E13c, E30a, E31b, E34

Latest application/Harvest interval
- HI apples, hops 14 d; blackcurrants 4 wk

Approval
- Accepted by BLRA for use on hops

---

404   pencycuron

A non-systemic urea fungicide for use on seed potatoes

See also imazalil + pencycuron

Products

1	Monceren DS	Bayer	12.5% w/w	DS	04160
2	Monceren Flowable	Bayer	250 g/l	FS	04907
3	Standon Pencycuron DP	Standon	12.5% w/w	DS	08774

Uses

Black scurf in **potatoes** *(tuber treatment)*. Stem canker in **potatoes** *(tuber treatment)*.

Efficacy
- Provides control of tuber-borne disease and gives some reduction of stem canker
- Apply to seed tubers in chitting trays, to bulk bins immediately before planting or in hopper at planting [1, 3]
- Apply to clean seed tubers into or out of store (follow agronomic guidelines in manufacturer's literature) [2]
- If rain interrupts planting cover tubers in hopper

Crop safety/Restrictions
- Maximum number of treatments 1 per batch of seed potatoes
- Treated tubers must be used only as seed and not for human or animal consumption
- Do not use on tubers previously treated with a dry powder seed treatment or hot water

Special precautions/Environmental safety
- Use of suitable dust mask is mandatory when applying dust, filling the hopper or riding on planter
- Do not use treated seed as food or feed
- Treated seed harmful to game and wildlife

Personal protective equipment/Label precautions
- A [1-3]; F [1, 3]
- U20b [1, 3]; U20a [2]; E03, E15, E30a, E32a, S01, S02, S03, S04b, S05, S06 [1-3]

Latest application/Harvest interval
- At planting

Approval
- May be applied by misting equipment mounted over roller table (see label for details) [2]

---

405   pendimethalin

A residual dinitroaniline herbicide for cereals and other crops

See also bentazone + pendimethalin
cyanazine + pendimethalin
isoproturon + pendimethalin

Products

1	Sovereign	Novartis	400 g/l	SC	08533
2	Stomp 400 SC	BASF	400 g/l	SC	10229
3	Stomp 400 SC	Cyanamid	400 g/l	SC	04183

Uses

Annual dicotyledons in **apples, blackberries, blackcurrants, carrots, cherries, gooseberries, hops, leeks, loganberries, onions, parsley, parsnips, pears, plums,** *protected lettuce* (off-label), **raspberries, strawberries, tayberries, transplanted brassicas** [1]. Annual dicotyledons in **herbs** (off-label), **leeks** (off-label), **onion sets** (off-label), **onions** (off-label), **outdoor leaf herbs** (off-label), **radicchio** (off-label), **runner beans** (off-label), **sweetcorn** (off-label) [1, 3]. Annual dicotyledons in **combining peas, durum wheat, fodder maize, potatoes, spring barley, sunflowers, triticale, winter barley, winter rye, winter wheat** [2, 3]. Annual dicotyledons in **evening primrose** (off-label), **farm forestry** (off-label), **lettuce** (off-label) [3]. Annual grasses in **apples, blackberries, blackcurrants, carrots, cherries, gooseberries, hops, leeks, loganberries, onions, parsley, parsnips, pears, plums, protected lettuce** (off-label), **raspberries, strawberries, tayberries, transplanted brassicas** [1]. Annual grasses in **herbs** (off-label), **leeks** (off-label), **onion sets** (off-label), **onions** (off-label), **outdoor leaf herbs** (off-label), **radicchio** (off-label), **runner beans** (off-label), **sweetcorn** (off-label) [1, 3]. Annual grasses in **combining peas, durum wheat, potatoes, spring barley, sunflowers, triticale, winter barley, winter rye, winter wheat** [2, 3]. Annual grasses in **evening primrose** (off-label), **farm forestry** (off-label), **lettuce** (off-label) [3]. Annual meadow grass in **combining peas, durum wheat, fodder maize, potatoes, spring barley, sunflowers, triticale, winter barley, winter rye, winter wheat** [2, 3]. Blackgrass in **combining peas, durum wheat, potatoes, spring barley, sunflowers, triticale, winter barley, winter rye, winter wheat** [2, 3]. Cleavers in **combining peas, durum wheat, potatoes, spring barley, sunflowers, triticale, winter barley, winter rye, winter wheat** [2, 3]. Cleavers in **winter field beans** (off-label) [3]. Speedwells in **combining peas, durum wheat, fodder maize, potatoes, spring barley, sunflowers, triticale, winter barley, winter rye, winter wheat** [2, 3]. Wild oats in **combining peas, durum wheat, potatoes, spring barley, sunflowers, triticale, winter barley, winter rye, winter wheat** [2, 3].

Efficacy
- Apply as soon as possible after drilling. Weeds are controlled as they germinate and emerged weeds will not be controlled by use of the product alone
- For effective blackgrass control apply not more than 2 d after final cultivation and before weed seeds germinate
- Tank mixes with approved formulations of isoproturon or chlorotoluron recommended for improved pre- and post-emergence control of blackgrass, with isoxaben for additional broad-leaved weeds [2, 3]
- Tank mixture with atrazine recommended for weed control in fodder maize [2, 3]
- Best results by application to fine firm, moist, clod-free seedbeds when rain follows treatment. Effectiveness reduced by prolonged dry weather after treatment
- Do not use on spring barley after end Mar (mid-Apr in Scotland) because dry conditions likely. Do not apply to dry seedbeds in spring unless rain imminent [2, 3]
- Effectiveness reduced on soils with more than 6% organic matter. Do not use where organic matter exceeds 10%
- Any trash, ash or straw should be incorporated evenly during seedbed preparation
- Do not disturb soil after treatment
- On peas drilled after end Mar (mid-Apr in Scotland) tank-mix with cyanazine [2, 3]
- Apply to potatoes as soon as possible after planting and ridging in tank-mix with cyanazine or metribuzin [2, 3]

Crop safety/Restrictions
- Maximum number of treatments 1 per crop or yr
- May be applied pre-emergence of cereal crops sown before 30 Nov provided seed covered by at least 32 mm soil, or post-emergence to early tillering stage (GS 23) [2, 3]
- Do not undersow treated crops
- Do not use on crops suffering stress due to disease, drought, waterlogging, poor seedbed conditions or chemical treatment or on soils where water may accumulate
- Apply to combining peas as soon as possible after sowing, not when plumule within 13 mm of surface [2, 3]
- Apply to potatoes up to 7 d before first shoot emerges [2, 3]
- In the event of crop failure specified crops may be sown after at least 2 mth following ploughing to 150 mm. See label for details
- After a dry season land must be ploughed to 150 mm before drilling ryegrass

- Apply to drilled crops as soon as possible after drilling but before crop and weed emergence
- Apply in top fruit, bush fruit and hops from autumn to early spring when crop dormant [1]
- In cane fruit apply to weed free soil from autumn to early spring, immediately after planting new crops and after cutting out canes in established crops [1]
- Apply in strawberries from autumn to early spring (not before Oct on newly planted bed). Do not apply pre-planting or during flower initiation period (post-harvest to mid-Sep) [1]
- Apply pre-emergence in drilled onions or leeks as tank-mixture with propachlor, not on Sands, Very Light, organic or peaty soils or when heavy rain forecast [1]
- Apply to brassicas after final plant-bed cultivation but before transplanting. Avoid unnecessary soil disturbance after application and take care not to introduce treated soil into the root zone when transplanting. Follow transplanting with specified post-planting treatments - see label [1]
- Do not use on protected crops or in greenhouses [1]

Special precautions/Environmental safety
- Dangerous to fish or other aquatic life. Do not contaminate surface waters or ditches with chemical or used container

Personal protective equipment/Label precautions
- U05a, U08, U13, U19, U20a, C03, E01, E13b, E30a, E31b [1-3]; E34 [2, 3]; E26 [1]

Latest application/Harvest interval
- Pre-emergence for spring barley, carrots, lettuce, fodder maize, parsnips, parsley, sage, peas, runner beans, potatoes, onions and leeks; before transplanting for brassicas; before main shoot and 3 tillers stage (GS 23) for winter cereals; before bud burst for blackcurrants, gooseberries, cane fruit; before flower trusses emerge for strawberries.
- HI evening primrose, outdoor parsley and sage 5 mth

Approval
- Off-label Approval unlimited for use in farm forestry (OLA 0226/94)[3]; unlimited for use on evening primrose (pre-emergence) (OLA 0319/92, 0660/92)[3]; unlimited for use on herbs and outdoor leaf herbs (OLA1727/97)[1]; unstipulated for use on winter field beans (OLA 1098/99)[3]; unstipulated for use on outdoor crops of lettuce, radicchio, scarole, lambs lettuce, cress grown under crop covers (OLA 0131/00)[3]; unstipulated for use on herbs and outdoor leaf herbs (OLA 0132/00)[3]; unstipulated for use on runner beans (OLA 0133/00)[3]; unstipulated for use on onions (OLA 0134/00)[3]; unstipulated for use on sweet corn grown under crop covers and mulches (OLA 0135/00)[3]; unstipulated for use on onions, onion sets, leeks (OLA 0136/00)[3]
- Accepted by BLRA for use on malting barley and hops

406   pendimethalin + simazine

A contact and residual herbicide for use in winter cereals

Products

1	Merit	BASF	300:100 g/l	SC	BASF
2	Merit	Cyanamid	300:100 g/l	SC	04976

Uses

Annual dicotyledons in **winter barley, winter wheat**. Meadow grasses in **winter barley, winter wheat**.

Efficacy
- Apply from pre-emergence to early tillering stage of crop
- Best results achieved on fine, firm, moist seedbed
- Do not use on soils with more than 10% organic matter
- Any straw residues should be incorporated before spraying

Crop safety/Restrictions
- Maximum number of treatments 1 per crop
- Do not use on durum wheat
- Ensure crop seed is evenly covered with at least 32 mm of settled soil
- Do not use on Sands, Very Light, stony or gravelly soils

- Do not apply to crops under stress or on waterlogged soils
- Avoid overlapping spray swathes

Special precautions/Environmental safety
- Irritating to eyes and skin
- Dangerous to fish or other aquatic life. Do not contaminate surface waters or ditches with chemical or used container
- LERAP Category B

Personal protective equipment/Label precautions
- A, C
- R04a, R04b, U05a, U08, U13, U19, U20a, C03, E01, E13b, E16a, E16b, E26, E30a, E31b, E34

Latest application/Harvest interval
- Before main shoot and 2 tillers (GS 21)

Approval
- Accepted by BLRA for use on malting barley
- Product registration number not available at time of printing [1]

---

407   pentanochlor

A contact anilide herbicide for various horticultural crops

See also chlorpropham + pentanochlor

Products

1	Atlas Solan 40	Nufarm Whyte	400 g/l	EC	07726
2	Croptex Bronze	Hortichem	400 g/l	EC	04087

Uses

Annual dicotyledons in *apples, carnations, celeriac, celery, cherries, conifer seedlings, currants, fennel, foxgloves, gooseberries, larkspur, parsley, pears, plums, roses, sweet williams, tomatoes, umbelliferous herbs* (off-label), *wallflowers* [1]. Annual dicotyledons in *anemones, annual flowers, carrots, chrysanthemums, freesias, nursery stock, parsnips, sweet peas* [1, 2]. Annual meadow grass in *apples, carnations, celeriac, celery, cherries, conifer seedlings, currants, fennel, foxgloves, gooseberries, larkspur, parsley, pears, plums, roses, sweet williams, tomatoes, umbelliferous herbs* (off-label), *wallflowers* [1]. Annual meadow grass in *anemones, annual flowers, carrots, chrysanthemums, freesias, nursery stock, parsnips, sweet peas* [1, 2].

Efficacy
- Best results achieved on young weed seedlings under warm, moist conditions
- Weeds most susceptible in cotyledon to 2-leaf stage, up to 3 cm high, redshank, fat-hen, fumitory and some others also controlled at later stages
- Some residual effect in early spring when adequate soil moisture and growing conditions good. Effectiveness reduced by very cold weather or drought

Crop safety/Restrictions
- Maximum number of treatments 1 or 2 per crop depending on dose applied. See label for details
- Apply pre-emergence in anemones, freesias, foxgloves, larkspur
- Not recommended for flower crops on light sandy soils
- Apply pre-emergence or after fully expanded cotyledon stage of carrots and related crops
- Any crop may be planted after 4 wk following ploughing and cultivation

Special precautions/Environmental safety
- Irritating to skin and eyes

Personal protective equipment/Label precautions
- A, C
- R04a, R04b, U05a, U08, U19, U20a, C03, E01, E15, E30a, E31b

Latest application/Harvest interval
- HI 28 d for parsnips, carrots

Approval

- Off-label Approval unlimited for use on chervil, coriander, dill, bronze and green fennel, wild celery leaf, lovage, sweet cicely (OLA 0465/92)[1]

---

408   permethrin

A broad spectrum, contact and ingested pyrethroid insecticide

See also fenitrothion + permethrin + resmethrin

Products

1	Darmycel Agarifume Smoke Generator	Sylvan	10% w/w	FU	07904
2	Fumite Permethrin Smoke	Hortichem	10% w/w	FU	00940
3	Permasect 10 EC	Mitchell Cotts	90 g/l	EC	03920
4	Permasect 25 EC	Mitchell Cotts	230 g/l	EC	01576

Uses

Aphids in **protected cucumbers, protected ornamentals, protected tomatoes** [3]. Aphids in **aubergines, chillies, cucumbers, ornamental plant production, tomatoes** [4]. Black pine beetle in **conifer seedlings** *(off-label)*, **cut logs/timber** *(off-label)*, **forestry** *(off-label)* [4]. Caterpillars in **ornamental plant production** [4]. Pine weevil in **conifer seedlings** *(off-label)*, **conifers** *(off-label)*, **cut logs/timber** *(off-label)*, **forestry** *(off-label)* [4]. Sciarid flies in **aubergines, mushrooms, protected celery, protected cucumbers, protected lettuce, protected ornamentals, protected peppers, protected tomatoes** [1]. Tomato fruitworm in **aubergines, chillies, cucumbers, tomatoes** [4]. Whitefly in **protected celery, protected lettuce, protected peppers** [2]. Whitefly in **protected cucumbers, protected ornamentals, protected tomatoes** [2, 3]. Whitefly in **aubergines** [2, 4]. Whitefly in **chillies, cucumbers, ornamental plant production, tomatoes** [4].

Efficacy

- Best results from glasshouse fumigation achieved by treating in late afternoon or evening [1, 2]
- Sprays give rapid knock-down effect with persistent protection on leaf surfaces [2-4]
- Apply as soon as pest or damage appears, or as otherwise recommended and repeat as necessary, usually at 14 d intervals [2-4]
- Number and timing of sprays vary with crop and pest. See label for details
- For whitefly or mushroom fly control use at least 3 fumigations at 5-7 d intervals [1, 2]
- Where glasshouse whitefly resistant to permethrin occurs control is unlikely to be satisfactory [1, 2]

Crop safety/Restrictions

- No of treatments not restricted
- Cut any open blooms before fumigating flower crops [1, 2]
- Do not treat rare or unusual plants without first testing on a small scale [1, 2]
- Do not fumigate young seedlings or plants being hardened off [1, 2]
- Allow 7 d after fumigating before introducing *Encarsia* or *Phytoseiulus* [1, 2]

Special precautions/Environmental safety

- Irritating to eyes and skin [3, 4]
- Irritating to eyes and respiratory system [1, 2]
- Harmful. May cause lung damage if swallowed [3]
- Flammable [3]
- Dangerous to bees. Do not apply to crops in flower or to those in which bees are actively foraging. Do not apply when flowering weeds are present [1-4]
- Extremely dangerous to fish or other aquatic life. Do not contaminate surface waters or ditches with chemical or used container [1-4]
- Domestic animals, birds and fish should be removed from the vicinity of buildings to be treated. Foodstuffs should be protected from smoke [1, 2]

Personal protective equipment/Label precautions

- D, E, H [2]; A [2-4]; C [3, 4]
- R04a [1-4]; R04c [1, 2]; R04b [3, 4]; R03, R04g, R07d [3]; U05a, U19 [1-4]; U20c [1, 2]; U04a, U08, U20b [3, 4]; C03, E01, E12c, E13a, E30a [1-4]; E29, E32a [1, 2]; E31a [3, 4]

Latest application/Harvest interval
- HI zero

Approval
- All approvals for plant protection uses of permethrin to be withdrawn during 2001 followed by 12-month use-up period, except forestry uses which will remain until end 2003
- Off-label Approval unlimited for use on forestry trees to control pine weevil - see OLA notice for restrictions (OLA 0135/92)[4]; unlimited for use on containerised forestry seedlings (OLA 0451/94)[4]; unlimited for use on forest saplings, felled trees, cut logs and timber (OLA 1210/97)[4]

Maximum Residue Level (mg residue/kg food)
- lettuces, herbs, celery, rhubarb, cereals (except maize) 2; pome fruits, stone fruits, grapes, kiwi fruit, Chinese cabbage, kale, spinach, beet leaves 1; citrus fruits, tomatoes, peppers, aubergines, beans (with pods), leeks, meat 0.5; cotton seed, maize 0.2; almond, celeriac, radishes, cucumbers, gherkins, courgettes, melons, squashes, watermelons, sweetcorn, cauliflowers, peas (with pods), peanuts, rape seed, mustard seed, hops 0.1; tree nuts (except almond), cane fruits, bilberries, cranberries, blackberries, loganberries, raspberries, wild berries, currants, miscellaneous fruit (except kiwi fruit), root and tuber vegetables (except celeriac, radishes), garlic, broccoli, kohlrabi, watercress, witloof, beans (without pods), peas (without pods), asparagus, cardoons, celery, globe artichokes, mushrooms, pulses, linseed, poppy seed, sesame seed, sunflower seed, soya bean, potatoes, milk, eggs 0.05

---

409 petroleum oil

An insecticidal and acaricidal hydrocarbon oil

Products

Hortichem Spraying Oil	Hortichem	710 g/l	EC

Uses

Mealy bugs in *protected cucumbers, protected grapevines, protected pot plants, protected tomatoes*. Red spider mites in *protected cucumbers, protected grapevines, protected pot plants, protected tomatoes*. Scale insects in *protected cucumbers, protected grapevines, protected pot plants, protected tomatoes*.

Efficacy
- Spray at 1% (0.5% on tender foliage) to wet plants thoroughly, particularly the underside of leaves, and repeat as necessary
- Apply under quick drying conditions but not in bright sun unless glass well shaded

Crop safety/Restrictions
- Treat grapevines before flowering
- On plants of unknown sensitivity test first on a small scale
- Mixtures with certain pesticides may damage crop plants. If mixing, spray a few plants to test for tolerance before treating larger areas
- Do not mix with sulfur or use sulfur sprays within 28 d of treatment

Special precautions/Environmental safety
- Harmful if swallowed. Irritating to eyes, skin and respiratory system

Personal protective equipment/Label precautions
- A, C
- M03, R03c, R04a, R04c, U05a, U08, U19, C03, E01, E30a, E34

Latest application/Harvest interval
- HI zero

Approval
- Product not controlled by Control of Pesticides Regulations because it acts by physical means only

410   phenmedipham

A contact carbamate herbicide for beet crops and strawberries

See also desmedipham + phenmedipham
ethofumesate + metamitron + phenmedipham
ethofumesate + phenmedipham
lenacil + phenmedipham

Products

1	Barclay Punter XL	Barclay	114 g/l	EC	08047
2	Beetup	United Phosphorus	114 g/l	EC	07520
3	Betanal E	Aventis	114 g/l	EC	07248
4	Betanal Flo	Aventis	160 g/l	SC	08898
5	Betosip	Sipcam	114 g/l	EC	06787
6	Dancer FL	Sipcam	160 g/l	SC	10048
7	Herbasan	Nufarm Whyte	160 g/l	SC	08553
8	Hickson Phenmedipham	Hickson & Welch	118 g/l	EC	02825
9	MSS Protrum G	Nufarm Whyte	114 g/l	EC	08342
10	Stefes Forte 2	Stefes	114 g/l	EC	08204
11	Stefes Medipham 2	Stefes	114 g/l	EC	08203
12	Tripart Beta	Tripart	118 g/l	EC	03111

Uses

Annual dicotyledons in **strawberries** [1, 2, 10, 11]. Annual dicotyledons in **fodder beet, mangels, sugar beet** [1, 2, 4-12]. Annual dicotyledons in **red beet** [1, 2, 4, 6, 7, 9-12]. Annual dicotyledons in **spinach beet** *(off-label)*, **spinach** *(off-label)* [4].

Efficacy

- Best results achieved by application to young seedling weeds, preferably cotyledon stage, under good growing conditions when low doses are effective
- 2-3 repeat applications at 7-10 d intervals using a low dose are recommended on mineral soils, 3-5 applications may be needed on organic soils
- Do not spray wet foliage or if rain imminent
- Addition of adjuvant oil may improve effectiveness on some weeds
- Various tank-mixtures with other beet herbicides recommended. See label for details
- Use of certain pre-emergence herbicides is recommended in combination with post-emergence treatment. See label for details

Crop safety/Restrictions

- Maximum number of treatments 1 per crop for spinach; see labels for other crops
- Apply to beet crops at any stage as low dose/low volume spray or from fully developed cotyledon stage with full rate. Apply to red beet after fully developed cotyledon stage
- At high temperatures (above 21°C) reduce rate and spray after 5 pm
- Do not apply immediately after frost or if frost expected
- Do not spray crops stressed by wind damage, nutrient deficiency, pest or disease attack etc. Do not roll or harrow for 7 d before or after treatment
- Apply to strawberries at any time when weeds in susceptible stage, except in period from start of flowering to picking
- Do not use on strawberries under cloches or polythene tunnels

Special precautions/Environmental safety

- Harmful if swallowed [2, 8, 9, 12]
- Irritating to eyes, skin and respiratory system [1-3, 5, 8-12]
- Harmful (dangerous [4, 6, 7]) to fish or other aquatic life. Do not contaminate surface waters or ditches with chemical or used container [1, 2, 5, 8-12]

Personal protective equipment/Label precautions

- H [8]; A, C [1-3, 5, 8-12]; P [12]
- M03 [2, 8, 9, 12]; R04a, R04b, R04c [1-3, 5, 8-12]; R03c [2, 8, 9, 12]; U05a [1-3, 5, 7-12]; U08 [1-5, 7-12]; U19 [1-12]; U20a [1, 3, 5, 8]; U20b [2, 4, 6, 7, 9-12]; U13 [2]; U09a [6]; C03 [1-3, 5, 7-12]; E13c [1-3, 5, 8-12]; E01 [1-3, 5, 7-12]; E31b [1-5, 7-12]; E30a [1-12]; E34 [1-3, 5, 8, 9]; E26 [2-6, 8, 9]; E13b [4, 6, 7]; E31c [6]

Latest application/Harvest interval
- Before crop leaves meet between rows for beet crops; before flowering for strawberries; before 5 leaf stage for spinach
- HI 5 d [10]

Approval
- Off-label unstipulated for use on outdoor spinach (OLA 1308/99)[4]; unstipulated for use on spinach beet (OLA 1319/99)[4]

411 phenothrin

A non-systemic contact and ingested pyrethroid insecticide

Products

Sumithrin 10 Sec	Sumitomo	103 g/l	EC	H3762

Uses

Flies in *livestock houses*.

Efficacy
- Spray walls and other structural surfaces of milking parlours, dairies, cow byres and other animal housing

Crop safety/Restrictions
- Do not apply directly to livestock
- Remove exposed milk and collect eggs before application. Protect milk machinery and containers from contamination

Special precautions/Environmental safety
- Irritating to eyes
- Wear approved respiratory equipment and eye protection (goggles) when applying with Microgen equipment
- Not to be sold or supplied to amateur users
- Dangerous to fish or other aquatic life. Do not contaminate surface waters or ditches with chemical or used container

Personal protective equipment/Label precautions
- A, C, D, E, F, H
- R04a, U05a, U08, U20b, C03, C06, C07, C10, E01, E05, E13b, E30a, E32a

Approval
- May be applied with Microgen ULV equipment

412 phenothrin + tetramethrin

A pyrethroid insecticide mixture for control of flying insects

Products

1	Deosan Fly Spray	DiverseyLever	0.066:0.028% w/w	AE	H5246
2	Killgerm ULV 500	Killgerm	36.8:18.4 g/l	UL	H4647

Uses

Flies in *agricultural premises* [1-3]. Grain storage mites in *grain stores* [2]. Mosquitoes in *agricultural premises* [1-3]. Wasps in *agricultural premises* [1-3].

Efficacy
- Close doors and windows and spray in all directions for 3-5 sec. Keep room closed for at least 10 min
- May be used in the presence of poultry and livestock

Crop safety/Restrictions
- Do not use space sprays containing pyrethrins or pyrethroid more than once per week in intensive or controlled environment animal houses in order to avoid development of resistance. If necessary, use a different control method or product [2]

Special precautions/Environmental safety

- Do not spray directly on food, livestock or poultry
- Remove exposed milk and collect eggs before application. Protect milk machinery and containers from contamination
- Flammable
- Dangerous to fish or other aquatic life. Do not contaminate surface waters or ditches with chemical or used container

Personal protective equipment/Label precautions

- A, H [2]
- M05, R03, R04g [2]; U19 [1-3]; U20a [1, 3]; U02, U09a, U16, U17, U20b [2]; C03 [1, 3]; C06, C07, C09 [2]; E01 [1-3]; E13b [1, 3]; E04, E05, E13a, E15, E30a, E32a [2]

Approval

- Product approved for ULV application. See label for details [2]

413 phenylmercury acetate

An organomercury fungicide seed treatment approvals for which were revoked with effect from 31 March 1992

414 phorate

A systemic organophosphorus insecticide with vapour-phase activity for potatoes

Products

Phorate 10G	United Phosphorus	10% w/w	GR	08007

Uses

Aphids in **potatoes**. Capsids in **potatoes**. Leafhoppers in **potatoes**. Wireworms in **potatoes** *(damage reduction)*.

Efficacy

- Must be applied as a soil incorporated treatment
- Natural degradation will reduce efficacy, but in intensive vegetable growing areas enhanced degradation by soil organisms may lead to reduced or unsatisfactory control. This risk can be reduced by not treating the same ground more than once every two yrs
- Control may be unsatisfactory where resistant strains of aphids occur. Repeat treatments are likely to result in lower levels of control

Crop safety/Restrictions

- Maximum number of treatments 1 per crop
- Product must only be applied as in-furrow or directed band application and must be incorporated
- Caibrate equipment before treatment. Use applicator to place granules in the planting furrow close to the seed tubers

Special precautions/Environmental safety

- This product contains an anticholinesterase organophosphorus compound. Do not use if under medical advice not to work with such compounds
- Toxic by inhalation or if swallowed
- Harmful in contact with skin
- Irritating to eyes and skin
- May cause sensitization by skin contact
- Keep in original container, tightly closed in a safe place, under lock and key
- Harmful to livestock
- Dangerous to game, wild birds and animals
- Dangerous to fish or other aquatic life. Do not contaminate surface waters or ditches with chemical or used container

Personal protective equipment/Label precautions

- A, B, C, D, E, H, J, K, M
- M01, M04, R02b, R02c, R03a, R04a, R04b, R04e, U02, U04a, U05a, U10, U11, U13, U19, U20a, C03; C02 (6 wk); E01, E10a, E13b, E26, E30b, E32a, E34; E06a (6 wk)

Withholding period
- Keep all livestock out of treated areas for at least 6 wk. Bury or remove spillages

Latest application/Harvest interval
- At sowing

Approval
- Approval for use on carrots and parsnips suspended in Dec 1997
- UK approvals for use of phorate on potatoes likely to be revoked in 2001 as a result of implementation of the MRL Directives

Maximum Residue Level (mg residue/kg food)
- milk and dairy produce 0.2; peanuts, tea, hops 0.1; all other produce 0.01

---

415   picloram

A persistent, translocated pyridine carboxylic acid herbicide for non-crop areas

See also 2,4-D + picloram
        bromacil + picloram

Products

Tordon 22K	Nomix-Chipman	240 g/l	SL	05790

Uses

Annual dicotyledons in **non-crop areas, non-crop grass**. Bracken in **non-crop areas, non-crop grass**. Japanese knotweed in **non-crop areas, non-crop grass**. Perennial dicotyledons in **non-crop areas, non-crop grass**. Woody weeds in **non-crop areas, non-crop grass**.

Efficacy
- May be applied at any time of year. Best results achieved by application as foliage spray in late winter to early spring
- For bracken control apply 2-4 wk before frond emergence
- Clovers are highly sensitive and eliminated at very low doses
- Persists in soil for up to 2 yr

Crop safety/Restrictions
- Maximum number of treatments 1 per yr
- Do not apply around desirable trees or shrubs where roots may absorb chemical
- Do not apply on slopes where chemical may be leached onto areas of desirable plants

Special precautions/Environmental safety
- Irritating to eyes
- Harmful to fish or other aquatic life. Do not contaminate surface waters or ditches with chemical or used container

Personal protective equipment/Label precautions
- R04a, U05a, U08, U20b, C03, E01, E07, E13c, E30a, E31b

Withholding period
- Keep livestock out of treated areas until foliage of any poisonous weeds such as ragwort has died and become unpalatable

---

416   pirimicarb

A carbamate insecticide for aphid control

See also deltamethrin + pirimicarb

Products

1	Aphox	Zeneca	50% w/w	WG	06633
2	Barclay Pirimisect	Barclay	50% w/w	WG	09057
3	Greencrop Glenroe	Greencrop	50% w/w	WG	09903
4	Landgold Pirimicarb 50	Landgold	50% w/w	WG	09018
5	Phantom	Bayer	50% w/w	WG	04519
6	Standon Pirimicarb 50	Standon	50% w/w	WG	08878

Uses

Aphids in *celeriac (off-label)*, *chicory (off-label)*, *endives (off-label)*, *fennel (off-label)*, *honesty (off-label)*, *horseradish (off-label)*, *kohlrabi (off-label)*, *marrows (off-label)*, *parsley root (off-label)*, *parsley (off-label)*, *plums (off-label)*, *protected courgettes (off-label)*, *protected gherkins (off-label)*, *radishes (off-label)*, *red beet (off-label)*, *salad brassicas, spinach beet (off-label)*, *spinach (off-label)*, *sweetcorn (off-label)* [1]. Aphids in *apples, barley, broad beans, broccoli, brussels sprouts, cabbages, calabrese, cauliflowers, durum wheat, field beans, oats, pears, peas, potatoes, rye, strawberries, sugar beet, swedes, triticale, turnips, wheat* [1-6]. Aphids in *cherries* [1-3, 5]. Aphids in *blackcurrants, carrots, celery, chinese cabbage, collards, dwarf beans, gooseberries, grassland, maize, oilseed rape, parsnips, raspberries, redcurrants, runner beans, sweetcorn* [1-3, 5, 6]. Aphids in *kale* [1, 2, 4-6]. Aphids in *cucumbers, forest nurseries, ornamentals, outdoor lettuce, peppers, protected carnations, protected chrysanthemums, protected cinerarias, protected cyclamen, protected lettuce, protected roses, tomatoes* [1, 3, 5]. Aphids in *lettuce, outdoor* [3]. Blackfly in *cherries* [6]. Cabbage aphid in *salad brassicas (off-label)* [1].

Efficacy

- Chemical has contact, fumigant and translaminar activity
- Best results achieved under warm, calm conditions when plants not wilting and spray does not dry too rapidly. Little vapour activity at temperatures below 15°C
- Apply as soon as aphids seen or warning issued and repeat as necessary
- Addition of non-ionic wetter recommended for use on brassicas
- Chemical has little effect on bees, ladybirds and other insects and is suitable for use in integrated control programmes on apples and pears
- On cucumbers and tomatoes a root drench is preferable to spraying when using predators in an integrated control programme
- Where aphids resistant to pirimicarb occur control is unlikely to be satisfactory

Crop safety/Restrictions

- Maximum number of treatments normally 2 per crop but not specified in some cases
- When treating ornamentals check safety by treating a small number of plants first

Special precautions/Environmental safety

- This product contains an anticholinesterase carbamate compound. Do not use if under medical advice not to work with such compounds
- Harmful if swallowed
- Harmful to livestock
- Dangerous to fish or other aquatic life. Do not contaminate surface waters or ditches with chemical or used container
- Spray equipment must only be used where the operator's normal working position is within a closed cab on a tractor or self-propelled sprayer when making air-assisted applications to apples or other top fruit

Personal protective equipment/Label precautions

- A, C, D, H, J, M [1-6]
- M03 [1-6]; M02 [1-3, 5, 6]; R03c, U05a, U08 [1-6]; U20a [2-6]; U20b [1]; C03, E01, E13b, E30a, E32a, E34 [1-6]; E06b [1-6] (7 d)

Withholding period

- Keep all livestock out of treated areas for at least 7 d. Bury or remove spillages

Latest application/Harvest interval

- HI oilseed rape, cereals, maize, sweetcorn, lettuce under glass 14 d (sweetcorn off-label 3 d); grassland 7 d; cucumbers, tomatoes and peppers under glass 2 d; protected courgettes and gherkins 24 h; other edible crops 3 d; flowers and ornamentals zero

Approval

- Approved for aerial application on cereals [1, 2, 5]. See notes in Section 1
- Off-label approval unlimited for use on honesty (OLA 0584/93)[1]; unlimited for use on a range of outdoor and protected vegetables - see OLA notices for details (OLA 1626/95)[1], (OLA 1634/95)[6694]; unlimited for use on protected courgettes, gherkins (OLA 1302/96)[1]; to Sep 2001 for use on plums (OLA 2178/96)[1]; to Feb 2001 for use on red beet, celeriac, fennel, kohlrabi (OLA 0328/96)[1]; unlimited for use on protected courgettes, gherkins (OLA 1203/

96)[6694]; unlimited for use on parsley and endives (OLA 1303/96)[1]; unlimited for use on chicory (OLA 1078/98)[1]; unstipulated for use on salad brassicas (see OLA notice for list) (OLA 1461/00)[1]
* Accepted by BLRA for use on malting barley

Maximum Residue Level (mg residue/kg food)
* brassicas 1.0; peas, oilseed rape 0.2; barley, potatoes 0.05

---

417   pirimiphos-methyl

A contact, fumigant and translaminar organophosphorus insecticide

Products

1 Actellic 2% Dust	Zeneca	2% w/w	DP	06931
2 Actellic D	Zeneca	250 g/l	EC	06930
3 Actellic Smoke Generator No 20	Zeneca	20 g a.i.	FU	06627
4 Fumite Pirimiphos Methyl Smoke	Hortichem	22.5% w/w	FU	00941

Uses

Ants in *aubergines, protected cucumbers, protected ornamentals, protected peppers, protected tomatoes* [4]. Aphids in *aubergines, protected cucumbers, protected ornamentals, protected peppers, protected tomatoes* [4]. Capsids in *aubergines, protected cucumbers, protected ornamentals, protected peppers, protected tomatoes* [4]. Earwigs in *aubergines, protected cucumbers, protected ornamentals, protected peppers, protected tomatoes* [4]. Flour beetles in *stored cereals* [1, 2]. Flour moths in *stored cereals* [1, 2]. Grain beetles in *stored cereals* [1, 2]. Grain storage mites in *stored cereals* [1, 2]. Grain storage pests in *grain stores* [2, 3]. Grain weevils in *stored cereals* [1, 2]. Leaf miners in *aubergines, protected cucumbers, protected ornamentals, protected peppers, protected tomatoes* [4]. Red spider mites in *aubergines, protected cucumbers, protected ornamentals, protected peppers, protected tomatoes* [4]. Sawflies in *aubergines, protected cucumbers, protected ornamentals, protected peppers, protected tomatoes* [4]. Thrips in *aubergines, protected cucumbers, protected ornamentals, protected peppers, protected tomatoes* [4]. Warehouse moth in *stored cereals* [1, 2]. Whitefly in *aubergines, protected cucumbers, protected ornamentals, protected peppers, protected tomatoes* [4].

Efficacy
* Chemical acts rapidly and has short persistence in plants but persists for long periods on inert surfaces
* Best results for protection of stored grain achieved by cleaning store thoroughly before use and employing a combination of pre-harvest and grain/seed treatments
* Disinfect empty grain stores by spraying surfaces and/or fumigation and treat grain by full or surface admixture. Treat well before harvest in late spring or early summer and repeat 6 wk later or just before harvest if heavily infested. See label for details of treatment and suitable application machinery
* Treatment volumes on structural surfaces should be adjusted according to surface porosity. See label for guidelines
* Treat inaccessible areas with smoke generating product used in conjunction with spray treatment of remainder of store
* Best results for admixture treatment obtained when grain stored at 15% moisture or less. Dry and cool moist grain coming into store but then treat as soon as possible, ideally as it is loaded
* Surface admixture can be highly effective on localised surface infestations but should not be relied on for long-term control unless application can be made to the full depth of the infestation
* For control of whitefly and other glasshouse pests apply as smoke or by thermal fogging. See label for details of techniques and suitable machines [4]
* Where insect pests resistant to pirimiphos-methyl occur control is unlikely to be satisfactory

Crop safety/Restrictions
- Maximum number of treatments 2 per grain store; 1 per batch of stored grain
- Grain treated by admixture as specified may be consumed by humans and livestock
- Do not mix with grain store disinfectants because of risk of chemical interaction [2]
- Do not fumigate glasshouses in bright sunshine or when foliage is wet or roots dry
- Do not fumigate young seedlings or plants being hardened off
- Do not fog open flowers of ornamentals without first consulting firm
- Do not fog mushrooms when wet as slight spotting may occur

Special precautions/Environmental safety
- This product contains an organophosphorus anticholinesterase compound. Do not use if under medical advice not to work with such compounds
- Irritating to eyes [2] and respiratory system [3, 4]
- Highly flammable [3]; flammable [2, 4]
- Ventilate fumigated or fogged spaces thoroughly before re-entry
- Unprotected persons must be kept out of fumigated areas within 3 h of ignition and for 4 h after treatment
- Do not harvest for human or animal consumption for at least 5 d after last application
- Extremely dangerous to bees. Do not apply to crops in flower or to those in which bees are actively foraging. Do not apply when flowering weeds are present [4]
- Harmful (extremely dangerous [4]) to fish or other aquatic life. Do not contaminate surface waters or ditches with chemical or used container [1-3]
- Wildlife must be excluded from buildings during treatment [2]
- Do not apply treated seed from the air [1, 2]

Personal protective equipment/Label precautions
- D [2, 3]; H [1-3]; C [2]; A [1, 2]
- M01 [1-4]; M05 [1-3]; R04c [3, 4]; R04a [2-4]; R07d [2, 4]; R07c [3]; U20a [4]; U19 [1-4]; U05a [1, 2, 4]; U08 [1, 4]; U04a, U20b [1-3]; U10 [1, 3]; U09a, U16, U17, U24 [2]; C02 [4] (5 d); C03 [2, 4]; C09 [2]; E12b, E13a [4]; E02 [3, 4] (4 h); E01 [2-4]; E30a [1-4]; E34 [1, 4]; E13c [1-3]; E26, E27, E31c [2]; S05, S07 [1, 2]

Withholding period
- Keep livestock out of treated areas/Keep livestock out of treated areas for at least xx weeks and until foliage of any poisonous weeds such as ragwort has died and become unpalatable [2]

Latest application/Harvest interval
- HI zero

Approval
- UK approvals for use of pirimiphos-methyl on tomatoes, aubergines, peppers likely to be revoked in 2001 as a result of implementation of the MRL Directives
- Accepted by BLRA for use in stores for malting barley

Maximum Residue Level (mg residue/kg food)
- cereals 5; mandarins, kiwi fruit, Brussels sprouts, mushrooms 2; citrus fruits (except mandarins), carrots, broccoli, cauliflower 1; other crops 0.05

---

418   prochloraz

A broad-spectrum protectant and eradicant conazole fungicide

See also carbendazim + prochloraz
cyproconazole + prochloraz
fenpropidin + prochloraz
fenpropimorph + prochloraz
fluquinconazole + prochloraz

Products

1	Alpha Mirage 40 EC	Makhteshim	400 g/l	EC	06770
2	Barclay Eyetak 450	Barclay	450 g/l	EC	09484
3	Prelude 20LF	Agrichem	200 g/l	LS	04371
4	Scotts Octave	Scotts	46% w/w	WP	09275
5	Sporgon 50WP	Sylvan	46% w/w	WP	08802
6	Sportak 45 EW	Aventis	450 g/l	EW	07996

Uses

Alternaria in *oilseed rape* [2, 6]. Botrytis in *ornamentals* [4]. Cobweb in *mushrooms* [5]. Dry bubble in *mushrooms* [5]. Eyespot in *spring wheat* [1]. Eyespot in *winter barley, winter wheat* [1, 2, 6]. Eyespot in *winter rye* [2, 6]. Fungus diseases in *container-grown stock, hardy ornamental nursery stock* [4]. Fusarium bulb rot in *flower bulbs (off-label)* [6]. Glume blotch in *winter wheat* [2, 6]. Grey mould in *oilseed rape* [6]. Leaf spot in *winter rye, winter wheat* [2, 6]. Light leaf spot in *spring oilseed rape, winter oilseed rape* [1]. Light leaf spot in *oilseed rape* [2, 6]. Net blotch in *spring barley, winter barley* [2, 6]. Penicillium rot in *flower bulbs (off-label)* [6]. Phoma in *spring oilseed rape, winter oilseed rape* [1]. Phoma in *oilseed rape* [2, 6]. Powdery mildew in *spring barley, spring wheat (protection), winter barley, winter rye* [2, 6]. Rhynchosporium in *spring barley, winter barley, winter rye* [2, 6]. Ring spot in *herbs (off-label), lettuce (off-label)* [4]. Sclerotinia stem rot in *oilseed rape* [2, 6]. Seed-borne diseases in *flax (seed treatment), linseed (seed treatment)* [3]. Wet bubble in *mushrooms* [5]. White leaf spot in *oilseed rape* [6].

Efficacy

- Spray cereals at first signs of disease. Protection of winter crops through season usually requires at least 2 treatments. See label for details of rates and timing. Treatment active against strains of eyespot resistant to benzimidazole fungicides
- Tank mixes with other fungicides recommended to improve control of rusts in wheat and barley. See label for details
- A period of at least 3 h without rain should follow spraying
- Can be used through most seed treatment machines if good even seed coverage is obtained. Check drill calibration before drilling treated seed [3]
- Apply as drench against soil diseases, as a spray against aerial diseases, as a dip for cuttings, or as a drench at propagation. Spray applications may be repeated at 10-14 d intervals. Under mist propagation use 7 d intervals [4]
- Apply to mushrooms as casing treatment or spray between flushes. Timing determined by anticipated disease occurrence [5]

Crop safety/Restrictions

- Maximum number of treatments 1 per batch for flax, linseed [3]; 2 at 120g or 3 at 60 g/100 sq m for mushrooms [5]; varies with dose and crop for other uses - see labels for details
- Do not treat linseed varieties Linda, Bolas, Karen, Laura, Mikael, Norlin, Moonraker, Abbey [3]

Special precautions/Environmental safety

- Harmful in contact with skin [2]
- Irritating to eyes
- Irritating to skin [1, 2, 4, 5]
- May cause sensitization by skin contact [1]
- Flammable [1]
- Dangerous to fish or other aquatic life. Do not contaminate surface waters or ditches with chemical or used container
- Do not use treated seed as food or feed [3]
- Treated seed harmful to game and wildlife [3]
- Product available in refillable container (see label for instructions) [3]

Personal protective equipment/Label precautions

- A, C [1-6]; H, J [5]
- M03 [2]; R04a [1-5]; R04e, R07d [1]; R04b [1, 2, 4, 5]; R03a [2]; U07, U11 [3]; U08 [1, 3]; U05a [1-6]; U20b [1-5]; U20a [6]; U09a [4, 5]; C03 [1, 2, 4, 5]; E03, E33, E36 [3]; E01, E13b, E30a [1-6]; E26 [1-3, 6]; E34 [2, 3, 5, 6]; E31b [1, 2, 4-6]; S01, S02, S03, S04b, S05, S06, S07 [3]

Latest application/Harvest interval

- Milky ripe stage (GS 77) for cereals; before drilling flax or linseed [3]
- HI 6 wk for oilseed rape and cereals; 2 d for mushrooms [5]

Approval

- Off-label approval to Mar 2002 for use on leafy herbs and lettuce. See notice for list of species (OLA 2002/99)[4]
- Accepted by BLRA for use on malting barley

419   prochloraz + propiconazole

A broad spectrum fungicide mixture for wheat and barley

Products

Bumper P	Makhteshim	400:90 g/l	EC	08548

Uses

Phoma leaf spot in *winter oilseed rape*. Rhynchosporium in *winter barley*. Septoria in *winter wheat*.

Efficacy
- Best results obtained from treatment when disease is active but not well established
- Treat wheat normally from flag leaf ligule just visible stage (GS 39) but earlier if there is a high risk of Septoria
- Treat barley from the first node detectable stage (GS 31)
- If disease pressure persists a second application may be necessary

Crop safety/Restrictions
- Maximum number of treatments 2 per crop

Special precautions/Environmental safety
- Irritating to eyes
- Dangerous to fish or other aquatic life. Do not contaminate surface waters or ditches with chemical or used container

Personal protective equipment/Label precautions
- A, C
- R04a, U05a, U08, U20b, C03, E01, E13b, E26, E30a, E31b, E34

Latest application/Harvest interval
- Before grain watery ripe (GS 71) for cereals; before most seeds green/brown mottled for winter oilseed rape
- HI 6 wk

Approval
- Accepted by BLRA for use on malting barley

Maximum Residue Level (mg residue/kg food)
- see propiconazole entry

420   prochloraz + tebuconazole

A broad spectrum systemic fungicide mixture for cereals

Products

Agate	Bayer	267:133 g/l	EC	08826

Uses

Brown rust in *barley, rye, wheat*. Eyespot in *barley, rye, wheat*. Glume blotch in *wheat*. Late ear diseases in *wheat*. Net blotch in *barley*. Powdery mildew in *barley, rye, wheat*. Rhynchosporium in *barley, rye*. Septoria leaf spot in *wheat*. Yellow rust in *barley, rye, wheat*.

Efficacy
- Best results achieved from applications at an early stage of disease development before infection spreads to new crop growth
- Adequate protection of winter cereals will usually require a programme of at least two treatments
- Optimum application timing is normally when disease first seen but varies with main target disease - see label
- To minimise possibility of development of resistance repeated applications should not be made against the same pathogen. Alternation with fungicides with a different mode of action is a preferred strategy

Crop safety/Restrictions
- Maximum total dose equivalent to two full dose treatments
- Occasionally transient leaf speckling may occur after treating wheat. Yield responses should not be affected

Special precautions/Environmental safety
- Harmful if swallowed and in contact with skin
- Risk of serious damage to eyes
- May cause sensitization by skin contact
- Dangerous to fish or other aquatic life. Do not contaminate surface waters or ditches with chemical or used container

Personal protective equipment/Label precautions
- A, C, H
- M03, R03a, R03c, R04d, R04e, U05a, U09b, U20a, C03, E01, E13b, E26, E30a, E31c, E34

Latest application/Harvest interval
- Up to and including inflorescence fully emerged (GS 59)
- HI 6 wk

Approval
- Accepted by BLRA for use on malting barley

---

421   prometryn

A contact and residual triazine herbicide for various field crops

Products

1	Alpha Prometryne 50 WP	Makhteshim	50% w/w	WP	04871
2	Gesagard	Novartis	50% w/w	WP	08410

Uses

Annual dicotyledons in *peas, transplanted leeks* [1]. Annual dicotyledons in *carrots, celery, early potatoes, parsley* [1, 2]. Annual dicotyledons in *celeriac* (off-label), *combining peas, drilled leeks* (off-label), *garlic* (off-label), *kohlrabi* (off-label), *onions* (off-label), *outdoor herbs* (off-label), *protected herbs* (off-label), *swedes* (off-label), *turnips* (off-label), *vining peas* [2]. Annual grasses in *peas, transplanted leeks* [1]. Annual grasses in *carrots, celery, early potatoes, parsley* [1, 2]. Annual grasses in *celeriac* (off-label), *combining peas, drilled leeks* (off-label), *garlic* (off-label), *kohlrabi* (off-label), *onions* (off-label), *outdoor herbs* (off-label), *protected herbs* (off-label), *swedes* (off-label), *turnips* (off-label), *vining peas* [2].

Efficacy
- Best results achieved by application to young seedling weeds up to 5 cm high (cotyledon stage for knotgrass, mayweed and corn marigold) on fine, moist seedbed when rain falls afterwards. Do not use on very cloddy soils
- On organic soils only contact action effective and repeat application may be needed (certain crops only)

Crop safety/Restrictions
- Maximum number of treatments 1 per crop for early potatoes, peas and transplanted leeks; 1 per crop for transplanted celery
- Apply to peas pre-emergence up to 3 d before crop expected to emerge
- Spring sown vining and drying peas may be treated. Damage may occur with Vedette or Printana, especially if emerging under adverse conditions. Do not treat forage peas
- Do not use on peas on Very Light soils, Sands, gravelly or stony soils
- Apply to early potatoes up to 10% emergence
- Apply to carrots, celery, parsley or coriander post-emergence after 2-rough leaf stage or after transplants established
- Apply to transplanted leeks or celery after transplants established. Do not use on drilled leeks
- Excessive rain after treatment may check crop
- In the event of crop failure only plant recommended crops within 8 wk

Special precautions/Environmental safety
- Harmful to fish or other aquatic life. Do not contaminate surface waters or ditches with chemical or used container [1]

Personal protective equipment/Label precautions
- U20b [1, 2]; C02 [1, 2] (6 wk); E13c [1]; E30a, E32a [1, 2]; E01, E15, E29, E34 [2]

Latest application/Harvest interval
- Pre-emergence for peas; 6 wk before harvest and before 10% emergence for early potatoes; before 10% crop emergence for swedes and turnips; 3 wk after planting out for onions and garlic.
- HI 6 wk

Approval
- Off-label approval unlimited for use on drilled leeks (OLA 2196/97)[2]; unlimited for use on outdoor and protected herbs, swedes, turnips, kohlrabi (OLA 2081/97)[2]; unlimited for use on outdoor onions (OLA 2080/97)[2]; to Feb 2001 for use on outdoor celeriac (OLA 1382/97)[2]

---

422 propachlor

A pre-emergence chloroacetanilide herbicide for various horticultural crops

See also chloridazon + propachlor

Products

1	Alpha Propachlor 50 SC	Makhteshim	500 g/l	SC	04873
2	Portman Brasson	Portman	480 g/l	SC	08158
3	Portman Propachlor 50 FL	Portman	500 g/l	SC	06892
4	Ramrod 20 Granular	Monsanto	20% w/w	GR	01687
5	Ramrod Flowable	Monsanto	480 g/l	SC	01688
6	Tripart Sentinel	Tripart	500 g/l	SC	03250
7	Tripart Sentinel 2	Tripart	480 g/l	SC	05140

Uses

Annual dicotyledons in *broccoli, brussels sprouts, cabbages, cauliflowers, kale, leeks, onions, swedes, turnips* [1-7]. Annual dicotyledons in *oilseed rape* [1, 3, 6, 7]. Annual dicotyledons in *strawberries* [1, 6]. Annual dicotyledons in *mustard, ornamentals* [2]. Annual dicotyledons in *fodder rape* [2, 4, 5]. Annual dicotyledons in *calabrese* [2, 4, 5, 7]. Annual dicotyledons in *sage, spring oilseed rape* [2, 5]. Annual dicotyledons in *herbaceous perennials, woody ornamentals* [4]. Annual dicotyledons in *blackcurrants* (off-label), *blueberries* (off-label), *brown mustard, chinese cabbage* (off-label), *gooseberries* (off-label), *kohlrabi* (off-label), *lettuce* (off-label), *outdoor leaf herbs* (off-label), *redcurrants* (off-label), *strawberries* (off-label), *white mustard, whitecurrants* (off-label) [5]. Annual dicotyledons in *rape* [7]. Annual grasses in *broccoli, brussels sprouts, cabbages, cauliflowers, kale, leeks, onions, swedes, turnips* [1-7]. Annual grasses in *oilseed rape* [1, 3, 6, 7]. Annual grasses in *strawberries* [1, 6]. Annual grasses in *mustard* [2]. Annual grasses in *ornamentals* [2, 4]. Annual grasses in *fodder rape* [2, 4, 5]. Annual grasses in *calabrese* [2, 4, 5, 7]. Annual grasses in *sage, spring oilseed rape* [2, 5]. Annual grasses in *herbaceous perennials, woody ornamentals* [4]. Annual grasses in *blackcurrants* (off-label), *blueberries* (off-label), *brown mustard, chinese cabbage* (off-label), *kohlrabi* (off-label), *lettuce* (off-label), *outdoor leaf herbs* (off-label), *redcurrants* (off-label), *strawberries* (off-label), *white mustard, whitecurrants* (off-label) [5]. Annual grasses in *rape* [7].

Efficacy
- Controls germinating (not emerged) weeds for 6-8 wk
- Best results achieved by application to fine, firm, moist seedbed free of established weeds in spring, summer and early autumn
- Effective results with granules depend on rainfall soon after application [4]
- Use higher rate on soils with more than 10% organic matter
- Recommended as tank-mix with glyphosate or chlorthal-dimethyl on onions, leeks, swedes, turnips; with chlorpropham on leeks, onions [4]. See label for details

Crop safety/Restrictions
- Maximum number of treatments 1 or 2 per crop, varies with product and crop - see label for details
- Maximum total dose equivalent to one full dose treatment [1]
- Apply in brassicas from drilling to time seed chits or after 3-4 true leaf stage but before weed emergence, in swedes and turnips pre-emergence only
- Apply in transplanted brassicas within 48 h of planting in warm weather. Plants must be hardened off and special care needed with block sown or modular propagated plants
- Apply in onions and leeks pre-emergence or from post-crook to young plant stage
- Apply to newly planted strawberries soon after transplanting, to weed-free soil in established crops in early spring before new weeds emerge
- Granules may be applied to onion, leek and brassica nurseries and to most flower crops after bedding out and hardening off
- Do not use pre-emergence of drilled wallflower seed
- Do not use on crops under glass or polythene
- Do not use under extremely wet, dry or other adverse growth conditions
- In the event of crop failure only replant recommended crops in treated soil

Special precautions/Environmental safety
- Harmful in contact with skin [1, 3]
- Harmful if swallowed [1, 2, 5]
- Irritating to skin [4] and eyes [1-3, 5-7]
- Irritating to respiratory system [7]
- May cause sensitization by skin contact [2, 3, 5, 7]

Personal protective equipment/Label precautions
- A [1-7]; C [1-3, 5-7]
- M03 [2, 3, 5]; R04b [1-7]; R04a [1-3, 5-7]; R03c [1, 2, 5]; R04e [2, 3, 5, 7]; R03a [1, 3]; R04c [7]; U02, U05a [1-7]; U20a [2-4, 7]; U20b [1, 5, 6]; U04a, U08, U13, U19 [1-3, 5-7]; C03 [1-7]; E32a [4]; E01, E15, E30a [1-7]; E26 [1, 3, 5-7]; E31b, E34 [1-3, 5-7]

Approval
- Off-label approval unlimited for use on outdoor lettuce, lamb's lettuce, frise, radicchio, cress, scarole (all under covers) (OLA 0364/96)[5]; unlimited for use on outdoor leaf herbs (see OLA notice for details) (OLA 0652/95, 0957/95)[5]; to Feb 2001 for use on gooseberries, blueberries, other Ribes species (OLA 1047/96)[5]; unlimited for use on Chinese cabbage (OLA 1572/97)[5]; unlimited for use on strawberries (OLA 1520/98)[5]; unlimited for use on outdoor kohl rabi (OLA 2929/98)[5]

---

423   propamocarb hydrochloride

A translocated protectant carbamate fungicide

See also chlorothalonil + propamocarb hydrochloride
mancozeb + propamocarb hydrochloride

Products

1 Filex	Scotts	722 g/l	SL	07631
2 Proplant	Fargro	722 g/l	SL	08572

Uses

Botrytis in **herbs** *(off-label)*, **lettuce** *(off-label)* [1]. Damping off and foot rot in **inert substrate cucumbers** *(off-label)*, **inert substrate peppers** *(off-label)*, **inert substrate tomatoes** *(off-label)*, **nft peppers** *(off-label)*, **nft tomatoes** *(off-label)* [1, 2]. Downy mildew in **herbs** *(off-label)*, **lettuce** *(off-label)*, **radishes** *(off-label)*, **watercress** *(off-label)* [1]. Downy mildew in **brassicas** [1, 2]. Phytophthora in **watercress** *(off-label)* [1]. Phytophthora in **aubergines, bedding plants, brassicas, container-grown ornamentals, cucumbers, leeks, nursery stock, ornamentals, peppers, pot plants, tomatoes, tulips** [1, 2]. Phytophthora in **flower bulbs** [2]. Pythium in **watercress** *(off-label)* [1]. Pythium in **aubergines, bedding plants, brassicas, cucumbers, flower bulbs, leeks, nursery stock, onions, ornamentals, peppers, pot plants, rockwool aubergines, rockwool cucumbers, rockwool peppers, rockwool tomatoes, tomatoes, tulips** [1, 2]. White blister in **horseradish** *(off-label)* [1]. White tip in **leeks** *(off-label)* [1].

Efficacy
- Chemical is absorbed through roots and translocated throughout plant
- Incorporate in compost before use or drench moist compost or soil before sowing, pricking out, striking cuttings or potting up
- Drench treatment can be repeated at 3-6 wk intervals
- Concentrated solution is corrosive to all metals other than stainless steel
- May also be applied in trickle irrigation systems
- To prevent root rot in tulip bulbs apply as dip for 20 min

Crop safety/Restrictions
- Maximum number of treatments 4 per crop for cucumbers, tomatoes, peppers, aubergines; 1 per crop for listed brassicas; 1 compost incorporation and/or 1 drench treatment for leeks, onions, tulips
- When applied over established seedlings rinse off foliage with water and do not apply under hot, dry conditions
- Do not treat seedlings with overhead drench
- On plants of unknown tolerance test first on a small scale

Special precautions/Environmental safety
- Store away from seeds and fertilisers

Personal protective equipment/Label precautions
- A, B, C, H, K, M [1, 2]
- U08, U20b, E15, E29, E30a, E32a [1, 2]; E26, E34 [2]

Latest application/Harvest interval
- Before transplanting for brassicas
- HI 14 d for cucumbers, tomatoes, peppers, aubergines; 4 wk for calabrese, cauliflower, sprouting broccoli, Brussels sprouts, Chinese cabbage; 19 wk for leeks, onions

Approval
- Off-label approval to Jul 2003 for use on leeks (OLA 2447/98)[1]; unstipulated for use on a range of outdoor and protected salad crops and herbs. See notice for details (OLA 1971/99; OLA 1972/99)[1]; unstipulated for use on outdoor and protected radishes (OLA 2030/99)[1]; unstipulated for use on NFT tomatoes and peppers, and tomatoes, cucumbers, peppers on inert substrate (OLA 2032/99)[1]; to Jan 2002 for use on water cress during propagation (OLA 2034/99)[1]; unstipulated for use on outdoor and protected horseradish (OLA 2036/99)[1]; unstipulated for use on tomatoes, peppers, cucumbers on inert substrates, and on tomatoes, peppers grown by nutrient film technique (OLA 2527/00)[2]

424   propaquizafop

A phenoxy alkanoic acid foliar acting grass herbicide

Products

1 Barclay Rebel II	Barclay	100 g/l	EC	08897
2 Falcon	Novartis	100 g/l	EC	09384
3 Landgold PQF 100	Landgold	100 g/l	EC	08976
4 Standon Propaquizafop	Standon	100 g/l	EC	09120

Uses

Annual grasses in *potatoes* [1]. Annual grasses in *bulb onions, carrots, combining peas, fodder beet, linseed, oilseed rape, parsnips, spring field beans, sugar beet, swedes, turnips, winter field beans* [1-4]. Annual grasses in *ware potatoes* [2]. Annual grasses in *farm forestry* [2-4]. Annual grasses in *maincrop potatoes* [3, 4]. Perennial grasses in *mustard, potatoes* [1]. Perennial grasses in *bulb onions, carrots, combining peas, fodder beet, linseed, oilseed rape, parsnips, spring field beans, sugar beet, swedes, turnips, winter field beans* [1-4]. Perennial grasses in *ware potatoes* [2]. Perennial grasses in *farm forestry* [2-4]. Perennial grasses in *maincrop potatoes* [3, 4]. Volunteer cereals in *red beet (off-label)* [2].

Efficacy
- Apply to emerged weeds when they are growing actively with adequate soil moisture
- Activity is slower under cool conditions

- Broad-leaved weeds and any weeds germinating after treatment are not controlled
- Annual meadow grass up to 3 leaves checked at low doses and severely checked at highest dose
- Spray barley cover crops when risk of wind blow has passed and before there is serious competition with the crop
- Various tank mixtures and sequences recommended for broader spectrum weed control in oilseed rape, peas and sugar beet. See label for details
- Severe couch infestations may require a second application at reduced dose when regrowth has 3-4 leaves unfolded

Crop safety/Restrictions
- Maximum total dose 2 l/ha per crop (or per yr for forestry)
- See label for earliest crop growth stages for treatment
- See label for list of tolerant tree species
- Application in high temperatures and/or low soil moisture content may cause chlorotic spotting especially on combining peas and field beans
- Overlaps at the highest dose can cause damage from early applications to carrots and parsnips
- Do not treat seed potatoes
- Products contain surfactants. Tank mixing with adjuvants not required or recommended
- An interval of 4 wk must elapse before redrilling a failed treated crop. Only broad-leaved crops may be redrilled

Special precautions/Environmental safety
- Irritating to eyes and skin
- Harmful to fish or other aquatic life. Do not contaminate surface waters or ditches with chemical or used container
- LERAP Category B [1, 3]

Personal protective equipment/Label precautions
- A, C [1-4]
- M05, R04a, R04b [1-4]; U08 [1, 3, 4]; U02, U05a, U14, U15, U19, U20b [1-4]; U09a [2]; C03 [1-4]; E16a, E16b [1, 3]; E01, E13c, E22b, E26, E30a, E31b [1-4]

Latest application/Harvest interval
- Before crop flower buds visible for winter oilseed rape, linseed, field beans; before 8 fully expanded leaf stage for spring oilseed rape, mustard; before weeds are covered by the crop for potatoes, sugar beet, fodder beet; when flower buds visible for peas
- HI early potatoes, carrots, parsnips, bulb onions 4 wk; peas 7 wk; sugar beet, fodder beet, maincrop potatoes, swedes, turnips 8 wk; field beans 14 wk

Approval
- Off-label approval unstipulated for use on red beet (OLA 2607/99)[2]

---

425   propham

A pre-sowing carbamate herbicide, all approvals for which were revoked in 1997

426　propiconazole

A systemic, curative and protectant conazole fungicide

See also *carbendazim + propiconazole*
cyproconazole + propiconazole
fenbuconazole + propiconazole
fenpropidin + propiconazole
fenpropidin + propiconazole + tebuconazole
fenpropimorph + propiconazole
prochloraz + propiconazole

Products

1	Barclay Bolt	Barclay	250 g/l	EC	08341
2	Bumper 250 EC	Makhteshim	250 g/l	EC	09039
3	Landgold Propiconazole	Landgold	250 g/l	EC	06291
4	Mantis	Novartis	250 g/l	EC	08423
5	Standon Propiconazole	Standon	250 g/l	EC	07037
6	Tilt	Novartis	250 g/l	EC	08456

Uses

Alternaria in *oilseed rape* [4, 6]. Brown rust in *barley, wheat* [1, 3-6]. Brown rust in *rye* [1, 4-6]. Brown rust in *spring barley, spring wheat, winter barley, winter rye, winter wheat* [2]. Cladosporium leaf blotch in *leeks* [6]. Cladosporium leaf spot in *garlic (off-label)*, *onions (off-label)*, *shallots (off-label)* [6]. Crown rust in *grass for ensiling, grass seed crops* [1, 2, 4-6]. Crown rust in *spring oats, winter oats* [2]. Drechslera leaf spot in *grass for ensiling, grass seed crops* [1, 2, 4-6]. Eyespot in *winter wheat (low levels only)* [1, 4, 5]. Eyespot in *winter barley (low levels only)* [1, 5]. Eyespot in *winter barley (with carbendazim), winter wheat (with carbendazim)* [3]. Fungus diseases in *honesty (off-label)* [6]. Light leaf spot in *oilseed rape* [1, 4-6]. Light leaf spot in *spring oilseed rape (reduction), winter oilseed rape (reduction)* [2]. Mildew in *grass for ensiling, grass seed crops* [1, 2, 4-6]. Mildew in *sugar beet* [1, 4-6]. Net blotch in *barley* [1, 3-6]. Powdery mildew in *barley, wheat* [1, 3-6]. Powdery mildew in *oats, rye* [1, 4-6]. Powdery mildew in *spring barley, spring oats, spring wheat, winter barley, winter oats, winter rye, winter wheat* [2]. Ramularia leaf spots in *sugar beet* [1, 4-6]. Ramularia leaf spots in *sugar beet (reduction)* [2]. Rhynchosporium in *grass for ensiling, grass seed crops* [1, 2, 4-6]. Rhynchosporium in *barley* [1, 3-6]. Rhynchosporium in *rye* [1, 4-6]. Rhynchosporium in *spring barley, winter barley, winter rye* [2]. Rust in *sugar beet* [1, 2, 4-6]. Rust in *leeks (qualified minor use)* [2]. Rust in *leeks* [2, 6]. Septoria in *wheat* [1, 3-6]. Septoria in *rye* [1, 4-6]. Septoria in *spring wheat, winter rye, winter wheat* [2]. Snow rot in *winter barley* [4]. Sooty moulds in *spring wheat, winter wheat* [2]. Sooty moulds in *wheat* [6]. White rust in *chrysanthemums (off-label)* [6]. Yellow rust in *barley, wheat* [1, 3-6]. Yellow rust in *spring barley, spring wheat, winter barley, winter wheat* [2].

Efficacy

- Best results achieved by applying at early stage of disease. Recommended spray programmes vary with crop, disease, season, soil type and product. See label for details

Crop safety/Restrictions

- Maximum number of treatments 4 per crop for wheat (up to 3 in yr of harvest); 3 per crop for barley, oats, rye, triticale (up to 2 in yr of harvest), leeks; 2 per crop or yr for oilseed rape, sugar beet, grass seed crops, honesty; 1 per yr on grass for ensiling
- On oilseed rape do not apply during flowering. Apply first treatment before flowering, the second afterwards
- Grass seed crops must not be treated in yr of harvest
- A minimum interval of 14 d must elapse between treatments on leeks and 21 d on sugar beet
- Avoid spraying crops under stress, e.g. during cold weather or periods of frost

Special precautions/Environmental safety

- Irritating to eyes and skin
- Dangerous to fish or other aquatic life. Do not contaminate surface waters or ditches with chemical or used container

Personal protective equipment/Label precautions
- A, C [1-6]
- R04a, R04b, U20a [1, 3-6]; U02, U05a, U19 [1-6]; U09a [3-6]; U08 [2-6]; U04a, U09b [1]; C03 [1-6]; C02 [3-6] (35 d cereals/grass seed/leeks; 28 d rape/silage/sugar beet); E01, E13b, E30a [1-6]; E31c [3-6]; E34 [2-6]; E31b [1, 2]; E26, S06 [2]

Latest application/Harvest interval
- Before 4 pairs of true leaves for honesty.
- HI cereals, grass seed crops, leeks 35 d; oilseed rape, sugar beet, grass for ensiling 28d; leeks 35 d

Approval
- Approved for aerial application on wheat, barley, rye, oats [1, 4-6]. See notes in Section 1
- Off-label approval unlimited for use only on *confirmed outbreaks* of white rust on chrysanthemums (OLA 1388/97)[6]; unlimited for use on outdoor crops of honesty undersown in cereals (OLA 1703/97)[6]; unlimited for use on outdoor crops of onions, garlic, shallots (OLA 2279/97)[6]
- Accepted by BLRA for use on malting barley

Maximum Residue Level (mg residue/kg food)
- grapes 0.5; apricots, peaches 0.2; bananas, tea, hops, ruminant liver 0.1; citrus fruits, tree nuts, pome fruits, strawberries, blackberries, dewberries, loganberries, raspberries, bilberries, cranberries, currants, gooseberries, wild berries, miscellaneous fruits (except bananas), root and tuber vegetables, bulb vegetables, tomatoes, aubergines, sweet corn, brassica vegetables, leaf vegetables and herbs, legume vegetables, asparagus, cardoons, fennel, leeks, rhubarb, fungi, pulses, peanuts, poppy seed, sesame seed, sunflower seed, soya beans, mustard seed, cotton seed, potatoes, cereals, meat (except ruminant liver) 0.05; milk and dairy produce 0.01

---

427   propiconazole + tebuconazole

A broad spectrum systemic fungicide mixture for cereals

Products

1 Cogito	Novartis	250:250 g/l	EC	08397
2 Endeavour	Bayer	250:250 g/l	EC	07385

Uses

Brown rust in **barley, wheat**. Glume blotch in **wheat**. Net blotch in **barley**. Powdery mildew in **barley, wheat**. Rhynchosporium in **barley**. Septoria leaf spot in **wheat**. Yellow rust in **barley, wheat**.

Efficacy
- Best disease control and yield benefit obtained when applied at early stage of disease development before infection has spread to new growth
- To protect the flag leaf and ear from Septoria diseases apply from flag leaf emergence to ear fully emerged (GS 37-59)
- Applications once foliar symptoms of *Septoria tritici* are already present on upper leaves will be less effective

Crop safety/Restrictions
- Maximum total dose 1 l/ha per crop
- Occasional slight temporary leaf speckling may occur on wheat but this has not been shown to reduce yield response or disease control
- Do not treat durum wheat

Special precautions/Environmental safety
- Irritating to eyes and skin. May cause sensitization by skin contact
- Dangerous to fish or other aquatic life. Do not contaminate surface waters or ditches with chemical or used container
- LERAP Category B

Personal protective equipment/Label precautions
- A, C, H
- R04a, R04b, R04e, U05a, U09b, U20a, C03, E01, E13b, E16a, E16b, E30a, E31c

Latest application/Harvest interval
- Before watery ripe stage (GS 71)

Approval
- Accepted by BLRA for use on malting barley

Maximum Residue Level (mg residue/kg food)
- see propiconazole entry

---

428 propyzamide

A residual amide herbicide for use in a wide range of crops

See also clopyralid + propyzamide

Products

1	Barclay Piza 400 FL	Barclay	400 g/l	SC	09633
2	Greencrop Saffron FL	Greencrop	400 g/l	SC	10244
3	Headland Judo	Headland	400 g/l	SC	08339
4	Kerb 50 W	SumiAgro	50% w/w	WP	10166
5	Kerb 50 W	Interfarm	50% w/w	WP	10200
6	Kerb 80 EDF	Interfarm	80% w/w	WG	10201
7	Kerb Flo	Interfarm	400 g/l	SC	10160
8	Kerb Flo	SumiAgro	400 g/l	SC	10157
9	Kerb Granules	PBI	4% w/w	GR	08917
10	Kerb Pro Flo	SumiAgro	400 g/l	SC	10158
11	Kerb Pro Granules	PBI	4% w/w	GR	09600
12	Landgold Propyzamide 50	Landgold	50% w/w	WP	09753
13	Menace 80 EDF	Interfarm	80% w/w	WG	10165
14	Mithras 80 EDF	Interfarm	80% w/w	WG	10164
15	Precis	PBI	400 g/l	SC	08678
16	Precis	SumiAgro	400 g/l	SC	10159
17	Redeem Flo	Interfarm	400 g/l	SC	10236
18	Standon Propyzamide 400 SC	Standon	400 g/l	SC	10255
19	Standon Propyzamide 50	Standon	50% w/w	WP	09054

Uses

Annual dicotyledons in **apples, blackberries, blackcurrants, clover seed crops, gooseberries, loganberries, lucerne, pears, plums, redcurrants, sugar beet seed crops, winter field beans, winter oilseed rape** [1-8, 12-19]. Annual dicotyledons in **raspberries** *(England only)*, **strawberries** [1-8, 13-18]. Annual dicotyledons in **fodder rape seed crops, kale seed crops, turnip seed crops** [1, 2, 4, 5, 7, 8, 12, 15, 16, 18, 19]. Annual dicotyledons in **lettuce, rhubarb** [1, 2, 4, 5, 7, 8, 15, 16, 18, 19]. Annual dicotyledons in **woody ornamentals** [1, 3-6, 9-11, 13, 14, 17]. Annual dicotyledons in **raspberries** [12, 19]. Annual dicotyledons in **ornamentals** [19]. Annual dicotyledons in **brassica seed crops, outdoor lettuce, rhubarb** *(outdoor)* [3, 6, 13, 14, 17]. Annual dicotyledons in **farm woodland** [4, 5]. Annual dicotyledons in **roses** [4, 5, 19]. Annual dicotyledons in **forestry, trees and shrubs** [4, 5, 9-11]. Annual grasses in **apples, blackberries, blackcurrants, clover seed crops, gooseberries, loganberries, lucerne, pears, plums, redcurrants, sugar beet seed crops, winter field beans, winter oilseed rape** [1-8, 12-19]. Annual grasses in **raspberries** *(England only)*, **strawberries** [1-8, 13-18]. Annual grasses in **fodder rape seed crops, kale seed crops, turnip seed crops** [1, 2, 4, 5, 7, 8, 12, 15, 16, 18, 19]. Annual grasses in **lettuce, rhubarb** [1, 2, 4, 5, 7, 8, 15, 16, 18, 19]. Annual grasses in **woody ornamentals** [1, 3-6, 9-11, 13, 14, 17]. Annual grasses in **forestry** [1, 4, 5, 9-11]. Annual grasses in **raspberries** [12, 19]. Annual grasses in **ornamentals** [19]. Annual grasses in **brassica seed crops, outdoor lettuce, rhubarb** *(outdoor)* [3, 6, 13, 14, 17]. Annual grasses in **farm woodland** [4, 5]. Annual grasses in **roses** [4, 5, 19]. Annual grasses in **trees and shrubs** [4, 5, 9-11]. Horsetails in **woody ornamentals** [10]. Horsetails in **forestry** [3, 6, 10, 13, 14, 17]. Perennial grasses in **apples, blackberries, blackcurrants, gooseberries, loganberries, pears, plums, redcurrants** [1-8, 12-19]. Perennial grasses in **raspberries** *(England only)* [1-8, 13-18]. Perennial grasses in **clover seed crops, fodder rape seed crops, kale seed crops, lucerne, sugar beet seed crops, turnip seed crops, winter field beans, winter oilseed rape** [1, 2, 4, 5, 7, 8, 12, 15, 16, 18, 19]. Perennial grasses in **strawberries** [1, 2, 4, 5, 7, 8, 15, 16, 18]. Perennial grasses in **lettuce, rhubarb** [1, 2, 4, 5, 7, 8, 15, 16, 18, 19]. Perennial grasses in **forestry, woody ornamentals** [1, 3-6, 9-11, 13, 14, 17]. Perennial grasses in **raspberries** [12, 19]. Perennial grasses in **ornamentals** [19]. Perennial grasses in **rhubarb** *(outdoor)* [3, 6, 13, 14, 17].

Perennial grasses in **farm woodland** [4, 5]. Perennial grasses in **roses** [4, 5, 19]. Perennial grasses in **trees and shrubs** [4, 5, 9-11]. Sedges in **woody ornamentals** [10]. Sedges in **forestry** [3, 6, 10, 13, 14, 17]. Volunteer cereals in **sugar beet seed crops, winter field beans, winter oilseed rape** [3, 6, 13, 14, 17]. Wild oats in **sugar beet seed crops, winter field beans, winter oilseed rape** [3, 6, 13, 14, 17].

Efficacy
- Active via root uptake. Weeds controlled from germination to young seedling stage, some species (including many grasses) also when established
- Best results achieved by winter application to fine, firm, moist soil. Rain is required after application if soil dry
- Do not use on soils with more than 10% organic matter except in forestry
- Excessive organic debris or ploughed-up turf may reduce efficacy
- For heavy couch infestations a repeat application may be needed in following winter
- In lettuce lightly incorporate in top 25 mm pre-drilling or irrigate on dry soil

Crop safety/Restrictions
- Maximum number of treatments 1 per crop or yr
- May be applied to listed edible crops only between 1 Oct and the date specified as the latest time of application
- Apply as soon as possible after 3-true leaf stage of oilseed rape (GS 1,3) and seed brassicas, after 4-leaf stage of sugar beet for seed, within 7 d after sowing but before emergence for field beans, after perennial crops established for at least 1 season, strawberries after 1 yr
- Only apply to strawberries on heavy soils, to field beans on medium and heavy soils, to established lucerne not less than 7 d after last cut. Use reduced dose on matted row strawberries
- Do not treat protected crops
- See label for lists of ornamental and forest species which may be treated
- Lettuce may be sown or planted immediately after treatment, with other crops period varies from 5 to 40 wk. See label for details
- Do not apply to the same land less than 9 mth after an earlier application

Special precautions/Environmental safety
- Do not harvest crops for human or animal consumption for at least 6 wk after last application

Personal protective equipment/Label precautions
- U20b [3, 17, 19]; U20c [1, 2, 4-8, 10, 12-16, 18]; U20a [9, 11]; C02 [2-9, 11-19] (6 wk); E32a [1, 3-17, 19]; E15, E30a [1-19]; E34 [3, 9, 11, 15, 16]; E01 [1, 9, 11, 12, 15, 16, 19]; E26 [2, 6-8, 13, 14, 17, 18]; E31b [2, 18]

Latest application/Harvest interval
- Labels vary but normally before 31 Dec for rhubarb, lucerne, strawberries and winter field beans; before 31 Jan for other crops
- HI normally 6 wk for edible crops but labels vary in detail

Approval
- UK approvals for use of propyzamide on strawberries, currants, gooseberries likely to be revoked in 2001 as a result of implementation of the MRL Directives
- May be applied through CDA equipment [7], through CDA equipment in forestry [4, 5]. See labels for details
- Accepted by BLRA for use on hops

Maximum Residue Level (mg residue/kg food)
- linseed, tea, meat 0.05; citrus fruit, tree nuts, pome fruit, stone fruit, berries and small fruit (except strawberries, currants, gooseberries), miscellaneous fruit, root and tuber vegetables, bulb vegetables, fruiting vegetables, brassica vegetables (except head cabbage), spinach, watercress, witloof, peas, stem vegetables, fungi, pulses, poppy seed, sesame seed, sunflower seed, soya bean, mustard seed, potatoes, cereals, eggs 0.02; milk and dairy produce 0.01

429   prosulfuron

A contact and residual sulfonyl urea herbicide available only in mixtures

See also bromoxynil + prosulfuron

430   pymetrozine

A novel azomethine insecticide

Products

1 Chess	Novartis	25% w/w	WP	09817
2 Plenum	Novartis	25% w/w	WB	09816

Uses

Aphids in **ornamentals, protected cucumbers, protected ornamentals** [1]. Aphids in **potatoes** [2].

Efficacy

- Pymetrozine acts by preventing feeding leading to death by starvation in 1-4 d. There is no immediate knockdown
- Aphids controlled include those resistant to organophosphorus and carbamate insecticides
- Best results achieved by starting spraying as soon as aphids seen in crop and repeating as necessary.
- To limit spread of persistent viruses such as potato leaf roll virus, apply from 90% crop emergence [2]

Crop safety/Restrictions

- Maximum total dose equivalent to two full dose treatments on potatoes [2], three full dose treatments on protected cucumbers, four full dose treatments on ornamentals [1]
- Consult processors before use on potatoes grown for processing [2]
- In cucumbers and ornamentals minimum spray interval is 1 wk [1]
- Follow pack instructions for handling water soluble bags [2]
- Check susceptibility of ornamentals on a small number of plants before large scale treatment. See label for details of species tested without crop damage [1]
- Product may leave visible spray deposits on leaves of some ornamentals [1]

Special precautions/Environmental safety

- Do not harvest edible crops for human or animal consumption for at least 3 d after last application [1] or 55 d after last application [2]
- High risk to bees. Do not apply to crops in flower or to those in which bees are actively foraging. Do not apply when flowering weeds are present
- Harmful to fish or other aquatic life. Do not contaminate surface waters or ditches with chemical or used container
- LERAP Category B [1]
- Avoid spraying within 6 m of field boundaries to reduce effects on non-target insects and other arthropods
- Limited evidence of safety to *Phytoseiulus, Amblyseius, Typhlodromus* and *Encarsia* [1]

Personal protective equipment/Label precautions

- A [1, 2]
- U22 [2]; U05a, U20c, C03 [1, 2]; C02 [1, 2] (55 d); E29 [2]; E01, E12a, E13c, E22b, E30a, E31a [1, 2]; E16a, E16b [1]

Latest application/Harvest interval

- HI protected cucmbers 3 d, potatoes 55 d

431   pyrethrins

A non-persistent, contact acting insecticide extracted from Pyrethrum

Products

1 Alfadex	Novartis A H	0.75 g/l	AL	08591
2 Dairy Fly Spray	B H & B	0.75 g/l	AL	H5579
3 Killgerm ULV 400	Killgerm	30 g/l	UL	H4838

Uses

Flies in *dairies, farm buildings, livestock houses, poultry houses*.

Efficacy

- For fly control close doors and windows and spray or apply fog as appropriate
- For best results outdoors spray during early morning or late afternoon and evening when conditions are still

Crop safety/Restrictions

- Do not allow spray to contact open food products or food preparing equipment or utensils
- Remove exposed milk and collect eggs before application
- Do not treat plants
- Do not use space sprays containing pyrethrins or pyrethroid more than once per week in intensive or controlled environment animal houses in order to avoid development of resistance. If necessary, use a different control method or product [3]

Special precautions/Environmental safety

- Harmful if swallowed [2]
- Irritating to eyes, skin [2] and respiratory system [1, 3]
- Do not apply directly to livestock
- Harmful to fish or other aquatic life. Do not contaminate surface waters or ditches with chemical or used container
- Exclude all persons and animals during treatment [2]

Personal protective equipment/Label precautions

- C, D, H [1, 3]; A [1-3]; B, E [2]
- R04c [1, 3]; R04a, R04b [1-3]; R03c [2]; U20a [1, 3]; U05a, U19 [1-3]; U09a, U20b [2]; C03 [1-3]; C04, C06, C07, C08, C09, C10 [2]; E31a [1, 3]; E01, E05, E13c, E30a [1-3]; E04, E32a [2]

Approval

- Products formulated for ULV application [2, 3]. May be applied through fogging machine or sprayer [2]. See label for details

432   pyrethrins + resmethrin

A contact insecticide for many glasshouse and horticultural crops

Products

Pynosect 30 Water Miscible	Mitchell Cotts	1.4:9.1% w/w	EC	01653

Uses

Aphids in *ornamental plant production* (protected crops), *protected cucumbers, protected tomatoes*. Phorid flies in *mushrooms* (off-label). Sciarid flies in *mushrooms* (off-label). Whitefly in *ornamental plant production* (protected crops), *protected cucumbers, protected tomatoes*.

Efficacy

- Apply when pest first seen and repeat as necessary, for whitefly every 3-4 d
- Apply as high volume spray. See label for details of suitable equipment
- Where glasshouse whitefly resistant to pyrethrins occur control is unlikely to be satisfactory

Crop safety/Restrictions

- Do not spray when temperature above 24°C
- Care should be taken in spraying `soft' cucumber plants grown during winter
- Do not treat crops suffering from drought or stress

- On ornamentals a small test spraying is recommended before treating whole crop
- Damage may occur to open flowers of ornamentals
- Do not apply to ferns

Special precautions/Environmental safety
- Harmful. May cause lung damage if swallowed
- Extremely dangerous to bees. Do not apply to crops in flower or to those in which bees are actively foraging. Do not apply when flowering weeds are present
- Extremely dangerous to fish or other aquatic life. Do not contaminate surface waters or ditches with chemical or used container

Personal protective equipment/Label precautions
- R03, R04g, U08, U19, U20b, E12b, E13a, E30a, E31b

Latest application/Harvest interval
- HI zero

Approval
- Off-label approval unlimited for use on mushrooms for control of sciarid and phorid flies (OLA 0505/95)

433   pyridate

A contact pyridazine herbicide for cereals, maize and brassicas

Products

Lentagran WP	Novartis	45% w/w	WB	08478

Uses

Black nightshade in **brussels sprouts, cabbages, maize, onions**. Cleavers in **brussels sprouts, cabbages, maize, onions**. Fat-hen in **brussels sprouts, cabbages, maize, onions**.

Efficacy
- Best results achieved by application to actively growing weeds at 6-8 leaf stage when temperatures are above 8°C before crop foliage forms canopy

Crop safety/Restrictions
- Maximum number of treatments 1 per crop
- Do not apply in mixture with or within 14 d of any other product which may result in dewaxing of crop foliage
- Apply to maize and sweetcorn after first-leaf stage. Do not use on cv. Meritos, Sunrise or Tainon 236
- Apply to cabbages and Brussels sprouts after 4 fully expanded leaf stage.   Allow 2 wk after transplanting before treating
- Do not use on crops suffering stress from frost, drought, disease or pest attack

Special precautions/Environmental safety
- Irritating to eyes and skin
- May cause sensitization by skin contact
- Harmful to fish or other aquatic life. Do not contaminate surface waters or ditches with chemical or used container

Personal protective equipment/Label precautions
- A, C
- R04a, R04b, R04e, U05a, U08, U20b, U22, C03, E01, E13c, E26, E30a

Latest application/Harvest interval
- Before 7 leaf stage for maize; before 5 true leaves for onions
- HI 6 wk for Brussels sprouts, cabbages

Approval
- Accepted by BLRA for use on malting barley

434　pyrifenox

A systemic pyridine fungicide for use on apples, pears and other fruit crops

Products

Dorado	Zeneca	200 g/l	EC	06657

Uses

Cobweb in *mushrooms* (off-label). Leaf spots in *blackcurrants, gooseberries*. Powdery mildew in *apples, blackcurrants, gooseberries, hardy ornamental nursery stock, ornamentals, pears*. Scab in *apples, pears*.

Efficacy

- Apply to apples and pears from bud-burst at 7-14 d intervals, depending on dose applied, until danger of scab infection ceases or extension growth completed
- Spray programmes on apples and pears which incorporate other suitable fungicides (such as Captan or Nimrod) will help to avoid development of tolerant strains
- Treat from early grape stage (blackcurrants and gooseberries) or just before blossom (strawberries) or at first signs of disease and repeat at 14 d intervals
- Treat ornamentals at first sign of disease and repeat at 7-14 d intervals

Crop safety/Restrictions

- Maximum number of treatments 5 per crop (blackcurrants and gooseberries)
- Maximum total dose equivalent to 8 full dose treatments
- Do not leave spray liquid in sprayer for long period

Special precautions/Environmental safety

- Harmful if swallowed. Irritating to eyes and skin
- Flammable
- Harmful to fish or other aquatic life. Do not contaminate surface waters or ditches with chemical or used container

Personal protective equipment/Label precautions

- A, C, H
- M03, R03c, R04a, R04b, R07d, U04b, U05a, U08, U14, U15, U19, U20a, C03; C02 (14 d); E01, E13c, E30a, E31c, E34

Latest application/Harvest interval

- HI 14 d (apples, pears, blackcurrants, gooseberries)

Approval

- Off-label approval unstipulated for use on mushrooms (OLA 1665/99)

435　pyrimethanil

An anilinopyrimidine fungicide for apples and strawberries

Products

Scala	Aventis	400 g/l	SC	07806

Uses

Botrytis in *aubergines* (off-label), *bilberries* (off-label), *blackcurrants* (off-label), *cane fruit* (off-label), *cranberries* (off-label), *gooseberries* (off-label), *grapevines* (off-label), *outdoor tomatoes* (off-label), *protected tomatoes* (off-label), *redcurrants* (off-label), *strawberries, whitecurrants* (off-label). Scab in *apples*.

Efficacy

- On apples a programme of sprays will give early season control of scab. Season long control can be achieved by continuing programme with other approved fungicides
- In strawberries product should be used as part of a programme of disease control treatments which should alternate with other materials to prevent or limit development of less sensitive strains of grey mould

Crop safety/Restrictions

- Maximum number of treatments 5 per yr for apples, 2 per yr for strawberries
- All varieties of apples and strawberries may be treated

- Treat apples from bud burst at 10-14 d intervals
- In strawberries start treatments at white bud to give maximum protection of flowers against grey mould and treat every 7-10 d. Product should not be used more than once in a 3 or 4 spray programme
- Product does not taint apples. Processors should be consulted before use on strawberries

Special precautions/Environmental safety
- Harmful to fish or other aquatic life. Do not contaminate surface waters or ditches with chemical or used container
- Product has negligible effect on hoverflies and lacewings. Limited evidence indicates some margin of safety to *Typhlodromus pyri*

Personal protective equipment/Label precautions
- U08, U20b, C03, E01, E13c, E26, E30a, E31b

Latest application/Harvest interval
- Before end of flowering for apples
- HI 1 d for strawberries; 3 d for aubergines, tomatoes

Approval
- Off-label approval unlimited for use on outdoor grapevines (OLA 1203/98); to March 20005 for use on aubergines, protected tomatoes (OLA 0509/99); to Jul 2003 for use on protected raspberries (OLA 1938/99); to Jul 2003 for use on a range of protected cane fruit crops. See notice for details (OLA 2182/99); unstipulated for use on currant crops, cranberries, bilberries, gooseberries (OLA 1939/99); to Jul 2003 for use on outdoor tomatoes (OLA 2735/99)

436 quinmerac

A residual herbicide available only in mixtures

See also metazachlor + quinmerac

437 quinoxyfen

A systemic protectant fungicide for cereals

See also cyproconazole + quinoxyfen
fenpropimorph + quinoxyfen

Products

1	Apres	Dow	500 g/l	SC	08881
2	Barclay Foran	Barclay	500 g/l	SC	09479
3	Erysto	Dow	250 g/l	SC	08697
4	Fortress	Dow	500 g/l	SC	08279
5	Standon Quinoxyfen 500	Standon	500 g/l	SC	08924

Uses

Powdery mildew in **spring barley, spring wheat, winter barley, winter wheat** [1-5]. Powdery mildew in **durum wheat, rye, spring oats, triticale, winter oats** [1, 3, 4].

Efficacy
- For best results treat at early stage of disease development before infection spreads to new crop growth. Further treatment may be necessary if disease pressure remains high
- Product not curative and will not control latent or established disease infections
- For broad spectrum control use in tank mixtures - see label
- Product rainfast after 1 h
- Systemic activity may be reduced in severe drought

Crop safety/Restrictions
- Maximum total dose equivalent to two full dose treatments
- Apply only in the spring from mid-tillering stage (GS 25)

Special precautions/Environmental safety
- Irritant. May cause sensitization by skin contact

- Dangerous to fish or other aquatic life. Do not contaminate surface waters or ditches with chemical or used container
- LERAP Category B

Personal protective equipment/Label precautions
- A, C, H [1-5]
- R04, R04e, U05a, U14, C03, E01, E13b, E16a, E16b, E34 [1-5]; E26 [2, 3]

Latest application/Harvest interval
- First awns visible stage (GS 49)

Approval
- Accepted by BLRA for use on malting barley

---

438   quintozene

A protectant soil applied chlorophenyl fungicide for horticultural crops

Products

1 Quintozene WP	Aventis Environ.	50% w/w	WP	09936
2 Quintozene WP	RP Amenity	50% w/w	WP	05404
3 Terraclor 20D	Hortichem	20% w/w	DP	06578
4 Terraclor Flo	Hortichem	497 g/l	SC	08666

Uses

Botrytis in **bedding plants, chrysanthemums, dahlias, fuchsias, lettuce, pelargoniums, pot plants, protected carnations, tomatoes** [3]. Botrytis in **outdoor lettuce, protected lettuce** [4]. Dollar spot in **turf** [1, 2]. Fusarium patch in **turf** [1, 2]. Red thread in **turf** [1, 2]. Rhizoctonia in **anemones, bedding plants, chrysanthemums, cucumbers, dahlias, flower bulbs, fuchsias, hyacinths, irises, lettuce, narcissi, pelargoniums, pot plants, protected carnations, tomatoes, tulips** [3]. Rhizoctonia in **outdoor lettuce, protected lettuce** [4]. Sclerotinia in **anemones, bedding plants, chicory** *(forcing)*, **chrysanthemums, corms, dahlias, flower bulbs, fuchsias, hyacinths, irises, lettuce, narcissi, pelargoniums, pot plants, protected carnations, tomatoes, tulips** [3]. Sclerotinia in **outdoor lettuce, protected lettuce** [4]. Wirestem in **brassicas** [3, 4].

Efficacy
- Apply to soil or compost and incorporate before planting or sowing [3, 4]
- For use in turf apply drenching spray when fungal growth active [1, 2]

Crop safety/Restrictions
- Maximum number of treatments 1 per crop for ornamental bulbs, brassicas, chicory, cucumber, lettuce, tomato, bedding and pot plants [3, 4]; 6 per yr for turf [1, 2]
- In general leave treated soil or compost for 2-3 wk before planting unrooted cuttings, 4 wk before sowing tomatoes or cucumbers and 2-3 d before planting other crops
- Do not allow soil to dry out leading to reduction of plant vigour [3, 4]

Special precautions/Environmental safety
- Irritating to skin, eyes and respiratory system [3]
- Dangerous to fish or other aquatic life. Do not contaminate surface waters or ditches with chemical or used container [4]

Personal protective equipment/Label precautions
- A [3, 4]
- R04a, R04b, R04c [3]; U19 [1-3]; U20b [1, 2, 4]; U05a, U08, U20a, C03 [3]; E30a, E32a [1-4]; E15, E34 [1-3]; E01 [3]; E13b, E26 [4]

Latest application/Harvest interval
- Before sowing or planting for listed crops other than turf

Approval
- Approvals for agricultural and garden uses of quintozene will be withdrawn during 2001 followed by a 12-month use-up period

Maximum Residue Level (mg residue/kg food)
- lettuce 3; bananas 1; potatoes 0.2; tomatoes, peppers, aubergines 0.1; cauliflowers, head cabbages 0.02; beans (with pods) 0.01

FOR FULL CONDITIONS OF USE ALWAYS READ THE PRODUCT LABEL

439   quizalofop-P-ethyl

An aryl phenoxypropionic acid post-emergence grass herbicide

Products

	1 CoPilot	Aventis	100 g/l	EC	08042
	2 Pilot D	Aventis	250 g/l	SC	08041
	3 Sceptre	Aventis	250 g/l	SC	08043

Uses

Annual grasses in *carrots, fodder beet, linseed, mangels, mustard, parsnips, red beet, spring oilseed rape, sugar beet, winter oilseed rape*. Couch in *carrots, fodder beet, linseed, mangels, mustard, parsnips, red beet, spring oilseed rape, sugar beet, winter oilseed rape*. Perennial grasses in *carrots, fodder beet, linseed, mangels, mustard, parsnips, red beet, spring oilseed rape, sugar beet, winter oilseed rape*. Volunteer cereals in *carrots, fodder beet, linseed, mangels, mustard, parsnips, red beet, spring oilseed rape, sugar beet, winter oilseed rape*.

Efficacy

- Apply to emerged weeds. Weeds emerging after treatment are not controlled. Effective on annual grasses from 2-leaf stage to fully tillered, on perennials from 4-6 leaf stage to before jointing
- Best results achieved by application to weeds growing actively in warm conditions with adequate soil moisture. Use split treatment to extend period of control
- Must be used with a recommended adjuvant - see label for details [3]
- Annual meadow-grass is not controlled
- Various spray programmes and tank-mixtures recommended to control mixed dicotyledon/grass weed populations. See label for details
- For effective couch control do not hoe beet crops within 21 d after spraying
- At least 2 h rain free period required for effective results

Crop safety/Restrictions

- Maximum number of treatments 2 as split dose on oilseed rape and mustard and 2 for couch control in all other crops. See labels for maximum total dose of individual products
- Apply to beet crops from 2 fully expanded leaves when weeds at appropriate stage and growing actively but before crop meets across rows
- Apply to oilseed rape and mustard from expanded cotyledon stage (GS 1,0), before crop covers larger weeds
- Do not spray crops under stress from any cause or in frosty weather
- Consult processor before using on carrot crops for processing
- An interval of at least 3 d must elapse between treatment and use of another herbicide on beet crops, 14 d on oilseed rape and mustard
- On all other recommended crops an interval of at least 3 wk must elapse after use of a broad-leaved herbicide
- In the event of crop failure broad-leaved crops may be resown at any time, cereals, onions, leeks or maize may be sown after 2-6 wk depending on dose used

Special precautions/Environmental safety

- Irritant. Risk of serious damage to eyes [1]
- Irritating to eyes and skin [2, 3]
- Flammable [1]
- Dangerous to fish or other aquatic life. Do not contaminate surface waters or ditches with chemical or used container

Personal protective equipment/Label precautions

- A, C [1-3]
- R04a, R04b [2, 3]; R04, R04d, R07d [1]; U09a, U20a [2, 3]; U05a, U11, U20b, C03 [1]; E13b, E26, E30a, E31b [1-3]; E01 [1]

Latest application/Harvest interval

- HI 16 wk for beet crops; 11 wk for oilseed rape, mustard linseed; 10 wk for parsnips; 9 wk for carrots

440   resmethrin

A contact acting pyrethroid insecticide available only in mixtures

See also fenitrothion + permethrin + resmethrin
pyrethrins + resmethrin

441   resmethrin + tetramethrin

A contact acting pyrethroid insecticide mixture

Products

Sorex Wasp Nest Destroyer	Sorex	0.10:0.05% w/w	AE	H4410

Uses

Wasps in *miscellaneous situations*.

Efficacy
- Product acts by lowering temperature in and around nest and producing a stupefying vapour in addition to its contact insecticidal effect. Spray only on surfaces
- Apply by directing jet at nest from up to 3 m away and approach nest gradually until jet can be directed into entrance hole. Continue spraying until nest saturated

Special precautions/Environmental safety
- For use only by trained operators and owners of commercial and agricultural premises
- Harmful by inhalation. Contains perchloroethylene
- Keep off skin. Ensure adequate ventilation
- Harmful to caged birds and pets
- Dangerous to bees. Do not apply to crops in flower or to those in which bees are actively foraging. Do not apply when flowering weeds are present
- Extremely dangerous to fish or other aquatic life. Remove fish tanks before spraying

Personal protective equipment/Label precautions
- A, C, H
- R03b, U14, U19, U20a, C03, E01, E10b, E12c, E13a, E15, E30a

442   rimsulfuron

A selective systemic sulfonylurea herbicide

Products

1	Landgold Rimsulfuron	Landgold	25% w/w	SG	08959
2	Me2 Aducksbackside	Me2	25% w/w	WG	10065
3	Standon Rimsulfuron	Standon	25% w/w	SG	09955
4	Titus	DuPont	25% w/w	SG	07908

Uses

Annual dicotyledons in *fodder maize, potatoes*. Volunteer oilseed rape in *fodder maize, potatoes*.

Efficacy
- Product should be used with a suitable adjuvant or a suitable herbicide tank-mix partner. See label for details
- Product acts by foliar action. Best results obtained from good spray cover of small actively growing weeds. Effectiveness is reduced in very dry conditions
- Split application provides control over a longer period and improves control of fat hen and polygonums
- Weed spectrum can be broadened by tank mixture with other herbicides. See label for details
- Susceptible weeds cease growth immediately and symptoms can be seen 10 d later

Crop safety/Restrictions
- Maximum number of treatments 2 per crop

- All varieties of ware potatoes and fodder maize may be treated, but variety restrictions of any tank-mix partner must be observed
- Crops should be treated in spring on potatoes up to 25 cm high and on maize up to the 4-collar stage
- Do not treat maize treated with organophosphorus insecticides
- Avoid high light intensity (full sunlight) and high temperatures on the day of spraying
- Do not treat during periods of substantial diurnal temperature fluctuation or when frost anticipated
- Do not apply to any crop stressed by drought, water-logging, low temperatures, pest or disease attack, nutrient or lime deficiency
- Only winter wheat should follow a treated crop in the same calendar yr. Only barley, wheat or maize should be sown in the spring of the yr following treatment. In the second autumn after treatment any crop except brassicas or oilseed rape may be drilled

Special precautions/Environmental safety

- Extremely dangerous to fish or other aquatic life. Do not contaminate surface waters or ditches with chemical or used container
- Herbicide is very active. Take particular care to avoid drift onto plants outside the target area
- Spraying equipment should not be drained or flushed onto land planted, or to be planted, with trees or crops other than potatoes or forage maize and should be thoroughly cleansed after use - see label for instructions

Personal protective equipment/Label precautions

- U08, U19 [1-4]; U20a [1, 2, 4]; U20b [3]; E13a, E30a, E32a [1-4]; E01, E34 [3]

Latest application/Harvest interval

- Before most advanced potato plants are 25 cm high; before 4-collar stage of fodder maize

---

443   rotenone

A natural, contact insecticide of low persistence

Products

1 Devcol Liquid Derris	Nehra	50 g/l	EC	06063
2 FS Liquid Derris	Ford Smith	50 g/l	EC	01213

Uses

Aphids in *flowers, ornamentals, protected crops, soft fruit, top fruit, vegetables*. Raspberry beetle in *cane fruit*. Sawflies in *gooseberries*. Slug sawflies in *pears, roses*.

Efficacy

- Apply as high volume spray when pest first seen and repeat as necessary
- Spray raspberries at first pink fruit, loganberries when most of blossom over, blackberries as first blossoms open. Repeat 2 wk later if necessary
- Spray to obtain thorough coverage, especially on undersurfaces of leaves

Special precautions/Environmental safety

- Dangerous to fish or other aquatic life. Do not contaminate surface waters or ditches with chemical or used container
- Flammable

Personal protective equipment/Label precautions

- R07d, U20a, C02 [1, 2]; E13b [2]; E30a, E31a, E34 [1, 2]; E15 [1]

Latest application/Harvest interval

- HI 1 d

---

444   sethoxydim

A post-emergence oxime grass weed herbicide

Products

Checkmate	Hortichem	193 g/l	EC	09122

Uses

Annual grasses in *bedding plants* (off-label), *combining peas, flax, flower bulbs* (off-label), *herbaceous perennials* (off-label), *linseed, lupins, mustard, sunflowers* (off-label), *trees and shrubs* (off-label), *vining peas, winter oilseed rape, woody ornamentals* (off-label). Awned canary grass in *spring oilseed rape, winter oilseed rape*. Barren brome in *field beans, flax, fodder beet, linseed, lupins, mangels, mustard, raspberries, spring oilseed rape, sugar beet*. Black bent in *field beans, fodder beet, mangels, potatoes, raspberries, sugar beet*. Blackgrass in *combining peas, field beans, flax, fodder beet, linseed, lupins, mangels, mustard, potatoes, raspberries, spring oilseed rape, sugar beet, vining peas, winter oilseed rape*. Couch in *field beans, fodder beet, mangels, potatoes, raspberries, sugar beet*. Creeping bent in *field beans, fodder beet, mangels, potatoes, raspberries, sugar beet*. Perennial grasses in *bedding plants* (off-label), *flower bulbs* (off-label), *herbaceous perennials* (off-label), *sunflowers* (off-label), *trees and shrubs* (off-label), *woody ornamentals* (off-label). Volunteer cereals in *combining peas, field beans, flax, fodder beet, linseed, lupins, mangels, mustard, potatoes, raspberries, spring oilseed rape, sugar beet, vining peas, winter oilseed rape*. Wild oats in *combining peas, field beans, flax, fodder beet, linseed, lupins, mangels, mustard, potatoes, raspberries, spring oilseed rape, sugar beet, vining peas*.

Efficacy

- Apply to emerged weeds in combination with Adder or Actipron adjuvant oil (except on peas). Weeds emerging after treatment are not controlled
- Best results when weeds growing actively. Action is faster under warm conditions
- Apply to annual grasses from 2-leaf stage to end of tillering, to couch when largest shoots at least 30 cm long. For effective couch control do not cultivate for 2 wk
- Annual meadow-grass is not controlled
- Apply when rain not expected for at least 2 h
- Various tank mixes recommended for grass/dicotyledon weed populations. See label for details of mixtures and sequential treatments
- Leave at least 7 d before applying another herbicide

Crop safety/Restrictions

- Maximum number of treatments 1 per crop or per yr
- See label for details of crop growth stage and timing
- Do not mix with adjuvant oil on peas. Check pea leaf-wax by Crystal Violet test and do not spray where wax insufficient or damaged
- Do not spray crops suffering damage from other herbicides
- In raspberries apply to inter-row or to base of canes only
- Brassicas, field beans and onions may be sown 3 d after treatment, grass and cereal crops after 4 d

Special precautions/Environmental safety

- Irritating to eyes and skin
- Harmful to fish or other aquatic life. Do not contaminate surface waters or ditches with chemical or used container

Personal protective equipment/Label precautions

- A, C
- R04a, R04b, U02, U05a, U08, U20b, C03, E01, E13c, E26, E30a, E31b

Latest application/Harvest interval

- HI linseed, lupins, mustard, oilseed rape 12 wk; combining peas 11 wk; fodder beet, mangels 9 wk; maincrop potatoes, sugar beet 8 wk; seed potatoes 7 wk; early potatoes, vining peas, raspberries 4 wk

445　simazine

A soil-acting triazine herbicide with restricted permitted uses

See also isoproturon + simazine
pendimethalin + simazine

Products

1 Alpha Simazine 50 SC	Makhteshim	500 g/l	SC	04801
2 Atlas Simazine	Nufarm Whyte	500 g/l	SC	07725
3 Gesatop	Novartis	500 g/l	SC	08412
4 Sipcam Simazine Flowable	Sipcam	500 g/l	SC	07622

Uses

Annual dicotyledons in **forestry plantations** [1]. Annual dicotyledons in **apples, blackcurrants, broad beans, cane fruit, field beans, gooseberries, hops, pears, redcurrants, strawberries** [1-4]. Annual dicotyledons in **forestry transplant lines** [1, 3, 4]. Annual dicotyledons in **rhubarb, whitecurrants, woody nursery stock** [2]. Annual dicotyledons in **roses** [2, 3]. Annual dicotyledons in **asparagus, maize** [2, 4]. Annual dicotyledons in **asparagus** *(off-label)*, **grapevines** *(off-label)*, **herbs** *(off-label)*, **mange-tout peas** *(off-label)*, **rhubarb** *(off-label)*, **runner beans** *(off-label)* [3]. Annual dicotyledons in **forest nursery beds** [3, 4]. Annual dicotyledons in **sweetcorn, trees and shrubs, woody ornamentals** [4]. Annual grasses in **forestry plantations** [1]. Annual grasses in **apples, blackcurrants, broad beans, cane fruit, field beans, gooseberries, hops, pears, redcurrants, strawberries** [1-4]. Annual grasses in **forestry transplant lines** [1, 3, 4]. Annual grasses in **rhubarb, whitecurrants, woody nursery stock** [2]. Annual grasses in **roses** [2, 3]. Annual grasses in **asparagus, maize** [2, 4]. Annual grasses in **asparagus** *(off-label)*, **herbs** *(off-label)*, **mange-tout peas** *(off-label)*, **rhubarb** *(off-label)*, **runner beans** *(off-label)* [3]. Annual grasses in **forest nursery beds** [3, 4]. Annual grasses in **sweetcorn, trees and shrubs, woody ornamentals** [4].

Efficacy

- Active via root uptake. Best results achieved by application to fine, firm, moist soil, free of established weeds, when rain falls after treatment
- Do not use on highly organic soils (soils with more than 10% organic matter [8032])
- Following repeated use of simazine or other triazine herbicides resistant strains of groundsel and some other annual weeds may develop

Crop safety/Restrictions

- Maximum number of treatments 1 per yr for sweetcorn, broad and field beans, asparagus, rhubarb, sage, strawberries, hops, bush and cane fruit, apples (lower rate), forest transplant lines; 1 every 2 yr for apples (higher rate); 2 per yr for mint
- See labels for maximum total dose of simazine and atrazine products for each situation
- Do not spray beans on sandy or gravelly soils or cultivate after treatment. Do not treat varieties Beryl, Feligreen or Rowena
- Apply in top, bush and cane fruit and woody ornamentals established at least 12 mth in Feb-Mar. Roses may be sprayed immediately after planting. See label for lists of resistant and susceptible species
- Apply to strawberries in Jul-Dec, not in spring. Do not treat spring-planted crops established less than 6 mth, or winter-planted crops less than 9 mth. Do not spray varieties Huxley Giant, Madame Moutot or Regina
- Apply to hops overall in Feb-Apr before weeds emerge. Newly planted sets must be covered with minimum of 50 mm soil
- Apply to forest nursery seedbed in second yr or to transplant lines after plants 5 cm tall. Do not treat Norway spruce (Christmas trees)
- To reduce soil run-off on gradients especially in orchard and forest plantations, users are advised to plant grass strips or leave 6 m wide strips between treated areas and surface waters
- On Sands, stony or gravelly soils there is risk of crop damage, especially with heavy rain
- Allow at least 7 mth before drilling or planting other crops, longer if weather dry
- Do not sow oats in autumn following spring application in maize

Special precautions/Environmental safety
- Dangerous to fish or other aquatic life and aquatic higher plants. Do not contaminate surface waters or ditches with chemical or used container
- LERAP Category B
- Use must be restricted to one product containing atrazine or simazine, and either to a single application at the maximum approved rate or (subject to any existing maximum permitted number of treatments) to several applications at lower doses up to the maximum approved rate for a single application

Personal protective equipment/Label precautions
- M [1, 2, 4]; A, C, D, H [1-4]; B [2, 4]
- R04e [3]; U20b [1, 2]; U20c [3, 4]; U23 [3]; E31b [1]; E16b [1, 2, 4]; E16a, E26, E30a [1-4]; E13b [1, 3, 4]; E31a [2-4]; E15 [2]

Latest application/Harvest interval
- Before emergence of spears for asparagus; 7 d after drilling for sweetcorn; before end of Feb (autumn planted), 10 d after drilling for spring planted field and broad beans; before end of Mar for top, bush and cane fruit; after harvest, before end of Nov for strawberries; after harvest, before 1 May for hops.
- HI mint and sage 4 mth

Approval
- Approvals for sale, supply and use of simazine on non-crop land were revoked in 1992 and 1993
- Off-label approval unlimited for use on edible-podded peas, runner beans (OLA 1723/97)[3]; unlimited for use on asparagus, outdoor rhubarb (OLA 1721/97)[3]; unlimited for use on a range of outdoor herbs (see notices for details) (OLA 1722/97)[3]
- Accepted by BLRA for use on hops

---

446   simazine + trietazine

A pre-emergence residual herbicide for peas, field and broad beans

Products

Remtal SC	Aventis	57.5:402.5 g/l	SC	07270

Uses

Annual dicotyledons in **broad beans, combining peas, edible podded peas** (off-label), **field beans, vining peas**.

Efficacy
- Chemical acts mainly via roots but some contact effect on cotyledon stage weeds. Gives residual control of germinating weeds for season
- Effectiveness increased by rain after spraying
- Weeds of intermediate susceptibility, including annual meadow grass and blackgrass, may be controlled if adequate rainfall occurs soon after spraying
- Control of deep germinating weeds on heavier soils may be incomplete
- Do not use on soils with more than 10% organic matter
- With repeated use of simazine or other triazine herbicides resistant strains of groundsel and some other annual weeds may develop

Crop safety/Restrictions
- Maximum number of treatments 1 per crop
- Apply to spring peas, winter or spring field and broad beans between drilling and 5% emergence
- On early-drilled peas apply when seed chitting, on later crops as soon as possible after drilling. Do not spray winter sown peas. Vedette peas may be checked by treatment
- Do not spray any forage pea varieties
- Crop seed should be covered by at least 25 mm of settled soil
- Do not use on sand or on gravelly or cloddy soils
- Other crops may be sown or planted 10 wk or more after spraying. With drilled brassicas allow a minimum of 14 wk

- In the event of crop failure only drill peas, field or broad beans without ploughing but do not respray
- To reduce soil run-off users are advised to plant grass strips or leave strips 6 m wide between treated areas and surface waters

Special precautions/Environmental safety
- Harmful if swallowed
- Dangerous to fish or other aquatic life and aquatic higher plants. Do not contaminate surface waters or ditches with chemical or used container
- LERAP Category B
- Use must be restricted to one product containing atrazine or simazine either to a single application at maximum approved rate or (subject to any existing maximum permitted number of treatments) to several applications at lower doses up to the maximum approved rate for single application

Personal protective equipment/Label precautions
- A, B, C, D, H, M
- M03, R03c, U05a, U20b, C03, E01, E13b, E16a, E16b, E26, E30a, E31b, E34

Latest application/Harvest interval
- Before 5% total crop emergence

Approval
- May be applied through CDA equipment (see label for details)

---

447   sodium chlorate

A non-selective inorganic herbicide for total vegetation control

Products

1 Atlacide Soluble Powder	Nomix-Chipman	58.2% w/w	SP	00125
2 Cooke's Weedclear	Nehra	50% w/w	SL	06512
3 Deosan Chlorate Weedkiller (Fire Suppressed)	DiverseyLever	55% w/w	SP	08521
4 Doff Sodium Chlorate Weedkiller	Doff Portland	53% w/w	SP	06049
5 TWK Total Weedkiller	Reabrook	89 g/l	SL	06393

Uses

Total vegetation control in *non-crop areas, paths and drives*.

Efficacy
- Active through foliar and root uptake
- Apply as overall spray at any time during growing season. Best results obtained from application to moist soil in spring or early summer
- Do not apply before heavy rain

Crop safety/Restrictions
- Maximum number of treatments 2 per yr for non-crop areas, paths and drives [4]
- Clothing, paper, plant debris etc become highly inflammable when dry if contaminated with sodium chlorate
- Fire risk has been reduced by inclusion of fire depressant but product should not be used in areas of exceptionally high fire risk e.g. oil installations, timber yards
- Treated ground must not be replanted for at least 6 mth after treatment [4]

Special precautions/Environmental safety
- Harmful if swallowed
- Oxidizing - contact with combustible material may cause fire
- Wash clothing thoroughly after use
- If clothes become contaminated do not stand near an open fire
- Must not be sold to persons under 18

Personal protective equipment/Label precautions
- L [1]; M [1-5]; A, H [2-5]
- M03, R03c [1-5]; R08 [1, 2, 4]; U20b [1, 4]; U02, U05a [1-3, 5]; U20a [3, 5]; U20c [2]; C03 [1-3, 5]; E15, E30a, E34 [1-5]; E32a [1, 3-5]; E01 [1-3, 5]; E07 [1, 3, 5]; E31a [2]

Withholding period
- Keep livestock out of treated areas until foliage of any poisonous weeds such as ragwort has died and become unpalatable

448   sodium chloride (commodity substance)

An inorganic salt for use as a sugar beet herbicide

Products
sodium chloride   various   -   SL

Uses
Polygonums in **sugar beet**. Volunteer potatoes in **sugar beet**.

Efficacy
- Best results obtained from good spray cover treatment in hot, humid weather
- Use as a follow-up treatment where earlier sprays have failed to control large volunteer potatoes or polygonums
- Acts by contact scorch and has no direct effect on daughter potato tubers
- Saturated spray solution should contain 0.1% w/w non-ionic wetter and be applied at 1000 l/ha

Crop safety/Restrictions
- Spray between three true leaf stage of crops and end Jul
- Crop scorch may occur after treatment

449   sodium hypochlorite (commodity substance)

An inorganic horticultural bactericide for use in mushrooms

Products
sodium hypochlorite   various   100% w/v

Uses
Bacterial blotch in **mushrooms**.

Crop safety/Restrictions
- Maximum concentration 315 mg/litre of water

Special precautions/Environmental safety
- Mixing and loading must only take place in a ventilated area
- Must only be used by suitably trained and competent operators
- Harmful to fish or other aquatic life. Do not contaminate surface waters or ditches with chemical or used container

Personal protective equipment/Label precautions
- A, C, H
- E13c

Latest application/Harvest interval
- HI 1 d

450   sodium monochloroacetate

A contact herbicide for various horticultural crops

Products
1 Atlas Somon	Nufarm Whyte	96% w/w	SP	07727
2 Croptex Steel	Hortichem	95% w/w	SP	02418

Uses
Annual dicotyledons in **apples, blackcurrants, gooseberries, pears, plums, redcurrants** [1]. Annual dicotyledons in **brussels sprouts, cabbages, kale, leeks, onions** [1, 2]. Annual dicotyledons in **apples** (directed spray), **blackcurrants** (directed spray), **broccoli** (off-label), **calabrese** (off-label), **cauliflowers** (off-label), **gooseberries** (directed spray), **kohlrabi** (off-

label), **pears** (directed spray), **plums** (directed spray), **redcurrants** (directed spray) [2]. Basal defoliation in **hops** [1]. Basal defoliation in **hops** (off-label) [2]. Sucker control in **blackberries** (off-label), **hybrid berries** (off-label), **loganberries** (off-label), **raspberries** (off-label) [2].

Efficacy
- Best results achieved by application to emerged weed seedlings up to young plant stage under good growing conditions
- Effectiveness reduced by rain within 12 h
- May be used for hop defoliation in tank-mixture with tar-oil [1], with Wayfarer adjuvant [2]. Apply as a directed spray to the base of the plant when they are up to 20 cm high, after training when main shoots are at least 100 cm high

Crop safety/Restrictions
- Maximum number of treatments 1 per crop or yr; 2 per yr for sucker control in cane fruit and hop defoliation
- Apply to brassicas from 2-4 leaf stage or after recovery from transplanting
- Do not spray cabbage that has begun to heart
- Apply to onions and leeks after crook stage but before 4-leaf stage
- Safety on brassicas, onions and leeks depends on presence of adequate leaf wax, check by crystal violet wax test. Do not add wetters, pesticides or nutrients to spray
- Apply in fruit crops established for at least 1 yr as a directed spray
- Do not spray if frost likely or if temperature likely to exceed 27°C
- Apply for sucker control in raspberries when canes 10-20 cm high with addition of Wayfarer adjuvant [2]

Special precautions/Environmental safety
- Harmful if swallowed, in contact with skin and by inhalation
- Irritating to respiratory system
- Irritating to respiratory system
- Risk of serious damage to eyes
- Corrosive, causes burns
- Harmful to bees. Do not apply to crops in flower or to those in which bees are actively foraging. Do not apply when flowering weeds are present

Personal protective equipment/Label precautions
- A, C, D, E [1, 2]
- R03a, R03b, R03c, R04c, R04d, R05, R05b, U05a, U08, U19, U20a [1, 2]; U10, U11 [1]; C03 [1, 2]; E15 [2]; E01, E05, E05a, E12e, E30a, E31a [1, 2]; E07, E34 [1]

Withholding period
- Keep livestock, especially poultry, out of treated areas for at least 2 wk

Latest application/Harvest interval
- Before 4 true leaf stage for onions and leeks.
- HI 21 d for cabbage, kale, Brussels sprouts; 28 d for cane fruit; 14 wk for hops

Approval
- Off-label approval unlimited for use on cane fruit (OLA 0115/93)[2], hops (OLA 0357/93)[2]; unlimited for use on hops to within 6 wk of harvest (OLA 0357/93)[2]; to Mar 2001 for use on kohlrabi (OLA 0615/96)[2]; unlimited for use on cauliflowers, broccoli, calabrese (OLA 2602/99)[2]
- Accepted by BLRA for use on hops

---

451   spiroxamine

A spiroketal amine fungicide for cereals

Products

1 Neon	Bayer	500 g/l	EW	08337
2 Torch	Bayer	500 g/l	EW	08336
3 Zenon	Bayer	800 g/l	EC	09193

Uses

Brown rust in **spring barley, spring rye, spring wheat, winter barley, winter rye, winter wheat** [1-3]. Powdery mildew in **spring barley, spring rye, spring wheat, winter barley,**

winter rye, winter wheat [1-3]. Rhynchosporium in **spring rye** *(reduction)*, **winter rye** *(reduction)* [1, 2]. Rhynchosporium in **spring barley** *(reduction)*, **winter barley** *(reduction)* [1-3]. Yellow rust in **spring barley, spring rye, spring wheat, winter barley, winter rye, winter wheat** [1-3].

Efficacy
- For best results treat at an early stage of disease development before infection spreads to new growth
- Products active against net blotch when applioed at first signs of disease but better control achieved with mixtures. See label for details
- To reduce the risk of development of resistance avoid repeat treatments on diseases such as powdery mildew. If necessary tank mix or alternate with other non-morpholine fungicides

Crop safety/Restrictions
- Maximum total dose equivalent to two full dose treatments

Special precautions/Environmental safety
- Harmful if swallowed
- Harmful in contact with skin [3]
- Irritating to skin [1, 2]
- Risk of serious damage to eyes
- May cause sensitization by skin contact
- Dangerous to fish or other aquatic life. Do not contaminate surface waters or ditches with chemical or used container
- LERAP Category B

Personal protective equipment/Label precautions
- A, C, H [1-3]
- M03 [1-3]; R04b [1, 2]; R03c, R04d, R04e [1-3]; R03a [3]; U05a, U11, U13, U14, U15, U20a, C03, E01, E13b, E16a, E16b, E26, E30a, E31b, E34 [1-3]

Latest application/Harvest interval
- Before caryopsis watery ripe (GS 71) for spring wheat, winter rye, winter wheat; ear emergence complete (GS 59) for spring barley, winter barley

Approval
- Spiroxamine included in Annex I under EC Directive 91/414
- Accepted by BLRA for use on malting barley

---

452   spiroxamine + tebuconazole

A broad spectrum systemic fungicide mixture for cereals

Products

1 Array	Bayer	400:100 g/l	EC	09352
2 Beam	Bayer	250:133 g/l	EW	08332
3 Sage	Bayer	250:133 g/l	EW	08334

Uses
Brown rust in **barley, rye, wheat** [1-3]. Fusarium ear blight in **wheat** [2, 3]. Glume blotch in **wheat** [2, 3]. Net blotch in **barley** [1-3]. Powdery mildew in **barley, rye, wheat** [1-3]. Rhynchosporium in **barley** [1-3]. Septoria leaf spot in **wheat** [1-3]. Sooty moulds in **wheat** [2, 3]. Yellow rust in **barley, rye, wheat** [1-3].

Efficacy
- Best results achieved from treatment at early stage of disease development before infection spreads to new growth
- To protect flag leaf and ear from Septoria apply from flag leaf emergence to ear fully emerged (GS 37-59). Earlier treatment may be necessary where there is high disease risk
- Control of rusts, powdery mildew, leaf blotch and net blotch may require second treatment 2-3 wk later

Crop safety/Restrictions
- Maximum total dose equivalent to two full dose treatments
- Do not use on durum wheat

FOR FULL CONDITIONS OF USE ALWAYS READ THE PRODUCT LABEL

- Some transient leaf speckling may occur on wheat but this has not been shown to reduce yield reponse or disease control

Special precautions/Environmental safety
- Harmful if swallowed
- Harmful in contact with skin [1]
- Irritating to skin. May cause sensitization by skin contact [2, 3]
- Risk of serious damage to eyes
- Dangerous to fish or other aquatic life. Do not contaminate surface waters or ditches with chemical or used container
- LERAP Category B

Personal protective equipment/Label precautions
- A, C, H [1-3]
- M03 [1-3]; R04b, R04e [2, 3]; R03c, R04d [1-3]; R03a [1]; U05a, U11, U13, U14, U15, U20b, C03, E01, E13b, E16a, E16b, E26, E30a, E31b, E34 [1-3]

Latest application/Harvest interval
- Before caryopsis watery ripe (GS 71) for rye, wheat; ear emergence complete (GS 59) for barley

Approval
- Spiroxamine included in Annex I under EC Directive 91/414
- Accepted by BLRA for use on malting barley

---

453   strychnine hydrochloride (commodity substance)

A vertebrate control agent for destruction of moles underground

Products

strychnine   various   -

Uses

Moles in **grassland** *(areas of restricted public access)*, **miscellaneous situations** *(areas of restricted public access)*.

Efficacy
- For use as poison bait against moles on commercial agricultural/horticultural land where public access restricted, on grassland associated with aircraft landing strips, horse paddocks, race and golf courses and other areas specifically approved by Agriculture Departments

Special precautions/Environmental safety
- Strychnine is subject to the Poisons Rules 1982 and the Poisons Act 1972. See notes in Section 1
- Must only be supplied to holders of an Authority to Purchase issued by appropriate Agricultural Department. Authorities to Purchase may only be issued to persons who satisfy the appropriate authority that they are trained and competent in its use
- Only to be supplied in original sealed pack in units up to 2 g
- Store in original container under lock and key and only on premises under control of holder of Authority to Purchase or a named individual
- A written COSHH assessment must be made before using
- Must be prepared for application with great care so that there is no contamination of the ground surface
- Any prepared bait remaining at the end of the day must be buried
- Operators must be supplied with a Section 6 (HSW) Safety Data Sheet before commencing work
- Other restrictions apply, see *Pesticides 2000*, Annex D

Personal protective equipment/Label precautions
- A

454   sulphuric acid (commodity substance)

A strong acid used as an agricultural desiccant

Products

sulphuric acid	various	77% w/w	SL

Uses

Haulm destruction in *potatoes*. Pre-harvest desiccation in *bulbs/corms, peas*.

Efficacy
- Apply with suitable equipment between 1 Mar and 15 Nov

Crop safety/Restrictions
- Maximum number of treatments 3 per crop for potatoes; 1 per crop for peas; 1 per yr for bulbs and corms
- Only 'sulphur burnt' sulfuric acid to be used

Special precautions/Environmental safety
- Sulfuric acid is subject to the Poisons Rules 1982 and the Poisons Act 1972. See notes in Section 1
- Must only be used by suitably trained operators competent in use of equipment for applying sulfuric acid
- Not to be applied using hand-held or pedestrian controlled applicators
- A written COSHH assessment must be made before use. Operators should observe OES set out in HSE guidance note EH40/90 or subsequent issues
- Operators must have liquid suitable for eye irrigation immediately available at all times throughout spraying operation
- Spray must not be deposited within 1 m of public footpaths
- Written notice of any intended spraying must be given to owners of neighbouring land and readable warning notices posted beforehand and left in place for 96 h afterwards. See PR 1990, Issue 12 for details
- Unprotected persons must be kept out of treated areas for at least 96 h after treatment
- Do not apply to crops in which bees are actively foraging. Do not apply when flowering weeds are present

Personal protective equipment/Label precautions
- B, C, D, H, K, M
- R05, R05b

Latest application/Harvest interval
- 15 Nov

455   sulphonated cod liver oil

An animal repellent

Products

Scuttle	Fine	800 g/l	EC	06232

Uses

Deer in *barley, cabbages, cauliflowers, forestry, lettuce, oats, oilseed rape, wheat*. Rabbits in *barley, cabbages, cauliflowers, forestry, lettuce, oats, oilseed rape, wheat*.

Efficacy
- Apply diluted as a spray before crop is being grazed or attacked or undiluted as a dip for forestry seedlings
- A rain-free period of about 5 h is required after application

Crop safety/Restrictions
- Before using on vegetables users should consider risk of taint and consult processor before using on crops for processing

Special precautions/Environmental safety
- Do not apply directly to animals

- Dangerous to bees. Do not apply to crops in flower or to those in which bees are actively foraging. Do not apply when flowering weeds are present.
- Dangerous to fish or other aquatic life. Do not contaminate surface waters or ditches with chemical or used container

Personal protective equipment/Label precautions
- A, C
- U05a, U20a, C03, E01, E12c, E13b, E30a, E31b

---

456   sulphur

A broad-spectrum inorganic protectant fungicide, foliar feed and acaricide

See also copper oxychloride + maneb + sulphur
copper sulphate + sulphur

Products

1	Ashlade Sulphur FL	Nufarm Whyte	720 g/l	SC	06478
2	Headland Sulphur	Headland	800 g/l	SC	03714
3	Headland Venus	Headland	80% w/w	WG	09572
4	Kumulus DF	BASF	80% w/w	SG	04707
5	Luxan Micro-Sulphur	Luxan	80% w/w	SG	06565
6	Microsul Flowable Sulphur	Stoller	960 g/l	SC	03907
7	Microthiol Special	PBI	80% w/w	MG	06268
8	Solfa	Nufarm Whyte	80% w/w	MG	06959
9	Stoller Flowable Sulphur	Stoller	720 g/l	SC	03760
10	Sulphur Flowable	United Phosphorus	800 g/l	SC	07526
11	Thiovit	Novartis	80% w/w	WG	08493
12	Tripart Imber	Tripart	720 g/l	SC	04050

Uses

Foliar feed in **barley, wheat** [4, 5, 7, 11, 12]. Foliar feed in **oilseed rape** [4, 5, 7, 8, 11, 12]. Foliar feed in **swedes** [5, 7, 8]. Foliar feed in **grassland** [5, 7, 8, 11]. Foliar feed in **sugar beet** [7]. Foliar feed in **oats** [7, 12]. Foliar feed in **brassicas, potatoes, soft fruit, top fruit, turf, turnips, vegetables** [7, 8]. Gall mite in **blackcurrants** [1, 4, 12]. Powdery mildew in **sugar beet** [1-12]. Powdery mildew in **hops** [1-4, 6, 9-12]. Powdery mildew in **apples, strawberries** [1, 2, 4, 6, 9, 10, 12]. Powdery mildew in **gooseberries** [1, 4, 12]. Powdery mildew in **grapevines** [1, 4, 6, 9, 12]. Powdery mildew in **aubergines** *(off-label)*, **herb crops** *(off-label)*, **parsnips** *(off-label)*, **protected peppers** *(off-label)*, **protected tomatoes** *(off-label)* [11]. Powdery mildew in **pears** [2, 10]. Powdery mildew in **swedes** [2, 3, 7, 9-12]. Powdery mildew in **oats** [6-9, 12]. Powdery mildew in **barley, wheat** [7]. Powdery mildew in **turnips** [7, 12]. Powdery mildew in **grassland** [9]. Scab in **apples** [2, 4, 6, 9, 10, 12]. Scab in **pears** [2, 6, 9, 10].

Efficacy

- Apply when disease first appears and repeat 2-3 wk later. Details of application rates and timing vary with crop, disease and product. See label for information
- Sulfur acts as foliar feed as well as fungicide and with some crops product labels vary in whether treatment recommended for disease control or growth promotion
- In grassland best at least 2 wk before cutting for hay or silage, 3 wk before grazing
- Treatment unlikely to be effective if disease already established in crop

Crop safety/Restrictions

- Maximum number of treatments 2 per crop for sugar beet (3 per yr [1, 12]), parsnips, swedes; 3 per yr for blackcurrants, gooseberries; variable on cereals - see label
- May be applied to cereals up to grain watery-ripe stage (GS 71) [1, 2, 6, 8, 9, 12], up to milky ripe stage (GS 75) [11], to first node detectable (GS 31) [5], at any time [4]
- Do not use on sulfur-shy apples (Beauty of Bath, Belle de Boskoop, Cox's Orange Pippin, Lanes Prince Albert, Lord Derby, Newton Wonder, Rival, Stirling Castle) or pears (Doyenne du Comice)
- Do not use on gooseberry cultivars Careless, Early Sulphur, Golden Drop, Leveller, Lord Derby, Roaring Lion, or Yellow Rough
- Do not use on apples or gooseberries when young, under stress or if frost imminent

- Do not use on fruit for processing, on grapevines during flowering or near harvest on grapes for wine-making
- Do not use on hops at or after burr stage
- Do not spray top or soft fruit with oil or within 30 d of an oil-containing spray

Special precautions/Environmental safety
- Highly flammable [11]

Personal protective equipment/Label precautions
- R07c [11]; U20a [1, 2]; U20b [6-10, 12]; U20c [3-5, 11]; E15, E30a [1-12]; E31b [2, 10]; E32a [3-9, 11, 12]; E01 [5, 7]; E31a [1]; E34 [5]; E26 [10]

Latest application/Harvest interval
- Late Jun for apples [1]; before 30 Sep (first wk of Aug [2]) in yr of harvest for sugar beet, swedes; up to and including fruit swell for gooseberries; up to burr stage for hops; not specified on other crops
- HI cutting grass for hay or silage 2 wk; grazing grassland 3 wk

Approval
- Approved for aerial application on sugar beet [11]. See notes in Section 1
- Off-label approval unlimited for use on protected tomatoes (OLA 0909/96)[5572], (OLA 1717/97)[11]; unlimited for use on sorrel, summer savory, tarragon (OLA 0910/96)[5572], (OLA 1716/97)[11]; unlimited for use on parsnips (OLA 0911/96)[5572], (OLA 1715/97)[11]; unlimited for use on protected aubergines (OLA 2079/98)[5572], (OLA 2080/98)[11]; unlimited for use on protected peppers (OLA 1714/97)[11]; unlimited for use on protected cucumbers (OLA 1713/97)[11]
- Accepted by BLRA for use on malting barley and on hops before burr stage

---

457   tar oils

Hydrocarbon and phenolic oils used as insecticidal and fungicidal winter washes

See also anthracene oil

Products

1	Sterilite Tar Oil Winter Wash 60% Stock Emulsion	Coventry Chemicals	636 g/l	EC	05061
2	Sterilite Tar Oil Winter Wash 80% Miscible Quality	Coventry Chemicals	800 g/l	EC	05062

Uses

Aphids in **apples, cane fruit, cherries, currants, gooseberries, nectarines, peaches, pears, plums**. Scale insects in **apples, cane fruit, cherries, currants, gooseberries, grapevines, nectarines, peaches, pears, plums**. Suckers in **apples**. Winter moth in **apples, cane fruit, cherries, currants, gooseberries, nectarines, peaches, pears, plums**.

Efficacy
- Spray kills hibernating insects and eggs as well as moss and lichens on trunk
- Spray winter wash over dormant branches and twigs. See labels for detailed timings
- Ensure that all parts of trees or bushes are completely wetted, especially cracks and crevices. Use higher concentration for cleaning up neglected orchard
- Treatment also partially controls winter and raspberry moths and reduces Botrytis on currants and powdery mildew on currants and gooseberries
- Combine with spring insecticide treatment for control of caterpillars
- When used in cold weather always add product to water and stir thoroughly before use
- Apply hop defoliant by spraying when bines 1.5-3 m high and direct spray downward at 45°

Crop safety/Restrictions
- Maximum number of treatments 2 per yr for hops
- Only spray fruit trees or bushes when fully dormant
- Do not spray hops if temperature is above 21°C or after cones have formed. Do not drench rootstocks
- Do not spray on windy, wet or frosty days

Special precautions/Environmental safety
- Harmful if swallowed. Irritating to eyes, skin and respiratory system
- Dangerous to fish or other aquatic life. Do not contaminate surface waters or ditches with chemical or used container

Personal protective equipment/Label precautions
- A, C, H
- M03, R03c, R04a, R04b, R04c, U05a, U08, U19, U20a, C03, E01, E13b, E26, E27, E30a, E31a, E34

Latest application/Harvest interval
- Before any signs of bud swelling for fruit crops

---

458   tau-fluvalinate

A contact pyrethroid insecticide for cereals and oilseed rape

Products

Mavrik	Novartis	240 g/l	EW	09697

Uses

Aphids in *oilseed rape, spring barley, spring wheat, winter barley, winter wheat*. Barley yellow dwarf virus vectors in *winter barley, winter wheat*. Pollen beetles in *oilseed rape*.

Efficacy
- For BYDV control on winter cereals follow local warnings or spray high risk crops in mid-Oct and make repeat application in late autumn/early winter if aphid activity persists
- For summer aphid control on cereals spray once when aphids present on two thirds of ears and increasing
- On oilseed rape treat peach potato aphids in autumn in response to local warning and repeat if necessary
- Best control of pollen beetle in oilseed rape obtained from treatment at green to yellow bud stage and repeat if necessary
- Good spray cover of target essential for best results

Crop safety/Restrictions
- Maximum total dose varies with crop. See label for details
- A minimum of 14 d must elapse between applications to cereals

Special precautions/Environmental safety
- Irritating to eyes and skin
- Extremely dangerous to fish or other aquatic life. Do not contaminate surface waters or ditches with chemical or used container
- LERAP Category A
- Presents low risk to many beneficial species when used as directed but some reduction in numbers may occur
- Avoid spraying oilseed rape within 6 m of field boundary to reduce effects on certain non-target species or other arthropods
- Must not be applied to cereals if any product containing a pyrethroid insecticide or dimethoate has been sprayed after the start of ear emergence (GS 51)

Personal protective equipment/Label precautions
- A, C, P
- R04a, R04b, U05a, U10, U11, U19, U20a, C03, E01, E13a, E16c, E16d, E22a, E26, E30a, E31b

Latest application/Harvest interval
- Before caryopsis watery ripe (GS 71) for barley; before flowering for oilseed rape; before kernel medium milk (GS 75) for wheat

Approval
- Accepted by BLRA for use on malting barley

459   tebuconazole

A systemic conazole fungicide for cereals and other field crops

See also carbendazim + tebuconazole
fenpropidin + propiconazole + tebuconazole
fenpropidin + tebuconazole
imidacloprid + tebuconazole + triazoxide
prochloraz + tebuconazole
propiconazole + tebuconazole
spiroxamine + tebuconazole

Products

1	Folicur	Bayer	250 g/l	EW	08691
2	Me2 Tebuconazole	Me2	250 g/l	EW	09751
3	Standon Tebuconazole	Standon	250 g/l	EC	09056

Uses

Alternaria in **brussels sprouts, cabbages, carrots** *(off-label)*, **horseradish** *(off-label)*, **kohlrabi** *(off-label)* [1]. Alternaria in **oilseed rape** [1-3]. Botrytis in **linseed** *(reduction)* [1-3]. Brown rust in **barley, rye, wheat** [1]. Brown rust in **spring barley, spring wheat, winter barley, winter rye, winter wheat** [2, 3]. Canker in **parsnips** *(off-label)* [1]. Chocolate spot in **field beans** [1-3]. Crown rust in **oats** *(reduction)* [1]. Fusarium ear blight in **wheat** [1]. Glume blotch in **wheat** [1]. Light leaf spot in **kohlrabi** *(off-label)* [1]. Light leaf spot in **brussels sprouts, cabbages, oilseed rape** [1-3]. Net blotch in **barley** [1]. Net blotch in **spring barley, winter barley** [2, 3]. Phoma leaf spot in **oilseed rape** [1]. Powdery mildew in **barley, herbs** *(off-label)*, **linseed, rye, wheat** [1]. Powdery mildew in **brussels sprouts, cabbages, oats, swedes, turnips** [1-3]. Powdery mildew in **spring barley, spring wheat, winter barley, winter rye, winter wheat** [2, 3]. Rhynchosporium in **barley, rye** [1]. Rhynchosporium in **spring barley, winter barley, winter rye** [2, 3]. Ring spot in **broccoli** *(off-label)*, **calabrese** *(off-label)*, **cauliflowers** *(off-label)*, **chinese cabbage** *(off-label)*, **collards** *(off-label)*, **kale** *(off-label)*, **kohlrabi** *(off-label)*, **oilseed rape** *(reduction)* [1]. Ring spot in **brussels sprouts, cabbages** [1-3]. Rust in **dwarf beans** *(off-label)*, **french beans** *(off-label)*, **herbs** *(off-label)*, **runner beans** *(off-label)* [1]. Rust in **leeks** [1, 2]. Rust in **field beans** [1-3]. Sclerotinia stem rot in **oilseed rape** [1-3]. Septoria diseases in **spring wheat, winter wheat** [2, 3]. Septoria leaf spot in **wheat** [1]. Sooty moulds in **wheat** [1]. Stem canker in **oilseed rape** [1]. White rot in **bulb onions** *(off-label)*, **garlic** *(off-label)*, **salad onions** *(off-label)*, **shallots** *(off-label)* [1]. Yellow rust in **barley, rye, wheat** [1]. Yellow rust in **spring barley, spring wheat, winter barley, winter rye, winter wheat** [2, 3].

Efficacy

- For best results apply at an early stage of disease development before infection spreads to new crop growth
- To protect flag leaf and ear from Septoria diseases apply from flag leaf emergence to ear fully emerged (GS 37-59). Earlier application may be necessary where there is a high risk of infection
- For light leaf spot control in oilseed rape apply in autumn/winter with a follow-up spray in spring/summer if required
- For control of most other diseases spray at first signs of infection with a follow-up spray 2-4 wk later if necessary. See label for details
- For disease control in Brussels sprouts and cabbages a 3-spray programme at 21-28 d intervals will give good control

Crop safety/Restrictions

- Maximum total dose equivalent to 1 full dose treatment on linseed; 2 full dose treatments on wheat, barley, rye, field beans, swedes, turnips; 2.25 full dose treatments on Brussels sprouts, cabbages; 2.5 full dose treatments on oilseed rape; 3 full dose treatments on leeks
- Some transient leaf speckling on wheat or leaf reddening/scorch on oats may occur but this has not been shown to reduce yield response to disease control
- Do not treat durum wheat
- Apply only to listed oat varieties (see label)
- Do not apply before swedes and turnips have a root diameter of 2.5 cm, or before start of button formation in Brussels sprouts or heart formation in cabbages

Special precautions/Environmental safety
- Harmful if swallowed. Risk of serious damage to eyes
- Irritating to skin
- Harmful to fish or other aquatic life and aquatic plants. Do not contaminate surface waters or ditches with chemical or used container

Personal protective equipment/Label precautions
- A, C, H [1-3]
- M03, R03c, R04b, R04d, U05a, U11, U20a, C03 [1-3]; E31b, E34 [1, 3]; E01, E13c, E26, E30a [1-3]; E31a [2]

Latest application/Harvest interval
- Before grain milky-ripe for cereals (GS 71); when most seed green-brown mottled for oilseed rape (GS 6,3); before brown capsule for linseed
- HI field beans, swedes, turnips 35 d; Brussels sprouts, cabbages 21 d; linseed 35 d; leeks 14 d

Approval
- Off-label approval unlimited for use on cauliflower, calabrese, broccoli, Chinese cabbage (OLA 2048/97)[1]; to Apr 2002 for use on bulb onions, garlic, shallots (OLA 2062/97)[1]; to July 2002 for use on outdoor parsnips, carrots and horseradish (OLA 1588/98)[1]; unlimited for use on outdoor leafy herbs - see notice for list of species (OLA 3020/98)[1]; unstipulated for use on kale, collards (OLA 0510/99)[1]; to Jul 2003 for use at higher dose on kale, collards (OLA 0511/99)[1]; unstipulated for use on outdoor salad onions (OLA 1447/99)[1]; unstipulated for use on outdoor kohl rabi (OLA 0072/00)[1]; unstipulated for use on runner beans, French beans, dwarf green beans (OLA 0073/00)[1]
- Accepted by BLRA for use on malting barley

460   tebuconazole + triadimenol

A broad spectrum systemic fungicide for cereals

Products

1	Garnet	Bayer	250:125 g/l	EC	06391
2	Silvacur	Bayer	250:125 g/l	EC	06387
3	Veto F	Bayer	225:75 g/l	EC	08057

Uses

Brown rust in **barley, rye, wheat** [1-3]. Crown rust in **oats** [1, 2]. Fusarium ear blight in **wheat** [1-3]. Glume blotch in **wheat** [1-3]. Leaf spot in **wheat** [1-3]. Net blotch in **barley** [1-3]. Powdery mildew in **oats** [1, 2]. Powdery mildew in **barley, rye, wheat** [1-3]. Rhynchosporium in **barley** [1-3]. Sooty moulds in **wheat** [1-3]. Yellow rust in **barley, rye, wheat** [1-3].

Efficacy
- For best results apply at an early stage of disease development before infection spreads to new crop growth
- To protect flag leaf and ear from Septoria diseases apply from flag leaf emergence to ear fully emerged (GS 37-59). Earlier application may be necessary where there is a high risk of infection
- For control of rust, powdery mildew, leaf and net blotch apply at first signs of disease with a second application 2-3 wk later if necessary

Crop safety/Restrictions
- Maximum total dose equivalent to two full dose treatments
- Do not use on durum wheat
- Use only on listed varieties of oats. See label [1, 2]
- Some transient leaf speckling may occur on wheat or leaf reddening/scorch on oats but this has not been shown to reduce yield response or disease control

Special precautions/Environmental safety
- Irritating to eyes
- Harmful to fish or other aquatic life. Do not contaminate surface waters or ditches with chemical or used container

Personal protective equipment/Label precautions
- A, C, H
- M03, R04a, U05a, U09b, U20a, C03, E01, E13c, E30a, E31c, E34

Latest application/Harvest interval
- Before grain milky-ripe (GS 71)

Approval
- Accepted by BLRA for use on malting barley

461   tebuconazole + triazoxide

A triazole and benzotriazine fungicide seed treatment for use in barley

Products

Raxil S	Bayer	20:20 g/l	FS	06974

Uses

Leaf stripe in *spring barley* (seed treatment), *winter barley* (seed treatment). Loose smut in *spring barley* (seed treatment), *winter barley* (seed treatment). Net blotch in *spring barley* (seed treatment), *winter barley* (seed treatment).

Efficacy
- Best applied through recommended seed treatment machines
- Evenness of seed cover improved by simultaneous application of equal volumes of product and water or dilution of product with an equal volume of water
- Diluted product must be used immediately
- Drill treated seed in the same season

Crop safety/Restrictions
- Maximum number of treatments 1 per batch of seed
- Slightly delayed and reduced emergence may occur but this is normally outgrown
- Any delay in field emergence, for whatever reason, may be accentuated by treatment
- Do not use on seed with more than 16% moisture content, or on sprouted, cracked or skinned seed

Special precautions/Environmental safety
- Harmful to fish or other aquatic life. Do not contaminate surface waters or ditches with chemical or used container
- Do not use treated seed as food or feed
- Treated seed harmful to game and wildlife
- Product also supplied in returnable containers. See label for guidance on handling, storage, protective clothing and precautions

Personal protective equipment/Label precautions
- A, H
- U20b, E03, E13c, E30a, E32a, S01, S02, S03, S04b, S05, S06, S07

Latest application/Harvest interval
- Before drilling

Approval
- Accepted by BLRA for use as a seed treatment on malting barley

462   tebufenpyrad

A pyrazole mitochondrial electron transport inhibitor aphicide and acaricide

Products

1 Masai	BASF	20% w/w	WB	10223
2 Masai	Cyanamid	20% w/w	WB	07452

Uses

Damson-hop aphid in *hops*. Red spider mites in *apples, pears*. Two-spotted spider mite in *hops, protected roses, strawberries*.

Efficacy
- Acts on eggs (except winter eggs) and all motile stages of spider mites up to adults
- Treat spider mites from 80% egg hatch but before mites become established
- For effective control total spray cover of the crop is required

- Product can be used in a programme to give season-long control of damson/hop aphids coupled with mite control
- Where aphids resistant to tebufenpyrad occur in hops control is unlikely to be satisfactory and repeat treatments may result in lower levels of control. Where possible use different active ingredients in a programme

Crop safety/Restrictions
- Maximum total dose equivalent to one full dose treatment on apples, pears, strawberries; 3 full dose treatments on hops
- Other mitochondrial electron transport inhibitor (METI) acaricides should not be applied to the same crop in the same calendar yr either separately or in mixture
- Do not treat apples before 90% petal fall
- Product has no effect on fruit quality or finish
- Small-scale testing of rose varieties to establish tolerance recommended before use
- Product not recommended for use in hand-held sprayers
- Inner liner of container must not be removed

Special precautions/Environmental safety
- Dangerous to fish or other aquatic life. Do not contaminate surface waters or ditches with chemical or used container
- High risk to bees. Do not apply to crops in flower or to those in which bees are actively foraging. Do not apply when flowering weeds are present
- LERAP Category B

Personal protective equipment/Label precautions
- A
- U02, U09a, U13, U14, U20b, U22, E12a, E13b, E16a, E30a, E32a; E17 (18 m)

Latest application/Harvest interval
- End of burr stage for hops
- HI apples 7d

Approval
- Accepted by BLRA for use on hops

---

463   tebutam

A pre-emergence amide herbicide used in brassica crops

Products

Comodor 600	Agrichem	600 g/l	EC	08398

Uses

Annual dicotyledons in *broccoli, brussels sprouts, cabbages, calabrese, cauliflowers, fodder rape, kale, swedes, turnips, winter oilseed rape*. Annual grasses in *broccoli, brussels sprouts, cabbages, calabrese, cauliflowers, fodder rape, kale, swedes, turnips, winter oilseed rape*. Volunteer cereals in *broccoli, brussels sprouts, cabbages, calabrese, cauliflowers, fodder rape, kale, swedes, turnips, winter oilseed rape*.

Efficacy
- Acts via roots and inhibits germination of weeds. Emerged weeds are not controlled
- Best results are achieved on seedbeds which are weed free and have a good tilth without clods
- Effectiveness requires adequate soil moisture. Under very dry conditions light rain or irrigation may be needed but heavy rain on light soils may result in loss of control
- On some heavy soils there may be a reduction in weed control. Do not use on soils with more than 10% organic matter
- Recommended for use in tank-mixture with propachlor and as a tank-mix or in sequence with trifluralin. Apply trifluralin tank-mix as a pre-plant incorporated treatment
- Emerged weeds can be controlled pre-crop emergence by tank mixtures with paraquat products. See label for details

Crop safety/Restrictions
- Maximum number of treatments 1 per crop
- Apply to brassica crops before drilling or planting in tank-mixture with trifluralin as a soil incorporated treatment. Do not use on brassica seedbeds

- Apply to oilseed rape or sown or transplanted brassicas shortly after drilling or planting but before weeds or crop germinate
- Transplants must be hardened off before treatment. Do not use post-planting on block or modular propagated brassicas
- Soil must be deep mouldboard ploughed before drilling or planting any following crop except brassicas
- In the event of crop failure only brassicas, oilseed rape, dwarf beans, maize, peas or potatoes may be grown in the same season

Special precautions/Environmental safety
- Irritating to eyes and skin
- Flammable
- Harmful to fish or other aquatic life. Do not contaminate surface waters or ditches with chemical or used container

Personal protective equipment/Label precautions
- A, C
- R04a, R04b, R07d, U05a, U08, U19, U20b, C03, E01, E13c, E26, E30a, E31c

Latest application/Harvest interval
- Pre-crop emergence for oilseed rape and other drilled named brassicas; 3 wk after transplanting for transplanted brassicas

---

464   tecnazene

A protectant chlorobenzene fungicide and potato sprout suppressant

See also carbendazim + tecnazene

Products

1	Hickstor 10 Granules	Hickson & Welch	10% w/w	GR	03121
2	Hickstor 5 Granules	Hickson & Welch	5% w/w	GR	03180
3	Hortag Tecnazene 10% Granules	Hortag	10% w/w	GR	03966
4	Hortag Tecnazene Double Dust	Hortag	6% w/w	DP	01072
5	Hortag Tecnazene Dust	Hortag	3% w/w	DP	01074
6	Hortag Tecnazene Potato Granules	Hortag	5% w/w	GR	01075
7	Hystore 10	Agrichem	10% w/w	GR	03581
8	Hytec	Agrichem	3% w/w	DP	01099
9	Hytec 6	Agrichem	6% w/w	DP	03580
10	New Hickstor 6	Hickson & Welch	6% w/w	DS	04221
11	New Hystore	Agrichem	5% w/w	GR	01485
12	Tripart Arena 10G	Hickson & Welch	10% w/w	GR	05603
13	Tripart Arena 5G	Hickson & Welch	5% w/w	GR	05604
14	Tripart New Arena 6	Hickson & Welch	6% w/w	DP	05813

Uses

Dry rot in **seed potatoes, ware potatoes**. Sprout suppression in **ware potatoes**.

Efficacy
- Apply dust or granules evenly to potatoes with appropriate equipment (see label for details) during loading. Works by vapour phase action so clamps or bins should be covered to avoid loss of vapour
- Best results achieved on dry, dirt-free potatoes. Fungicidal activity best at 12-14°C, sprouting control best at 9-12°C. For best sprouting control maintain temperature at 10-14°C for first 14 d, then reduce to 7°C
- Under suitable storage conditions sprouting prevented for 4-6 mth
- Excessive ventilation shortens control period
- Sprouting not controlled in tubers which have already broken dormancy
- Treatment can be used in sequence with chlorpropham treatment [7]
- Treatment gives some control of skin spot, gangrene and silver scurf

Crop safety/Restrictions
- Maximum number of treatments 1 per batch of potatoes
- Air tubers for 6 wk before planting; chitting should have commenced
- Treated tubers must not be removed for sale or processing, including washing, for at least 6 wk after application
- To prevent contamination of other crops remove all traces of treated potatoes and soil after the store has been emptied

Special precautions/Environmental safety
- Dangerous to fish or other aquatic life. Do not contaminate surface waters or ditches with chemical or used container [1, 2, 12, 13]
- Do not remove from store for sale or processing (including washing) for at least 6 wk after application

Personal protective equipment/Label precautions
- U20c [1-6, 10, 12-14]; U19 [2-5, 7-14]; U20a [8, 9, 11]; U20b [7]; C02 [8, 9, 11]; E30a, E32a [1-14]; E15 [3-11, 14]; E31a, E34 [8, 9, 11]; E13b [1, 2, 12, 13]

Latest application/Harvest interval
- 6 wk before removal from store for sale or processing

Approval
- All UK approvals for products containing tecnazene were revoked in Jan 2000. Existing stocks may continue to be used until 31 Jan 2002

465   tecnazene + thiabendazole

A fungicide and potato sprout suppressant

Products

Hytec Super	Agrichem	6:1.8% w/w	DS	01100

Uses

Dry rot in *seed potatoes, ware potatoes*. Gangrene in *seed potatoes, ware potatoes*. Silver scurf in *seed potatoes, ware potatoes*. Skin spot in *seed potatoes, ware potatoes*. Sprout suppression in *ware potatoes*.

Efficacy
- Apply dust evenly with suitable dusting machine (not by hand) as tubers enter store and cover as soon as possible
- Treat tubers as soon as possible after lifting, no longer than 14 d
- Under suitable storage conditions sprouting prevented for 3-6 mth. Effectiveness may be decreased with excessive ventilation or inadequate covering
- Best results achieved on dry, dirt-free tubers
- Sprouting not controlled on tubers that have broken dormancy

Crop safety/Restrictions
- Maximum number of treatments 1 per batch
- Treated tubers must not be removed for sale or processing, including washing, for at least 6 wk after application
- Air tubers for 6 wk before planting, chitting should have commenced

Special precautions/Environmental safety
- Harmful to fish or other aquatic life. Do not contaminate surface waters or ditches with chemical or used container

Personal protective equipment/Label precautions
- A, C, D, H
- U19, U20a, E13c, E30a, E32a, E34

Latest application/Harvest interval
- 6 wk before removal from store for sale or processing (including washing)

Approval
- All UK approvals for products containing tecnazene were revoked in Jan 2000. Existing stocks may continue to be used until 31 Jan 2002

466   teflubenzuron

A benzoylurea insecticide for use on ornamentals

Products

1 Nemolt	Fargro	150 g/l	SC	10226
2 Nemolt	Fargro	150 g/l	SC	07012

Uses

Browntail moth in **ornamental plant production**. Caterpillars in **ornamental plant production**. Whitefly in **ornamental plant production**.

Efficacy
- Product acts as larval stomach poison interfering with moulting process leading to cessation of feeding and larval death
- Apply as soon as first stage larvae seen. This will often coincide the peak of moth flight

Crop safety/Restrictions
- Maximum number of treatments 3 per yr
- Test specific varieties before carrying out extensive treatments

Special precautions/Environmental safety
- Limited evidence suggests some margin of safety to *Encarsia formosa*. Effects on other parasites and predators not fully tested

Personal protective equipment/Label precautions
- A, C
- U20c, E15, E26, E30a, E32a

467   tefluthrin

A soil acting pyrethroid insecticide seed treatment

Products

1 Evict	Bayer	100 g/l	CS	08731
2 Force ST	Bayer	200 g/l	LS	09713

Uses

Bean seed flies in **bulb onions** *(off-label - seed treatment)*, **herbs** *(off-label - seed treatment)*, **leeks** *(off-label - seed treatment)*, **salad onions** *(off-label - seed treatment)*, **spinach beet** *(off-label - seed treatment)*, **spinach** *(off-label - seed treatment)* [2]. Carrot fly in **outdoor carrots** *(off-label - seed treatment)*, **outdoor parsnips** *(off-label - seed treatment)* [2]. Millipedes in **fodder beet** *(seed treatment)*, **sugar beet** *(seed treatment)* [2]. Onion fly in **bulb onions** *(off-label - seed treatment)*, **leeks** *(off-label - seed treatment)*, **salad onions** *(off-label - seed treatment)* [2]. Pygmy beetle in **fodder beet** *(seed treatment)*, **sugar beet** *(seed treatment)* [2]. Springtails in **fodder beet** *(seed treatment)*, **sugar beet** *(seed treatment)* [2]. Symphylids in **fodder beet** *(seed treatment)*, **sugar beet** *(seed treatment)* [2]. Wheat bulb fly in **barley, wheat** [1]. Wireworms in **barley, oats, wheat** [1].

Efficacy
- Apply during process of pelleting seed. Consult manufacturer for details of specialist equipment required [2]
- Apply to cereal seed through a suitable liquid seed treater calibrated to achieve even coverage [1]
- Micro-capsule formulation allows slow release to provide a protection zone around treated seed during establishment [1]
- Product is non-systemic and may not protect against wireworm attack at the soil surface [1]

Crop safety/Restrictions
- Maximum number of treatments 1 per batch of seed
- Sow treated seed as soon as possible. Do not store treated seed from one drilling season to next
- Must be co-applied with suitable water-based fungicide seed treatment to protect against seed and soil borne diseases [1]
- Co-application with other seed treatments may block or damage application equipment [1]

- Do not use on seed that is sprouted, cracked, damaged or over 16% moisture content [1]
- Where wireworm levels exceed 1.25 million per ha, increased crop damage may occur

Special precautions/Environmental safety
- Irritant. May cause sensitization by skin contact
- Powered visor respirator must be worn when bagging treated seed [1]
- Can cause a transient tingling or numbing sensation to exposed skin. Avoid skin contact with product and treated seed
- Do not handle seed unnecessarily
- Extremely dangerous to fish or other aquatic life. Do not contaminate surface waters or ditches with chemical or used container
- Do not use treated seed as food or feed
- Keep treated seed secure from people, domestic stock/pets and wildlife at all times during storage and use [1]
- Treated seed harmful to game and wild life. Bury spillages
- Do not reuse sacks or containers that have been used for treated seed for food or feed
- Do not apply treated seed from the air

Personal protective equipment/Label precautions
- A [1]; C, D, H [1, 2]; B [2]
- R04 [1]; R04e [1, 2]; U07, U14 [1]; U05a [1, 2]; U02, U04a, U08, U20a [2]; C03 [1, 2]; E26, E32a, E36 [1]; E01, E13a [1, 2]; E30a, E31a [2]; S03 [1]; S01, S02, S04b, S05, S06, S07 [1, 2]

Withholding period
- Keep livestock out of areas drilled with treated seed for at least 80 d [2]

Latest application/Harvest interval
- Before drilling seed

Approval
- Off-label approval unstipulated for use on outdoor carrots and parsnips (OLA 0873/00)[2]
- Accepted by BLRA for use a seed treatment on malting barley
- Approval expiry 31 Jul 2001 [1]

---

468   terbacil

A residual uracil herbicide for use in asparagus and herbs

Products

Sinbar	DuPont	80% w/w	WP	01956

Uses

Annual dicotyledons in **asparagus, outdoor herbs** *(off-label)*, **protected herbs** *(off-label)*. Annual grasses in **asparagus, outdoor herbs** *(off-label)*, **protected herbs** *(off-label)*. Bent grasses in **asparagus, outdoor herbs** *(off-label)*, **protected herbs** *(off-label)*. Couch in **asparagus, outdoor herbs** *(off-label)*, **protected herbs** *(off-label)*.

Efficacy
- Has some foliar activity but uptake mainly through roots
- Adequate rainfall needed to ensure penetration to weed root zone
- Best results achieved by application in early spring, before mid-Apr

Crop safety/Restrictions
- Maximum number of treatments 1 per yr
- Apply to asparagus established at least 2 yr. Do not apply after first spears emerge
- Do not use on Sands, Loamy Sand or gravels or soils with less than 1% organic matter
- Do not use for at least 2 yr prior to grubbing and replanting

Personal protective equipment/Label precautions
- U08, U19, U20a, E15, E30a, E31a

Latest application/Harvest interval
- Before spears emerge for asparagus; pre-emergence of new growth in spring for herbs

Approval
- Off-label approval unlimited for use in range of herbs, see off-label approval notice for details (OLA 0082/93)

469   terbuthylazine

A triazine herbicide available only in mixtures

See also bromoxynil + terbuthylazine
cyanazine + terbuthylazine
diflufenican + terbuthylazine

470   terbuthylazine + terbutryn

A pre-emergence herbicide for peas, beans, lupins and potatoes

Products

1 Batallion	Makhteshim	150:350 g/l	SC	08305
2 Opogard	Novartis	150:350 g/l	SC	08427

Uses

Annual dicotyledons in *edible podded peas* [1]. Annual dicotyledons in *broad beans, combining peas, field beans, potatoes, vining peas* [1, 2]. Annual dicotyledons in *lupins* [2]. Annual grasses in *edible podded peas* [1]. Annual grasses in *broad beans, combining peas, field beans, potatoes, vining peas* [1, 2]. Annual grasses in *lupins* [2].

Efficacy

- Active via root uptake and with foliar activity on cotyledon stage weeds
- Best results achieved by application to fine, firm, moist seedbed, preferably at weed emergence, when rain falls after spraying
- Effectiveness may be reduced by excessive rain, drought or cold
- Residual control lasts for up to 8 wk on mineral soils. Effectiveness reduced on highly organic soils and subsequent use of post-emergence treatment recommended
- Do not cultivate after treatment
- Do not use on soils which are very cloddy or have more than 10% organic matter

Crop safety/Restrictions

- Maximum number of treatments 1 per crop
- Apply to spring sown peas, beans and lupins as soon as possible after drilling
- Crop seed must be covered by at least 25 mm of settled soil
- Vedette and Printana peas may be damaged by treatment. Do not use on forage peas
- Heavy rain after application may cause damage to peas on light soils
- Do not treat peas, beans or lupins on soils lighter than loamy fine sand, potatoes on soils lighter than loamy sand or on very stony soils or lupins on silty clay soil
- Subsequent crops may be sown or planted after 12 wk (14 wk if prolonged drought)

Special precautions/Environmental safety

- Harmful if swallowed
- Harmful to fish or other aquatic life. Do not contaminate surface waters or ditches with chemical or used container

Personal protective equipment/Label precautions

- A, C [2]
- M03 [1]; R03c [1, 2]; U02, U13, U14, U15 [1]; U05a, U08, U19, U20a, C03, E01, E13c, E30a, E31b, E34 [1, 2]

Latest application/Harvest interval

- 3 d before emergence for peas, beans and lupins; before 10% of plants emerged for potatoes

471   terbutryn

A residual triazine herbicide for cereals and aquatic weed control

See also fomesafen + terbutryn
terbuthylazine + terbutryn

Products

1	Alpha Terbutryn 50 SC	Makhteshim	500 g/l	SC	04809
2	Clarosan	Scotts	1% w/w	GR	09394
3	Prebane	Novartis	490 g/l	SC	08432

Uses

Annual dicotyledons in *winter barley, winter oats, winter wheat* [1, 3]. Annual grasses in *winter oats* [3]. Annual meadow grass in *winter barley, winter oats, winter wheat* [1, 3]. Aquatic weeds in *areas of water* [2]. Blackgrass in *winter barley, winter wheat* [1, 3]. Perennial ryegrass in *winter barley, winter wheat* [1, 3]. Rough meadow grass in *winter barley, winter oats, winter wheat* [1, 3].

Efficacy

- Apply in autumn after drilling but before crop emergence. Best results given on fine, firm seedbed in which any trash or ash dispersed. Do not use on soils with more than 10% organic matter [1, 3]
- Effectiveness reduced by long, dry period after spraying or by waterlogging [1, 3]
- Effectiveness reduced by excessive surface trash and an impermeable surface layer [1, 3]
- Do not use on soils with more than 10% organic matter [1, 3]
- Do not harrow after treatment [1, 3]
- For aquatic weed control apply granules to water surface with suitable equipment, when growth active but before heavy infestation develops, usually Apr-May, sometimes to Aug [2]
- Use only in static or sluggishly moving water. The flow in moving water should be stopped for at least 7 d after treatment otherwise weed control may be reduced. Treat small ponds before start of new growth [2]
- Effectiveness reduced in water with peaty bottom [2]

Crop safety/Restrictions

- Maximum number of treatments 1 per crop
- Do not use on crops drilled after 30 Nov [1, 3]
- Do not use on spring barley cultivars sown in autumn [1, 3]
- Do not treat winter oats with dose higher than that recommeded otherwise serious crop damage may result. Winter oats in unconsolidated seedbeds may also be damaged [1, 3]
- Do not treat winter barley on Sands or Very Light soils at recommended dose [1, 3]
- Only use in programme with tri-allate in winter barley on Medium and heavier soils [1, 3]
- Before spraying direct-drilled crops ensure seed well covered and drill slots closed [1, 3]
- Risk of crop damage on stony or gravelly soils especially if heavy rain falls soon after treatment [1, 3]

Special precautions/Environmental safety

- Do not use treated water for irrigation purposes within 7 d of treatment, or dump surplus herbicide in water or ditch bottoms [2]
- For maximum concentration in water see label [2]
- Harmful to fish or other aquatic life. Do not contaminate surface waters or ditches with chemical or used container [1, 3]
- If dense weed growth in watercourses to be controlled without de-oxygenation treat in sections of about 400 m at a time with intervals of at least 14 d [2]

Personal protective equipment/Label precautions

- U20a [1, 3]; U20c [2]; E13c, E31b [1, 3]; E30a [1-3]; E26 [3]; E19, E21, E29, E32a [2]

Latest application/Harvest interval

- Aug for areas of water [2]; before crop emergence for cereals [1, 3]

Approval

- Approved for aquatic weed control [2]. See notes in Section 1 on use of herbicides in or near water
- Accepted by BLRA for use on malting barley [1, 3]

472   terbutryn + trietazine

A residual triazine herbicide for peas, potatoes and field beans

Products

Senate	Aventis	250:250 g/l	SC	07279

Uses

Annual dicotyledons in *peas, potatoes, spring field beans*. Annual meadow grass in *peas, potatoes, spring field beans*.

Efficacy

- Product acts mainly through roots of germinating seedlings but also has some contact effect on cotyledon stage weeds
- Weeds germinating from deeper than 2.5 cm may not be completely controlled
- Weed control is improved by light rain after spraying but reduced by excessive rain, drought or cold
- Persistence may be reduced on soils with high organic matter content

Crop safety/Restrictions

- Maximum number of treatments 1 per crop
- Apply between drilling and 5% emergence (peas), pre-emergence (field beans) between sowing and 10% emergence (potatoes), provided weeds not beyond cotyledon stage
- May be used on spring sown vining, combining or forage peas and field beans covered by not less than 3 cm soil
- Do not spray any winter sown peas or beans sown before Jan
- May be used on early and maincrop potatoes, including seed crops
- On stony or gravelly soils there is risk of crop damage, especially if heavy rain falls soon after treatment. Frost after application may also check crop
- Do not use on soils classed as Sands
- Plough or cultivate to 15 cm before sowing or planting another crop. Any crop may be sown or planted after 12 wk (14 wk with drilled brassicas)
- If sprayed pea crop fails redrill without ploughing but do not respray

Special precautions/Environmental safety

- Harmful if swallowed
- Harmful to fish or other aquatic life. Do not contaminate surface waters or ditches with chemical or used container

Personal protective equipment/Label precautions

- A, C
- M03, R03c, U05a, U08, U20b, C03, E01, E13c, E26, E30a, E31b, E34

Latest application/Harvest interval

- Pre-emergence of crop for field beans; 5% emergence for peas; 10% emergence for potatoes

473   terbutryn + trifluralin

A residual, pre-emergence herbicide for winter cereals

Products

Alpha Terbalin 35 SC	Makhteshim	150:200 g/l	SC	04792

Uses

Annual dicotyledons in *winter barley, winter wheat*. Annual grasses in *winter barley, winter wheat*. Annual meadow grass in *winter barley, winter wheat*. Blackgrass in *winter barley, winter wheat*. Chickweed in *winter barley, winter wheat*. Mayweeds in *winter barley, winter wheat*. Speedwells in *winter barley, winter wheat*.

Efficacy

- Apply as soon as possible after drilling, before weed or crop emergence when soil moist
- Best results achieved by application to fine, firm, moist seedbed when adequate rain falls after application. Do not use on crops drilled after 30 Nov

- Do not use on uncultivated stubbles or soils with trash or more than 10% organic matter. Do not harrow after treatment
- Effectiveness reduced in mild, wet winters, especially in south-west

Crop safety/Restrictions
- Maximum number of treatments 1 per crop
- Crops should only be treated if drilled 25-35 mm deep
- Do not use on sandy soils (Coarse Sandy Loam - Coarse Sand) or on spring-drilled crops
- Do not treat undersown crops or crops to be undersown
- Crops suffering stress due to waterlogging, deficiency, poor seedbed preparation or pest problems may be damaged
- Any crop may be sown following harvest of treated crop. In the event of crop failure only resow with wheat or barley

Special precautions/Environmental safety
- Irritating to skin, eyes and respiratory system
- Harmful to fish or other aquatic life. Do not contaminate surface waters or ditches with chemical or used container

Personal protective equipment/Label precautions
- A, C
- R04a, R04b, U05a, U08, U13, U19, U20b, C03, E01, E13c, E30a, E31b

Latest application/Harvest interval
- Pre-emergence of crop

Approval
- Accepted by BLRA for use on malting barley

---

474   tetraconazole

A systemic, protectant and curative triazole fungicide for cereals

See also chlorothalonil + tetraconazole

Products

1 Eminent	Monsanto	125 g/l	ME	09410
2 Juggler	Sipcam	100 g/l	EC	09391

Uses

Brown rust in **spring barley, spring wheat, winter barley, winter wheat**. Crown rust in **spring oats, winter oats**. Powdery mildew in **spring barley, spring wheat, winter barley, winter wheat**. Septoria leaf spot in **spring wheat, winter wheat**. Sooty moulds in **spring wheat, winter wheat**. Yellow rust in **spring barley, spring wheat, winter barley, winter wheat**.

Efficacy
- Apply at any time from the tillering stage up to the latest times indicated for each crop
- Best results achieved from applications before disease becomes established
- If disease pressure persists a second treatment may be required to prevent late season attacks
- Best control of ear diseases in wheat obtained from application at or after the ear completely emerged stage

Crop safety/Restrictions
- Maximum total dose equivalent to three full dose aplications on wheat and two full dose applications on other cereals

Special precautions/Environmental safety
- Harmful if swallowed
- Irritating to eyes and skin
- Harmful to fish or other aquatic life. Do not contaminate surface waters or ditches with chemical or used container
- LERAP Category B

Personal protective equipment/Label precautions
- A, C, H
- M03, R03c, R04a, R04b, U05a, U19, U20a, C03, E01, E13c, E16a, E16b, E26, E30a, E31b

Latest application/Harvest interval
- Before end of ear emergence (GS 59) for oats; before end of flowering (GS 69) for barley; before grain watery ripe (GS 71) for wheat

475   tetradifon

A bridged-diphenyl acaricide for use in horticultural crops

Products

Tedion V-18 EC	Hortichem	80 g/l	EC	03820

Uses

Red spider mites in **apples, apricots** (protected crops), **blackberries, blackcurrants, cherries, gooseberries, grapevines, hops, loganberries, nectarines** (protected crops), **peaches** (protected crops), **pears, plums, protected beans, protected cucumbers, protected flowers, protected grapevines, protected melons, protected ornamentals, protected peppers, protected tomatoes, raspberries, strawberries**.

Efficacy
- Chemical active against summer eggs and all stages of red spider larvae but does not kill adult mites
- Apply as soon as first mites seen and repeat as required. See label for details of timing
- May be used in conjunction with biological control
- Resistant strains of red spider mite have developed in some areas

Crop safety/Restrictions
- Some rose cultivars are slightly susceptible. Do not treat cissus, dahlia, ficus, kalanchoe or primula

Special precautions/Environmental safety
- Harmful in contact with skin and if swallowed. Irritating to eyes and skin
- Flammable

Personal protective equipment/Label precautions
- A, C
- M03, R03a, R03c, R04a, R04b, U05a, U08, U20a, C03, E01, E15, E30a, E31a, E34

Approval
- Accepted by BLRA for use on hops

476   tetramethrin

A contact acting pyrethroid insecticide

See also phenothrin + tetramethrin
resmethrin + tetramethrin

Products

Killgerm Py-Kill W	Killgerm	10.2% w/v	EC	H4632

Uses

Flies in **agricultural premises, livestock houses**.

Efficacy
- Dilute in accordance with directions and apply as space or surface spray

Crop safety/Restrictions
- Do not apply directly on food or livestock
- Remove exposed milk before application. Protect milk machinery and containers from contamination
- Do not use space sprays containing pyrethrins or pyrethroid more than once per wk in intensive or controlled environment animal houses in order to avoid development of resistance. If necessary, use a different control method or product

Special precautions/Environmental safety
- Harmful to fish or other aquatic life. Do not contaminate surface waters or ditches with chemical or used container

Personal protective equipment/Label precautions
- R07d, U08, U19, U20c, C06, C10, E05, E13c, E30a, E31a

FOR FULL CONDITIONS OF USE ALWAYS READ THE PRODUCT LABEL

477   thiabendazole

A systemic, curative and protectant benzimidazole (MBC) fungicide

See also carboxin + imazalil + thiabendazole
imazalil + thiabendazole
tecnazene + thiabendazole

Products

1	Hykeep	Agrichem	2% ww	DS	06744
2	Storite Clear Liquid	Banks	220 g/l	LS	08982
3	Storite Flowable	Banks	450 g/l	FS	08703
4	Tecto Superflowable Turf Fungicide	Vitax	500 g/l	SC	09258

Uses

Dollar spot in **managed amenity turf** [4]. Dry rot in **ware potatoes** *(post-harvest)* [1]. Dry rot in **seed potatoes** *(tuber treatment - post-harvest)*, **ware potatoes** *(tuber treatment - post-harvest)* [2, 3]. Dutch elm disease in **elm trees** *(off-label - injection)* [2]. Fungus diseases in **asparagus** *(off-label)* [2]. Fusarium basal rot in **narcissi** [2]. Fusarium patch in **managed amenity turf** [4]. Gangrene in **ware potatoes** *(post-harvest)* [1]. Gangrene in **seed potatoes** *(tuber treatment - post-harvest)*, **ware potatoes** *(tuber treatment - post-harvest)* [2, 3]. Red thread in **managed amenity turf** [4]. Silver scurf in **ware potatoes** *(post-harvest)* [1]. Silver scurf in **seed potatoes** *(tuber treatment - post-harvest)*, **ware potatoes** *(tuber treatment - post-harvest)* [2, 3]. Skin spot in **ware potatoes** *(post-harvest)* [1]. Skin spot in **seed potatoes** *(tuber treatment - post-harvest)*, **ware potatoes** *(tuber treatment - post-harvest)* [2, 3].

Efficacy

* Apply post-harvest treatment to potatoes using suitable equipment within 24 h of lifting. See label for details [1-3]
* For turf disease apply as a preventive spray or promptly when disease appears. Repeat as necessary at 14 d (Fusarium Patch) or 21 d (Dollar Spot) intervals [4]
* Apply to turf as spray after mowing and do not mow for at least 48 h. Best results on red thread obtained in combination with fertilizer [4]
* Apply to bulbs as dip. Ensure bulbs are clean [2]

Crop safety/Restrictions

* Maximum number of treatments 1 per batch for seed potato treatments; 2 per yr for bulbs [2]; 1 per batch as dip and module drench, or 10 per crop for drench in field grown asparagus [2]

Special precautions/Environmental safety

* Harmful to fish or other aquatic life. Do not contaminate surface waters or ditches with chemical or used container
* Treated seed potatoes must not be used for food or feed [2, 3]

Personal protective equipment/Label precautions

* D [1, 2]; A, C, H, M [1, 2, 4]; B, K [2]
* U19, U20c [1-3]; U20b [4]; E29, E32a [1]; E01 [1-3]; E13c, E30a [1-4]; E26, E31a [2-4]; E15 [4]; S01, S05 [2]

Latest application/Harvest interval

* Before planting for seed potatoes; 21 d before removal from store for sale, processing or consumption for ware potatoes.
* HI asparagus 6 mth [2]

Approval

* Product formulated for application with ULV equipment (see label for details) [2]
* Off-label Approval unlimited for use on asparagus (OLA 0525/95)[2]; unlimited for use on elm trees (OLA 0990/95)[2]

Maximum Residue Level (mg residue/kg food)

* citrus fruits 6; pome fruit, strawberries, broccoli, potatoes 5; bananas 3; tea, hops, meat, eggs 0.1; other crops 0.05

478   thiabendazole + thiram

A fungicide seed dressing mixture for field and vegetable crops

Products

1 Hy-TL	Agrichem	225:300 g/l	FS	06246
2 sHYlin	Agrichem	258:255 g/l	FS	10030

Uses

Ascochyta in **broad beans** *(seed treatment)*, **field beans** *(seed treatment)*, **peas** *(seed treatment)* [1]. Botrytis root rot in **flax** *(seed treatment)*, **linseed** *(seed treatment)* [2]. Damping off in **broad beans** *(seed treatment)*, **field beans** *(seed treatment)*, **peas** *(seed treatment)* [1]. Damping off in **carrots** *(seed treatment)*, **flax** *(seed treatment)*, **linseed** *(seed treatment)*, **lupins** *(seed treatment)*, **parsnips** *(seed treatment)* [2]. Fusarium root rot in **flax** *(seed treatment)*, **linseed** *(seed treatment)* [2].

Efficacy
- Dress seed as near to sowing as possible
- Dilution may be needed with particularly absorbent types of seed. If diluted material used, seed may require drying before storage

Crop safety/Restrictions
- Maximum number of treatments 1 per batch
- Seed to be treated should be of satisfactory quality and moisture content

Special precautions/Environmental safety
- Harmful if swallowed and in contact with skin
- Irritating to eyes, skin and respiratory system
- Dangerous to fish or other aquatic life. Do not contaminate surface waters or ditches with chemical or used container
- Do not use treated seed as food or feed
- Treated seed harmful to game and wildlife

Personal protective equipment/Label precautions
- A, C, D, H, M [1, 2]
- M03, R03a, R03c, R04a, R04b, R04c [1, 2]; U20b [1]; U05a, U08, U20a, C03, E01, E03, E13b, E26, E30a, E31c, E34, S01, S02, S03, S04b, S05, S06, S07 [1, 2]

Latest application/Harvest interval
- Before drilling

Maximum Residue Level (mg residue/kg food)
- see thiabendazole entry

479   thifensulfuron-methyl

A translocated sulfonylurea herbicide

See also carfentrazone-ethyl + thifensulfuron-methyl
flupyrsulfuron-methyl + thifensulfuron-methyl
metsulfuron-methyl + thifensulfuron-methyl

Products

Prospect	Nufarm Whyte	75% w/w	WB	06541

Uses

Docks in **established grassland, land temporarily removed from production** *(in green cover)*.

Efficacy
- Best results achieved from application to young green dock foliage when growing actively
- Only broad-leaved docks (*Rumex obtusifolius*) are controlled; curled docks (*Rumex crispus*) are resistant
- Apply 7-10 d before grazing and do not graze for 7 d afterwards
- Docks with developing or mature seed heads should be topped and the regrowth treated later

- Established docks with large tap roots may require follow-up treatment
- High populations or poached grassland will require further treatment in following yr
- Ensure good spray coverage and apply to dry foliage

Crop safety/Restrictions
- Maximum number of treatments 1 per calender yr. Must only be applied from 1 Feb in yr of harvest
- Do not treat new leys in year of sowing
- Do not treat where nutrient imbalances, drought, waterlogging, low temperatures, lime deficiency, pest or disease attack have reduced sward vigour
- Do not roll or harrow within 7 d of spraying
- Product may cause a check to both sward and clover which is usually outgrown

Special precautions/Environmental safety
- Extremely dangerous to aquatic higher plants. Do not contaminate surface waters or ditches with chemical or used container
- LERAP Category B
- Take extreme care to avoid drift onto broad-leaved plants outside the target area or onto surface waters or ditches, or land intended for cropping
- Spraying equipment should not be drained or flushed onto land planted, or to be planted, with trees or crops other than cereals and should be thoroughly cleansed after use - see label for instructions
- Only grass or cereals may be sown within 4 wk of treatment

Personal protective equipment/Label precautions
- U08, U19, U20b, E14a, E16a, E16b, E30a, E32a; E07 (7 d)

Withholding period
- Keep livestock out of treated areas for at least 7 d following treatment

Latest application/Harvest interval
- Before 1 Aug

Approval
- Approval expiry 31 Dec 2001; replaced by [9389]

---

480   thifensulfuron-methyl + tribenuron-methyl

A mixture of two sulfonylurea herbicides for cereals

Products

Calibre	DuPont	50:25% w/w	WG	07795

Uses

Annual dicotyledons in **barley, wheat**. Charlock in **barley, wheat**. Chickweed in **barley, wheat**. Mayweeds in **barley, wheat**.

Efficacy
- Apply after 1 Feb when weeds are small and actively growing
- Ensure good spray cover of the weeds
- Ensure that weeds present are those that are susceptible. See label
- Follow label mixing instructions
- Effectiveness reduced by rain within 4 h of treatment

Crop safety/Restrictions
- Maximum number of treatments 1 per crop
- May be sprayed on all varieties of wheat and barley from 3 leaf stage (GS 13) up to and including flag leaf fully emerged (GS 39)
- Igri winter barley may suffer damage during period of rapid growth and must not be treated before leaf sheath erect stage (GS 30)
- Do not apply to cereals undersown with grass, clover or other legumes
- Do not apply within 7 d of rolling
- Various tank mixtures recommended to broaden weed control spectrum. Other mixtures are specifically excluded. See label

- Do not apply in sequence or in tank mixture with a product containing any other sulfonyl urea except as directed on label
- Do not apply to any crop suffering from stress or not actively growing
- Take particular care to avoid damage by drift onto broad-leaved plants outside the target area or onto surface waters or ditches
- Only cereals, field beans or oilseed rape may be sown in the same calendar year as harvest of a treated crop. In the event of failure of a treated crop sow only a cereal crop within 3 mth of product application

Special precautions/Environmental safety
- Extremely dangerous to fish or other aquatic life. Do not contaminate surface waters or ditches with chemical or used container
- LERAP Category B
- Spraying equipment should not be drained or flushed onto land planted, or to be planted, with trees or crops other than cereals and should be thoroughly cleansed after use - see label for instructions

Personal protective equipment/Label precautions
- U08, U19, U20a, E13a, E16a, E16b, E30a, E32a

Latest application/Harvest interval
- Before flag leaf ligule first visible (GS 39)

Approval
- Accepted by BLRA for use on malting barley

---

481 thiodicarb

A carbamate insecticide with molluscicide uses

Products

1	Genesis	RP Agric.	4% w/w	RB	06168
2	Genesis	Aventis	4% w/w	RB	09993
3	Genesis ST	RP Agric.	4% w/w	RB	08211
4	Judge	RP Agric.	4% w/w	RB	08163
5	Judge	Aventis	4% w/w	RB	10000
6	Me2 Exodus	Me2	4.0% w/w	RB	09786

Uses

Slugs in *barley, durum wheat, oats, oilseed rape, potatoes, triticale, wheat* [1, 2, 4-6]. Slugs in *barley* (seed admixture), *durum wheat* (seed admixture), *oats* (seed admixture), *triticale* (seed admixture), *wheat* (seed admixture) [3].

Efficacy
- Apply bait as broadcast treatment or admixed with seed
- Additional broadcast treatments may be needed after drilling admixed seed

Crop safety/Restrictions
- Maximum number of treatments 1 per crop (admixture), 3 per crop (broadcast)
- Do not treat grain with more than 16% moisture content or allow treated seed to rise above this level
- Sow admixed seed as soon as possible after treatment
- May only be applied to oilseed rape and potatoes as broadcast application
- Must not be admixed with cereal seed already admixed with any other product containing thiodicarb

Special precautions/Environmental safety
- Product contains an anticholinesterase carbamate compound. Do not use if under medical advice not to work with such compounds
- Harmful if swallowed. Irritating to eyes. May cause sensitization by skin contact
- Harmful to game, wild birds and animals
- Dangerous to fish or other aquatic life. Do not contaminate surface waters or ditches with chemical or used container
- Do not use treated seed as food or feed

Personal protective equipment/Label precautions
- A, H [1-6]; J [1, 2, 4-6]; D [1-3, 6]
- M02, M04, R03c, R04a, R04e, U05a, U13, U20b, C03, E01, E13b, E30a, E32a, E34 [1-6]; E10b [1, 2, 4-6]; S01, S02, S03, S04b, S05, S06, S07 [1-3, 6]

Latest application/Harvest interval
- Before drilling when admixed with barley, durum wheat, oats, triticale, wheat; before first node detectable (GS 31) for barley, durum wheat, oats, triticale, wheat; before stem extension (GS 2,0) for oilseed rape
- HI potatoes 3 wk

Approval
- Accepted by BLRA for use on malting barley

482   thiophanate-methyl

A carbendazim precursor fungicide with protectant and curative activity

See also gamma-HCH + thiophanate-methyl

Products

1	Cercobin Liquid	Aventis	500 g/l	SC	09984
2	Mildothane Liquid	Hortichem	500 g/l	SC	06211
3	Mildothane Liquid	Aventis	500 g/l	SC	10002
4	Mildothane Turf Liquid	Aventis Environ.	500 g/l	SC	09935
5	Mildothane Turf Liquid	RP Amenity	500 g/l	SC	05331

Uses

Botrytis in **winter oilseed rape** [1]. Canker in **apples, pears** [2, 3]. Dollar spot in **managed amenity turf** [4, 5]. Fusarium patch in **managed amenity turf** [4, 5]. Light leaf spot in **winter oilseed rape** [1]. Powdery mildew in **apples, pears** [2, 3]. Red thread in **managed amenity turf** [4, 5]. Scab in **apples, pears** [2, 3]. Sclerotinia stem rot in **winter oilseed rape** [1]. Storage rots in **apples, pears** [2, 3]. Wormcast formation in **managed amenity turf** [4, 5].

Efficacy
- Timing of foliar sprays varies with crop variety and disease problem (see label for details) [2, 3]
- Spray treatments can reduce wood canker on apples [2, 3]
- Apply a post-blossom spray or as a post-harvest dip fruit to reduce storage rot diseases [2, 3]
- Apply to turf during period of active growth and do not mow for 48 h [4, 5]

Crop safety/Restrictions
- Maximum number of treatments (including applications of products containing carbendazim or benomyl) 12 per crop as pre-harvest foliar spray on apples and pears, 1 per batch as post-harvest dip [2, 3]
- Spraying apples from blossom to fruitlet stage may increase russeting on prone cultivars [2, 3]
- Do not use on fruit trees at petal fall or early fruitlet stage when preceded by a protectant such as captan [2, 3]
- Do not mix with MCPB herbicides or copper. See label for compatible mixtures [4, 5]

Special precautions/Environmental safety
- To dispose of tip empty solution into an approved soakaway, not into drains, open ditches or soakaways

Personal protective equipment/Label precautions
- A, C, H, M [1-5]; B, K [2, 3]
- U20b [1-3]; E15, E30a [1-5]; E31a [1, 4, 5]; E26 [1-3]; E32a [2, 3]

Latest application/Harvest interval
- HI apples, pears 7 d; winter oilseed rape 21 d

Approval
- Following implementation of Directive 98/82/EC, approval for use of thiophanate-methyl on several crops was revoked in 1999

483   thiram

A protectant dithiocarbamate fungicide

See also carbendazim + cymoxanil + oxadixyl + thiram
carboxin + thiram
metalaxyl + thiabendazole + thiram
metalaxyl + thiram
thiabendazole + thiram

Products

1	Agrichem Flowable Thiram	Agrichem	600 g/l	FS	06245
2	Thiraflo	Uniroyal	480 g/l	FS	09496
3	Unicrop Thianosan DG	Unicrop	80% w/w	WG	05454

Uses

Botrytis in **chrysanthemums, freesias, ornamentals** (except hydrangea), **outdoor lettuce, protected lettuce, raspberries, strawberries, tomatoes** [3]. Botrytis fruit rot in **apples** [3]. Cane spot in **raspberries** [3]. Damping off in **broad beans** (seed treatment), **cabbages** (seed treatment), **carrots** (seed treatment), **cauliflowers** (seed treatment), **dwarf beans** (seed treatment), **field beans** (seed treatment), **grass seed** (seed treatment), **leeks** (seed treatment), **lettuce** (seed treatment), **maize** (seed treatment), **oilseed rape** (seed treatment), **onions** (seed treatment), **peas** (seed treatment), **radishes** (seed treatment), **runner beans** (seed treatment), **soya beans** (off-label - seed treatment), **turnips** (seed treatment) [1]. Damping off in **combining peas** (seed treatment), **vining peas** (seed treatment) [2]. Downy mildew in **protected lettuce** [3]. Fire in **tulips** [3]. Gloeosporium in **apples** [3]. Rust in **blackcurrants, carnations, chrysanthemums** [3]. Scab in **apples, pears** [3]. Seed-borne diseases in **carrots** (seed soak), **celery** (seed soak), **fodder beet** (seed soak), **mangels** (seed soak), **parsley** (seed soak), **red beet** (seed soak), **sugar beet** (seed soak) [1]. Spur blight in **raspberries** [3].

Efficacy

- Spray before onset of disease and repeat every 7-14 d. Spray interval varies with crop and disease. See label for details [3]
- Apply as seed treatment for protection against damping off [1, 2]
- Co-application of 175 ml water per 100 kg of seed likely to improve evenness seed coverage [2]
- Do not spray when rain imminent [3]
- For use on tulips, chrysanthemums and carnations add non-ionic wetter [3]

Crop safety/Restrictions

- Maximum number of treatments 3 per crop for protected winter lettuce (thiram based products only [3]; 2 per crop if sequence of thiram and other EBDC fungicides used) [3]; 2 per crop for protected summer lettuce [3]; 1 per batch of seed for seed treatments [1, 2]
- Do not apply to hydrangeas [3]
- Notify processor before dusting or spraying crops for processing [3]
- Do not dip roots of forestry transplants [3]
- Do not treat seed of tomatoes, peppers or aubergines [1]
- Do not treat pea seed with more than 15% moisture content and do not treat cracked or sprouted seed [2]

Special precautions/Environmental safety

- Harmful in contact with skin and if swallowed [1]
- Irritating to eyes, skin and respiratory system
- Dangerous to fish or other aquatic life. Do not contaminate surface waters or ditches with chemical or used container
- Do not use treated seed as food or feed [1, 2]
- Treated seed harmful to game and wildlife [1, 2]

Personal protective equipment/Label precautions

- A [1-3]
- M03 [1, 2]; R04a, R04b, R04c [1, 3]; R03a, R03c [1]; U19 [3]; U05a, U08, U20b [1-3]; U02 [1]; C03 [1-3]; E13b, E32a [3]; E01, E30a [1-3]; E03, E15, E31c, E34 [1]; E26, E31a [2]; S01, S02, S03, S04b, S05, S07 [1, 2]; S06 [2]

Latest application/Harvest interval

- 21 d after planting out or 21 d before harvest, whichever is earlier, for protected winter lettuce; 14 d after planting out or 21 d before harvest, whichever is earlier, for protected summer lettuce.
- HI protected lettuce 21 d; outdoor lettuce 14 d; apples, pears, blackcurrants, raspberries, strawberries, tomatoes 7 d

Approval

- Off-label approval to Mar 2002 for seed treatment of soya beans (OLA 0372/97)[1]

484   tolclofos-methyl

A protectant organophosphorus fungicide for soil-borne diseases

Products

1 Basilex	Scotts	50% w/w	WP	07494
2 Rizolex	AgrEvo	10% w/w	DS	07271
3 Rizolex	Hortichem	10% w/w	DS	09673
4 Rizolex Flowable	AgrEvo	500 g/l	FS	07273
5 Rizolex Flowable	Hortichem	500 g/l	FS	09358

Uses

Black scurf and stem canker in *potatoes (tuber treatment)* [2, 3, 5]. Black scurf and stem canker in *potatoes (seed tuber treatment)* [4]. Bottom rot in *lettuce, protected lettuce* [1]. Damping off in *seedlings of ornamentals* [1]. Damping off and wirestem in *leaf brassicas* [1]. Foot rot in *ornamentals, seedlings of ornamentals* [1]. Rhizoctonia in *protected celery* *(off-label)*, *protected radishes (off-label)* [1]. Rhizoctonia in *seed potatoes (off-label - tuber treatment)* [5]. Root rot in *ornamentals, seedlings of ornamentals* [1].

Efficacy

- Apply dust to seed potatoes during hopeer loading [2, 3]
- Apply flowable formulation to clean tubers with suitable misting equipment over a roller table. Spray as potatoes taken into store (first earlies) or as taken out of store (second earlies, maincrop, crops for seed) pre-chitting [4, 5]
- Do not mix flowable formulation with any other product [4, 5]
- To control rhizoctonia in vegetables and ornamentals apply as drench before sowing, pricking out or planting or incorporate into compost [1]
- On established seedlings and pot plants apply as drench and rinse off foliage [1]

Crop safety/Restrictions

- Maximum number of treatments 1 per batch for seed potatoes; 1 per crop for pre-transplanted lettuce and brassicas; 1 at each stage of growth (ie sowing, pricking out, potting) to a maximum of 3, for ornamentals
- Only to be used with automatic planters [2, 3]
- Not recommended for use on seed potatoes where hot water treatment used or to be used [2, 3]
- Do not apply as overhead drench to vegetables or ornamentals when hot and sunny [1]
- Do not use on heathers [1]

Special precautions/Environmental safety

- Tolclofos-methyl is an atypical organophosphorus compound which has weak anticholinesterase activity. Do not use if under medical advice not to work with such compounds
- Irritating to eyes, skin and respiratory system [2-5]
- Dangerous (harmful [1-3, 5]) to fish or other aquatic life. Do not contaminate surface waters or ditches with chemical or used container [4]
- LERAP Category B [1]
- Treated tubers to be used as seed only, not for food or feed [2-5]

Personal protective equipment/Label precautions

- A, H [1-5]; D, E [2-5]; M [1-3]; J [2, 3]; C [1, 4, 5]
- M01 [1]; R04a, R04b, R04c [2-5]; U19 [1-5]; U05a [2-4]; U20b [1-3, 5]; U20a [4]; U16 [5]; C03 [2-4]; E30a [1-5]; E01 [2-4]; E13c [1-3, 5]; E32a [1-3]; E13b [4]; E31a [4, 5]; E16a, E16b [1]

Latest application/Harvest interval
- At planting of seed potatoes [2-5]; before transplanting for lettuce and brassicas [1]

Approval
- May be applied by misting equipment mounted over roller table. See label for details [4, 5]
- Off-label approval unstipulated for use from pre-emergence to second leaf stage on protected radishes (OLA 2006/99)[1]; unstipulated for use on protected radishes pre-emergence (OLA 2007/99)[1]; unlimited for use on protected celery (OLA 2009/99)[1]
- Approval expiry 31 Dec 2001 [4]; replaced by [5]

---

485   tolylfluanid

A multi-site protectant fungicide

Products

Elvaron Multi	Bayer	50.5% w/w	WG	10080

Uses

Botrytis in **blackberries, blackcurrants, gooseberries, loganberries, raspberries, redcurrants, strawberries, whitecurrants**.

Efficacy
- Best results in soft fruit achieved by good spray coverage of flowers and young fruitlets
- To achieve required coverage of flowers and young fruitlets spray using a hand lance, drop arm attachments or a special strawberry boom with leaf rolling bar
- Best results using high volume directed over the crop rows

Crop safety/Restrictions
- Maximum total dose equivalent to 2.5 full dose treatments on currants, gooseberries, and four full dose treatments on other soft fruit crops

Special precautions/Environmental safety
- Irritant. May cause sensitization by skin contact
- Dangerous to fish or other aquatic life. Do not contaminate surface waters or ditches with chemical or used container
- LERAP Category B
- Use best available application technique that minimises off-target drift to reduce effects on non-target insects or other arthropods

Personal protective equipment/Label precautions
- A, H, M
- R04, R04e, U02, U05a, U20b, C03, E01, E13b, E16a, E16b, E22b, E30a, E32a

Latest application/Harvest interval
- HI blackberries, loganberries, raspberries, strawberries 14 d; currants, gooseberries 21 d

---

486   tralkoxydim

A foliar applied oxime herbicide for grass weed control in cereals.

Products

1	Grasp	Zeneca	250 g/l	SC	06675
2	Greencrop Gweedore	Greencrop	250 g/l	SC	09882
3	Landgold Tralkoxydim	Landgold	250 g/l	SC	08604
4	Standon Tralkoxydim	Standon	250 g/l	SC	09579

Uses

Awned canary grass in **spring barley** *(qualified minor use)*, **spring wheat** *(qualified minor use)* [1]. Blackgrass in **autumn sown spring barley, autumn sown spring wheat** [1]. Blackgrass in **durum wheat, spring barley, spring wheat, triticale, winter barley, winter rye, winter wheat** [1-4]. Rough meadow grass in **autumn sown spring barley, autumn sown spring wheat, durum wheat, spring wheat, triticale, winter barley, winter rye, winter wheat** [1]. Ryegrass in **autumn sown spring barley, autumn sown spring wheat, durum wheat, spring barley, spring wheat, triticale, winter barley, winter rye, winter wheat** [1]. Wild oats in **autumn sown spring barley, autumn sown spring wheat** [1]. Wild

---

oats in **durum wheat, spring barley, spring wheat, triticale, winter barley, winter rye, winter wheat** [1-4]. Yorkshire fog in **autumn sown spring barley** *(qualified minor use)*, **autumn sown spring wheat** *(qualified minor use)*, **durum wheat** *(qualified minor use)*, **triticale** *(qualified minor use)*, **winter barley** *(qualified minor use)*, **winter rye** *(qualified minor use)*, **winter wheat** *(qualified minor use)* [1].

Efficacy
- Product leaf-absorbed and translocated rapidly to growing points. Best results achieved when weeds growing actively in competitive crops under warm humid conditions with adequate soil moisture
- Activity not dependent on soil type. Weeds germinating after application will not be controlled
- Best control of wild oats obtained from 2 leaf to 1st node detectable stage of weeds, and of blackgrass up to 3 tillers
- Authorised adjuvant must always be added. See label

Crop safety/Restrictions
- Maximum number of treatments 1 on spring sown crops; 2 on winter sown crops
- Apply to winter cereals from 2 leaves unfolded up to and including flag leaf sheath extending (GS 12-41). If necessary winter cereals may be sprayed twice: once in autumn and once in spring
- Apply to spring cereals from end of tillering up to and including flag leaf sheath extending (GS 29-41)
- Do not spray undersown crops or crops to be undersown
- Do not spray when foliage wet or covered in ice or crop otherwise under stress
- Do not spray if a protracted period of cold weather forecast
- Do not spray crops under stress from chemical treatment, grazing, pest attack, mineral deficiency or low fertility
- Do not roll or harrow within 1 wk of spraying
- Do not tank-mix with phenoxy hormone or sulfonylurea herbicides or apply these chemicals within 15 d before and 5 d after treatment

Special precautions/Environmental safety
- Irritating to eyes
- May cause sensitization by skin contact

Personal protective equipment/Label precautions
- A, C, H [1-4]
- M03 [1, 4]; R04a, R04e, U02, U04a, U05a, U08, U20b, C03, E01, E15, E26, E30a, E31b [1-4]; E34 [1, 4]

Latest application/Harvest interval
- Before booting (GS 41)

Approval
- Accepted by BLRA for use on malting barley

---

487   triadimefon

A systemic conazole fungicide with curative and protectant action

Products
Bayleton	Bayer	25% w/w	WP	00221

Uses

Brown rust in **barley, wheat**. Crown rust in **grassland**. Powdery mildew in **apples, barley, barley** *(off-label - research/breeding)*, **blackberries** *(off-label)*, **blackcurrants** *(off-label)*, **brussels sprouts, brussels sprouts** *(off-label - research/breeding)*, **cabbages, fodder beet, gooseberries** *(off-label)*, **grapevines** *(off-label)*, **grassland, hops, loganberries** *(off-label)*, **oats, parsnips, peas** *(off-label - research/breeding)*, **raspberries** *(off-label)*, **rye, strawberries** *(off-label)*, **sugar beet, swedes, turnips, wheat, wheat** *(off-label - research/breeding)*. Rhynchosporium in **grassland**. Rust in **barley** *(off-label - research/breeding)*, **brussels sprouts** *(off-label - research/breeding)*, **leeks, peas** *(off-label - research/breeding)*, **wheat** *(off-label - research/breeding)*. Snow rot in **barley**. Yellow rust in **barley, wheat**.

Efficacy
- Apply at first sign of disease and repeat as necessary, applications to established infections are less effective. Spray programme varies with crop and disease. See label
- Applications to control mildew will also reduce crown rust in oats, snow rot in winter barley and rust in beet

Crop safety/Restrictions
- Maximum number of treatments 2 per crop for wheat, oats, rye, spring barley, grassland, sugar beet, fodder beet, turnips, swedes, parsnips; 3 per crop for winter barley (1 in autumn), Brussels sprouts, cabbages, cane fruit and herbs; 4 per crop for leeks; 6 per yr for grapevines; 8 per yr for hops; 12 per yr for apples
- Continued use of fungicides from the same group in cereals may lead to reduced effectiveness against mildew

Special precautions/Environmental safety
- Harmful to fish or other aquatic life. Do not contaminate surface waters or ditches with chemical or used container

Personal protective equipment/Label precautions
- U20a, C02, E06b, E13c, E30a, E32a

Withholding period
- Keep all livestock out of treated areas for 21 d

Latest application/Harvest interval
- Before grain milky ripe (GS 71) for cereals.
- HI sugar beet, fodder beet, Brussels sprouts, cabbages, turnips, swedes, parsnips, leeks, apples, hops, cane fruit, strawberries, blackcurrants, gooseberries 14 d; grassland, herbs 21 d; grapes 6 wk

Approval
- Approved for aerial application on sugar beet, brassicas, swedes, turnips, cereals. See notes in Section 1
- Off-label approval unlimited for use on blackcurrants, gooseberries, strawberries, outdoor grapes, raspberries, loganberries, blackberries, other Rubus hybrids (OLA 0024/95); unlimited for use on protected peas, wheat, barley, Brussels sprouts (for research/breeding) (OLA 0158/97)
- Accepted by BLRA for use on malting barley and hops

---

488   triadimenol

A systemic conazole fungicide for cereals, beet and brassicas

See also fuberidazole + imidacloprid + triadimenol
fuberidazole + triadimenol
tebuconazole + triadimenol

Products

Bayfidan	Bayer	250 g/l	DC	02672

Uses

Alternaria in **brussels sprouts, cabbages**. Fungus diseases in **carrots** (off-label), **horseradish** (off-label), **parsley** (off-label), **parsnips** (off-label), **salsify** (off-label). Light leaf spot in **brussels sprouts, cabbages**. Powdery mildew in **barley, brussels sprouts, cabbages, fodder beet, oats, rye, sugar beet, swedes, turnips, wheat**. Rhynchosporium in **barley, oats, rye, wheat**. Ring spot in **brussels sprouts, cabbages**. Rust in **barley, oats, rye, wheat**. Septoria in **wheat**. Snow rot in **barley, oats, rye, wheat**.

Efficacy
- Apply sprays at first signs of disease and ensure good cover
- When applied for mildew and rust control in wheat product also gives good reduction of *Septoria tritici* but for specific protection against *S. tritici* and *S. nodorum* use tank-mix with chlorothalonil
- Treatment also reduces crown rust in oats and rust in beet
- Continued use of fungicides from same group can result in reduced effectiveness against powdery mildew. See label for recommended tank mixtures

Crop safety/Restrictions
- Maximum number of treatments 2 per crop for wheat, oats, spring barley, rye, sugar beet, fodder beet, turnips, swedes, carrots, parsnips; 3 per crop (only 2 in spring) for winter barley, cabbages, Brussels sprouts

Special precautions/Environmental safety
- Harmful if swallowed. Irritating to eyes
- Dangerous to fish or other aquatic life. Do not contaminate surface waters or ditches with chemical or used container

Personal protective equipment/Label precautions
- A, C
- M03, R03c, R04a, U05a, U08, U10, U11, U13, U20a, C03, E01, E13b, E30a, E31b, E34

Latest application/Harvest interval
- Before grain milky ripe (GS 71) for cereals.
- HI beet crops, brassicas 14 d; carrots, parsnips 21 d

Approval
- Off-label approval unlimited for use on outdoor crops of carrots, parsnips, parsley root, salsify, horseradish (OLA0836/95)
- Accepted by BLRA for use on malting barley

---

489   tri-allate

A soil-acting thiocarbamate herbicide for grass weed control

Products

Avadex Excel 15G	Monsanto	15% w/w	GR	07117

Uses

Blackgrass in *barley, durum wheat, field beans, fodder beet, forage legumes, mangels, peas, red beet, sugar beet, triticale, winter rye, winter wheat*. Meadow grasses in *barley, durum wheat, field beans, fodder beet, forage legumes, mangels, peas, red beet, sugar beet, triticale, winter rye, winter wheat*. Wild oats in *barley, durum wheat, field beans, fodder beet, forage legumes, mangels, peas, red beet, sugar beet, triticale, winter rye, winter wheat*.

Efficacy
- Incorporate or apply to surface pre-emergence (post-emergence application possible in winter cereals up to 2-leaf stage of wild oats)
- Do not use on soils with more than 10% organic matter
- Wild oats controlled up to 2-leaf stage
- If applied to dry soil, rainfall needed for full effectiveness, especially with granules on surface. Do not use if top 5-8 cm bone dry
- Do not apply with spinning disc granule applicator; see label for suitable types
- Do not apply to cloddy seedbeds
- Use sequential treatments to improve control of barren brome and annual dicotyledons (see label for details)

Crop safety/Restrictions
- Consolidate loose, puffy seedbeds before drilling to avoid chemical contact with seed
- Drill cereals well below treated layer of soil (see label for safe drilling depths)
- Do not use on direct-drilled crops or undersow grasses into treated crops
- Do not sow oats or grasses within 1 yr of treatment

Special precautions/Environmental safety
- Irritating to eyes and skin
- May cause sensitization by skin contact
- Harmful to fish or other aquatic life. Do not contaminate surface waters or ditches with chemical or used container

Personal protective equipment/Label precautions
- A, C, H
- R04a, R04b, R04e, U02, U05a, U20a, C03, E01, E13c, E30a, E32a

Latest application/Harvest interval
- Pre-drilling for beet crops; before crop emergence for field beans, spring barley, peas, forage legumes; before first node detectable stage (GS 31) for winter wheat, winter barley, durum wheat, triticale, winter rye

Approval
- Approved for aerial application on wheat, barley, beans, peas. See notes in Section 1
- Accepted by BLRA for use on malting barley

---

490   triasulfuron

A sulfonylurea herbicide for annual broad-leaved weed control in cereals

See also bromoxynil + ioxynil + triasulfuron

Products

Lo-Gran	Novartis	20% w/w	WG	08421

Uses

Annual dicotyledons in **barley, durum wheat, rye, triticale, wheat, winter oats**. Charlock in **barley, durum wheat, rye, triticale, wheat, winter oats**. Chickweed in **barley, durum wheat, rye, triticale, wheat**. Mayweeds in **barley, durum wheat, rye, triticale, wheat, winter oats**.

Efficacy
- Best results achieved on small, actively growing weeds up to 6 leaf or 50 mm growth stage (up to flower bud for charlock, chickweed and mayweed)
- Some species, although not controlled remain stunted and uncompetitive with crop

Crop safety/Restrictions
- Maximum number of treatments 1 per crop
- Product must be applied only after 1 Feb from 1 leaf stage of crop to before 3rd node detectable (GS 11-33)
- Do not apply to undersown crops or those due to be undersown
- Do not spray in windy weather. Avoid drift onto neighbouring crops
- Do not spray during period of frosty weather, when frost imminent or onto crops under stress from frost, waterlogging or drought
- Do not spray in tank mixture, or in sequence, with a product containing any other sulfonylurea
- See label for restrictions on succeeding crops

Special precautions/Environmental safety
- Extremely dangerous to aquatic higher plants. Do not contaminate surface waters or ditches with chemical or used container
- LERAP Category B
- Ensure spraying equipment is washed thoroughly according to specific instructions. Do not allow washings to drain onto land intended for cropping or growing crops

Personal protective equipment/Label precautions
- A, H
- U20a, E16a, E16b, E30a, E32a

Latest application/Harvest interval
- Before 3rd node detectable (GS 33)

Approval
- Triasulfuron included in Annex I under EC Directive 91/414
- Accepted by BLRA for use on malting barley

---

491   triazamate

A carbamoyl triazole insecticide for sugar beet and specified brassicas

Products

1 Aztec	BASF	140 g/l	EW	10211
2 Aztec	Cyanamid	140 g/l	EW	07817

Uses

Aphids in **brussels sprouts, cabbages** *(not savoys)* [1]. Aphids in **sugar beet** *(including myzus persicae)* [1, 2].

Efficacy

- Treat as soon as aphids seen in crop or immediately after official warnings are issued. Repeat as necessary
- Products are systemic and act through contact and stomach action. Ensure foliage well covered with spray by using higher volumes in dense crops
- Controls aphids resistant to other chemical groups
- To reduce risk of development of resistance consider use of products with alternative modes of action in intensive pest control programmes. If tank mixtures used, do not reduce dose of either component

Crop safety/Restrictions

- Maximum number of treatments on sugar beet 3 per crop
- Maximum total dose on brassicas equivalent to three full dose treatments
- Label warning that damage may result unless all recommendations carefully followed
- Products must be used with Swirl adjuvant
- Consult processor before use on crops for processing

Special precautions/Environmental safety

- Harmful by inhalation and if swallowed
- Extremely dangerous to fish or other aquatic life. Do not contaminate surface waters or ditches with chemical or used container
- LERAP Category B
- Product not harmful to bees when used as directed but spraying in late evening/early morning or in dull weather recommended to avoid unnecessary stress on foraging bees

Personal protective equipment/Label precautions

- A, H [1, 2]
- M03, M06, R03b, R03c, U05a, U19 [1, 2]; U02, U09a, U13, U14, U20b [1]; C03, E01, E13a, E16a, E16b, E26, E34 [1, 2]; E27, E30a, E32a [1]

Latest application/Harvest interval

- HI 28 d

---

492   triazoxide

A benzotriazine fungicide available only in mixtures

See also imidacloprid + tebuconazole + triazoxide
tebuconazole + triazoxide

---

493   tribenuron-methyl

A foliar acting sulfonylurea herbicide with some root activity for use in cereals

See also metsulfuron-methyl + tribenuron-methyl
thifensulfuron-methyl + tribenuron-methyl

Products

1 Quantum	DuPont	50% w/w	TB	06270
2 Quantum 75 DF	DuPont	75% w/w	WG	09340

Uses

Annual dicotyledons in **durum wheat, oats, spring barley, spring wheat, triticale, winter barley, winter rye, winter wheat**. Charlock in **durum wheat, oats, spring barley, spring wheat, triticale, winter barley, winter rye, winter wheat**. Chickweed in **durum wheat, oats, spring barley, spring wheat, triticale, winter barley, winter rye, winter wheat**. Mayweeds in **durum wheat, oats, spring barley, spring wheat, triticale, winter barley, winter rye, winter wheat**.

Efficacy
- Best control achieved when weeds small and actively growing
- Good spray cover must be achieved since larger weeds often become less susceptible
- Susceptible weeds cease growth almost immediately after treatment and symptoms can be seen in about 2 wk
- Products can be used on all soil types
- Weed control may be reduced when conditions very dry
- See label for details of technique to be used for dissolving tablets in spray tank
- When tank mixing ensure product fully dispersed before adding other products

Crop safety/Restrictions
- Maximum number of treatments 1 per crop
- Do not spray in tank mixture, or in sequence, with a product containing any other sulfonylurea except as directed on label
- Apply in autumn or in spring (after 1 Feb) from 3 leaf stage of crop up to and including flag leaf fully emerged (GS 13-39)
- Do not apply to crops undersown with grass, clover or other broad-leaved crops
- Do not apply to any crop suffering stress from any cause or not actively growing
- Do not apply within 7 d of rolling
- Special care must be taken to avoid damage by drift onto nearby broad-leaved crops, surface waters or ditches
- In the event of crop failure sow only a cereal within 3 mth of application. See label for other restrictions on subsequent cropping

Special precautions/Environmental safety
- Irritant. May cause sensitization by skin contact
- Extremely dangerous to fish or other aquatic life. Do not contaminate surface waters or ditches with chemical or used container [2]
- Spraying equipment should not be drained or flushed onto land planted, or to be planted, with trees or crops other than cereals and should be thoroughly cleansed after use - see label for instructions

Personal protective equipment/Label precautions
- A [1, 2]
- R04e [1, 2]; U20a [1]; U05a, U08 [1, 2]; U20b [2]; C03 [1, 2]; E15 [1]; E01, E30a, E32a [1, 2]; E13a [2]

Latest application/Harvest interval
- Up to and including flag leaf ligule/collar just visible (GS 39)

Approval
- Accepted by BLRA for use on malting barley

---

494   triclopyr

An aryloxyalkanoic acid herbicide for perennial and woody weed control

See also clopyralid + fluroxypyr + triclopyr
clopyralid + triclopyr
fluroxypyr + triclopyr

Products

1	Chipman Garlon 4	Nomix-Chipman	480 g/l	EC	06016
2	Convoy	Nufarm Whyte	480 g/l	EC	09185
3	Garlon 2	Zeneca	240 g/l	EC	06616
4	Garlon 4	Dow	480 g/l	EC	05090
5	Timbrel	Dow	480 g/l	EC	05815
6	Triptic 48 EC	United Phosphorus	480 g/l	EC	09294

Uses

Brambles in *forestry, land not intended to bear vegetation* [1, 2, 4-6]. Brambles in *established grassland, land not intended for cropping, non-crop areas* (directed treatment) [3]. Broom in *forestry, land not intended to bear vegetation* [1, 2, 4-6]. Broom in *established grassland, land not intended for cropping, non-crop areas* (directed *treatment*) [3]. Brush clearance in **industrial sites** [1, 2, 4-6]. Docks in **forestry, land not intended to bear vegetation** [1, 2, 4-6]. Docks in **established grassland, land not intended for cropping, non-crop areas** [3]. Gorse in **forestry, land not intended to bear vegetation** [1, 2, 4-6]. Gorse in **established grassland, land not intended for cropping, non-crop areas** (*directed treatment*) [3]. Hard rush in **established grassland, land not intended for cropping, non-crop areas** [3]. Perennial dicotyledons in **forestry, industrial sites, land not intended to bear vegetation** [1, 2, 4-6]. Perennial dicotyledons in **established grassland, land not intended for cropping, non-crop areas** [3]. Rhododendrons in **forestry, land not intended to bear vegetation** [1, 2, 4-6]. Scrub clearance in **industrial sites** [1, 2, 4-6]. Scrub clearance in **land not intended for cropping, non-crop areas** (*directed treatment*) [3]. Stinging nettle in **forestry, land not intended to bear vegetation** [1, 2, 4-6]. Stinging nettle in **established grassland, land not intended for cropping, non-crop areas** [3]. Woody weeds in **forestry, industrial sites, land not intended to bear vegetation** [1, 2, 4-6]. Woody weeds in **established grassland, land not intended for cropping, non-crop areas** (*directed treatment*) [3].

Efficacy
- Uses on land not intended for cropping include grassland of no agricultural interest such as roadside verges, railway and motorway embankments
- Apply in grassland as spot treatment or overall foliage spray when weeds in active growth in spring or summer. Details of dose and timing vary with species. See label [3, 4]
- Apply to woody weeds as summer foliage, winter shoot, basal bark, cut stump or tree injection treatment [1, 4, 5]
- Apply foliage spray in water when leaves fully expanded but not senescent
- Apply winter shoot, basal bark or cut stump sprays in paraffin or diesel oil. Dose and timing vary with species. See label for details
- Inject undiluted or 1:1 dilution into cuts spaced every 7.5 cm round trunk [1, 4, 5]
- Do not spray in drought, in very hot or cold conditions
- Control may be reduced if rain falls within 2 h of application
- Control of rhododendron can be variable. If higher than 1.8 m cut stump treatment recommended. A follow-up shoot treatment may be required

Crop safety/Restrictions
- Maximum number of treatments 1 per yr on non-crop land (as directed spray) [3]; 2 per yr on established grassland (including land not intended for cropping) and forestry
- See label for maximum concentrations when applying in oil, water or via watering can [3]
- Clover will be killed or severely checked by application in grassland. Do not apply to grass leys less than 1 yr old [3]
- Do not allow spray to drift onto agricultural or horticultural crops, amenity plantings, gardens, ponds, lakes or water courses. Vapour drift may occur under hot conditions
- Do not drill kale, swedes, turnips, grass or mixtures containing clover within 6 wk of treatment. Allow at least 6 wk before planting trees

Special precautions/Environmental safety
- Harmful in contact with skin and if swallowed [1, 2, 4-6]
- May cause lung damage if swallowed [4, 5]
- Irritating to skin. May cause sensitization by skin contact
- Irritating to eyes [3]
- Flammable [3]
- Not to be used on food crops
- Dangerous to fish or other aquatic life. Do not contaminate surface waters or ditches with chemical or used container
- Do not apply through hand held rotary atomisers
- LERAP Category B
- Not to be applied in or near water
- Do not apply from tractor-mounted sprayer within 250 m of susceptible crops, ponds, lakes or watercourses

Personal protective equipment/Label precautions
- A, C, H, M [1-6]
- M05, R04g [4, 5]; R04b, R04e [1-6]; R03a, R03c [1, 2, 4-6]; R04a, R07d [3]; U02, U05a, U08, U20b [1-6]; U19 [3]; C01, C03 [1-6]; E31b [1-5]; E01, E13b, E16a, E16b, E23, E30a, E34 [1-6]; E07 [1-6] (7 d); E26 [3]

Withholding period
- Keep livestock out of treated areas for at least 7 d and until foliage of any poisonous weeds such as buttercups or ragwort has died and become unpalatable

Latest application/Harvest interval
- 6 wk before replanting; 7 d before grazing

---

495 tridemorph

A systemic, eradicant and protectant morpholine fungicide available only in mixtures

See also cyproconazole + tridemorph

---

496 trietazine

A triazine herbicide available only in mixtures

See also simazine + trietazine

---

497 trifloxystrobin

A protectant strobilurin fungicide for cereals

Products

Twist	Novartis	125 g/l	EC	10125

Uses

Brown rust in **winter wheat**. Net blotch in **spring barley, winter barley**. Rhynchosporium in **spring barley, winter barley**.

Efficacy
- Should be used protectively before disease is established in crop. Further treatment may be necessary if disease attack prolonged
- Where wheat powdery mildew is a common problem, or where strobilurin resistance has been reported, mixtures with a different mode of action mildewicide recommended

Crop safety/Restrictions
- Maximum number of treatments 2 per crop per yr
- To minimise likelihood of resistance do not include more than two treatments of products containing a strobilurin fungicide in the spray programme

Special precautions/Environmental safety
- Irritating to eyes
- May cause sensitization by skin contact
- Do not harvest for human or animal consumption for at least 35 days after last application
- Dangerous to fish or other aquatic life. Do not contaminate surface waters or ditches with chemical or used container
- LERAP Category B

Personal protective equipment/Label precautions
- A, C
- R04a, R04e, U02, U05a, U09a, U19, U20b, C03; C02 (35 d); E01, E13b, E16a, E16b, E26, E29, E30a, E31b

Latest application/Harvest interval
- HI 35 d

Approval
- Accepted by BLRA for use on malting barley

---

498 trifluralin

A soil-incorporated dinitroaniline herbicide for use in various crops

See also clodinafop-propargyl + trifluralin
diflufenican + trifluralin
isoxaben + trifluralin
terbutryn + trifluralin

Products

1	Alpha Trifluralin 48 EC	Makhteshim	480 g/l	EC	07406
2	Ashlade Trifluralin	Nufarm Whyte	480 g/l	EC	08303
3	Ashlade Trimaran	Nufarm Whyte	480 g/l	EC	06228
4	Atlas Trifluralin	Nufarm Whyte	480 g/l	EC	08498
5	DAPT Trifluralin 48 EC	DAPT	480 g/l	EC	07906
6	MSS Trifluralin 48 EC	Nufarm Whyte	480 g/l	EC	07753
7	Nufarm Triflur	Nufarm Whyte	480 g/l	EC	08311
8	Portman Trifluralin	Portman	480 g/l	EC	05751
9	Treflan	Dow	480 g/l	EC	05817
10	Tripart Trifluralin 48 EC	Tripart	480 g/l	EC	02215
11	Whyte Trifluralin	Nufarm Whyte	480 g/l	EC	09286

Uses

Annual dicotyledons in **broad beans, broccoli, brussels sprouts, cabbages, carrots, cauliflowers, french beans, kale, lettuce, parsnips, raspberries, runner beans, strawberries, swedes, turnips, winter barley, winter wheat** [1-11]. Annual dicotyledons in **oilseed rape** [1, 2, 4-11]. Annual dicotyledons in **mustard** [1, 3, 5-10]. Annual dicotyledons in **linseed** [1, 5-7, 11]. Annual dicotyledons in **field beans** [1, 5, 6, 8]. Annual dicotyledons in **calabrese** [2-4, 8-11]. Annual dicotyledons in **navy beans** [2-4, 9, 10]. Annual dicotyledons in **spring oilseed rape, winter oilseed rape** [3]. Annual dicotyledons in **sugar beet** [3, 4, 7-10]. Annual dicotyledons in **parsley** [3, 4, 9, 10]. Annual dicotyledons in **spring linseed, winter linseed** [3, 9]. Annual dicotyledons in **celeriac** (off-label), **combining peas** (off-label), **evening primrose** (off-label), **gold-of-pleasure** (off-label), **kohlrabi** (off-label), **nursery fruit trees and bushes** (off-label), **ornamentals** (off-label), **outdoor herbs** (off-label), **radish seed crops** (off-label), **soft fruit** (off-label), **soya beans** (off-label), **sunflowers** (off-label) [9]. Annual grasses in **broad beans, broccoli, brussels sprouts, cabbages, carrots, cauliflowers, french beans, kale, lettuce, parsnips, raspberries, runner beans, strawberries, swedes, turnips, winter barley, winter wheat** [1-11]. Annual grasses in **oilseed rape** [1, 2, 4-11]. Annual grasses in **mustard** [1, 3, 5-10]. Annual grasses in **linseed** [1, 5-7, 11]. Annual grasses in **field beans** [1, 5, 6, 8]. Annual grasses in **calabrese** [2-4, 8-11]. Annual grasses in **navy beans** [2-4, 9, 10]. Annual grasses in **spring oilseed rape, winter oilseed rape** [3]. Annual grasses in **sugar beet** [3, 4, 7-10]. Annual grasses in **parsley** [3, 4, 9, 10]. Annual grasses in **spring linseed, winter linseed** [3, 9]. Annual grasses in **celeriac** (off-label), **combining peas** (off-label), **evening primrose** (off-label), **gold-of-pleasure** (off-label), **kohlrabi** (off-label), **nursery fruit trees and bushes** (off-label), **ornamentals** (off-label), **outdoor herbs** (off-label), **radish seed crops** (off-label), **soft fruit** (off-label), **soya beans** (off-label), **sunflowers** (off-label) [9].

Efficacy

- Acts on germinating weeds and requires soil incorporation to 5 cm (10 cm for crops to be grown on ridges) within 30 min of spraying (2 h [10], except in mixtures on cereals). See label for details of suitable application equipment
- Best results achieved by application to fine, firm seedbed, free of clods, crop residues and established weeds
- Do not use on sand, fen soil or soils with more than 10% organic matter
- In winter cereals normally applied as surface treatment without incorporation in tank-mixture with other herbicides to increase spectrum of control. See label for details
- Follow-up herbicide treatment recommended with some crops. See label for details

Crop safety/Restrictions

- Maximum number of treatments 1 per crop
- Apply and incorporate at any time during 2 wk before sowing or planting

- Do not apply to brassica plant raising beds
- Transplants should be hardened off prior to transplanting
- Apply in sugar beet after plants 10 cm high with 4-8 leaves and harrow into soil
- Apply in cereals after drilling up to and including 3 leaf stage (GS 13)
- Minimum interval between application and drilling or planting may be up to 12 mth. See label for details

Special precautions/Environmental safety
- Harmful if swallowed [11]
- May cause lung damage if swallowed [9, 11]
- Irritating to skin [7] and eyes [1, 2, 4-6, 8, 10]
- Risk of serious damage to eyes [7]
- Irritating to respiratory system [2, 4, 8]
- May cause lung damage if swallowed [11]
- Flammable
- Harmful to fish or other aquatic life. Do not contaminate surface waters or ditches with chemical or used container

Personal protective equipment/Label precautions
- A, C [1, 2, 4-8]
- M05 [9, 11]; R07d [1-6, 8-11]; R04b [1, 2, 4-8, 10]; R04a [1, 2, 4-6, 8, 10]; R04c [2, 4, 8]; R04g [9, 11]; R04d [7]; R03c [11]; U08, U13 [1-11]; U05a [1, 2, 4-8, 10]; U20a [3, 4, 6, 10]; U20b [1, 2, 5, 7-9, 11]; U11 [7]; C03 [1, 2, 4-8, 10]; E30a, E31b [1-11]; E26 [1, 2, 5, 6, 8-11]; E01 [1-8, 10]; E13c [1-7, 9-11]; E27 [1, 5-7, 9]; E28 [9, 11]; E29 [1, 5, 6]; E34 [7, 11]

Latest application/Harvest interval
- Varies between products; check labels. Normally before 4 leaves unfolded (GS 13) for cereals; up to 6, 8 or 10 leaves for sugar beet; pre-sowing/planting for other crops

Approval
- Off-label approval unlimited for use on evening primrose (OLA 1080/92)[9], gold-of-pleasure, radish seed crops, sunflowers, soya beans, combining peas, nursery fruit trees and bushes, ornamentals, outdoor herbs (OLA 0074/93)[9]; unlimited for use on outdoor kohlrabi (OLA 2341/98)[9]; unlimited for use on combining peas, sunflower, linseed, celeriac, ornamentals (OLA 1560/98)[9]
- Accepted by BLRA for use on malting barley

---

499   triflusulfuron-methyl

A sulfonyl urea herbicide for sugar beet

Products

1 Debut	DuPont	50% w/w	WG	07804
2 Landgold TFS 50	Landgold	50% w/w	WG	08941
3 Standon Triflusulfuron	Standon	50% w/w	WG	09487

Uses
Annual dicotyledons in **red beet** *(off-label)* [1]. Annual dicotyledons in **sugar beet** [1-3]. Annual dicotyledons in **fodder beet** [1, 3].

Efficacy
- Product should be used with a recommended adjuvant or a suitable herbicide tank-mix partner - see label for details
- Product acts by foliar action. Best results obtained from good spray cover of small actively growing weeds
- Susceptible weeds cease growth immediately and symptoms can be seen 5-10 d later
- Best results achieved from a programme of up to 4 treatments starting when first weeds have emerged with subsequent applications every 5-14 d when new weed flushes at cotyledon stage
- Weed spectrum can be broadened by tank mixture with other herbicides. See label for details
- Product may be applied overall or via band sprayer

Crop safety/Restrictions
- Maximum number of treatments 4 per crop
- All varieties of sugar beet (and fodder beet [1, 3]) may be treated from early cotyledon stage until the leaves begin to meet between the rows
- Do not apply to any crop stressed by drought, water-logging, low temperatures, pest or disease attack, nutrient or lime deficiency
- Only winter cereals should follow a treated crop in the same calendar yr. Any crop may be sown in the next calendar yr
- After failure of a treated crop, sow only spring barley, linseed or sugar beet within 4 mth of spraying unless prohibited by tank-mix partner

Special precautions/Environmental safety
- Irritant. May cause sensitization by skin contact
- Extremely dangerous to fish or other aquatic life. Do not contaminate surface waters or ditches with chemical or used container
- LERAP Category B
- Triflusulfuron-methyl is very active. Take particular care to avoid drift onto plants outside the target area
- Spraying equipment should not be drained or flushed onto land planted, or to be planted, with trees or crops other than sugar beet and should be thoroughly cleansed after use - see label for instructions

Personal protective equipment/Label precautions
- A [1-3]
- R04, R04e [1-3]; U20a [1, 2]; U08, U19 [1-3]; U20b [3]; E26 [1, 2]; E13a, E16a, E30a, E32a [1-3]

Latest application/Harvest interval
- Before crop leaves meet between rows

Approval
- Off-label approval to Oct 2003 for use on outdoor red beet (OLA 2351/98)[1]

---

500 triforine

A locally systemic fungicide with protectant and curative activity

See also bupirimate + triforine

Products

Fairy Ring Destroyer	Vitax	190 g/l	EC	05541

Uses

Fairy rings in **turf**.

Efficacy
- For fairy ring control apply as high volume spray or drench as soon as infection noted and repeat twice at 14 d intervals

Crop safety/Restrictions
- Maximum number of treatments 3 per yr for turf

Special precautions/Environmental safety
- Harmful in contact with skin, irritating to eyes

Personal protective equipment/Label precautions
- A, C
- M03, R03a, R04a, U02, U05a, U08, U20b, C03, E01, E13c, E26, E30a, E31a, E34

Maximum Residue Level (mg residue/kg food)
- hops 30; pome fruit, cherries, currants, gooseberries 2; plums 1; cucumbers, gherkins, courgettes 0.5; tea, wheat, rye, barley, oats, triticale 0.1; citrus fruit, tree nuts, cane fruit, bilberries, cranberries, wild berries, miscellaneous fruit, root and tuber vegetables, sweet corn, lettuce, spinach beet, watercress, witloof, chervil, chives, celery leaves, cardoons, fennel, rhubarb, fungi, pulses, oilseeds, potatoes, sorghum, maize, buckwheat, millet, rice, meat, milk and dairy produce, eggs 0.05

501   trinexapac-ethyl

A novel cyclohexanecarboxylate plant growth regulator for cereals, turf and amenity grassland

Products

1 Moddus	Novartis	250 g/l	EC	08801
2 Shortcut	Scotts	25% w/w	WB	09254

Uses

Growth retardation in *amenity grass, amenity turf* [2]. Lodging control in *durum wheat, ryegrass seed crops, spring barley, spring oats, spring rye, triticale, winter barley, winter oats, winter rye, winter wheat* [1].

Efficacy

- On wheat apply as single treatment between leaf sheath erect stage (GS 30) and flag leaf fully emerged (GS 39) [1]
- On barley, rye, triticale and durum wheat apply as single treatment between leaf sheath erect stage (GS 30) and second node detectable (GS 32), or on winter barley at higher dose between flag leaf just visible (GS 37) and flag leaf fully emerged (GS 39) [1]
- On oats and ryegrass seed crops apply between leaf sheath erect stage (GS 30) and first node detectable stage (GS 32) [1]
- Best results on turf achieved from application to actively growing weed free turf grass that is adequately fertilised and watered and is not under stress [2]
- Product rainfast after 12 h [2]

Crop safety/Restrictions

- Maximum total dose equivalent to one full dose on cereals, ryegrass seed crops [1]
- Maximum number of treatments on turf 5 per yr [2]
- Do not apply if rain or frost expected or if crop wet
- Only use on crops at risk of lodging [1]
- Treatment may cause ears to remain erect through to harvest [1]
- Turf under stress when treated may show signs of damage. Do not apply within 12 h of mowing [2]
- Not recommended for closely mown fine turf [2]

Special precautions/Environmental safety

- May cause sensitization by skin contact [1]
- Not to be used on food crops [2]
- Harmful to fish or other aquatic life. Do not contaminate surface waters or ditches with chemical or used container [1, 2]
- Avoid drift outside target area
- Do not compost or mulch clippings [2]

Personal protective equipment/Label precautions

- C [1]; A [1, 2]
- R04e, U05a, U15, U20c [1]; U20b [2]; C03 [1]; C01 [2]; E01 [1]; E13c, E26, E30a, E31a [1, 2]

Latest application/Harvest interval

- Before flag leaf sheath extending stage (GS 41)
- Before 2nd node detectable (GS 32) for oats, ryegrass seed crops; before 3rd node detectable (GS 33) for durum wheat, spring barley, triticale, rye; before flag leaf sheath extending (GS 41) for winter barley, winter wheat

Approval

- Accepted by BLRA for use on malting barley

502   triticonazole

A conazole fungicide available only in mixtures

See also imazalil + triticonazole

---

503   Verticillium lecanii

A fungal parasite of aphids and whitefly

Products

1	Mycotal	Koppert	16.1% w/w	WP	04782
2	Vertalec	Koppert	20% w/w	WP	04781

Uses

Aphids in *aubergines, cuttings, protected beans, protected chrysanthemums, protected cucumbers, protected peppers, protected roses, protected tomatoes* [2]. Whitefly in *aubergines, cucumbers, peppers, protected beans, protected cut flowers, protected lettuce, protected ornamentals, protected tomatoes* [1].

Efficacy

- Apply spore powder as spray as part of biological control programme
- Spray during late afternoon and early evening directing spray onto underside of leaves and to growing points
- Best results require minimum 80% relative humidity and 18°C within the crop canopy

Crop safety/Restrictions

- Maximum number of treatments 3 per crop [1]; 2 per crop at high dose for chrysanthemums, 1 per crop for other crops, 12 per crop at low dose, low volume [2]
- Never use in tank mixture
- A fungicide may not be used within 3 d of treatment. Pesticides containing captan, chlorothalonil, dichlofluanid, triforine, maneb, thiram or tolylfluanid may not be used on the same crop
- Keep in a refrigerated store at 2-6°C

Special precautions/Environmental safety

- Product does not affect natural predators or parasites

Personal protective equipment/Label precautions

- U20b, E15, E26, E29, E30a, E32a

---

504   vinclozolin

A protectant dicarboximide fungicide

See also carbendazim + vinclozolin

Products

1	Barclay Flotilla	Barclay	500 g/l	SC	07905
2	Landgold Vinclozolin SC	Landgold	500 g/l	SC	06459
3	Ronilan FL	BASF	500 g/l	SC	02960
4	Standon Vinclozolin	Standon	500 g/l	SC	07836

Uses

Alternaria in *spring oilseed rape, winter oilseed rape* [1]. Alternaria in *oilseed rape* [2-4]. Ascochyta in *combining peas* [2-4]. Blossom wilt in *apples* [3]. Botrytis in *spring oilseed rape, winter oilseed rape* [1]. Botrytis in *combining peas, dwarf beans, navy beans, runner beans, vining peas* [1-4]. Botrytis in *oilseed rape* [2-4]. Chocolate spot in *broad beans, field beans* [1-4]. Mycosphaerella in *combining peas* [2-4]. Sclerotinia stem rot in *spring oilseed rape, winter oilseed rape* [1]. Sclerotinia stem rot in *oilseed rape* [2-4].

Efficacy

- Timing of sprays varies with crop and disease. See label for details
- May be used at reduced rate on peas and field beans in tank-mix with chlorothalonil. Mixture with chlorothalonil essential on combining peas and oilseed rape [3]
- Where dicarboximide resistant strains have developed product may not be effective
- Do not spray if crop wet or if rain or frost expected

Crop safety/Restrictions

- Maximum number of treatments 2 per crop
- Do not treat mange-tout varieties of peas

Special precautions/Environmental safety
- Irritant. May cause sensitization by skin contact
- Irritating to skin [2]
- Product presents a minimal hazard to bees when used as directed but consider informing local bee-keepers if intending to spray crops in flower
- Harmful to fish and aquatic life. Do not contaminate surface waters or ditches with chemical or used container
- Operator must use a vehicle fitted with a cab and forced air filtration unit with a pesticide filter complying with HSE Guidance Note PM 74 or to an equally effective standard

Personal protective equipment/Label precautions
- A, C, H, K [1-4]; M [1-3]
- R04e [1-4]; R04b [2]; R04 [4]; U05a, U19, U20a [1-4]; U08 [1-3]; U09a [4]; C03 [1-4]; E31c [3]; E01, E13c, E30a [1-4]; E26 [2, 4]; E31b [1, 2, 4]; E34 [4]

Latest application/Harvest interval
- Before end of petal fall for apples
- HI beans, peas 2 wk; oilseed rape 7 wk

Maximum Residue Level (mg residue/kg food)
- hops 40; grapes, cane fruits, lettuces, celery 5; tomatoes, peppers, aubergines 3; apricots, peaches, nectarines, Chinese cabbage, witloof, beans (with pods), peas (with pods) 2; pome fruits, bulb vegetables, cucumbers, gherkins, courgettes, melons, squashes, watermelons, rape seed 1; cherries 0.5; tea 0.1; citrus fruits, tree nuts, bilberries, cranberries, gooseberries, wild berries, miscellaneous fruit (except kiwi fruit), beetroot, celeriac, Jerusalem artichokes, parsnips, parsley root, salsify, sweet potatoes, turnips, yams, sweetcorn, broccoli, cauliflowers, Brussels sprouts, head cabbages, kale, kohlrabi, spinach, beet leaves, watercress, herbs, stem vegetables (except celery), mushrooms, oilseed (except rape seed), potatoes, cereals, animal products 0.05

---

505   warfarin

A hydroxycoumarin rodenticide

Products

1	Grey Squirrel Liquid Concentrate	Killgerm	0.5% w/w	CB	06455
2	Sakarat Ready-to-Use (Cut Wheat Base)	Killgerm	0.025% w/w	RB	H6807
3	Sakarat Ready-to-Use (Whole Wheat)	Killgerm	0.025% w/w	RB	H6808
4	Sakarat X	Killgerm	0.05% w/w	RB	H6809
5	Sewarin Extra	Killgerm	0.05% w/w	RB	H6810
6	Sewarin P	Killgerm	0.025% w/w	RB	H6811
7	Sewercide Cut Wheat Rat Bait	Killgerm	0.05% w/w	RB	H6805
8	Sewercide Whole Wheat Rat Bait	Killgerm	0.05% w/w	RB	H6806
9	Sorex Warfarin 250 ppm Rat Bait	Sorex	0.025% w/w	RB	07371
10	Sorex Warfarin 500 ppm Rat Bait	Sorex	0.05% w/w	RB	07372
11	Sorex Warfarin Sewer Bait	Sorex	0.05% w/w	RB	07373
12	Warfarin 0.5% Concentrate	B H & B	0.5%	CB	H6815
13	Warfarin 0.5% Concentrate	B H & B	0.5%	CB	02325
14	Warfarin Ready Mixed Bait	B H & B	0.025%	RB	H6816
15	Warfarin Ready Mixed Bait	B H & B	0.025%	RB	02333

Uses

Grey squirrels in **agricultural premises, forestry, industrial sites** [1]. Mice in **agricultural premises** [9-11, 13, 15]. Rats in **agricultural premises** [2-15].

Efficacy

- For rodent control place ready-to-use or prepared baits at many points wherever rats active. Out of doors shelter bait from weather
- Inspect baits frequently and replace or top up as long as evidence of feeding. Do not underbait
- For grey squirrel control mix with whole wheat and leave to stand for 2-3 h before use
- Use bait in specially constructed hoppers and inspect every 2-3 d. Replace as necessary

Crop safety/Restrictions

- For use only between 15 Mar and 15 Aug for tree protection [1]

Special precautions/Environmental safety

- For use only by local authorities, professional operators providing a pest control service and persons occupying industrial, agricultural or horticultural premises
- Prevent access to baits by children and animals, especially cats, dogs and pigs
- Rodent bodies must be searched for and burned or buried, not placed in refuse bins or rubbish tips. Remains of bait and containers must be removed after treatment and burned or buried
- Bait must not be used where food, feed or water could become contaminated
- The use of warfarin to control grey squirrels is illegal unless the provisions of the Grey Squirrels Order 1973 are observed. See label for list of counties in which bait may not be used [1]
- Must not used outdoors at all in Scotland, nor in areas of England or Wales where pine martens occur naturally [1]

Personal protective equipment/Label precautions

- A, C, D, E, H [2-8]
- M05 [9-11]; M03 [2, 3, 5, 6]; U13 [1-15]; U20b [7-15]; U20a [1-6]; E30a [1-15]; E31a [12-15]; E32a [1-11]; V04a [1, 4, 7-15]; V02 [1-10, 12-15]; V01a, V03a [1, 4, 7-10, 12-15]; V05 [1]; V01b, V03b, V04b [2, 3, 5, 6]

Approval

- Product not approved for use in N Ireland [1]

---

506   zeta-cypermethrin

A contact and stomach acting pyrethroid insecticide

Products

1 Fury 10 EW	FMC	100 g/l	EW	10268
2 Minuet EW	FMC	100 g/l	EW	09605

Uses

Aphids in **spring barley, spring wheat, winter barley, winter wheat**. Barley yellow dwarf virus vectors in **spring barley, spring wheat, winter barley, winter wheat**. Cabbage stem flea beetle in **spring oilseed rape, winter oilseed rape**. Cutworms in **potatoes, sugar beet**. Pea and bean weevils in **combining peas, spring field beans, vining peas, winter field beans**. Pea aphid in **combining peas, vining peas**. Pea moth in **combining peas, vining peas**. Pod midge in **spring oilseed rape, winter oilseed rape**. Pollen beetles in **spring oilseed rape, winter oilseed rape**. Rape winter stem weevil in **spring oilseed rape, winter oilseed rape**. Seed weevil in **spring oilseed rape, winter oilseed rape**.

Efficacy

- On winter cereals spray when aphids first found in the autumn for BYDV control. A second spray may be required on late drilled crops or in mild conditions
- For summer aphids on cereals spray when treatment threshold reached
- For listed pests in other crops spray when feeding damage first seen or when treatment threshold reached. Under high infestation pressure a second treatment may be necessary
- Best results for pod midge and seed weevil control in oilseed rape obtained from treatment after pod set but before 80% petal fall

- Pea moth treatments should be applied according to ADAS/PGRO warnings or when economic thresholds reached as indicated by pheromone traps
- Treatments for cutworms should be made at egg hatch and repeated no sooner than 10 d later

Crop safety/Restrictions
- Maximum number of treatments 2 per crop
- Consult processors before use on crops for processing

Special precautions/Environmental safety
- Harmful if swallowed. May cause sensitization by skin contact
- Extremely dangerous to fish or other aquatic life. Do not contaminate surface waters or ditches with chemical or used container
- LERAP Category A

Personal protective equipment/Label precautions
- A, C, H
- M03, R03c, R04e, U05a, U08, U14, U15, U19, U20b, C03, E01, E13a, E16c, E16d, E22a, E22b, E30a, E31b, E34

Latest application/Harvest interval
- Before end of flowering for oilseed rape; before flowering completed (GS 69) for cereals
- HI potatoes, field beans 14 d; sugar beet 60 d. No information for other crops

Approval
- Following implementation of Directive 98/82/EC, approval for use of zeta-cypermethrin on numerous crops was revoked in 1999

---

507   zinc phosphide

A phosphine generating rodenticide

Products

1	Grovex Zinc Phosphide	Killgerm	100%	CB	H6800
2	RCR Zinc Phosphide	Killgerm	100%	CB	H6801
3	ZP Rodent Pellets	Antec	2.0% w/w	GB	07814

Uses

Mice in **agricultural premises** [1, 2]. Mice in **farm buildings** [3]. Rats in **agricultural premises** [1, 2]. Rats in **farm buildings** [3].

Efficacy
- Use to prepare baits either by dry or wet baiting as directed

Special precautions/Environmental safety
- Zinc phosphide is subject to the Poisons Rules 1982 and the Poisons Act 1972. See notes in Section 1
- For use only by local authorities, professional operators providing a pest control service and persons occupying industrial, agricultural or horticultural premises
- Toxic if swallowed
- Spontaneously inflammable in contact with acid
- Keep in original container, tightly closed, in a safe place, under lock and key
- Wear respirator if mixing baits in a confined space
- Prevent access to baits by children and domestic animals, especially cats, dogs and pigs
- Search for and burn or bury all rodent bodies. Do not place in refuse bins or on rubbish tips
- Do not prepare or use baits where food or water could be contaminated
- Remove all remains of baits and bait containers after treatment and burn or bury
- Wash out all mixing equipment thoroughly at the end of every operation

Personal protective equipment/Label precautions
- A [1-3]; D, H [1, 2]
- M03 [3]; R02c [1, 2]; U13, U20a [1-3]; U02, U04a, U05a, C03 [1, 2]; E15 [3]; E30b, E32a [1-3]; E01, E34 [1, 2]; V03b [3]; V01a, V02, V04a [1-3]; V03a [1, 2]

---

508   zineb

A protectant dithiocarbamate fungicide for many horticultural crops

See also ferbam + maneb + zineb

Products

Unicrop Zineb   Unicrop   70 % w/w   WP   02279

Uses

Blight in **potatoes**. Botrytis in **anemones**. Downy mildew in **hops, outdoor lettuce, protected lettuce**. Fire in **tulips**. Rust in **carnations**.

Efficacy

- Best results obtained by treatment in settled weather conditions. Do not spray if rain imminent
- Follow local blight warnings to ensure accuracy of first treatment on potatoes
- Recommended timing and spray volume varies with crop and disease. See label for details
- For root rot control in tomatoes apply as a trench watering at about the time of set of 5th truss
- Addition of wetting agent recommended for tulips and anemones

Crop safety/Restrictions

- Maximum number of treatments on protected lettuce (including zineb, maneb, mancozeb, other EBDC fungicides or thiram) 2 per crop post-planting up to 2 wk later, and none thereafter. If thiram-based products are used post-planting on crops that will mature from Nov to Mar, 3 treatments are permitted within 3 wk of planting out
- Do not apply pre-picking to blackcurrants intended for canning

Special precautions/Environmental safety

- Irritating to eyes, skin and respiratory system

Personal protective equipment/Label precautions

- A
- R04a, R04b, R04c, U05a, U08, U19, U20a, C03, E01, E15, E30a, E32a

Latest application/Harvest interval

- HI blackcurrants 4 wk; protected lettuce 3 wk; outdoor lettuce 2 wk; other edible outdoor crops 1 wk; other edible glasshouse crops 2 d

Approval

- Approved for aerial application on potatoes. See notes in Section 1
- Accepted by BLRA for use on hops before burr stage

Maximum Residue Level (mg residue/kg food)

- hops 25; lettuces, herbs 5; oranges, apricots, peaches, nectarines, grapes, strawberries, peppers, aubergines 2; cherries, plums 1; garlic, onions, shallots, cucumbers 0.5; celeriac, witloof 0.2; tree nuts, oilseeds (except rape seed), tea 0.1; bilberries, cranberries, wild berries, blackberries, loganberries, miscellaneous fruit, root and tuber vegetables, horseradish, Jerusalem artichokes, parsley root, sweet potatoes, swedes, turnips, yams, spinach, beet leaves, watercress, asparagus, cardoons, fennel, globe artichokes, rhubarb, mushrooms, potatoes, maize, rice, animal products 0.05

509   ziram

A dithiocarbamate bird and animal repellent

Products

AAprotect   Unicrop   32% w/w   PA   03784

Uses

Birds in **field crops, forestry, ornamentals, top fruit**. Deer in **field crops, forestry, ornamentals, top fruit**. Hares in **field crops, forestry, ornamentals, top fruit**. Rabbits in **field crops, forestry, ornamentals, top fruit**.

Efficacy
- Apply undiluted to main stems up to knee height to protect against browsing animals at any time of yr or spray 1:1 dilution on stems and branches in dormant season
- Use dilute spray on fully dormant fruit buds to protect against bullfinches
- Only apply to dry stems, branches or buds
- Use of diluted spray can give limited protection to field crops in areas of high risk during establishment period

Crop safety/Restrictions
- Do not spray elongating shoots or buds about to open
- Do not apply concentrated spray to foliage, fruit buds or field crops

Special precautions/Environmental safety
- Irritating to eyes, skin and respiratory system

Personal protective equipment/Label precautions
- A, C
- R04a, R04b, R04c, U05a, U08, U19, U20b, C03; C02 (8 wk); E01, E13c, E30a, E31c

Latest application/Harvest interval
- HI edible crops 8 wk

SECTION 5
APPENDICES

Appendix 1
Suppliers of Pesticides and Adjuvants

AgrEvo: AgrEvo UK Ltd
See Aventis CropScience UK Ltd

AgrEvo Environ.: AgrEvo Environmental
Health
See Aventis CropScience UK Ltd

Agrichem: Agrichem (International) Ltd
Industrial Estate
Station Road
Whittlesey
Cambs
PE7 2EY
Tel: (01733) 204019
Fax: (01733) 204162

Agricola: Agricola Ltd
Abbey Farm
Snape
Saxmundham
Suffolk
IP17 1RQ
Tel: (01728) 688078

AgriGuard: AgriGuard Ltd
75 Shandon Park
Phibsboro
Dublin 7
Ireland
Tel: (+353) 1 845 1599
Fax: (+353) 1 845 1620
Email: brian@agritech.iol.ie

Allied Colloids: Allied Colloids Limited
See Ciba Specialty Chemicals

Antec: Antec International
Windham Road
Chilton Industrial Estate
Sudbury
Suffolk
CO10 2XD
Tel: (01787) 377305
Fax: (01787) 310846
Email: antec_international@compuserve.com
Web: http://www.antecint.com

Aquaspersions: Aquaspersions Ltd
Beacon Hill Road
Halifax
W. Yorks.
HX3 6AQ
Tel: (01422) 386200
Fax: (01422) 386239

Aventis: Aventis CropScience UK Ltd
Fyfield Road
Ongar
Essex
CM5 0HW
Tel: (01227) 301301
Fax: (01227) 362610
Email: ian.cockram@aventis.com
Web: http://www.aventis.com

Aventis Environ.: Aventis Evironmental
Science (Turf & Amenity)
Fyfield Road
Ongar
Essex
CM5 0HW
Tel: (01277) 301301
Fax: (01277) 362610
Email: ian.cockram@aventis.com
Web: http://www.aventis.co.uk

B H & B: Battle Hayward & Bower Ltd
Victoria Chemical Works
Crofton Drive
Allenby Road Industrial Estate
Lincoln
LN3 4NP
Tel: (01522) 529206
Fax: (01522) 538960

Banks: Banks Agriculture Ltd
Cattle Market Chase
Wisbech
Cambs.
PE13 1RE
Tel: (01767) 680351
Fax: (01767) 6902215

Barclay: Barclay Chemicals (UK) Ltd
  Barclay House
  Lilmar Industrial Estate
  Santry
  Dublin 9
  Ireland
  Tel:(+353) 1 842 5755
  Fax: (+353) 1 842 5381

Barrettine: Barrettine Environmental Health
  Barrettine Works
  St. Ivel Way
  Warmley
  Bristol
  BS15 5TY
  Tel: (0117) 967 2222
  Fax: (0117) 961 4122

BASF: BASF plc.
  Agricultural Divison
  PO Box 4
  Earl Road, Cheadle Hulme
  Cheadle
  Cheshire
  SK8 6QG
  Tel: (0161) 485 6222
  Fax: (0161) 485 2229

Batsons: Joseph Batsons Ltd
  Dudley Road
  Tipton
  W. Midlands
  DY4 8EH
  Tel: (0121) 522 0100
  Fax: (0121) 522 0115

Bayer: Bayer plc.
  Crop Protection Business Group
  Eastern Way
  Bury St Edmunds
  Suffolk
  IP32 7AH
  Tel: (01284) 763200
  Fax: (01284) 702810
  Email: crop.protection@bayer.co.uk

Biowise: Biowise
  Hoyle Depot
  Graffham
  Petworth
  W. Sussex
  GU28 0LR
  Tel: (01798) 867574
  Fax: (01798) 867574
  Email: woodbugs@newscientist.net

Brian Jones: Brian Jones and Associates Ltd
  Fluorocarbon Building
  Caxton Hill
  Hertford
  Herts.
  SG13 7NH
  Tel: (01992) 553065
  Fax: (01992) 551873

Cardel: Cardel Agro SAS
  Europarc du Chene
  11, Rue Pascal
  69500 Bron
  France
  Tel: (01653) 617000

Chiltern: Chiltern Farm Chemicals Ltd
  11 High Street
  Thornborough
  Buckingham
  MK18 2DF
  Tel: (01280) 822400
  Fax: (01280) 822082

CMI: CMI Ltd (incorporating Collingham Marketing)
  United House
  113 High Street
  Collingham
  Newark
  Notts.
  NG23 7NG
  Tel: (01636) 892078
  Fax: (01636) 893037
  Email: collmark@fsbdial.co.uk

Coalite: Coalite Chemicals
  PO Box 152
  Buttermilk Lane
  Bolsover
  Chesterfield
  Derbyshire
  S44 6AZ
  Tel: (01246) 826816
  Fax: (01246) 240309

Coventry Chemicals: Coventry Chemicals Ltd
  Woodhams Road
  Siskin Drive
  Coventry
  CV3 4FX
  Tel: (024) 7663 9739
  Fax: (024) 7663 9717

Cyanamid: Cyanamid Agriculture UK
  See BASF plc

DAPT: DAPT Agrochemicals Limited
14 Monks Walk
Southfleet
Gravesend
Kent
DA13 9NZ
Tel: (01474) 834448
Fax: (01474) 834449

Dax: Dax Products Ltd
P O Box 119
76 Cyprus Road
Nottingham
NG3 5NA
Tel: (0115) 926 9996
Fax: (0115) 966 1173

De Sangosse: De Sangosse (UK) SA
PO Box 135
Market Weighton
York
YO43 4YY
Tel: (01430) 872525
Fax: (01430) 873123

Deosan: Deosan Ltd
See DiverseyLever

Dewco-Lloyd: Dewco-Lloyd Ltd
Cyder House
Ixworth
Suffolk
IP31 2HT
Tel: (01359) 230555
Fax: (01359) 232553

DiverseyLever: DiverseyLever Ltd
Weston Favell Centre
Northampton
NN3 8PD
Tel: (01604) 783505
Fax: (01604) 783506

Doff Portland: Doff Portland Ltd
Aerial Way
Hucknall
Nottingham
NG15 6DW
Tel: (0115) 963 2842
Fax: (0115) 963 8657
Email: info@doff.co.uk
Web: http://www.doff.co.uk

Dow: Dow AgroSciences
Latchmore Court
Brand Street
Hitchin
Herts.
SG5 1NH
Tel: (01462) 457272
Fax: (01462) 426605

DuPont: DuPont (UK) Ltd
Agricultural Products Department
Wedgwood Way
Stevenage
Herts.
SG1 4QN
Tel: (01438) 734000
Fax: (01438) 734452

Elliott: Thomas Elliott Ltd
143A High Street
Edenbridge
Kent
TN8 5AX
Tel: (01732) 866566
Fax: (01732) 864709

Euroagkem: Euroagkem Ltd
Byemoor Farm
Melmerby
Leyburn
N. Yorks
DL8 4TW
Tel: (01969) 640655
Fax: (01969) 640633

Fargro: Fargro Ltd
Toddington Lane
Littlehampton
Sussex
BN17 7PP
Tel: (01903) 721591
Fax: (01903) 730737
Email: promos-fargro@btinternet.com
Web: http://www.fargro.co.uk

FCC: Farmers Crop Chemicals Ltd
Thorn Farm
Evesham Road
Inkberrow
Worcs.
WR7 4LJ
Tel: (01386) 793401
Fax: (01386) 793184

Fine: Fine Agrochemicals Ltd
Hill End House
Whittington
Worcester
WR5 2RL
Tel: (01905) 361800
Fax: (01905) 361810
Email: enquire@fine-agrochemicals.com

FMC: FMC Corporation (UK) Ltd
Wolseley Road
Woburn Road Industrial Estate
Kempston
Bedford
MK42 7EF
Tel: (01234) 841177
Fax: (01234) 841097
Web: http://www.fmcapg.co.uk

Ford Smith: Ford Smith & Co. Ltd
Lyndean Industrial Estate
Felixstowe Road
Abbey Wood
London
SE2 9SG
Tel: (020) 8310 8127
Fax: (020) 8310 9563

Graincare: Graincare (Colchester) Ltd
17 Woodlands
Colchester
Essex
CO4 3JA
Tel: (01206) 862436
Fax: (01206) 862436

Greencrop: Greencrop Technology Ltd
Burren House
2 Cowbrook Court
Glossop
Derbyshire
SK13 8SL
Tel: (01457) 856001
Fax: (01457) 857137

Greenhill: Rawdon Works
Rawdon Road
Moira
Swadlincote
Derbyshire
DE12 6DA
Tel: (01283) 551500
Fax: (01283) 550413

Growing Success: Growing Success Organics Ltd
Wessex House
1-3 Hilltop Business Park
Devizes Road
Salisbury
Wilts.
SP3 4UF
Tel: (01722) 337744
Fax: (01722) 333177
Email: growing@wessexhortgrp.demon.co.uk

Headland: Headland Agrochemicals Ltd
Norfolk House
Gt. Chesterford Court
Gt. Chesterford
Saffron Walden
Essex
CB10 1PF
Tel: (01799) 530146
Fax: (01799) 530229

Heatherington: J.V. Heatherington Farm & Garden Supplies
29 Main Street
Glenary
Crumlin
Co. Antrim
BT29 4LN
Tel: (01849) 422227

Helm: Helm Great Britain Chemicals Ltd
Wimbledon Bridge House
1 Hartfield Road
London
SW19 3RU
Tel: (020) 8544 9000
Fax: (020) 8544 1011
Web: http://www.helmag.com

Hickson & Welch: Hickson & Welch Ltd
Wheldon Road
Castleford
W. Yorks.
WF10 2JT
Tel: (01977) 556565
Fax: (01977) 550910

Hortag: Hortag Chemicals Ltd
Salisbury Road
Downton
Wilts.
SP5 3JJ
Tel: (01725) 512822
Fax: (01725) 512840

Hortichem: Hortichem Ltd
1b, Mills Way
Boscombe Down Business Park
Amesbury
Wilts.
SP4 7RX
Tel: (01980) 676500
Fax: (01980) 626555
Email: hortichem@hortichem.co.uk

I T Agro: I T Agro Ltd
805 Salisbury House
31 Finsbury Circus
London
EC2M 5SQ
Tel: (020) 7628 2040

Interagro: Interagro (UK) Ltd
Sworders Barn
Sworders Yard
North Street
Bishop's Stortford
Herts.
CM23 2LD
Tel: (01279) 501995
Fax: (01279) 501996
Email: info@interagro.co.uk
Web: http://www.interagro.co.uk

Interfarm: Interfarm (UK) Ltd
Kinghams's Place
36 Newgate Street
Doddington
March
Cambs.
PE15 0SR
Tel: (01354) 741414
Fax: (01354) 741004

Intracrop: Intracrop
Byemoor Farm
Melmerby
Leyburn
N. Yorks.
DL8 4TW
Tel: (01969) 640655
Fax: (01969) 640633

Irish Drugs: Irish Drugs Ltd
Burnfoot
Lifford
Co. Donegal
Ireland
Tel: (+353) 77 68103/4
Fax: (+353) 77 68311

Isagro: Isagro Sp.A
See Sipcam UK Ltd

K&S Fumigation: K&S Fumigation Services
Ltd
Shirley Farmhouse
Moor Lane
Woodchurch
Ashford
Kent
TN26 3SS
Tel: (01233) 758252
Fax: (01233) 758343

Killgerm: Killgerm Chemicals Ltd
115 Wakefield Road
Flushdyke
Ossett
W. Yorks.
WF5 9AR
Tel: (01924) 268400
Fax: (01924) 264757
Email: info@killgerm.com
Web: http://www.killgerm.com

Koppert: Koppert (UK) Ltd
Homefield Road
Faircrouch Lane
Haverhill
Suffolk
CB9 8QP
Tel: (01440) 704488
Fax: (01440) 704487

Lambson: Lambson Agricultural Services
Cinder Lane
Castleford
W. Yorks.
WF10 1LU
Tel: (01977) 510511
Fax: (01977) 603049

Landgold: Landgold & Co. Ltd
PO Box 829
Charles House
Charles Street
St. Helier
Jersey
JE4 0UE
Tel: (01534) 768446
Fax: (01534) 732843

Landseer: Landseer Ltd
Landseer House
9 Rous Chase
Galleywood
Chelmsford
Essex
CM2 8QF
Tel: (01245) 357109
Fax: (01245) 494165

Luxan: Luxan (UK) Ltd
Sysonby Lodge
Nottingham Road
Melton Mowbray
Leics.
LE13 0NU
Tel: (01664) 566372
Fax: (01664) 480137
Email: enquiry@luxan.co.uk
Web: http://www.luxan.co.uk

Makhteshim: Makhteshim-Agan (UK) Ltd
Unit 16
Thatcham Business Village
Colthrop Way
Thatcham
Berks.
RG19 4LW
Tel: (01635) 860555
Fax: (01865) 861555

Mandops: Mandops (UK) Ltd
36 Leigh Road
Eastleigh
Hants.
SO50 9DT
Tel: (023) 8064 1826
Fax: (023) 8062 9106
Email: enquiries@mandops.co.uk
Web: http://www.mandops.co.uk

Maxicrop: Maxicrop International Ltd
Weldon Road
Corby
Northants.
NN17 5US
Tel: (01536) 402182
Fax: (01536) 204254
Email: info@maxicrop.co.uk

Me2: Me2 Crop Protection Ltd
Tower House
Fishergate
York
YO10 4HA
Tel: (01904) 567331
Fax: (01904) 567301
Email: sales@me2cpl.co.uk

Microcide: Microcide Ltd
Shepherds Grove
Stanton
Bury St. Edmunds
Suffolk
IP31 2AR
Tel: (01359) 251077
Fax: (01359) 251545

Mitchell Cotts: Mitchell Cotts Chemicals Ltd
PO Box 6
Steanard Lane
Mirfield
W. Yorks.
WF14 8QB
Tel: (01924) 493861
Fax: (01924) 490972

Monsanto: Monsanto (UK) Ltd
The Maris Centre
Hauxton Road
Trumpington
Cambridge
CB2 2LQ
Tel: (01223) 849540
Fax: (01223) 849414
Email: technical.helpline.uk@monsanto.com
Web: http://www.monsanto.com

MTM Agrochem.: MTM Agrochemicals Ltd
See United Phosphorus Ltd

Nehra: Nehra Cookes Chemicals Ltd
16 Chiltern Close
Warren Wood
Arnold
Nottingham
NG5 9PX
Tel: (0115) 973 5999
Fax: (0115) 973 6700

Newman: Newman Agrochemicals Ltd
Swaffham Bulbeck
Cambridge
CB5 0LU
Tel: (01223) 811215
Fax: (01223) 812725

Nickerson Seeds: Nickerson Seeds Ltd
JNRC
Rothwell
Lincs.
LN7 6DT
Tel: (01472) 371661
Email: Nickerson.Seeds@farmline.com

Nomix-Chipman: Nomix-Chipman Ltd
Portland Building
Portland Street
Staple Hill
Bristol
BS16 4PS
Tel: (0117) 957 4574
Fax: (0117) 956 3461

Novartis: Novartis Crop Protection UK Ltd
Whittlesford
Cambridge
CB2 4QT
Tel: (01223) 833621
Fax: (01223) 835211
Web: http://www.novartiscrop.co.uk

Novartis A H: Novartis Animal Health UK Ltd
Whittlesford
Cambridge
CB2 4XW
Tel: (01223) 833634
Fax: (01223) 836526

Novartis BCM: Novartis BCM Ltd
Telstar Nursery
Holland Road
Little Clacton
Essex
CO16 9QG
Tel: (01255) 863200
Fax: (01255) 863206
Email: bcm.novartis@cp.novartis.com
Web: http://www.novartis-agri.co.uk

Nufarm Whyte: Nufarm Whyte Agriculture Ltd
Denaby Lane Industrial Estate
Denaby Lane
Old Denaby
Doncaster
S. Yorks.
DN12 4LQ
Tel: (01709) 772200
Fax: (01709) 772201

PBI: pbi Agrochemicals Ltd
See SumiAgro (UK) Ltd

Portman: Portman Agrochemicals Ltd
Apex House
Grand Arcade
Tally-Ho Corner
North Finchley
London
N12 0EH
Tel: (020) 8446 8383
Fax: (020) 8445 6045

Powaspray: Powaspray Ltd
7 Browntoft Lane
Donington
Spalding
Lincs.
PE11 4TQ
Tel: (01775) 821031
Fax: (01775) 821034

Email: powaspray@compuserve.com

Quadrangle: Quadrangle Agrochemicals
Crook Farm
North Deighton
Wetherby
Yorks.
LS22 5HW
Tel: (01937) 584228
Fax: (01937) 580937

Reabrook: Reabrook Ltd
Stanhope Road
Swadlincote
Burton-on-Trent
Staffs.
DE11 9BE
Tel: (01283) 221044
Fax: (01283) 225731

Rentokil: Rentokil Initial UK Ltd
Felcourt
East Grinstead
W. Sussex
RH19 2JY
Tel: (01342) 833022
Fax: (01342) 326229

Rigby Taylor: Rigby Taylor Ltd
Rigby Taylor House
Garside Street
Bolton
Lancs.
BL1 4AE
Tel: (01204) 394888
Fax: (01204) 385276

Roebuck Eyot: Roebuck Eyot Ltd
7a Hatfield Way
South Church Enterprise Park
Bishop Auckland
Co. Durham
DL14 6XF
Tel: (01388) 772233
Fax: (01388) 775233
Email: sales@roebuck-eyot.co.uk
Web: http://www.roebuck-eyot.co.uk

Rohm & Haas: Rohm & Haas (UK) Ltd
Lennig House
2 Masons Avenue
Croydon
Surrey
CR9 3NB
Tel: (020) 8774 5300
Fax: (020) 8774 5301

RP Agric.: Rhone-Poulenc Agriculture Ltd
See Aventis CropScience UK Ltd

RP Amenity: Rhone-Poulenc Amenity
See Aventis CropScience UK Ltd

Scotts: The Scotts Company (UK) Ltd
Paper Mill Lane
Bramford
Ipswich
Suffolk
IP8 4BZ
Tel: (01473) 830492
Fax: (01473) 830386

Service Chemicals: Service Chemicals Ltd
Lanchester Way
Royal Oak Industrial Estate
Daventry
Northants.
NN11 5PH
Tel: (01327) 704444
Fax: (01327) 71154

Sinclair: William Sinclair Horticulture Ltd
Firth Road
Lincoln
LN6 7AH
Tel: (01522) 537561
Fax: (01522) 513609

Sipcam: Sipcam UK Ltd
Sheraton House
Castle Park
Cambridge
CB3 0AX
Tel: (01223) 370030
Fax: (01223) 354026

Sorex: Sorex Ltd
St Michael's Industrial Estate
Hale Road
Widnes
Cheshire
WA8 8TJ
Tel: (0151) 420 7151
Fax: (0151) 495 1163
Email: sorex@sorex.com

Sphere: Sphere Laboratories (London) Ltd
The Yews
Main Street
Chilton
Oxon.
OX11 0RZ
Tel: (01235) 831802
Fax: (01235) 833896

Standon: Standon Chemicals Ltd
48 Grosvenor Square
London
W1X 9LA
Tel: (020) 7493 8648
Fax: (020) 7493 4219

Stefes: Stefes UK Ltd
See Aventis CropScience UK Ltd

Stoller: Stoller Chemical Ltd
53 Bradley Hall Trading Estate
Bradley Lane
Standish
Lancs.
WN6 0XQ
Tel: (01257) 427722
Fax: (01257) 427888

SumiAgro: SumiAgro (UK) Ltd
Merlin House
Falconry Court
Bakers Lane
Epping
Essex
CM16 5DQ
Tel: (01992) 563700
Fax: (01992) 563800
Email: sumiagro@sumiagro.co.uk

Sumitomo: Sumitomo Corporation (UK) PLC
Vintners' Place
68 Upper Thames Street
London
EC4V 3BJ
Tel: (020) 7246 3796
Fax: (020) 7246 3933

Sylvan: Sylvan Spawn Limited
Broadway
Yaxley
Peterborough
PE7 3EJ
Tel: (01733) 240412
Fax: (01733) 245020

Tomen: Tomen (UK) plc
Tomen House
13 Charles Street
London
SW1Y 4QT
Tel: (020) 7321 6621
Fax: (020) 7321 6624
Email: meadsa@ldn.tomen.co.uk

Tripart: Tripart Farm Chemicals Ltd
The Grove
Cambridge Road
Godmanchester
Huntingdon
Cambs.
PE18 8BW
Tel: (01480) 417951
Fax: (01480) 417651

Truchem: Truchem Ltd
Brook House
30 Larwood Grove
Sherwood
Nottingham
NG5 3JD
Tel: (0115) 926 0762
Fax: (0115) 967 1153

Unicrop: Universal Crop Protection Ltd
Park House
Maidenhead Road
Cookham
Berks.
SL6 9DS
Tel: (01628) 526083
Fax: (01628) 810457

Uniroyal: Uniroyal Chemical Ltd
Kennet House
4 Langley Quay
Slough
Berks.
SL3 6EH
Tel: (01753) 603000
Fax: (01753) 603077

United Phosphorus: United Phosphorus Ltd
Chadwick House
Birchwood Park
Warrington
Cheshire
WA3 6AE
Tel: (01925) 819999
Fax: (01925) 817425
Email: chris@upluk6.demon.co.uk

Vitax: Vitax Ltd
Owen Street
Coalville
Leicester
LE67 3DE
Tel: (01530) 510060
Fax: (01530) 510299

Wheatley: Wheatley Chemical Co.
Denaby Lane Industrial Estate
Denaby Lane
Old Denaby
Doncaster
S. Yorks.
DN12 4LQ
Tel: (01709) 772200
Fax: (01709) 772201

Whyte Chemicals: Whyte Chemicals Ltd
Marlborough House
298 Regents Park Road
Finchley
London
N3 2UA
Tel: (020) 8346 5946
Fax: (020) 8349 4589

Zeneca: Zeneca Crop Protection
Fernhurst
Haslemere
Surrey
GU27 3JE
Tel: (01428) 657222
Fax: (01428) 657330
Email: agassist@aguk.zeneca.com
Web: http://www.zeneca-crop.co.uk

Appendix 2
Useful Contacts

BASIS Ltd
Bank Chambers
34 St John Street
Ashbourne
Derbyshire DE6 1GH
Tel: (01335) 343945/346138
Fax: (01335) 346488

**Brewers' and Licensed Retailers'
Association (BLRA)**
42 Portman Square
London W1H 0BB
Tel: (020) 7486 4831
Fax: (020) 7935 3991

British Crop Protection Council (BCPC)
49 Downing Street
Farnham
Surrey GU9 7PH
Tel: (01252) 733072
Fax: (01252) 727194

BCPC Publications Sales
Bear Farm
Binfield
Bracknell
Berks. RG42 5QE
Tel: (0118) 934 2727
Fax: (0188) 934 1998

British Beekeepers' Association
National Agricultural Centre
Stoneleigh
Kenilworth
Warwickshire CV8 2LZ
Tel: (024) 7669 6679
Fax: (024) 7669 0682

British Pest Control Association
3 St James Court
Friar Gate
Derby DE1 1ZU
Tel: (01332) 294288
Fax: (01332) 295904

**Crop Protection Association Ltd (previously
British Agrochemicals Association Ltd)**
4 Lincoln Court
Lincoln Road
Peterborough
Cambs. PE1 2RP
Tel: (01733) 349225
Fax: (01733) 562523

Department of Agriculture Northern Ireland
Pesticides Section
Dundonald House
Upper Newtownards Road
Belfast BT4 3SB
Tel: (028) 9052 4704
Fax: (028) 9052 4266

Department of the Environment
Water Directorate
Romney House
43 Marsham Street
London SW1P 3PY
Tel: (020) 7276 8220
Fax: (020) 7276 8639

Department of Health
Food Safety and Public Health Branch
Skipton House
80 London Road
London SE1 6LW
Tel: (020) 7972 5033
Fax: (020) 7972 5137

**European Crop Protection Association
(ECPA)**
Avenue E van Nieuwenhuyse 6
B-1160 Brussels
Belgium
Tel: (+32) 2 663 1550
Fax: (+32) 2 663 1560

**Environment Agency (previously The
National Rivers Authority)**
Rio House
Waterside Drive
Aztec West
Almondsbury
Bristol BS12 4UD
Tel: (01454) 624400
Fax: (01454) 624409

Farmers' Union of Wales
Llys Amaeth
Queen's Square
Aberystwyth
Dyfed SY23 2EA
Tel: (01970) 612755
Fax: (01970) 624369

Forestry Commission
231 Corstorphine Road
Edinburgh EH12 7AT
Tel: (0131) 334 0303
Fax: (0131) 334 3047

Global Crop Protection Federation
Avenue Louise 143
B-1050 Brussels
Belgium
Tel: (+32) 2 542 0410
Fax: (+32) 2 542 0419

Health and Safety Executive
Information Centre
Broad Lane
Sheffield
Yorkshire S3 7HQ
Tel: (0114) 289 2345/6
Fax: (0114) 289 2333

Health and Safety Executive
Biocides & Pesticides Assessment Unit
Magdalen House
Bootle
Merseyside L20 3QZ
Tel: (0151) 951 4000

Health and Safety Executive – Books
PO Box 1999
Sudbury
Suffolk CO10 6FS
Tel: (01787) 881165
Fax: (01787) 313995

**Lantra National Training Organisation Ltd
(previously ATB-LandBase)**
National Agricultural Centre
Stoneleigh
Kenilworth
Warwickshire CV8 2UG
Tel: (024) 7669 6996
Fax: (024) 7669 6732

Ministry of Agriculture Fisheries and Food
Food and Veterinary Science Division
Nobel House
17 Smith Square
London SW1P 3JR
Tel: (020) 7238 5526
Fax: (020) 7238 6129

**National Association of Agricultural
Contractors (NAAC)**
Samuelson House
Paxton Road
Orton Centre
Peterborough PE2 5LT
Tel: (01733) 362920
Fax: (01733) 362921

National Farmers' Union
Agriculture House
164 Shaftesbury Avenue
London WC2H 8HL
Tel: (020) 7331 7200
Fax: (020) 7331 7313

National Poisons Information Service
Guys' and St Thomas' Hospital Trust
London SE14 5ER
Tel: (020) 7635 9191

The Royal Hospitals
Belfast BT12 6BA
Tel: (028) 9024 0503

City Hospital NHS Trust
Birmingham B18 7QH
Tel: (0121) 507 5588/5589

Llandough Hospital
Penarth CF64 2XX
Tel: (029) 2070 9901

Royal Infirmary
Edinburgh
Tel: (0131) 536 2300

The General Infirmary
Leeds LS1 3EX
Tel: (0113) 243 0715

Royal Victoria Infirmary
Newcastle NE1 4LP
Tel: (0191) 232 5131

**National Proficiency Tests
Council (NPTC)**
National Agricultural Centre
Stoneleigh
Kenilworth
Warwickshire CV8 2LG
Tel: (024) 7669 6553
Fax: (024) 7669 6128

National Turfgrass Council
Hunter's Lodge
Dr Brown's Road
Minchinhampton
Glos. GL6 9BT
Tel: (01453) 883588
Fax: (01453) 731449

Pesticides Safety Directorate
Mallard House
King's Pool
3 Peasholme Green
York YO1 2PX
Tel: (01904) 640500
Fax: (01904) 455733

Processors and Growers Research Organisation
The Research Station
Great North Road
Thornhaugh
Peterborough
Cambs. PE8 6HJ
Tel: (01780) 782585
Fax: (01780) 783993

Scottish Beekeepers' Association
North Trinity House
114 Trinity Road
Edinburgh EH5 3JZ
Tel: (0131) 552 5341

Scottish Environment Protection Agency (SEPA)
Erskine Court
The Castle Business Park
Stirling FK9 4TR
Tel: (01786) 457 700
Fax: (01786) 446 885

Stationery Office (previously HMSO)
Publications Centre
PO Box 276
London SW8 5DT
Tel: (020) 7873 9090 (orders)
(020) 7873 0011 (enquiries)
Fax: (020) 7873 8200

UK Agricultural Supply Trade Association (UKASTA)
3 Whitehall Court
London SW1A 2EQ
Tel: (020) 7930 3611
Fax: (020) 7930 3952

Ulster Beekeepers' Association
57 Liberty Road
Carrickfergus
Co. Antrim BT38 9DJ
Tel: (01960) 362998

Welsh Beekeepers' Association
Trem y Clawdd
Fron Isaf
Chirk
Wrexham
Clwyd LL14 5AH
Tel/Fax: (01691) 773300

Appendix 3
Keys to Crop and Weed Growth Stages

Decimal Code for the Growth Stages of Cereals

Illustrations of these growth stages can be found in the reference indicated below and in some company product manuals.

0 Germination
- 00 Dryseed
- 03 Imbibition complete
- 05 Radicle emerged from caryopsis
- 07 Coleoptile emerged from caryopsis
- 09 Leaf at coleoptile tip

1 Seedling growth
- 10 First leaf through coleoptile
- 11 First leaf unfolded
- 12 2 leaves unfolded
- 13 3 leaves unfolded
- 14 4 leaves unfolded
- 15 5 leaves unfolded
- 16 6 leaves unfolded
- 17 7 leaves unfolded
- 18 8 leaves unfolded
- 19 9 or more leaves unfolded

2 Tillering
- 20 Main shoot only
- 21 Main shoot and 1 tiller
- 22 Main shoot and 2 tillers
- 23 Main shoot and 3 tillers
- 24 Main shoot and 4 tillers
- 25 Main shoot and 5 tillers
- 26 Main shoot and 6 tillers
- 27 Main shoot and 7 tillers
- 28 Main shoot and 8 tillers
- 29 Main shoot and 9 or more tillers

3 Stem elongation
- 30 Ear at 1 cm
- 31 1st node detectable
- 32 2nd node detectable
- 33 3rd node detectable
- 34 4th node detectable
- 35 5th node detectable
- 36 6th node detectable
- 37 Flag leaf just visible
- 39 Flag leaf ligule/collar just visible

4 Booting
- 41 Flag leaf sheath extending
- 43 Boots just visibly swollen
- 45 Boots swollen
- 47 Flag leaf sheath opening
- 49 First awns visible

5 Inflorescence
- 51 First spikelet of inflorescence just visible
- 52 $^1/_4$ of inflorescence emerged
- 55 $^1/_2$ of inflorescence emerged
- 57 $^3/_4$ of inflorescence emerged
- 59 Emergence of inflorescence completed

6 Anthesis
- 60 61 } Beginning of anthesis
- 64 65 } Anthesis half way
- 68 69 } Anthesis complete

7 Milk development
- 71 Caryopsis watery ripe
- 73 Early milk
- 75 Medium milk
- 77 Late milk

8 Dough development
- 83 Early dough
- 85 Soft dough
- 87 Hard dough

9 Ripening
- 91 Caryopsis hard (difficult to divide by thumb-nail)
- 92 Caryopsis hard (can no longer be dented by thumb-nail)
- 93 Caryopsis loosening in daytime

(From Tottman, 1987. *Annals of Applied Biology*, **110**, 441–454)

Stages in Development of Oilseed Rape

Illustrations of these growth stages can be found in the reference indicated below and in some company product manuals.

0   Germination and emergence

1   Leaf production
1,0   Both cotyledons unfolded and green
1,1   First true leaf
1,2   Second true leaf
1,3   Third true leaf
1,4   Fourth true leaf
1,5   Fifth true leaf
1,10   About tenth true leaf
1,15   About fifteenth true leaf

2   Stem extension
2,0   No internodes ('rosette')
2,5   About five internodes

3   Flower bud development
3,0   Only leaf buds present
3,1   Flower buds present but enclosed by leaves
3,3   Flower buds visible from above ('green bud')
3,5   Flower buds raised above leaves
3,6   First flower stalks extending
3,7   First flower buds yellow ('yellow bud')

4   Flowering
4,0   First flower opened
4,1   10% all buds opened
4,3   30% all buds opened
4,5   50% all buds opened

5   Pod development
5,3   30% potential pods
5,5   50% potential pods
5,7   70% potential pods
5,9   All potential pods

6   Seed development
6,1   Seeds expanding
6,2   Most seeds translucent but full size
6,3   Most seeds green
6,4   Most seeds green-brown mottled
6,5   Most seeds brown
6,6   Most seeds dark brown
6,7   Most seeds black but soft
6,8   Most seeds black and hard
6,9   All seeds black and hard

7   Leaf senescence

8   Stem senescence
8,1   Most stem green
8,5   Half stem green
8,9   Little stem green

9   Pod senescence
9,1   Most pods green
9,5   Half pods green
9,9   Few pods green

(From Sylvester-Bradley, 1985. *Aspects of Applied Biology*, **10**, 395–400)

Stages in Development of Peas
Illustrations of these growth stages can be found in the reference indicated below and in some company product manuals.

0 Germination and emergence
000 Dry seed
001 Imbibed seed
002 Radicle apparent
003 Plumule and radicle apparent
004 Emergence

1 Vegetative stage
101 First node (leaf with one pair leaflets, no tendril)
102 Second node (leaf with one pair leaflets, simple tendril)
103 Third node (leaf with one pair leaflets, complex tendril)
•
•
l0x X nodes (leaf with more than one pair leaflets, complex tendril)
•
•
10n Last recorded node

2 Reproductive stage (main stem)
201 Enclosed buds
202 Visible buds
203 First open flower
204 Pod set (small immature pod)
205 Flat pod
206 Pod swell (seeds small, immature)
207 Podfill
208 Pod green, wrinkled
209 Pod yellow, wrinkled (seeds rubbery)
210 Dry seed

3 Senescence stage
301 Desiccant application stage. Lower pods dry and brown, middle yellow, upper green. Overall moisture content of seed less than 45%
302 Pre-harvest stage. Lower and middle pods dry and brown, upper yellow. Overall moisture content of seed less than 30%
303 Dry harvest stage. All pods dry and brown, seed dry

(From Knott, 1987. *Annals of Applied Biology*, **111**, 233–244)

Stages in Development of Faba Beans

Illustrations of these growth stages can be found in the reference indicated below and in some company product manuals.

0 Germination and emergence
- 000 Dry seed
- 001 Imbibed seed
- 002 Radicle apparent
- 003 Plumule and radicle apparent
- 004 Emergence
- 005 First leaf unfolding
- 006 First leaf unfolded

1 Vegetative stage
- 101 First node
- 102 Second node
- 103 Third node
- •
- •
- l0x X nodes
- •
- •
- 10n N, last recorded node

2 Reproductive stage (main stem)
- 201 Flower buds visible
- 203 First open flowers
- 204 First pod set
- 205 Pods fully formed, green
- 207 Pod fill, pods green
- 209 Seed rubbery, pods pliable, turning black
- 210 Seed dry and hard, pods dry and black

3 Pod senescence
- 301 10% pods dry and black
- •
- •
- 305 50% pods dry and black
- •
- •
- 308 80% pods dry and black, some upper pods green
- 309 90% pods dry and black, most seed dry. Desiccation stage.
- 310 All pods dry and black, seed hard. Pre-harvest (glyphosate application stage)

4 Stem senescence
- 401 10% stem brown/black
- •
- •
- 405 50% stem brown/black
- •
- •
- 409 90% stem brown/black
- 410 All stems brown/black. All pods dry and black, seed hard.

(From Knott, 1990. *Annals of Applied Biology*, **116**, 391–404)

Stages in Development of Potato

Illustrations of these growth stages can be found in the reference indicated below and in some company product manuals.

0 Seed germination and seedling emergence
000 Dry seed
001 Imbibed seed
002 Radicle apparent
003 Elongation of hypocotyl
004 Seedling emergence
005 Cotyledons unfolded

1 Tuber dormancy
100 Innate dormancy (no sprout development under favourable conditions)
150 Enforced dormancy (sprout development inhibited by environmental conditions)

2 Tuber sprouting
200 Dormancy break, sprout development visible
21x Sprout with 1 node
22x Sprout with 2 nodes
•
•
29x Sprout with 9 nodes
21x(2) Second generation sprout with 1 node
22x(2) Second generation sprout with 2 nodes
•
•
29x(2) Second generation sprout with 9 nodes

Where x = 1, sprout <2 mm;
2, 2–5 mm; 3, 5–20 mm;
4, 20–30 mm; 5, 50–100 mm;
6, 100–150 mm long

3 Emergence and shoot expansion
300 Main stem emergence
301 Node 1
302 Node 2
•

•
319 Node 19
Second order branch
321 Node 1
•
•
Nth order branch
3N1 Node 1
•
•
3N9 Node 9

4 Flowering
Primary flower
400 No flowers
410 Appearance of flower bud
420 Flower unopen
430 Flower open
440 Flower closed
450 Berry swelling
460 Mature berry
Second order flowers
410(2) Appearance of flower bud
420(2) Flower unopen
430(2) Flower open
440(2) Flower closed
450(2) Berry swelling
460(2) Mature berry

5 Tuber development
500 No stolons
510 Stolon initials
520 Stolon elongation
530 Tuber initiation
540 Tuber bulking (> 10 mm diam)
550 Skin set
560 Stolon development

6 Senescence
600 Onset of yellowing
650 Half leaves yellow
670 Yellowing of stems
690 Completely dead

(From Jefferies & Lawson, 1991. *Annals of Applied Biology*, **119**, 387–389)

Stages in Development of Linseed

Illustrations of these growth stages can be found in the reference indicated below and in some company product manuals.

0 Seed germination and seedling emergence
00 Dry seed
01 Imbibed seed
02 Radicle apparent
04 Hypocotyl extending
05 Emergence
07 Cotyledon unfolding from seed case
09 Cotyledons unfolded and fully expanded

1 Vegetative stage (of main stem)
10 True leaves visible
12 First pair of true leaves fully expanded
13 Third pair of true leaves fully expanded
1n n leaf fully expanded

2 Basal branching
21 One branch
22 Two branches
23 Three branches
2n n branches

3 Flower bud development (on main stem)
31 Enclosed bud visible in leaf axils
33 Bud extending from axil
35 Corymb formed
37 Buds enclosed but petals visible
39 First flower open

4 Flowering (whole plant)
41 10% of flowers open
43 30% of flowers open
45 50% of flowers open
49 End of flowering

5 Capsule formation (whole plant)
51 10% of capsules formed
53 30% of capsules formed
55 50% of capsules formed
59 End of capsule formation

6 Capsule senescence (on most advanced plant)
61 Capsules expanding
63 Capsules green and full size
65 Capsules turning yellow
67 Capsules all yellow brown but soft
69 Capsules brown, dry and senesced

7 Stem senescence (whole plant)
71 Stems mostly green below panicle
73 Most stems 30% brown
75 Most stems 50% brown
77 Stems 75% brown
79 Stems completely brown

8 Stems rotting (retting)
81 Other tissue rotting
85 Vascular tissue easily removed
89 Stems completely collapsed

9 Seed development (whole plant)
91 Seeds expanding
92 Seeds white but full size
93 Most seeds turning ivory yellow
94 Most seeds turning brown
95 All seeds brown and hard
98 Some seeds shed from capsule
99 Most seeds shed from capsule

(From Freer, 1991. *Aspects of Applied Biology*, **28**, 33–40)

Stages in Development of Annual Grass Weeds
Illustrations of these growth stages can be found in the reference indicated below and in some company product manuals.

0 Germination and emergence
00 Dry seed
01 Start of imbibition
03 Imbibition complete
05 Radicle emerged from caryopsis
07 Coleoptile emerged from caryopsis
09 Leaf just at coleoptile tip

1 Seedling growth
10 First leaf through coleoptile
11 First leaf unfolded
12 2 leaves unfolded
13 3 leaves unfolded
14 4 leaves unfolded
15 5 leaves unfolded
16 6 leaves unfolded
17 7 leaves unfolded
18 8 leaves unfolded
19 9 or more leaves unfolded

2 Tillering
20 Main shoot only
21 Main shoot and 1 tiller
22 Main shoot and 2 tillers
23 Main shoot and 3 tillers
24 Main shoot and 4 tillers
25 Main shoot and 5 tillers
26 Main shoot and 6 tillers
27 Main shoot and 7 tillers
28 Main shoot and 8 tillers
29 Main shoot and 9 or more tillers

3 Stem elongation
31 First node detectable
32 2nd node detectable
33 3rd node detectable
34 4th node detectable
35 5th node detectable
36 6th node detectable
37 Flag leaf just visible
39 Flag leaf ligule just visible

4 Booting
41 Flag leaf sheath extending
43 Boots just visibly swollen
45 Boots swollen
47 Flag leaf sheath opening
49 First awns visible

5 Inflorescence emergence
51 First spikelet of inflorescence just visible
53 $1/4$ of inflorescence emerged
55 $1/2$ of inflorescence emerged
57 $3/4$ of inflorescence emerged
59 Emergence of inflorescence completed

6 Anthesis
61 Beginning of anthesis
65 Anthesis half-way
69 Anthesis complete

(From Lawson & Read, 1992. *Annals of Applied Biology*, **12**, 211–214)

Growth Stages of Annual Broad-leaved Weeds
Preferred Descriptive Phrases

Illustrations of these growth stages can be found in the reference indicated below and in some company product manuals.

Pre-emergence
Early cotyledons
Expanded cotyledons
One expanded true leaf
Two expanded true leaves
Four expanded true leaves
Six expanded true leaves
Plants up to 25 mm across/high
Plants up to 50 mm across/high
Plants up to 100 mm across/high
Plants up to 150 mm across/high
Plants up to 250 mm across/high
Flower buds visible
Plant flowering
Plant senescent

(From Lutman & Tucker, 1987. *Annals of Applied Biology*, **110**, 683–687)

Appendix 4
A. Key to Label Requirements for Personal Protective Equipment (PPE)

COSHH requires that exposure of anyone, including members of the public, who may be affected by work involving substances hazardous to health be either prevented or, where this is not reasonably practicable, adequately controlled. Engineering or other control measures should be used in preference to personal protective equipment (PPE). However, in the case of pesticides, spray operators are clearly most at risk from exposure and PPE will usually be needed in addition to engineering controls to achieve adequate control of exposure. Even if COSHH does not apply (because the product in use is not classed as hazardous), employers may still have responsibilities under the Protective Equipment at Work Regulations 1992.

European Council Directive 86/686/EEC defines the requirements for the use of PPE and these are enacted in UK under the Personal Protective Equipment (EC Directive) Regulations 1992. The Regulations require that any PPE made or imported after 1 July 1995 must meet the requirements set out in the Directive and be CE-marked. Older PPE may continue to be used provided it gives adequate protection and is properly maintained.

Where a product label specifies the use of PPE, the requirements are listed in the profiles under the heading **Personal Protective Equipment/label precautions**, using letter codes to denote the protective items, according to the list below. Often PPE requirements are different for specified operations, e.g. handling the concentrate, cleaning equipment etc., but it is not possible to list them separately. The lists of PPE are therefore an indication of what the user may require to have available to use the product in different ways. **When making a COSHH assessment it is therefore essential that the product label is consulted for information on the particular use that is being assessed.**

A   Suitable protective gloves (the product label should be consulted for any specific requirements about the material of which the gloves should be made)
B   Rubber gauntlet gloves
C   Face-shield
D   Approved respiratory protective equipment
E   Goggles
F   Dust mask
G   Full face-piece respirator
H   Coverall
J   Hood
K   Rubber apron
L   Waterproof coat
M   Rubber boots
N   Waterproof jacket and trousers
P   Suitable protective clothing

Appendix 4
B. Key to Numbered Label Precautions

The series of numbers listed in the profiles refer to the numbered precautions below. Where the generalised wording includes a phrase such as "... for xx days" the specific requirement for each pesticide is given in the **Special precautions/Environmental safety** section of each profile.

Medical Advice

M01 This product contains an anticholinesterase organophosphorus compound. DO NOT USE if under medical advice NOT to work with such compounds

M02 This product contains an anticholinesterase carbamate compound. DO NOT USE if under medical advice NOT to work with such compounds

M03 If you feel unwell, seek medical advice (show the label where possible)

M04 In case of accident or if you feel unwell, seek medical advice immediately (show the label where possible)

M05 If swallowed, seek medical advice immediately and show this container or label

M06 This product contains an anticholinesterase carbamoyl triazole compound. DO NOT USE if under medical advice NOT to work with such compounds

Risk Phrases

R01 Very toxic
R01a Very toxic: In contact with skin
R01b Very toxic: By inhalation
R01c Very toxic: If swallowed
R02 Toxic
R02a Toxic: In contact with skin
R02b Toxic: By inhalation
R02c Toxic: If swallowed
R03 Harmful
R03a Harmful: In contact with skin
R03b Harmful: By inhalation
R03c Harmful: If swallowed
R04 Irritant
R04a Irritating to eyes
R04b Irritating to skin
R04c Irritating to respiratory system
R04d Risk of serious damage to eyes
R04e May cause sensitization by skin contact
R04f May cause sensitization by inhalation
R04g May cause lung damage if swallowed
R05 Corrosive
R05a Causes severe burns
R05b Causes burns
R06 Danger of serious damage to health by prolonged exposure
R07a Extremely flammable liquefied gas
R07b Extremely flammable
R07c Highly flammable
R07d Flammable
R08 Oxidising agent. Contact with combustible material may cause fire
R09 Explosive when mixed with oxidising substances

VERY TOXIC

TOXIC

HARMFUL

IRRITANT

CORROSIVE

User Safety Precautions

U01 To be used only by operators instructed or trained in the use of chemical/product/type of produce and familiar with the precautionary measures to be observed

U02 Wash all protective clothing thoroughly after use, especially the inside of gloves/Avoid excessive contamination of coveralls and launder regularly

U03 Wash splashes off gloves immediately
U04a Take off immediately all contaminated clothing
U04b Take off immediately all contaminated clothing and wash underlying skin. Wash clothes before re-use
U05a When using do not eat, drink or smoke
U05b When using do not eat, drink, smoke or use naked lights
U06 Handle with care and mix only in a closed container
U07 Open container only as directed
U08 Wash concentrate/dust from skin or eyes immediately
U09a Wash any contamination/splashes/dust/powder/concentrate from skin or eyes immediately
U09b Wash any contamination/splashes/dust/powder/concentrate from eyes immediately
U10 After contact with skin or eyes wash immediately with plenty of water
U11 In case of contact with eyes rinse immediately with plenty of water and seek medical advice
U12 In case of contact with skin rinse immediately with plenty of water and seek medical advice
U13 Avoid all contact by mouth
U14 Avoid all contact with skin
U15 Avoid all contact with eyes
U16 Ensure adequate ventilation in confined spaces
U17 Ventilate treated areas thoroughly when smoke has cleared/Ventilate treated rooms thoroughly before occupying
U18 Extinguish all naked flames, including pilot lights, when applying the fumigant/dust/liquid/ product
U19 Do not breathe dust/fog/fumes/gas/smoke/spray mist/vapour. Avoid working in spray mist
U20a Wash hands and exposed skin before eating, drinking or smoking and after work
U20b Wash hands and exposed skin before meals and after work
U20c Wash hands before meals and after work
U21 Before entering treated crops, cover exposed skin areas, particularly arms and legs
U22 Do not touch sachet with wet hands or gloves/Do not touch water soluble bag directly
U23 This product must not be applied using knapsack sprayers/This product must not be applied by hand-held equipment
U24 Do not handle grain unnecessarily

Consumer Safety Precautions

C01 Not to be used on food crops
C02 Do not harvest for human or animal consumption for at least xx days/weeks after last application
C03 Keep away from food, drink and animal feeding-stuffs
C04 Do not apply to surfaces on which food/feed is stored, prepared or eaten
C05 Remove/cover all foodstuffs before application
C06 Remove exposed milk before application
C07 Collect eggs before application
C08 Protect food preparing equipment and eating utensils from contamination during application
C09 Cover water storage tanks before application
C10 Protect exposed water/feed/milk machinery/milk containers from contamination

Environmental Safety Precautions (Public, Livestock, Wildlife etc.)

E01 Keep out of reach of children
E02 Keep unprotected persons out of treated areas for at least xx hours/days
E03 Label treated seed with the appropriate precautions, using the printed sacks, labels or bag tags supplied
E04 Remove all pets/livestock/fish tanks before treatment/spraying
E05 Do not apply directly to livestock/poultry
E05a Keep poultry out of treated areas for at least xx days/weeks
E06a Dangerous to livestock. Keep all livestock out of treated areas/away from treated water for at least xx days/weeks. Bury or remove spillages
E06b Harmful to livestock. Keep all livestock out of treated areas/away from treated water for at least xx days/weeks. Bury or remove spillages
E07 Keep livestock out of treated areas/Keep livestock out of treated areas for at least xx

weeks and until foliage of any poisonous weeds such as ragwort has died and become unpalatable

E08   Do not feed treated straw or haulm to livestock within x days of spraying

E09   Do not use on crops if the straw is to be used as animal feed/bedding

E10a   Dangerous to game, wild birds and animals

E10b   Harmful to game, wild birds and animals

E11   Harmful to livestock. Paraquat may be harmful to hares; stubbles must be sprayed early in the day

E12a   High risk to bees. Do not apply to crops in flower or to those in which bees are actively foraging. Do not apply when flowering weeds are present

E12b   Extremely dangerous to bees. Do not apply to crops in flower or to those in which bees are actively foraging. Do not apply when flowering weeds are present

E12c   Dangerous to bees. Do not apply to crops in flower or to those in which bees are actively foraging. Do not apply when flowering weeds are present

E12d   Dangerous to bees. Do not apply to crops in flower or to those in which bees are actively foraging except as directed on [crop]. Do not apply when flowering weeds are present

E12e   Harmful to bees. Do not apply to crops in flower or to those in which bees are actively foraging. Do not apply when flowering weeds are present

E13a   Extremely dangerous to fish or other aquatic life. Do not contaminate surface waters or ditches with chemical or used container

E13b   Dangerous to fish or other aquatic life. Do not contaminate surface waters or ditches with chemical or used container

E13c   Harmful to fish or other aquatic life. Do not contaminate surface waters or ditches with chemical or used container

E14a   Extremely dangerous to aquatic higher plants. Do not contaminate surface waters or ditches with chemical or used container

E14b   Dangerous to aquatic higher plants. Do not contaminate surface waters or ditches with chemical or used container

E15   Do not contaminate surface waters or ditches with chemical or used container

E16   Do not allow direct spray/granule applications from vehicle mounted/drawn hydraulic sprayers to fall within 6 m of surface waters or ditches/Do not allow direct spray/granule applications from hand-held sprayers to fall within 2 m of surface waters or ditches. Direct spray away from water

E16a   Do not allow direct spray from ground crop sprayers to fall within 5 m of the top of the bank of a static or flowing waterbody, unless a Local Environment Risk Assessment for Pesticides (LERAP) permits a narrower buffer zone, or within 1 m of the top of a ditch which is dry at the time of application. Direct spray away from water

E16b   Do not allow direct spray from hand-held sprayers to fall within 1 m of the top of the bank of a static or flowing waterbody. Direct spray away from water

E16c   Do not allow direct spray from ground crop sprayers to fall within 5 m of the top of the bank of a static or flowing waterbody, or within 1 m of the top of a ditch which is dry at the time of application. Direct spray away from water. This product is not eligible for buffer zone reduction under the LERAP scheme.

E16d   Do not allow direct spray from hand-held sprayers to fall within 1 m of the top of the bank of a static or flowing waterbody. Direct spray away from water. This product is not eligible for buffer zone reduction under the LERAP scheme.

E16e   Do not allow direct spray from ground crop sprayers to fall within 5 m of the top of the bank of a static or flowing water body or within 1 m from the top of any ditch which is dry at the time of application. Spray from hand-held sprayers must not in any case be allowed to fall within 1 m of the top of the bank of a static or flowing water body. Always directed spray away from water. The LERAP scheme does not extend to adjuvants. This product is therefore not eligible for a reduced buffer zone under the LERAP scheme.

E17   Do not allow direct spray from broadcast air-assisted sprayers to fall within xx m of surface waters or ditches. Direct spray away from water

E18   Do not spray from the air within 250 m horizontal distance of surface waters or ditches

E19   Do not dump surplus herbicide in water or ditch bottoms

E20   Prevent any surface run-off from entering storm drains

E21   Do not use treated water for irrigation purposes within xx days/weeks of treatment

E22a	High risk to non-target insects or other arthropods. Do not spray within 6 m of the field boundary
E22b	Risk to certain non-target insects or other arthropods. See directions for use
E23	Avoid damage by drift onto susceptible crops or water courses
E24	Store away from seeds, fertilizers, fungicides and insecticides
E25	Store well away from corms, bulbs, tubers and seeds
E26	Store away from frost
E27	Store away from heat
E28	Flammable. Do not store near heat or open flame
E29	Store under cool, dry conditions
E30a	Store/keep in original container, tightly closed, in a safe place
E30b	Store/keep in original container, tightly closed, in a safe place, under lock and key
E31a	Wash out container thoroughly and dispose of safely
E31b	Wash out container thoroughly, empty washings into spray tank and dispose of safely
E31c	Rinse container thoroughly by using an integrated pressure rinsing device or manually rinsing three times. Add washings to sprayer at time of filling and dispose of container safely
E32a	Empty container completely and dispose of safely
E32b	Empty container completely and dispose of it in the specified manner
E33	Return empty container as instructed by supplier
E34	Do not re-use this container for any purpose
E35	Do not burn this container
E36	Do not rinse out the container
E37	Do not use with any pesticide which is to be applied in or near water

Sack Label Precautions for Treated Seed

S01	Do not handle treated seed unnecessarily
S02	Do not use treated seed as food or feed
S03	Keep treated seed secure from people, domestic stock/pets and wildlife at all times during storage and use
S04a	Bury or remove spillages
S04b	Harmful to game and wild life. Bury spillages/Bury or remove spillages
S04c	Dangerous to birds, game and other wildlife. Treated seed should not be left on the soil surface. Bury spillages
S05	Do not re-use sack for food or feed/Do not re-use sacks or containers that have been used for treated seed for food or feed
S06	Wash hands and exposed skin before meals and after work
S07	Do not apply treated seed from the air
S08	Treated seed must not be applied as a broadcast application

Vertebrate/Rodent Control Product Precautions

V01a	Prevent access to baits/powder by children, birds and other animals, particularly cats, dogs, pigs and poultry
V01b	Prevent access to bait/gel/dust by children, birds and non-target animals, particularly dogs, cats, pigs, poultry
V02	Do not prepare/use/lay baits/dust/spray where food/feed/water could become contaminated
V03a	Remove all remains of bait, tracking powder or bait containers after use and burn or bury
V03b	Remove all remains of bait and bait containers/exposed dust/after treatment (except where used in sewers) and dispose of safely (e.g. burn/bury). Do not dispose of in refuse sacks or on open rubbish tips
V04a	Search for and burn or bury all rodent bodies. Do not place in refuse bins or on rubbish tips
V04b	Search for rodent bodies (except where used in sewers) and dispose of safely (e.g. burn/bury). Do not dispose of in refuse sacks or on open rubbish tips
V04c	Dispose of safely any rodent bodies and remains of bait and bait containers that are recovered after treatment (e.g. burn/bury). Do not dispose of in refuse sacks or on open rubbish tips
V05	Use bait containers clearly marked POISON at all surface baiting points

Appendix 5
Key to Abbreviations and Acronyms

The abbreviations of formulation types in the following list are used in Section 4 (Pesticide Profiles) and are derived from the Catalogue of Pesticide Formulation Types and International Coding System (GCPF Technical Monograph 2, 4th edition, April 1999)

1 Formulation Types

AE	Aerosol generator
AL	Other liquids to be applied undiluted
BB	Block bait
CB	Bait concentrate
CG	Encapsulated granule (controlled release)
CR	Crystals
CS	Capsule suspension
DC	Dispersible concentrate
DP	Dustable powder
DS	Powder for dry seed treatment
EC	Emulsifiable concentrate
ES	Emulsion for seed treatment
EW	Oil in water emulsion
FG	Fine granules
FP	Smoke cartridge
FS	Flowable concentrate for seed treatment
FT	Smoke tablet
FU	Smoke generator
FW	Smoke pellets
GA	Gas
GB	Granular bait
GE	Gas-generating product
GG	Macrogranules
GP	Flo-dust
GR	Granules
GS	Grease
HN	Hot fogging concentrate
KK	Combi-pack (solid/liquid)
KL	Combi-pack (liquid/liquid)
KN	Cold-fogging concentrate
KP	Combi-pack (solid/solid)
LA	Lacquer
LI	Liquid, unspecified
LS	Solution for seed treatment
ME	Microemulsion
MG	Microgranules
OL	Oil miscible liquid
PA	Paste
PC	Gel or paste concentrate
PS	Seed coated with a pesticide
PT	Pellet
RB	Ready-to-use bait
RH	Ready-to-use spray in hand-operated sprayer
SA	Sand
SC	Suspension concentrate (= flowable)
SE	Suspo-emulsion
SG	Water soluble granules
SL	Soluble concentrate

SP	Water soluble powder
SS	Water soluble powder for seed treatment
SU	Ultra low-volume suspension
TB	Tablets
TC	Technical material
TP	Tracking powder
UL	Ultra-low-volume liquid
VP	Vapour releasing product
WB	Water soluble bags
WG	Water dispersible granules
WP	Wettable powder
WS	Water dispersible powder for slurry treatment of seed
XX	Other formulations
ZZ	Not Applicable

2 Other Abbreviations and Acronyms

ACP	Advisory Committee on Pesticides
ACTS	Advisory Committee on Toxic Substances
ADAS	Agricultural Development and Advisory Service
a.i.	active ingredient
BAA	British Agrochemicals Association
BLRA	Brewers' and Licensed Retailers' Association
CDA	controlled droplet application
cm	centimetre(s)
COPR	Control of Pesticides Regulations 1986
COSHH	Control of Substances Hazardous to Health Regulations
CPA	Crop Protection Association
d	day(s)
EA	Environment Agency
EBDC	ethylene-bis-dithiocarbamate fungicide
FEPA	Food and Environment Protection Act 1985
g	gram(s)
GCPF	Global Crop Protection Federation
GS	growth stage (unless in formulation column)
h	hour(s)
ha	hectare(s)
HBN	hydroxybenzonitrile herbicide
HI	harvest interval
HMIP	HM Inspectorate of Pollution
HSE	Health and Safety Executive
ICM	Integrated Crop Management
IPM	Integrated Pest Management
kg	kilogram(s)
l	litre(s)
LERAP	Local Environmental Risk Assessments for Pesticides
m	metre(s)
MAFF	Ministry of Agriculture, Fisheries and Food
MBC	methyl benzimidacarbamate fungicide
MEL	Maximum Exposure Limit
min	minute(s)
mm	millimetre(s)
MRL	Maximum Residue Level
mth	month(s)
NA	notice of approval
NFU	National Farmers Union
OES	Occupational Exposure Standard
OLA	off-label approval
PM	Pesticides Monitor

PPE	personal protective equipment
PPPR	Plant Protection Products Regulations
PR	Pesticides Register
PSD	Pesticides Safety Directorate
SOLA	specific off-label approval
ULV	ultra-low volume
w/v	weight/volume
w/w	weight/weight
wk	week(s)
yr	year(s)

Index of Proprietary Names of Products

The references are to entry numbers, not to pages. Adjuvant names are referred to as 'Adj' and are listed separately in Section 2

The references are to entry numbers, not to pages

The references are to entry numbers, not to pages

The references are to entry numbers, not to pages

The references are to entry numbers, not to pages

The references are to entry numbers, not to pages

The references are to entry numbers, not to pages

The references are to entry numbers, not to pages

USING PESTICIDES

A complete guide to safe effective spraying

Using Pesticides is a practical yet comprehensive guide to safe, effective spraying. It is designed for sprayer operators, their managers and supervisors in the agricultural, horticultural and amenity industries.

This publication is presented in an easy to use binder, divided into 12 easy to follow sections:

- Definitions
- The Planned Approach
- Using The Product Label
- Keeping People Safe
- Applying Pesticides
- Environmental Issues

- Transporting Pesticides
- The Pesticide Store
- Disposal
- Keeping Records
- Emergency Procedures
- Further Information

This revised edition is in line with the "Green" and "Orange" Codes, buffer zones, LERAP and the Groundwater Regulations. It also reflects current ICM and quality assurance matters, plus nozzle selection guidance.

This guide is essential for all those involved in any way with pesticides.

188 pages.   ISBN 1 901396 01 0
£18.50. UK delivery included. Overseas delivery extra.
Credit cards accepted.

BCPC Publications Sales, Bear Farm, Binfield, Bracknell, Berks, RG42 5QE, UK.
Tel: +44 (0) 118 934 2727. Fax: +44 (0) 118 934 1998.
Email: publications@bcpc.org   Web: www.bcpc.org

BRITISH
CROP
PROTECTION
COUNCIL

The UK Pesticide Guide is also available on CD-ROM as

The e-UK Pesticide Guide

The e-UK Pesticide Guide contains all the information found in this book and much more, making searching for information quick and easy.

The 2001 edition of
The e-UK Pesticide Guide
has been made even more user friendly.

The 2001 edition has a number of new features including:

- Improved crop search facilities
- Improved save, download and printing of search results
- Updated background information

In addition it has all the functionality of the previous edition:

More user-friendly

- Search by keywords in Straight Search and Search Wizard
- Simple Straight Search – by product number or name
- Simple Search Wizard – enhanced with Boolean operators
- Personal notes can be added to a saved results page
- Results output in a standardised format
- Symbols to provide recognition at a glance of key features and hazards

Related information

- Comprehensive Glossary
- Internet links

Other features

- Undo feature to allow back-tracking through search history
- Networking can be initiated over the phone
- On screen help

Free demonstration available via the web at:
http://www.bcpc.org/publications or
http://www.cabi.org/publishing/

The e-UK Pesticide Guide (CD-ROM)

Edited by R Whitehead

Release date: January 2001

ISBN: 0 85199 552 7

To order your copy today, contact BCPC Publishing sales on
0118 934 2727

CABI *Publishing*
A division of CAB *International*

THE UK PESTICIDE GUIDE 2001
RE-ORDER FORM

Surname _____ Initials _____ Dr/Mr/Mrs/Ms

Company/organization _____

Address _____

_____

Postcode _____ Country _____

Telephone _____ Fax _____

Email _____

Date _____ Signed _____

Please send me _____ more copies of *The UK Pesticide Guide 2001*

☐ Payment enclosed, cheques made payable to BCPE Ltd.

☐ Please debit my Master Card/Eurocard/Visa Card/American Express account

by the sum of _____ Name of cardholder _____

Name of issuing bank _____

Card no _____ Expiry date _____

Address of cardholder (if different from above) _____

_____

_____

OR TELEPHONE YOUR ORDER NOW AND ASK FOR THE SALES DEPARTMENT

List price £24.50
Discount for bulk sales

No. of copies	Discount
100+	30%
50–99	25%
10–49	15%
1–9	No discount

2002 EDITION

☐ Please send me advance price details for the 2002 Edition of *The UK Pesticide Guide* as soon as they are available.

Also available on CD-ROM as *The e-UK Pesticide Guide*. Details can be obtained from the address below and on the web at www.bcpc.org or www.cabi.org.

Please detach and return to:
British Crop Protection Enterprises Ltd.
BCPC Publications Sales
Bear Farm, Binfield, Bracknell, Berkshire RG42 5QE, UK
Telephone: 0118 934 2727
Fax: 0118 934 1998